Studies in Computational Intelligence

Volume 1056

Series Editor

Janusz Kacprzyk, Polish Academy of Sciences, Warsaw, Poland

The series "Studies in Computational Intelligence" (SCI) publishes new developments and advances in the various areas of computational intelligence—quickly and with a high quality. The intent is to cover the theory, applications, and design methods of computational intelligence, as embedded in the fields of engineering, computer science, physics and life sciences, as well as the methodologies behind them. The series contains monographs, lecture notes and edited volumes in computational intelligence spanning the areas of neural networks, connectionist systems, genetic algorithms, evolutionary computation, artificial intelligence, cellular automata, self-organizing systems, soft computing, fuzzy systems, and hybrid intelligent systems. Of particular value to both the contributors and the readership are the short publication timeframe and the world-wide distribution, which enable both wide and rapid dissemination of research output.

Indexed by SCOPUS, DBLP, WTI Frankfurt eG, zbMATH, SCImago.

All books published in the series are submitted for consideration in Web of Science.

Muhammad Alshurideh ·
Barween Hikmat Al Kurdi · Ra'ed Masa'deh ·
Haitham M. Alzoubi · Said Salloum
Editors

The Effect of Information Technology on Business and Marketing Intelligence Systems

Volume 3

Springer

Editors
Muhammad Alshurideh (ID)
Department of Management, College
of Business Administration
University of Sharjah
Sharjah, United Arab Emirates

Department of Marketing, School
of Business
The University of Jordan
Amman, Jordan

Ra'ed Masa'deh (ID)
Management Information Systems
Department, School of Business
University of Jordan
Aqaba, Jordan

Said Salloum (ID)
School of Comupting, Science
and Engineering
University of Salford
Salford, England

Barween Hikmat Al Kurdi (ID)
Department of Marketing, Faculty
of Economics and Administrative Sciences
The Hashemite University
Zarqa, Jordan

Haitham M. Alzoubi (ID)
Skyline University College
Sharjah, United Arab Emirates

ISSN 1860-949X   ISSN 1860-9503 (electronic)
Studies in Computational Intelligence
ISBN 978-3-031-12381-8   ISBN 978-3-031-12382-5 (eBook)
https://doi.org/10.1007/978-3-031-12382-5

Contents

Business and Data Analytics

Innovation, Entrepreneurship and Leadership

Knowledge Management

Machine Learning, IOT, BIG DATA, Block Chain and AI

Machine Learning Techniques for Stock Market Predictions: A Case of Mexican Stocks

Aqila Rafiuddin, Jesus Cuauhtemoc Tellez Gaytan, Gouher Ahmed, and Muhammad Alshurideh

Abstract Prediction of stock markets prices in a financial time series is very difficult and challenging task due to the nature characterized by dynamic, volatile, sentimental and chaotic behavior of the investors. The quantitative nature of data of the stocks gives a pace for the advanced simulation techniques, machine learning models a branch of Artificial intelligence learning process utilizes their predictive algorithms and forecasts the prices for future periods. With high productivity in the machine learning area applied to the prediction of financial market prices, objective methods are required for a consistent analysis of the most relevant existing literature on financial markets. Research revealed that most commonly used models for prediction of stocks are support vector machine (SVM) models and artificial neural networks (ANN). This research work is from the Mexican stock market with a comparison of SVM and ANN.

Keywords Artificial intelligence · Machine learning · Deep learning · Neural networks · Stock markets

A. Rafiuddin · J. C. T. Gaytan
Fair Centre for Financial Access Inclusion and Research, Tecnologico de Monterrey, San Luis Potosi, Mexico
e-mail: cuauhtemoc.tellez@tec.mx

J. C. T. Gaytan
Tecnologico de Monterrey, San Luis Potosi, Mexico

G. Ahmed (✉)
Professor of Strategic Leadership and International Business, School of Business, Skyline University College, P.O. Box: 1797, Sharjah, UAE
e-mail: gouher@usa.net

M. Alshurideh
Department of Marketing, School of Business, The University of Jordan, Amman, Jordan
e-mail: m.alshurideh@ju.edu.jo; malshurideh@sharjah.ac.ae

Department of Management, College of Business Administration, University of Sharjah, Sharjah, UAE

1   Introduction

According to the efficient market hypothesis (EMH) by markets are expected to follow random pathways and making it more unpredictable leading to a continuous quest for a model and the most important and challenging problems involving time series (Atsalakis & Valavanis, 2009; Chen et al., 2017) highlighted ANN and SVM, with data pre-processing. The research work by Laboissiere et al. (2015) who continued with pattern recognition for financial market prediction in brazillian market. According, there is a need for profitable system that attracts lot of attention from the business and academia (Weng et al., 2017). Due to non stationary in nature its difficult to predict the stock market prices (Tay & Cao, n.d.; Zhang et al., 2017). They are dynamic, chaotic, noisy, non-linear series (Bezerra & Albuquerque, 2017; Göçken et al., 2016; Kumar et al., 2016), that are influenced by the general economy, characteristics of the industries, politics, and even the psychology of investors (Chen et al., 2017; Zhong & Enke, 2017). An evidence of the efficiency of financial markets is summarized by a model capable of consistently generating returns above the market indices and represents EMH and enable large profits with financial operations (Kumar & Thenmozhi, 2014a). According to Wang et al. (2012), initial techniques were moving averages, autoregressive models, discriminating analyses, and correlations but artificial intelligence is considered to be an ideal and promising area of the research in the prediction of time series. ANN techniques are designed to address chaotic data, randomness, and non-linearity (Zhong & Enke, 2017). Computational system with the technological advances made it possible to analyse large historical price databases by Chiang et al. (2016). The intense computational use of intelligent predictive models is commonly considered as machine learning techniques.

2   Machine Learning

Fundamental analysis information of financial statements for the evaluation of the company helps the investors in decision making through specific components and returns (Foster et al., 1984; Lev & Thiagarajan, 1993; Ou & Penman, 1989). Based on the past financial statements the intrinsic value of a company can be estimated and compared to market prices for investments. There is a deviation of market prices from the original value due to under reaction or overreaction of customers on the available information (Desai et al., 2004; Kothari et al., 2005). With the revision in beliefs and sentiments of the investors, earnings announcements, future dividends (Collins & Kothari, 1989), new information released affects the market participants to predict the future change in the prices (Bernard et al., 1997). For developing effective market trading strategies (Leung et al., 2000a). there is an immense need for accurate predictions of movement of stock price indexes Stock market prediction is regarded as a challenging task of the financial time series prediction process since the stock market is essentially dynamic, nonlinear, complicated, nonparametric, and chaotic in nature (Abu-Mostafa & Atiya n.d.). Integration of Machine learning techniques with artificial intelligence systems seek to extract patterns learned from historical data in a process known as training or learning to subsequently make predictions about new data and integrate (Xiao et al., 2013). Empirical studies using machine learning commonly have two main phases. The first one addresses the selection of relevant variables and models for the prediction, separating a portion of the data for the training and validation of the models, thus optimizing them. The second phase applies the optimized models to the data intended for testing, thus measuring predictive performance. The basic techniques used in the literature include the following: artificial neural networks (ANNs), support vector machines (SVMs), Stock returns based on accounting variables are mispriced at time as a violation of efficient markets due to the varying time risk factors (Fama, 1991; Hirshleifer et al., 2012).

The layers of basic processing units of the neural networks are interconnected, attributing weights for each connection (Lahmiri & Boukadoum, 2015), which are adjusted in the learning process of the network (Kumar & Thenmozhi, 2014b), in the first training phase optimization of not only the interconnections between the layers of neurons but also the parameters of the transfer functions between one layer and another, thus minimizing the errors. Finally, the last layer of the neural network is responsible for summing all the signals from the previous layer into just one output signal—the network's response to certain input data.

Whereas neural networks seek to minimize the errors of their empirical responses in the training stage, an SVM seeks to minimize the upper threshold of the error of its classifications. Huang et al. (2005) took the training samples and transforms them from their original dimension space to another space, with a greater number of dimensions, in which a linear separation is approximated by a hyperplane (Kara et al., 2011). This algorithm, which is commonly used to classify data based on input variables in the model, seeks to minimize the margin of the classification hyperplane during the training stage of the model. The transformation of the space of original dimensions to the space in which the classifications are performed is done with the assistance of kernel functions, from estimated parameterization in the training of the model, as detailed by Pai and Lin (2005).

ANN and SVM have been successfully used for modeling and predicting financial time series. Although ANN can be one of the very useful tools in time series prediction, several studies showed that ANN had some limitations in learning the patterns because stock market data has tremendous noise, non-stationary characteristics, and complex dimensionality. ANN often exhibit inconsistent and unpredictable performance on noisy data (Huang et al., 2005; Kim, 2003; Kim & Han, 2000; Kumar & Thenmozhi, 2014b; Lahmiri & Boukadoum, 2015; Manish & Thenmozhi, 2005). Therefore, predicting stock price movements is quite difficult.

Thus, the literature regarding financial market prediction using machine learning is vast, which hinders revisions, systematizations of models and techniques, and searches for material to determine the state of the art. Tools are needed to objectively and quantitatively select the most relevant works for a literature review covering the most influential articles. Thus, the intention of this work is to present methods for the selection of the main advances in machine learning applied to financial market prediction in the Mexican market clarifies a knowledge flow that literature brings a summary of the best procedures followed by the literature on applying machine learning to financial time series forecasting.

ANN is applied predict accurately the stock price return and the direction of its movement. ANN has been demonstrated to provide promising results in predict the stock price return (Avci, 2007; Egeli et al., 2003; Karaatlı et al., 2005; Kimoto et al., 1990; Olson & Mossman, 2003; White, 1990; Yoon & Swales, 1991). Leung et al. (2000b) examined various prediction models based on multivariate classification techniques and compared them with a number of parametric and nonparametric models which forecast the direction of the index return. Empirical experimentation suggested that the classification models (discriminant analysis, logit, probit and probabilistic neural network) outperform the level estimation models (adaptive exponential smoothing, vector auto regression with Kalman filter updating, multivariate transfer function and multilayered feed forward neural network) in terms of predicting the direction of the stock market movement and maximizing returns from investment trading.

Recently the support vector machines (SVM), has been also successfully applied to predict stock price index and its movements. Kim (2003) used SVM to predict the direction of daily stock price change in the Korea composite stock price index (KOS-I). The experimental results showed that SVM outperform random forest, neural network and other traditional models (Huang et al., 2005), in their study, investigated the predictability of financial movement direction with SVM by predicting the weekly movement direction of NIKKEI 225 Index. Support vector machine (SVM) is a very specific type of learning algorithms characterized by the capacity control of the decision function, the use of the kernel functions and the sparsity of the solution. Established on the unique theory of the structural risk minimization principle to estimate a function by minimizing an upper bound of the generalization error, SVM is shown to be very resistant to the over-1tting problem, eventually achieving a high generalization performance. Another key property of SVM is that training SVM is equivalent to solving a linearly constrained quadratic programming problem so that the solution of SVM is always unique and globally optimal, unlike neural networks training which requires nonlinear optimization with the danger of getting stuck at local minima. Some applications of SVM to financial forecasting problems have been reported recently in most cases, the degree of accuracy and the acceptability of certain forecasts are measured by the estimates' deviations from the observed values. For the practitioners in financial market, forecasting methods based on minimizing forecast error may not be adequate to meet their objectives. In other words, trading driven by a certain forecast with a small forecast error may not be as pro1table as trading guided by an accurate prediction of the direction of movement.

3 Objective

This research objective is to predict stock price movements based on three machine learning techniques: Support Vector Machines, Artificial Neural Networks, and Deep Neural Networks. The Machine Learning approach can be equally applied to forecast continuous and categorical variables. If the predictive variable is a continuous one it is said to be a regression based approach. Otherwise, when the predictive variable is a categorical one it is referred to a classification problem. This research considers stock price movements as a categorical variable since it is aimed to predict an "up" or "down" movement.

4 Methodology

The predictive model considers as target variables the daily closing prices of the main Mexican Stock Index (IPC) and seven representative Mexican companies: Grupo Alsea (ALSEA), America Movil (AMXL), Grupo Bimbo (BIMBOA), Cemex (CEMEXCPO), Grupo Carso (GCARSOA1), Grupo Mexico (GMEXICO), and Grupo Maseca (GRUMA). Data ranges from September 2013 up to December 2019 which acounts for 1,424 observations and were downloaded from the Bloomberg Terminal. Daily closing prices were transformed to log-returns and the categorical variable was built by the following conditional: if the log-return is positive then it is an upward movement (1), else if the log-return is negative then it is a downward movement (0). Whole estimations were performed in R v.3.5.1.

Based on Kara et al. (2011), the predictive model considers technical indicators as predictors such as: Simple Moving Average (SMAVG), Weighted Moving Average (WMAVG), Stochastic Slow (SLOWD), Stochastic Fast (FASTK), Relative Strength Index (RSI), Moving Average Convergence Divergence (MACD), Larry Williams Percent Range (WLPR), and Commodity Channel Index (CMCI). Predictors were not transformed to log-returns. Once, data was subsetted into training (80%) and testing (20%) sets. The first step was to perform a Principal Component Analysis to extract the main influential features where it was used the correlation matrix to scale the features. Figure 1 shows the PCA biplot of the IPC where it is observed that RSI, SLOW and FASTD are the most representative features along the PC1 axis. Meanwhile SMAVG and DIFF are the most representative features along the PC2 exists.

Once main drivers were identified that mostly explain the categorical variable ("movement"), they were normalized under the minmax strategy in such a way to speed the convergence process. Then, it was estimated a Support Vector Machine model considering a linear kernel. Next, a Neural Network was estimated with one hidden layer, three neurons, and a logistic activation function. Finally, the Deep Learning approach was considered by increasing the number of hidden layers in the

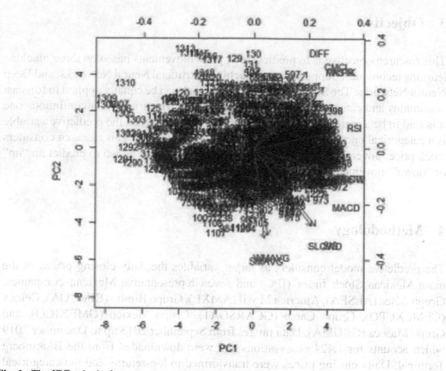

Fig. 1 The IPC principal component analysis

Neural Network. In both techniques it was estimated the confusion matrix to predict (classify) true positives, true negatives, false positives, and false negatives.

Findings Table 1 Summarizes the Support Vector Machine (SVM) and Neural Network (NN) market prediction accuracy levels. The NN columns are read as the number of hidden layers and neurons in the network. For example, (5,4,2) states three hidden layers, each one with 5, 4, and 2 neurons, respectively.

In most cases the NN model outperformed the SVM. But within the NN approach not all the Deep Learning estimations could outperform the single (3)-NN hidden layer. The greatest accuracy level was obtained on Gruma with a (5,4,2)-NN hidden layers. In this case, it is argued that an upward or downward movement can be predicted with a 75.1% accuracy level. The accuracy level was calculated from the confusion matrix. For example, a predicted movement with an accuracy level of 70.8% for GMexico under the SVM approach was obtained as (Fig. 2).

Other studies like (Huang et al., 2005) which got an accuracy level by the SVM model of 75% applied to NIKKEI 225 weekly prices. Kara et al. (2011) obtained a SVM accuracy level of 71.52%. Patel et al. (2015) estimated an accuracy level of 86.69% under the NN approach and an 88.69% accuracy level by the SVM model applied to the S&P Bombay Stock Exchange SENSEX.

Table 1 SVM and NN estimations

Index/company	SVM (%)	NN			
		(3)	(3,2)	(4,3,2)	(5,4,2)
IPC	68.0	69.2% (94.8)	67.5% (94.28)	68.5% (91.13)	68.9% (94.28)
Alsea	69.4	73.4% (94.40)	73.4% (94.90)	72.5% (93.58)	71.5% (93.40)
Amxl	71.1	72.5% (74.70)	74.8% (95.74)	72.1% (94.37)	70.8% (93.93)
Bimbo	70.4	71.5% (92.06)	67.5% (91.19)	70.05% (90.83)	71.8% (89.70)
Comex	71.1	69.2% (102.13)	66.9% (110.24)	70.8% (97.63)	72.5% (95.84)
GCarso	72.5	69.2% (102.13)	66.9% (110.24)	70.8% (97.63)	72.5% (95.84)
GMexico	70.8	73.4% (94.03)	74.4% (93.98)	72.8% (93.44)	72.5% (93.96)
Gruma	69.0	71.1% (110.55)	70.8% (108.90)	70.5% (110.13)	75.1% (95.93)

The Error estimation in parenthesis

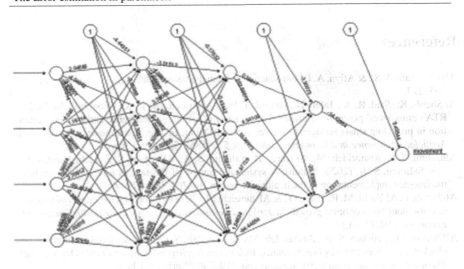

Error: 95.930901 Steps: 9909

Fig. 2 Gruma NN deep learning estimation

5 Practical Implications

The importance of this research is for trading and investment strategy purposes that could be automated given an uncertainty environment (Al-Jarrah et al., 2012; Assad & Alshurideh, 2020a, b; Shah et al., 2020, 2021). Specifically, it is referred to Algorithmic Trading where behavioral patterns are analyzed, modeled, and predicted under mathematical approaches such as Machine Learning techniques (Al Shebli et al., 2021; Alhashmi et al., 2020; AlShamsi et al., 2021; Nuseir et al., 2021; Yousuf et al., 2021).

6 Originality

Most of research on Artificial Intelligence has been done on developed financial markets where liquidity and number of listed companies are higher than financial emergency markets such as the Mexican. It was shown for an emergency financial market that "market prediction" accuracy levels are similar to those obtained in other research works done on mode developed financial markets.

References

Abu-Mostafa, Y. S., & Atiya, A. F., Introduction to financial forecasting. *Applied Intelligence, 6*(3), 205–213.

Al Shebli, K., Said, R. A., Taleb, N., Ghazal, T. M., Alshurideh, M. T., & Alzoubi, H. M. (2021). RTA's employees' perceptions toward the efficiency of artificial intelligence and big data utilization in providing smart services to the residents of Dubai. In *The International Conference on Artificial Intelligence and Computer Vision* (pp. 573–585). Springer, Cham.

Alhashmi, S. F., Alshurideh, M., Al Kurdi, B., Salloum, S. A., Alhashmi, S. F. S., Alshurideh, M., .. & Salloum, S. A. (2020, March). A systematic review of the factors affecting the artificial intelligence implementation in the health care sector. In *AICV* (pp. 37–49).

Al-Jarrah, I., Al-Zu'bi, M. F., Jaara, O., & Alshurideh, M. (2012). Evaluating the impact of financial development on economic growth in Jordan. *International Research Journal of Finance and Economics, 94*, 123–139.

AlShamsi, M., Salloum, S. A., Alshurideh, M., & Abdallah, S. (2021). Artificial intelligence and blockchain for transparency in governance. In *Artificial Intelligence for Sustainable Development: Theory, Practice and Future Applications* (pp. 219–230). Springer, Cham.

Assad, N. F., & Alshurideh, M. T. (2020a). Financial reporting quality, audit quality, and investment efficiency: Evidence from GCC economies. *WAFFEN-UND Kostumkd. Journal, 11*(3), 194–208.

Assad, N. F., & Alshurideh, M. T. (2020b). Investment in context of financial reporting quality: A systematic review. *WAFFEN-UND Kostumkd. Journal, 11*(3), 255–286.

Atsalakis, G. S., & Valavanis, K. P. (2009). Surveying stock market forecasting techniques–Part II: Soft computing methods. *Expert Systems with Applications, 36*(3), 5932–5941.

Avci, E. (2007). Forecasting daily and sessional returns of the Ise-100 index with neural network models. *Dogus Universitesi Dergisi, 2*(8), 128–142.

Bernard, V., Thomas, J., & Wahlen, J. (1997). Accounting-based stock price anomalies: Separating market inefficiencies from risk. *Contemporary Accounting Research, 14*(2), 89–136.

Bezerra, P. C. S., & Albuquerque, P. H. M. (2017). Volatility forecasting via SVR—GARCH with mixture of Gaussian kernels. *Computational Management Science, 14*(2), 179–196.

Chen, H., Xiao, K., Sun, J., & Wu, S. (2017). A double-layer neural network framework for high-frequency forecasting. *ACM Transactions on Management Information systems (TMIS), 7*(4), 11:2–11:17.

Chiang, W.-C., Enke, D., Wu, T., & Wang, R. (2016). An adaptive stock index trading decision support system. *Expert Systems with Applications, 59*(1), 195–207.

Collins, D. W., & Kothari, S. (1989). An analysis of intertemporal and cross-sectional determinants of earnings response coefficients. *Journal of Accounting and Economics, 11*(2–3), 143–181.

Desai, H., Rajgopal, S., & Venkatachalam, M. (2004). Value-glamour and accruals mispricing: One anomaly or two?. *The Accounting Review, 79*, 355–385.

Egeli, B., Ozturan, M., & Badur, B. (2003). Stock market prediction using artificial neural networks. In *Hawaii International Conference on Business* (June 2003) 14.

Fama, E. F. (1991). Efficient capital markets: II. *The Journal of Finance, 46*(5), 1575–1617.

Foster, G., Olsen, C., & Shevlin, T. (1984). Earnings releases, anomalies, and the behavior of security returns. *Accounting Review*, 574–603.

Göçken, M., Özçalıcı, M., Boru, A., & Dosdoğru, A. T. (2016). Integrating metaheuristics and artificial neural networks for improved stock price prediction. *Expert Systems with Applications, 44*, 320–331.

Hirshleifer, D., Low, A., & Teoh, S. (2012). Are overconfident CEOs better innovators? *Journal of Finance, 67*(4), 1457–98.

Huang, W., Nakamori, Y., & Wang, S.-Y. (2005). Forecasting stock market movement direction with support vector machine. *Computers & Operations Research, 32*(10), 2513–2522.

Kara, Y., Boyacioglu, M. A., & Baykan, O. K. (2011). Predicting direction of stock price index movement using artificial neural networks and support vector machines: The sample of the Istanbul Stock Exchange. *Expert Systems with Applications, 38*(5), 5311–5319.

Karaatlı, M., Güngör, I., Demir, Y., & Kalaycı, S. (2005). Hisse Senedi Fiyat Hareketlerinin Yapay Sinir Ağları Yöntemi ile Tahmin Edilmesi. *Yönetim Ve Ekonomi Araştırmaları Dergisi, 3*(3), 38–48.

Kim, K. (2003). Financial time series forecasting using support vector machines. *Neurocomputing, 55*(1–2), 307–319.

Kimoto, T., Asakawa, K., Yoda, M., & Takeoka, M. (1990). Stock market prediction system with modular neural network. *Proceedings of the International Joint Conference on Neural Networks*, pp. 1 ± 6. San Diego, CA.

Kim, K.-J., & Han, I. (2000). Genetic algorithms approach to feature discretization in artificial neural networks for the prediction of stock price index. *Expert Systems with Applications, 19*(2), 125–132.

Kothari, S. P., Leone, A. J., & Wasley, C. E. (2005). Performance-matched discretionary accruals. *Journal of Accounting and Economics, 39*, 163–197.

Kumar, M., & Thenmozhi, M. (2014a). International Journal of Banking. *Accounting and Finance, 5*(3), 284–308.

Kumar, M., & Thenmozhi, M. (2014b). Forecasting stock index returns using ARIMA-SVM, ARIMA-ANN, and ARIMA-random forest hybrid models. *International Journal of Banking, Accounting and Finance, 5*(3), 284–308.

Kumar, D., Meghwani, S. S., & Thakur, M. (2016). Proximal support vector machine based hybrid prediction models for trend forecasting in financial markets. *Journal of Computational Science, 17*(1), 1–13.

Laboissiere, L. A., Fernandes, R. A., & Lage, G. G. (2015). Maximum and minimum stock price forecasting of Brazilian power distribution companies based on artificial neural networks. *Applied Soft Computing, 35*(1), 66–74.

Lahmiri, S., & Boukadoum, M. (2015). An ensemble system based on hybrid EGARCH-ANN with different distributional assumptions to predict S&P 500 intraday volatility. *Fluctuation and Noise Letters, 14*(1), 1550001.

Leung, M. T., Daouk, H., & Chen, A.-S. (2000a). Forecasting stock indices: A comparison of classification and level estimation models. *International Journal of Forecasting, 16*(2), 173–190.

Leung, Y., Mei, C. L., & Zhang, W. X. (200b). Statistical tests for spatial nonstationarity based on the geographically weighted regression model. *Environment and Planning A, 32*, 9–32.

Lev, B., & Thiagarajan, S. R. (1993). Fundamental information analysis. *Journal of Accounting Research, 31*, 190–215.

Manish, K., & Thenmozhi, M. (2005). Forecasting stock index movement: A comparison of support vector machines and random forest. In *Proceedings of ninth Indian institute of capital markets conference, Mumbai, India.*

Nuseir, M. T., Al Kurdi, B. H., Alshurideh, M. T., & Alzoubi, H. M. (2021). Gender discrimination at workplace: Do artificial intelligence (AI) and machine learning (ML) have opinions about it. In *The International Conference on Artificial Intelligence and Computer Vision* (pp. 301–316). Springer, Cham.

Olson, D., & Mossman, C. (2003). Neural network forecasts of Canadian stock returns using accounting ratios. *International Journal of Forecasting, 19*(3), 453–465.

Ou, J. A., & Penman, S. H. (1989). Financial statement analysis and the prediction of stock returns. *Journal of Accounting and Economics, 11*(4), 295–329.

Pai, P.-F., & Lin, C.-S. (2005). A hybrid ARIMA and support vector machines model in stock price forecasting. *Omega, 33*(6), 497–505.

Patel, J., Shah, S., Thakkar, P., & Kotecha, K. (2015). Predicting stock and stock price index movement using trend deterministic data preparation and machine learning techniques. *Expert Systems with Applications, 42*(1), 259–268.

Shah, S. F., Alshurideh, M., Al Kurdi, B., & Salloum, S. A. (2020). The Impact of the behavioral factors on investment decision-making: a systemic review on financial institutions. In *International Conference on Advanced Intelligent Systems and Informatics* (pp. 100–112). Springer, Cham.

Shah, S. F., Alshurideh, M. T., Al-Dmour, A., & Al-Dmour, R. (2021). Understanding the influences of cognitive biases on financial decision making during normal and COVID-19 pandemic situation in the United Arab Emirates. *The Effect of Coronavirus Disease (COVID-19) on Business Intelligence, 334*, 257–274.

Tay, F. E., & Cao, L., Application of support vector machines in financial time series forecasting. *Omega, 29*(4), 309–317.

Wang, J.-J., Wang, J.-Z., Zhang, Z.-G., & Guo, S.-P. (2012). Stock index forecasting based on a hybrid model. *Omega, 40*(6), 758–766.

Weng, B., Ahmed, M. A., & Megahed, F. M. (2017). Stock market one-day ahead movement prediction using disparate data sources. *Expert Systems with Applications, 79*(1), 153–163.

White, E. N. (1990). The Stock market boom and crash of 1929 revisited. *The Journal of Economic Perspectives, 4*(2), 67–83.

Xiao, Y., Xiao, J., Lu, F., & Wang, S. (2013). Ensemble ANNs-PSO-GA approach for day-ahead stock e-exchange prices forecasting. *International Journal of Computational Intelligence Systems, 6*(1), 96–114.

Yoon, Y. & Swales, G. (1991). Predicting stock price performance: a neural network approach. *Proceedings of the Twenty-Fourth Annual Hawaii International Conference on System Sciences,* vol. 4, pp. 156–162. https://doi.org/10.1109/HICSS.1991.184055.

Yousuf, H., Zainal, A. Y., Alshurideh, M., & Salloum, S. A. (2021). Artificial intelligence models in power system analysis. In *Artificial Intelligence for Sustainable Development: Theory, Practice and Future Applications* (pp. 231–242). Springer, Cham.

Zhang, N., Lin, A., & Shang, P. (2017). Multidimensional k-nearest neighbor model based on EEMD for financial time series forcasting. Ph*ysica, A 477*(1), 161–173, 309.

Zhong, X., & Enke, D. L. (2017). Forecasting daily stock market return using dimensionality reduction. *Expert Systems with Applications, 67*, 126–139, Elsevier, Jan 2017. The definitive version is available at https://doi.org/10.1016/j.eswa.2016.09.027.

Machine Learning Price Prediction During and Before COVID-19 and Consumer Buying Behavior

Tauqeer Faiz⬛, Rakan Aldmour⬛, Gouher Ahmed,
Muhammad Alshurideh⬛, and Ch. Paramaiah

Abstract In the unprecedented situation of COVID-19, the global economy has turned upside down. This has led to sudden and unprecedented pressures on products demand and price forecasting. The study utilized regression techniques to predict product prices during and before COVID-19 using multiple influencing factors such as increase of COVID-19 positive cases on daily basis, number of deaths on a particular day, and government restrictions level. The data was gathered from worldometers website and combined with local store on sales based on the date. The results were eye opening as the product sold in the months of Mach, April, May and June 2020 were different than last year. This means the customers buying habits were totally altered due to many reasons such as job loss, wages reduction due to remote working, or promotions. Moreover, these products prices were directly proportional to increase of new COVID-19 cases, rise of daily deaths and government restriction levels imposed during the pandemic. The study uses machine learning data mining algorithms such as Logistic regression (LR), Decision Tree, Random Forest and K-Nearest Neighbor. Decision Tree and Random Forest works best in the pandemic situation to predict

T. Faiz (✉)
School of Information Technology, Skyline University College, University City of Sharjah, Sharjah, UAE
e-mail: tauqeer.khan@skylineuniversity.ac.ae

R. Aldmour
School of Computing and Digital Technologies, Staffordshire University, Stoke-on-Trent, UK

G. Ahmed
School of Business, Skyline University College, University City of Sharjah, Sharjah, UAE

M. Alshurideh
Department of Marketing, School of Business, The University of Jordan, Amman, Jordan
e-mail: m.alshurideh@ju.edu.jo; malshurideh@sharjah.ac.ae

Department of Management, College of Business Administration, University of Sharjah, Sharjah, UAE

Ch. Paramaiah
School of Business, Skyline University College, Sharjah, UAE

© The Author(s), under exclusive license to Springer Nature Switzerland AG 2023   1845
M. Alshurideh et al. (eds.), *The Effect of Information Technology on Business and Marketing Intelligence Systems*, Studies in Computational Intelligence 1056,
https://doi.org/10.1007/978-3-031-12382-5_101

product price as compared to Logistic Regression and KNN. However, different outcomes were recorded when comparing the sales during pandemic and before pandemic.

Keywords Price prediction · Regression analysis · Covid-19 sales · logistic regression · Decision tree · Random forest · k-nearest neighbor

1 Introduction

The COVID-19 pandemic has turned the global economy upside down (Akour et al., 2021; Alshurideh et al., 2021; Nuseir et al., 2021a). The COVID-19 crisis is exerting sudden and unprecedented pressures on products demand and price forecasting (Al-Dmour et al., 2021a; Shah et al., 2021; Taryam et al., 2020). Many industries, starting from overseas travel to consumer goods have dropped in demand. Consumers are contracting for enormous discounts and asking for renegotiations (Abbas, 2020; Al-Dmour et al., 2021b; Aljumah et al., 2021). Dubai Economy and Trade licenses authority has confirmed increase of 179% issuance of online shopping trade licenses as compared to last year. Companies have to put their energy and resources to re-invent the ways to look into the market demand and price forecasting during this unprecedented and uncertain times to expand their business (Abdelnour et al., 2020; Al-Jarrah et al., 2012; Alzoubi & Inairat, 2020).

The aim of this study is to find out most selling products during the pandemic, see the change in product price over the period of four months, predict prices of most selling items in normal days as compared to products which have been sold during the pandemic.

2 Research Background

This research project compares the effect on product prices before and during the pandemic. Small to medium enterprises (SME's) deploy resources to understand the market situation based on collected data to uncover the best deals for consumers (Al Suwaidi et al., 2020; Al-Gasaymeh et al., 2020; Assad & Alshurideh, 2020). Product prices are dependent on various other industries e.g. oil, real estate and other interlinked industries. US crude oil has gone negative for the first time in the history. An overflow of unwanted oil in the market caused US Oil price to plunge to almost − $40 a barrel (Ambrose, 2020). This focuses on looking at product demand during the pandemic and predict product prices and compare this result with the data available before the pandemic. Ministry of human resources and emiratization (the "MoHRE") has declared it legal to reduce employee salary until the COVID-19 crisis prevails (Barr, 2020). Organizations have reduced employee's salaries ranging from 20 to

35% in the UAE. There is no proper statistics on actual salary reduction as it varies from company to company. This makes consumer's buying power very limited and stress upon daily or weekly re-pricing of products.

Therefore, this research investigates "Re-pricing of products" during the COVID-19 pandemic as price prediction or forecasting has been investigated in best case situation so far. Nevertheless, predicting product prices in a crisis like COVID-19 has not been described in the literature. This article reports the internal and external factors involved in Re-pricing of products during the pandemic or emergency like COVID-19 and sales comparison will be made before the pandemic.

3 Aims and Objectives

The study will focus on analysing data mining techniques used to predict product prices during COVID-19, design and develop model based on the requirements, test the model accuracy, and deploy the model to the industry. Finally, this research will compare the results of findings before and during the pandemic. The overall research objective is to apply regression data mining technique to study the effects of prices on certain products during and before COVID. To accomplish this goal, the report will focus on the following major objectives:

RO1[1]: Obtain data and apply data mining techniques such as Logistic Regression, Decision Tree, Random Forest and K-Nearest Neighbour for price prediction during and before the emergency situation like COVID-19.

RO2: Determine how much of the variation in the dependent variable can be explained by the independent variables. In other words, this defines the strength of the relationship.

RO3: Build and Assess model accuracy based on the findings and suggest which model performs the best based on the given data.

RO4: Discuss the outcomes of different models.

Regression analysis is most widely used for forecasting, and this requires an expert understanding and readiness. Previous work on price forecasting contains all the required information to pursue with this report. However, researchers have focused on biological or health issues during H1N1 and Swine pandemics rather impact on pricing in unprecedented situation like COVID-19. Online collaboration required at each development phase with the SMEs to consider all important factors of pricing.

---

[1] Research Objectives.

4   Literature Review

The purpose of this section is to discuss published work on price forecasting during the pandemic. Keeping this information in mind, different repositories of research articles were interrogated. There is adequate amount of research done in price forecasting, however, forecasting during the pandemic has never been carried out, which clearly indicates the research gap.

Support vector regression (SVR) against Long short-term memory (LSTM) neural network algorithms was used by Bakir et al. (2018). In their comparison study of SVR and LSTM, they introduced a framework to predict mobile phone prices in European shops. According to Bakir et al. (2018), better prediction performance was obtained with the use of multivariate models of SVR and LSTM neural networks. In fact, the Root mean squared error (RMSE) of €33.43 was recorded with SVR model but accurate prediction results were obtained using multivariate LSTM RNN model for next days' cost with an RMSE of €23.640. However, LSTM and SVR time series models work best in price prediction, but the authors have not shared any information of external factors effecting the product prices. In case of mobile phones, the stock can last for years but this will really change the discussion when items expiry date is taken into consideration.

Researchers studied the dynamic price forecasting in Simultaneous Online Art Auctions and investigated two questions: The first question referred to the function of price dynamics in predicting SOA and how it varied from the individual-auction case. The second question probed why the inclusion of price dynamics resulted in superior prediction, and they examined the role of bidder competition and its relation to price dynamics (Dass et al., 2010). The writers used functional data analysis modelling approach which is an emerging statistical methodology that works on functions observations for example price curves in online auctions. (Chen & Fan, 2010; Dass et al., 2010; Elgerd, 1982) had proposed FDA-based an alternate model to the Bass model for price forecasting in 2008. According to Dass et al. (2010) and (Elgerd, 1982), FDA belongs to Fuzzy Logic and Neural Network family as it mimics the continuity processing performed by the human brain. In their research contribution, they developed two main dynamic prediction models, DFM-I and DFM-II for price forecasting. In DFM-I, it incorporated both the price and the price velocity information until the time of prediction, whereas the second model (DFM-II) considered only the price path till the time of prediction. Both these models included static pre-auction information for example the opening price or the item characteristics). DFM-I outperformed all competing models in terms of predictive accuracy. Forecasting price dynamics in particular the price velocity has a major impact on forecasting price, statement made in their first conclusion. Though the context of "An effective pricing framework during COVID-19" is very much aligned with "Dynamic Price Forecasting in SOA", however, the most important factor of re-pricing during the pandemic situation and how other factors impact the product prices still remains buried. This article will primarily focus on the relationship of external factors on predicting prices during the virus situation.

The literature work discussed of Elgerd (1982)various stock price prediction techniques. Opinion mining or Sentiment Analysis in other words, was used to predict stock prices by using historical data. The prediction model training was based on messages which was grouped in to five classifications such as "Strong Buy", "Buy", "Hold", "Sell" and "Strong Sell". All other irrelevant data was discarded, and the purpose of this method was to check how human mood can be used to predict the stock. Support Vector Machine (SVM) has been used a lot for high dimensional data for time-series analysis due to its performance and the authors of this article tested this model for its popularity.

Lab of tropical crops information technology China (Mucedola, 2021) predicted price of vegetables using Back propagation (BP) neural network algorithm. Back Propagation NN has multilayer feed forward networks being trained by the error back propagation algorithm. The BP model contains input layer, hidden layer and output layer, with key attributes of which is that information transfers forward, and the error back propagation. The study revolved around the green pepper price prediction, which is produced locally, and its price fluctuated in recent years. The price is affected due to seasonal production, May to October each year is considered as off-season and busy season remains from November to April. Due to spring festival, the demand of green pepper becomes high during Mach and April. The article also indicates the change of price due to high temperature and rain fall. The price drop due to these external factors are negligible in the current scenario. In order to determine "Re-pricing" due to pandemic, it is necessary to analyze all the relevant impact factors to forecast the product prices. This research article will focus on predicting prices during the pandemic.

OSEMN framework covers all the steps from gathering raw data all the way up to analyzing data and building dashboards or reports (Faiz, 2019). This model contains the framework as (O)—Obtain raw data, (S)—Scrubbing or cleaning data, (E)—Exploring or Visualizing data to search patterns or trends, (M)—Modelling data for prediction, (N)—Interpret the results. Any project related to machine learning or data analytics always follow the given steps. "An effective Pricing framework during COVID-19" will be using the similar techniques discussed in "Scrubbing data for medium-sized enterprises" to pre-process data to make it fit for use. Data pre-processing techniques, finding outliers, and removing irrelevant information will be followed according to the framework mentioned in this article.

The airfare price prediction using machine learning techniques provides the best time for customer to purchase an air ticket before it gets increase (Elgerd, 1982). According to this technique, the most informative features such as departure time, arrival time, free luggage, days left until departure, number of transits, holidays, overnight flights and day of the week play a vital role in ticket prices (Elgerd, 1982). Eight different machine learning regression models were selected which includes, multilayer perception (MLP), generalized regression neural network, extreme learning machine (ELM), random forest regression tree, regression tree, bagging regression tree, regression SVM and linear regression. The historical data was gathered of a particular airline, and experimental models have shown that machine learning prediction models brought satisfactory results. They concluded that

"Regression Tree", "Random Forest Regression", and "Bagging Regression Tree" machine learning models brought accurate results while "Random Forest Regression" and "Regression Tree" outperformed considering the execution time.

There are numerous research articles based on price prediction using machine learning algorithms. According to Elgerd (1982), Regression Tree, Random Forest and Bagging Regression works the best to forecast airfare prices. The LSTM and SVR were compared (Elgerd, 1982) against the mobile phone prices prediction and they obtained best accurate results with these models. In a nutshell, articles reviewed in this research have used SVR, Random Forest and LSTM brought accurate results as compared to Nearest Neighbour and other similar machine learning algorithms. However, there is no single machine learning technique which works best with all kinds of data. This publication recommends that some other popular regression analysis techniques such as Multinomial Logistic Regression, Decision Tree, Random Forest and KNN should be used and evaluated for large amounts of data with multiple categorical fields. As the price prediction during the pandemic or emergency situation has not been explored before, so, certain parameters which affects the pricing were included such as promotions, new COVID cases, new daily deaths and restriction levels. Worldometers website provides covid information on daily basis of all the countries and the sales information were obtained from a well-established supermarket. The two data sets were combined together based on date column. There were almost 50,000 data rows which were sufficient to be used for model building. The data mining model accuracies were improved by discarding the irrelevant and multicollinear columns from the data and the results were compared against pandemic vs non-pandemic situation.

5 Methodology

This section focuses on data analytics project lifecycle and actual implementation of the problem discussed in aims and objectives of Sect. 3. Due to the unprecedented time of COVID-19, ecommerce has observed price fluctuations in different products. As mentioned in Sect. 1, customer's purchasing habits are shifted to selection based due to its limited buying power. Most of the business entities relies on price prediction framework to run their businesses. This data science project research is illustrated in the following Fig. 2, Research Methodology. As the past data will be collected and analyzed, the research will be a quantitative analysis. The whole research process will involve finding the most selling products in the month of March, April, May, and June and predicting the product prices during and before COVID-19. Regression data mining techniques such as Logistic Regression, Decision Tree, Random Forest and KNN will be applied, and the model outcomes will be compared and generalized. Descriptive analysis answers the "what happened" by summarizing past data, usually in the form of dashboards. In this research first we will create a dashboard to see most selling products using Microsoft Power Business Intelligence tool or Python coding can also help to show graph for most selling products. The data will be imported to Power BI tool and a dashboard will be designed based on the requirements. Following reports will be generated using this tool:

- Top 5 most selling products (Sect. 3—Aims and Objectives), Frequency of change of Unit_price during the COVID-19 when there is increase of new cases diagnosed as COVID positive (Sect. 3—Aims and Objectives), Frequency of change of Unit_price of the most selling products before the COVID-19
- Filtering data to required months (i.e. March, April, May and June 2020)

The second type of analysis is the Predictive Analysis, and it answer the question "what is likely to happen". This technique utilizes previous data to make predictions about the future sales. Predictive analytics relies on statistical modelling to forecast. The accuracy of prediction depends upon the data quality and sufficient training dataset.

In this supervised machine learning problem, "Logistic Regression", "Random Forest", "Decision Tree" and "K-Nearest Neighbour" will be applied and their results will be recorded. The focus of this study is on data mining regression techniques, so the research approach of this study is quantitative. Secondary data (data from business or data resided in computerized databases) will be collected from the e-commerce store and will be used in this research. The Worldometers website provides COVID-19 cases on daily basis. A CSV file related to COVID cases is downloaded and will be merged with ecommerce file. The reason to merge COVID new cases and deaths information is to find a relationship among product Unit_price and daily new cases or new deaths. The purpose of this research is to explore the effect of product prices when new cases or deaths are increased, and the research approach is quantitative. Another most interesting feature of Government imposing restrictions to be followed during the pandemic will also be merged with the excel sheet.

The objective of this research is to understand change of certain product prices when the COVID positive cases are increased. In this study, the data science project methodology has been followed. There are several steps in data mining price prediction. In order to implementing these methods, Python machine learning libraries such as pandas, scikit learn, matplotlib, seaborn and sklearn will be used. Microsoft Excel and Power BI is also required for column addition and dashboard creation. All outliers will be checked and removed to pursue the analysis following the techniques mentioned in article "Multiple approaches on scrubbing data for medium sized enterprises" (Faiz, 2019).

According to Beklemysheva (2019), Tziridis et al. (2017), AlHamad et al. (2021), Nuseir et al. (2021b), Salloum et al. (2020), Ghazal et al. (2021), Alshurideh et al. (2020), Al Kurdi et al. (2021) implementing machine learning solutions using python is time saving as python offers number of frameworks and libraries. Python provides extensive libraries for machine learning such as data analysis, visualizing and provides machine learning algorithms. Model building is based on some certain steps as shown in Fig. 1.

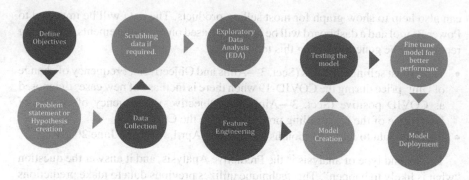

Fig. 1 Supervised learning—model building

6 Findings and Analysis

This section presents the results, and analysis relating to the prices of selected products during and before COVID-19 which is the aim of this research. Data Science project lifecycle methods is followed before building any model as mentioned in model building Sect. 5. Some fancy python coding is used to build the model. Let us understand the pandemic data used in this research first.

The dataset is obtained from a renowned E-Commerce store in UAE and is used for research purposes only. The company name will not be revealed due to non-disclosure agreement. The obtained data consists of 51,308 individual data rows. In this dataset, New_cases, New_deaths and Restriction_level columns are manually added using Microsoft Excel. The COVID-19 cases information is updated on regular bases on worldometers.info site. The site allows to download more than 200 countries daily COVID cases as a CSV file. The file was downloaded from the website and only UAE COVID cases were filtered from the file and merged to the sales report using Microsoft Excel according to the "Order_date".

There are 17 columns in the dataset, which are described below and a sample is shown in Fig. 2:

- **Order ID**—Customer order is given a unique id and it is auto-generated by the system at the time of purchase.
- **Order date**—It lists the date of purchase in 'yyyy-mm-dd' format.

```
# Loading Excel Sheet data - Sales during COVID-19
data = pd.read_excel('/content/covid_sales.xlsx', sheet_name='covid_sales')
data.head()
```

	Order_ID	Order_date	Day	Month	Product_category	Product	Type	Unit_price	Quantity	Promotion	Customer_ID	Customer_name	City	Country	New_cases	New_deaths	Restriction_level
0	AE-2020-209	2020-03-01	1	3	General	Mask	N95	450	46	1	DS-00209	Dixon Childs	Abu Dhabi	UAE	2	0	Level-1
1	AE-2020-2487	2020-03-01	1	3	General	Thermometer	Contactless	350	76	1	LL-002487	Rose Connell	Ajman	UAE	2	0	Level-1
2	AE-2020-1064	2020-03-01	1	3	General	Mask	N95	450	51	1	KA-001064	Deleon Pisteka	Abu Dhabi	UAE	2	0	Level-1
3	AE-2020-366	2020-03-01	1	3	General	Sanitizer	1000ML	55	64	1	AN-00366	Mack Hoffman	Dubai	UAE	2	0	Level-1
4	AE-2020-402	2020-03-01	1	3	General	Sanitizer	50ML	10	8	1	ER-00402	Drake Macallister	Dubai	UAE	2	0	Level-1

```
[112] data.shape
(51298, 17)
```

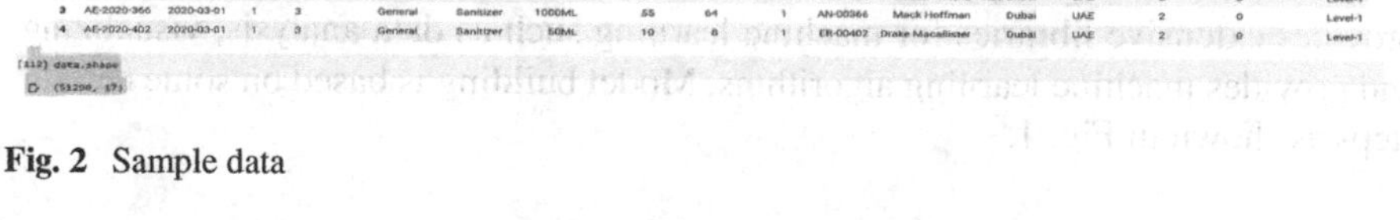

Fig. 2 Sample data

- **Day**—Day is obtained through Order date column and display the day of the month e.g. 1 represent first day of the month. The column has range of values from 1–31.
- **Month**—This column converts Order date to month and only month information is retained. The proposed dataset has month of March, April, May and June values.
- **Product Category**—It holds the product categories and has 4 distinct values. (1) General, (2) Fashion, (3) Fruits & Vegetables, (4) Electronics
- **Product**—One of the most useful columns and it consists of names of the products. The column has 34 unique products in the list.
- **Type**—Product is divided into further categories e.g. Mask has N95 and Surgical types.
- **Unit_price**—Actual price of the product and it's a numerical field.
- **Quantity**—Number of items ordered by the customer.
- **Promotion**—This tells whether the product is on promotion, 0 means product is sold on normal price, 1 means product is sold on offer price.
- **Customer ID**—This field contains customer id information.
- **Customer Name**—The column saves customer name.
- **City**—Name of the city where the product is delivered.
- **Country**—The column has single value as data is based on United Arab Emirates.
- **New_cases**—This field is inserted manually to establish relationship between number of products sold when the cases were increased. The data is obtained from worldometer.com website.
- **New_deaths**—This field lists the number of deaths on a particular day recorded during the COVID-19. This field is also taken from worldometer.com and merged based on the dates.
- **Restriction Level**—This field may affect the Unit_price as the Government of UAE was imposing different restrictions alerts on the public to contain the spread of COVID-19. The values are populated in the excel sheet based on the Ministry of Health and Emergency Crisis & Disasters Committee for COVID-19 pandemic announcements on different dates. The data is obtained from UAE ministry of health website (mohap.gov.ae). This field has five distinct alert levels which includes:
  - **Level-1** allows normal activities with precautions; **Level-2** requires physical distancing and restrictions on leisure and social activities.
  - **Level-3** applies restrictions on many activities including at workplaces and socially, to address a high risk of transmission; **Level-4** Extreme precautions to limit community transmission and outbreaks, while allowing some activities to resume whereas **Level-5** Drastic measures to contain the spread of the virus and save lives.

From the above datapoints, "Unit_price" will be dependent (target) variable and "Day", "Month", "Product", "Type", "Quantity", "Promotion", "New_cases", "New_deaths" and "Restriction_level" are the independent attributes.

First, let us understand the pricing changes during the COVID-19. Comparison is made on pure on-demand product pricing during the month of March, April, May

Fig. 3 Quantity sold

Month	Quantity			
Product	3	4	5	6
Dragon_fruit	624	1211	2153	1533
Gloves	3807	2285	3984	1620
Mask	7140	6367	7366	10706
Sanitizer	10874	10282	12732	13674
Thermometer	2990	2372	3950	4613
Vitamin_C	1523	7190	8356	8365
Water_melon	1284	181	265	87

and June 2020 as this was the peak time of COVID-19 pandemic. Figure 3 Sales Report offers several useful information to simplify the analysis report.

The significant change of "Unit_price" reflects that the price of Sanitizer was shifted 24 times during these four months, similarly, Vitamin_C had been modified 16 times, Mask prices 13 times, Gloves 4, Thermometer 6, Green_leaves 5, Hair_clipper 2 and Samsung Mobile 2 times. Next important feature of interest in the given dataset is "Quantity" sold during the pandemic. The Fig. 3, discovers that Sanitizers are at the top with around 47 K units sold, Mask holds the second place with 37 K, Vitamin_C is the third most demanding product, Thermometer, Green_leaves and Gloves are also among the top selling products category during this time.

The following two figures (Fig. 4—Change of Masks and Sanitizers Prices) are evident to see the change of Mask and Sanitizer prices during the month of March, April, May and June. The Vitamin_C, Thermometer, Hair_clipper and Gloves depicts the similar kind of behavior in these four months.

Sanitizer, Mask, Vitamin_C, Contactless Thermometer, Gloves, and Green_leaves are most selling products found in the given dataset as shown in the Fig. 5—Top Selling Products in COVID-19.

Monthly quantity sold of different products statistics is shown below. Dragon_fruit is included in this study because this is the top season for Dragon_fruit to be sold in the market. However, the demand of Dragon_fruit has gone down which can be proved when considering the data before pandemic.

Another useful graph that shows the distribution of quantitative data to facilitate comparisons between variables is box plot. The box depicts the quartiles of the dataset whereas the whiskers reflect the rest of the distribution. The box plot also known as box and whisker chart is a way of displaying the distribution of data based on the minimum, first quartile, median, third quartile and maximum.

Outliers can be seen in case of Vitamin_C and Thermometer in Fig. 4 Boxplot. However, this will not be removed as the Unit_price has been changed with respect to different factors. A violin plot is like box plot, except that violin plots show probability density of the data at different values. Some box plots and violin charts pertaining to the dataset are shown in Figs. 6 and 7.

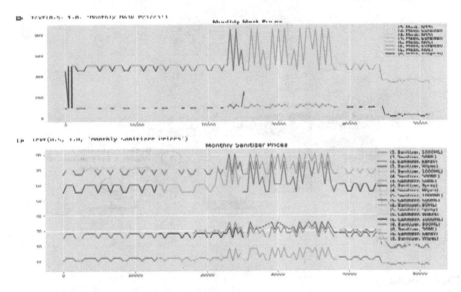

Fig. 4 Change of masks and sanitizers prices

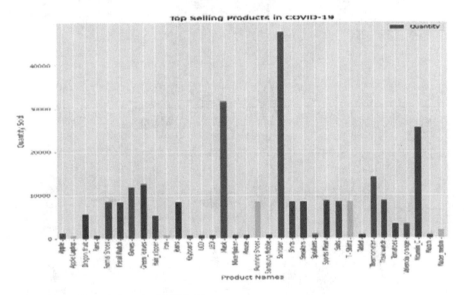

Fig. 5 Top selling products in COVID-19

Unit_price of Mask and Thermometer seems to be spread out in the above violin graph. The value of Unit_price ranges from 5 to almost 850.

Product Mask sub-category N95 and Surgical Masks prices are varying with respect to time, Similarly, Vitamin_C 20Tab bottle and Contactless Thermometer prices have changed slightly as shown in the Fig. 8 Violin Chart.

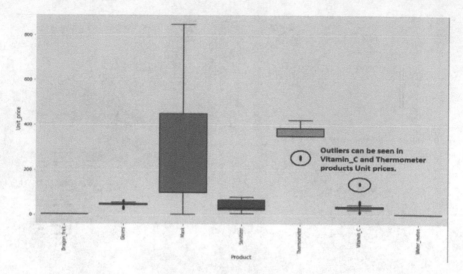

Fig. 6 Boxplot

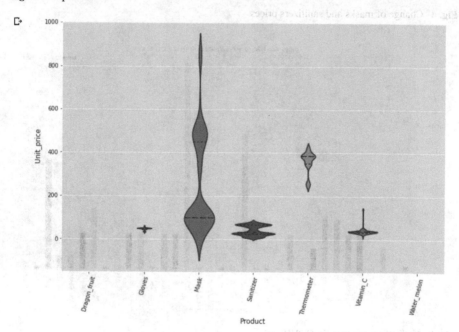

Fig. 7 Product prices during COVID

Unit price of Mask and Thermometer seems to be spread out in the above violin graph. The value of Unit price ranges from 0 to almost 850.

Product price sub examples Mask and surgical Masks prices are varying with respect to time, similarly Vitamin C, Dragon fruit, and Container less Thermometer prices have changed slightly as shown in the Fig. 8 Violin Charts.

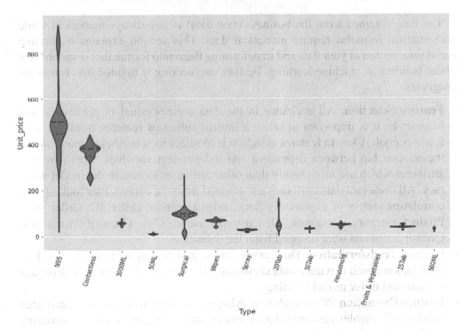

Fig. 8 Violin chart—product sub-categories

The Swarm plot Fig. 9, displays the different unique values of the Unit_price of selected products. According to this graph, 12 different masks prices are recorded, and other products prices are self-explanatory.

```
[ ] sns.swarmplot(x=data.Unit_price, y=data.Product, data=data)
```

```
<matplotlib.axes._subplots.AxesSubplot at 0x7f6b6c668b00>
```

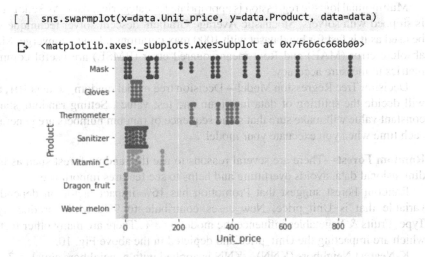

Fig. 9 Swarm plot during COVID-19

The data obtained from the business (raw data) is not always numerical while mathematical formulas require numerical data. This section explains on finding useful information in your data and transforming them into format that is suitable for model building in machine learning. Feature engineering is divided into following categories:

- **Feature Selection**: All attributes in the data are not equal in preference and magnitude. It is important to select a limited subset of features from the large features pool. A key to features selection is to select such attributes which shows the relationship between dependent and independent variables. There are few attributes which are more handy than other attributes to ensure the model accuracy. All irrelevant data attributes are dropped from the dataset after looking into correlation matrix of exploratory data analysis section. Order_ID, Order_date, Product_category, Customer_ID, Customer_name, City, Country, Month and Quantity columns were dropped from the dataset.
- **Feature Transformation**: This covers scaling and dealing with null values. Log functions are used to reduce data skewness. The data used in this research project is clean and fit for model building.
- **Feature Extraction**: When dealing with huge data, it requires processing and large number of variables consumes lot of processing power and memory. Therefore, duplicate data should be removed, or dimensionality of such variables should be removed.

As of now, there is a strong correlation among dependent and independent variables and it's time to build the logistic regression models in SKLearn. Logistic Regression model in machine learning is useful for discrete dependent variables as well as categorical variables.

Multinomial logistic regression is appropriate for categorical data. As Scikit-learn is shipped with solvers, Stochastic Average Gradient descent solver technique will be used as it is fast for big dataset with 1000 max iterations taken to converge. Mean absolute error (MAE) and Root mean squared error (RMSE) are useful common metrics to measure accuracy.

Decision Tree Regression Model—Decision tree model random_state is 101, this will decide the splitting of data into train and test values. Setting random_state a constant value will make sure that same sequence of random numbers are generated each time when you execute your model.

Random Forest—There are several reasons to use the random forest such as high dimensional data, avoids overfitting and helps to see features importance.

Random Forest suggest that Promotion has 16% impact factor on dependent variable that is Unit_price, New_cases contribute to 9%, while product type Type_Fruits & Vegetables influence the model by 8%. There are many other factors which are impacting the Unit_price and depicted in the above Fig. 10.

K-Nearest Neighbors (KNN)—KNN is applied with n_neighbors equal to 3.

Finally, the accuracy graph shown below Fig. 11, depicts the model's accuracies. According to the sample data fed into these models, the Decision Tree and Random

```
feat_importances = pd.Series(model_rf.feature_importances_, index=x.columns)
feat_importances.nlargest(25).plot(kind='barh',figsize=(10,10))
```

<matplotlib.axes._subplots.AxesSubplot at 0x7effbd06e518>

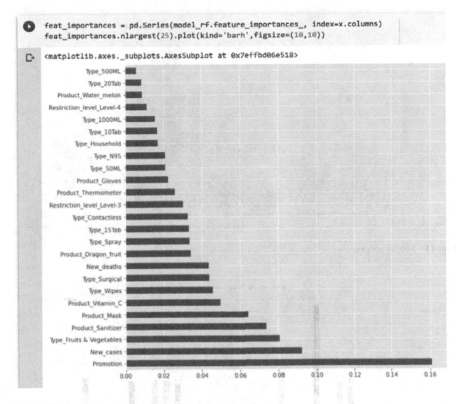

Fig. 10 Feature importance

Forest brings the highest accuracy level of 99%, whereas KNN is 83% and the least accurate model is the Logistic Regression in the current scenario.

It's time to understand data before pandemic. Similar approach is followed to understand the data before COVID-19. The top 5 most selling products before the pandemic are Green_leaves, Jeans, Titak watch, Sports Wear, and Formal shoes as shown in Fig. 12. To compare the similar product prices sold during the COVID-19, all other products are dropped for the analysis and the price distribution before COVID-19 is shown in (Fig. 13).

The product "Mask" of "N95" and "Surgical Mask" types prices are shown in the graph for the months of March, April, May and June 2019. The values displayed in the Fig. 15 below are stable throughout these three months and no noticeable changes are recorded.

The correlation matrix in Fig. 14, highlights the positive and negative correlation between quantity and moth variables.

The Swarm plot in Fig. 15 indicates the different prices of products in the given dataset which appears to be constant. The OLS matrix in Fig. 14 below lists the R-squared value and P value which is appropriate for this dataset.

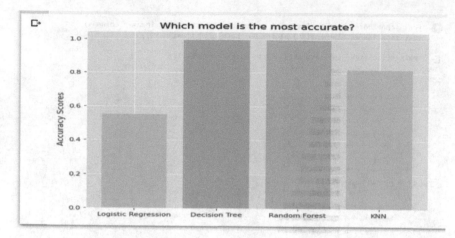

Fig. 11 Model accuracies

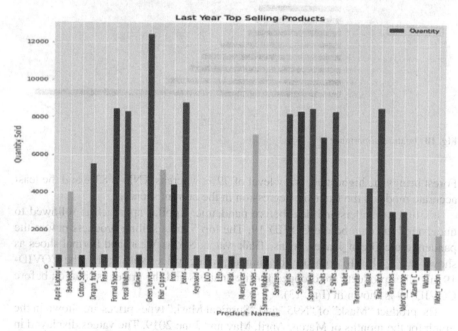

Fig. 12 Last year top selling products

The data is split into training and test sizes, 70% of the data will be used for training purpose and remaining 30% would be considered for testing.

Models' accuracy Scores—After passing this data to Logistic Regression, Decision Tree, Random Forest and KNN, model accuracies were recorded as follows.

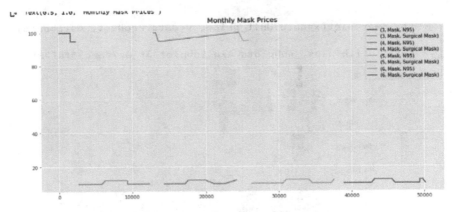

Fig. 13 Price distribution before COVID-19

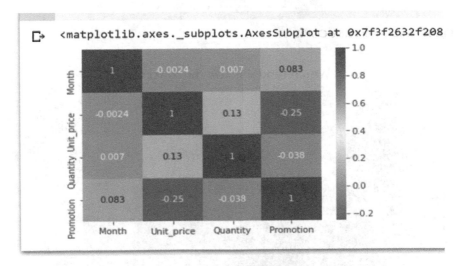

Fig. 14 Correlation matrix

All the four models are predicting almost the same and their predictions scores are identical as shown in the Fig. 16.

7 Results Comparison and Recommendations

In this study, logistic regression, decision tree, random forest, and knn regression models were applied to the collected dataset, and the following result comparisons are made.

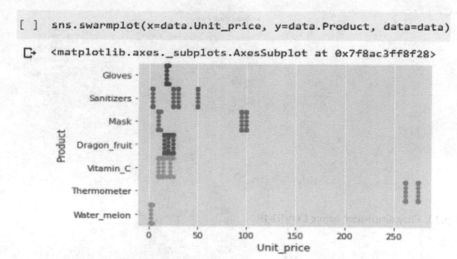

```
[ ]  sns.swarmplot(x=data.Unit_price, y=data.Product, data=data)

[>   <matplotlib.axes._subplots.AxesSubplot at 0x7f8ac3ff8f28>
```

Fig. 15 Swarm plot before COVID-19

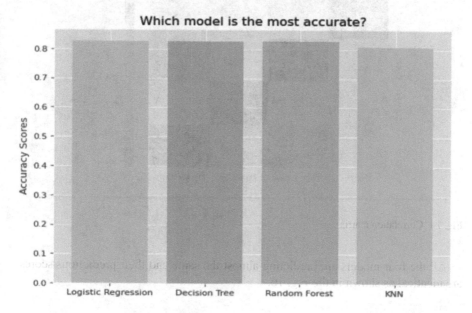

Fig. 16 Model accuracies before COVID-19

Starting with the features used in both datasets (during and before covid-19) discussed in Sect. 4, there were 17 features used during the pandemic and 14 features were used before covid-19. Addition of these extra four columns during the pandemic is explained and supported by arguments in previous sections. The New_cases, New_deaths, Restriction_level, Promotion, Quantity, Months and Promotion are the most important features influencing the Unit_price. Both datasets share different outcomes in terms of most selling products. During the COVID-19, the most selling products were "Mask", "Gloves", "Sanitizers", "Contactless Thermometers", "Vitamin_C", and "Green leaves", whereas before the Covid-19 factor, the top selling products were "Green leaves", "Jeans", "Titak watch", "Sportswear" and "Formal shoes". Comparing and contrasting these two different outcomes support the initial argument mentioned in Sect. 1 that the consumers buying habits were changed and customers were buying the products as per the situation. As there were no physical activities outside the house, the sales of footwear, clothing and other leisure products were gone down drastically.

During the pandemic, some products prices were dependent on certain conditions. The Swarm plots shown above are evident that the Mask and Sanitizer prices were changed due to certain conditions while it remained constant before the pandemic. According to the regression models used in last section, prices of mask, sanitizers, gloves, vitamin_c and thermometer were changed (swarm plot) due to the following impact factors:

- Increase in covid-19 positive cases
- Number of daily deaths increased during the pandemic
- When Government was imposing restrictions to follow health and safety precautions
- Product demand was increased

The two correlation matrixes in Table 1 tabulate the relationship between dependent and independent variables. There is a slight relationship between dependent and independent variables however, multicollinearity problem exist in the dataset as independent variables are correlated as well. New_deaths and New_cases are strongly correlated independent variables in this scenario. As the focus of this analysis is to find the effect of COVID-19 impact factors on Unit_price of different products, so discarding these attributes will not serve the purpose. A similar kind of dilemma has been noticed in the dataset used for before COVID-19.

Table 1 Correlation matrixes (During & Before COVID-19)

| Correlation Matrix – During COVID-19 | Correlation Matrix – Before COVID-19 |

Table 2 Features importance

	During COVID-19		Before COVID-19	
	Random forest		Random forest	
1	Promotion	16%	Promotion	29%
2	New_cases	10%	Dragon_fruit	14%
3	Fruits & Vegetables	8%	Fruit & Vegetable	8%
Decision tree			*Decision tree*	
1	New_cases	24%	Dragon_fruit	24%
2	Fruit & Vegetables	24%	Thermometer	16%
3	Promotion	18%	Promotion	15%

The top three important features which are influencing the model are listed in the following Table 2 based on the two datasets. According to the Random Forest, the top feature impacting the model is Promotion field during the COVID-19 and before the COVID-19, whereas Decision Tree has two variants at the top level which is New_cases in the COVID-19 situation and Dragon_fruit before the pandemic. The second most feature influencing the model is New_case and Dragon_fruit in Random Forest with the values of 10% and 14% during and before COVID respectively. The Decision Tree's second most impacting features are Fruits & Vegetable and Thermometer during and before the COVID-19 with distribution of 24% and 16% accordingly. Promotion comes at number three with the values of 18% and 15% influencing the DT model during and before the pandemic. Here, only top 3 features are discussed.

Four regression techniques were applied to the two datasets as discussed in Sect. 4. The datasets were split into training and testing purposes. The training set data was set to 70% to build the model while remaining 30% data was used for model evaluation. All categorical variables were converted to dummy variables using panda's built-in library. Table 3 displays the accuracy scores obtained when applying Logistic Regression, Decision Tree, Random Forest and KNN to the data tables. Accuracy evaluation metric is used to determine how good our model is, based on dependent and independent variables. During the COVID-19 situation, Logistic regression accuracy scores are 55.1% which means the model has been failed to produce the accurate predictions. KNN accuracy scores are 81.8% which is fairly good. On the other hand, Decision Tree and Random Forest predicted 99% accurate results and standout from the rest.

Before COVID-19, the models' predictions are almost same as the number dependent variables used in the dataset was fairly simple and straightforward. That is why the differences are not significant in these four models. According to the Table 3, first three models (LR, DT and RF) have similar 82.5% accuracies while KNN is reduced to 1.9% and was recorded 80.6% accurate predictions. Figure 17—Accuracy scores signifies the comparison accuracy scores of the mentioned four models.

Table 3 Accuracy comparison

S. no	During COVID-19—accuracy results		BeforeCOVID-19—accuracy results
1	Logistic regression	55.1%	82.5%
2	Decision Tree	99.2%	82.5%
3	Random Forest	99.3%	82.5%
4	KNN	81.8%	80.6%

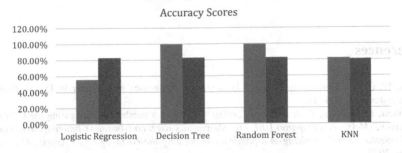

Fig. 17 Accuracy scores

There could be many factors that may have contributed to this Logistic Regression inaccuracies such as it requires more data, weak relationship, or multicollinearity.

8 Conclusion

This research focused on regression analysis during the COVID-19 pandemic. It predicts the unit price of most selling products during the COVID-19 and compares the prediction with the last year data i.e. before COVID-19. According to the findings obtained in Sect. 6, Masks, Sanitizers, Gloves, Vitamin, Contactless Thermometers are the most selling products, and it makes sense as well. During the COVID-19, daily routine activities were badly affected, and people were forced to practice social distancing. That is why, people were purchasing the items which were highly recommended by doctors and government officials.

Decision Tree and Random Forest showed the maximum accuracy score of 99.2% during the pandemic. Which means these two machine learning algorithms are reliable for e-commerce price prediction based on multiple independent variables. The second accurate model was K-Nearest Neighbor with the accuracy score of 81.8% which is again acceptable results for this kind of data during the COVID-19. Since, Logistic Regression gave the least accuracy score of 55.1%.

The accuracy score of Logistic Regression, Decision Tree and Random Forest was the same as 82.5%, while accuracy score of KNN model was 1.9% less than the rest. However, all the model's accuracy score is above 80% which makes all of the somewhat acceptable for this situation. The reason for this was that the Unit Price of products stayed constant throughout the four months. Which is why all these models behaving the same.

The future augmentation to this study is to collect data from multiple e-commerce stores from different emirates (states) and pass this data through models currently used in this study, mentioned in Sect. 6. The relationship between dependent and independent variables should be re-evaluated to see the patterns. Moreover, models accuracy should also be compared to form a generalized opinion of these findings.

References

Abbas, W. (2020). Coronavirus impact: E-commerce businesses grow faster in Dubai. *Dubai: Khaleejtimes, 7*(3), 1–18.

Abdelnour, A., Babbitz, T., & Moss, S. (2020). "Pricing in a pandemic: Navigating the COVID-19 crisis," *Mckinsey Company) URL https//www. mckinsey. com/business-functions/marketing-and-sales/our-insights/pricing-in-a-pandemic-navigating-the-covid-19-crisis Assess. May*, vol. 20, p. 2020.

Akour, I., Alshurideh, M., Al Kurdi, B., Al Ali, A., & Salloum, S. (2021). Using machine learning algorithms to predict people's intention to use mobile learning platforms during the COVID-19 pandemic: Machine learning approach. *JMIR Medical Education, 7*(1), 1–17.

Al Kurdi, B. A., Alshurideh, M., Nuseir, M., Aburayya, A., & Salloum, S. A. (2021, March). The effects of subjective norm on the intention to use social media networks: an exploratory study using PLS-SEM and machine learning approach. In *International Conference on Advanced Machine Learning Technologies and Applications* (pp. 581–592). Springer, Cham.

Al Suwaidi, F., Alshurideh, M., Al Kurdi, B., & Salloum, S. A. (2020). The impact of innovation management in SMEs performance: A systematic review. In *International Conference on Advanced Intelligent Systems and Informatics* (pp. 720–730). Springer, Cham.

Al-Dmour, A., Al-Dmour, H., Al-Barghuthi, R., Al-Dmour, R., & Alshurideh, M. T. (2021a). Factors influencing the adoption of E-payment during pandemic outbreak (COVID-19): Empirical evidence. *The Effect of Coronavirus Disease (COVID-19) on Business Intelligence, 334*, 133–154.

Al-Dmour, R., AlShaar, F., Al-Dmour, H., Masa'deh, R., & Alshurideh, M. T. (2021b). The effect of service recovery justices strategies on online customer engagement via the role of "Customer Satisfaction" During the Covid-19 Pandemic: An empirical study. *The Effect of Coronavirus Disease (COVID-19) on Business Intelligence, 334*, 325–346.

Al-Gasaymeh, A., Almahadin, A., Alshurideh, M., Al-Zoubid, N., & Alzoubi, H. (2020). The role of economic freedom in economic growth: Evidence from the MENA region. *International Journal of Innovation Create Chang, 13*(10), 759–774.

AlHamad, M., Akour, I., Alshurideh, M., Al-Hamad, A., Kurdi, B., & Alzoubi, H. (2021). Predicting the intention to use google glass: A comparative approach using machine learning models and PLS-SEM. *International Journal of Data and Network Science, 5*(3), 311–320.

Al-Jarrah, I., Al-Zu'bi, M. F., Jaara, O., & Alshurideh, M. (2012). Evaluating the impact of financial development on economic growth in Jordan. *International Research Journal of Finance and Economics, 94*, 123–139.

Aljumah, A., Nuseir, M. T., & Alshurideh, M. T. (2021). The impact of social media marketing communications on consumer response during the COVID-19: Does the brand equity of a university matter. *The Effect of Coronavirus Disease (COVID-19) on Business Intelligence*, 367–384.

Alshurideh, M., Al Kurdi, B., Salloum, S. A., Arpaci, I., & Al-Emran, M. (2020). Predicting the actual use of m-learning systems: a comparative approach using PLS-SEM and machine learning algorithms. *Interactive Learning Environments*, 1–15.

Alshurideh, M. T., Kurdi, B. A., AlHamad, A. Q., Salloum, S. A., Alkurdi, S., Dehghan, A., .. & Masa'deh, R. E. (2021). Factors affecting the use of smart mobile examination platforms by universities' postgraduate students during the COVID 19 pandemic: An empirical study. In *Informatics*, 8 2), 1–21. Multidisciplinary Digital Publishing Institute.

Alzoubi, H. M., & Inairat, M. (2020). Do perceived service value, quality, price fairness and service recovery shape customer satisfaction and delight? A practical study in the service telecommunication context. *Uncertain Supply Chain Management, 8*(3), 579–588.

Ambrose, J. (2020). Over a barrel: How oil prices dropped below zero. *Guard, 6*(2), 300–310.

Assad, N. F., & Alshurideh, M. T. (2020). Investment in context of financial reporting quality: A systematic review. *WAFFEN-UND Kostumkd. Journal, 11*(3), 255–286.

Bakir, H., Chniti, G., & Zaher, H. (2018). E-Commerce price forecasting using LSTM neural networks. *International Journal of Machtch Learning Computer, 8*(2), 169–174.

Barr, G. (2020). New ministerial resolution relating to employment and COVID-19, vol. 7, no. 2, pp. 1–12, 2020.

Beklemysheva, A. (2019). Why use python for AI and machine learning. *Machine Learning, 4*(2), 1–18.

Chen, J., & Fan, X. (2010). Effectiveness analysis of promotions in supermarket Chain. In *2010 Second International Conference on Communication Systems, Networks and Applications*, vol. 2, pp. 42–43.

Dass, M., Jank, W., & Shmueli, G. (2010). Dynamic price forecasting in simultaneous online art auctions. In *Marketing Intelligent Systems using Soft Computing*, Springer, pp. 417–445.

Elgerd, O. I. (1982). *Electric energy systems theory: An introduction*. McGraw-Hill Book Company.

Faiz, T. (2019). Multi-approaches on scrubbing data for medium-sized enterprises. In *2019 International Conference on Digitization (ICD)*, 2019, pp. 75–86.

Ghazal, T. M., Hasan, M. K., Alshurideh, M. T., Alzoubi, H. M., Ahmad, M., Akbar, S. S., .. & Akour, I. A. (2021). IoT for smart cities: Machine learning approaches in smart healthcare—A review. *Future Internet, 13*(8), 218, 1–19.

Mucedola, M. P. (2021). *Impact assessment and forecast through 2019–2025 for enery, food & feverage, retail & e-commerce*. Dublin: Research and Market.

Nuseir, M. T., El-Refae, G. A., & Aljumah, A. (2021a). The e-learning of students and university's brand image (Post COVID-19): How successfully Al-Ain University have embraced the paradigm shift in digital learning. In *The Effect of Coronavirus Disease (COVID-19) on Business Intelligence* (pp. 171–187). Springer, Cham.

Nuseir, M. T., Al Kurdi, B. H., Alshurideh, M. T., & Alzoubi, H. M. (2021b). Gender Discrimination at Workplace: Do Artificial Intelligence (AI) and Machine Learning (ML) Have Opinions About It. *In The International Conference on Artificial Intelligence and Computer Vision* (pp. 301–316). Springer, Cham.

Salloum, S. A., Alshurideh, M., Elnagar, A., & Shaalan, K. (2020, April). Machine learning and deep learning techniques for cybersecurity: a review. In *The International Conference on Artificial Intelligence and Computer Vision* (pp. 50–57). Springer, Cham.

Shah, S. F., Alshurideh, M. T., Al-Dmour, A., & Al-Dmour, R. (2021). Understanding the influences of cognitive biases on financial decision making during normal and COVID-19 pandemic situation in the United Arab Emirates. *The Effect of Coronavirus Disease (COVID-19) on Business Intelligence, 334*, 257–274.

Taryam, M., Alawadhi, D., Aburayya, A., Albaqa'een, A., Alfarsi, A., Makki, I., .. & Salloum, S. A. (2020). Effectiveness of not quarantining passengers after having a negative COVID-19 PCR test at arrival to Dubai airports. *Systematic Reviews in Pharmacy, 11*(11), 1384–1395.

Tziridis, K., Kalampokas, T., Papakostas, G. A., & Diamantaras, K. I. (2017). Airfare prices prediction using machine learning techniques. In *2017 25th European Signal Processing Conference (EUSIPCO)*, 2017, pp. 1036–1039.

Secure CoAP Application Layer Protocol for the Internet of Things Using Hermitian Curves

Raja Masadeh [ID], Omar Almomani [ID], Esra Masadeh, and Ra'ed Masa'deh [ID]

Abstract Internet of Things (IoT) is a modern technology that allows devices to sense status and take suitable reactions without human interference; this leads machines to interact directly with another machine, called machine to machine (m2m) communication. Many network technologies were involved in making the IoT idea work well. Devices in IoT are connected through the wireless network. Thus, different protocols on different network layers are used in IoT. One of the primary routing protocols used on the application layer is Constrained Application Protocol (CoAP) which is a new web application transfer protocol. Where it is better suited to constrained environments than HTTP for IoT. In this study, Hermitian-based Cryptosystem (HBC) is applied on the CoAP protocol in order to increase the security of IoT. The Hermitian curve is an algebraic geometric curve capable of generating a vast number of points. Simulation results are presented, showing significant performance improvements of AG codes over Hermitian codes against Reed-Solomon (RS) codes and Elliptic Curve Cryptography (ECC). This study proved that HBC outperformed the RS and ECC in terms of several points and key length, which leads to increase security.

R. Masadeh (✉)
Department of Computer Science, College of Information Technology, The World Islamic Sciences and Education University, Amman, Jordan
e-mail: raja.masadeh@wise.edu.jo

O. Almomani
Department of Computer Network and Information Systems, College of Information Technology, The World Islamic Sciences and Education University, Amman, Jordan
e-mail: Omar.almomani@wise.edu.jo

E. Masadeh
An Independent Researcher of Information Technology, Amman, Jordan

R. Masa'deh
Department of Management Information Systems, College of Business, The University of Jordan, Amman, Jordan
e-mail: r.masadeh@ju.edu.jo

Keywords Internet o Things (IoT) · CoAP · Security · Hermitian curve · Hermitian based Cryptosystem (HBC)

1 Introduction

Internet of Things (IoT) is a modern concept, appears in the latest few years. It consists of two terms: Internet and Things. The idea behind this concept is to connect any item in the world using the Internet. Such items or Thing are not only smart devices (Yassein & Aljawarneh, 2017; Yassein et al., 2016). A milk bottle in your fridge that is supplied by a sensor might be connected with the shopping list on your mobile phone. In case that the milk bottle is empty the sensor sends a message to your phone to add milk as one of the items on the shopping list. Also, the coffee machine may be connected to the Internet with the alarm on your smart phone, in order to get the coffee ready when you are awake. In this case the mobile sends a message to the coffee machine to turn on and get the coffee ready.

Since lifestyle become more complicated and connected technology take the most important responsibility to make such communication become true using the most important concept "Internet of Things". As shown in Fig. 1, Cisco (Kalyani et al., 2015) had done an analysis about the number of devices connected to the internet over years with respect to world population. The study shows in 2015 the population of the world was about 7.2 billion and about 25 billion devices were connected to the internet, which means each person in average used about 3.47 devices. However, a huge growth is expected in 2020. The study expected the population to become about 7.6 billion, however about 50 billion devices expected to be connected to the internet, which is in average about 6.58 device for each Pearson.

The first appearance of IoT was in 1999, by a group of network specialists who utilized the preexisting sciences such as radio frequency identification (RFID) and emerging sensing technologies (Evans, 2011; Hart & Martinez, 2015; Kalyani et al., 2015). Moreover, IoT include sensor networks that use sensors to read the status of a specific device in specific time and send the information to a sink node to take the appropriate action. The flow of the information should choose the best path between nodes, here appears the responsibility of the routing protocol used (Yassein & Aljawarneh, 2017; Yassein et al., 2016). Many routing protocols are used in IoT such as 6LowPAN-IPV6 over IEEE 802.15.4, Routing Protocol for Low Power and Lossy Networks (RPL) and Constrained Application Protocol (CoAP).

CoAP is a routing protocol that stands on the application layer. It aims to take place of HTTP protocol. This protocol was designed to meet low power devices with low overhead and multicast support. CoAP was also build as an alternative protocol to connect low power devices to the internet.

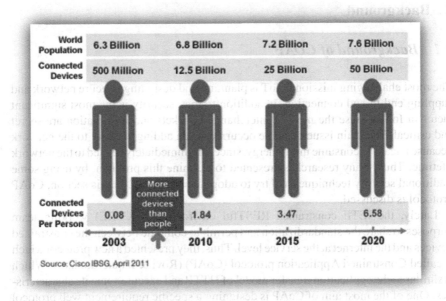

Fig. 1 Cisco analysis of IoT (Kalyani et al., 2015)

The main issue that faced the CoAP is the security. Many researches are suggested many solutions in order to solve this problem. Such as the data encryption standard (DES) which was the most familiar and used over than 30 years but now it has a main drawback which is easy to crack by any attacker; and the advanced encryption standard (AES). While, there is another algorithm is not used in CoAP till now which is called Hermitian curves (Rovi & Izquierdo, n.d.). Which is algebraic geometry (AG) curve that plays a main role in coding theory and cryptography. Moreover, the main goal of using AG is to create an effective coding and secured system.

The testing environment will use Contiki-Cooja 3.0 simulator. The difficulty of this simulator that it is not documented. However, all IoT protocols are well programmed which make the environment easier for the researchers than start from scratch. The Contiki-Cooja is written in two programming languages JAVA and C. the windows of the simulator are written in JAVA while the nodes inside and the flow of scenarios is written in C.

The reset of this paper is organized as follows: Sect. 2 contains the background in detail for both CoAP and Hermitian curve. Section 3 presents the related work. While Sect. 4 illustrates the proposed algorithm. Section 5 presents the simulation results. Finally, Sect. 6 draws the conclusion.

2 Background

2.1 Background of COAP

The most challenging mission in IoT is planning and designing a secure network and suppling end-to-end connection. In addition to that, security is the most significant factor in IoT because the most of interchanged packets and information are secret and critical. The main issue could be occurred when adding security to the network because it leads to consume more energy, since it is immediately related to the network lifetime. Thus, many researches presented to examine this problem, by using some traditional security techniques and try to adopt them into IoT. In this section, CoAP protocol is discussed.

Lately, the IETF constrained RESTful environments (CoRE) working team purposes to share the standardization and permit to combine between the constrained devices and the internet at the service level. Thus, they presented a new protocol which is called Constrained Application protocol (CoAP) (Rovi & Izquierdo, n.d.). Which is similar to the application transfer model of HTTP and execute some its characteristics. One of the most aim of CoAP is designing a specific requirement web protocol that goals to utilize in 6lowPAN, taking into account the structuring automation, power and other M2M (machine to machine) applications which are used in smart towns (Kushalnagar et al., 2007). Thus, these applications can be managed by using CoAP protocol which leads to rise the significance of web services during the daily lifetime for people. However, CoAP use UDP connection protocol between client and server by default to guarantee datagram arrival even through connectionless (Kerasiotis et al., 2010). The features of CoAP that recognize from other protocols that make it more suitable for constrained network and devices are presented in the following (Kovatsch et al., 2011):

- Web protocol that fulfills M2M constrained requirements which is consolidated protocol that is designed for constrained nodes. Moreover, it is suitable and applicable on RAM of 8–12 K and 8–16 bits microcontroller nodes.
- Simple caching and proxy abilities, where this protocol props caching, and the last responses could be cached in case the proxy is utilized in order to reply later such as the work of agents when they in the sleep mode.
- UDP binding that supports reliability using the multicast and unicast requests. In other words, CoAP has the choice to transmit a large number of packets by using UDP with reliability guarantee using Back-off algorithm. In addition to that, supporting the multicast requests but the reliability is not guarantee.
- The variety of message exchanging, which means this protocol permits to read, create, delete, modify and update a resource on any node because it depends and based on REST architecture.
- Content type and URI support. That means CoAP supports all the huge media types.

As mentioned above, CoAP is similar to the application transfer model of HTTP and execute some of its features. Thus, the standard methods which are completely supported by CoAP such as POST, PUT, DELETE and GET methods which are close to HTTP methods. In addition to that, these methods capable to monitor and effect the resources. These methods are presented as follows (Shelby et al., 2014):

- POST, this type of methods is used when a new subordinate resource is required in order to be differentiated under a present resource URI.
- PUT, this method is used within updating or creating a resource stable with URI that is defined in advance.
- DELETE, this type is used in case of deleting a URI specified resource.
- GET, this method is utilized in order to gain any resource information which is known by using URI.

As shown in Fig. 2 CoAP has client—server interaction model is similar to HTTP (Levä et al., 2014). When the sender (client) sends a request for any action, the receiver (server) replies with response message that includes the resource representation and the response code. However, CoAP functions the roles of clients and servers at the same time. CoAP achieves M2M requirements through supporting asynchronous operations over the UDP. This operation is done by using four exchange messages which are CON, Non-Con, ACK and REST as presented in Fig. 2.

2.2 Hermitian Curves

The AG curve is by depicting the number of points which accepted the curve equation which is called point on the curve. Thus, this curve is called either a projective curve because j represents the x-axis, k represents the y-axis and s represents the z-axis, which means it represents three—axis (j, k, s) or affine curve in case represents two-axis (j, k).

The projective Hermitian curve which is defined over a finite filed $(2\ m)$ where $b = \sqrt{f}$ and $f = 2\ m$ is a finite filed length (Alzubi, 2015), the proactive Hermitian curve is mathematically modeled by the following equation (Mohammad, 2017):

$$H(j, k, s) = j^{b+1} + k^b w + ks^b$$

If all partial derivatives of the curve do not disappear at this point on the Hermitian curve, $H(j, k, s) = 0$, the Hermitian curve is smooth, or the projective points must be nonsingular. The curve is considered to be nonsingular or smooth if all of the points that fulfill it are nonsingular (Carrasco & Johnston, 2008).

These three partial derivatives make the Hermitian curve smooth or nonsingular (Mohammad, 2017):

$$\frac{\partial H(j, k, s)}{\partial j} = (b + 1)j^b = j^b$$

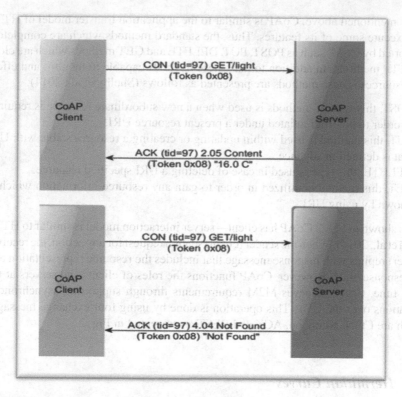

Fig. 2 CoAP transaction model (Levä et al., 2014)

$$\frac{\partial H(j,k,s)}{\partial k} = bk^{b-1}s + s^b = s^b$$

$$\frac{\partial H(j,k,s)}{\partial s} = k^b + bks^{b-1} = k^b$$

The only point where all three partial derivatives disappear is (0, 0, 0). However, this is not a point in projective space, the curve is smooth, and all points are nonsingular (Carrasco & Johnston, 2008).

The affine Hermitian curve, which is defined over finite fields (2b), is mathematically represented by the following Eq. 2 (Mohammad, 2017):

$$H(j,k,s) = j^{b+1}k^b + k$$

H (j, k, 1), which s-axis equal 1, when substation it in Eq. 1 (Mohammad, 2017):

$$H(j,k,1) = j^{b+1}k^b1 + k1^b$$

where $1^b = 1$, that will get Eq. 2.

If all partial derivatives of the curve do not disappear at this point, the Hermitian curve should be smooth, or the affine points must be nonsingular, recalling that the Hermitian curve should be smooth or the affine points must be nonsingular. The curve is considered to be nonsingular or smooth if all points that fulfill it are nonsingular (Carrasco & Johnston, 2008).

These three partial derivatives (Mohammad, 2017) smooth, or nonsingular points on the affine Hermitian curve $(j, k, 1)$:

$$\frac{\partial H(j, k, 1)}{\partial j} = (b + 1)j^b = j^b$$

$$\frac{\partial H(j, k, 1)}{\partial k} = bk^{b-1}1 + 1 = 1^b$$

The only position on the projective Hermitian curve where all three partial derivatives disappear is $(0, 0, 1)$, and this point does not satisfy Eq. (3) (Mohammad, 2017).

$$H(0, 0) = 0^{b+1} + 0^b + k$$

$$\frac{\partial H(j, k, 1)}{\partial 1} = k^b + bk1^{b-1} = k^b$$

The first partial derivatives (Mohammad, 2017):

$$\frac{\partial H(j, k, 1)}{\partial j} = (k + 1)0^b = 0^b$$

The second partial derivatives (Mohammad, 2017):

$$\frac{\partial H(j, k, 1)}{\partial k} = b0^{b-1}1 + 1 = 1^b = 1, \text{where} 1^b = 1$$

The third partial derivatives (Mohammad, 2017):

$$\frac{\partial H(j, k, 1)}{\partial 1} = 0^b + 0k1^{b-1} = 0^b$$

Over finite fields (2^b), the number of locations that fulfilled Hermitian curve H in Eq. (3) (Mohammad, 2017):

1. Compute $r = \sqrt{f} \, \epsilon f = 2^b$.
2. Compute number of points $\#p = r^3 + 1$.
3. It's worth noting that we added one to step 2: the point at infinity. Hermitian curves, like algebra curves, have a specific point at infinity $(j, k, 0)$.

3 Literature Review

Many researchers have recognized that IoT encompasses many concerns that need to be investigated in many ways, with a particular focus on security. As a result, many solutions have been proposed to address this problem. In Raza et al. (2012), the researchers suggested using the datagram transport layer security (DTLS) protocol with CoAP in order to supply end-to-end security in the internet of things. While the DTLS wasn't designed for constrained devices, it was designed for the internet, which is a heavyweight protocol that has a long header to suit a singular IEEE 802.15.4 maximum transmission unit (MTU). Thus, 6LoWPAN is utilized to pressure the header of the IP layer. According to that, the researchers suggested mechanisms to employ the compression abilities of 6LoWPAN to compress the header and the message of DTLS. The researchers defined many compression techniques that are related to their contribution, such as record header, handshake header, client hello message, and server hello message. One of the outcomes shown for the record header of the DTLS could minimize the number of bits by 62%. However, the researcher's study did not prove that compression does not impact security. According to that, the researchers (Kothmayr, 2011; Kothmayr et al., 2013) suggested a new solution based on RSA which energy consumption is high.

The researchers (Raza et al., 2013) proposed another resolution by using DTLS. The suggested solution was a combination between DTLS and CoAP for the internet. Moreover, it includes four main components; DTLS, CoAP, DTLS header compression (using 6LowPAN), and a CoAP-DTLS integration module that was improved to permit the application to access CoAP automatically. This solution is evaluated by using Contiki OS, which uses actual sensors. The evaluation results show the essential gains in the network response time and power consumption by decreasing the packet size. Thus, they could reduce the number of fragments or avert them by utilizing compression in case the payload was a little above the threshold.

The researchers (Brachmann et al., 2011) illustrated two main issues that required to be resolved to secure the communication links; secure group communications that aim by using DTLS to establish a secure connection for the collection of devices, and end-to-end security purpose is to perform a fully secure communication among HTTP and CoAP by using 6LoWPAN border router and DTLS-PSK protocol.

In Alghamdi et al. (2013), the researchers presented an analysis of two well-known protocols that could be used to increase security in CoAP networks. These protocols are DTLS and IPSec. The analysis was based on the architecture of X.805 standard that supplies a top-down method to do the following; edit, predict and discover the security vulnerabilities in the network. In addition, the researchers mentioned that these protocols weren't designed and transacted for constrained devices. In comparison, the researchers in Ukil et al. (2014) suggested a lightweight approach by using an advanced encryption standard (AES) which is 128 bits symmetric key algorithm, which came up with the Auth-Lite approach that becomes the lightweight security for CoAP. However, the suggested process is applicable in vehicle tracking systems but may or may not be efficient and valuable for other applications.

In Bhattacharyya et al. (2015), many solutions were presented by the researchers depending on the idea of session security. The researchers used the Lightweight Establishment of Secure Session (LESS) algorithm in CoAP and the DTLS-PSK channel encryption method. Theory communication and message exchange LESS algorithm is used to protect the session key since it uses the AES-CCM encryption technique. Moreover, the algorithm separates the key to ensure secure communication. The DTLS algorithm includes six main steps; pre-sharing secret, session initiation, server challenge, client response and challenge, client authentication, and server authentication. The analysis of one study shows that DTLS is more reliable since it uses six steps rather than request/response in LESS, and each step in DTLS may have multiple messages. However, the researchers' solution is applied for unicast security only, not to multicast security issues.

4 Proposed Algorithm

In this section, a proposed algorithm is described, which its main goal being to secure communications between different objects in IoT networks that are connected with each other. The Hermitian-based Cryptosystem (HBC) method (Mohammad, 2017), which is based on AG code over Hermitian curve, is employed in this study. Hermitian curve can create a huge number of points in a curve which plays a significant role in generating encryption keys. In addition, that its main employment in this technique is to generate private and public keys based on (Mohammad, 2017). Moreover, HBC can perform data encryption in addition to error correction, which is another advantage compared with other algebraic geometric code methods.

As shown in Fig. 3, HBC (Hermitian Based Cryptosystem) is applied to CoAP protocol located in the application layer. HBC is based on an algebraic geometric curve, Hermitian Curves (HC). Moreover, it inherits all advantages and characteristics of HC, whose main significant characteristic is the number of points that can be generated. Thus, it plays the primary role in generating a massive size of encryption keys and defining the size of the plaintext message. Moreover, HBC has three main functions; keys generation, encryption, and decryption. Keys generation includes public and private keys; a public key is employed to encrypt a message, while a private key is used to decrypt the same message. In HBC, three types of private keys, Q, I, and G (Mohammad, 2017), are used to compute the public key by the following Eq. 4 (Mohammad, 2017), as illustrated in Fig. 4.

$$Kpub = G * I * Q$$

As shown in Fig. 5, the main inputs for the encryption algorithm are plaintext which denotes by M and a public key that indicates to $Kpub$, which is selected

Fig. 3 General model of
suggested work

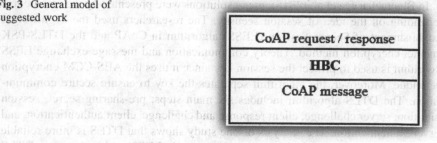

Fig. 4 Keys generation
(Mohammad, 2017)

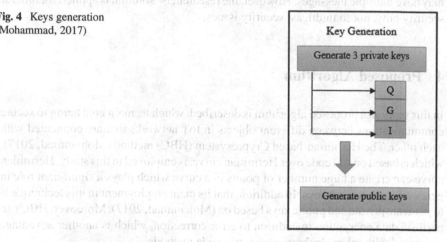

randomly from many keys that are generated by Eq. 4. In addition to that; by applying
the public key on the message that will be sent, ciphertext (C) is created as presented
in Eq. 5 (Mohammad, 2017).

$$C = M * Kpub$$

In turn, as presented in Fig. 6, the decryption algorithm has ciphertext (C) as
an input that is transmitted from the sender to the receiver. As shown in Fig. 6,
this algorithm has three main steps; compute D, Z, and P. Where D is calculated
by applying Eq. 6 (Mohammad, 2017) to eliminate the effect of the permutation
matrix form the C. While Z indicates to error detection and correction on D. Finally,
Plaintext denotes as P, which is the original message sent by the sender after the
decryption process.

$$D = C \times Q - 1$$

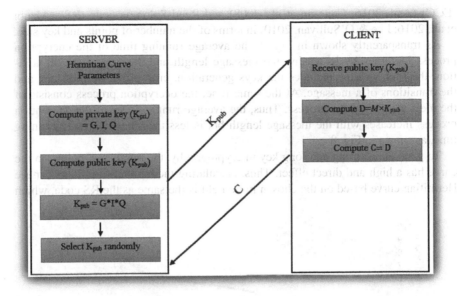

Fig. 5 Encryption algorithm in CoAP (Mohammad, 2017)

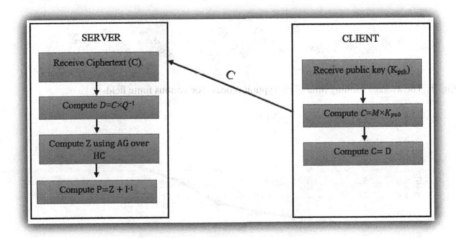

Fig. 6 Decryption algorithm in CoAP (Mohammad, 2017)

5 Simulation Results

This study presents the average running time results of the encryption and decryption processes for various message lengths. Moreover, the proposed work is compared with other cryptographies applied on CoAP, such as Elliptic Curve Cryptography

(ECC) (Asif, 2016; Ganesan et al., 2008) and Reed-Solomon code (RS) (Alzubi et al., 2016; Lee & O'Sullivan, 2010) in terms of the number of points and key size.

As transparently shown in Fig. 7, the average running time of the encryption process increases by increasing the message length and the finite field. In addition, the running time includes the keys generation, encrypting the message, and the transitions of a message. At the same time, the decryption process consists of the message decryption process. Thus, the average running time of the decryption process increases with the message length but is less than the encryption running time, as shown in Fig. 8.

To generate a strong and long key in cryptography, the number of points on the curve has a high and direct effect. Thus, calculating the number of points over the Hermitian curve based on the chosen finite field is the same as the RS code, which

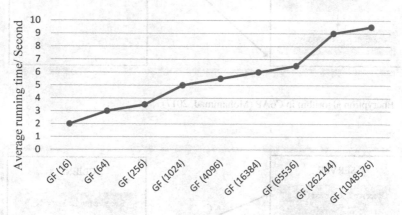

Fig. 7 The average running time of encryption process for various finite field

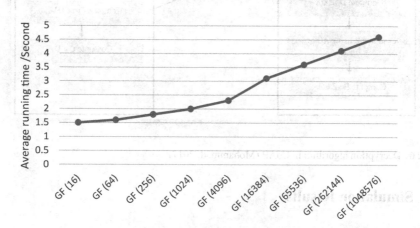

Fig. 8 The average running time of decryption process for various finite field

contrasts ECC. The number of points over HC depends on the selected finite field, while EC is computed by applying the following Eq. 7 (Blake et al., 1998). However, RS is calculated by using Eq. 8 (Johnston & Carrasco, 2005).

$$\neq P = (w - 1)(w - 2)\sqrt{q} + 1 + q$$

$$\neq P = q2 - 1$$

where the number of points denotes as P, w indicates to the code length and q indicates to the characteristic of the finite field for instance filed.

It is evident in Table 1 that HC is more secure than EC and RS because it can generate many points compared to them. In other words, the number of points that satisfied the HC are significantly increased, especially when choosing a higher field compared to the slightly raising in both EC and RS, as shown in Fig. 9. Thus, applying the Hermitian curve to COAP will increase the security. For instance, GF (2^{20}) can generate 1,073,741,824 points which satisfies HC, while EC can generate 1,050,625 points and RS can generate 262,143 points.

Depending on Fig. 10, it is obvious that the public key length of HC is larger than the other key length of both EC and RS, which leads to being more secure and more complicated to decrypt the message. Thus, two main reasons to increase the complex message decryption; are the extensive length of the public key and each point represented in non-binary.

Table 1 Number of points for Hermitian curve compared to RS and EC for various GF (2^b)	$GF(2^b)$	Number of points for Hermitian	Number of points for Elliptic	Number of points for RS code
	$GF(2^2)$	9	9	3
	$GF(2^4)$	65	25	15
	$GF(2^6)$	513	81	63
	$GF(2^8)$	4097	289	255
	$GF(2^{10})$	32,769	1097	1023
	$GF(2^{12})$	262,145	4225	4095
	$GF(2^{14})$	2,097,153	16,384	16,383
	$GF(2^{16})$	16,777,217	66,149	65,535
	$GF(2^{18})$	134,217,729	263,169	262,143
	$GF(2^{20})$	1,073,741,825	1,050,625	262,143

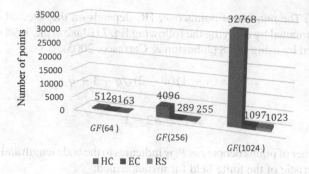

Fig. 9 Number of points that satisfy HC, EC and RS

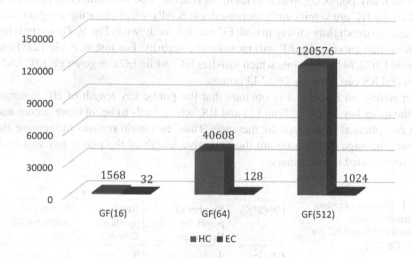

Fig. 10 The public key length for HC, EC and RS

6 Conclusion

This study proposed a new solution based on AG codes over the Hermitian curve to solve the central issue of IoT, which is security, by applying this algorithm to the CoAP protocol. The Hermitian curve can generate a maximum number of points on the curve that plays a crucial role in creating the encryption key compared to the other types of AG curves, such as the EC and RS. This work proved that the suggested solution outperforms RS code and ECC in terms of key length and the number of points which indicates the number of generation keys. However, the drawback of this algorithm is the power consumption.

As a future direction, a new solution will suggest decreasing the power consumption, preserving the security, or increasing it by employing metaheuristic optimization algorithms such as (Alzaqebah et al., 2019; Masadeh et al., 2019).

References

Alghamdi, T. A., Lasebae, A., & Aiash, M. (2013). Security analysis of the constrained application protocol in the Internet of Things. In *Future Generation Communication Technology (FGCT), 2013 Second International Conference* (pp. 163–168). IEEE.

Alzaqebah, A., Al-Sayyed, R., & Masadeh, R. (2019). Task scheduling based on modified grey wolf optimizer in cloud computing environment. In *2019 2nd International Conference on new Trends in Computing Sciences (ICTCS)* (pp. 1–6). IEEE.

Alzubi, O. A. (2015, September). Performance Evaluation of AG Block Turbo Codes over Fading Channels Using BPSK. In *Proceedings of the The International Conference on Engineering & MIS 2015* (p. 36). ACM.

Alzubi, O. A., Chen, T. M., Alzubi, J. A., Rashaideh, H., & Al-Najdawi, N. (2016). Secure channel coding schemes based on algebraic-geometric codes over hermitian curves. *Journal of Universal Computer Science, 22*(4), 552–566.

Asif, R. A. (2016). *Efficient computation for hyper elliptic curve based cryptography.* University of Windsor.

Bhattacharyya, A., Bose, T., Bandyopadhyay, S., Ukil, A., & Pal, A. (2015). LESS: Lightweight Establishment of Secure Session: A Cross-Layer Approach Using CoAP and DTLS-PSK Channel Encryption. In *Advanced Information Networking and Applications Workshops (WAINA), 2015 IEEE 29th International Conference on* (pp. 682–687). IEEE.

Blake, I., Heegard, C., Hoholdt, T., & Wei, V. (1998). Algebraic-geometry codes. *IEEE Transactions on Information Theory, 44*(6), 2596–2618.

Brachmann, M., Garcia-Morchon, O., & Kirsche, M. (2011). *Security for practical coap applications: Issues and solution approaches.* Universitt Stuttgart.

Carrasco, R. A., & Johnston, M. (2008). *Non-binary error control coding for wireless communication and data storage.* John Wiley & Sons.

Evans, D. (2011). The internet of things how the next evolution of the internet is changing everything. *Cisco Internet Business Solutions Group (IBSG).*

Ganesan, R., Gobi, M., & Vivekanandan, K. (2008). Elliptic and hyper elliptic curve cryptography over finite field Fp. *Manager's Journal on Software Engineering, 3*(2), 43.

Hart, J. K., & Martinez, K. (2015). Toward an environmental Internet of Things. *Earth and Space Science, 2*(5), 194–200.

Johnston, M., & Carrasco, R. A. (2005). Construction and performance of algebraic–geometric codes over AWGN and fading channels. *IEE Proceedings-Communications, 152*(5), 713–722.

Kalyani, V., Gaur, P., & Vats, S. (2015). IoT:'machine to machine'application A future vision. *Journal of Management Engineering and Information Technology (JMEIT), 2*(4), 15–20.

Kerasiotis, F., Prayati, A., Antonopoulos, C., Koulamas, C., & Papadopoulos, G. (2010). Battery lifetime prediction model for a WSN platform. In *2010 Fourth International Conference on Sensor Technologies and Applications* (pp. 525–530). IEEE.

Kothmayr, T. (2011). A security architecture for wireless sensor networks based on DTLS. Master's Thesis in the Software Engineering Elite Graduate Program at the University of Augsburg.

Kothmayr, T., Schmitt, C., Hu, W., Brünig, M., & Carle, G. (2013). DTLS based security and two-way authentication for the Internet of Things. *Ad Hoc Networks, 11*(8), 2710–2723.

Kovatsch, M., Duquennoy, S., & Dunkels, A. (2011). A low-power CoAP for Contiki. In *Mobile Adhoc and Sensor Systems (MASS), 2011 IEEE 8th International Conference on* (pp. 855–860). IEEE.

Kushalnagar, N., Montenegro, G., & Schumacher, C. (2007). IPv6 over low-power wireless personal area networks (6LoWPANs): Overview, assumptions, problem statement, and goals. *RFC, 4919*, 10.

Lee, K., & O'Sullivan, M. E. (2010). Algebraic soft-decision decoding of Hermitian codes. *IEEE Transactions on Information Theory, 56*(6), 2587–2600.

Levä, T., Mazhelis, O., & Suomi, H. (2014). Comparing the cost-efficiency of CoAP and HTTP in Web of Things applications. *Decision Support Systems, 63*, 23–38.

Masadeh, R., Mahafzah, B. A., & Sharieh, A. (2019). *Sea lion optimization algorithm. Sea, 10*(5), 388.
Mohammad, M. A. (2017). *"Cryptosystem design based on hermitian curves" (dissertation)*, Al-Balqa Applied University.
Raza, S., Shafagh, H., Hewage, K., Hummen, R., & Voigt, T. (2013). Lithe: Lightweight secure CoAP for the internet of things. *IEEE Sensors Journal, 13*(10), 3711–3720.
Raza, S., Trabalza, D., & Voigt, T. (2012). 6LoWPAN compressed DTLS for CoAP. In *2012 IEEE 8th International Conference on Distributed Computing in Sensor Systems* (pp. 287–289). IEEE.
Rovi, C., & Izquierdo, M., *Maximal curves covered by the Hermitian curve*.
Shelby, Z., Hartke, K., & Bormann, C. (2014). *The constrained application protocol (CoAP)*.
Ukil, A., Bandyopadhyay, S., Bhattacharyya, A., Pal, A., & Bose, T. (2014). Lightweight security scheme for IoT applications using CoAP. *International Journal of Pervasive Computing and Communications, 10*(4), 372–392.
Yassein, M. B., & Aljawarneh, S. (2017). A new elastic trickle timer algorithm for Internet of Things. *Journal of Network and Computer Applications, 89*, 38–47.
Yassein, M. B., Aljawarneh, S., & Ghaleb, B. (2016). A new dynamic trickle algorithm for low power and lossy networks. In *2016 International Conference on Engineering & MIS (ICEMIS)* (pp. 1–6). IEEE.

Effectiveness of Introducing Artificial Intelligence in the Curricula and Teaching Methods

Hani Yousef Jarrah⊙, Saud Alwaely⊙, Saddam Rateb Darawsheh⊙, Muhammad Alshurideh⊙, and Anwar Saud Al-Shaar

Abstract This research examined educators' perceptions of the effectiveness of introducing artificial intelligence (A.I.) and expert systems into the curricula and teaching methods of higher educational institutions in the United Arab Emirates. Ninety-three faculty members in various departments of UAE colleges of education completed a questionnaire assessing the impacts of Artificial Intelligence on students' creativity, problem-solving skills, data were analyzed to identify differences based on gender, educational qualification, and experience. The results of t-tests and ANOVA analysis revealed that participants had overall positive attitudes toward introducing A.I. and expert systems into the curricula and teaching methods. Although no differences were found based on gender or experience, statistically significant differences were associated with academic qualifications, whereby participants with master's

H. Y. Jarrah · S. Alwaely
College of Education, Humanities and Social Sciences, Al Ain University, Abu Dhabi, UAE
e-mail: Hani.jarrah@aau.ac.ae

S. Alwaely
e-mail: Suad.alwaely@aau.ac.ae

S. Alwaely
The Hashemite University, Zarqa, Jordan

S. R. Darawsheh (✉)
Department of Administrative Sciences, The Applied College, (Imam Abdulrahman Bin Faisal University), P.O. Box: 1982, Dammam, Saudi Arabia
e-mail: srdarawsehe@iau.edu.sa

M. Alshurideh
Department of Management, College of Business Administration, University of Sharjah, Sharjah, UAE
e-mail: malshurideh@sharjah.ac.ae; m.alshurideh@ju.edu.jo

Marketing Department, School of Business, The University of Jordan, Amman, Jordan

A. S. Al-Shaar
Department of Self Development, Deanship of Preparatory Year and Supporting Studies (Imam Abdulrahman Bin Faisal University), P.O. Box: 1982, Dammam 43212, Saudi Arabia

degrees gave higher ratings than those with doctorates. Based on the findings, introducing A.I. and expert systems into higher education curricula and teaching methods is recommended.

Keywords Artificial intelligence · Expert systems · Teaching methods

1 Introduction

Artificial intelligence (A.I.) is an efficacious mechanism to deal with complex conditions (Claudé & Combe, 2018) and it is among the most rapidly developing multimedia educational technologies (Salloum et al., 2020a; Tripathi & Al Shahri, 2016). AI has two types: strong and weak. The strong one is supposed to emulate the function of brains of the humans but it is not created yet while the weak one is common in life of people and it comprises Natural Language Processing (NLP), Machine Learning (ML), Speech recognition, Machine vision, and Expert Systems (ES) (Al Claudé & Combe, 2018; Salloum et al., 2020b; Shebli et al., 2021).

The rate of use of Artificial intelligence and expert systems is proportional to the increasing number of courseware inventions among educational colleges (Amarneh et al., 2021; Nuseir et al., 2021). Adaptive learning courseware can be referred to as the technology requiring students to master the same objectives of learning. The adaptive software engine is responsible for determining timing and order of the content and it evaluates performance of the student based on a number of factors and after that the student is guided through the course content (Alhashmi et al., 2020; AlShamsi et al., 2021). Adaptive courseware is existed for many years ago but nowadays it is used in all of the education stages and professional development. And also, it is expanding to other branches like foreign languages, economics, physiology, anatomy, and business (Gebhardt, 2018; Yousuf et al., 2021).

Intelligent tutoring systems (ITSs) is a special genre of e-learning systems, where the operation of ITSs is based on making an emulation of the human teacher in the process of teaching and learning. The personal and portable characteristic of e-learning that appeared with the coming of the mobile computing system, put a key stone to the beginning of mobile intelligent tutoring. The aim of the Mobile intelligent systems (MITSs) is to adapt with daily routine of the mobile learners without any bad effect on their other activities. We should note that there are limited capabilities in improving the efficacy and potency of learning in case of handheld terminals. While in the non-portable ITS rivals, MITSs is considered agent-based systems as their tasks are achieved by agents (Glavinic et al., 2007).

Over the past 2–3 decades, the development of tools for multimedia technology-enhanced student learning has accelerated quickly to enhance the interactive capabilities of students as well as educators. Multimedia tools have immensely benefitted educational institutions across fields, fostering learners' problem-solving abilities and communication and thereby augmenting their creativity, imagination, and cognition, as well as enhancing teachers' planning by providing insights into student aptitudes and learning processes.

NLP abilities and social intelligence embedded in A.I. give machines the ability to read and understand human languages as well as to adapt to different learning styles, aptitudes, and rates. Expert systems, which are a type of A.I. comprising software programs, have human influence approaches to problem-solving.

The following items: Human–computer interaction interface, interpreter, working memory, reasoning machine, and knowledge base constitute the essential structure of any expert system. The expertise of the expert system including rules and facts is stored in the knowledge base. Acquiring a new knowledge, storing and expressing knowledge in a trend that the computer can accomplish, should be enabled by the knowledge base in the stage of establishing a knowledge base. The facts in the working memory are matched with the knowledge by the reasoning machine and obtains new information. The storage unit should furthermore store the intermediate information gotten during processing. Explaining the outcome of the inference engine output, taking into consideration to interpret the cause and correctness of the conclusion is the responsibility of the interpreter (Tan, 2017).

It is worthy to mention that in the latest several years NLP has belonged to the most challenging areas. Analyzing of the education web pages and learning systems based on internet are in a need to the essential role played by the NLP techniques based on semantics and syntax of the natural languages (Salem, 2000).

The application of expert systems offers many advantages that support the educational process (Ghadirli & Rastgarpour, 2013). The chief benefit of an expert system is its ability to act as a "smart" teacher; thus, enabling independent learning outside of the classroom. The expert system also interacts with the learner by asking a set of questions and providing its conclusions in the form of diagnoses or suggested treatments. A similar process to the humanistic approach is to solve complex problems. A.I. and expert systems operate at a constant scientific and consultative level that is stable (Clement, 1992). This technology provides students with enjoyment that differs from traditional methods and introduces students' happiness and creativity in education, helping them to imagine all that is new.

The main objective of this study was to assess different perceptions on the effectiveness of A.I. expert system implementation on the curricula and teaching method in education colleges, a case of Colleges of Education in the United Arab Emirates.

Accurately, this research measured perceptions of faculty members at Colleges of Education in the United Arab Emirates (UAE) regarding the effectiveness of introducing A.I. and expert systems in the curricula and teaching methods on the students' creativity and the problem-solving ability.

The research questions relevant for this study include (1) how the Artificial Intelligence expert systems implementation has influenced the curricula and teaching methods of education colleges? (2) How does the implementation of the Artificial Intelligence systems in education colleges affect the creativity and the student's ability to solve problems?

2   Literature Review

2.1   The Meaning of A.I. and How It Works in Educational Colleges

There are various perspectives to view A.I., if we take for instance the business perspective, we can define A.I. as the combination of robust methodologies and tools used in solving the problems of business. And if we turn into the research perspective, A.I. is the interest of making computers replace humans in doing things that they do better nowadays (Brown & O'Leary, 1995) or AI can be defined as the mechanism interested in the creation and study of computer systems which shows some intelligence forms. We mean by intelligence as the system which can draw and reason valuable conclusions, learn new duties and conceptions about the world surrounding us (Rao et al., 1999).

A.I. refers to the aptitude exhibited by machines or software (Claudé & Combe, 2018), and the field of A.I. researches how to create computers and computer programs capable of intelligent behavior. Baker and Smith (2019) defined A.I. rather broadly as "computers which perform cognitive tasks, usually associated with human minds, particularly learning and problem-solving" (p. 10). As they explained, A.I. describes a range of technologies and methods, including machine learning, natural language processing (NLP), data mining, neural networks, and algorithms. Artificial intelligence systems execute tasks that usually need human intelligence, such as visual insight, speech recognition, decision-making, and translation between languages (Jewandah, 2018).

Accordingly, the A.I. field draws upon multiple disciplines, including computer science, information engineering, mathematics, linguistics, psychology, and philosophy. Based on the simulation of the human mind, A.I. technology has belonged to the science of algorithms and intelligent techniques based on replicating and understanding the archetype of the human psyche. The human brain sends signals to the body, which in turn controls all social processes and activities that occur daily, such as feeling, sensing, thinking, and movement. A.I. was introduced to emulate these processes and has had some success.

A.I. and expert systems have published several studies on the development across diverse areas such as medicine, chemistry, physics, engineering, education, training, manufacturing, and management (Tan, 2017). Furthermore, applications of AI nowadays expanded to finance, consumer products, medicine, and manufacturing taking into consideration that implementing of an adequate A.I. technique in an application should always be competitive. Knowledge-based systems (KBS) and expert systems (ES) are contributing deeply in numerous areas of decision making in industries such as production planning, advanced manufacturing processes, plant layout, materials, and purchasing (Rao et al., 1999). Claudé and Combe (2018) highlighted the effectiveness of A.I. as a tool for managing complex situations. A.I. and its branches of expert systems can address complex problems in record time, which accelerates the process of accumulating new knowledge by presenting the information in a manner tailored to learner's needs, thereby enhancing their reasoning skills and ability to conclude. Additionally, the A.I make the content easily memorizable by the students and make the work more comfortable for the teachers who prefer using them.

Jewandah (2018) emphasized that improvement and development in the A.I. industry would increase productivity for organizations at a lower cost And also, his results indicated that integration of the adaptive learning courseware as low-stakes assignments improves the outcomes of the students with a relatively modicum effort on both sides of the students and the instructor.

The inferencing system and the knowledge base are the two major parts of any AI-based education software. The knowledge base comes from procedures, theories, facts, and relationships which represent knowledge of the real world about people, events, objects, places, etc. while the thinking mechanism or the inference system is the technique of employing the knowledge base to resolve the complicated problems (Salem, 2000).

2.2   The Role of A.I. in the Achievement of Educational Goals

Information and communication technology (ICT) became an important part in the system of education for either completely replacing the teacher or just helping him in realizing a traditional lecture. E-learning is achieved between education on one hand and ICT on the other hand (Glavinic et al., 2007). Hence, the implementation of multimedia technology in higher education has been recognized as critical to maintain educational quality and achieve global competitiveness (Duncan, 2010; Popenici & Kerr, 2017a, 2017b; Warschauer, 2004).

In the future, the expert system may compromise applications which are more interactive, mobile, and intelligent that may blur the distinction between agents and expert system (Grove, 2000).

Mobile learning or m-learning is a subclass of e-learning and currently it is attracting high attention as it is the upcoming type of learning but the development of this type of ought to be achieved through evolutions of technologies related to wireless infrastructure like UMTS, GSM, and WiMAX. Utilizing of the previous infrastructure will help in making a network of mobile devices dedicated to the process of learning (Glavinic et al., 2007).

Accordingly, countries with highly developed higher education systems are increasingly incorporating multimedia and mobile technologies as intrinsic elements of new "connected learning" networked economies. However, educators have only recently begun to explore the potential pedagogical opportunities that A.I. applications afford for supporting learners throughout their education (Zawacki-Richter et al., 2019). Many teachers are yet to understand the benefits of I.A. in schools. They have a limited understanding of its pedagogical uses and have many questions regarding how to integrate A.I. as a supportive resource (Hinojo-Lucena et al., 2019; Zawacki-Richter et al., 2019). A.I. can impede students' imagination and learning self-efficiency if not applied to the university curriculum in a creative and balanced manner. As a result, the students may complete courses and graduate without enhancing their level of performance, development, and innovation.

Expert systems, such as hardware, software, and knowledge engineering, have a statistically significant impact on the quality of decision-making. This was confirmed by Rafeeq (2015), who established that A.I. applications could provide substantial assistance to employees in the completion of their tasks because of their superior ability to perform tasks perceived as painful.

Al-Bishtawi et al. (2015) underscored the importance of expert systems in banks in facilitating electronic audit procedures such as swiftness in task execution and obtaining necessary data and information. A.I. enhanced the efficiency and quality of the audit by saving effort, time, and costs allocated to the implementation of the audit procedures and plan.

2.3 Expert Systems as Educational Tools

The technology of expert system firstly was created in 1960 but the origin of development in this branch of research go back to the concept of production system two decades earlier. In the current days, there are a considerable number of expert system obtainable on the internet containing applications in science, medicine, and industry (Grove, 2000).

Expert systems—the most notable applications of A.I.—transfer the logical processes of the human mind and present these consistent processes through a machine. Khanna et al. (2010) defined expert systems as a computer application that resolves efficiently the complicated issues that need vast human experience; it simulates rational thinking using precise knowledge and interfaces. Additionally, these systems employ human expertise to solve the problems that usually require human intelligence.

Expert systems have an immense capacity to carry out tasks, have access to rare human knowledge and expertise, and seek to preserve it and facilitate their use in specific areas by applying multiple perspectives to identify appropriate solutions to complex problems (Al-Maroof et al., 2021; Wan Ishak et al., 2010). Expert systems can provide solutions, answer questions, or provide advice at similar to or higher levels than that provided by human experts in the same field (Leo et al., 2021; Lucas & Van Der Gaag, 1991). In this way, users can exploit their insights and experiences using the analytical and data retrieval powers of the system (Al-Hamad et al., 2021; Keen, 1980). The teaching processes employed by expert systems include dynamic and rational training, and they play a critical role in enhancing teaching so that both teachers and students become interested and engaged in the learning process (Nazir et al., 2021; Sultan et al., 2021). These processes can then be used in educational curricula to promote and improve students' performance by fostering creative skills such as thinking, imagination, and cognition (Alshurideh, 2014; Alshurideh et al., 2021).

The study of Stella and Madhu (2017) emphasized the promotion of an expert system in the Nigerian educational system, finding that the tool provided great assistance for repeated study and understanding. Organizations can lose more of their useful problem-solving and analytical capabilities if A.I. technology and expert systems get neglected. Hassan (2010) identified the uses and efficacy of A.I. in libraries, finding that librarians' insufficient knowledge of artificial intelligence affected their understanding of any type of practical application in libraries and how to extract knowledge and use it in information bases in expert systems in libraries.

Using of internet grows continuously and additionally, the wireless communication supports the portability of internet access. Hence, the internet-based expert systems are available for accessing everywhere in the world at the needed locations (Grove, 2000).

2.4 Expert Systems as Educational Tools

The success or failure of educational technology use is highly dependent on teachers' attitudes toward technology. Teachers' technology self-efficacy is a critical factor in attitudes toward educational technology and the propensity to utilize it. Some studies have indicated that men tend to have higher computer self-efficacy than women (Koch et al., 2008; Ong & Lai, 2006; Sieverding & Koch, 2009) which can affect their attitudes towards its usefulness for education. The study of Wong et al. (2012) found that the effect of computer teaching efficacy on perceived value and attitude toward computer use was significant for female student teachers but not for their male counterparts.

Several studies have applied technology acceptance models (TAM) to assess teachers' perceptions of the usefulness, ease of use, and attitudes toward technology use and education in the classroom (Wong et al., 2012). The study of Wong et al. (2012) found that their TAM model accounted for 36.8% of the variance in intention to use computers among student teachers, and significant differences were identified based on computer teaching efficacy and gender.

Regarding the use of educational technology in Southeast Asia and Serbia, respectively, other researchers have used different measures to evaluate teachers' attitudes toward technology. Some studies have reported more positive attitudes towards technology among males than females (Chou et al., 2011; Durndell & Haag, 2002; Durndell et al., 2000; Kay, 2009; Kesici et al., 2009; Ong & Lai, 2006).

Cai et al. (2017) meta-analysis of gendered attitudes toward technology use revealed that males tend to hold more favorable attitudes toward technology use than females. However, some studies have found the opposite (Chou et al., 2011; Johnson, 2011; Price, 2006), and others indicate no significant differences (Imhof et al., 2007; Teo et al., 2015, 2016). For example, the studies of (Teo et al., 2015, 2016) examined pre-service teachers' perceptions and found no significant differences based on gender.

Table 1 Distribution of the study sample

Variable	Categories	Frequency	Percentage
Gender	Male	55	59.1
	Female	38	40.9
	Total	93	100.0
Educational qualification	M.AMA	40	43.0
	Ph.D	53	57.0
	Total	93	100.0
Experience	<5 years	10	10.8
	5–10 years	44	47.3
	>10 years	39	41.9
	Total	1093	10.8100

3 Method and Procedures

3.1 Population and Sample

This study used the survey method to collect quantitative data using questionnaires. Random sampling identified 93 faculty members from the college of education in 5 UAE Universities. To gather the whole sample from the population of the five UAE universities using the simple sampling technique, 15–17 faculty members were selected from each university. Data collection occurs during the second semester of the academic year 2018/2019. Table 1 presents a summary of the participants according to gender, educational qualification, and years of experience.

3.2 Instrument

Data was collected using a two-part tool. The first part collected primary socio-demographic data (gender, educational qualification, and experience). The second part comprised 25 questions measuring faculty's perceptions of the effectiveness of the introduction of A.I. and expert systems in the curricula and teaching methods of university education colleges. A five-point Likert scale scored the responses. The case ranged from 5 = completely agree to 1 = disagree.

The questionnaire was presented to a group of arbitrators to ensure its reliability. After an initial review, some paragraphs got deleted, new sections added, and others modified. The Cronbach's Alpha coefficient used to verify the scale's stability and internal consistency calculated resulted in a high value of 0.932.

3.3 Statistical Processing

Frequencies and percentages of personal information, arithmetic averages, and standard deviations of questionnaires are used for developing questionnaire paragraphs. An independent-sample t-test is applied to variables with two categories and analysis of variance (ANOVA) is used in analyzing variables with more than two groups.

4 Results

Table 2 presents the means and standard deviations of the questionnaire responses. Regarding RQ1 ("How do faculty assess the effectiveness of introducing A.I. and expert systems in the curricula and teaching methods of university education colleges?"), the findings indicate above-neutral attitudes among faculty with means ranging between 3.48 and 4.26. The lowest scoring items related to A.I.'s ability to help in developing students' perception and imagination (item 4), giving students the ability to manage difficult and complex situations (item 7), and to creatively solve problems presented in the absence of complete information (item 5).

Regarding RQ2 ("Does the effectiveness of introducing A.I. and expert systems in the curricula and teaching methods of university education colleges vary according to variables such as gender, educational qualification, and experience?"), independent-sample t-tests and ANOVA were applied to measure differences between means depending on the gender, educational qualification, and experience. Table 3 presents the results of a t-test analyzing differences according to gender. Although female faculty considered A.I. to be more useful and practical than their male counterparts, there was no statistically significant difference between assessments based on gender.

Table 4 summarizes the results of the t-test assessing differences in faculty perceptions according to academic qualifications. There was a statistically significant difference in faculty's understanding of A.I., whereby participants with master's degrees more commonly assessed the use of such technology as useful and practical ($t = 2.457, p < 0.05$).

The degree of freedom has calculated using the formula of (n–1) where n is the sample size. Therefore, there are 39 degrees of freedom for *MA* and 52 degrees of freedom for a Ph.D.

As shown by Table 5, an ANOVA found no statistical difference between participants' perceptions of AI-based on teaching experience.

5 Discussion and Recommendations

This study used a questionnaire to assess UAE faculty members' attitudes towards the usefulness and effectiveness of A.I. in enhancing learning. Although overall, the

Table 2 Summary of questionnaire results

No	Item	Mean	SD
1	Introducing artificial intelligence (A.I.) and expert systems into the university curricula increases students' creativity and performance	4.26	0.71
2	Introducing A.I. and expert systems into the university curricula enhances students' ability to think creatively	4.24	0.79
3	Introducing A.I. and expert systems into the curriculum helps students to respond quickly and creatively to new situations and circumstances	4.05	0.77
4	Introducing A.I. and expert systems into the curriculum develops students' perception and imagination	3.48	0.97
5	Introducing A.I. and expert systems into the curriculum helps students to solve in a creative manner the problems presented in the absence of complete information	3.61	1.11
6	Introducing A.I. and expert systems into university curricula increases the students' ability to discover knowledge creatively	3.82	1.06
7	A.I. and expert systems play a vital role in giving students the ability to manage difficult and complex situations	3.49	1.13
8	A.I. and expert systems in the university curricula increase students' pleasure during the lecture	3.94	0.94
9	A.I. in the university curricula increases students' ability to learn and understand based on past experiences	3.87	0.92
10	Introducing A.I. and expert systems increase the students' ability to develop creatively	3.75	1.00
11	A.I. and expert systems in the university curricula increase students' ability to understand and perceive visual matters	3.86	0.90
12	A.I. and expert systems increase the students' ability to innovate and renew creatively	4.00	0.94
13	Introducing A.I. and expert systems in the university curricula enhances students' creative skills	3.80	1.03
14	A.I. and expert systems in the university curricula creatively improve students' decision-making	3.95	0.88
15	A.I. and expert systems in the curriculum teach students the skill of interpretation	3.81	0.89
16	A.I. and expert systems help students to develop their performance	3.68	0.93
17	Introducing A.I. and expert systems in the university curriculum help students to creatively apply knowledge	3.67	0.78
18	A.I. and expert systems in the curriculum enable students to creatively present information	3.78	0.86
19	A.I. and expert systems in the curriculum help students to creatively carry out their assigned tasks	3.97	0.97
20	A.I. and expert systems in the curriculum help students to understand the information quickly	4.02	0.90
21	A.I. and expert systems help students to obtain timely and accurate information	4.14	0.90

(continued)

Table 2 (continued)

No	Item	Mean	SD
22	Introducing A.I. and expert systems in the university curricula enable students to keep abreast of technological developments	3.96	1.03
23	A.I. and expert systems help students to choose the best alternative	4.12	0.78
24	Introducing A.I. and expert systems in the university curricula increase productivity of the student	4.12	0.95
25	Introducing A.I. and expert systems in the university curricula enhances student intelligence	3.89	0.90

Table 3 Test Results for independent sample *t-test* to detect variances by gender variable

Gender	N	Mean	SD	t	df	P
Male	55	3.870	0.554	−0.423	54	0.673
Female	38	3.921	0.601		37	

Table 4 Differences in perception based on academic qualifications

Degree	N	Mean	SD	T	Df
MA	40	4.055	0.477	2.475	39
Ph.D	53	3.767	0.609		52

Table 5 Results of an ANOVA measuring difference based on experiences

Variance source	Sum of squares	Df	Mean square	F	p
Within groups	1.535	2	0.767	2.425	0.094
Between groups	28.482	90	0.316		
Total	30.016	92	1.083		

participants reported that A.I. and Expert systems to be a useful tool in fostering students' innovation and creative thinking, the lowest scores were related to the role of A.I. and expert systems in developing students' perception and imagination (item 4), providing students with the ability to deal with complex cases (item 7). Faculty also had somewhat less confidence in the capacity of A.I. to assist problem-solving in the absence of complete information (item 5). Claudé and Combe (2018) described A.I. as a useful tool for managing complex situations; however, finding of this study indicates the limits of A.I. and the continued need for teachers to help in guiding students to address complex problems and obtain the necessary information to solve them. A study on the use of intelligent information systems in biomedical education highlighted the importance of students' use of search engines, bibliographic websites and library resources to support their research (Aparicio et al., 2018). Zawacki-Richter et al. (2019) meta-analysis demonstrated that the faculty used A.I. tools to provide teaching content to students, it was generally combined with support by giving feedback and hints on how to solve questions related to the content, as well as detecting the difficulties/errors of students (Alsharari & Alshurideh, 2021; Alshurideh et al., 2019a, 2021).

The lack of significant difference in faculty perceptions of A.I. according to gender aligns with previous findings regarding the impact of gender on attitudes toward technology (North & Noyes, 2002; Teo et al., 2015, 2016; Imhof et al., 2007; Alshurideh et al., 2019a, b, c; Kurdi et al., 2020, 2020a, b)

In the context of the Arab Gulf region, there is a limited research on educators' attitudes toward A.I.; however, we can compare the results with findings on attitudes toward other educational technology tools. Al-Emran et al. (2016) found no statistical difference between male and female educators' perspectives, which they attributed to equal levels of use and experience. However, a large-scale study of educators at King Saud University. Alwraikat and Al Tokhaim (2014) found that female instructors' attitudes, towards mobile learning, were significantly more positive than that of male instructors. The study of Al-Hunaiyyan et al. (2017), which studied the impact of instructors age and gender differences in the acceptance of mobile learning, found that the younger instructors were more likely to accept the mobile learning compare to the elderly instructors. Additionally, the study of Alwraikat and Al Tokhaim (2014) study findings affirmed that female instructors were more favorable to mobile learning compared to male counterparts.

The lack of difference based on experience suggests that the participants are using this technology with a high degree of mastery regardless of experience. Therefore, they view the same impact on students after the inclusion in the teaching methods and curriculum. Teachers' performance and interaction are at the same level, despite their experience. Other studies in the Arab Gulf region similarly identified no differences in attitudes toward mobile learning based on the faculty members' expertise (Al-Emran et al., 2016; Eltahir, 2019). Gebhardt (2018), in the study on Adaptive Learning Courseware as a Tool to Build Foundational Content Mastery: Evidence from Principles of Microeconomics, also found that the use of the courseware promoted the student's mastery of concepts on the principles of Microeconomics.

Based on the findings, the researchers offer two main recommendations:

(1) A.I. and expert systems introduced into the curricula and teaching methods of university education college. The two have a significant impact on students' creativity and excellence return.

(2) Further research, similar to this study, must be conducted on applications of A.I. in higher education systems in developing countries, and different perspectives should have been explored. The reason for choosing this study is most of the developing countries are yet to embrace e-learning due to some limiting factors such as cultures, economy, and the level of technological development. For example, this study focused on teachers' perceptions of learning-facing tools, i.e., software that responds to students' needs and is used to receive and understand new information (Baker & Smith, 2019). Future research should examine

precisely how faculty members use A.I. to support students' learning and the level of acceptance of A.I. in educational colleges in developing countries.

This study has several limitations. First, the sample was limited to the college of education faculty rather than encompassing faculty across departments. Future studies can examine perceptions of A.I. educational technology across disciplines to evaluate if any differences can be related to educators' area of expertise. Future research can also incorporate additional variables such as age to measure teachers' perceptions, as well as measures of teacher self-efficacy and mastery of A.I. and expert systems.

Declaration of Conflicting Interest The author(s) declared no potential conflicts of interest concerning the research, authorship, and publication of this article

Funding The author(s) received no financial support for the research, authorship, and publication of this article.

References

Al-Bishtawi, S. H., & Al-Buqami, M. A. (2015). The impact of the application of expert systems in commercial banks on electronic auditing procedures, from the viewpoint of public accountants: A comparative study in the Hashemite Kingdom of Jordan and Saudi Arabia. *Jordan Journal of Business Administration, 11*(1), 117–151.

Al-Emran, M., Elsherif, H. M., & Shaalan, K. (2016). Investigating attitudes towards the use of mobile learning in higher education. *Computers in Human Behavior, 56*, 93–102.

Al-Hamad, M. Q., Mbaidin, H. O., Alhamad, A. Q. M., Alshurideh, M. T., Kurdi, B. H. A., & Al-Hamad, N. Q. (2021). Investigating students' behavioral intention to use mobile learning in higher education in UAE during Coronavirus-19 pandemic. *International Journal of Data and Network Science, 5*(3). https://doi.org/10.5267/j.ijdns.2021.6.001.

Al-Hunaiyyan, A., Alhajri, R., & Al-Sharhan, S. (2017). Instructors age and gender differences in the acceptance of mobile learning. *International Journal of Interactive Mobile Technologies, 11*(4), 4–16.

Al-Maroof, R., Ayoubi, K., Alhumaid, K., Aburayya, A., Alshurideh, M., Alfaisal, R., & Salloum, S. (2021). The acceptance of social media video for knowledge acquisition, sharing and application: A comparative study among YouYube users and TikTok users' for medical purposes. *International Journal of Data and Network Science, 5*(3). https://doi.org/10.5267/j.ijdns.2021.6.013.

Al Shebli, K., Said, R. A., Taleb, N., Ghazal, T. M., Alshurideh, M. T., & Alzoubi, H. M. (2021). RTA's Employees' perceptions toward the efficiency of artificial intelligence and big data utilization in providing smart services to the residents of Dubai. *The International Conference on Artificial Intelligence and Computer Vision*, 573–585.

Alhashmi, S. F. S., Alshurideh, M., Al Kurdi, B., & Salloum, S. A. (2020). A systematic review of the factors affecting the artificial intelligence implementation in the health care sector. In *Advances in Intelligent Systems and Computing: Vol. 1153 AISC* (pp. 37–49). https://doi.org/10.1007/978-3-030-44289-7_4.

AlShamsi, M., Salloum, S. A., Alshurideh, M., & Abdallah, S. (2021). Artificial intelligence and blockchain for transparency in governance. In *Studies in Computational Intelligence* (Vol. 912, pp. 219–230). https://doi.org/10.1007/978-3-030-51920-9_11.

Alsharari, N. M., & Alshurideh, M. T. (2021). Student retention in higher education: the role of creativity, emotional intelligence and learner autonomy. *International Journal of Educational Management, 35*(1). https://doi.org/10.1108/IJEM-12-2019-0421.

Alshurideh, M., Al Kurdi, B., & Salloum, S. (2019a). Examining the main mobile learning system drivers' effects: a mix empirical examination of both the expectation-confirmation model (ECM) and the technology acceptance model (TAM). *International Conference on Advanced Intelligent Systems and Informatics*, 406–417.

Alshurideh, M., Salloum, S. A., Al Kurdi, B., & Al-Emran, M. (2019b). Factors affecting the social networks acceptance: An empirical study using PLS-SEM approach. *PervasiveHealth: Pervasive Computing Technologies for Healthcare, Part F1479*. https://doi.org/10.1145/3316615.3316720.

Alshurideh, M. T., Al Kurdi, B., AlHamad, A. Q., Salloum, S. A., Alkurdi, S., Dehghan, A., Abuhashesh, M., & Masa'deh, R. (2021). Factors affecting the use of smart mobile examination platforms by universities' postgraduate students during the COVID-19 pandemic: An empirical study. *Informatics, 8*(2). https://doi.org/10.3390/informatics8020032.

Alshurideh, M. (2014). The factors predicting students' satisfaction with universities' healthcare clinics' services. *Dirasat. Administrative Sciences, 41*(2), 451–464.

Alshurideh, Muhammad, Salloum, S. A., Al Kurdi, B., Monem, A. A., & Shaalan, K. (2019c). Understanding the quality determinants that influence the intention to use the mobile learning platforms: A practical study. *International Journal of Interactive Mobile Technologies, 13*(11).

Alwraikat, M., & Al Tokhaim, H. (2014). Exploring the potential of mobile learning use among faculty members. *International Journal of Interactive Mobile Technologies, 8*(3), 4–10.

Amarneh, B. M., Alshurideh, M. T., Al Kurdi, B. H., & Obeidat, Z. (2021). The Impact of COVID-19 on E-learning: Advantages and challenges. *The International Conference on Artificial Intelligence and Computer Vision*, 75–89.

Aparicio, F., Morales-Botello, M. L., Rubio, M., Hernando, A., Muñoz, R., López-Fernández, H., Glez-Peña, D., Fdez-Riverola, F., de la Villa, M., & Maña, M. (2018). Perceptions of the use of intelligent information access systems in university level active learning activities among teachers of biomedical subjects. *International Journal of Medical Informatics, 112*, 21–33.

Baker, T., & Smith, L. (2019). *Educ-AI-tion rebooted? Exploring the future of artificial intelligence in schools and colleges. Retrieved from Nesta Foundation.*

Brown, C. E., & O'Leary, D. E. (1995). Introduction to artificial intelligence and expert systems. *International Journal of Intelligent Systems.*

Cai, Z., Fan, X., & Du, J. (2017). Gender and attitudes toward technology use: A meta-analysis. *Computers & Education, 105*, 1–13.

Chou, C., Wu, H.-C., & Chen, C.-H. (2011). Re-visiting college students' attitudes toward the Internet-based on a 6-T model: Gender and grade level difference. *Computers & Education, 56*(4), 939–947.

Claudé, M., & Combe, D. (2018). *The roles of artificial intelligence and humans in decision making: Towards augmented humans?: A focus on knowledge-intensive firms..*

Clement, R. P. (1992). Learning expert systems by being corrected. *International Journal of Man-Machine Studies, 36*(4), 617–637.

Duncan, S. C. (2010). Gamers as designers: A framework for investigating design in gaming affinity spaces. *E-Learning and Digital Media, 7*(1), 21–34.

Durndell, A., & Haag, Z. (2002). Computer self efficacy, computer anxiety, attitudes towards the Internet and reported experience with the Internet, by gender, in an East European sample. *Computers in Human Behavior, 18*(5), 521–535.

Durndell, A., Haag, Z., & Laithwaite, H. (2000). Computer self efficacy and gender: A cross cultural study of Scotland and Romania. *Personality and Individual Differences, 28*(6), 1037–1044.

Eltahir, M. E. (2019). E-learning in developing countries: Is it a panacea? A case study of Sudan. *IEEE Access, 7*, 97784–97792.

Gebhardt, K. (2018). Adaptive learning courseware as a tool to build foundational content mastery: Evidence from principles of microeconomics. *Current Issues in Emerging ELearning, 5*(1), 7–19.

Ghadirli, H. M., & Rastgarpour, M. (2013). An adaptive and intelligent tutor by expert systems for mobile devices. ArXiv Preprint ArXiv:1304.4619.

Glavinic, V., Rosic, M., & Zelic, M. (2007). Agents in m-learning systems based on intelligent tutoring. *International Conference on Universal Access in Human-Computer Interaction*, 578–587.

Grove, R. (2000). Internet-based expert systems. *Expert Systems, 17*(3), 129–135.

Hassan, A. (2010). *Using artificial intelligence applications in university libraries: Designing an expert reference system for the University of Khartoum Library*. University of Khartoum.

Hinojo-Lucena, F.-J., Aznar-Díaz, I., Cáceres-Reche, M.-P., & Romero-Rodríguez, J.-M. (2019). Artificial intelligence in higher education: A bibliometric study on its impact in the scientific literature. *Education Sciences, 9*(1), 1–9.

Imhof, M., Vollmeyer, R., & Beierlein, C. (2007). Computer use and the gender gap: The issue of access, use, motivation, and performance. *Computers in Human Behavior, 23*(6), 2823–2837.

Jewandah, S. (2018). How artificial intelligence is changing the banking sector–A case study of top four commercial Indian banks. *International Journal of Management, Technology and Engineering, 8*(7), 525–530.

Johnson, R. D. (2011). Gender differences in e-learning: Communication, social presence, and learning outcomes. *Journal of Organizational and End User Computing (JOEUC), 23*(1), 79–94.

Kay, R. H. (2009). Examining gender differences in attitudes toward interactive classroom communications systems (ICCS). *Computers & Education, 52*(4), 730–740.

Keen, P. G. W. (1980). *MIS research: Reference disciplines and a cumulative tradition*.

Kesici, S., Sahin, I., & Akturk, A. O. (2009). Analysis of cognitive learning strategies and computer attitudes, according to college students' gender and locus of control. *Computers in Human Behavior, 25*(2), 529–534.

Khanna, S., Kaushik, A., & Barnela, M. (2010). Expert systems advances in education. *Proceedings of the National Conference on Computational Instrumentation NCCI-2010. CSIO*, 109–112.

Koch, S. C., Müller, S. M., & Sieverding, M. (2008). Women and computers. Effects of stereotype threat on attribution of failure. *Computers & Education, 51*(4), 1795–1803.

Kurdi, B. A., Alshurideh, M., & Salloum, S. A. (2020a). Investigating a theoretical framework for e-learning technology acceptance. *International Journal of Electrical and Computer Engineering, 10*(6). https://doi.org/10.11591/IJECE.V10I6.PP6484-6496.

Kurdi, B. A., Alshurideh, M., Salloum, S. A., Obeidat, Z. M., & Al-dweeri, R. M. (2020b). An empirical investigation into examination of factors influencing university students' behavior towards elearning acceptance using SEM approach. *International Journal of Interactive Mobile Technologies, 14*(2). https://doi.org/10.3991/ijim.v14i02.11115.

Leo, S., Alsharari, N. M., Abbas, J., & Alshurideh, M. T. (2021). From offline to online learning: A qualitative study of challenges and opportunities as a response to the COVID-19 Pandemic in the UAE higher education context. In *Studies in Systems, Decision and Control* (Vol. 334). https://doi.org/10.1007/978-3-030-67151-8_12.

Lucas, P., & Van Der Gaag, L. (1991). *Principles of expert systems* (Vol. 13). Addison-Wesley Wokingham.

Nazir, M. I. J., Rahaman, S., Chunawala, S., & AlHamad, A. Q. M. (2021). Perceived factors affecting students academic performance. Nazir, J., Rahaman, S., Chunawala, S., Ahmed, G., Alzoubi, H., Alshurideh, M., & AlHamad, A. (2022) *Perceived* factors affecting students academic performance. *Academy of Strategic Management Journal, 21*(Special Issue 4), 1–15., *21*(Special Issue 4), 1–15.

North, A. S., & Noyes, J. M. (2002). Gender influences on children's computer attitudes and cognitions. *Computers in Human Behavior, 18*(2), 135–150.

Nuseir, M. T., Al Kurdi, B. H., Alshurideh, M. T., & Alzoubi, H. M. (2021). Gender discrimination at workplace: Do artificial intelligence (AI) and machine learning (ML) have opinions about It. *The International Conference on Artificial Intelligence and Computer Vision*, 301–316.

Ong, C.-S., & Lai, J.-Y. (2006). Gender differences in perceptions and relationships among dominants of e-learning acceptance. *Computers in Human Behavior, 22*(5), 816–829.

Popenici, S. A. D., & Kerr, S. (2017a). Exploring the impact of artificial intelligence on teaching and learning in higher education. *Research and Practice in Technology Enhanced Learning, 12*(1), 22.

Popenici, S. A. D., & Kerr, S. (2017b). Exploring the impact of artificial intelligence on teaching and learning in higher education. *Research and Practice in Technology Enhanced Learning, 12*(1), 1–13.

Price, L. (2006). Gender differences and similarities in online courses: Challenging stereotypical views of women. *Journal of Computer Assisted Learning, 22*(5), 349–359.

Rafeeq, A. (2015). *The use of artificial intelligence applications in the management of the organization's activities A case study of a group of economic institutions.* Um El Bouaghi, Algeria.

Rao, S. S., Nahm, A., Shi, Z., Deng, X., & Syamil, A. (1999). Artificial intelligence and expert systems applications in new product development—A survey. *Journal of Intelligent Manufacturing, 10*(3), 231–244.

Salem, A.-B. M. (2000). The potential role of artificial intelligence technology in education. *Proceedings of the International Conference on Technology in Mathematics Education (July 5–7, 2000, Beirut, Lebanon)*, 178–185.

Salloum, S. A., Alshurideh, M., Elnagar, A., & Shaalan, K. (2020a). Machine learning and deep learning techniques for cybersecurity: A review. In *Advances in Intelligent Systems and Computing: Vol. 1153 AISC.* https://doi.org/10.1007/978-3-030-44289-7_5.

Salloum, S. A., Alshurideh, M., Elnagar, A., & Shaalan, K. (2020b). Mining in educational data: Review and future directions. In *Advances in Intelligent Systems and Computing: Vol. 1153 AISC.* https://doi.org/10.1007/978-3-030-44289-7_9.

Sieverding, M., & Koch, S. C. (2009). (Self-) Evaluation of computer competence: How gender matters. *Computers & Education, 52*(3), 696–701.

Stella, N. N., & Madhu, B. K. (2017). Impact of expert system as tools for efficient teaching and learning process in educational system in Nigeria. *International Journal on Recent and Innovation Trends in Computing and Communication, 5*(11), 129–133.

Sultan, R. A., Alqallaf, A. K., Alzarooni, S. A., Alrahma, N. H., AlAli, M. A., & Alshurideh, M. T. (2021). How students influence faculty satisfaction with online courses and do the age of faculty matter. *The International Conference on Artificial Intelligence and Computer Vision*, 823–837.

Tan, H. (2017). A brief history and technical review of the expert system research. *IOP Conference Series: Materials Science and Engineering, 242*(1), 1–5.

Teo, T., Fan, X., & Du, J. (2015). Technology acceptance among pre-service teachers: Does gender matter? *Australasian Journal of Educational Technology, 31*(3), 235–251.

Teo, T., Milutinović, V., & Zhou, M. (2016). Modelling Serbian pre-service teachers' attitudes towards computer use: A SEM and MIMIC approach. *Computers & Education, 94*, 77–88.

Tripathi, S., & Al Shahri, M. (2016). Omani community in digital age: A study of Omani women using back channel media to empower themselves for frontline entrepreneurship. *International Journal of Social, Behavioral, Educational, Economic, Business and Industrial Engineering, 10*(6), 1748–1753.

Wan Ishak, W. H., Ku-Mahamud, K. R., & Md Norwawi, N. (2010). Conceptual framework for intelligent decision support system in emergency management. In *International Conference on Arts, Social Sciences and Technology 2010 (ICAST2010), 24–26 February 2010, Gurney Hotel, Penang.* https://repo.uum.edu.my/id/eprint/3468/.

Warschauer*, M. (2004). The rhetoric and reality of aid: Promoting educational technology in Egypt. *Globalisation, Societies and Education, 2*(3), 377–390.

Wong, K.-T., Teo, T., & Russo, S. (2012). Influence of gender and computer teaching efficacy on computer acceptance among Malaysian student teachers: An extended technology acceptance model. *Australasian Journal of Educational Technology, 28*(7), 1190–1207.

Yousuf, H., Zainal, A. Y., Alshurideh, M., & Salloum, S. A. (2021). Artificial intelligence models in power system analysis. In *Artificial Intelligence for Sustainable Development: Theory, Practice and Future Applications* (pp. 231–242). Springer.

Zawacki-Richter, O., Marín, V. I., Bond, M., & Gouverneur, F. (2019). Systematic review of research on artificial intelligence applications in higher education–where are the educators? *International Journal of Educational Technology in Higher Education, 16*(1), 1–27.

A Roadmap for SMEs to Adopt an AI Based Cyber Threat Intelligence

Abhilash J. Varma, Nasser Taleb, Raed A. Said, Taher M. Ghazal⑩,
Munir Ahmad, Haitham M. Alzoubi⑩, and Muhammad Alshurideh⑩

Abstract Cybersecurity has started to become the most significant concern among organizations as the number of threats and criminal activities in the past decade has increased exponentially. Cybercriminals and their attacking techniques have become increasingly sophisticated over the past couple of years. Conventional security measures will no longer be able to detect and mitigate the propagation of such advanced attacking trends. More and more hackers have started focusing on Small and medium-sized enterprises (SMEs) taking advantage of their limited resources. Therefore, SMEs will have to quickly adopt Artificial Intelligence (AI) based cybersecurity system in their infrastructure to defend themselves effectively and efficiently. It is currently forecasted that by 2021, 75% of all organizations will use AI and Machine learning (ML) applications in their security architecture to protect against all cyber threats. In this paper, the researchers identify the various challenges faced by SMEs in

A. J. Varma · N. Taleb · R. A. Said
Canadian University Dubai, Dubai, UAE

T. M. Ghazal
Faculty of Information Science and Technology, Center for Cyber Security, Universiti Kebansaan
Malaysia (UKM), 43600 Bangi, Selangor, Malaysia
e-mail: taher.ghazal@skylineuniversity.ac.ae

School of Information Technology, Skyline University College, Sharjah, UAE

M. Ahmad
School of Computer Science, National College of Business Administration and Economics,
Lahore 54000, Pakistan

H. M. Alzoubi (✉)
School of Business, Skyline University College, Sharjah, UAE
e-mail: haitham.alzubi@skylineuniversity.ac.ae

M. Alshurideh
Department of Marketing, School of Business, University of Jordan, Amman, Jordan
e-mail: m.alshurideh@ju.edu.jo; malshurideh@sharjah.ac.ae

Department of Management, College of Business Administration, University of Sharjah, Sharjah,
UAE

adopting an AI based cybersecurity due to their knowledge gap and lack of expertise. The researcher intends to provide a good background on AI, Cyber Threat Intelligence (CTI) and highlight some of the significant benefits provided by an AI based CTI system. A simple roadmap is developed using a qualitative research methodology to help SMEs effectively implement an AI based Cyber Threat Intelligent system in their infrastructure.

Keywords Cybersecurity · Artificial intelligence · Machine learning · Cyber threat intelligence · Deep learning · AI · ML

1 Introduction

Cyberattacks have increased exponentially over the globe. The world-famous investor Warren Buffet sees cyber risk as one of the gravest threats to humanity. In the past decade, we have seen some of the largest data breaches, national level hacking activities, political manipulations, and the use of botnets to bring down major telecom industries (Alzoubi et al., 2021a, b). These are compromising our private, professional, and national existence. No industry or organization is now completely safe from a cyber threat and therefore, the security professionals are always expected to stay current and up to date on the latest attack trends and vulnerabilities used by hackers (Kashif et al., 2021). They must detect, analyze and protect the organization in real-time (Alshurideh, 2022; Alshurideh et al., 2022a, b). Alkhalil et al. (2021) mentioned that the global cost of cybercrimes is expected to reach USD 6 trillion by 2021; 43% of the total cyberattacks target small businesses; $3.9 million is the average cost of a data breach for small to medium size businesses; organizations usually take nearly 6 months to detect a breach; the number of connected IoT devices will reach 75 billion by the year 2025 (Alnuaimi et al., 2021a, b).

Cyber threat intelligence is a process that proactively and iteratively searches various systems, databases, networks (Farouk, 2021), and other resources to detect and educate itself about the changing cybersecurity threats that can evade existing security controls. It enables the cybersecurity professionals to quickly recognize indicators of cyberattacks, analyze the attack methods and respond in a timely manner (Ali Alzoubi, 2021a, b). The cyber threat intelligence helps organizations to isolate and remedy these advanced threats before a cyber threat occurs. Even though various cyber threats follow different methods of cyberattack, they have a similar life cycle starting with the victim reconnaissance to performing malicious activities on the victim's network\devices (Mondol, 2021). The primary purpose of Cyber threat intelligence is to detect all weak points in the existing security solution and thereby take the necessary actions to safeguard the organization (Al Ali, 2021).

Over the years, Cyber Threat Intelligence has evolved from small ad-hoc tasks to a much more powerful program with their own dedicated staff, tools and processes that can support the entire organization (Radwan & Farouk, 2021). Most organizations nowadays not only consume cyber threat intelligence but also produce them.

This shows the growing maturity, popularity and professionalization in this field (Lee, 2020). Without the help of advanced data mining techniques, ML and AI, it is practically impossible for cyber analysts to stay on top of the latest attack trends, analyzing the current attack logs, generating intelligent reports that can be used to share and report on cyber security (Al Kurdi et al., 2021; Alhashmi et al., 2020; Salloum et al., 2020). AI and ML systems can significantly support cyber analysts in early detection as well as providing timely recommendations for mitigating various threats (AlShamsi et al., 2021; Nuseir et al., 2021; Yousuf et al., 2021).

In this research, the researcher will be reviewing some of the existing literatures on AI and CTI, discussing their core processes, functions and development lifecycle (Akour et al., 2021; Almaazmi et al., 2020; Alshurideh et al., 2020a, b, c). One of the key issues identified in the literature review is that SMEs are currently lagging in adopting this technology in their infrastructure due to several challenges such as limitations in budget, expertise, knowledge, etc. This problem can be overcome if there is a roadmap that can help SMEs to plan and successfully implement an AI based CTI system (Al Al Suwaidi et al., 2021; Alzoubi et al., 2021a, b; Shebli et al., 2021). Thus, the aim of this research is to develop a simple, effective, customizable roadmap for SMEs. The effectiveness of the roadmap was evaluated using a qualitative analysis and based on this result, an updated, customized, and focused roadmap is finally presented by the researcher (AlHamad et al., 2022). The main advantage of this research would be that it will provide enough knowledge and the confidence for SMEs to go ahead with their investments in AI based cybersecurity and how to use intelligent AI based systems to protect the organization from the vast number of cyber threats in the modern technological environment (Lee et al., 2022a, b). The research also covers some of the main challenges and risks that are usually faced by SMEs as part of their adoption of AI based CTI system (Miller, 2021).

2 Literature Review

In recent years, there has been a significant increase in the number of cyberattacks faced by organizations. Traditional firewall and antivirus based cyber defense tools is now primitive and are no longer able to effectively block cyber threats or keep up with the rapidly developing threat vectors. Cyber attackers are developing new sophisticated AI based smart malwares that can understand the target system, learn their environment, evade detection, and make intelligent decisions making it more and more complex and challenging for the organizations to detect and defend against various cyber threats (Alzoubi, 2021a, b). It is practically impossible for cyber analysts and forensic investigators to keep up with the advanced number of threats, the amount of data to be analyzed and the speed of processes to actively respond to all cyber threats in real-time (Alhamad et al., 2021). Existing literature shows that even though there has been a significant increase in the number of organizations that have started to adopt AI based Cyber (Alzoubi & Aziz, 2021a, b). Threat Intelligence for their cyber defense, a vast majority of cyber security professionals still lack a deep understanding of Cyber Threat Intelligence, how AI based algorithms work and most importantly how to secure and harden an AI based CTI system (Ali et al., 2021). This is mostly because AI based CTI systems are not mature yet and is still an evolving technology (Alshurideh et al., 2020a, b, c). Large enterprises have the luxury of higher budget to invest in the latest innovative solutions, infrastructure, resources and for acquiring the required knowledge and expertise. Small and Medium sized Enterprises (SMEs) do not have this luxury and often struggle to invest and implement such new technologies in their infrastructure (Alshraideh et al., 2017; AlShurideh et al., 2019; Ghannajeh et al., 2015). SMEs therefore become most vulnerable to these advanced cyber threats as they are not kept up to date with the emerging cyber defense mechanisms (Alzoubi et al., 2021a, b).

Lee (2020) based on their latest CTI Survey, the biggest challenges faced by most organizations on the successful implementation of CTI are difficulties in integrating CTI with existing systems, the overall cost of implementation and continual improvements, lack of trained cyber security analysts and limited management support. Lidestri (2018) surveyed about 603 IT and IT Security professionals working for various US organizations who have already deployed or are planning to deploy an AI based CTI program in their organization. In this survey, while 71% of AI users voted that the AI technology will increase the speed of analyzing threats and providing a deeper security, only 60% of the current non-AI users believed this to be the primary benefit. 69% of the respondents believe that incorporating AI in Cybersecurity will increase the speed of analyzing threats while 64% believe it will help to quickly contain the infected endpoint (Ghazal et al., 2021). Cyber threat intelligence will increase the requirement for organizations to retain and employ more talented cyber analysts with AI exposure. This will bring a positive cybersecurity posture. Based on the survey, 68% of AI users said AI will improve the productivity of the IT staff while only 60% of non-AI users agreed on this. In one hand, 69% of AI users said that AI and ML technology will significantly improve the effectiveness of various application security activities, while on the other hand, only 59% of the non-AI users believed in this (Alzoubi & Yanamandra, 2020). The AI security experts\users believe that AI adaptation will help reduce the complexities in various security architecture. 56% of AI users responded that the adoption of AI based CTI system will decrease the overall complexity of the organizations security architecture while only 40% of non-AI users shared this view (Lee et al., 2022a, b). As the AI maturity level increases in the organization, the cyber analysts become more capable in understanding and identifying areas where AI implementation would be most beneficial (Aasriya, 2021). The stability, knowledge and expertise of AI based threat detection is drastically expected to improve as more and more organizations start to incorporate it and invest in this technology (Alkalha et al., 2012; Alnuaimi et al., 2021a, b; Altamony et al., 2012; Zu'bi et al., 2012). While non-AI users believe that an AI based CTI system will help them detect 41% of the previously undetected zero-day exploits, the AI users responded that they are able to detect 63% of these exploits with the help of AI (Lidestri, 2018).

With the boom of the Internet of Things (IoT) and cloud computing, the SMEs have become heavily dependent on the cyber world and therefore need to adopt the latest cyber forensic investigation techniques on exchanged and stored data (Singh & Singh, 2021). The growing attack surface includes amateur threats, such as phishing, sophisticated distributed denial of service attacks and skilled nation-state actors. Prevention is nearly impossible. Advanced persistent threats show that hackers are patient (Lee & Ahmed, 2021). Given enough time, attackers will be able to get in since the cost of an attack is low and automated probing will eventually find a weakness. ML, Data mining and AI can quickly and efficiently detect any network abnormalities, including static and dynamic malware attacks. They can quickly analyze, learn and act intelligently against advanced cyber threats (Hanaysha et al., 2021a, b).

3 Background Information

3.1 Artificial Intelligence (AI)

AI was explained as the science and development of intelligent systems and programs that can understand human intelligence (McCarthy, 2007). Artificial intelligent systems are technologies that can gather or read data from their various inputs, learn from their environment and make autonomous decisions based on the intelligence acquired through their experience and knowledge (Hanaysha et al., 2021a, b). They can adapt to different situations and act independently without any human intervention. AI should be able to automate most of your repeatable tasks and in the long run make intelligent decisions on behalf of the cyber analysts (Ahmed et al., 2021; M. Alshurideh et al., 2020a, b, c; Harahsheh et al., 2021; Naqvi et al., 2021). This will include functions like planning, reasoning, training, problem solving, etc. It's a highly effective tool that can continually learn from internal, as well as external data and become a vital part of the cyber security infrastructure. According to Davies (2020), AI systems can include capabilities like Cognitive computing, ML, DL, Natural Language Processing (NLP), Computer Vision, Speech, RPA, etc. as depicted in Fig. 1.

Fig. 1 Artificial intelligence

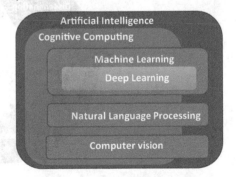

Cognitive Computing works with multiple subsets of AI such as ML, DL, NLP, Computer vision, etc. to simulate the human thought process and provide recommendations to help humans make better decisions. **Machine Learning** is a subset of AI and it enables machines to interact with data, learn from them and probably make changes to the algorithm in response to the data received without the need to follow explicitly programmed instructions (Alshurideh et al., 2022a, b). It includes deep learning, supervised algorithms and unsupervised algorithms to support predictions, analytics and data mining. These algorithms are used to recognize patterns and anomalies in data. Deep learning is a subset of ML. **Natural Language Processing** is a subset of AI that enables machines to work with text and languages, extracting their meaning and generating texts that are natural and grammatically correct (Ali et al., 2022).

3.2 Cyber Threat Intelligence (CTI)

Cyber threat intelligence is the final product that is disseminated after all cyber information is collected, processed and analyzed through a rigorous procedure called the intelligence cycle. It enables organizations to collect the correct threat intelligence and create a secure plan to detect, respond, prevent and mitigate various types of cyber threats and strengthen the organizations defense (Alzoubi et al., 2020a, b, c, d). The data should provide accurate useable intelligence. Figure 2, shows the various phases in the cycle.

According to Lee (2020), the Intelligence cycle starts with the Planning phase where the requirements are identified. This includes the specific questions and concerns that are to be addressed by the CTI program. Even though the requirements have a generic nature to all organizations, the specifications are unique to each

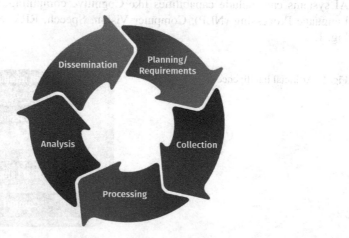

Fig. 2 The intelligence cycle. *Source* 2020 SANS CTI Survey

organization (Ahmad et al., 2021; AlMehrzi et al., 2020; Alzoubi et al., 2020a, b, c, d; Hayajneh et al., 2021). They are updated in an ad hoc manner based on the past incidents or upcoming trends in the industry. Specify who will be consuming the finished product—will it be used as an input to another system, will it be sent to Cyber analysts with technical expertise or will it be sent to top executives for their broad overview of the current cyber trends (Mehmood, 2021).

These requirements are then passed on to the Collection phase where they identify and evaluate the sources of intelligence that will help answer the requirements in an efficient manner. In addition to the inhouse data that is collected from the different departments\groups, information is also gathered from commercial threat feeds coming from CTI-specific vendors, generic security vendors, open-source threat feeds and forensics data (Alzoubi et al., 2020a, b, c, d). Information should also be periodically evaluated to ensure that it is current, effective and usable (Aburayya et al., 2020; AlShehhi et al., 2020; Svoboda et al., 2021). A data source that may have been critical in the past might no longer be needed, and new data sources might need to be identified as the organization and the threat landscape change (Alzoubi & Aziz, 2021a, b).

This data is then processed in the Processing phase to a format that is usable for the analysis phase (Alzoubi et al., 2022). Thousands of log events are generated every day by each system, these need to be collected, filtered and processed by automated systems to make any sense from the data. In the Analysis phase, the data is synthesized to support the requirements that were defined in the first phase. It searches for any potential security events and alerts the respective teams (Alzoubi & Yanamandra, 2020).

The final step is the Dissemination phase where the intelligence gathered is disseminated to the intended audience outlined in the planning phase. The data should go to the right recipient in the right format so that they are able to use it effectively. Timeliness and relevance are critical to the effectiveness of CTI dissemination. Cyber Threat Intelligence is primarily presented in the form of reports or brief summaries via emails, spreadsheets, power point presentations, etc. It is also used as an input to various security tools to generate alerts in an automated fashion (Alzoubi & Ahmed, 2019).

3.3   The Benefits of an AI Based CTI

Organizations generate and consume huge amounts of raw CTI data but it's incredibly time consuming and cumbersome to convert into intelligence. Most of these data is used for analyzing the application and its security (Alzoubi et al., 2020a, b, c, d). AI and ML programs can be the most beneficial for such tasks as they can analyze thousands of records\events, learn their behavior and raise an alert to the cyber analyst in case a vulnerability is detected (Alzoubi et al., 2020a, b, c, d). Semi-automation of cyber security tasks and data processing will be the golden standard followed by industries as the AI based CTI program maturity improves (Joghee et al., 2020).

3.4 The Core Functions of an AI Based CTI System

Alnuaimi et al., (2021a, b) proposed the core functions of an AI based cyber threat intelligence as shown in Figure 3. *Identification* of threats and the vulnerabilities is key to understanding the risk a potential attack can have on the organization. AI and ML algorithms can *discover* abnormalities in behavior patterns, flag it immediately and alert the cyber security analyst of a potential breach or threat. AI can use its data sets to *detect* cyber threats by identifying common threat characteristics. It is important to *investigate* events to gather maximum amount of information about the attack, the attack vector, its status, exploits and vulnerabilities (Aziz & Aftab, 2021). *Analysis* would enable the security analyst to identify the vulnerabilities, the devices that were compromised, the elements that failed to prevent the attack and how it happened (Alzoubi & Ahmed, 2019). AI would enable the security analysts to respond in a timely manner and *prevent* any form of progression. AI would be able to immediately *respond* to threats and isolate the infected equipment and stop the threat from progressing any further. Deep learning and ML enable AI to learn the behavior of attacks and thereby *predict* when, how and where the attack will begin. This will enable the security analysts to take appropriate actions to better prevent or mitigate an attack before it happens (Akhtar et al., 2021). AI enables the organization to *continuously monitor* the cyberspace to discover new threats vectors as well as defend against known and unknown threats.

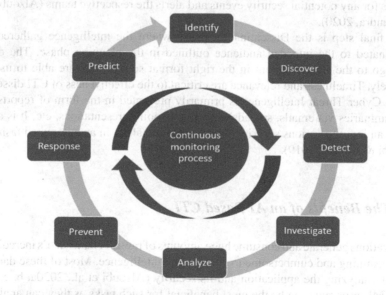

Fig. 3 Core functions of an AI based cyber threat intelligence

3.5 Designing an AI Based CTI system—The AI Development Life Cycle

Darraj et al. (2019) married the traditional SDLC with AI specific design and tech-nology. Figure 4 depicts the 9 phases in the AI development life cycle: Planning, Analysis, Design, Implementation, Training, Optimization & Validation, Testing & Integration, Deployment and Maintenance. In the Planning phase, we need to clearly specify the Cyber threat intelligence requirements, its scope, security, access control, privacy and data encryption along with governance, laws and compliance (Emerita, 2021). In the Analysis phase, various cybersecurity AI tools are analyzed to iden-tify the right tools, algorithms and data sets that are optimal for the organization. Here we will be able to decide which tasks can be automated and which needs a human intervention. In the Design phase, the AI based CTI solution is designed for the organization using the above identified AI tools covering ML, natural language processing, deep neural nets and cognitive computing (Khan, 2021). You will specify what will be the outcome and who will be using the system. In Phase four, the designed AI-CTI solution is implemented. In the Training phase, the designated data sets are used to train the AI-CTI algorithms for building its intelligence that will later be used for automation and taking intelligent decisions (Guergov & Radwan, 2021). In phase six, the cyber security analyst should optimize and validate the AI algorithm and data, ensure the cybersecurity hardening and patching are met and the system is functioning as intended. In phase seven, AI-CTI solution is integrated with the various systems in the organization and tested. The cyber security analyst starts testing the system ensuring all security and privacy requirements are completely met and confirm all PII, PHI, PCI/SOX and FTI data are appropriately protected (Obaid, 2021). In phase eight, the solution is deployed to production environment and becomes completely active. Here the cybersecurity analyst will continue hardening the system, patching and continuously monitor the AI-CTI solution. Phase nine is continual operation and maintenance of the AI-CTI system. The system, as any other application, needs to be continuously monitored and patched to ensure it meets expectations and requirements of the organization. We must make sure the AI system is behaving and functioning as planned. Threat forecasting and risk assessment should be performed at each phase considering the indicators of attack, compromise and interest. All vulnerabilities should be identified and remediated at each phase (Alnazer et al., 2017a, b).

3.6 Incorporating AI into Cyber Threat Intelligence

Learning can be defined as the process of acquiring knowledge or skill sets and when this knowledge is applied in a decision-making process it is called intelli-gence. An AI system must first learn before it can apply intelligence. To start off, AI systems can be used to perform daily repetitive tasks such as scanning, auditing, analyzing and reviewing huge volumes of data, logs and reports. It can perform

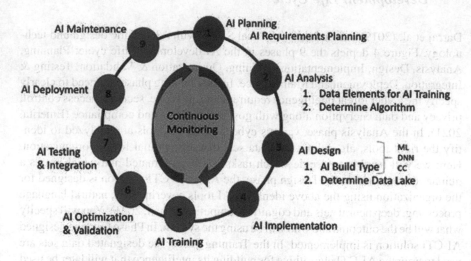

AI-DLC (Development Life Cycle)

AI Maintenance

AI Deployment

AI Testing
& Integration

AI Optimization
& Validation

AI Training

AI Planning
AI Requirements Planning

AI Analysis
1. Data Elements for AI Training
2. Determine Algorithm

AI Design
1. AI Build Type
2. Determine Data Lake

ML
DNN
CC

AI Implementation

Continuous
Monitoring

Fig. 4 An AI development life cycle

event correlation and generate intelligent reports which can then be further analyzed by the cyber analysts to strengthen the cyber defense system. AI allows to free up security operators so that they can now focus on more advanced tasks that require deeper thought processes (Alnazer et al., 2017a, b). There are several AI enabled tools and applications such as Neural Networks for intrusion detection, JASK security platform for detecting cyberthreats at its early stage, Cylance's security solution for predictive threat prevention, etc. to support cyber defense (Hamadneh et al., 2021). Adopting an AI based CTI system would improve the cyber resilience of the organization making its cyber defense strategy to be more proactive than reactive. The most significant feature of an AI based CTI system is its ability to adapt and learn. Overtime the system would be able to detect and respond to both known as well as unknown threats. Truong et al. (2020) categorized the most common applications of AI based cybersecurity into: Malware detection, Network intrusion and Phishing/Spam detection as illustrated in Fig. 5.

4   Research Design and Methodology

To address the research problem, the researcher has come up with a simple roadmap for SMEs that can help them to easily and effectively incorporate an AI based CTI system in their cyber security architecture. This roadmap was shared with some of the subject matter experts who are currently working in the cyber security team for various reputable organizations. This research is limited to the development of a

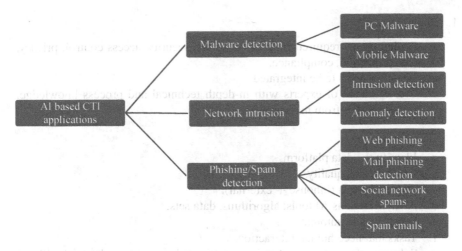

Fig. 5 Common AI based CTI applications

roadmap for SMEs for the successful adoption of an AI based CTI system as part of their cyber defense.

A qualitative research primarily deals with data which is verbal in the participants own written or spoken words based on his beliefs and understanding of the phenomenon (Bless et al., 2000). It tries to identify the problems that might be experienced by the participants with respect to the subject. The qualitative approach is appropriate for this study because the data collected and used focuses on the participants' subjective experiences in the field of organizational cyber security and the way they interpret them.

To evaluate the effectiveness of this roadmap we will be using a qualitative research method where the newly developed framework, as shown in Fig. 6 will be shared with various IT professionals. These participating IT professionals would be requested to evaluate the effectiveness, efficiency and the simplicity of the roadmap. The feedbacks from each of the participants are noted and used to develop our final roadmap for SMEs for the successful adoption of an AI based CTI system.

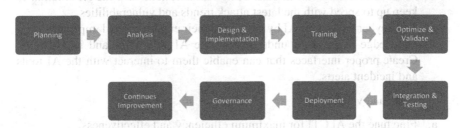

Fig. 6 A roadmap for SMEs for adopting an AI based CTI

1. Planning

   a. Setting up the requirements, scope, goals, security, access control, privacy, data protection, compliance.
   b. List of systems to be integrated.
   c. Create a team of experts with in-depth technical and process knowledge. Include SMEs from third party partners.

2. Analysis

   a. Identify the data platform.
   b. Evaluate data quality.
   c. Data collection (inhouse + external).
   d. Analyze various AI tools, algorithms, data sets.
   e. What tasks to automate.
   f. Tasks that need human interaction.
   g. Selecting the use cases that is easy to implement, provides a significant benefit, high impact to the cyber defense, high quality, complete, up-to-date data set that is readily available, has a with a very high impact.
   h. Look at some of the common use cases like malware detection, network intrusion detections and Spam\phishing detection.

3. Design and implement

   a. Designing the AI solution using the selected AI tools, algms, data sets identified from step 2.
   b. Specify what will be the outcome, who will be using it.
   c. Implement the solution in development environment.

4. Training

   a. Use pre-defined data sets to train AI to build intelligence.
   b. Exchanging and collaborating with other threat researchers and security professionals through various open-source threat intelligence platforms like Open threat exchange, Facebook threat exchange or IBM X-Force exchange is very critical for improving the efficiency and effectiveness of the AI algorithm in detecting new threats. Such collaborations help the AI solution to keep up to speed with the latest attack trends and vulnerabilities.
   c. Train cyber analyst for AI by upskilling their employees and improving their knowledge in the logic underpinning the AI algorithms and its behavior. Create proper interfaces that can enable them to interact with the AI tools and incident alerts.

5. Optimize and validate

   a. Fine tune the AI CTI for maximum efficiency and effectiveness.
   b. Ensure the AI CTI is functioning as intended.
   c. Continue patching and hardening.
   d. Assess the performance wrt initial requirements.

6. Integration and Testing

    a. Deploy Security orchestration, automation and response (SOAR) technologies to improve incident alert triage quality, defining a standardized incident response workflow, improving the security and operations management as well as reducing the onboarding time for a cyber analyst.

    b. Use SIEM to integrate the silos.

    c. Security and privacy requirements are met.

    d. Data is properly protected.

    e. Test the system thoroughly.

7. Deployment

    a. Deploy to production.

    b. Continue patching and hardening.

    c. Ensure the AI CTI is functioning as intended.

    d. Begin continues monitoring.

8. Governance

    a. A clear, transparent governance process needs to be implemented and adhered to monitor the performance of the AI-CTI solution. The controls should include checks on the roles and responsibilities, AI algorithm behavior abnormalities, risk tolerance, output verification and key performance indicators to measure the success of the program.

    b. Control processes to monitor the performance of AI CTI.

    c. BCP if the AI CTI stops working or goes rogue.

    d. It gives you the information you need to reduce Mean Time to Detect and Mean Time to Respond (MTTD and MTTR)—with a quicker, more decisive escalation process.

    e. It enables SOCs to assess and refine their IR processes, continually.

9. Continues Improvement

    a. Look at adding other tasks into the AI CTI.

The roadmap was sent via email to:

a. Senior Cyber Advisor working for a manufacturing company.

b. SIEM—Subject Matter Expert working for a reputed Information Technology and Service provider.

c. Director of Network, Infrastructure and Security working in the hospitality industry.

5 Findings and Results

5.1 *Comments Received from the Senior Cyber Advisor*

The road plan looks fab. It is really very impressive and efficient from what I understand. Great Job. I have some suggestions hoping they will be of some help.

Continual Improvement: I was expecting it from the planning phase itself. I was happy to see a heading in the end. I would recommend that every phase should have a clause of continual improvement. For instance, the processes defined need to be changed and improved with changing time. Our design needs to continuously evolve to better our product and its security. In the heading 9 itself we can say something of this sort in a couple of lines maybe?

Reviews and Approvals: Can we add some reviews and approvals? For instance, I would be more comfortable saying "reviewed and approved pre-defined data set" than 'pre-defined data set'.

One subheading, maybe a one liner somewhere about 'Segregation of Duties'?

5.2 *Comments Received from the SIEM—Subject Matter Expert*

Vulnerability assessment and Penetration testing should be used in the Planning phase for a better understanding of the current security infrastructure. This can be used to customize the solution for the customer. Getting to know the type of industry and customizing the solution accordingly for EG: Bank (SWIFT, PCIDSS), Healthcare (HIPPA etc.). Knowing the Number of users in an organization is important for the right sizing of Tools and devices. Use cases have to be custom tailored as per industry type of the organization. eg: create use cases for SWIFT application for banks etc. Providing Analyst with a Database for them to input information based on customer feedback after an incident is raised. Creating automated Templates in SOAR based on the incident triggered and for low-risk incidents automating the action using SOAR. For eg: IP is blacklisted, automatic action of blocking IP. Data should always be encrypted and make sure the main hardware is not connected to internet. Testing of the system should include PT or VA activity and detection of the threat. Tasks based on customer requirements and required less analysis by the Engineer needs to be automated. Further automation can be achieved but should have a final approval from Analyst before performing the actions.

5.3 Comments Received from the Director of Network, Infrastructure and Security

In the Planning phase, the company's mission, vision and executive management directions has to be analyzed. Based on these directions, the current security infrastructure must be reviewed and their setbacks to be noted. The AI based enhancements should be able to overcome these setbacks to gain the confidence and buy-in from the executive management.

Future research on various AI based cyberattacks by nefarious actors, their prevention and mitigation methods can further bridge the knowledge gap and secure the defense against such cyberattacks. The application of AI is a relatively new trend in combating cybercrimes and is still evolving. Further research should be conducted to improve the maturity of AI based cyber threat intelligence, their design, architecture and implementation. International Government bodies should also research and study how best they can legally cooperate and fight cybercrimes. The roadmap was evaluated and developed specifically for an SME; therefore, larger organizations might have added requirements to be addressed in the roadmap.

Considering the generic audience with backgrounds from different industries and having their unique requirement, we have restricted ourselves from discussing any vendor application or solution in detail. This gives us the opportunity to create a general guideline for SMEs that want to develop their own AI based CTI application as well as helping them negotiate with third-party service providers to make an educated and detailed blueprint to support their strategic decision on how they want to implement an AI based CTI system in their infrastructure.

6 Conclusion

Cybercrimes are increasing day by day and the attackers are using technologically advanced, complex and sophisticated threat models to evade detection. Adaptation of the various AI technologies have several benefits and can be used to detect and prevent attacks before it takes place. It is imperative that cybersecurity analysts have a deep and clear understanding of the AI algorithms, their functions and various possible applications to make the most of the cyber threat intelligence. The cyber analysts should ensure the AI systems are secured, hardened and continuously monitored to check and confirm they are behaving as expected, meeting all security/privacy requirements. More and more SMEs are acknowledging the need for an AI based Cyber Threat Intelligence to provide a secure and reliable cyberspace. It is critical that SMEs work together with larger organizations, Government, and other not-for-profit organizations to improve the current overall maturity of AI programs so that they become more dependable and less prone to going rogue.

The researcher believes that SMEs should start investing and adopting the latest AI based cyber security systems in their infrastructure so that they are all well prepared and understand the nuances of future AI-powered exploitations and other attacking trends. This will also help in the continued research, development and maturing of the AI technology. With the help of this article, cyber security leaders would have gained the knowledge and confidence required for the successful implementation of an AI based cybersecurity system as part of its cyber defense and resilience program. Organizations can start off with the basic automation of daily repetitive tasks such as analyzing security logs, errors and reports for the Cyber analyst. Eventually, the AI applications scope can be increased to a broader set of more complicated cases to significantly increase the speed, efficiency, and effectiveness of their cyber defense.

7 Recommendation and Future Research

Even though there are several benefits in incorporating AI in CTI, very few SMEs have successfully implemented this in their infrastructure. There are several factors contributing to this, such as lack of expertise, lack of thorough knowledge about the algorithm behavior, its stability, cost of implementation, lack of awareness about the evolving cybercrimes, false positives, marketing hypes, etc. Due to these, SMEs often struggle to come up with a strong roadmap that would enable them to implement an efficient AI based CTI system. With the help of the qualitative analysis that was completed in this research, the researcher proposes the final roadmap, Fig. 7, for SMEs to develop their own customized AI based cybersecurity in their current security infrastructure.

Planning—In the Planning phase, based on the company's mission, vision and executive management direction, analyze the security infrastructure of the company using vulnerability assessment and penetration testing to understand the current risks. The AI based CTI system should be an enabler for the SME to achieve their strategic goals and should be able to mitigate the identified risks to gain the confidence and buy-in from the executive management. create a team of subject matter experts that include inhouse members with technical expertise, deep process knowledge as well as third-party partners. The third-party partners should be well experienced working with SMEs so that they can provide innovative solutions that are effective, optimized for SME requirements and are budget friendly. Specify the list of requirements, objectives and scope of the AI based CTI system. The specifications should also include information regarding the security, access control, privacy, data protection, encryption and compliance requirements. You should also list out all existing applications

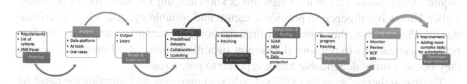

Fig. 7 Roadmap for AI implementation

and systems in the organization so that we do not miss out any critical application during rollout.

Analysis—In this phase, you need to analyze the data sets, data platform, AI tools and algorithms available in the market to identify the best ones that meet your organizations requirements. AI systems are heavily depended on the data that is fed to the system and is popularly publicized to be Garbage in garbage out systems. Identify the datasets that will be used by the AI based CTI system and evaluate them. The data sets used should be of the highest quality, up-to-date and complete. Specify which tasks should be automated by the AI based CTI system and which processes require human intervention. By the end of this phase, you should have selected the use cases that is easy to implement, provides significant benefits, has a high impact in the organizations cyber defense strategy. Select use cases that already have a high quality, complete, up-to-date data set that is readily available. The use cases have to be custom tailored based on the industry type of the SME. Some of the popular use cases are:

- Use of ML for analyzing huge amounts of data and identifying threats before it is exploited. ML can observe anomalies much quicker and with lot more accuracy than humans can and thereby enabling us to predict future threats.
- AI and ML can be used to track and detect any phishing attempts and remediate much faster than humans can. AI-ML can scan through every email in its network to detect phishing campaigns, block them or tag them immediately.
- AI-CTI systems can effectively combine threat intelligence received from various external sources such as discussion boards in the dark web, hacking patterns used, zero-day vulnerabilities, latest attacking trends, etc. This information can be used by the system to proactively determine when and how a cyber-attack might make its way to the organization. This will give the cyber advisors enough lead time to close the vulnerability or mitigate the risk.
- AI can be used to automatically learn the network topology, the traffic pattern and suggest security improvements. AI algorithms can be used to scour through large volumes of data, analyze them and identify anomalies. The AI algorithm can be trained to take some predefined steps in the event of an attack through inputs from the subject matter experts and overtime it will learn these response pattern and make its own decisions.
- ML can be used to learn and create a behavior pattern for your devices. AI can use this pattern to detect any anomalies on the behavior from these devices such as unusual amounts of downloads, uploads, financial transactions, shipments, data access, sudden change in typing speed, etc. Such activities will then trigger the AI algorithm to immediately flag those devices and block them. AI based antivirus software's can detect unusual behavior from programs rather than depending on traditional methods such as matching signatures of known malwares.

Design & Implementation—Design the AI solution using the selected data sets, platform, AI algorithms and tools. Specify what is the expected output from the system and who will be using the system. The solution is then implemented in a development environment.

AI Training—Start building the AI intelligence using reviewed and approved data sets that were pre-defined as per our algorithm requirements. Exchanging and collaborating with other threat researchers and security professionals through various open-source threat intelligence platforms like Open threat exchange, Facebook threat exchange or IBM X-Force exchange is very critical for improving the efficiency and effectiveness of the AI algorithm in detecting new threats. Such collaborations help the AI solution to keep up to speed with the latest attack trends and vulnerabilities. Train cyber analyst for AI by upskilling their employees and improving their knowledge in the logic underpinning the AI algorithms and its behavior. Create proper interfaces that can enable them to interact with the AI tools and incident alerts. Provide the cyber analyst with a database for them to input information based on the customer feedback after an incident is raised.

Optimize & Validate—Optimize the AI based CTI system through fine tuning, patching and hardening to maximize the effectiveness and efficiency of the system. Deploy Security orchestration, automation and response (SOAR) technologies to improve incident alert triage quality, defining a standardized incident response workflow as well as improving the security and operations management. Create automated templates in SOAR for low-risk incidents with the actions to be taken such blacklisting IP, blocking IP, etc. Access and validate the performance of the system with respect to the initial requirements that were specified, ensuring they are functioning as intended.

Integration & Testing—Start integrating the system with the various applications that are part of the scope of this project. Applications such as SIEM can be used to integrate various security systems in the organization working in silos and then SIEM can be integrated to our AI based CTI to automate the tasks. The main hardware should not be connected to the internet. Test the system thoroughly to ensure the security and privacy requirements are met and data is appropriately protected with encryption. Use penetration testing and vulnerability assessment to detect the threats.

Deployment—The AI based CTI solution is deployed in the production environment. Continuously monitor the performance of the system, patching and hardening the system as and when required.

Governance—A clear, transparent governance process needs to be implemented and adhered to monitor the performance of the AI-CTI solution. The controls should include checks on the roles and responsibilities, segregation of duties (Mehmood et al. 2019), AI algorithm behavior abnormalities, risk tolerance, output verification and key performance indicators to measure the success of the program. It should allow the cyber analysts to assess and refine their Incident Response processes, Mean Time to Detect and Mean Time to Respond (MTTD and MTTR) matrix. The data sets and the initial requirements specified in the planning phase must be periodically re-evaluated to ensure the quality and effectiveness of implemented AI based CTI system. We should also look at documenting a business continuity plan (BCP) in case the algorithm goes rogue.

Continues Improvement—Once the organization and the cyber advisors are confident with the performance and the logic controlling the AI based CTI system, we should start looking at expanding the scope of tasks to be automated and include more complex tasks to maximize the productivity of the program. Moreover, SMEs should continuously monitor and review each process so that the products efficiently and effectiveness improves over time.

Future research on various AI based cyberattacks by nefarious actors, their prevention and mitigation methods can further bridge the knowledge gap and secure the defense against such cyberattacks. The application of AI is a relatively new trend in combating cybercrimes and is still evolving. Further research should be conducted to improve the maturity of AI based cyber threat intelligence, their design, architecture and implementation. International Government bodies should also research and study how best they can legally cooperate and fight cybercrimes. The roadmap was evaluated and developed specifically for an SME; therefore larger organizations might have added requirements to be addressed in the roadmap.

Considering the generic audience with backgrounds from different industries and having their unique requirement, we have restricted ourselves from discussing any vendor application or solution in detail. This gives us the opportunity to create a general guideline for SMEs that want to develop their own AI based CTI application as well as helping them negotiate with third-party service providers to make an educated and detailed blueprint to support their strategic decision on how they want to implement an AI based CTI system in their infrastructure.

References

Aasriya, N. Al. (2021). *International Journal of Technology, Innovation and Management (IJTIM), 1*(Special Issue 1), 90–104.

Aburayya, A., Alshurideh, M., Al Marzouqi, A., Al Diabat, O., Alfarsi, A., Suson, R., Salloum, S. A., Alawadhi, D., & Alzarouni, A. (2020). Critical success factors affecting the implementation of tqm in public hospitals: A case study in UAE Hospitals. *Systematic Reviews in Pharmacy, 11*(10). https://doi.org/10.31838/srp.2020.10.39.

Ahmad, A., Alshurideh, M. T., Al Kurdi, B. H., & Salloum, S. A. (2021). Factors impacts organization digital transformation and organization decision making during Covid19 Pandemic. In *Studies in Systems, Decision and Control* (Vol. 334). https://doi.org/10.1007/978-3-030-671 51-8_6.

Ahmed, A., Alshurideh, M., Al Kurdi, B., & Salloum, S. A. (2021). Digital transformation and organizational operational decision making: A systematic review. In *Advances in Intelligent Systems and Computing: Vol. 1261 AISC* (Issue September). Springer International Publishing. https://doi.org/10.1007/978-3-030-58669-0_63.

Akhtar, A., Akhtar, S., Bakhtawar, B., Kashif, A. A., Aziz, N., & Javeid, M. S. (2021). COVID-19 detection from CBC using machine learning techniques. *International Journal of Technology, Innovation and Management (IJTIM), 1*(2), 65–78. https://doi.org/10.54489/ijtim.v1i2.22.

Akour, I., Alshurideh, M., Al Kurdi, B., Al Ali, A., & Salloum, S. (2021). Using machine learning algorithms to predict people's intention to use mobile learning platforms during the COVID-19 pandemic: machine learning approach. *JMIR Medical Education, 7*(1), 1–17.

Al Ali, A. (2021). The impact of information sharing and quality assurance on customer service at UAE banking sector. *International Journal of Technology, Innovation and Management (IJTIM), 1*(1), 1–17. https://doi.org/10.54489/ijtim.v1i1.10.

Al Kurdi, B., Alshurideh, M., Nuseir, M., Aburayya, A., & Salloum, S. A. (2021). The effects of subjective norm on the intention to use social media networks: An exploratory study using

PLS-SEM and machine learning approach. *Advanced Machine Learning Technologies and Applications: Proceedings of AMLTA, 2021*, 581–592.

Al Shebli, K., Said, R. A., Taleb, N., Ghazal, T. M., Alshurideh, M. T., & Alzoubi, H. M. (2021). RTA's Employees' perceptions toward the efficiency of artificial intelligence and big data utilization in providing smart services to the residents of Dubai. *The International Conference on Artificial Intelligence and Computer Vision*, 573–585.

Al Suwaidi, F., Alshurideh, M., Al Kurdi, B., & Salloum, S. A. (2021). The impact of innovation management in SMEs performance: A systematic review. In *Advances in Intelligent Systems and Computing*, Vol. 1261 AISC. https://doi.org/10.1007/978-3-030-58669-0_64.

AlHamad, A., Alshurideh, M., Alomari, K., Kurdi, B. A., Alzoubi, H., Hamouche, S., & Al-Hawary, S. (2022). The effect of electronic human resources management on organizational health of telecommunications companies in Jordan. *International Journal of Data and Network Science, 6*(2), 429–438. https://doi.org/10.5267/j.ijdns.2021.12.011.

Alhamad, A. Q. M., Akour, I., Alshurideh, M., Al-Hamad, A. Q., Kurdi, B. A., & Alzoubi, H. (2021). Predicting the intention to use google glass: A comparative approach using machine learning models and PLS-SEM. *International Journal of Data and Network Science, 5*(3), 311–320. https://doi.org/10.5267/j.ijdns.2021.6.002.

Alhashmi, S. F. S., Alshurideh, M., Al Kurdi, B., & Salloum, S. A. (2020). A systematic review of the factors affecting the artificial intelligence implementation in the health care sector. In *Advances in Intelligent Systems and Computing*, Vol. 1153 AISC. https://doi.org/10.1007/978-3-030-44289-7_4.

Ali, N., Ahmed, A., Anum, L., Ghazal, T. M., Abbas, S., Khan, M. A., … & Ahmad, M. (2021). Modelling supply chain information collaboration empowered with machine learning technique. *Intelligent Automation and Soft Computing, 30*(1), 243–257.

Ali, N., M. Ghazal, T., Ahmed, A., Abbas, S., A. Khan, M., Alzoubi, H., Farooq, U., Ahmad, M., & Adnan Khan, M. (2022). Fusion-based supply chain collaboration using machine learning techniques. *Intelligent Automation & Soft Computing, 31*(3), 1671–1687. https://doi.org/10.32604/iasc.2022.019892.

Alkalha, Z., Al-Zu'bi, Z., Al-Dmour, H., Alshurideh, M., & Masa'deh, R. (2012). Investigating the effects of human resource policies on organizational performance: An empirical study on commercial banks operating in Jordan. *European Journal of Economics, Finance and Administrative Sciences, 51*(1), 44–64.

Alkhalil, Z., Hewage, C., Nawaf, L., & Khan, I. (2021). Phishing attacks: A recent comprehensive study and a new anatomy. *Frontiers in Computer Science, 3*, 6.

Almaazmi, J., Alshurideh, M., Al Kurdi, B., & Salloum, S. A. (2020). The effect of digital transformation on product innovation: A critical review. *International Conference on Advanced Intelligent Systems and Informatics*, 731–741.

AlMehrzi, A., Alshurideh, M., & Al Kurdi, B. (2020). Investigation of the key internal factors influencing knowledge management, employment, and Organisational performance: A qualitative study of the UAE hospitality sector. *International Journal of Innovation, Creativity and Change, 14*(1), 1369–1394.

Alnazer, N. N., Alnuaimi, M. A., & Alzoubi, H. M. (2017a). Analysing the appropriate cognitive styles and its effect on strategic innovation in Jordanian universities. *International Journal of Business Excellence, 13*(1), 127–140.

Alnuaimi, M., Alzoubi, H. M., Ajelat, D., & Alzoubi, A. A. (2021a). Towards intelligent organisations: An empirical investigation of learning orientation's role in technical innovation. *International Journal of Innovation and Learning, 29*(2), 207–221.

AlShamsi, M., Salloum, S. A., Alshurideh, M., & Abdallah, S. (2021). Artificial intelligence and blockchain for transparency in governance. In *Artificial Intelligence for Sustainable Development: Theory, Practice and Future Applications* (pp. 219–230). Springer.

AlShehhi, H., Alshurideh, M., Al Kurdi, B., & Salloum, S. A. (2020). The impact of ethical leadership on employees performance: A systematic review. *International Conference on Advanced Intelligent Systems and Informatics*, 417–426.

Alshraideh, A., Al-Lozi, M., & Alshurideh, M. (2017). The impact of training strategy on organizational loyalty via the mediating variables of organizational satisfaction and organizational performance: An empirical study on Jordanian agricultural credit corporation staff. *Journal of Social Sciences (COES&RJ-JSS)*, *6*, 383–394.

Alshurideh, M. T., Al Kurdi, B., Alzoubi, H. M., Sahawneh, N., & Al-kassem, A. H. (2022a). Fuzzy assisted human resource management for supply chain management issues. *Annals of Operations Research, 24*(1), 1–19.

Alshurideh, M. (2022). Does electronic customer relationship management (E-CRM) affect service quality at private hospitals in Jordan? *Uncertain Supply Chain Management, 10*(2), 1–8.

AlShurideh, M., Alsharari, N. M., & Al Kurdi, B. (2019). Supply chain integration and customer relationship management in the airline logistics. *Theoretical Economics Letters, 9*(02), 392–414.

Alshurideh, M., Gasaymeh, A., Ahmed, G., Alzoubi, H., & Kurd, B. A. (2020a). Loyalty program effectiveness: Theoretical reviews and practical proofs. *Uncertain Supply Chain Management,* *8*(3). https://doi.org/10.5267/j.uscm.2020a.2.003.

Alshurideh, M, Al Kurdi, B., Alzoubi, H., Ghazal, T., Said, R., AlHamad, A., Hamadneh, S., Sahawneh, N., & Al-kassem, A. (2022b). Fuzzy assisted human resource management for supply chain management issues. *Annals of Operations Research*, 1–19.

Alshurideh, Muhammad, Al Kurdi, B., Salloum, S. A., Arpaci, I., & Al-Emran, M. (2020b). Predicting the actual use of m-learning systems: A comparative approach using PLS-SEM and machine learning algorithms. *Interactive Learning Environments*, 1–15.

Alshurideh, M., Gasaymeh, A., Ahmed, G., Alzoubi, H., & Kurd, B. A. (2020c). Loyalty program effectiveness: Theoretical reviews and practical proofs. *Uncertain Supply Chain Management,* *8*(3), 599–612. https://doi.org/10.5267/j.uscm.2020.2.003.

Altamony, H., Masa'deh, R. M. T., Alshurideh, M., & Obeidat, B. Y. (2012). Information systems for competitive advantage: Implementation of an organisational strategic management process. *Innovation and Sustainable Competitive Advantage: From Regional Development to World Economies—Proceedings of the 18th International Business Information Management Association Conference, 1*.

Alzoubi, H., & Ahmed, G. (2019). Do total quality management (TQM) practices Improve Organisational success? A case study of electronics industry in the UAE. *International Journal of Economics and Business Research, 17*(4), 459–472.

Alzoubi, Ali. (2021a). The impact of process quality and quality control on organizational competitiveness at 5-star hotels in Dubai. *International Journal of Technology, Innovation and Management (IJTIM), 1*(1), 54–68. https://doi.org/10.54489/ijtim.v1i1.14.

Alzoubi, Asem. (2021b). Renewable Green hydrogen energy impact on sustainability performance. *International Journal of Computations, Information and Manufacturing (IJCIM), 1*(1), 94–110. https://doi.org/10.54489/ijcim.v1i1.46.

Alzoubi, H., Alshurideh, M., Kurdi, B. A., & Inairat, M. (2020a). Do perceived service value, quality, price fairness and service recovery shape customer satisfaction and delight? A practical study in the service telecommunication context. *Uncertain Supply Chain Management, 8*(3). https://doi.org/10.5267/j.uscm.2020a.2.005.

Alzoubi, H., & Yanamandra, R. (2020). Investigating the mediating role of information sharing strategy on agile supply chain in supply chain performance. *Uncertain Supply Chain Management, 8*(2), 273–284.

Alzoubi, H. M., Vij, M., Vij, A., & Hanaysha, J. R. (2021a). What leads guests to satisfaction and Loyalty in UAE Five-Star Hotels? AHP analysis to service quality dimensions. *enlightening tourism. A Pathmaking Journal, 11*(1), 102–135.

Alzoubi, H., Ahmed, G., Al-Gasaymeh, A., & Kurdi, B. (2020b). Empirical study on sustainable supply chain strategies and its impact on competitive priorities: The mediating role of supply chain collaboration. *Management Science Letters, 10*(3), 703–708.

Alzoubi, H., Alshurideh, M., Kurdi, B. A., Akour, I., & Azi, R. (2022). Does BLE technology contribute towards improving marketing strategies, customers' satisfaction and loyalty? The role of open innovation. *International Journal of Data and Network Science, 6*(2), 449–460. https://doi.org/10.5267/j.ijdns.2021.12.009.

Alzoubi, H., Alshurideh, M., Kurdi, B. A., & Inairat, M. (2020c). Do perceived service value, quality, price fairness and service recovery shape customer satisfaction and delight? A practical study in the service telecommunication context. *Uncertain Supply Chain Management, 8*(3), 579–588. https://doi.org/10.5267/j.uscm.2020.2.005.

Alzoubi, H. M., Ahmed, G., Al-Gasaymeh, A., & Al Kurdi, B. (2020d). Empirical study on sustainable supply chain strategies and its impact on competitive priorities: The mediating role of supply chain collaboration. *Management Science Letters, 10*(3), 703–708. https://doi.org/10.5267/j.msl.2019.9.008.

Alzoubi, H. M., & Aziz, R. (2021a). Does emotional intelligence contribute to quality of strategic decisions? The mediating role of open innovation. *Journal of Open Innovation: Technology, Market, and Complexity, 7*(2), 130. https://doi.org/10.3390/joitmc7020130.

Alzoubi, H. M., & Yanamandra, R. (2020). Investigating the mediating role of information sharing strategy on agile supply chain. *Uncertain Supply Chain Management, 8*(2), 273–284. https://doi.org/10.5267/j.uscm.2019.12.004.

Alzoubi, H. M., Alshurideh, M., & Ghazal, T. M. (2021b). Integrating BLE beacon technology with intelligent information systems IIS for operations' performance: A managerial perspective. *The International Conference on Artificial Intelligence and Computer Vision*, 527–538.

Alzoubi, H. M., & Aziz, R. (2021b). Does emotional intelligence contribute to quality of strategic decisions? *The Mediating Role of Open Innovation*.

Aziz, N., & Aftab, S. (2021). Data mining framework for nutrition ranking Nauman Aziz. *International Journal of Technology, Innovation and Management, 1*(1), 90–100.

Bless, C., Higson-Smith, C., & Kagee, A. (2000). Fundamentals of social research methods. *An African Perspective, 3.*

Darraj, E., Sample, C., & Justice, C. (2019). Artificial intelligence cybersecurity framework: Preparing for the here and now with ai. *ECCWS 2019 18th European Conference on Cyber Warfare and Security*, 132.

Davies, A. (2020). *AI Software Development life cycle: Explained.* DevTeam.Space.

Emerita, A. (2021). Convergence between blockchain and the internet of things Alma Emerita. *International Journal of Technology, Innovation and Management (IJTIM), 1*(1), 35–56.

Farouk, M. (2021). The universal artificial intelligence efforts to face coronavirus COVID-19. *International Journal of Computations, Information and Manufacturing (IJCIM), 1*(1), 77–93. https://doi.org/10.54489/ijcim.v1i1.47.

Ghannajeh, A. M., AlShurideh, M., Zu'bi, M. F., Abuhamad, A., Rumman, G. A., Suifan, T., & Akhorshaideh, A. H. O. (2015). A qualitative analysis of product innovation in Jordan's pharmaceutical sector. *European Scientific Journal, 11*(4), 474–503.

Ghazal, T. M., Hasan, M. K., Alshurideh, M. T., Alzoubi, H. M., Ahmad, M., Akbar, S. S., Al Kurdi, B., & Akour, I. A. (2021). IoT for smart cities: Machine learning approaches in smart healthcare—a review. *Future Internet, 13*(8), 218. https://doi.org/10.3390/fi13080218.

Guergov, S., & Radwan, N. (2021). Blockchain convergence: Analysis of issues affecting IoT, AI and blockchain. *International Journal of Computations, Information and Manufacturing (IJCIM), 1*(1), 1–17. https://doi.org/10.54489/ijcim.v1i1.48.

Hamadneh, S., Pedersen, O., Alshurideh, M., Kurdi, B. Al, & Alzoubi, H. (2021). An investigation of the role of supply chain visibility into the scottish blood supply chain. *Journal of Legal, Ethical and Regulatory Issues, 24*(Special Issue 1), 1–12.

Hanaysha, J. R., Al Shaikh, M. E., & Alzoubi, H. M. (2021a). Importance of marketing mix elements in determining consumer purchase decision in the retail market. *Internation Journal of Service Science, 12*(6), 56–72.

Hanaysha, J. R., Al-Shaikh, M. E., Joghee, S., & Alzoubi, H. (2021b). Impact of innovation capabilities on business sustainability in small and medium enterprises. *FIIB Business Review*, 1–12.https://doi.org/10.1177/23197145211042232.

Harahsheh, A. A., Houssien, A. M. A., & Alshurideh, M. T. (2021). The effect of transformational leadership on achieving effective decisions in the presence of psychological capital as an intermediate variable in Private Jordanian. In *The Effect of Coronavirus Disease (COVID-19) on Business Intelligence* (pp. 243–221). Springer Nature.

Hayajneh, N., Suifan, T., Obeidat, B., Abuhashesh, M., Alshurideh, M., & Masa'deh, R. (2021). The relationship between organizational changes and job satisfaction through the mediating role of job stress in the Jordanian telecommunication sector. *Management Science Letters, 11*(1), 315–326.

Joghee, S., Alzoubi, H. M., & Dubey, A. R. (2020). Decisions effectiveness of FDI investment biases at real estate industry: Empirical evidence from Dubai smart city projects. *International Journal of Scientific and Technology Research, 9*(3), 3499–3503.

Kashif, A. A., Bakhtawar, B., Akhtar, A., Akhtar, S., Aziz, N., & Javeid, M. S. (2021). Treatment response prediction in hepatitis C Patients using machine learning techniques. *International Journal of Technology, Innovation and Management (IJTIM), 1*(2), 79–89. https://doi.org/10.54489/ijtim.v1i2.24.

Khan, M. A. (2021). Challenges facing the application of iot in medicine and healthcare. *International Journal of Computations, Information and Manufacturing (IJCIM), 1*(1), 39–55. https://doi.org/10.54489/ijcim.v1i1.32.

Lee, C., & Ahmed, G. (2021). Improving IoT privacy, data protection and security concerns. *International Journal of Technology, Innovation and Management (IJTIM), 1*(1), 18–33. https://doi.org/10.54489/ijtim.v1i1.12.

Lee, K. L., Azmi, N. A. N., Hanaysha, J. R., Alzoubi, H. M., & Alshurideh, M. T. (2022a). The effect of digital supply chain on organizational performance: An empirical study in Malaysia manufacturing industry. *Uncertain Supply Chain Management, 10*(2), 495–510. https://doi.org/10.5267/j.uscm.2021.12.002.

Lee, K. L., Romzi, P. N., Hanaysha, J. R., Alzoubi, H. M., & Alshurideh, M. (2022b). Investigating the impact of benefits and challenges of IOT adoption on supply chain performance and organizational performance: An empirical study in Malaysia. *Uncertain Supply Chain Management, 10*(2), 537–550. https://doi.org/10.5267/j.uscm.2021.11.009.

Lee, R. M. (2020). *2020 SANS cyber threat intelligence (CTI) survey.* Sans.Org.

Lidestri, N. (2018). The impact of artificial intelligence in cybersecurity. *ProQuest Dissertations and Theses, 6*(2), 709–717.

McCarthy, J. (2007). *What is artificial intelligence?* (pp. 1–15). Stanford University. http://faculty.otterbein.edu/dstucki/inst4200/whatisai.pdf.

Mehmood, T. (2021). Does information technology competencies and fleet management. *International Journal of Technology, Innovation and Management, 1*(1), 14–41.

Mehmood, T., Alzoubi, H. M., Alshurideh, M., Al-Gasaymeh, A., & Ahmed, G. (2019). Schumpeterian entrepreneurship theory: Evolution and relevance. *Academy of Entrepreneurship Journal, 25*(4), 1–10.

Miller, D. (2021). The best practice of teach computer science students to use paper prototyping. *International Journal of Technology, Innovation and Management (IJTIM), 1*(2), 42–63. https://doi.org/10.54489/ijtim.v1i2.17.

Mondol, E. P. (2021). The impact of block chain and smart inventory system on supply chain performance at retail industry. *International Journal of Computations, Information and Manufacturing (IJCIM), 1*(1), 56–76. https://doi.org/10.54489/ijcim.v1i1.30.

Naqvi, R., Soomro, T. R., Alzoubi, H. M., Ghazal, T. M., & Alshurideh, M. T. (2021). The nexus between big data and decision-making: a study of big data techniques and technologies. *The International Conference on Artificial Intelligence and Computer Vision*, 838–853.

Nuseir, M. T., Al Kurdi, B. H., Alshurideh, M. T., & Alzoubi, H. M. (2021). Gender discrimination at workplace: Do artificial intelligence (AI) and machine learning (ML) have opinions about it. *The International Conference on Artificial Intelligence and Computer Vision*, 301–316.

Obaid, A. J. (2021). Assessment of smart home assistants as an IoT. *International Journal of Computations, Information and Manufacturing (IJCIM), 1*(1), 18–36. https://doi.org/10.54489/ijcim.v1i1.34.

Radwan, N., & Farouk, M. (2021). The growth of internet of things (IoT) in the management of healthcare issues and healthcare policy development. *International Journal of Technology, Innovation and Management (IJTIM), 1*(1), 69–84. https://doi.org/10.54489/ijtim.v1i1.8.

Salloum, S. A., Alshurideh, M., Elnagar, A., & Shaalan, K. (2020). Machine learning and deep learning techniques for cybersecurity: A review. *Joint European-US Workshop on Applications of Invariance in Computer Vision*, 50–57.

Singh, R., & Singh, P. K. (2021). Integrating blockchain technology with IoT. *CEUR Workshop Proceedings, 2786*(1), 81–82.

Svoboda, P., Ghazal, T. M., Afifi, M. A. M., Kalra, D., Alshurideh, M. T., & Alzoubi, H. M. (2021). Information systems integration to enhance operational customer relationship management in the pharmaceutical industry. *The International Conference on Artificial Intelligence and Computer Vision*, 553–572.

Truong, T. C., Diep, Q. B., & Zelinka, I. (2020). Artificial intelligence in the cyber domain: Offense and defense. *Symmetry, 12*(3), 410.

Yousuf, H., Zainal, A. Y., Alshurideh, M., & Salloum, S. A. (2021). Artificial intelligence models in power system analysis. In *Artificial Intelligence for Sustainable Development: Theory, Practice and Future Applications* (pp. 231–242). Springer.

Zu'bi, Z., Al-Lozi, M., Dahiyat, S., Alshurideh, M., & Al Majali, A. (2012). Examining the effects of quality management practices on product variety. *European Journal of Economics, Finance and Administrative Sciences, 51*(1), 123–139.

NoSQL: Future of BigData Analytics Characteristics and Comparison with RDBMS

Muhammad Arshad, M. Nawaz Brohi, Tariq Rahim Soomro,
Taher M. Ghazal ⓘ, Haitham M. Alzoubi ⓘ, and Muhammad Alshurideh ⓘ

Abstract The growth of digital world has become so fast and the data volume turned to more composite in relations to size from terabyte and higher, the variation of data including structured, unstructured and hybrid, pace has gone so high in growth of data. This is called 'Big Data' as a worldwide phenomenon. Typically, this is measured as the collection of data from multiple sources has become so huge that it can't be well exploited or maintained by using regular methods used for data management: e.g., RDBMS the conventional relational database management systems or search engines used conventionally. The experts have been working to solve handle these problems, the traditional relational database management systems are modified by precisely

M. Arshad
Department of Computer Science, SZABIST Dubai, Dubai, UAE

M. N. Brohi
Bath Spa University RAK Campus, RAK, UAE
e-mail: mnbrohi@bathspa.ae

T. R. Soomro
Institute of Business Management, CCSIS, Karachi, Sind, Pakistan
e-mail: tariq.soomro@iobm.edu.pk

T. M. Ghazal
School of Information Technology, Skyline University College, Sharjah, UAE
e-mail: taher.ghazal@skylineuniversity.ac.ae

Center for Cyber Security, Faculty of Information Science and Technology, Universiti Kebansaan
Malaysia (UKM), Selangor, Malaysia

H. M. Alzoubi (✉)
School of Business, Skyline University College, Sharjah, UAE
e-mail: haitham.alzubi@skylineuniversity.ac.ae

M. Alshurideh
Department of Marketing, School of Business, University of Jordan, Amman, Jordan
e-mail: malshurideh@sharjah.ac.ae

Department of Management, College of Business Administration, University of Sharjah, Sharjah,
UAE

M. Alshurideh et al. (eds.), *The Effect of Information Technology on Business
and Marketing Intelligence Systems*, Studies in Computational Intelligence 1056,
https://doi.org/10.1007/978-3-031-12382-5_106

designed a set of substitutes DBMS's; which are search based systems, New SQL and NoSQL. This goal of this study is to provide characteristics, classification and evaluation of database management system called NoSQL in Big Data Analytics. The study is planned to help people, specifically the companies to gain. The current computation power also falls less against the massive extended storage volume. The fast growth of unstructured data requires a complete model change for the new age to meet the incoming challenges and observe the progress of new proficient techniques for data engineering. The explanation of BigData, RDBMS and NoSQL have also been described in this study to put light on difference of data and requirement of new technology. This study will highlight the differences, capabilities of both RDBMS and NoSQL database management systems in accordance with BigData. With the help of qualitative and quantitative approach the study has emphasized on the limitation of RDBMS and requirement of new technology NoSQL used for unstructured BigData used for analytics.

1 Introduction

NoSQL, for "Not Only SQL", attributes to a diverse and progressively more recognizable collection of non-relational data management systems; where SQL is not being used for manipulation of data as well as databases are not built primarily on tables (Akhtar et al., 2021; Al Ali, 2021; Alhamad et al., 2021, 2022; Ali et al., 2021). While working with massive data whose architecture does not necessitate a relational model NoSQL database management system are preferable. These structures are disseminated, non-relational databases planned for all-encompassing wide range data storage as well as for densely-parallel data operations and processing over a great amount of commodity servers. Non- SQL languages and systems are also employed by them to interrelate with data and information (although several latest APIs that renovate SQL inquiries to the system's indigenous query language or tool). NoSQL database systems came up next to chief Internet corporations, like Facebook, Amazon and google which experienced difficulties while working with massive amounts of data and information with usual RDBMS systems were not able to deal with. Multiple processes can be carried out by them like, investigative and prophetic analytics, ETL-style the transformation of data, and OLTP non-mission-critical (like handling transactions between organizations and long duration transactions). Unlike the conventional DBMSs and data warehouses, these systems as inspired by Web 2.0 applications are planned to reach out to the maximum number of those users who are doing updates and the reads as well (Ali et al., 2022; Alnazer et al., 2017; Alnuaimi et al., 2021; Alsharari, 2021; Alshurideh et al., 2022).

A relational database system which is planned and intended to offer ACID (Atomicity, Consistency, Isolation, Durability) properties, traditional SQL-based OLAP in Big Data domain and real-time OLTP (Online Transaction) are known as NewSQL system. Utilizing NoSQL-approach features like column-based information storage and scattered styles this system has covered up all the limitations encountered by conformist RDBMS system. Other novel features introduced by this system are in-memory processing, symmetric multiprocessing (SMP) or Massively Parallel Processing.

The data that is too much comprehensive, ambiguous, readily changing is hard to be tackled using the conventional methods as per the observation of analytics. These days every research institution, businesses, and governments are generating extraordinary amounts of data which is too complex as well. Hunting of required information from such enormous data is very crucial for the organizations. It's a great challenge to extract the meaningful insight out of the bulk of data swiftly. That is why analytics has become inextricably important to understand the significance of Big Data to perk up their business performance and boost their market share. In the past few years, the ways to deal with the variety, velocity, and volume of big data has improved to a great extent. The immense rise in data size necessitates rapid analytics with each new inquiry by the application user. This situation has escorted the technologists to introduce a new DBMS system that can triumph over this processing holdup at Database side. As the structural design of RDBMS has restrictions to handle such huge data and carry out analytics. NoSQL architecture is specifically designed to deal with such speed breakers. Thanks to the flexible and adaptable architecture of NoSQL massive volumes of information can be processed rapidly.

In this paper, we will discuss the conventional RDBMS features as well as its limits to manage the huge data. Furthermore, NoSQL Databases will be discussed along with their types and distinctive characteristics to deal with Big Data will be explained. The application areas where NoSQL databases can incorporate will also be discussed. By means of the industry's experience with NoSQL, we will also attempt to elucidate the problems that can be encountered using later mentioned systems for Big Data. Finally, there is going to be a comparison between the two systems about how they contend with normal data to Big Data.

2   Literature Review

From the several data-models, the model that has been surpassing all others is a relational model from the start of early 80 s, with achievements like Oracle databases, MySQL and Microsoft SQL Serveralso known as Relational Database Management System (RDBMS). The models mentioned before are all designed on the relational model. The main reason to build RDBMS was to provide data processing to businesses and from that time till now RDBMS is proving as the best tool for information storage whether the information is personal data, financial statements, transaction processing and so on (Aziz & Aftab, 2021; Cruz, 2021; Eli, 2021; Farouk, 2021; Ghazal et al., 2021a, 2021b, 2021c, 2021d).

2.1 Big Data

As time went by, the provisions for data kept on growing. This data has seen a revolution from structured to unstructured in form and from Megabytes, Gigabytes, Terabytes to Petabytes in size this ever-changing data has caused people to consider about a different solution to supervise this big amount of data. As the data became big, comprehensive, multifaceted, structured or amorphous, and diverse, it needed noteworthy consideration and concentration. Vast amounts of data are being produced at a very swift rate from a variety of distinct prospective areas, systematic tools, and the internet, especially the world known social media, mentioning a few of them. This type of data was termed as Big data. Big data points to those datasets the size of which are past the ability of typical database software techniques to capture, process and manage, store, and examine (Al-Khayyal et al., 2021; Al Batayneh et al., 2021; Alshurideh, 2022; Alshurideh et al., 2022). This characterization is deliberately subjective because it is evident that as technology progresses over the passage of time, the volume of datasets that become licensed as big data will also enlarge (Al Guergov & Radwan, 2021; Hamadneh et al., 2021; Hanaysha et al., 2021a, 2021b; Joghee et al., 2020; Naqvi et al., 2021; Shebli et al., 2021).

4v's can explain the nature of Big data easily. 4v's include Volume, Variety, Velocity, Variability.

2.1.1 Volume

The magnitude of produced and stored data, its importance, value and potential insight is determined by its size. Means whether the data is worthy enough to be termed as Big data or not.

2.1.2 Variety

Variety is the kind, category, and the architecture of the data. This assists experts who scrutinize it to efficiently employ the consequential insight. Big data it the collection of unstructured type data including videos, audios, images and text; also, it uses data fusion to fill in the missing pieces.

2.1.3 Velocity

It can be described as, the pace of data production and processing in the system to comply with the business requirements, tests, and obstacles that stretch out in the technique of escalation, improvement, and lastly the growth. Big data is usually obtainable in up to date format.

2.1.4 Variability

Irregularity and unpredictability of the data set can obstruct the course of action to hold and handle it (Fig. 1).

While considering big data, one problem is its growth, but the other bigger issue is the dire need to manage and store not just the structured data but also the unstructured data as well as the pictures, videos and files. An analytical case in point is that the relational model cannot deal with the data traffic that the social media sites like Facebook and Twitter produce, also it is not the type of data they want to store. Now, for the conventional data processing systems, this significant velocity of the increase in the volume of the data poses a solemn challenge (Alshurideh et al., 2020; Alzoubi, 2021a, 2021b; Alzoubi et al., 2022a; Alzoubi et al., 2020b).

Newly, though, in many cases, the utilization of relational databases escorted to troubles both due to discrepancy and glitches in the design of data and restraints of parallel scalability among various servers and massive size of data. The two main tendencies that brought the mentioned issues into the consideration of international software community are.

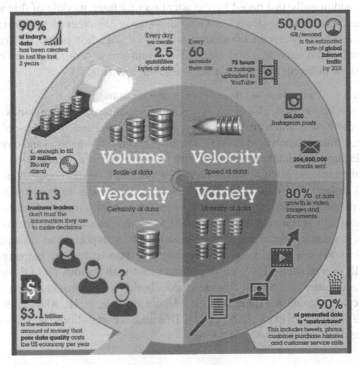

Fig. 1 BigData 4 V explanation ("Big Data," IBM, [Online]. Available: http://www.ibmbigdat ahub.com/infographic/extracting-business-value-4-vs-big-data)

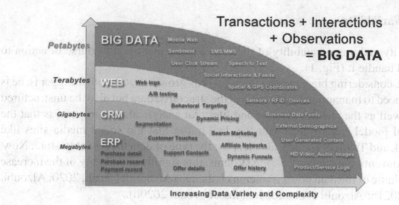

1. The massive increase of the quantity of data produced by sensors, systems and users, additionally fast paced by the attention of a huge module of this volume on big disseminated systems like Google, Amazon and other cloud services.
2. The escalating interconnection and intricacy of data fast paced by Web 2.0, Internet, social networks and exposed and uniform access to sources of data consists of different systems at a very large number (Fig. 2).

Big Data = Transaction + Observation + Interactions.

Due to this very reason, a lot of emerging companies took up different kinds of non-relational databases, which are also known as NoSQL databases and the application run arise e.g. Yahoo which used PNUTTS to fulfill enormously parallel and physically worldwide dispersed database system to run their web based applications.

Since their release NoSQL ("Not Only SQL") systems are extensively accepted in several realms. The main idea behind NoSQL systems is to hold up applications not properly served by relational systems, specifically those which involved in managing and processing BigData. NoSQL system can be classified as graph databases, document stores and key-value stores. It is important to mention here that, there is not one specific query language like standard query language used in RDBMS or a typical APIs used to communicate with various NoSQL systems. Normally, customers are required to use custom build APIs at programming level to communicate (Alzoubi & Ahmed, 2019; Alzoubi & Aziz, 2021; Alzoubi & Yanamandra, 2020; Alzoubi et al., 2020, 2021). This makes portability to be reduced and necessitates code at system-specific level.

2.2 Characteristics of RDBMS

The data organized in relational databases is in the form of tables, which are made up of columns and rows. In order to remove ambiguities during queries, these tables cannot have duplicate rows and every table has been assigned a primary key to a column that distinctively recognize each and every row that is known as record. For example, Fig. 3 demonstrates that Product_ID in the product table is the primary key. Author_ID Column used as Foreign Key in the Product_Book table, which is a child table, Author_ID column is used to reference Author table which is a parent table. The keys used like Foreign and performance, that scan be described as supplementary table relationships may possibly be mandatory while retrieval of data.

Multiple tale inheritances are utilized by libraries' database to hoard familiar traits in a common table knows as Product table (please refer to Fig. 3), every single one of unique attributes is kept in specific type product tables. This said method is far additional well-organized in contrast to concrete table inheritance where for every product category a new table is fashioned, and the queries used are custom-made for specific products. But, as explained to get all the significant characteristics of an established solution multiple table inheritance necessitate many joined operations.

The RDBMS make it certain to have Database ACID properties as the primary prerequisite. For databases, ACID properties act as the vital concept. The abbreviation globally knows as Atomicity, Consistency, Isolation, and Durability.

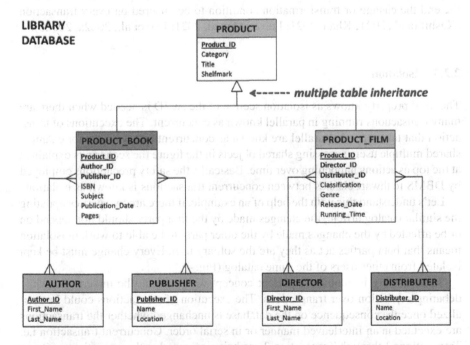

Fig. 3 Relational Database Schema [17]

The ACID makes certain arrangements to business to keep these purchase of sweater dealings from overlaying one another thus the merchant is kept safe from the flawed register and account balances.

2.2.1 Atomicity

Atomicity is the first ACID property and it is best explained by the phrase "all or nothing" to understand this let's consider an example; When a database gets an update, either all of it is available or none of the updates happen to be accessible to anybody past the application or user executing the update. The above-mentioned action performed on database is called a transaction and it is either assigned or canceled. In other words, only a part of an update cannot be put into the database, you get whole of it or none.

2.2.2 Consistency

ACID property of consistency makes sure that if there is a change to values in an instance then there will be a consistent change to other all values in that specific instance. The constraint of consistency is a base on data and it assists in the system as precondition, the condition after the execution knows as post-condition and at the end the change or transformation condition to be ensured on every transaction (Kashif et al., 2021; Khan, 2021; Lee & Ahmed, 2021; Lee et al., 2022a, 2022b).

2.2.3 Isolation

The third property knows as isolation section of the ACID is needed when there are many transactions running in parallel known as concurrent. The executions of transaction that take place in parallel are known as concurrent transactions, for example shared multiple users accessing shared objects in the figure the scenario is explained at the top as actions happening over time. Basically, the safety precautions employed by DBMS to thwart clashes between concurrent transactions is known as isolation.

Let's understand this with the help of an example, if there are two parties updating the similar catalog article, the changes made by the first party should not depend on or be affected by the changes made by the other party to be able to work in isolation means that both parties act as they are the solitary user. Every change must be kept isolated from other users of the same catalog (Fig. 4).

Serializability is another important concept which should be understood while debating separation over transactions. The execution of transactions could be serialized once the consequence on the database is unchanged whether the transactions are executed in an interleaved manner or in serial order. Concurrent transaction i.e. Transactions 1 through Transaction 3 are being executed at the same time as it can be seen in the figure. An important point to keep in mind here is that in serialized

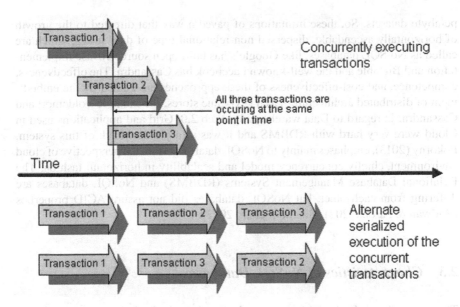

Fig. 4 Isolation Relational Database Property ("Database ACID Properties," [Online]. Available: https://www.servicearchitecture.com/articles/database/acid_properties.html.)

execution it is not compulsory that transactions started first will be the ones automatically completed before the finishing the other transactions in the sequential execution (Mehmood, 2021; Mehmood et al., 2019; Miller, 2021; Mondol, 2021; Obaid, 2021).

2.2.4 Durability

Durability is that ACID property which attends to the need of keeping a record of committed transactions. These updates are not supposed to be missed in any case as it's a very critical thing. It's the systems capability to recover the completed transactions in system or storage media failure. The durability features are as follows:

- The recovery of recently committed transactions in case of database failure
- The recovery of recently committed transactions in case of application failure
- The recovery of recently committed transactions in case of CPU failure
- The recovery of recently committed transactions in case of storage.

The restrictions of RDBMS's are to deal with amorphous, diverse, heterogeneous, massive amounts of data. For RDBMS retailer it is a huge challenge because of its architecture. This test to manage the big data has compelled them to devise a new technology that can handle the said amounts of data and information.

SQL-Like centralized databases have been pushed towards their perimeters by computational processing and storage requirements of applications like Big Data used for Analytics, Social Networking and Business Intelligence having large than petabyte datasets. So, these limitations of paved a way that directed to the growth of horizontally ascendable, dispersed non-relational type of data stores, which are called as No-SQL databases, like Google's has built open source HBase implementation and Bigtable and the well-known Facebook has Cassandra. The effectiveness, competence, and cost-effectiveness of these approaches are gained by the embodiment of distributed architecture-based key-value stores, for example Voldemort and Cassandra. In regard to Data warehousing, Web 2.0, Grid and applications used in Cloud were very hard with RDBMS and it was a major drawback of this system. Pokorny (2013), emphases mainly to NoSQL databases from the perspective of cloud environment, chiefly concurrency model and scalability in horizontal fashion. The Relational Database Management Systems (RDBMS) and NoSQL databases are differing from each other, but NoSQL databases did not assure ACID properties (Radwan & Farouk, 2021; Shamout et al., 2022).

2.3   Characteristics of NoSQL Databases

The conventional database systems are designed on the basic idea of execution of transactions in the manner to keep the data veracity and reliability. This keeps the data consistent while managing it. The features of transactions are also known as ACID (Atomicity, Consistency, Isolation, and Durability) as we have already discussed. Though, developing a system compliant with ACID has made known to be a trouble. CAP-theorem has been observed, i.e. clashes arose among distributed systems diverse sides of high availability that are not completely resolvable.

2.3.1   Strong Consistency

On updates to the data set the version of data seen by the clients is totally same e. g. through the method of two-phase commit protocol (XA transactions), and ACID.

2.3.2   High Availability

In the case, if a few of the machines in a cluster are not working, all the clients still can always find a copy of the requested data. Down machines do not create a problem in this matter.

2.3.3   Partition-Tolerance

The goal of the entire system always is to maintains its characteristics and features even while being positioned on various servers at the same time being transparent to

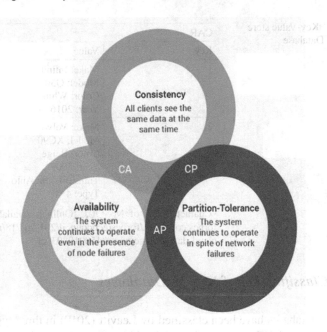

Fig. 5 Characteristics of NoSQL Database

the client. According to the CAP-Theorem, at the same time, out of three only two dissimilar aspects of scaling out can be attained entirely (see Fig. 5).

To attain improved Availability and Partitioning a lot of the NoSQL databases mentioned above have lessened the needs for Consistency. This step laid the way to develop systems globally called as a BASE (Basically Available, Soft-state, Eventually consistent). NoSQL databases have been classified according to the CAP theorem by Han, J. They had compared different NoSQL databases by executing multiple different criteria (Al Ali et al., 2021; Alzoubi et al., 2021; Batayneh et al., 2021).

Main Usages of NoSQL Database can be categorized as (1) Huge-scale and wide data calculation and processing (processing in parallel in the distributed systems); (2) Embedded IR (general machine-to-machine data search and reclamation); (3) Investigative analytics done on unstructured and structured data (knows as expert level); (4) Huge size data storage (unstructured, semi-structured, small-packet structured) (Afifi et al., 2020).

They prove valuable as well for machine-to-machine communication for information and data retrieval, recovery and exchange, for dispensing large number of executions, to the extinct ACID restrictions can be made soft, or the way is to apply them on application side not on DBMS side. In conclusion, when we are to deal with semi-structured or hybrid data these systems act as very good probing analytics, nonetheless to get to the lowermost of intellect, the researcher should be a skillful mathematician working in accordance with an expert programmer (Ghazal, 2021; Ghazal et al., 2021a, 2021b).

Table 1 Key-Value store
NoSQL Database

CAR	
Key	Value
1	Make: Infiniti Model: Q50 Color: White Year: 2016
2	Make: Volvo Model: XC90 Color: Beige Year: 2009 Transmission: Auto Type: 4*4

"Characteristics of NoSQL," [Online]. Available: https://www.
forbes.com/sites/forbestechcouncil/2017/09/18/renaissance-in-
cloud-datamanagement/#4509108030cf

2.4 Classification of NoSQL Databases

NoSQL databases have been classified by Leavitt (2010) in three types: Key-value
stores e.g.SimpleDB column-oriented databases—e.g. Cassandra, HBase, Big Table
and document-based stores—e.g. CouchDB, MongoDB. In this segment, according
to the suitability of different kinds of tasks we categorize NoSQL Databases into
four basic categories,

(1) Key-Value stores.
(2) Document databases (or stores).
(3) Wide-Column stores.
(4) Graph databases.

2.4.1 Key-Value Stores

Classically, in these DBMS the data objects are stored as alpha-numeric identifiers
(keys) and related values in plain, standalone tables (also known as —hash tables‖).
The values could possibly be as simple as text strings or could be more complex like
lists and sets. Data searches can usually one can perform data searches only with the
use of keys, not values, and they are restricted to precise matches. See Table 1.

2.4.2 Document Databases

As the name indicates, document databases and the idea were derived from Lotus
Notes, these mainly are designed and intended to store and manage the different
kind of documents. Customary data exchange systems like JSON (Javascript Option
Notation), XML, or BSON (documents are encoded by Binary JSON). Contrasting to

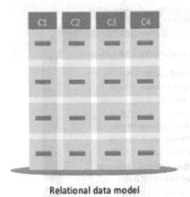 

Relational data model   **Document data model**

Highly-structured table organization with rigidly-defined data formats and record structure.   Collection of complex documents with arbitrary, nested data formats and varying "record" format.

Fig. 6 Document Type Database vs Relational Database (Slide Share, [Online]. Available: https://www.slideshare.net/cloudstack/vbacd-july-2012-apache-hadoopnow-and-beyond. [Accessed 12 February 2018].)

the uncomplicated key-value stores illustrated above, in the value column of document databases structured and unstructured data is present—particularly attribute name/value pairs. Hundreds of these attributes can dwell into a single column, also from row to row the type and number of attributes recorded can differ. In document databases, the values and keys are totally searchable which is in contrast with simple key-value stores (Ghazal et al., 2013, 2021c; Kalra et al., 2020) (Fig. 6).

2.4.3 Wide-Column (or Column-Family) Stores (BigTable-Implementations)

Wide-Column (or Column-Family) stores (after this WC/CF) are just like document databases. To house multiple attributes for each key they utilize a column based distributed data structure. Whilst several WC/CF stores comprise a Key-Value DNA (for example the Cassandra Dynamo-inspired), the majority are designed like Google's Bigtable, that is petabyte scale internal system based on distributed storage for data. This system is developed by Google for its famous search engine and additional products like Finance by Google and Google Earth. In general, the capability is not only to reproduce Google's storage structure BigTable, but Google's file system which is distributed in architecture (GFS) and its processing framework which is parallel MapReduce too. Similar scenario is with Hadoop, which comprises the file system called the Hadoop File System (HDFS, based on GFS) + Hbase (a

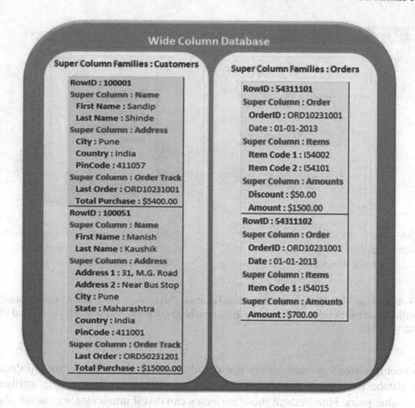

Fig. 7 NoSQL Wide Column Database ("Graph Databases: NOSQL and Neo4j," infoQ, [Online]. Available: http://www.infoq.com/articles/graphnosql-neo4j [Accessed 26 March 2018].)

Bigtable-style storage system) + MapReduce (Khan et al., 2021; Lee et al., 2021) (Fig. 7).

2.4.4 Graph Databases

Relational databases have been replaced by graph databases with more organized relational graphs having key value pairing interconnected to each other. These are like object-oriented databases because of the graph is illustrated as object-oriented network of graph nodes, (objects in concept), relationship of node knows as edges and properties (the object characteristics stated as key-value pairs). The four NoSQL forms conferred here are those that are related with relations. Among other NoSQL DMS, these types are considered more human-friendly because they focus on the visual depiction of data and information (Fig. 8).

Many big organizations that deal with big data have now adopted NoSQL. Following is the table that shows a few of these big businesses (Table 2).

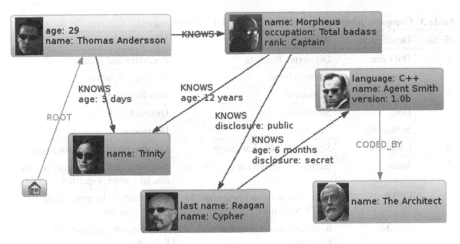

Fig. 8 NoSQL Graph Database Store ("Graph Databases: NOSQL and Neo4j," infoQ, [Online]. Available: http://www.infoq.com/articles/graphnosql-neo4j. [Accessed 26 March 2018].)

Table 2 NoSQL type used by Companies

Company name	NoSQL name	NoSQL storage type
eBay	Cassandra MongoDB	Column Document
Facebook	Cassandra	Column
Lots of Words	CouchDB	Document
MongoHQ	MongoDB	Document
Mozilla	Hbase	Column
Netflix	Hbase Cassandra	Column Column
Twitter	Cassandra	Column

Big businesses because of their high data storage demands have converted to NoSQL and its experts are also in the favorable light now.

2.5   Comparison Between RDBMS and NoSQL

In this study as the characteristics of both RDBMS and NoSQL has been described, the comparison between RDBMS and NoSQL has been analyzed in detail. Which shows the major differences and capabilities of both systems according to customer needs (Table 3).

Table 3 Comparison between RDBMS and NoSQL (Author Created)

S. No	Description	RDBMS	NoSQL
1	Data size	Gigabyte, Terabyte	Petabytes and greater
2	Schema	Static Schema	Dynamic Schema
3	Type	Relational	Non-Relational
4	Data	Structured	Unstructured
5	Scalability	Vertical	Horizontal
6	Language	Standard Query Language	Un-structured Query language
7	Joins	Helpful to design complex queries	No Joins, don't have interface to design complex queries
7	OLTP	Recommended and best used for OLTP Systems	Less likely to be considered as OLTP systems
8	Flexible	Rigid Schema, relationship bounded	Very flexible and no rigid Schemas
9	Auto elasticity	Require down time in most cases	Automatic, no outage required
10	Consistency	Strong consistency supported	Consistency varies per solution; some solutions have tunable consistency
11	Scale	Scale well vertically	Scale well horizontally
12	Transaction	ACID	CAP Theorem

"Characteristics of NoSQL," [Online]. Available: https://www.forbes.com/sites/forbestechcouncil/2017/09/18/renaissance-in-cloud-datamanagement/#4509108030cf

3 Material and Methods

In 1980's the primary cohort of commercial systems come into sight by Teradata Corporation and in the same time, the necessity surfaced for well-defined systems to determine the ability of DBMS dealing with very big quantities of data. Motivated by vendor's desires to weigh the commercial systems, in the starting of 90's the Transaction Processing Performance Council designed a series of data warehouse end-to-end benchmarks. Likely systems have been developed by TPC-H and PC-R at the beginning of 2000 (the details are all accessible from the TPC website2). With some update on a venture data warehouse, these benchmarks are limited to a data size of a terabyte, highlighting single and multi-user performance of complex SQL query processing abilities. Even before this, academia had started developing micro-benchmarks like EXRT and XMark benchmarks for XML-related DBMS technologies and the OO7, the Wisconsin benchmark, and BUCKY benchmarks for object-oriented DBMSs, (Matloob et al., 2021; Naqvi et al., 2021).

With the passage of time, the volume of data kept on growing from megabytes to petabytes in size and from simple data models (a few tables with a small number of relationships) to complex ones (big tables with many complex relationships). This change in the demand for data needs has led TPC to act in response. In the dawn of 2000's TPC-DS developed its next generation decision support benchmark. Its foundation is based on the SQL programming language, but it consists of several big data elements, like exceedingly large system sizes and data. Even bthough the existing limit is 100 terabytes; the schema and data generator can be expanded to petabytes. Quite composite analytical data queries are also contained by it using sophisticated and complicated SQL structures and a synchronized update model.

3.1 Adaptation of NoSQL

The term NoSQL was invented in 1998. Lots of people assume NoSQL is a deprecating term fashioned to jab at SQL but in actual, the term NoSQL stands for Not Only SQL. Putting forth the idea that both these technologies SQL and NoSQL can exist together in their own specific place. For the previous few years, NoSQL technology has been heard and seen in the news most likely because of the reason that as many of the Web 2.0 leaders have taken the NoSQL technology. Facebook, Twitter, Digg, Amazon, LinkedIn and Google all these companies use NoSQL in one way or another.

The main factors behind the adaptation of NoSQL includes flexibility of data, no rigid schema and scalability.

3.2 Questionnaire Development

As discussed earlier technology is changing extensively, and data is becoming more crucial to any organization. The accessing and manipulation of data is much more necessary than saving it to storage. Accessing and storing of data to storage obviously is time consuming. The RDBMS's due to their architecture must consider the data types, relations and other hidden processes involved in execution and storing the data to disk and the same when accessing the data from storage. This adds up the time which is very crucial for the applications to respond. The massive data growth requires additional storage on the fly. This is difficult to manage in RDBMS environment (Rehman et al., 2021; Suleman et al., 2021).

On the other hand, accessing and storing of heterogenous type of data in NoSQL environment is very fast. The flexibility of multitype data such as text, images, videos and documents are managed very efficiently. The schema free and no predefined architecture gives a strong advantage over RDBMS. Dealing with enormous growth of data is very easy in NoSQL environment as it gives flexible scalable architecture by adding more and more additional servers called shards in running environment.

Qualitative Study has been done on previous researches for the same topic. This has helped enormously to get the basic idea about the database vendors and organizations requirements and improvements which have been done to accomplish the day-to-day challenges. This also has shed light on the technology enhancements in this specific field within last 2 decades.

3.3 Data Collection

A questionnaire survey has been conducted as part of quantitative method. To get the current situation in dealing with bigdata and analytics a questionnaire has been developed and spread in IT field. The targeted people were IT company CEO's, CTO,'s, Database Administrators which are dealing with day to day management challenges and the network professionals which are designing networks to deal with bigdata. The results are then analyzed by the respondents and have helped in getting the result and findings for this study. As the restriction of direct access to all the respondents the study focused on various methods to collect data which include professional groups, emails, printed copies to professionals in contact. The results are then analyzed and discussed in Chap. 4.

4 Results and Findings

As the world already has adopted the NoSQL Database, to analyze the levels and reasons of moving the software developing companies towards it a survey has been conducted. The survey has been conducted to analyze and judge the necessity of NoSQL with respect to business needs, Data management problems, flexibility or low-cost DBMS adaptation.

The survey has been conducted among database administrators, IT company CEO's, technology decision makers, software developers and IT experts. The survey has been conducted considering those people doing business in RDBMS and Big Data.

The main points of the survey are given below.

- About half of the more than 250 respondents pointed to the fact that they have worked on NoSQL projects in past couple of years. The companies which have secured large software projects more than 50% of their projects are to deal with Big Data for their clients. The reasons of adaption of NoSQL is described by them as:
- 49% referred to inflexible schemas to be the major reason for their migration to NoSQL technology from the relational database system. A prime reason for switching to NoSQL is the deficiency of scalability and high latency/low performance when dealing with Big Data.
- Overall, 40% were of the viewpoint that NoSQL is very essential and significant to their daily operations, and it is continuing to become more important.

- The management of Big Data in NoSQL is much easier than in RDBMS due to the pre-defined architecture and limitation. As the capturing of multi structural data is greeted in NoSQL. The type of data is high rank in considering the NoSQL database to be used (Figs. 9, 10).

As per the above survey results explains according to software professionals the major factors which take part in deciding about NoSQL and RDBMS are BigData, unstructured data and the management of the same. This means when the experts have to deal with massive amount of unstructured data, they are more likely to adopt NoSQL (Fig. 11).

- Another question regarding the performance of NoSQL using BigData shows the experts agreed on having better performance than RDBMS.

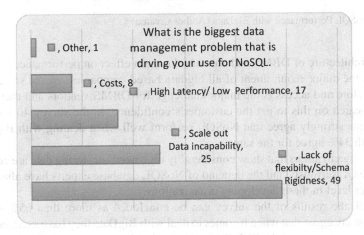

Fig. 9 Problems to driving towards NoSQL (Created by authors)

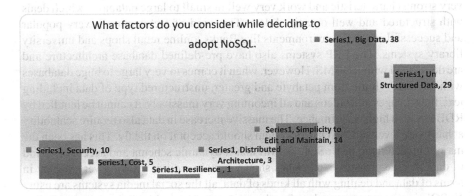

Fig. 10 Factor to decide NoSQL (Created by authors)

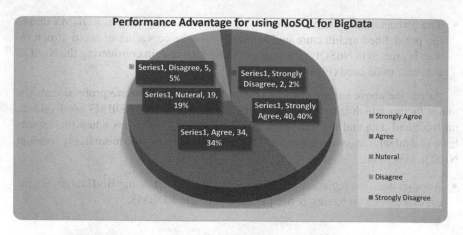

Fig. 11 NoSQL Performance with BigData (Author Created)

The architecture of DBMS and data has the main effect on performance. Performance is the major requirement of all bigdata based companies. At the same time fast data store and access is the major challenge for DBMS vendors and they keep doing research on this to get the customer's confidence. In the survey 40% of the respondents strongly agree that NoSQL perform well when dealing with BigData along with 34% agree for the same.

Those organizations that show considerably high needs of storing data are considering NoSQL seriously and the demand of NoSQL database experts have also risen to a higher level in these developing organizations.

Overall, the results of the survey can be concluded as more than 65% of the respondent are agree that when it comes to deal with BigData they have used, or their first choice is to go with NoSQL. The reasons that have emerged from the survey answers are particularly highlighted as pre-defined schema in RDBMS. This has very strong characteristic and work very well in small to large databases which deals with structured and well-organized data. This kind of architecture is very popular and successful in OLTP environments like Banks, online retail shops and university library systems. The ERP systems also have pre-defined database architecture and are doing well with RDBMS. However, when it comes to very large to huge databases with the data volume from petabyte and greater, unstructured type of data including text, files, images and videos and all incoming very massively, it cannot be handled by RDBMS with high performance. The massive increase in data also require scalability at hardware level and the DBMS system should accept it on the fly. This has been the major characteristic of NoSQL DBMS that it dynamic schema and accept any kind of data. As the previous studies has shown social media is growing very widely in terms of data and dealing with all kinds of data, all the social media systems are using NoSQL in one way or another. So, the other top factors highlighted as scalability, performance when dealing with BigData, simple in maintaining the system.

5 Conclusion

Storage and processing requirements of some applications like Analytics for Big Data, Business Intelligence and social networking which is growing rapidly over peta-byte datasets have forced RDBMS to their limits. This has directed to the development of technology that is horizontally scalable, dispersed non-relational database named as No-SQL. The study speculate about the primarily usages of NoSQL Databeses: The larger scale data processing system (parallel processing over distributed systems), (machine-to-machine data look-up & recovery); Analytics on semi-structured data (professional level); Huge capacity data storage (structured, semi-structured, unstructured) NoSQL is a huge and growing field, for the purposes of this study, characteristics (benefits and features of NoSQL DBMS); classification (the four categories with their features); and the comparison and assessment (with a table on basis of few characteristics- strategy, integrity, attributes, distribution) of different kinds of NoSQL databases. The study has also shown the difference between DBMS and NoSQL with present state and reason of acceptance of NoSQL databases. This study with motivation has provided an autonomous understating about the weakness and strengths of NoSQL databases which are supporting the applications that are dealing with large volume of data. The study has concluded the applications dealing with BigData performs well in NoSQL environment. Still the requirement varies from solution to solution. NoSQL will be emerging as the solution for BigData analytics in future.

5.1 Future Work

As the technology has been changing rapidly, the business is more relying on analytics. The analytics is based on data and the data is growing massively. The fast processing of this data is the major need of every organization. This is forcing technology leaders to put more efforts in making the DBMS more efficient in dealing with BigData. NoSQL has fulfilled the requirement to some level. However, there are still challenges that need more efforts from DBMS vendors to get confidence of customers. The main challenges are like data security and more compatible environments to most available development systems. Hope this study will help the users to understand the different DBMS architectures, BigData and its requirements and how to deal with this by understanding the nature. And will help users to guide to make decision while choosing DBMS according to their need and DBMS capabilities.

References

Afifi, M. A. M., Kalra, D., Ghazal, T. M., & Mago, B. (2020). Information Technology Ethics and Professional Responsibilities. *International Journal of Advanced Science and Technology, 29*, 11336–11343.

Akhtar, A., Akhtar, S., Bakhtawar, B., Kashif, A.A., Aziz, N., Javeid, M.S., (2021). COVID-19 Detection from CBC using machine learning techniques. *International Journal of Innovation and Technology Management* 1, 65–78. https://doi.org/10.54489/ijtim.v1i2.22

Al-Khayyal, A., Alshurideh, M., Al Kurdi, B., Salloum, S.A., (2021). Factors influencing electronic service quality on electronic loyalty in online shopping context: data analysis approach, in: Enabling AI Applications in Data Science. Springer, pp. 367–378.

Al Ali, A., (2021). The Impact of Information Sharing and Quality Assurance on Customer Service at UAE Banking Sector. *International Journal of Innovation and Technology Management* 1, 01–17. https://doi.org/10.54489/ijtim.v1i1.10

Al Batayneh, R.M., Taleb, N., Said, R.A., Alshurideh, M.T., Ghazal, T.M., Alzoubi, H.M., (2021). IT Governance Framework and Smart Services Integration for Future Development of Dubai Infrastructure Utilizing AI and Big Data, Its Reflection on the Citizens Standard of Living, In: *The International Conference on Artificial Intelligence and Computer Vision*. pp. 235–247. Springer.

Al Shebli, K., Said, R.A., Taleb, N., Ghazal, T.M., Alshurideh, M.T., Alzoubi, H.M., (2021). RTA's employees' perceptions toward the efficiency of artificial intelligence and big data utilization in providing smart services to the residents of Dubai, In: *The International Conference On Artificial Intelligence And Computer Vision*. pp. 573–585. Springer.

AlHamad, A., Alshurideh, M., Alomari, K., Kurdi, B., Alzoubi, H., Hamouche, S., & Al-Hawary, S. (2022). The effect of electronic human resources management on organizational health of telecommuni-cations companies in Jordan. *International Journal of Data and Network Science, 6*, 429–438.

Alhamad, A.Q.M., Akour, I., Alshurideh, M., Al-Hamad, A.Q., Kurdi, B.A., Alzoubi, H., (2021). Predicting the intention to use google glass: A comparative approach using machine learning models and PLS-SEM. *International Journal of Data and Network Science* 5. https://doi.org/10.5267/j.ijdns.2021.6.002

Ali, N., Ahmed, A., Anum, L., Ghazal, T.M., Abbas, S., Khan, M.A., Alzoubi, H.M., Ahmad, M., (2021). Modelling supply chain information collaboration empowered with machine learning technique. *Intelligent Automation and Soft Computing* 30, 243–257. https://doi.org/10.32604/iasc.2021.018983

Ali, N., M. Ghazal, T., Ahmed, A., Abbas, S., A. Khan, M., Alzoubi, H., Farooq, U., Ahmad, M., Adnan Khan, M., (2022). Fusion-Based supply chain collaboration using machine learning techniques. *Intelligent Automation and Soft Computing* 31, 1671–1687. https://doi.org/10.32604/iasc.2022.019892

Alnazer, N. N., Alnuaimi, M. A., & Alzoubi, H. M. (2017). Analysing the appropriate cognitive styles and its effect on strategic innovation in Jordanian universities. *International Journal of Business Excellence, 13*, 127–140. https://doi.org/10.1504/IJBEX.2017.085799

Alnuaimi, M., Alzoubi, H. M., Ajelat, D., & Alzoubi, A. A. (2021). Towards intelligent organisations: An empirical investigation of learning orientation's role in technical innovation. *International Journal of Innovation and Learning, 29*, 207–221. https://doi.org/10.1504/IJIL.2021.112996

Alsharari, N. (2021). Integrating blockchain technology with internet of things to efficiency. *International Journal of Innovation and Technology Management, 1*, 1–13.

Alshurideh, M. (2022). Does electronic customer relationship management (E-CRM) affect service quality at private hospitals in Jordan? *Uncertain Supply Chain Manag., 10*, 1–8.

Alshurideh, M., Al Kurdi, B., Alzoubi, H., Ghazal, T., Said, R., AlHamad, A., Hamadneh, S., Sahawneh, N., Al-kassem, A., (2022a). Fuzzy assisted human resource management for supply chain management issues. *Annals of Operations Research* 1–19.

Alshurideh, M., Gasaymeh, A., Ahmed, G., Alzoubi, H., Kurd, B.A., (2020). Loyalty program effectiveness: Theoretical reviews and practical proofs. Uncertain Supply Chain Management 8. https://doi.org/10.5267/j.uscm.2020.2.003

Alshurideh, M.T., Al Kurdi, B., Alzoubi, H.M., Ghazal, T.M., Said, R.A., AlHamad, A.Q., Hamadneh, S., Sahawneh, N., Al-kassem, A.H., (2022b). Fuzzy assisted human resource management for supply chain management issues. *Annals of Operations Research* 1–19.

Alzoubi, Ali, 2021a. The Impact of Process Quality and Quality Control on Organizational Competitiveness at 5-star hotels in Dubai. *International Journal of Innovation and Technology Management* 1, 54–68. https://doi.org/10.54489/ijtim.v1i1.14

Alzoubi, Asem, 2021b. Renewable Green hydrogen energy impact on sustainability performance. *International Journal of Computer Integrated Manufacturing* 1, 94–110. https://doi.org/10.54489/ijcim.v1i1.46

Alzoubi, H., & Ahmed, G. (2019). Do TQM practices improve organisational success? A case study of electronics industry in the UAE. *International Journal of Economics and Business Research, 17*, 459–472. https://doi.org/10.1504/IJEBR.2019.099975

Alzoubi, H., Ahmed, G., Al-Gasaymeh, A., & Kurdi, B. (2020a). Empirical study on sustainable supply chain strategies and its impact on competitive priorities: The mediating role of supply chain collaboration. *Management Science Letters, 10*, 703–708.

Alzoubi, H., Alshurideh, M., Kurdi, B. A., & Inairat, M. (2020b). Do perceived service value, quality, price fairness and service recovery shape customer satisfaction and delight? A practical study in the service telecommunication context. *Uncertain Supply Chain Management, 8*, 579–588. https://doi.org/10.5267/j.uscm.2020.2.005

Alzoubi, H., Alshurideh, M., Kurdi, B., Akour, I., & Aziz, R. (2022). Does BLE technology contribute towards improving marketing strategies, customers' satisfaction and loyalty? The role of open innovation. *International Journal of Data and Network Science, 6*, 449–460.

Alzoubi, Haitham M, Alshurideh, M., Ghazal, T.M., (2021a). Integrating BLE beacon technology with intelligent information systems iis for operations' performance: a managerial perspective, In: *The International Conference on Artificial Intelligence and Computer Vision*. pp. 527–538.

Alzoubi, H. M., & Aziz, R. (2021). Does emotional intelligence contribute to quality of strategic decisions? *The Mediating Role of Open Innovation*. https://doi.org/10.3390/joitmc7020130

Alzoubi, Haitham M., Vij, M., Vij, A., Hanaysha, J.R., (2021b). What leads guests to satisfaction and loyalty in UAE five-star hotels? AHP analysis to service quality dimensions. *Enlightening Tour*. 11, 102–135. https://doi.org/10.33776/et.v11i1.5056

Alzoubi, H. M., & Yanamandra, R. (2020). Investigating the mediating role of information sharing strategy on agile supply chain. *Uncertain Supply Chain Manag., 8*, 273–284. https://doi.org/10.5267/j.uscm.2019.12.004

Aziz, N., & Aftab, S. (2021). Data mining framework for nutrition ranking: methodology: SPSS modeller. *International Journal of Innovation and Technology Management, 1*, 85–95.

Cruz, A. (2021). Convergence between blockchain and the internet of things. *International Journal of Innovation and Technology Management, 1*, 35–56.

Eli, T. (2021). StudentsPerspectives on the Use of Innovative and Interactive Teaching Methods at the University of Nouakchott Al Aasriya, Mauritania: English Department as a Case Study. *International Journal of Innovation and Technology Management, 1*, 90–104.

Farouk, M., 2021. The Universal Artificial Intelligence Efforts to Face Coronavirus COVID-19. International Journal of Computer Integrated Manufacturing 1, 77–93. https://doi.org/10.54489/ijcim.v1i1.47

Ghazal, T., Soomro, T. R., & Shaalan, K. (2013). Integration of project management maturity (PMM) based on capability maturity model integration (CMMI). *European Journal of Scientific Research, 99*, 418–428.

Ghazal, T.M., (2021). Positioning of UAV base stations using 5G and beyond networks for IoMT applications. *Arabian Journal for Science and Engineering* 1–12.

Ghazal, T. M., Anam, M., Hasan, M. K., Hussain, M., Farooq, M. S., Ali, H. M., Ahmad, M., & Soomro, T. R. (2021a). Hep-Pred: Hepatitis C staging prediction using fine gaussian SVM. *Computers, Materials and Continua, 69*, 191–203.

Ghazal, T. M., Hasan, M. K., Alshurideh, M. T., Alzoubi, H. M., Ahmad, M., Akbar, S. S., Al Kurdi, B., & Akour, I. A. (2021b). IoT for smart cities: machine learning approaches in smart Healthcare—a review. *Futur. Internet, 13*, 218. https://doi.org/10.3390/fi13080218

Ghazal, Taher M, Hussain, M.Z., Said, R.A., Nadeem, A., Hasan, M.K., Ahmad, M., Khan, M.A., Naseem, M.T., (2021b). Performances of K-means clustering algorithm with different distance metrics.

Ghazal, Taher M, Said, R.A., Taleb, N., (2021c). Internet of vehicles and autonomous systems with AI for medical things. Soft Computing 1–13.

Guergov, S., Radwan, N., (2021). Blockchain convergence: analysis of issues affecting IoT, AI and Blockchain. *International Journal of Computer Integrated Manufacturing* 1, 1–17. https://doi.org/10.54489/ijcim.v1i1.48

Hamadneh, S., Pedersen, O., & Al Kurdi, B. (2021). An investigation of the role of supply chain visibility into the scottish bood supply chain. *Journal of Legal, Ethical and Regulatory Issues, 24*, 1–12.

Hanaysha, J.R., Al-Shaikh, M.E., Joghee, S., Alzoubi, H., (2021a). Impact of innovation capabilities on business sustainability in small and medium enterprises. FIIB Business Review 1–12. https://doi.org/10.1177/23197145211042232

Hanaysha, J. R., Al Shaikh, M. E., & Alzoubi, H. M. (2021b). Importance of marketing mix elements in determining consumer purchase decision in the retail market. *International Journal Services Science Management Engineering Technology, 12*, 56–72.

Joghee, S., Alzoubi, H. M., & Dubey, A. R. (2020). Decisions effectiveness of FDI investment biases at real estate industry: Empirical evidence from Dubai smart city projects. *International Journal of Scientific & Technology Research, 9*, 3499–3503.

Kalra, D., Ghazal, T. M., & Afifi, M. A. M. (2020). Integration of collaboration systems in hospitality management as a comprehensive solution. *International Journal of Advanced Science and Technology, 29*, 3155–3173.

Kashif, A.A., Bakhtawar, B., Akhtar, A., Akhtar, S., Aziz, N., Javeid, M.S., (2021). Treatment response prediction in hepatitis c patients using machine learning techniques. *International Journal of Innovation and Technology Management*. 1, 79–89. https://doi.org/10.54489/ijtim.v1i2.24

Khan, M.A., (2021). Challenges facing the application of IoT in medicine and healthcare. *International Journal of Computer Integrated Manufacturing*. 1, 39–55. https://doi.org/10.54489/ijcim.v1i1.32

Khan, M.F., Ghazal, T.M., Said, R.A., Fatima, A., Abbas, S., Khan, M A, Issa, G.F., Ahmad, M., Khan, Muhammad Adnan, (2021a). An IoMT-Enabled smart healthcare model to monitor elderly people using machine learning technique. *Computational Intelligence and Neuroscience.*

Khan, Q.-T.-A., Ghazal, T.M., Abbas, S., Khan, W.A., Khan, M.A., Said, R.A., Ahmad, M., Asif, M., (2021b). Modeling habit patterns using conditional reflexes in agency.

Leavitt, N., (2010). Will NoSQL databases live up to their promise? Computer (Long. Beach. Calif). 43, 12–14.

Lee, C., Ahmed, G., (2021). Improving IoT privacy, data protection and security concerns. *International Journal of Innovation and Technology Management*. 1, 18–33. https://doi.org/10.54489/ijtim.v1i1.12

Lee, K., Azmi, N., Hanaysha, J., Alzoubi, H., & Alshurideh, M. (2022a). The effect of digital supply chain on organizational performance: An empirical study in Malaysia manufacturing industry. *Uncertain Supply Chain Manag., 10*, 495–510.

Lee, K., Romzi, P., Hanaysha, J., Alzoubi, H., & Alshurideh, M. (2022b). Investigating the impact of benefits and challenges of IOT adoption on supply chain performance and organizational performance: An empirical study in Malaysia. *Uncertain Supply Chain Management, 10*, 537–550.

Lee, S.-W., Hussain, S., Issa, G. F., Abbas, S., Ghazal, T. M., Sohail, T., Ahmad, M., & Khan, M. A. (2021). Multi-Dimensional trust quantification by artificial agents through evidential fuzzy multi-criteria decision making. *IEEE Access, 9,* 159399–159412.

Matloob, F., Ghazal, T.M., Taleb, N., Aftab, S., Ahmad, M., Khan, M.A., Abbas, S., Soomro, T.R., (2021). Software defect prediction using ensemble learning: A systematic literature review. IEEE Access.

Mehmood, T. (2021). Does information technology competencies and fleet management practices lead to effective service delivery? empirical evidence from e-commerce industry. *International Journal of Innovation and Technology Management, 1,* 14–41.

Mehmood, T., Alzoubi, H.M., Ahmed, G., (2019). Schumpeterian entrepreneurship theory: evolution and relevance. *Academy of Entrepreneurship Journal.* 25.

Miller, D., (2021). The best practice of teach computer science students to use paper prototyping. *International Journal of Innovation and Technology Management.* 1, 42–63. https://doi.org/10.54489/ijtim.v1i2.17

Mondol, E.P., (2021). The impact of block chain and smart inventory system on supply chain performance at retail industry. *International journal of computer integrated manufacturing* 1, 56–76. https://doi.org/10.54489/ijcim.v1i1.30

Naqvi, R., Soomro, T.R., Alzoubi, H.M., Ghazal, T.M., Alshurideh, M.T., (2021). The nexus between big data and decision-making: a study of big data techniques and technologies, In: *The International Conference on Artificial Intelligence and Computer Vision.* pp. 838–853.

Obaid, A.J., (2021). Assessment of smart home assistants as an IoT. *International Journal of Computer Integrated Manufacturing.* 1, 18–36. https://doi.org/10.54489/ijcim.v1i1.34

Pokorny, J. (2013). NoSQL databases: A step to database scalability in web environment. *International Journal of Web Information Systems., 9,* 69–82.

Radwan, N., Farouk, M., (2021). The growth of internet of things (IoT) in the management of healthcare issues and healthcare policy development. *International Journal of Innovation and Technology Management.* 1, 69–84. https://doi.org/10.54489/ijtim.v1i1.8

Rehman, E., Khan, M. A., Soomro, T. R., Taleb, N., Afifi, M. A., & Ghazal, T. M. (2021). Using blockchain to ensure trust between donor agencies and ngos in under-developed countries. *Computers, 10,* 98.

Shamout, M., Ben-Abdallah, B., Alshurideh, M., Alzoubi, H., Al Kurdi, B., & Hamadneh, S. (2022). A conceptual model for the adoption of autonomous robots in supply chain and logistics industry. *Uncertain Supply Chain Management., 10,* 1–16.

Suleman, M., Soomro, T.R., Ghazal, T.M., Alshurideh, M., (2021). Combating against potentially harmful mobile apps, In: the international conference on artificial intelligence and computer vision, pp. 154–173 Springer.

Internet of Things Connected Wireless Sensor Networks for Smart Cities

Taher M. Ghazal ⓘ, **Mohammad Kamrul Hasan, Haitham M. Alzoubi** ⓘ, **Muhammad Alshurideh** ⓘ, **Munir Ahmad, and Syed Shehryar Akbar**

Abstract The Smart City is the most complete and covered framework that meets the need of different project facets related to the smart city. It allows the cities to use the urban network and raise their economic power, unique solutions of technology, and build the most efficient systems. Smart City is the advanced developmental product of the smart economy and information technology. It relies upon wireless networking, broadcast networking, internet mesh networking, telecommunication network, and the end-to-end sensor network in which the Internet of Things (IoT) is the core. The IoT serves as the core for integrating the wide variety of sensors in each day's objects and interconnects the sensors through the internet using specific protocols for exchanging the communications and information that lead to location tracking, management, monitoring, and intelligent achievement recognition. It not only supports one city but also interconnects it with the other smart cities. This paper

T. M. Ghazal · M. K. Hasan
Center for Cyber Security, Faculty of Information Science and Technology, Universiti Kebansaan Malaysia (UKM), 43600 Bangi, Selangor, Malaysia
e-mail: Taher.ghazal@skylineuniversity.ac.ae

M. K. Hasan
e-mail: mkhasan@ukm.edu.my

T. M. Ghazal
Skyline University College, Sharjah, UAE

H. M. Alzoubi (✉)
School of Business, Skyline University College, Sharjah, UAE
e-mail: haitham.alzubi@skylineuniversity.ac.ae

M. Alshurideh
Department of Marketing, School of Business, University of Jordan, Amman, Jordan
e-mail: malshurideh@sharjah.ac.ae

Department of Management, College of Business Administration, University of Sharjah, Sharjah, UAE

M. Ahmad · S. S. Akbar
National College of Business Administration & Economics, Lahore 54000, Pakistan
e-mail: munir@ncbae.edu.pk

© The Author(s), under exclusive license to Springer Nature Switzerland AG 2023   1953
M. Alshurideh et al. (eds.), *The Effect of Information Technology on Business and Marketing Intelligence Systems*, Studies in Computational Intelligence 1056,
https://doi.org/10.1007/978-3-031-12382-5_107

aims to explore the use of IoT-based machine learning approaches that help develop a smart city.

Keywords IoT · Machine learning approaches · Smart City · Wireless sensor network

1 Introduction

Internet of Things (IoT) has been one of the leading technologies of the rapidly blooming trends within the environment and smart cities (Alshurideh et al., 2022). According to the sources, the term IoT was first defined in 1999 as the basis of infrastructure for the information society. The IoT or Internet of Things is the system that includes buildings, modern vehicles, physical devices, and the most important electronic devices that we use daily and are interconnected with each other on the internet to enable the collection and exchange of data (Ali et al., 2020). The Internet of Things allows users to control their devices through remote access across the network's already built infrastructure. It then enables the user to an effective and direct computer-controlled system implementation within the physical world (Ghazal et al., 2020). This is how it allows the rise in the precision, reliability, speed, and efficiency of the system and reduces human beings' intervention. When the IoT is connected with the actuators and the sensors, it is converted into the cyber-physical system that is the smart city establishment (Saha et al., 2017).

From the last decade, the Internet of Things technology has played the most significant role not only in smart cities but also in smart homes, smart factories, and various smart grids, etc. The IoT operations here rely upon different sensors, video cameras, and mobile devices within the cosmopolitan city to share the data through the internet (Minoli et al., 2017). The most important assisted element of this Internet of Things (IoT) technology is the Wireless Sensor Network (WSN) which is the core of the IoT approach. The WSN has numerous sensor nodes that allow data management within the smart city circle (Ali et al., 2021). Moreover, every node of the wireless sensor network is utilized for a particular purpose and is operated with the battery's help that consumes the energy to process the data in every framework communication of smart city (Aljumah et al., 2021; MAlshurideh & Shaltoni, 2014; Alyammahi et al., 2020; Hasan et al., 2020; Shakeel et al., 2019; Sweiss et al., 2021).

At the current moment, Smart cities and Smart homes have the infrastructure that requires extensive control and monitoring of the information collected. This process has been made simple through the wireless sensor network that organizes the data through sensor nodes formulation as the self-sufficient model (Alzoubi & Aziz, 2021). Within a few years, the smart city's monitoring is extensively utilizing WSN and its use has been extended to home automation, environmental monitoring, and healthcare monitoring (Zanella et al., 2014). According to the results of a recent survey, it has been showing that the IoT technology has earned quite a significance due to its use in the construction of smart buildings that have inheritance property to make them more sustainable and allows conservation of energy (Zygiaris, 2013).

The smart city's management framework utilizes the IoT devices used for various purposes, including designing nonstandard cooling and heating that connects the lighting which is dissimilar, and the fire safety system for a central management application to build up the standards. The IoT can change the water consumption ways in urban areas by implementing smart meters that enhance information integrity and leak detection (Ali et al., 2022). Additionally, it allows the assistance in strengthening different companies production and helps them in monitoring the loss in revenue owing to the mismanagement of time to process the information (Vlacheas et al., 2013).

Finally, complete content and organizational editing before formatting. Please take note of the following items when proofreading spelling and grammar

Furthermore, such smart meters can also be used to throw light over the technology necessary for the user by establishing real-time access to the residents. It allows monitoring of the consumer-facing gateways to collect data about water supply management and its use (Ejaz et al., 2017).

The IoT is also helpful in checking the smart traffic signals by adjusting the traffic and shifts during the holidays and keeping the cars moving without any manual interference. It allows the city authorities to collect information regarding traffic from vehicles, traffic cameras, road sensors, and monitors the accidents of traffic without any hurdle (Alzoubi et al., 2021). This puts the drivers on caution against the coordination and incidents of driving. Such types of prospective are quite infinite and the expected outcomes are generous (Sanchez et al., 2014). By taking the help of the IoT system, public transport is easily monitored and stays unobstructed. It keeps an eye on the roadblocks either due to heavy weather or the equipment breakdown (Guergov & Radwan, 2021; Khan, 2021; Mondol, 2021). It offers real-time data for the transport experts to carry out the emergency plans. It guarantees constant access to the network, providing full security and quick public transport through assisted devices and smart cameras at the transport covers and the other areas (Perera et al., 2014).

The smart cities now carry many surveillance cameras that have smart access and they monitor the control of traffic for the safety concern incidents within public streets. The software for video monitoring utilizes the services of IoT. It connects every camera of the system within the sensor to sense and measure data based on cloud-assisted IoT through computing and edge analytics using the WSN channel. The IoT's machine learning approach allows the completion of the investigation study by sending the recording of video to belonging people who can explain the issue and keep the people under protection (Talari et al., 2017).

The basic structure of the cloud-assisted wireless sensor network includes the sensor node. These sensor nodes are made up of the sensor arrays, System-On-Chip (SoC), distribution units, wireless communication interface, and the power supply (Alnuaimi et al., 2021; Alshurideh et al., 2019; AlShurideh et al., 2019; Joghee et al., 2021). The wireless communication interface of this sensor allows each node to work separately.

2   Iot-Based Wsn for Reducing Energy Consumption

There are two approaches: Hierarchical strategic making (HCSM) and the Dynamic Stochastic Optimization Technique (DSOT) system assisted by IoT that helps optimize the energy within the Wireless Sensor Network to monitor smart city to manage the data. One study has been conducted by Meenakshi et al., has combined the two approaches within the smart city for helping in reducing maintenance cost, consumption of energy (Akhtaruzzaman et al., 2020), and the environmental impact to lead towards efficiency with adoption of distinct elements within a single framework. The proposed models utilizing the two technologies achieved a better performance ratio (95.3%), higher prediction ratio (96.54%), higher reliability ratio (92.4%) in comparison with the existing approaches. The study showed better environmental impact with low cost and low consumption of emergency by combining HCSM and DSOT. Moreover, the IoT based Wireless sensor network has helped enhance network lifetime and the node battery life (Sundhari & Jaikumar, 2020).

Many other studies have utilized machine learning approaches to reduce energy consumption. According to the study conducted by Karan Nair et al., the Bluetooth Low Energy (BLE) method was utilized to optimize the consumption of energy within wireless sensor networks of IoT (Nair et al., 2015). The study utilized the hybrid topology for reducing the cost as well as consumption of energy (Alshurideh et al., 2012). It has been mentioned that the BLE approach can be used in operating utilized network topology whereas the hybridized star and mesh topologies use the multi-hop mesh topology (Alzoubi et al., 2022; Obaid, 2021).

A study carried out by Xue Wang et al. (2009) started algorithms of parallel sensor deployment optimization and the parallel particle swarm optimization (PSDO— PPSO) to mark the optimized energy tracking within WSN. The later algorithm, Parallel particle swarm optimization is utilized within-cluster head to enhance the area of coverage and reduce the power of communication within each cluster. On the other side, the former algorithm, the Parallel sensor deployment optimization algorithm, has been utilized to reduce the tracking of a target and energy consumption (Ahmed et al., 2020). Both approaches have found significant improvement in the reduction of energy consumption.

RERUM (Reliable, Resilient, and Secure IoT for Smart City) utilize framework to secure the smart city and provide security using IoT (Pöhls et al., 2014). The RERUM algorithm efficiently sensed the environment in a timely and trustworthy manner and is found best in improving the smart city's architecture. The RERUM has been integrated with IoT to enhance IoT reliability by providing security and protection to a smart city's applications (Aburayya et al., 2020; Al Batayneh et al., 2021; Al Shebli et al., 2021; Alshurideh et al., 2021; Alzoubi et al., 2020). The RERUM framework offers the application scenario and attacker model to achieve protection within a smart city.

One other method, named Media Based Surveillance system (EAMSuS) has been suggested by Vasileios et al., for the inculcation of the quick algorithm within the IoT smart city system (Memos et al., 2018). This framework provides the daily security system that assures the most secured and light transmission of the rapid media sharing within the smart city (AlHamad et al., 2021; Alzoubi, 2021a, b; Farouk, 2021). The study achieved the low consumption of memory at IoT integrating WSN assistance in cybercrime social cloud for management of the smart city.

3 Iot Based Water Supply System

Water is the most demanding thing in the industries in urban cities due to economic sprouting. Such demands are often not met because of the short water supply and the high-water wastage culture. The water is wasted due to many reasons including majorly human error and then pipeline leakages or errors due to operator's negligence (AlHamad et al., 2022; Alzoubi & Ahmed, 2019). Water supply is quite irregular and water wastage can be managed with PLC use that can efficiently control the water supply system. Monitoring industries' water resources can properly control water theft. The supply chain of water in the urban areas is directly related to the water supply and its consumption (Joghee et al., 2020).

The water quality control distribution system must be by the water quality control, its continuity, water theft prevention, and monitoring technological processes. It must also cater to restrictions imposed concerning water availability, water tanks' storage capacity, and the wide water use (Hanaysha et al., 2021a, b). The supply system is made up of storage tanks, filtering units, pump stations, and the distribution network through dispatching units or pipelines. One structural system SCADA carries more than one PC major station that is the communication center linked through IoT with the other PLCs that are integrated within the pumping station. PLC also controls the technological process and the data collection and reports a similar operational hub for the analysis (Verma et al., 2015).

3.1 Current Water Supply System

The existing supply system is full of water wastage and theft. It is irregular but inefficient and doesn't meet the restrictions applied to it considering water tanks' storage capacity, water availability, and wide water usage. The existing system is widely dependent upon manpower and the person chose performs the job manually. Such higher dependency upon manpower makes the system highly prone to human error and is inefficient. This system allows a higher chance of personal water theft by simply attaching a house pump with a supply system and it leads to the wastage of the resources (Alzoubi & Yanamandra, 2020) This is why the system is full of unethical and irregular water distribution.

3.2 Automated Water Supply System

In the automatic supply system, the equipment is installed at the pumping station and is controlled with the help of IoT related PLC (Programmable Logical Controller) equipment that uses electrical and hydraulic parameters. The pumping function module present in PLC carries an optimization tool that allows a schedule for water distribution. It prevents water theft, creates water demanding statistics, has proximity sensors for detecting water and plans, and maintains water wastage reduction.

If the water levels detected are low, the sensor in the equipment signals PLC to initiate the pump station motor. The sensors' current status is displayed on the PC and the user graphical interface is created with SCADA software. Module for optimization helps in the reduction of water resources wastage based on machine learning algorithms (Alzoubi et al., 2019). Such algorithms reduce electrical energy cost by managing the schedule for pumping and predicting the times of engine on and off. It also helps in maintenance planning because of loading. The PLC is the central part of the automated water supply system that manages motor and pumping units and data is displayed on SCADA software in real-time for the user (Saha et al., 2017).

4 Iot Based Traffic Management System

The smart cities are now using the IoT in their traffic system. The problem of congestion and unforeseen events has been reduced by the integration of metal loops in the road that has data points that send a report about such events. The metal loop has sensors on it and street cameras allow the information transfer to major operation rooms (Alzoubi et al., 2019). The operation room staff analyze the data with Big Data Analysis's help and organize it under priority by utilizing a machine learning approach. The operation room then transmits the relevant information to the traffic lights integrated with IoT to change traffic direction away from the site of the incident or make room for other traffic in congestion. The digital display boards notify the traffic about the speed light and measures required to be taken for current road conditions (Misbahuddin et al., 2015).

The emergency vehicles are also caught in the congestion and have fatal consequences. This problem is also resolved by marking such vehicles with infrared (IR) beamers that provide IR signals. This method allows the release of signals that are taken by road cameras and they send it to the main room where the data is categorized and transmits orders to switch traffic lights immediately (Alzoubi, 2021a, 2021b; Miller, 2021).

The other most notable traffic problem is the parking space needed. It is also dealt with the IoT at the parking space. The counter for parking has sensors that show whether there is an empty parking slot. Such sensors need very little power because they are used only if any parking is offered (Mehmood et al., 2019). Moreover, they also require less bandwidth and every sensor has its routing capabilities which allow the traveling of data to network doorway by combining such networks. This is why a network gateway is not required in parking sensor limits. The sensor network nodes will create a mesh and use the data of neighboring nodes to reach a network gateway (Alsharari, 2021). This final entrance will transfer information to a cloud application or computer and enables drivers to be seen in real-time. It allows safe streets, reduces congestion on the road and lower rates of the accident. It reduces the money spent on the roads fixation and provides an excellent boost to the economy (Saha et al., 2017).

5 Iot Based Deep Reinforcement Learning

Deep learning is the most effective machine-based learning approach which offers purpose classification, estimate, and forecast functions. Reinforcement learning provides a decision-making process and optimal control in which the software agents allow optimal actions policy learning on the states set in the environment (LeCun et al., 2015). The deep learning model is used for approximation of action values in applications where states number is quite large. The systems that merge reinforcement and deep learning are gaining considerable attention in smart cities' different applications (Ghazvini et al., 2020; Memon et al., 2020).

According to research carried out by Nemati et al., a deep reinforcement learning algorithm is used for learning the actionable policies to provide the medicines optimal dose such as heparin (Nemati et al., 2016). The researchers utilized the dosage trials sample dataset and its outcomes from various electronic medical records (Alnazer et al., 2017). They used Hidden Markov Model for the Q-network and state estimation with the neuron's two layers. The dosage agent learns the optimal policy by enhancing the entire reward, that is the overall time fraction when the patients are within the therapeutic range of activated Partial Thromboplastin Time (aPTT).

The machine-based approach is also applied in the vehicle image classification. Researchers proposed the Convolutional Neural Network (CNN) model combined with a learning component that guides where to look in car key parts image. The focused image created by the CNN model is measured with the reinforcement learning agent reward of learning identification of following visual attention part in the image. A similar effort is reported by Caicedo et al., where the focused attention on the object localization is reported through the use of a deep reinforcement learning approach (Caicedo & Lazebnik, 2015).

Li et al. have used the deep reinforcement learning approach in controlling the time of traffic signal to provide good signal time (Li et al., 2016). The proposed model has the four layers stacked auto-encoder neural network that works to estimate Q-function. The two actions defined include: Change lanes and allow other traffic to move through the intersection or stay in the current traffic lane. The reward function is represented by the difference in absolute value among opposite lanes length (Akhtar et al., 2021; Aziz & Aftab, 2021; Mehmood, 2021). Such a result shows that the model works well than the conventional approach.

Resource management is one other task that utilizes DRL. Mao et al., formulates job scheduling problem with the various resources demand as the deep reinforcement learning procedure. In this method, the primary aim is the reduction of the slowdown of jobs. The reciprocal job duration is defined as a reward function (Mao et al., 2016). One other application where deep reinforcement learning plays the most critical role is the innate understanding of the language for text-based games. The researchers used the long short term memory networks for the training of agents with the useful text descriptions presentation along with the deep Q network for approximating the Q functions (Radwan & Farouk, 2021). Some other fields, such as energy management, have also included the DRL for improving the use of energy.

Indoor localization.

Only one group of researchers have utilized the DRL for the indoor localization services and the most promising approach found in the Relative Signal Strength (RSS) (Hasan et al., 2019; Islam et al., 2017). However, it has come up with some challenges in deployment, including device diversity and fingerprint annotation. Different machine-based learning approaches have been studied including transfer learning, SVM, neural networks, Bayesian-based filtering, and KNN. It is showed that for the Bluetooth low energy RSS fingerprinting based positioning applications, the estimation to decide that if the device is within the room provides quite reliable results (Islam et al., 2020). The researchers report the indoor based positioning results reading. They have studied three methods of accuracy including Centroid positioning, three border positioning, and the least square estimation. There are four BLE stations in the 6×8 sq. meter classroom testing area, and the LSE algorithm is applied that showed accurate positions compared with other methods (Nurelmadina et al., 2021). However, all three algorithms showed complete satisfaction (Kajioka et al., 2014).

Museums are best for utilizing BLE to offer location awareness as the building does not permit changes due to its protection rules. Researchers have established the method for making an interactive display of culture in the museum by BLE examples combined with the image recognition wearable device. This device provides localization to receive the signals of BLE for identifying the room where it is present. The device also recognizes artwork with the image processing facility. A combination of artwork identifier and closest beacon identifier is installed in the processing center to earn relevant cultural material (Al Ali, 2021; Eli, 2021; Kashif et al., 2021; Lee & Ahmed, 2021).

Wang et al. (2015) provide a system named DeepFi that uses a deep learning method for locating the indoor positions based on channel state information. This system includes the online localization and offline training phase. In the offline training stage, the researchers used deep learning for training total weights as the fingerprints upon the basis of before kept CSI. Improved localization accuracy of 20% was reported in the laboratory and living room settings. As their CSI approach use is only restricted to the WiFi network, not every network interface card supports the measurements from wide channels.

Joined semi-supervised learning, deep learning, and the extreme learning machine to the unlabeled data for study classification phases and feature extraction performance for the indoor localization (Gu et al., 2015). They generated the accurate classification and higher abstract features by combining the semi-supervised learning and deep learning network in which the learning machine can raise the learning process speed. The results showed that deep learning enhances fingerprint accuracy by 1.3% for similar dataset training compared to the shallow learning method. The unlabeled data also showed a positive effect over accuracy. When the approach is compared with other methods, the approach has raised the accuracy level to 10%.

In another study conducted by Zhang et al. (2016), the group utilized the WiFi localization approach by utilizing deep neural networks (DNN). The four-layered deep learning model was utilized for the extraction of features from the WiFi RSS data (Wang et al., 2016). Within the approach, the researchers used Backpropagation and Stacked Denoising Autoencoder for the preparation phases. In the level of online positioning, DNN based location is additionally advanced by part of HMM. The experiments indicate that the neurons and hidden layers pose an effect directly upon localization accuracy. Raising the number of layers provides better results but when the network is deep, the results degrade. According to their results, when three hidden layers each layer carrying 200 neurons is used, the model showed optimum accuracy.

One other researcher has also utilized an artificial neural network for the localization of WiFi fingerprints. They used the localization approach, which used this artificial neural network and the clustering method based on affinity propagation. With help of affinity propagation clustering, the ANN model training is quick whereas memory overhead reduces. It also showed better positioning accuracy in comparison with baseline methods (Ding et al., 2013).

One other study has used the deep belief network as the localization approach based on ultra-wideband signals fingerprinting within indoor atmospheric conditions. The channel impulse response parameters were utilized to provide the dataset for the fingerprints. When the model was compared with the other methods, the DBN has shown better and improved localization accuracy (Luo & Gao, 2016). This work was further expanded by Zhang et al. (2016), where they used a deep learning model along with the regression model for learning the discriminative features automatically from wireless signals. The softmax regression algorithm is used for performing the activity recognition and device-free localization. It has been reported that the method enhances the accuracy of localization up to 10% when compared with the other methods.

6   Iot Based Underwater and Wireless Sensor Network

The IoUT is defined as the chain of smart interconnected underwater objects that are now defined as the IoT labeled underwater and wireless sensor networks (I-UWSNs). The smart objects have different sensor types such as autonomous underwater vehicles, ships, buoys, watchman nodes, etc. That is why, the I-UWSN is expected to carry various practical applications, including monitoring the environment, disaster prevention, and underwater exploitation (Alshurideh, 2022; Cruz, 2021; T. Ghazal et al., 2021a, 2021b; Hamadneh et al., 2021a, 2021b; Shamout et al., 2022).

There are different ways of implying sensor nodes within I-UWSN for these purposes. The nodes are either deployed randomly or in the tree structure or grid form. For providing reliable communication, every sensor has communicational, computational, and intelligence capabilities in dealing with smart city environments. Different routing protocols are utilized for providing effective and reliable communication in the smart underwater environment. One of them is the self-organized underwater wireless sensor network that utilizes tree topology to deploy nodes. The dynamic address is assigned to every node for hierarchical, hop count, and identification level. For new node addition at each level, it only requires a parent address. It also resolves the problem of node isolation as well as the close loop (Alshurideh et al., 2020; Lee, Azmi, et al., 2022a, 2022b; Lee, Romzi, et al., 2022a, 2022b).

6.1 Topology Controlled Protocols

Another reliable routing protocol is the topology efficient discovery that can reduce the time overhead and accurately evaluating the reliability of the link. This protocol relies upon the time window that is allocated for transmission within which the time slot has been defined for the nodes for transmission of one or different packets.

One other work measures the reliability of the link. The model supports multiple-input multiple-output (MIMO), code division multiple access (CDMA), and Binary Phase Shift Key (BPSK) communication system. Geethu et al., used the depth-based routing protocol as mentioned above, which improves data reliability. This is an efficient approach for handling dynamic networks (Diamant et al., 2017).

7 Conclusion

Modernization is at the peak at this time as rural to the urban movement of population has been greatly observed, shaping people's lifestyle. More smart cities are in process that is expected to accommodate such a large number of figures. Several developments are undergoing in different countries regarding Smart City. The smart city is the shorthand for the technology-based services within society, which means the ICT (information and communication technology) is crucial for the implementation. The justified use of IoT based machine learning approaches has been proved valuable in various effective ways, including IoT-based WSN for reducing energy consumption. IoT based Water Supply System, IoT based traffic management System, and the IoT based Underwater and Wireless sensor network.

References

Aburayya, A., Alshurideh, M., Al Marzouqi, A., Al Diabat, O., Alfarsi, A., Suson, R., Bash, M., & Salloum, S. A. (2020). An empirical examination of the effect of TQM practices on hospital service quality: An assessment study in uae hospitals. *Systematic Reviews in Pharmacy, 11*(9). https://doi.org/10.31838/srp.2020.9.51

Ahmed, Z. E., Hasan, M. K., Saeed, R. A., Hassan, R., Islam, S., Mokhtar, R. A., Khan, S., & Akhtaruzzaman, M. (2020). Optimizing energy consumption for cloud internet of things. *Frontiers of Physics, 8*, 358. https://doi.org/10.3389/Fphy

Akhtar, A., Akhtar, S., Bakhtawar, B., Kashif, A. A., Aziz, N., & Javeid, M. S. (2021). COVID-19 Detection from CBC using Machine Learning Techniques. *International Journal of Technology, Innovation and Management (IJTIM), 1*(2), 65–78. https://doi.org/10.54489/ijtim.v1i2.22

Akhtaruzzaman, M., Hasan, M. K., Kabir, S. R., Abdullah, S. N. H. S., Sadeq, M. J., & Hossain, E. (2020). HSIC bottleneck based distributed deep learning model for load forecasting in smart grid with a comprehensive survey. *IEEE Access.*

Al Ali, A. (2021). The impact of information sharing and quality assurance on customer service at uae banking sector. *International Journal of Technology, Innovation and Management (IJTIM), 1*(1), 01–17. https://doi.org/10.54489/ijtim.v1i1.10.

Al Batayneh, R. M., Taleb, N., Said, R. A., Alshurideh, M. T., Ghazal, T. M., & Alzoubi, H. M. (2021). IT governance framework and smart services integration for future development of Dubai infrastructure utilizing ai and big data, its reflection on the citizens standard of living. *The International Conference on Artificial Intelligence and Computer Vision*, 235–247.

Al Shebli, K., Said, R. A., Taleb, N., Ghazal, T. M., Alshurideh, M. T., & Alzoubi, H. M. (2021). RTA's employees' perceptions toward the efficiency of artificial intelligence and big data utilization in providing smart services to the residents of Dubai. *The International Conference on Artificial Intelligence and Computer Vision*, 573–585.

AlHamad, A., Alshurideh, M., Alomari, K., Kurdi, B., Alzoubi, H., Hamouche, S., & Al-Hawary, S. (2022). The effect of electronic human resources management on organizational health of telecommuni-cations companies in Jordan. *International Journal of Data and Network Science, 6*(2), 429–438.

Alhamad, A. Q. M., Akour, I., Alshurideh, M., Al-Hamad, A. Q., Kurdi, B. A., & Alzoubi, H. (2021). Predicting the intention to use google glass: A comparative approach using machine learning models and PLS-SEM. *International Journal of Data and Network Science, 5*(3). https://doi.org/10.5267/j.ijdns.2021.6.002

Ali, N., Ahmed, A., Anum, L., Ghazal, T. M., Abbas, S., Khan, M. A., Alzoubi, H. M., & Ahmad, M. (2021). Modelling supply chain information collaboration empowered with machine learning technique. *Intelligent Automation and Soft Computing, 30*(1), 243–257. https://doi.org/10.32604/iasc.2021.018983

Ali, N., M. Ghazal, T., Ahmed, A., Abbas, S., A. Khan, M., Alzoubi, H., Farooq, U., Ahmad, M., & Adnan Khan, M. (2022). Fusion-Based Supply Chain Collaboration Using Machine Learning Techniques. *Intelligent Automation & Soft Computing, 31*(3), 1671–1687. https://doi.org/10.32604/iasc.2022.019892

Ali, T., Irfan, M., Shaf, A., Saeed Alwadie, A., Sajid, A., Awais, M., & Aamir, M. (2020). A secure communication in IoT enabled underwater and wireless sensor network for smart cities. *Sensors, 20*(15), 4309.

Aljumah, A., Nuseir, M. T., & Alshurideh, M. T. (2021). The impact of social media marketing communications on consumer response during the COVID-19: Does the brand equity of a university matter. *The Effect of Coronavirus Disease (COVID-19) on Business Intelligence, 334*, 384–367.

Alnazer, N. N., Alnuaimi, M. A., & Alzoubi, H. M. (2017). Analysing the appropriate cognitive styles and its effect on strategic innovation in Jordanian universities. *International Journal of Business Excellence, 13*(1), 127–140. https://doi.org/10.1504/IJBEX.2017.085799

Alnuaimi, M., Alzoubi, H. M., Ajelat, D., & Alzoubi, A. A. (2021). Towards intelligent organisations: An empirical investigation of learning orientation's role in technical innovation. *International Journal of Innovation and Learning, 29*(2), 207–221. https://doi.org/10.1504/IJIL.2021.112996

Alsharari, N. (2021). Integrating blockchain technology with internet of things to efficiency. *International Journal of Technology, Innovation and Management (IJTIM), 1*(2), 1–13.

Alshurideh, M. (2022). Does electronic customer relationship management (E-CRM) affect service quality at private hospitals in Jordan? *Uncertain Supply Chain Management, 10*(2), 1–8.

AlShurideh, M., Alsharari, N. M., & Al Kurdi, B. (2019). Supply chain integration and customer relationship management in the airline logistics. *Theoretical Economics Letters, 9*(02), 392–414.

Alshurideh, M., Gasaymeh, A., Ahmed, G., Alzoubi, H., & Kurd, B. A. (2020). Loyalty program effectiveness: Theoretical reviews and practical proofs. *Uncertain Supply Chain Management, 8*(3). https://doi.org/10.5267/j.uscm.2020.2.003

Alshurideh, M. T., Al Kurdi, B., Alzoubi, H. M., Ghazal, T. M., Said, R. A., AlHamad, A. Q., Hamadneh, S., Sahawneh, N., & Al-kassem, A. H. (2022). Fuzzy assisted human resource management for supply chain management issues. *Annals of Operations Research*, 1–19.

Alshurideh, M. T., Kurdi, B. A., AlHamad, A. Q., Salloum, S. A., Alkurdi, S., Dehghan, A., Abuhashesh, M., & Masa'deh, R. (2021). Factors affecting the use of smart mobile examination platforms by universities' postgraduate students during the COVID 19 pandemic: An empirical study. *Informatics, 8*(2), 32.

Alshurideh, M. T., & Shaltoni, A. M. (2014). Marketing communications role in shaping consumer awareness of cause-related marketing campaigns. *International Journal of Marketing Studies, 6*(2), 163.

Alshurideh, M., & Masa'deh, R. M. d. T., & Alkurdi, B. (2012). The effect of customer satisfaction upon customer retention in the Jordanian mobile market: An empirical investigation. *European Journal of Economics, Finance and Administrative Sciences, 47*(47), 69–78.

Alyammahi, A., Alshurideh, M., Kurdi, B. Al, & Salloum, S. A. (2020). The impacts of communication ethics on workplace decision making and productivity. In: *International Conference on Advanced Intelligent Systems and Informatics*, 488–500.

Alzoubi, Ali. (2021a). The impact of process quality and quality control on organizational competitiveness at 5-star hotels in Dubai. *International Journal of Technology, Innovation and Management (IJTIM), 1*(1), 54–68. https://doi.org/10.54489/ijtim.v1i1.14

Alzoubi, Asem. (2021b). Renewable Green hydrogen energy impact on sustainability performance. *International Journal of Computations, Information and Manufacturing (IJCIM), 1*(1), 94–110. https://doi.org/10.54489/ijcim.v1i1.46

Alzoubi, H., Ahmed, G., Al-Gasaymeh, A., & Alkurdi, B. (2019). Empirical study on sustainable supply chain strategies and its impact on competitive priorities: the mediating role of supply chain collaboration. *Management Science Letters, 10*(3), 703–708.

Alzoubi, H., Alshurideh, M., Kurdi, B. A., & Inairat, M. (2020). Do perceived service value, quality, price fairness and service recovery shape customer satisfaction and delight? A practical study in the service telecommunication context. *Uncertain Supply Chain Management, 8*(3). https://doi.org/10.5267/j.uscm.2020.2.005

Alzoubi, H. M., & Aziz, R. (2021). Does emotional intelligence contribute to quality of strategic decisions? the mediating role of open innovation. *Journal of Open Innovation: Technology, Market, and Complexity, 7*(2), 130. https://doi.org/10.3390/joitmc7020130

Alzoubi, H. M., Vij, M., Vij, A., & Hanaysha, J. R. (2021). What leads guests to satisfaction and loyalty in UAE five-star hotels? AHP analysis to service quality dimensions. *Enlightening Tourism, 11*(1), 102–135. https://doi.org/10.33776/et.v11i1.5056

Alzoubi, H. M., & Yanamandra, R. (2020). Investigating the mediating role of information sharing strategy on agile supply chain. *Uncertain Supply Chain Management, 8*(2), 273–284. https://doi.org/10.5267/j.uscm.2019.12.004

Alzoubi, H., Alshurideh, M., Kurdi, B., Akour, I., & Aziz, R. (2022). Does BLE technology contribute towards improving marketing strategies, customers' satisfaction and loyalty? The role of open innovation. *International Journal of Data and Network Science, 6*(2), 449–460.

Alzoubi, H., & Ahmed, G. (2019). Do TQM practices improve organisational success? A case study of electronics industry in the UAE. *International Journal of Economics and Business Research, 17*(4), 459–472. https://doi.org/10.1504/IJEBR.2019.099975

Aziz, N., & Aftab, S. (2021). Data mining framework for nutrition ranking: methodology: SPSS modeller. *International Journal of Technology, Innovation and Management (IJTIM), 1*(1), 85–95.

Caicedo, J. C., & Lazebnik, S. (2015). Active object localization with deep reinforcement learning. In: *Proceedings of the IEEE International Conference on Computer Vision,* 2488–2496.

Cruz, A. (2021). Convergence between blockchain and the internet of things. *International Journal of Technology, Innovation and Management (IJTIM), 1*(1), 35–56.

Diamant, R., Francescon, R., & Zorzi, M. (2017). Topology-efficient discovery: A topology discovery algorithm for underwater acoustic networks. *IEEE Journal of Oceanic Engineering, 43*(4), 1200–1214.

Ding, G., Tan, Z., Zhang, J., & Zhang, L. (2013). Fingerprinting localization based on affinity propagation clustering and artificial neural networks. *IEEE Wireless Communications and Networking Conference (WCNC), 2013,* 2317–2322.

Ejaz, W., Naeem, M., Shahid, A., Anpalagan, A., & Jo, M. (2017). Efficient energy management for the internet of things in smart cities. *IEEE Communications Magazine, 55*(1), 84–91.

Eli, T. (2021). Students perspectives on the use of innovative and interactive teaching methods at the university of nouakchott al aasriya, mauritania: english department as a case study. *International Journal of Technology, Innovation and Management (IJTIM), 1*(2), 90–104.

Farouk, M. (2021). The Universal Artificial Intelligence Efforts to Face Coronavirus COVID-19. *International Journal of Computations, Information and Manufacturing (IJCIM), 1*(1), 77–93. https://doi.org/10.54489/ijcim.v1i1.47

Ghazal, T., Alshurideh, M., & Alzoubi, H. (2021a). Blockchain-Enabled Internet of Things (IoT) Platforms for Pharmaceutical and Biomedical Research. In: *The International Conference on Artificial Intelligence and Computer Vision,* 589–600.

Ghazal, T. M., Hasan, M. K., Alshurideh, M. T., Alzoubi, H. M., Ahmad, M., Akbar, S. S., Al Kurdi, B., & Akour, I. A. (2021b). IoT for smart cities: machine learning approaches in smart Healthcare—a review. *Future Internet, 13*(8), 218. https://doi.org/10.3390/fi13080218

Ghazal, T. M., Hasan, M. K., Hassan, R., Islam, S., Abdullah, S., Afifi, M. A., & Kalra, D. (2020). Security vulnerabilities, attacks, threats and the proposed countermeasures for the Internet of things applications. *Solid State Technology, 63*(1s), 2513–2521.

Ghazvini, A., Abdullah, S. N. H. S., Hasan, M. K., Kasim, D. Z. A., & Bin. (2020). Crime spatiotemporal prediction with fused objective function in time delay neural network. *IEEE Access, 8,* 115167–115183.

Gu, Y., Chen, Y., Liu, J., & Jiang, X. (2015). Semi-supervised deep extreme learning machine for Wi-Fi based localization. *Neurocomputing, 166,* 282–293.

Guergov, S., & Radwan, N. (2021). Blockchain Convergence: analysis of issues affecting IoT, AI and blockchain. *International Journal of Computations, Information and Manufacturing (IJCIM), 1*(1), 1–17. https://doi.org/10.54489/ijcim.v1i1.48

Hamadneh, S., Pedersen, O., Alshurideh, M., Kurdi, B. A., & Alzoubi, H. M. (2021a). An investigation of the role of supply chain visibility into the scottish blood supply chain. *Journal of Legal, Ethical and Regulatory Issues, 24*(Special Issue 1).

Hamadneh, Samer, Pedersen, O., & Al Kurdi, B. (2021b). An investigation of the role of supply chain visibility into the scottish bood supply chain. *Journal of Legal, Ethical and Regulatory Issues, 24*(Special Issue 1), 1–12.

Hanaysha, J. R., Al-Shaikh, M. E., Joghee, S., & Alzoubi, H. (2021a). Impact of innovation capabilities on business sustainability in small and medium enterprises. *FIIB Business Review,* 1–12. https://doi.org/10.1177/23197145211042232

Hanaysha, J. R., Al Shaikh, M. E., & Alzoubi, H. M. (2021b). Importance of marketing mix elements in determining consumer purchase decision in the retail market. *International Journal of Service Science, Management, Engineering, and Technology (IJSSMET)*, *12*(6), 56–72.

Hasan, M. K., Ahmed, M. M., Hashim, A. H. A., Razzaque, A., Islam, S., & Pandey, B. (2020). A novel artificial intelligence based timing synchronization scheme for smart grid applications. *Wireless Personal Communications*, *114*(2), 1067–1084.

Hasan, M. K., Ismail, A. F., Islam, S., Hashim, W., Ahmed, M. M., & Memon, I. (2019). A novel HGBBDSA-CTI approach for subcarrier allocation in heterogeneous network. *Telecommunication Systems*, *70*(2), 245–262.

Islam, S., Hashim, A.-H.A., Habaebi, M. H., & Hasan, M. K. (2017). Design and implementation of a multihoming-based scheme to support mobility management in NEMO. *Wireless Personal Communications*, *95*(2), 457–473.

Islam, S., Khalifa, O. O., Hashim, A.-H.A., Hasan, M. K., Razzaque, M. A., & Pandey, B. (2020). Design and evaluation of a multihoming-based mobility management scheme to support inter technology handoff in PNEMO. *Wireless Personal Communications*, *114*(2), 1133–1153.

Joghee, S., Alzoubi, H. M., Alshurideh, M., & Al Kurdi, B. (2021). The role of business intelligence systems on green supply chain management: empirical analysis of FMCG in the UAE. In: *The International Conference on Artificial Intelligence and Computer Vision*, 539–552.

Joghee, S., Alzoubi, H. M., & Dubey, A. R. (2020). Decisions effectiveness of FDI investment biases at real estate industry: Empirical evidence from Dubai smart city projects. *International Journal of Scientific and Technology Research*, *9*(3), 3499–3503.

Kajioka, S., Mori, T., Uchiya, T., Takumi, I., & Matsuo, H. (2014). Experiment of indoor position presumption based on RSSI of Bluetooth LE beacon. In: *2014 IEEE 3rd Global Conference on Consumer Electronics (GCCE)*, 337–339.

Kashif, A. A., Bakhtawar, B., Akhtar, A., Akhtar, S., Aziz, N., & Javeid, M. S. (2021). Treatment response prediction in hepatitis C patients using machine learning techniques. *International Journal of Technology, Innovation and Management (IJTIM)*, *1*(2), 79–89. https://doi.org/10.54489/ijtim.v1i2.24

Khan, M. A. (2021). Challenges facing the application of iot in medicine and healthcare. *International Journal of Computations, Information and Manufacturing (IJCIM)*, *1*(1), 39–55. https://doi.org/10.54489/ijcim.v1i1.32

LeCun, Y., Bengio, Y., & Hinton, G. (2015). Deep learning. *Nature*, *521*(7553), 436–444.

Lee, C., & Ahmed, G. (2021). Improving IoT privacy, data protection and security concerns. *International Journal of Technology, Innovation and Management (IJTIM)*, *1*(1), 18–33. https://doi.org/10.54489/ijtim.v1i1.12

Lee, K., Azmi, N., Hanaysha, J., Alzoubi, H., & Alshurideh, M. (2022a). The effect of digital supply chain on organizational performance: An empirical study in Malaysia manufacturing industry. *Uncertain Supply Chain Management*, *10*(2), 495–510.

Lee, K., Romzi, P., Hanaysha, J., Alzoubi, H., & Alshurideh, M. (2022b). Investigating the impact of benefits and challenges of IOT adoption on supply chain performance and organizational performance: An empirical study in Malaysia. *Uncertain Supply Chain Management*, *10*(2), 537–550.

Li, L., Lv, Y., & Wang, F.-Y. (2016). Traffic signal timing via deep reinforcement learning. *IEEE/CAA Journal of Automatica Sinica*, *3*(3), 247–254.

Luo, J., & Gao, H. (2016). Deep belief networks for fingerprinting indoor localization using ultrawideband technology. *International Journal of Distributed Sensor Networks*, *12*(1), 5840916.

Mao, H., Alizadeh, M., Menache, I., & Kandula, S. (2016). Resource management with deep reinforcement learning.In: *Proceedings of the 15th ACM Workshop on Hot Topics in Networks*, 50–56.

Mehmood, T. (2021). Does information technology competencies and fleet management practices lead to effective service delivery? empirical evidence from E-commerce industry. *International Journal of Technology, Innovation and Management (IJTIM)*, *1*(2), 14–41.

Mehmood, T., Alzoubi, H. M., Alshurideh, M., Al-Gasaymeh, A., & Ahmed, G. (2019). Schumpeterian entrepreneurship theory: evolution and relevance. *Academy of Entrepreneurship Journal, 25*(4), 1–10.

Memon, I., Shaikh, R. A., Hasan, M. K., Hassan, R., Haq, A. U., & Zainol, K. A. (2020). Protect mobile travelers information in sensitive region based on fuzzy logic in IoT Ttechnology. *Security and Communication Networks, 2020.*

Memos, V. A., Psannis, K. E., Ishibashi, Y., Kim, B.-G., & Gupta, B. B. (2018). An efficient algorithm for media-based surveillance system (EAMSuS) in IoT smart city framework. *Future Generation Computer Systems, 83*, 619–628.

Miller, D. (2021). The Best Practice of Teach Computer Science Students to Use Paper Prototyping. *International Journal of Technology, Innovation and Management (IJTIM), 1*(2), 42–63. https://doi.org/10.54489/ijtim.v1i2.17

Minoli, D., Sohraby, K., & Occhiogrosso, B. (2017). IoT considerations, requirements, and architectures for smart buildings—Energy optimization and next-generation building management systems. *IEEE Internet of Things Journal, 4*(1), 269–283.

Misbahuddin, S., Zubairi, J. A., Saggaf, A., Basuni, J., Sulaiman, A., Al-Sofi, A., & others. (2015). IoT based dynamic road traffic management for smart cities. In: *2015 12th International Conference on High-Capacity Optical Networks and Enabling/Emerging Technologies (HONET)*, 1–5.

Mondol, E. P. (2021). The impact of block chain and smart inventory system on supply chain performance at retail industry. *International Journal of Computations, Information and Manufacturing (IJCIM), 1*(1), 56–76. https://doi.org/10.54489/ijcim.v1i1.30

Nair, K., Kulkarni, J., Warde, M., Dave, Z., Rawalgaonkar, V., Gore, G., & Joshi, J. (2015). Optimizing power consumption in iot based wireless sensor networks using bluetooth low energy. *International Conference on Green Computing and Internet of Things (ICGCIoT), 2015*, 589–593.

Nemati, S., Ghassemi, M. M., & Clifford, G. D. (2016). Optimal medication dosing from suboptimal clinical examples: A deep reinforcement learning approach. In: *2016 38th Annual International Conference of the IEEE Engineering in Medicine and Biology Society (EMBC)*, 2978–2981.

Nurelmadina, N., Hasan, M. K., Memon, I., Saeed, R. A., Zainol Ariffin, K. A., Ali, E. S., Mokhtar, R. A., Islam, S., Hossain, E., Hassan, M., et al. (2021). A systematic review on cognitive radio in low power wide area network for industrial IoT applications. *Sustainability, 13*(1), 338.

Obaid, A. J. (2021). Assessment of smart home assistants as an IoT. *International Journal of Computations, Information and Manufacturing (IJCIM), 1*(1), 18–36. https://doi.org/10.54489/ijcim.v1i1.34

Perera, C., Zaslavsky, A., Christen, P., & Georgakopoulos, D. (2014). Sensing as a service model for smart cities supported by internet of things. *Transactions on Emerging Telecommunications Technologies, 25*(1), 81–93.

Pöhls, H. C., Angelakis, V., Suppan, S., Fischer, K., Oikonomou, G., Tragos, E. Z., Rodriguez, R. D., & Mouroutis, T. (2014). RERUM: Building a reliable IoT upon privacy-and security-enabled smart objects. *IEEE Wireless Communications and Networking Conference Workshops (WCNCW), 2014*, 122–127.

Radwan, N., & Farouk, M. (2021). The growth of internet of things (IoT) in the management of healthcare issues and healthcare policy development. *International Journal of Technology, Innovation and Management (IJTIM), 1*(1), 69–84. https://doi.org/10.54489/ijtim.v1i1.8

Saha, H. N., Auddy, S., Chatterjee, A., Pal, S., Sarkar, S., Singh, R., Singh, A. K., Sharan, P., Banerjee, S., Sarkar, R., & others. (2017). IoT solutions for smart cities. *2017 8th Annual Industrial Automation and Electromechanical Engineering Conference (IEMECON)*, 74–80.

Sanchez, L., Muñoz, L., Galache, J. A., Sotres, P., Santana, J. R., Gutierrez, V., Ramdhany, R., Gluhak, A., Krco, S., Theodoridis, E., et al. (2014). SmartSantander: IoT experimentation over a smart city testbed. *Computer Networks, 61*, 217–238.

Shakeel, P. M., El Tobely, T. E., Al-Feel, H., Manogaran, G., & Baskar, S. (2019). Neural network based brain tumor detection using wireless infrared imaging sensor. *IEEE Access, 7*, 5577–5588.

Shamout, M., Ben-Abdallah, B., Alshurideh, M., Alzoubi, H., Al Kurdi, B., & Hamadneh, S. (2022). A conceptual model for the adoption of autonomous robots in supply chain and logistics industry. *Uncertain Supply Chain Management, 10*, 1–16.

Sundhari, R. P. M., & Jaikumar, K. (2020). IoT assisted hierarchical computation strategic making (HCSM) and dynamic stochastic optimization technique (DSOT) for energy optimization in wireless sensor networks for smart city monitoring. *Computer Communications, 150*, 226–234.

Sweiss, N., Obeidat, Z. M., Al-Dweeri, R. M., Ahmad, M. K., & A., M. Obeidat, A., & Alshurideh, M. (2021). The moderating role of perceived company effort in mitigating customer misconduct within online brand communities (OBC). *Journal of Marketing Communications*. https://doi.org/10.1080/13527266.2021.1931942

Talari, S., Shafie-Khah, M., Siano, P., Loia, V., Tommasetti, A., & Catalão, J. P. S. (2017). A review of smart cities based on the internet of things concept. *Energies, 10*(4), 421.

Verma, P., Kumar, A., Rathod, N., Jain, P., Mallikarjun, S., Subramanian, R., Amrutur, B., Kumar, M. S. M., & Sundaresan, R. (2015). Towards an IoT based water management system for a campus. In: *2015 IEEE First International Smart Cities Conference (ISC2)*, 1–6.

Vlacheas, P., Giaffreda, R., Stavroulaki, V., Kelaidonis, D., Foteinos, V., Poulios, G., Demestichas, P., Somov, A., Biswas, A. R., & Moessner, K. (2013). Enabling smart cities through a cognitive management framework for the internet of things. *IEEE Communications Magazine, 51*(6), 102–111.

Wang, J., Zhang, X., Gao, Q., Yue, H., & Wang, H. (2016). Device-free wireless localization and activity recognition: A deep learning approach. *IEEE Transactions on Vehicular Technology, 66*(7), 6258–6267.

Wang, X., Ma, J., Wang, S., & Bi, D. (2009). Distributed energy optimization for target tracking in wireless sensor networks. *IEEE Transactions on Mobile Computing, 9*(1), 73–86.

Wang, X., Gao, L., Mao, S., & Pandey, S. (2015). DeepFi: Deep learning for indoor fingerprinting using channel state information. *IEEE Wireless Communications and Networking Conference (WCNC), 2015*, 1666–1671.

Zanella, A., Bui, N., Castellani, A., Vangelista, L., & Zorzi, M. (2014). Internet of things for smart cities. *IEEE Internet of Things Journal, 1*(1), 22–32.

Zhang, X., Wang, J., Gao, Q., Ma, X., & Wang, H. (2016). Device-free wireless localization and activity recognition with deep learning. *IEEE International Conference on Pervasive Computing and Communication Workshops (PerCom Workshops), 2016*, 1–5.

Zygiaris, S. (2013). Smart city reference model: Assisting planners to conceptualize the building of smart city innovation ecosystems. *Journal of the Knowledge Economy, 4*(2), 217–231.

Machine Learning Approaches for Sustainable Cities Using Internet of Things

Taher M. Ghazal [ORCID], **Mohammad Kamrul Hasan, Munir Ahmad,**
Haitham M. Alzoubi [ORCID]**, and Muhammad Alshurideh** [ORCID]

Abstract As these tech-based cities continue to emerge, so do the experts dive deeper into the research on the various Internet of Things methods, and the most plausible machine learning techniques. What is Machine Learning? What role does it play in IoT-based platforms? How does it contribute to the continuous advancement of smart cities? What is IoT? How has it changed the lives of urban households residing in smart cities? How does IoT interact with machine learning to ensure more efficient cities? This research paper delves into the various existing literature in a bid to respond to these questions as effectively as possible. It looks the research that have handled issues of IoT and machine learning. The search criteria are set out in the methodology section that has inclusion and exclusion criteria for the articles considered in this paper. There is systematic evidence showing that IOT and machine learning can be used for the management of the future cities. The results suggest that IoT and machine learning are at the core of the development, maintenance, and

T. M. Ghazal · M. K. Hasan
Center for Cyber Security, Faculty of Information Science and Technology, Universiti Kebansaan Malaysia (UKM), 43600 Bangi, Selangor, Malaysia
e-mail: taher.ghazal@skylineuniversity.ac.ae

T. M. Ghazal
School of Information Technology, Skyline University College, Sharjah, UAE

M. Ahmad
School of Computer Science, National College of Business Administration & Economics, Lahore, Pakistan

H. M. Alzoubi (✉) · H. M. Alzoubi (✉)
School of Business, Skyline University College, Sharjah, UAE
e-mail: haitham.alzubi@skylineniversity.ac.ae

M. Alshurideh
Department of Marketing, School of Business, University of Jordan, Amman, Jordan
e-mail: malshurideh@sharjah.ac.ae

Department of Management, College of Business Administration, University of Sharjah, Amman, UAE

M. Alshurideh et al. (eds.), *The Effect of Information Technology on Business and Marketing Intelligence Systems*, Studies in Computational Intelligence 1056, https://doi.org/10.1007/978-3-031-12382-5_108

sustainability of smart cities. Further studies are required on how IOT can be securely used by analysis some of the security vulnerabilities faced by these systems.

Keywords IOT · Machine Learning · Digital Cities

1   Introduction

Humanity has advanced its technologies to the extent that they have become an integral part of our lives (Al Kurdi et al., 2020; Alshurideh et al., 2019; Kurdi et al., 2020). A few decades ago, the idea of having a city whose main operations and services are run and operated by modern technology seemed like something out of a sci-fi movie. Today, however, smart cities are mushrooming in different countries across the world, including Singapore, the UAE, Japan, Italy, the US, and the UK. A few decades ago, the idea of a city that utilizes different electronic sensors and methods to gather and analyze data seemed outlandish (Ismail et al., 2019). Today, however, the idea of smart cities is fast growing in popularity, as countries like Japan, the UK and Italy lead in the digitization of their City Asset Management Systems. Cities like Tokyo and Milan have been remarkably digitized in terms of traffic control, power plant operations, transportation systems, water supply networks, and waste management, and so on (Habibzadeh et al., 2019). With technology becoming not only more advanced but also accessible to the masses, smart cities are going to be part of the urban planning agenda in many countries. Since the concept of smart cities is a novel idea, there is no consensus on what the exact definition of the term 'smart city' is (). However, experts' scholars, and pundits agree on the key characteristics of smart cities. For instance, they all agree that smart city applications utilize the information and data gathered in many processes, such as the management of assets, services, and resources (Chhaya et al., 2018). In all these cities, the primary reason for the use of technology in the management of resources is to improve the quality and efficiency of services (Al Alshurideh, 2022; Alzoubi & Aziz, 2021; Batayneh et al., 2021; Hasan et al., 2022). This research paper seeks to explore the role of internet of things (IoT), electronic sensors and machine learning (ML) in the development, improvement, and maintenance of sustainable smart cities.

2   Research Methods

A.   *Research Methods and Materilas*

The work was done by use of the secondary sources through a systematic review of the published papers. This method was widely used by authors lately as seen in the following paper (Ahmad et al., 2021; Al Khayyal et al., 2021; Almazrouei et al., 2021; Alshamsi et al., 2021; Aisha Alshamsi et al., 2020; AlShehhi et al., 2021; Assad & Alshurideh, 2020; Mehrez et al., 2021). Only credible papers that have been published within acceptable journals were selected for this work. Papers that have been peer reviewed and published in the IEEE journal were considered.

B. *Time Frame*

The researcher considered only the papers that have been published in the past 10 years based on their relevance to the research. The timeframe used in the references to achieve this research was to include papers published in 2010 to the once published in 2020.

C. *Search Key Words*

The key words that were considered here include IOT, Internet of Things, Machine learning, IOT architectures, learning-based algorithms, IOT security, Secure communication, and IOT taxonomy.

D. *Inclusion Criteria*

The materials must be available within the IEEE websites. The information must be available within three clicks and that was to ensure that the selected website meets the useability criteria. The top five websites when searching using the phrases selected and that was to ensure the researcher uses the once that are most visited. Work from peer reviewed journals to ensure high quality of the articles selected.

E. *Exclusion Criteria*

The type of website where the materials was located was a major exclusion criterion. Websites that do not support academics content, are political in nature, supported by organizations with biases were excluded from the list and the researcher would not select articles based on those sites. The sponsored websites as well as advertisement were excluded when it comes to website selection. The website providers can make payments to make their site appear within the search engines results, the above has a potential of biases hence must be excluded in the website selection. Duplicate work was excluded to reduce the time needed to carry out the research.

F. *Research Queations*
i. What is the role of internet of things and electronic census in the development and maintenance of smart cities?
ii. What role do the different machine learning approaches play in the improvement of the effectiveness of smart City applications?
iii. How will the internet of things revolutionize the lives of the thousands of households living in smart cities?

3 Literature Review

Broadly, the term smart city refers to an urban center or area that applies different electronic sensors and methods to gather data from devices, buildings, citizens, and city assets (Sayakkara et al., 2019). These smart city applications analyze such data and use it to ensure more efficient and timely services, and real-time responses to various arising issues. In a smart city, the applications based on internet of things (IoT) gather and process data, which is used in the management and monitoring of information systems (IS), libraries, schools' hospitals, services, traffic transportation systems, water supply network, waste management systems, power plants, crime detection, and other community services (Shin, 2019). In smart cities, the internet of things (IoT) offers a platform of interaction, where the city authorities can directly interact with the residents, as well as the city infrastructure. Specifically, the platform on which smart cities run is based on the interaction between information and technology (Kortesniemi et al., 2019). In other words, in smart cities, information communication technology (ICT) is applied in the enhancement of performance, quality and interactivity of all urban services (Park et al., 2018). It is also used as a tool for controlling resource consumption and reducing costs, by improving the contact between the citizenry and the authorities. In smart cities the smart applications are used in managing the urban flows and sharing real-time responses.

Infrastructure Development, while these variables may vary from one context to another, there are some primary defining ones, including infrastructure development. Smart cities prioritize the optimization of infrastructural development to improve and support social, economic, cultural and urban development (Hahm et al., 2016). This explains why smart cities are associated with improved communication channels to connect such essential services as entertainment, housing, business, and telecommunications, among others. All these services can be linked through advanced technologies that enable a city to develop and grow.

Competitive Environment: Smart cities endeavor to create a highly competitive environment through information and communication technologies (ICT) and planning with the aim to expand urban sectors, which enhances the growth and development of businesses and the enhancement of the city's social economic performance (Park et al., 2018). In addition to promoting healthy competition, smart cities are sustainable and inclusive because their main strategic element is to achieve sustainability to find participation drivers, use renewable energies, and create better consumption habits, all with the aim to preserve the natural resources and to take care of the urban environment.

Protection and safety for the residents and ICT incorporation. The peace of mind of the residents goes together with new and emerging technologies. Security cameras are located all over the urban territory to ensure that the people's lives are safeguarded (Bairagi et al., 2016). These cameras and surveillance systems are accompanied by superior public lighting, and an effective system of surveillance (Ismagilova et al., 2020). Additionally, smart city authorities run processes for identity verification to minimize irregularities in the various dealings and settings (Alzoubi et al., 2021) Most importantly, the city residents get to receive instant responses to emergency calls regarding any accidents or domestic incidents, or anything related to the health of the citizens.

Ecological concerns: In the modern world, the environment keeps suffering due to unlimited human activity and other factors, which may be directly associated with health hazards. Smart cities have effective environmental leaders, whose main concern is to create recycling and ecological mechanisms to enhance the quality of air and the general surrounding, and to protect them from further contamination (Rayan et al., 2019). One of the main aims of the smart cities, therefore, is to reduce the carbon footprint of the entire city to support its sustainability goals. When polluting gases are reduced, green spaces such as public parks, will be better preserved and less contaminated (Kouliaridis et al., 2020). As mentioned earlier, recycling is one of the citizens' daily routines. Smart cities have minimal or no problems associated with water collection and supply.

Participation and Open Innovative Processes. Overall, it has been established that smart cities use Information Communication Technologies (ICT) to adapt and innovate and increase their capacity to respond more promptly and effectively to the ever changing circumstances of the city by enhancing its intelligence (Erdin et al., 2015). Also, these cities use ICT to engage and interact effectively with the residents in decision-making and governance using a participation and open innovation processes. They also engage with the local people with the aim to improve the city's collective intelligence, as well as that of its institutions and organizations through a governance (Kreso et al., 2021). Here, the emphasis is on the participation of citizens and the core design of the cities with government and public efforts working together (Jamshidi et al., 2020). Lastly, smart cities apply ICT to ensure more efficient utilization of public infrastructure, including built environments, roads, and other physical assets, through data analytics and artificial intelligence (AI) to support a robust economic, cultural, and social development.

4 The Internet of Things (IoT) in Smart City Development and Maintenance

The internet of things (IoT), a key component of smart cities, refers to the interconnectivity of various physical devices through an internet platform. The interconnected digital devices can be controlled remotely using different telecommunication devices, including mobile phones (Wang et al., 2016). The internet of things (IoT), in more straight-forward terms, is the connection between diverse physical devices linked to and in constant communication with each other over an online based platform. A good example of a common application of internet of things (IoT) is the collection and analysis of weather fluctuations and changes (Zou & Wang, 2016). One of the most significant contributions of the internet of things to the development and improvement of smart cities is that it offers low-cost management of assets. For instance, the use of internet of things (IoT) eliminates or mitigates the need for human labor in various services (Rabbachin et al., 2015). Also, the internet of things offers fresh insights that can further revolutionize how the city functions. Research has demonstrated that the internet of things (IoT), through its intricate data analysis capabilities, has been able to unveil new insights such as establishing relationship between industrial production and the weather (Almeida et al., 2015). Such insights and relationships demonstrate the ability of internet of things (IoT) and other aspects of technology to improve how cities function.

4.1 Smart Energy

One of the most critical areas of research that is related to the internet of things (IoT) is smart energy because it is important in ensuring the reduction of overall power consumption. Smart energy provides high quality and highly affordable ecologically friendly energy (Gnad et al., 2019). The concept of smart energy includes a broad multiplicity of energy and operational measures, which include smart leak monitoring, smart energy applications, and renewable energy resources, and so on (Fernando et al., 2020) Deploying a smart energy grid or using smart energy, implies a critical re-engineering of the services within the city. The smart energy grid is one of the most critical applications of the concept (Sayakkara et al., 2019). The smart energy grid includes a broad diversity of high-speed time series data in the monitoring of devices. To manage this kind of data, experts brought in a method to analyze and manage the time series information so as to make them organized on demand (Abawajy et al., 2018). Additionally, the smart energy grid is expected to grow in complexity into the foreseeable future(Ahmed et al., 2021; Al Naqbia et al., 2020; Al Suwaidi et al., 2021; Almaazmi et al., 2020). Therefore, experts propose a simulation system for testing the novel concepts and optimizing approaches and forecasting or projecting future consumption. Another critical application of smart energy grids is seen in leak monitoring systems (AlHamad et al., 2022; Alzoubi et al., 2022; Shamout et al., 2022). The main function of these systems is to model a gas or water management system that has the capacity to optimize the consumption of energy resources (Kaur & Sharma, 2015).

4.2 Smart Energy

Mobility is one of the most essential and critical aspects of any urban area. Through the internet of things (IoT) city authorities can enhance the quality of life for people. Smart mobility can be evaluated from three dimensions. The first dimension is traffic control. The optimization of the flow of traffic through the analysis of sensor data is one of the key elements of the city's mobility (Hosseinian Far et al., 2020) For effective control of traffic, data will be gathered from road cameras, cars, and counter sensors installed on the city's roads. The second dimension of smart mobility in cities is the use of autonomous vehicles. The internet of things (IoT) is expected to have numerous effects on how cars are operated. The most critical issue is about how the internet of things can enhance other vehicle services (Alzoubi et al., 2020a, 2020b). Wireless connections and internet of things (IoT) sensors make it easy to monitor the performance of vehicles and to create self-driving vehicles. With the data gathered from the automobiles, the most congested roads can be predicted enabling the city and traffic authorities to make decisions that will significantly decrease congestion on the popular roads (Alshurideh et al., 2022; Lee, Azmi, et al., 2022a, 2022b; Lee, Romzi, et al., 2022a, 2022b). Self-driving automobiles can enhance the safety of passengers because they have the capacity to monitor and detect the driving of other motorists on the road (Zhang et al., 2017).

Regarding smart mobility, the last dimension is that off public transportation. The internet of things (IoT) can improve the system of public transportation by offering precise routing and location information to a city's smart system of transportation (Din et al., 2019). Also, it has the capacity to assist the city's passengers in making more informed decisions in their schedules, and to decrease the amount of time wasted. There are various perspectives on how to construct smart city public transportation systems (Joghee et al., 2020). The transportation systems in smart cities need to manage various types of data, like traffic and vehicle location data (Alhamad et al., 2021; Ali et al., 2022; Hanaysha et al., et al., 2021). Public transportation systems in smart cities should be designed in such a way that they are real-time oriented. This will help them make more informed decisions and prepare appropriate responses in real time. Also, they can use historical data analysis through the various machine learning methods (Gupta & Gupta, 2018).

4.3 Smart Citizens

The use of internet of things in smart cities covers a wide range of aspects in human lives, including the monitoring of crimes, the management of the environment, as well as the monitoring of social and public health (Park, 2021) All these components are fundamental and vital for human life within the urban settings. Additionally, making significant progress in technology is sure to improve the security status of the city (Alzoubi & Ahmed, 2019; Alzoubi & Yanamandra, 2020). Close monitoring and surveillance dedicated to crime will also have significant contribution to the citizens overall public and social health (Alnazer et al., 2017).

4.4 Urban Planning

Urban planning is yet another critical feature in smart cities because it helps them attain long-term goals and decisions. Since the environment and the city play critical roles in human life, achieving quality decisions is critical in this context. By Gathering data from citizens, devices, and the various sources in the city, it becomes possible to make effective decisions regarding the city's future. Urban planning refers to the process of making decisions that affect the city's functionality, design, and infrastructure. The internet of things (IoT) is highly resourceful in this context because through the smart analysis of a cities data the authorities can predict those parts of the city that will be more populated or crowded in the foreseeable future and to come up with solutions for the possible problems (Alzoubi & Aziz, 2021; Hamadneh et al., 2021; Hanaysha et al., 2021). A combination of effective urban planning and the internet of things (IoT) will have a major impact on planning future improvements to infrastructure (Syam & Sharma, 2018).

5 Machine Learning Approaches, IoT and Data Analytics for Smart Cities

Machine learning (ML) is one of the most celebrated modern innovations that has assisted human beings enhance not only numerous professional and industrial processes, but also improve everyday living. However, the term 'machine learning' has not been assigned universally accepted definition, but it is generally agreed that machine learning is a subclass of the broader field of artificial intelligence, which concerns itself with using statistical methods to create intelligent computer systems with the capacity to learn from historical data and databases. Presently, machine learning (ML) is widely applied in diverse industries and fields. For instance, machine learning is used in medical diagnosis, classification, prediction, image processing, regression, learning, and association and so on (Ali et al., 2021a, 2021b; Alnuaimi et al., 2021; Alzoubi et al., 2021). The intelligent computer systems based on machine learning algorithms have the immense capacity to learn from historical experiences and past data without supervision. Machine learning applications produce results and make projections based on past experiences of similar nature. Machine learning is considered one of the most remarkable breakthroughs in the broader spectrum of artificial intelligence. While machine learning comes with some frightening implications, its applications are some of the ways through which technology can ultimately improve human lives.

5.1 Machine Learning Internet of Things and Data Analytics for Smart Cities

The transformation of big data analytics, along with the internet of things (IoT) machine learning (ML) has helped various places around the world realized the idea of a Smart City. The idea underlying the concept of a smart city is to ensure effective rendering of services to the citizens and people dwelling in the city through advanced data analytics and technologies (Sharma & Arya, 2015). In this regard, the data analytics focuses on the data gathered by the various internet of things (IoT) systems and electronic sensors. The extent to which a city is smart can be determined by initiatives on environmental issues, technology-driven infrastructure, smart public transport systems, and the utilization of technology in reduction of crimes, and the use of technology to manage wastes and water, and so on (Alshurideh et al., 2020; Alzoubi et al., 2020a, 2020b; Ghazal et al., 2021). In the context of smart cities, one of the machine learning (ML) techniques that is considered highly effective in gaining insights from the various gathered data is deep learning. Deep learning can be used to make sense of the patterns from the data collected, and to predict or to classify the data.

With rapid and widespread urbanization efforts and the unprecedented rates of population growth in cities, smart cities are becoming highly favored options, as they can offer a better lifestyle to the citizens and residents by applying innovative technological advancements. The smart cities highly depend on actuators and sensors to gather large volumes of information and data. This data is analyzed and applied in the extraction of potentially beneficial decision-making information. However, because of some limited capacities, only a small percentage of the data collected in the smart city is used because of various reasons (Mehmood et al., 2019). For instance, the data generated or collected may be highly diverse and noisy. Machine learning and artificial intelligence are known for their capacity to handle high volumes of error-prone or messy data. Advanced machine learning applies algorithms that can exploit the availability of both labelled and unlabeled data to provide effective personalized services and resource management in smart cities. The processed data or information can be used in numerous daily applications including prevention of pollution, management of healthcare, and designation of efficient transportation, stronger security measures, and better management of energy (Aziz & Aftab, 2021; Cruz, 2021; Radwan & Farouk, 2021). Academia, research and practice have demonstrated that the application of machine learning (ML) in the leveraging of internet of things (IoT-based techniques) have consistently proven to be more sensitive, accurate, cost-effective, and timesaving (Malche, 2019).

Machine learning has proven to be a proficient and powerful technology data has revolutionized how big data is analyzed. In smart cities, huge amounts of data are gathered from smart devices and sensors around the city (Al-Fuqaha, 2018). Deep learning and machine learning collectively have the potential to significantly enhance the quality of life through the application of diverse domains (Al Ali, 2021; Alzoubi, 2021a; Lee & Ahmed, 2021). Deep studies have been and are still being done in this area to reveal and explore more useful applications of deep and machine learning. Machine learning provides the computer systems the ability to learn automatically by accessing information and data provided and applying it in its self-learning process. This gives the computer systems and the internet of things (IoT) by extension, the ability to improve from experience without any programming efforts by humans (Akhtar et al., 2021; Eli, 2021; Kashif et al., 2021). Coined by Arthur Samuel in 1959, the term 'machine learning' has become a central area of focus in the planning of large smart cities as urbanization keeps growing by the day. With the data overload from the smart devices and sensors in smart cities individual reviewing may not be possible which makes it necessary to have a system that can not only learn but improve by itself from historical experiences (Napiah et al., 2018).

Machine learning requires a dynamic general and continuous learning mechanism as the smart city undergoes evolution in its operating environments and applications. Therefore, there is a need to explore the capacity and potential of machine learning and large amounts of data in the creation of personalized services in smart cities (Chilamkurti, 2018). In a nutshell, machine learning, and deep learning specifically play a critical role in ensuring improved efficiency in smart cities. The concept of machine learning is undergoing rapid evolution as an outcome of the improvement of the various algorithms, improved computer networks, upgraded approaches of capturing data, innovative sensors, and heightened interest in customization (Elhoseny, 2018). The primary objective of machine learning algorithms is to effectively interpret the information and data that has never been observed before, and to make projections beyond the samples used in training, such as real-life data (Alsharari, 2021; Mehmood, 2021; Miller, 2021). The precision and accuracy of the machine learning algorithms can be further improved by increasing the volume of training data given so as to strengthen or to reinforce their learning capacities.

5.2 Machine Learning Approaches for Smart Cities

Machine learning approaches are widely categorized into three main classes, according to the nature of the learning feedback or signal available in a system of learning. These three categories include the supervised, unsupervised, and semi-supervised approaches. The semi-supervised approaches refer to a combination of the supervised and unsupervised types. The semi-supervised algorithm works by applying huge amounts of input data with only a small section of it labelled as 'training data.' In a similar approach, referred to as reinforcement, learning offers feedback, which guides the computer programs in its interactions with a highly dynamic environment. The unsupervised algorithms work through self-organizing and recognizing concealed patterns in input data that is unlabeled to form neural networks (NN) through unsupervised algorithms of machine learning (Guergov & Radwan, 2021; Khan, 2021; Mondol, 2021). It has the capacity to process data and information without any error signal to assess and evaluate possible solutions. Unsupervised learning is important because to annotate large sets of data like speech recognition can be quite expensive and cannot be evaluated. Clustering is a classic example of unsupervised learning (Alzoubi, 2021b; Farouk, 2021; Obaid, 2021).

6   Performance Evaluation

A detailed review of the existing literature suggests that IoT and artificial intelligence are some of the most powerful forces on the list of technologies revolutionizing the modern idea of a city. Particularly, research studies suggest that machine learning is a critical component in the IoT program, when a city wants to achieve better communication technology. Among others, the key findings from the research papers reviewed here include:

i. From Kortesniemi et al. (2019) the work characterized the machine learning problem in cognitive radios and showed the value of artificial intelligence when it comes to attaining real cognitive commination systems. The machine must be well advanced to levels where they can aid in decision making. The decision levels algorithm developed for the machine will help in classifying different observation models. It possible to have unsupervised algorithms that can be secure for the IOT. The challenges experienced within the learning machines algorithms should be known and addressed.

ii. Overall, the application of machine learning in the efforts to leverage IoT based methods, have consistently demonstrated to be more time saving, accurate, cost-effective and precise (Ismail et al., 2019). The ability of machine learning to process heavy amounts of data at once and with remarkable accuracy makes it a primary component in successful IoT based systems in smart cities. Particularly, the findings of most scholarly research studies suggest that busy systems as traffic and transportation systems need superior machine learning approaches because accidents can be fatal and costly.

iii. Research findings of Chhaya et al. (2018) study suggested that one of the most important aspect of machine learning is the unique ability to process huge volumes of data, and to allow machines to learn from past events. By so doing, machine learning gives researchers the opportunity to reevaluate standards in several sectors hence promoting

iv. Continuous improvement.

v. The internet of things will offer remarkable contribution on the developments of smart cities and maintenance from references (Wang et al., 2016; Zou & Wang, 2016). IOT will offer connections between the physical devices that are linked to it and the online platforms that are used for city planning and maintenance. The internet of things is now used in smart city systems such as those for the collection and analysis of weather and that has a potential to other areas such as city health status.

vi. Machine learning applications can be adopted to special wire sensor networks which can be incorporated in the management of smart cities (Elhoseny, 2018). It will be possible to have special networks that can be used for the transformation of the big data analytics together with internet of things and machine learning. Once integrated the smart cities can benefit through effective rendering of services to the residents. Data would be gathered by different IOTs and machine learning will offer the computer systems the capabilities to learn automatically and apply the experiences within the learning process of managing the cities. They will be continuous improvements on the way the cities are going to run and managed.

vii. From Syam and Sharma (2018), IOT facilitates integration between physical objects and sensors that are responsible for providing commination between them without having any human interventions. Despite their good algorithms they are faced with different security vulnerabilities that should be addressed to have temper proof systems. Lightweight schemes are the most favored systems to solve the security challenges of IOT. Other new measures should be sought since the security vulnerabilities keep on changing.

viii. Also, Sharma and Arya (2015), findings firmly suggest that one of the sectors that heavily relies on IoT and machine intelligence is healthcare. In addition to healthcare, machine learning improves the quality of life by revolutionizing transport, security, law enforcement and crime management.

ix. From Nia (2021), machine learning, and the internet of things change how people respond to disasters and emergencies by gathering analyzing and disseminating important decision-making information. IOT and machine learning are tools that can be incorporated towards managing the cities security systems.

x. Based on the findings from Bairagi et al. (2016), IoT and Machine intelligence will alter the socio-political climate of the smart cities because the relationship between the government and the citizenry will be strengthened by inclusive and participatory decision making. The IOT will rely on information faster and Machine learning utilized to make the required analytics.

7 Conclusion

In conclusion, IoT, electronic sensors and machine learning are revolutionizing our idea of mega smart cities. These AI-based concepts have revolutionized how people go about their everyday lives. As smart cities continue to emerge, the word is going to change in a broad number of ways. First, ICT is going to reduce the number of people needed in the running of such services as water and waste management, security and surveillance, transport sectors, traffic management and industrial operations, among others. This paper has effectively responded to the research questions because it has demonstrated the centrality of IoT, electronic sensors and machine learning in the designing and running of smart cities, such as Tokyo. These concepts contribute immensely to the fast and reliable conveyance of information and data among the relevant people and departments, for effective and well-informed decision making. Machine learning's biggest contributions include its almost limitless capacity to gather and process multiple volumes of data at remarkable speeds, and the reality that it allows machine to learn and contribute to the improvement of the city systems. Overall, IoT and Machine learning are gradually becoming seamlessly interweaved with the everyday lives of the citizens living in smart cities.

References

Abawajy, J., Huda, S., Sharmeen, S., Hassan, M. M., & Almogren, A. (2018). Identifying cyber threats to mobile-IoT applications in edge computing paradigm. *Future Generation Computer Systems, 89*, 525–538. https://doi.org/10.1016/J.FUTURE.2018.06.053

Ahmad, A., Alshurideh, M. T., Al Kurdi, B. H., & Alzoubi, H. M. (2021). Digital Strategies: A Systematic Literature Review. *The International Conference on Artificial Intelligence and Computer Vision*, 807–822.

Ahmed, A., Alshurideh, M., Al Kurdi, B., & Salloum, S. A. (2021). Digital transformation and organizational operational decision making: a systematic review. In *Advances in Intelligent Systems and Computing: Vol. 1261 AISC*. https://doi.org/10.1007/978-3-030-58669-0_63

Akhtar, A., Akhtar, S., Bakhtawar, B., Kashif, A. A., Aziz, N., & Javeid, M. S. (2021). COVID-19 Detection from CBC using machine learning techniques. *International Journal of Technology, Innovation and Management (IJTIM), 1*(2), 65–78. https://doi.org/10.54489/ijtim.v1i2.22

Al-Fuqaha, M. M., & A. (2018). Enabling cognitive smart cities using big data and machine learning: approaches and challenges. *IEEE Communications Magazine, 56*(2), 94–101.

Al Ali, A. (2021). The impact of information sharing and quality assurance on customer service at uae banking sector. *International Journal of Technology, Innovation and Management (IJTIM), 1*(1), 01–17. https://doi.org/10.54489/ijtim.v1i1.10

Al Batayneh, R. M., Taleb, N., Said, R. A., Alshurideh, M. T., Ghazal, T. M., & Alzoubi, H. M. (2021). IT governance framework and smart services integration for future development of Dubai infrastructure utilizing ai and big data, its reflection on the citizens standard of living. *The International Conference on Artificial Intelligence and Computer Vision*, 235–247.

Al Khayyal, A. O., Alshurideh, M., Al Kurdi, B., & Salloum, S. A. (2021). Women empowerment in UAE: A systematic review. In *Advances in Intelligent Systems and Computing: Vol. 1261 AISC*. https://doi.org/10.1007/978-3-030-58669-0_66

Al Kurdi, B., Alshurideh, M., & Salloum, S. A. (2020). Investigating a theoretical framework for e-learning technology acceptance. *International Journal of Electrical and Computer Engineering (IJECE), 10*(6), 6484–6496.

Al Naqbia, E., Alshuridehb, M., AlHamadc, A., & Al, B. (2020). The impact of innovation on firm performance: A systematic review. *International Journal Innovative Creat Change, 14*(5), 31 58.

Al Suwaidi, F., Alshurideh, M., Al Kurdi, B., & Salloum, S. A. (2021). The impact of innovation management in SMEs performance: A systematic review. In *Advances in Intelligent Systems and Computing: Vol. 1261 AISC*. https://doi.org/10.1007/978-3-030-58669-0_64

AlHamad, A., Alshurideh, M., Alomari, K., Kurdi, B. A., Alzoubi, H., Hamouche, S., & Al-Hawary, S. (2022). The effect of electronic human resources management on organizational health of telecommuni-cations companies in Jordan. *International Journal of Data and Network Science, 6*(2), 429–438. https://doi.org/10.5267/j.ijdns.2021.12.011

Alhamad, A. Q. M., Akour, I., Alshurideh, M., Al-Hamad, A. Q., Kurdi, B. A., & Alzoubi, H. (2021). Predicting the intention to use google glass: A comparative approach using machine learning models and PLS-SEM. *International Journal of Data and Network Science, 5*(3), 311–320. https://doi.org/10.5267/j.ijdns.2021.6.002

Ali, N., Ahmed, A., Anum, L., Ghazal, T. M., Abbas, S., Khan, M. A., Alzoubi, H. M., & Ahmad, M. (2021a). Modelling supply chain information collaboration empowered with machine learning technique. *Intelligent Automation and Soft Computing, 30*(1), 243–257. https://doi.org/10.32604/iasc.2021.018983

Ali, W., Din, I. U., Almogren, A., Guizani, M., & Zuair, M. (2021b). A lightweight Privacy-aware IoT-based metering scheme for smart industrial ecosystems. *IEEE Transactions on Industrial Informatics, 17*(9), 6134–6143. https://doi.org/10.1109/TII.2020.2984366

Ali, N., M. Ghazal, T., Ahmed, A., Abbas, S., A. Khan, M., Alzoubi, H., Farooq, U., Ahmad, M., & Adnan Khan, M. (2022). Fusion-based supply chain collaboration using machine learning techniques. *Intelligent Automation & Soft Computing, 31*(3), 1671–1687. https://doi.org/10.32604/iasc.2022.019892

Almaazmi, J., Alshurideh, M., Al Kurdi, B., & Salloum, S. A. (2020). The effect of digital transformation on product innovation: A critical review.In: *International Conference on Advanced Intelligent Systems and Informatics*, 731–741.

Almazrouei, F. A., Alshurideh, M., Al Kurdi, B., & Salloum, S. A. (2021). Social media impact on business: a systematic review. In *Advances in Intelligent Systems and Computing: Vol. 1261 AISC*. https://doi.org/10.1007/978-3-030-58669-0_62

Almeida, V. A. F., Doneda, D., & Monteiro, M. (2015). Governance challenges for the internet of things. *IEEE Internet Computing, 19*(4), 56–59. https://doi.org/10.1109/MIC.2015.86

Alnazer, N. N., Alnuaimi, M. A., & Alzoubi, H. M. (2017). Analysing the appropriate cognitive styles and its effect on strategic innovation in Jordanian universities. *International Journal of Business Excellence, 13*(1), 127–140. https://doi.org/10.1504/IJBEX.2017.085799

Alnuaimi, M., Alzoubi, H. M., Ajelat, D., & Alzoubi, A. A. (2021). Towards intelligent organisations: An empirical investigation of learning orientation's role in technical innovation. *International Journal of Innovation and Learning, 29*(2), 207–221. https://doi.org/10.1504/IJIL.2021.112996

Alshamsi, A., Alshurideh, M., Kurdi, B. A., & Salloum, S. A. (2021). The influence of service quality on customer retention: a systematic review in the higher education. In *Advances in Intelligent Systems and Computing: Vol. 1261 AISC*. https://doi.org/10.1007/978-3-030-58669-0_37

Alshamsi, Aisha, Alshurideh, M., Al Kurdi, B., & Salloum, S. A. (2020). The influence of service quality on customer retention: a systematic review in the higher education. *International* In: *Conference on Advanced Intelligent Systems and Informatics*, 404–416.

Alsharari, N. (2021). Integrating blockchain technology with internet of things to efficiency. *International Journal of Technology, Innovation and Management (IJTIM), 1*(2), 1–13.

AlShehhi, H., Alshurideh, M., Kurdi, B. A., & Salloum, S. A. (2021). The impact of ethical leadership on employees performance: a systematic review. In *Advances in Intelligent Systems and Computing: Vol. 1261 AISC*. https://doi.org/10.1007/978-3-030-58669-0_38

Alshurideh, M. (2022). Does electronic customer relationship management (E-CRM) affect service quality at private hospitals in Jordan? *Uncertain Supply Chain Management, 10*(2), 1–8.

Alshurideh, M., Al Kurdi, B., & Salloum, S. (2019). Examining the main mobile learning system drivers' effects: a mix empirical examination of both the expectation-confirmation model (ECM) and the technology acceptance model (TAM). In: *International Conference on Advanced Intelligent Systems and Informatics*, 406–417.

Alshurideh, M. T., Al Kurdi, B., Alzoubi, H. M., Ghazal, T. M., Said, R. A., AlHamad, A. Q., Hamadneh, S., Sahawneh, N., & Al-kassem, A. H. (2022). Fuzzy assisted human resource management for supply chain management issues. *Annals of Operations Research*, 1–19.

Alshurideh, M., Gasaymeh, A., Ahmed, G., Alzoubi, H., & Kurd, B. A. (2020). Loyalty program effectiveness: Theoretical reviews and practical proofs. *Uncertain Supply Chain Management, 8*(3), 599–612. https://doi.org/10.5267/j.uscm.2020.2.003

Alzoubi, H., & Ahmed, G. (2019). Do total quality management (TQM) practices improve organisational success? A case study of electronics industry in the UAE. *International Journal of Economics and Business Research, 17*(4), 459–472.

Alzoubi, Ali. (2021a). The impact of process quality and quality control on organizational competitiveness at 5-star hotels in Dubai. *International Journal of Technology, Innovation and Management (IJTIM), 1*(1), 54–68. https://doi.org/10.54489/ijtim.v1i1.14

Alzoubi, Asem. (2021b). Renewable Green hydrogen energy impact on sustainability performance. *International Journal of Computations, Information and Manufacturing (IJCIM), 1*(1), 94–110. https://doi.org/10.54489/ijcim.v1i1.46

Alzoubi, H., Alshurideh, M., Kurdi, B. A., Akour, I., & Azi, R. (2022). Does BLE technology contribute towards improving marketing strategies, customers' satisfaction and loyalty? The role of open innovation. *International Journal of Data and Network Science, 6*(2), 449–460. https://doi.org/10.5267/j.ijdns.2021.12.009

Alzoubi, H., Alshurideh, M., Kurdi, B. A., & Inairat, M. (2020a). Do perceived service value, quality, price fairness and service recovery shape customer satisfaction and delight? A practical study in the service telecommunication context. *Uncertain Supply Chain Management, 8*(3), 579–588. https://doi.org/10.5267/j.uscm.2020.2.005

Alzoubi, H. M., Ahmed, G., Al-Gasaymeh, A., & Al Kurdi, B. (2020b). Empirical study on sustainable supply chain strategies and its impact on competitive priorities: The mediating role of supply chain collaboration. *Management Science Letters, 10*(3), 703–708. https://doi.org/10.5267/j.msl.2019.9.008

Alzoubi, H. M., & Aziz, R. (2021). Does emotional intelligence contribute to quality of strategic decisions? the mediating role of open innovation. *Journal of Open Innovation: Technology, Market, and Complexity, 7*(2), 130. https://doi.org/10.3390/joitmc7020130

Alzoubi, H. M., Vij, M., Vij, A., & Hanaysha, J. R. (2021). What leads guests to satisfaction and loyalty in UAE five-star hotels? AHP analysis to service quality dimensions. *Enlightening Tourism, 11*(1), 102–135. https://doi.org/10.33776/et.v11i1.5056

Alzoubi, H. M., & Yanamandra, R. (2020). Investigating the mediating role of information sharing strategy on agile supply chain. *Uncertain Supply Chain Management, 8*(2), 273–284. https://doi.org/10.5267/j.uscm.2019.12.004

Assad, N. F., & Alshurideh, M. T. (2020). Investment in context of financial reporting quality: A systematic review. *WAFFEN-UND Kostumkd. J, 11*(3), 255–286.

Aziz, N., & Aftab, S. (2021). Data mining framework for nutrition ranking: methodology: SPSS modeller. *International Journal of Technology, Innovation and Management (IJTIM), 1*(1), 85–95.

Bairagi, A. K., Khondoker, R., & Islam, R. (2016). An efficient steganographic approach for protecting communication in the Internet of Things (IoT) critical infrastructures. *Information Security Journal, 25*(4–6), 197–212. https://doi.org/10.1080/19393555.2016.1206640

Chhaya, L., Sharma, P., Kumar, A., & Bhagwatikar, G. (2018). IoT-Based implementation of field area network using smart grid communication infrastructure. *Smart Cities 2018, Vol. 1, Pages 176–189, 1*(1), 176–189. https://doi.org/10.3390/SMARTCITIES1010011

Chilamkurti, A. A., & N. (2018). Deep learning: the frontier for distributed attack detection in Fog-to-Things computing. *IEEE Communications Magazine, 56*(2), 169–175.

Cruz, A. (2021). Convergence between blockchain and the internet of things. *International Journal of Technology, Innovation and Management (IJTIM), 1*(1), 35–56.

Din, I. U., Guizani, M., Rodrigues, J. J. P. C., Hassan, S., & Korotaev, V. V. (2019). Machine learning in the internet of things: designed techniques for smart cities. *Future Generation Computer Systems, 100*, 826–843. https://doi.org/10.1016/J.FUTURE.2019.04.017

Eli, T. (2021). Students perspectives on the use of innovative and interactive teaching methods at the university of nouakchott al aasriya, mauritania: english department as a case study. *International Journal of Technology, Innovation and Management (IJTIM), 1*(2), 90–104.

Erdin, E., Zachor, C., & Gunes, M. H. (2015). How to find hidden users: a survey of attacks on anonymity networks. *IEEE Communications Surveys and Tutorials, 17*(4), 2296–2316. https://doi.org/10.1109/COMST.2015.2453434

Farouk, M. (2021). The universal artificial intelligence efforts to face coronavirus COVID-19. *International Journal of Computations, Information and Manufacturing (IJCIM), 1*(1), 77–93. https://doi.org/10.54489/ijcim.v1i1.47

Fernando, D. W., Komninos, N., & Chen, T. (2020). A study on the evolution of ransomware detection using machine learning and deep learning techniques. *IoT 2020, Vol. 1, Pages 551–604, 1*(2), 551–604. https://doi.org/10.3390/IOT1020030

Ghazal, T. M., Hasan, M. K., Alshurideh, M. T., Alzoubi, H. M., Ahmad, M., Akbar, S. S., Al Kurdi, B., & Akour, I. A. (2021). IoT for smart cities: machine learning approaches in smart Healthcare—a review. *Future Internet, 13*(8), 218. https://doi.org/10.3390/fi13080218

Gnad, D. R. E., Krautter, J., & Tahoori, M. B. (2019). Leaky Noise: New Side-Channel Attack Vectors in Mixed-Signal IoT Devices. *IACR Transactions on Cryptographic Hardware and Embedded Systems, 2019*(3), 305–339. https://doi.org/10.13154/TCHES.V2019.I3.305-339

Guergov, S., & Radwan, N. (2021). Blockchain convergence: analysis of issues affecting IoT, AI and Blockchain. *International Journal of Computations, Information and Manufacturing (IJCIM), 1*(1), 1–17. https://doi.org/10.54489/ijcim.v1i1.48

Gupta, S., & Gupta, B. B. (2018). Robust injection point-based framework for modern applications against XSS vulnerabilities in online social networks. *International Journal of Information and Computer Security, 10*(2–3), 170–200. https://doi.org/10.1504/IJICS.2018.091455

Habibzadeh, H., Kaptan, C., Soyata, T., Kantarci, B., & Boukerche, A. (2019). Smart city system design: A comprehensive study of the application and data planes. *ACM Computing Surveys, 52*(2). https://doi.org/10.1145/3309545

Hahm, O., Baccelli, E., Petersen, H., & Tsiftes, N. (2016). Operating systems for low-end devices in the internet of things: a survey. *IEEE Internet of Things Journal, 3*(5), 720–734. https://doi.org/10.1109/JIOT.2015.2505901

Hamadneh, S., Pedersen, O., & Al Kurdi, B. (2021). An investigation of the role of supply chain visibility into the scottish bood supply chain. *Journal of Legal, Ethical and Regulatory Issues, 24*(Special Issue 1), 1–12.

Hanaysha, J. R., Al-Shaikh, M. E., Joghee, S., & Alzoubi, H. (2021a). Impact of innovation capabilities on business sustainability in small and medium enterprises. *FIIB Business Review*, 1–12. https://doi.org/10.1177/23197145211042232

Hanaysha, J. R., Al Shaikh, M. E., & Alzoubi, H. M. (2021b). Importance of marketing mix elements in determining consumer purchase decision in the retail market. *International Journal of Service Science, Management, Engineering, and Technology (IJSSMET), 12*(6), 56–72.

Hasan, O., McColl, J., Pfefferkorn, T., Hamadneh, S., Alshurideh, M., & Kurdi, B. (2022). Consumer attitudes towards the use of autonomous vehicles: Evidence from United Kingdom taxi services. *International Journal of Data and Network Science, 6*(2), 537–550.

Hosseinian Far, A., Montasari, R., Hill, R., Montasari, F., & Jahankhani, H. (2020). Internet of things devices: digital forensic process and data reduction. *International Journal of Electronic Security and Digital Forensics, 12*(1), 1. https://doi.org/10.1504/IJESDF.2020.10030308

Ismagilova, E., Hughes, L., Rana, N. P., & Dwivedi, Y. K. (2020). Security, privacy and risks within smart cities: literature review and development of a smart city interaction framework. *Information Systems Frontiers*, 1–22. https://doi.org/10.1007/S10796-020-10044-1/TABLES/2

Ismail, A., Saad, M., & Abbas, R. (2019). Cyber security in internet of things. *Review of Computer Engineering Studies, 5*(1), 17–22. https://doi.org/10.18280/rces.050104

Jamshidi, M., Esnaashari, M., Ghasemi, S., Qader, N. N., & Meybodi, M. R. (2020). DSLA: Defending against selective forwarding attack in wireless sensor networks using learning automaton. *IEIE Transactions on Smart Processing and Computing, 9*(1), 58–74. https://doi.org/10.5573/IEIESPC.2020.9.1.058

Joghee, S., Alzoubi, H. M., & Dubey, A. R. (2020). Decisions effectiveness of FDI investment biases at real estate industry: Empirical evidence from Dubai smart city projects. *International Journal of Scientific and Technology Research, 9*(3), 3499–3503.

Kashif, A. A., Bakhtawar, B., Akhtar, A., Akhtar, S., Aziz, N., & Javeid, M. S. (2021). Treatment response prediction in hepatitis c patients using machine learning techniques. *International Journal of Technology, Innovation and Management (IJTIM), 1*(2), 79–89. https://doi.org/10.54489/ijtim.v1i2.24

Kaur, P., & Sharma, S. (2015). Spyware detection in android using hybridization of description analysis, permission mapping and interface analysis. *Procedia Computer Science, 46*, 794–803. https://doi.org/10.1016/J.PROCS.2015.02.148

Khan, M. A. (2021). Challenges facing the application of iot in medicine and healthcare. *International Journal of Computations, Information and Manufacturing (IJCIM), 1*(1), 39–55. https://doi.org/10.54489/ijcim.v1i1.32

Kortesniemi, Y., Lagutin, D., Elo, T., & Fotiou, N. (2019). Improving the privacy of IoT with decentralised identifiers (DIDs). *Journal of Computer Networks and Communications, 2019*. https://doi.org/10.1155/2019/8706760

Kouliaridis, V., Barmpatsalou, K., Kambourakis, G., & Chen, S. (2020). A survey on mobile malware detection techniques. *IEICE Transactions on Information and Systems, E103D*(2), 204–211. https://doi.org/10.1587/TRANSINF.2019INI0003

Kreso, I., Kapo, A., & Turulja, L. (2021). Data mining privacy preserving: research agenda. *Wiley Interdisciplinary Reviews: Data Mining and Knowledge Discovery, 11*(1), e1392. https://doi.org/10.1002/WIDM.1392

Kurdi, B. A., Alshurideh, M., Salloum, S. A., Obeidat, Z. M., & Al-dweeri, R. M. (2020). An empirical investigation into examination of factors influencing university students' behavior towards elearning acceptance using SEM approach. *International Journal of Interactive Mobile Technologies, 14*(2). https://doi.org/10.3991/ijim.v14i02.11115

Lee, C., & Ahmed, G. (2021). Improving IoT Privacy, Data Protection and Security Concerns. *International Journal of Technology, Innovation and Management (IJTIM), 1*(1), 18–33.

Lee, K. L., Azmi, N. A. N., Hanaysha, J. R., Alzoubi, H. M., & Alshurideh, M. T. (2022a). The effect of digital supply chain on organizational performance: An empirical study in Malaysia manufacturing industry. *Uncertain Supply Chain Management, 10*(2), 495–510. https://doi.org/10.5267/j.uscm.2021.12.002

Lee, K. L., Romzi, P. N., Hanaysha, J. R., Alzoubi, H. M., & Alshurideh, M. (2022b). Investigating the impact of benefits and challenges of IOT adoption on supply chain performance and organizational performance: An empirical study in Malaysia. *Uncertain Supply Chain Management, 10*(2), 537–550. https://doi.org/10.5267/j.uscm.2021.11.009

Elhoseny, M., & X. Y. and M. G. (2018). Special issue on machine learning applications for self-organized wireless sensor networks. *Neural Computing and Applications, 30*(1), 1–8.

Mehmood, T. (2021). Does information technology competencies and fleet management practices lead to effective service delivery? empirical evidence from e-commerce industry. *International Journal of Technology, Innovation and Management (IJTIM), 1*(2), 14–41.

Mehmood, T., Alzoubi, H. M., & Ahmed, G. (2019). Schumpeterian entrepreneurship theory: evolution and relevance. *Academy of Entrepreneurship Journal, 25*(4).

Mehrez, A. A. A., Alshurideh, M., Kurdi, B. A., & Salloum, S. A. (2021). Internal factors affect knowledge management and firm performance: a systematic review. In *Advances in Intelligent Systems and Computing: Vol. 1261 AISC*. https://doi.org/10.1007/978-3-030-58669-0_57

Miller, D. (2021). The best practice of teach computer science students to use paper prototyping. *International Journal of Technology, Innovation and Management (IJTIM), 1*(2), 42–63. https://doi.org/10.54489/ijtim.v1i2.17

Mondol, E. P. (2021). The impact of block chain and smart inventory system on supply chain performance at retail industry. *International Journal of Computations, Information and Manufacturing (IJCIM), 1*(1), 56–76. https://doi.org/10.54489/ijcim.v1i1.30

Napiah, M. N., Idris, M. Y. I., Bin, R., & R., & Ahmedy, I. (2018). Compression header analyzer intrusion detection system (CHA-IDS) for 6LoWPAN communication protocol. *IEEE Access, 6*, 16623–16638.

Nia, J. W. & R. M. (2021). Location prediction using GPS trackers: Can machine learning help locate the missing people with dementia? 00035, 2019. *Internet of Things, 13*, 100035.

Obaid, A. J. (2021). Assessment of smart home assistants as an IoT. *International Journal of Computations, Information and Manufacturing (IJCIM), 1*(1), 18–36. https://doi.org/10.54489/ijcim.v1i1.34

Park, E., del Pobil, A. P., & Kwon, S. J. (2018). The role of internet of things (IoT) in smart cities: technology Roadmap-oriented approaches. *Sustainability 2018, Vol. 10, Page 1388, 10*(5), 1388. https://doi.org/10.3390/SU10051388

Park, J. (2021). A review on inheritance tax and gift tax focusing on deductions and coherence with property-related taxes. *The Korean Association of Space and Environment Research, 76*, 140–170. https://doi.org/10.19097/KASER.2021.31.2.140

Rabbachin, A., Conti, A., & Win, M. Z. (2015). Wireless network intrinsic secrecy. *IEEE/ACM Transactions on Networking, 23*(1), 56–69. https://doi.org/10.1109/TNET.2013.2297339

Radwan, N., & Farouk, M. (2021). The growth of internet of things (IoT) in the management of healthcare issues and healthcare policy development. *International Journal of Technology, Innovation and Management (IJTIM), 1*(1), 69–84. https://doi.org/10.54489/ijtim.v1i1.8

Rayan, Z., Alfonse, M., & Salem, A. B. M. (2019). Machine Learning approaches in smart health. *Procedia Computer Science, 154*, 361–368. https://doi.org/10.1016/J.PROCS.2019.06.052

Sayakkara, A., Le-Khac, N. A., & Scanlon, M. (2019). Leveraging electromagnetic Side-channel analysis for the investigation of IoT devices. *Digital Investigation, 29*, S94–S103. https://doi.org/10.1016/J.DIIN.2019.04.012

Shamout, M., Ben-Abdallah, R., Alshurideh, M., Alzoubi, H., Kurdi, B. A., & Hamadneh, S. (2022). A conceptual model for the adoption of autonomous robots in supply chain and logistics industry. *Uncertain Supply Chain Management, 10*(2), 577–592. https://doi.org/10.5267/j.uscm.2021.11.006

Sharma, R. S., & Arya, R. (2015). Analysis and optimization of energy of sensor node using ACO in wireless sensor network. *Procedia Computer Science, 45*, 681–686.

Shin, Y. (2019). A VM-Based detection framework against remote code execution attacks for closed source network devices. *Applied Sciences 2019, Vol. 9, Page 1294, 9*(7), 1294. https://doi.org/10.3390/APP9071294

Syam, N., & Sharma, N. (2018). Waiting for a sales renaissance in the fourth industrial revolution: Machine learning and artificial intelligence in sales research and practice. *Industrial Marketing Management, 69*, 135–146.

Malche T., P. M. & R. K. (2019). Environmental monitoring system for smart city based on secure internet of things (IoT) architecture. *Wireless Personal Communications, 107*(4), 2143–2172.

Wang, Q., Dai, H. N., Li, X., Wang, H., & Xiao, H. (2016). On modeling eavesdropping attacks in underwater acoustic sensor networks. *Sensors 2016, Vol. 16, Page 721, 16*(5), 721. https://doi.org/10.3390/S16050721

Zhang, P., Nagarajan, S. G., & Nevat, I. (2017). Secure location of things (SLOT): Mitigating localization spoofing attacks in the internet of things. *IEEE Internet of Things Journal, 4*(6), 2199–2206. https://doi.org/10.1109/JIOT.2017.2753579

Zou, Y., & Wang, G. (2016). Intercept behavior analysis of industrial wireless sensor networks in the presence of eavesdropping attack. *IEEE Transactions on Industrial Informatics, 12*(2), 780–787. https://doi.org/10.1109/TII.2015.2399691

DDoS Intrusion Detection with Ensemble Stream Mining for IoT Smart Sensing Devices

Taher M. Ghazal⬤, Nidal A. Al-Dmour, Raed A. Said, Alireza Omidvar,
Urooj Yousuf Khan, Tariq Rahim Soomro, Haitham M. Alzoubi⬤,
Muhammad Alshurideh⬤, Tamer Mohamed Abdellatif,
Abdullah Moubayed, and Liaqat Ali

Abstract Security threats in the Smart City Systems are becoming a challenge. These Smart City Systems, generating Big Data, are a revolutionizing application of the Internet of Things(IoT). Data Stream Mining, which is an efficient way of

T. M. Ghazal
Center for Cyber Security, Faculty of Information Science and Technology, University Kebansaan Malaysia (UKM), Putrajaya, Malaysia
e-mail: taher.ghazal@skylineuniversity.ac.ae

School of Information Technology, Skyline University College, Sharjah, UAE

N. A. Al-Dmour
Department of Computer Engineering, College of Engineering, Mutah University, Mu'tah, Jordan

R. A. Said
Faculty of Management, Canadian University Dubai, Dubai, UAE

A. Omidvar
Engineering Management Department (TUSRC), Tehran Urban and Suburban Railway Co, Tehran, Iran

U. Y. Khan · T. R. Soomro
Department of Computer Science, IoBM, Karachi, Pakistan

H. M. Alzoubi (✉)
School of Business, Skyline University College, Sharjah, UAE
e-mail: haitham_zubi@skylineuniversity.ac.ae

M. Alshurideh
Department of Marketing, School of Business, University of Jordan, Amman, Jordan
e-mail: malshurideh@sharjah.ac.ae

Department of Management, College of Business Administration, University of Sharjah, Sharjah, UAE

T. M. Abdellatif
Faculty of Engineering, Applied Science and Technology, Canadian University Dubai, Dubai, UAE
e-mail: tamer.mohamed@cud.ac.ae

handling Big Data, is now of great concern. The acquired information is computationally expensive to process in terms of efficiency and runtime. Detection of suspicious activities on decentralized servers, generating and computing massive data streams requires time. Moreover, several stakeholders should be engaged to train the heterogenous malware data streams in the level of service application. Small experiments can be performed on the functionality of Batch ML on IoT datasets with available heap size resources. Among these candidate datasets, a little contribution has been already represented on the Mirai Attack. This research aims at the study of Data Stream Mining algorithms. Owing to the accuracy and interferences of the measurement, these algorithms are able to handle the non-hierarchical and unbalanced datasets similar to the Mirai Attacks. No single method can solely improve these critical standpoints. Thus, an Ensemble technique should be implemented. According to our study, a pool of meta or selective classifiers that interact based on the temporal Data Mining swiftly can outperform others. The maintainability and security concerns of such applications can be best fulfilled in meta-heuristics with the one-time scanning network approach for the recognition of the most frequent attacking pattern with the on-the-fly scheme. These are implemented in Create, Read, Update and Delete (CRUD) operations of the Big Data Systems.

Keywords Data stream mining · Ensemble active learning · Prequential learning · Mirai dataset · Security and privacy · Smart city · Wireless sensors · And internet of things

1 Introduction

The human mind is routinely sensing the reality of things through five sense organs. However, none of these senses can singularly result in the perception of objects in contexts such as planning, investigating, composing music, playing chess, or journalism while a mystic unconscious mind is incorporated in the aggregation of all those senses and builds the perception of the objects (Akhtar et al., 2021; Al Ali, 2021; AlHamad et al., 2021, 2022; Ali et al., 2021). Perhaps the human mind's neurons, evolving over thousands of decades, are capable of tree mining and associative rule reasoning ability. How a learning machine can similarly train from high volume streams, processed with distributed structure in parallel with other measurements, is an unanswered question. The combination of state-of-the-art sensor technology with

A. Moubayed
Engineering & Computer Engineering Department, University of Western Ontario, Ontario, ON, Canada
e-mail: amoubaye@uwo.ca

L. Ali
College of Engineering and Information Technology, University of Science and Technology of Fujairah, Fujairah, UAE
e-mail: l.ali@ustf.ac.ae

Table 1 Sensor network characteristics

Details	Wireless sensor network data
Processing architecture	Distributed
Data type	Dynamic
Memory usage	Restricted
Processing time	Restricted
Computing ability	Low
Energy	Restricted
Data length	Restricted
Response time	Real time
Updating	High
Data flow	Continuous

IoT in which is rapidly penetrating human life is inevitable. In 2020, the number of connected things on the Internet exceeded 50 billion. Data Streams are typically dynamic such as time-series format and their memory usage and processing time are restricted to hardware and database servers' limitations (Ali et al., 2022; Alnazer et al., 2017; Alnuaimi et al., 2021; Alsharari, 2021; Alshurideh et al., 2022). Regarding the unbalance attribute, high missing and noisy percentual, the dataset exhibits a low ability for the computation therefore any structuring or classification for such data should be updated in on-the-sky responding time. Sensor Network characteristics are shown in Table 1.

Nowadays, tread events can be handled when more sensor nodes are connected through cluster-based routings. Therefore, there are principal protocols in the field of planning Wireless Sensors. These protocols are better in energy saving, result in a longer network lifecycle and higher network efficiency. Wireless sensors that are small, adoptable, cost-effective, energy-efficient, and secure play an important role in the management of security, monitoring, and privacy issues of IoT Systems (Alshurideh et al., 2020; Ali Alzoubi, 2021; Alzoubi, 2021a; H Alzoubi et al., 2020, 2022; Ghazal et al., 2021). Since sensors are relatively cheap and small therefore there is a restriction on their hardware resources like memory and CPU to process the detection log files and sensors should maintain the energy consumption and needed to operate on the simplest form of processors. Furthermore, the environment doesn't always allow us to have a constant sensor functionality in comparison to the wired sensor networks. Hence, there is an essentiality for the implementation of real-time Intrusion Detection Systems on groups with the least cost and energy consumption, which is working on the basis of a decentralized detection approach.

The importance of security is not hidden from any stakeholder on the Internet of Things. Today Artificial Intelligence is handling security more accurately and professionally. It fills the gaps of Internet of Things Systems against the form of robots that hijack their information for their own sake into other levels and increase the risk of stakeholders. Wireless Sensor Networks are working in hostile environments where fraud issues can easily happen in the physical layer (Sensor). IoT systems in smart cities especially in applications such as electric power, health care systems, financial sector, and environmental disaster prediction (flood, tornado) can be the purpose of the attacker according to the level of damage they tend to bring to the public good. Similar to the aspect of term Internet advertisement and ransomware Web usage, in the Internet of Things Systems, malicious botnets are now threatening the privacy of the Users. Distributed Denial of Service (DDoS) attack is known as the most serious attack that transmits lots of data by taking up the bandwidth traffic. According to the reports published in the course of the pandemic on the DDoS attacks, there are 929,000 DDoS attacks in a 31-day (April–May 2020) with more than 2.8 TBPS bandwidth, and more than a thousand botnets are disrupting legitimate services like Amazon, eBay, Netflix, etc. or even public services (Alzoubi & Ahmed, 2019; Alzoubi & Aziz, 2021; Alzoubi & Yanamandra, 2020; Alzoubi et al., 2020, 2021). Unlike network layer (DoS (that occurs in the hardware such as I/O or Sockets, in the Distributed DoS (DDoS) attacks are performed from a legitimate request usually by a HTTP get command on a machined network. In contexts like Smart Cities and the Internet of Things, a middle-ware layer can be constructed to partly solve the above issue with located features' API architecture by semantic analysis of the code. It can be further extended not only by the Application level but also from the viewpoint of repository level which migrates the central cloud framework to Distributed Clouds (Fog Computing). An example of the Fog nodes clustering based on the Actors of different layer and applications are shown in Fig. 1.

A survey was conducted on the cyber criminal's behaviors and intension in order to develop a framework for IoT devices, applications, and Big Data Systems that require immediate awareness such as healthcare and public and financial services, who are threatened and lost control of their contents. The functional blocks and maps send the clustering results to actors by the implementation of the actors' model. This allows the users to contribute to the holistic and work for the sake of their prenatal network layers in the IoT. The attacks on the health, financial and public sector were reported that with the increasing demand on the performance of data mining the less traffic between the end cloud and the user can be gained. Table 2 presents the most important form of security and breach of the Internet of Things framework.

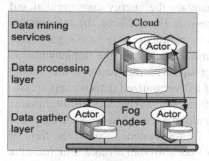 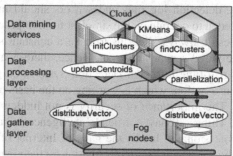

Fig. 1 Fog computing in the distributed architecture of IoT

Table 2 Design requirements of security scheme	Security requirement	Integrity
		Confidentiality
		Availability
		Authentication
		Access control
		Prevent node replication or addition

In our proposed framework a packet payload has been utilized to discover if an activity is benign or attacked based on a network packet. After the training of the classification model (Intrusive user experiment) with attributes (IP and MAC address), Channel, Socket characteristics were imported to the data stream mining ensemble of different algorithms. A simulation was performed with density and weight-based ensemble in the MOA (Massive Online Analysis) software. The results were compared with the accuracy and false-positive rates for the batch dataset of the DoS attack. Secondly, detection of a network intrusion in the server even at the Application level by the implementation of a Test Case scenario is elaborated. Here, different CRUD operations, feature locations are acting on a defined number of Test Cases, although applied filtering, high level of penetration, and big sample size stand as a generalized and wide strict rule for the Internet of Things.

The paper is organized as follows. Section 2 provides a summary of the results and research pathway of the recent publications and introduce the implementation of Data Stream Mining and IoT systems in DDoS attacks of different applications. Section 3 presents the dataset and the designed experiment with the proposed Ensemble. Section 4 presents the results of the Mirai case study and the accompanying risk on the CRUD processes. Section 5 concludes idea of each of the Ensemble techniques and the future research for smart cities.

2 Literature Review

The application layer DDoS Intrusion Detection form a new trend in Industry and professional issues and are now of great interest scientifically and is practiced in many algorithms such as Markov-based wavelet singularities detection model, Flexibility induced Random Walk Graphs, and Novel method Bayes-based Entropy minimization Clustering in temporal data Mining. Temporal effects caused lots of computational costs and should not be ignored as well as other effects. One main advancement for the prediction of the Big Data System of IoT system was the application of very fast algorithms for the implementation in the AI, especially in the networking and service development cases which need the fulfillment of the requirement of security planning. Full implementation can be explained according to the CENELEC Standards procedure of software development which includes, System planning, software architecture planning, and Coding, Software Interface planning and static testing, software/hardware compatibility, and dynamic testing like Black-Box, simulators, and documentation. IoT systems' inventions always extending to new tools and methods and in between the usage of data science in the area of mechanical, civil and mechatronics engineering are among the most wanted specialization in the field of technological development such as smart aircraft and satellites, high-speed hyperloop driverless System, smart cities and health care system. The very fast algorithms can also be implemented at best for smart cities for usages such as indoor application, edge computing, subways surveillances systems, smart commercial buildings, and also for the specialized sustainability measures where mounting of the temperature, humidity, and pollution level can be of great importance (Aziz & Aftab, 2021; Cruz, 2021; Eli, 2021; Farouk, 2021; Ghazal et al., 2021; Hamadneh et al., 2021; Nuseir et al., 2021).

The field of Stream Data Mining is very wide and it has become a very challenging task among computer science researchers. Even though Data Stream Mining is thoroughly analyzed in the literature, the concept of the method can be defined by major steps. It is started by the splitting of the incoming data into windows frame with fixed-width according to the hourly, daily, etc. bases. Then the summary of each window is used as input for the local clustering of the time-series. Two clustering algorithms are performed in this case one for the local clustering and the other one for the comparison and building of the proximity matrix. From this step, the data that is updated in an online manner is used for the clustering of the time-series, which is required by the users. Figure 2 is demonstrating the mining path in the data stream (Al Guergov & Radwan, 2021; Hamadneh et al., 2021a, 2021b; Hanaysha et al., 2021a, 2021b; Joghee et al., 2020; Salloum et al., 2020; Shebli et al., 2021).

According to the proposed diagram Data Stream Mining requires clustering (micro clustering) according to one of the following summarizations' algorithms: Growing

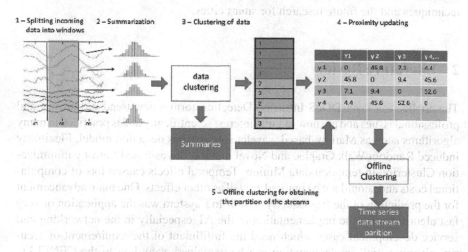

Fig. 2 Analyzing work plan

Neural Gas, hierarchical, Partitioning, density-based, and grid-based streaming method.

In a context such as Smart Cities, the memory and computational resources of the data streams are constrained by the hardware and software requirements, while the attractivity of the very fast decision tree is its temporal processing part of the dataset. The field of Stream Data Mining is very wide and it has become a very challenging task among computer science researchers. Detection of patterns, faulty and noisy data has been also widely discussed for the identification of patterns, planning of the IoT system safety, data repository, and security. There are normally two steps of cached-based meta- classification and main classification which can be organized in the mode. The principal idea behind the very fast decision trees (or Forest) is that it learns on the total macro view of an object and receive the knowledge from each singular Tree. The applications can be further extended in Jigsaw puzzles, Electrocardiogram Patterns and dynamic stock Market Data.

Incrementally Optimized Very Fast Decision Forest as an extension to the VFDT is reviewed on the Simon Fong Work on temporal data. Their work was mainly focusing on the iOVDFT algorithm and was published in two papers in 2017 and in 2018. This method was implemented to better evaluate the effect of the equal timeframe windows over a single day measurement of the sensors. After the implementation of the required MOA (Massive Online Analysis) component the chosen dataset on Intel Research Lab in Berkeley was evaluated (Part of the Weka Application). There are normally two steps of cached-based meta- classification and main classification which can be organized in the colorful visualization mode. Figure 3 represents the output of the Clustering module in two dimensions by MOA.

Obviously, for Sensor datasets, some attributes are not usually normalized, and their final cluster cannot visually distinguish as the above Generator based Visualization. This framework has no automatic way of normalizing the data and for this reason, the managing of the sensor from the data source is necessary when data classification is conducted using iOVDFT. But this makes it possible to look at the concept of Drift visually for the detection of the anomalies or abrupt changes in the Stream. It is anyhow possible to split the attribute with a supervised filter through different algorithms like Entropy and Gain ratio or using Weka classifiers for the attribute selection. Sometimes, MOA algorithm is used in the iOVDFT such as Tree Clustering, Randomized RBF, or KM Stream partitioning. However, the choice of an algorithm should be experimented with and must consider the complexity of the case. This transformation was also utilized in the DBSCAN, in a density-based streaming of the time-series. The chosen dataset for this case study has been set to measure the indoor environmental parameters based on the network of sensors positioned at the Intel Berkeley Research lab (Kashif et al., 2021; Khan, 2021; Lee & Ahmed, 2021; Lee et al., 2022a, 2022b). Several sensors are measuring the temperature, humidity, and light in the indoor environment and broadcast their data to the Raspberry pi Board for recording and processing. The sensors were covered the public area of the Lab and according to the following figure, Server Rack and Lab area, as well as rooms, were excluded from the measurements. The data communication is performed by the LAN network. The recorded temperature by each sensor has been analyzed by

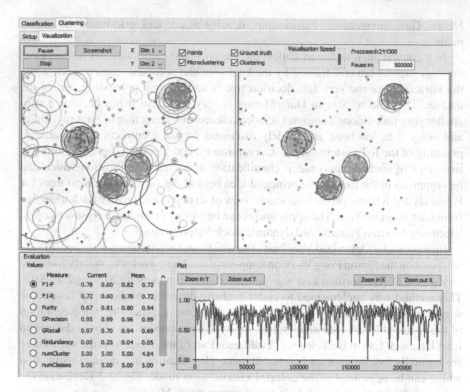

Fig. 3 RBF cluster generator visualization on MOA with 5 distinct colorful partitions

testing a set of 54 time-series each one made by 65,000 observations. In this dataset, epoch number was a sequence increasing monotones of each sensor i.e., readings from a similar number of epochs were read at the same time from different sensors. There is some missing value in both epochs regarding power blackout and missing mote (Sensors) dataset because of certain sensor malfunctions or dead battery. The measurements in the case study are in the unit of degrees Celsius for the temperature, the humidity was corrected in relation to the temperature, and light was measured in Lux. Voltage in volts, between 2 and 3 supplying by a lithium-ion battery and the changes of battery voltage is strongly correlated with temperature. One important achievement in the iOVFDT method is the consideration of the temporal aspects during the pattern recognition and in the prediction of similarity between time-series the ordering of the events was not neglected. Some time series could contain rich information such as hidden patterns that seemed to be treated in an unordered manner and were outranked. The same author has compared the CPU runtime of important data mining methods namely in another article.

From the above results (Fig. 4) it is clear for sake of disposition of distortion and latency there is a need to perform stream mining in a two-step data stream mining algorithm. The cached-based meta-classifier stands for the sliding window are updated in the process of knowledge discovery, and the second (main) classifier is to verify whether the data is nominated for the knowledge discovery. The sliding timeframe should be adjusted in a way that covers the rate of data change for the prediction of the best sensor. Incremental learning, like IBK, is the better choice for the meta-classifier compared to the ones that need to search the full database— it is not efficient and the models cannot predict but with sufficient power, proper clustering can be achieved. They have also concluded how one HT (Hoeffding Tree) can be improved with the noise and misclassified recall along their CPU Time. The main limitation of this work is the restriction of the data stream in one single data stream in full processing. This means that if the even and odd measurements should be treated in different ways, the method should be adopted from the multiple data stream to discover the subsequences in the sequences of the multiple time series in single/parallel for the cashed-based classification. In a signaling study approach, acceleration and energy of the particles are given parameters of the partial differential equation model, thus, the system time and discrete frequency cycle are important. From the continuous segmentation like windowing, segmentation and discretizing representation like self-organizing map, signal shape language and transformation types like Laplace and Kalman filter or Fast Fourier Transform would be implemented which are out of the scope of Data Mining (Mehmood, 2021; Mehmood et al., 2019; Miller, 2021; Mondol, 2021; Obaid, 2021).

In research, published in relation to RGCE (Realistic Global Cyber Environment), a network traffic dataset was selected for the simulation of real-time DDoS attacks. RGCE is a project-based in Finland for simulation of the closed Internet environment provided by the RGCE project. One of the RGCE services is to provide real IP addresses and geolocations. While the virtual server operating on normal traffic with

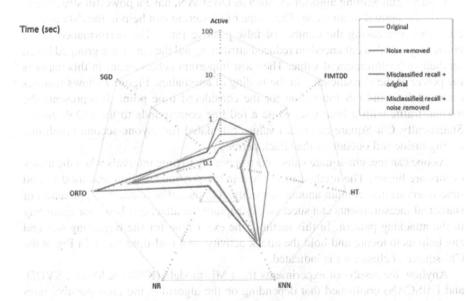

Fig. 4 Runtime measurement in different data stream mining

included text and gif format pictures on the homepage, the legitimate service is performed via HTTPS protocol (SSL/TLS) as a part of the program infrastructure. Here within, the raw network data were extracted in the Libpcap format (This format is a combination of captured Network Data). The captured dataset contains mainly HTTPS network traffic while test hosting communicated with the end-users only via an encrypted protocol.

In this configuration, unencrypted network traffic (HTTP) in the dataset is the first try for the clients (bots) and hosting server before a secure channel was created. Figure 5 represents the important region RNA of the RGCE service provided for the DDoS attacks which were extracted from the project in Feature Vectors captured in Libpcap format can have different scales. Therefore, all of the attributes namely size of packets, size of TCP, time interval since the last previous attack, and percentages of different TCP flags were standardized to the range between 0 and 1 according to a linear alteration on the Max–Min normalization principle. At least two parameters namely the clustering neighborhood size ε and the minimum number of records a cluster needed should be specified in the DBSCAN. The results showed that false-positive rates are quite low, especially with the PCA reduced feature. Network anomalies are the aggregate of remaining clusterless after the finishing of the simulation. An application of DBSCAN is shown in Fig. 6, where Nmin is equal to 3 and clustering neighborhood size of 0.25 (radius of each circle). As been discussed modeling of un-attacking behavior was performed with the help of density-based clustering by noise algorithms for discovering the outliers (DBSCAN). DBSCAN is considered a strong density-based clustering algorithm for detecting outliers and finding clusters in the dataset (Radwan & Farouk, 2021; Shamout et al., 2022).

Data Stream Mining algorithms, such as DBSCAN, have a powerful structure to explain the normal or attacked. The usage of clustering can help in the data mining efficiency, decreasing the number of false-positive rates. The performance of the clustering was also enhanced on reduced attributes, and the data were grouped based on their ordinal/numerical value. The most important achievement in this paper is the power of the classification in the finding of anomalies. Figure 7 shows features obtained for windows timeframe for the considered time point. It represents the normal traffic with a blue line, while a red line corresponds to the DDoS attack. Statistically, Chi-Square computed values calculated for anyone-second timeframe during inside and outside of the attack period.

As one can see, chi-square values corresponding to time intervals when the attack occurs are higher. The method mentioned in the previous section was used to find time intervals that contain anomalous traffic. On the other hand, the application of statistical measurements can successfully capture the attacker's behavior according to the attacking pattern. In this method, the exact time for the beginning and end can help us to locate and hold the attack activity on a real-time basis. In Fig. 8 the Chi-square of clustering is indicated.

Anyhow the results of experiments from ML models (KNN, K-Means, SVDD, and DBSCAN) confirmed that depending on the algorithm the false-positive rates can be varied slightly to be between 1 to 6% (false positive rate) (Afifi et al., 2020; Ali et al., 2021; Alzoubi et al., 2021a, 2021b). The main distraction of the mentioned

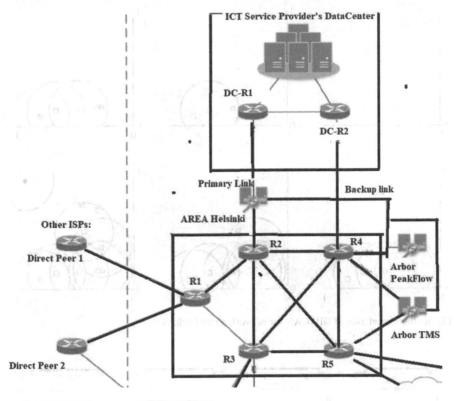

Fig. 5 RNA map (arbor networks peakflow)

methodology was that the target class was considered normal or attacked. In a real test case, there can be a range of different attacks happening at the same time; DOS, Probing (Port scanning), U2R (Buffer Overflow), and R2L (Password Guessing) could be mentioned. A multi-binomial model can be used to predict network intrusion. Additionally, nominal and numerical attributes must be identified as well as both the accuracy and precision have been checked. It was clear that density-based algorithm outperforms other approaches with regard to accuracy and false alarms. The false alarm can be explained low probably for model building. The results of different ML are presented in Table 3.

In the literature there are many DDoS attacks discussed the Ransomware and attacks in form of DDoS, for instance in a study dataset gained on Bitcoins total transactions from 2009 to 2018 were selected to perform clustering heuristics on data temporal analysis and graph mapping. The analysis revealed a significant repeated pattern between families of Ransomware by the techniques such as COSINE, TDA (Topological Data Analysis), and DBSCAN. The reliability of the analysis was confirmed for very little or too many Bitcoins addresses in the relevant clusters and

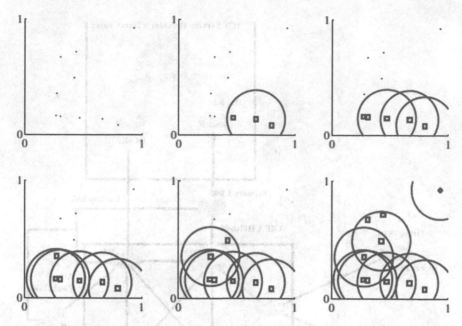

Fig. 6 Implementation of DBSCAN for network normal behavior

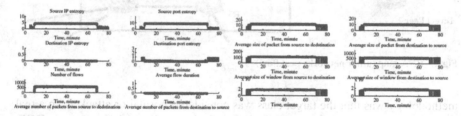

Fig. 7 Convolution between important attributes (timeseries)

their validity by statistical confidence level and compared with recent different scientific papers and studies. Figure 9 shows an example graph of 10 addresses (accounts) in a bitcoin chain block with 7 transactions.

This article aimed to use the Dataset on the Bitcoin transaction by seven different constitutes as a case. Although applied filtering, the high level of penetration, and the big sample size stand as a generalized and wide strict rule for the Internet of Things, we insist on concluding the results through more realistic initiatives and incentives for the better Internet of Things on the thematic too. The attribute of the selected dataset using 24 h transactions from 2009 to 2018 of the bitcoins testing through different features selection by mappers cluster. The dataset was filtered out by some heuristics and the selection was based on two parameters namely inclusion and threshold parameters. Other than the two mentioned above parameters we

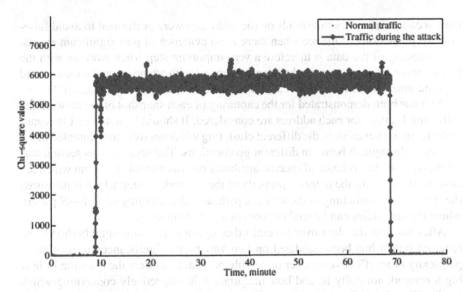

Fig. 8 Network Chi-square in regard to DBSCAN, (red line anomaly condition)

Table 3 Comparison of the results by different ML

Algorithm	TPR	FPR	Accuracy
K-means	100	0.488	99.9951
KNN	100	0.209	99.9979
SVDD	100	6.063	99.939
SOM	100	0.488	99.9951
DBSCAN	100	0.07	99.9993

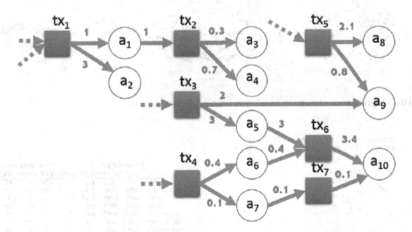

Fig. 9 BitCoin network (edge number is the number of Coins)

have seen that quantiles thresholds on the addresses were performed to avoid false-positive predictions of malice when there is no evidence of past significant events. Preprocessing of the data is therefore a very important step when working with the dataset more sensitive to the imbalance—temporal with attention on the metrics and hyperparameters (Al Batayneh et al., 2021; Ghazal, 2021; Ghazal et al., 2021).

As it has been demonstrated for the capturing of each snapshot of the network the following features for each address are considered. It should be noted that in regard to the temporal behavior of the different clustering windows were implemented in the dataset to distinguish between different geolocations. The result of this section was for this specific importance of specific attributes such as weight and count which are designed to quantify the merging property of the network or loop which is measured the splitting and switching of the network path, and also merging all to 1—N pattern where the currencies can be sold or converted to flat currencies.

After running the data over several other options in 2-dim neighborhood, the required output has been validated on two variables t-significances in relation to perplexity or an effective number of neighbors, which answers the question of how big a network normally is, and how important it is subjectively concerning which family (Fig. 10). Distances and sizes of the cluster are not conveying any concept, but the group of addresses means that DMALocker and Crypxxx are locating in the center while CryptoLocker and CryptoWall are often together in many groups. The most common application is the cluster purity for every Ransom with all addresses from the same family. The Cerber, CryptoWall, and CryptoLocker, have gained around 40 percent of purity where the most improved number of clusters were in the amount of k = 12,500 (Fig. 11).

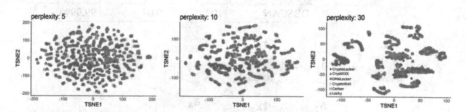

Fig. 10 T-significance clustering in relation with perplexity

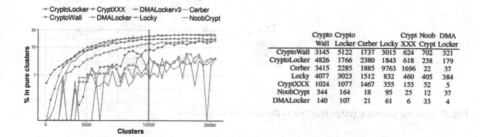

	Crypto Wall	Crypto Locker	Cerber	Locky	Crypt XXX	Noob Crypt	DMA Locker
CryptoWall	3145	5122	1737	3015	624	702	321
CryptoLocker	4826	1766	2380	1843	618	238	179
Cerber	3415	2285	1885	9763	1696	22	37
Locky	4077	3023	1512	832	460	405	384
CryptXXX	1024	1077	1467	355	155	52	5
NoobCrypt	344	164	18	95	25	12	37
DMALocker	140	107	21	61	6	33	4

Fig. 11 Percent of purity and no. of shared clusters of the ransom families

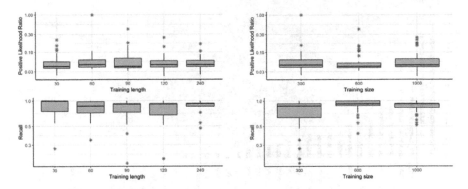

Fig. 12 Effect of training size and length on the positive likelihood ratio and recall

The co-clustering families were shown in the last tables were also supported for CryptXXX, from the Symantec warns security company that on 2016 as Trojan Trojan.CryptoLocker.AN, therefore it can advantageous to search for similar patterns of the Ransom families. There is a daily more than 500 K transaction on the Bitcoin alone and the number of false-positive rates can happen to the dataset. One of the most important achievements in this report is the class rebalancing to different Ransomware addresses. On the other hand, regarding the temporal characteristics of the dataset, the concept of sliding windows was introduced to the 30, 60, 90, 120, and 240 for the training of the model (Fig. 12). The effect of training size is indicated in the following graph. Finally, Accuracy and Recall were both along with the maximum likelihood function were improved by this setup (Ghazal et al., 2013, 2021; Ghazal et al., 2021;).

3   Research Method

In this research, we have firstly conducted a network anomaly detection experiment from KDDCUP 99, LAN activity log of a military environment. We aimed to predict the probabilities of malicious activities (intrusions). The class attribute of the model was binary with 38 attributes of numerical type and 3 attributes were of the nominal character. The attributes are categorized into four categories including inbound and outbound connection features which were basically:

- footprints of each network from source to host and vice versa (e.g. protocol, service type, duration, status of service, and data bitrate).
- Content Access Level Features such as the number of times a user access system directory entrance, administer passwords trials, access control, and user privileges,
- Network connection Time-dependent Features indicating the number of connections to the host in the last 2 s considering the error flag on the same port number.

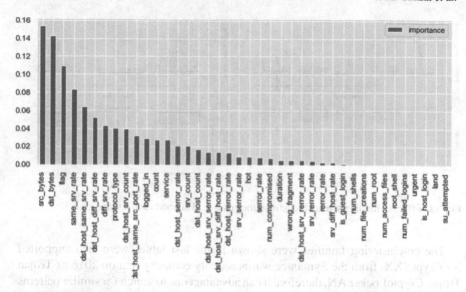

Fig. 13 Feature selection

- Destination Traffic Features, which are a representation of a connection to the same destination IP address of the host while activating the error flag. The feature selection method was performed by Random Forest in Fig. 13.

This experiment was firstly performed with the 4NN batch processing. Later on, by the evaluation of the model when the cost–benefit analysis would be considered the accuracy was 98.8% FP average rate and 1.4% of FN rate with the above 4NN. For avoiding the overfitting error, attributes were filtered on the hybridized subset selection and wrapper techniques models by using model information gain and SVM model. Both were purposely used to improve the model accuracy.

This prediction power resulted in approximately 1000 false positive alarms in network 50 K instances, where 6, 9, and 23 numbers of nominal and numerical attributes were selected. The following table shows the efficiency results of the SVM model. Table 4 shows the efficiency results of the SVM model.

Results of our experiments from batched ML models (SMO, KNN, ANN, and Logistic regression) confirmed that depending on the batch size, and other hyper parametrical, the F-Score model can be varied to be between 94 to 98% (with 2 to

Table 4 Batch processing results on the KDDCUP 99 network anomaly detection with 6 attributes

TP rate	FP rate	Precision	Recall	F-Measure	MCC	ROC area	PRC area	
0.954	0.135	0.89	0.954	0.921	0.825	0.909	0.873	Normal
0.865	0.046	0.942	0.865	0.902	0.825	0.909	0.878	Anomaly
0.912	0.094	0.914	0.912	0.912	0.825	0.909	0.8	Avg

Table 5 Batch processing results on the KDDCUP 99 network anomaly detection with 23 attributes

TP rate	FP rate	Precision	Recall	F-Measure	MCC	ROC area	PRC area	
0.981	0.11	0.911	0.981	0.945	0.879	0.969	0.956	Normal
0.89	0.019	0.977	0.89	0.931	0.879	0.969	0.975	Anomaly
0.939	0.067	0.941	0.939	0.938	0.879	0.969	0.965	Avg

Table 6 Batch processing results on the KDDCUP 99 network anomaly detection with 9 attributes

TP rate	FP rate	Precision	Recall	F-Measure	MCC	ROC area	PRC area	
0.984	0.111	0.911	0.984	0.946	0.882	0.975	0.971	Normal
0.889	0.016	0.98	0.889	0.932	0.882	0.975	0.976	Anomaly
0.94	0.067	0.943	0.94	0.94	0.882	0.975	0.973	Avg

4% false rate). Interestingly, in case 4NN algorithm with a batch size of 50 outperformed Bootstrapped LogitBoost, logistic regression, and Bayes Naïve. One of the achievements was the runtime performance of 0.25 s on Logitboost on KDDCUP 99 dataset with a high number of classes that are not distributed independently. Repeating the same setup for filtered SVM, TP and FP rates were promoted around 4 percent, but the runtimes were 58.93 and 13.98 s with 23 and 9 attributes respectively. These results are shown in the following tables (Tables 5 and 6)

This proved that each packet of data cannot implicitly explain the state of containing packet. This fact was also reported in the literature on the network anomaly detection on KDDCUP 99 dataset where temporal dependence is present. Nevertheless, the 97% True Positive rate accompanied by 1.5% False negative motivated us to switch to other datasets for the designing of the data streaming experiment on the Mirai DDoS network activity, a type of Ensemble detection strategy based on two learners. The stability learner stands for the ground truth on the past information and an evaluator that updates, replaces, or removes instances according to the most recent data streams forms a plasticity region around the classification problem. In an era of anomaly detection, Ensemble approaches were chosen for the concept drift in this study, which was DDM, EDDM, and Hinckley page (Kalra et al., 2020; Khan et al., 2021a, 2021b). The data was further classified and evaluated by Adoptive Hoeffding Tree, Leveraging Bagging, no-change, and/or majority classifiers with prequential Interleaved Test and Train samples. Regarding the temporal dependence of KDDCUP 99, usage of no-change and/or majority classifiers was proposed in the literature but in this experiment, the Kappa Statistics based on Random classifier by prequential cannot be misleading, measurements are compared based on the Classification analysis on Damped Statistic while computational costs and proposed Concept Drift Strategy resulted in relatively higher accuracy and recall. For sake of availability in real-time, the CRUD operations were performed on MongoDB DBS. The objective of this experiment is to work with a temporal type of attributes by application of stream data mining with fewer false-positive rates. We aim to make comparable results as we believe that fewer false-positive rates with lower precision will be achieved in comparison with the Mirai dataset.

3.1 Dataset Description

The selected database is IoT devices' Mirai attack which is the Network Mirror data of 10 min of activity in a simulated controlled environment connected to an access point which is designed to have a victim and an attacker device. More than 700 k data packets are recognized and captured by having a binomial value for the class attribute of the model (value 0 when there is no attack and value 1 for a network attack) (Lee et al., 2021; Matloob et al., 2021; Naqvi et al., 2021). When a fraud packet is received the behavioral attributes can be extracted. Two types of data are contained in this packet which are statistics or bandwidth behavior of a small window. These temporal analyses could be on the general information of the packet senders and also on the traffic between the sender and receivers. The first type is defined as Source MAC and IP address (SrcMAC-IP). The second type is the traffic between source and destination (either IPs or TCP/UDP Sockets). The structure of the dataset embedded in MongoDB can represent that there are 20 classes of nominal on the features as well as numerical attributes (see Fig. 14).

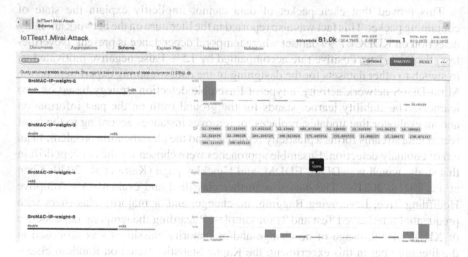

Fig. 14 Big data system implementation on MongoDB

3.2   Feature Extraction

The attributes harmonized through damped statistics such as packet id number, Linear Sum and Squared Sum can be further extended to explain the signaling behavior of the packets in regard to their context and traffic. Every packet was implemented in the feature extractor to create 23 features for the description of outbound bandwidth (8 features), inbound and outbound bandwidth (8 features), packet rate (4 features), and inter-packet delays of the outbound bandwidth (3 features). The packet size of outbound bandwidth statistic measures was divided to mean and standard deviations aggregated by SrcMAC-IP, SrcIP, Channel, and Socket. The packet size of inbound and outbound bandwidth statistic measures was divided into Magnitude, Radius, Approximate Covariance, and Correlation Coefficient aggregated by Channel and Socket. Packet count by the weight of SrcMAC-IP, SrcIP, Channel, and Socket, and packet's jitter by the weight, mean, and standard deviation of the Channel. For the consideration of timestamp, these values are extracted by the same feature with five windows of length 100, 500, 1500, 10,000, and 60,000 ms, which sums the total attributes into 115 plus one class attribute on the basis of their weight and decay factors while they are extracted. For the network administration, the IP address is like the postal code, and each channel is a mailbox. On the other phrase, residents (services) need to access their messages through their mailbox. The number of a channel cannot solely identify the path in which the packet would be sent while the UDP/TCP isn't defined. Mirai bot through ports number 23 and 2323 uses TCP/IP to map packets to services. These ports are belonged to Telnet service enter the command and control by legitimate access. The botnet can firstly disable the Telnet and search the ports by sending SYN, ACK, RST, etc. flags on large scales. One of the millions of inbound traffic could be accompanied by more than 1 GB ack flag and is known as TCP stomp flood. These bots are not anyway easy to handle but using CNC control on TCP/23 and looking for the NetFlow on this channel is also a good idea. The chart below (Fig. 15) is one out of 115 attribute representations of Channel chart 100-ms versus 1-min windows in the past as it has been aggregated by Channel of 10-s windows in the past.

3.3   Evaluations Sets

Contrary to the batched Learning which is leveraged by the random arrangement of train and test sets, in data stream Mining the challenge of streaming posed another type of evaluation. Among the recently developed approaches for the evaluation of data streams, the hold-out trainset in stream mining is used when cross-validation computational costs are big enough that make splitting of the data into train and test efficient. The problem in our case was that network vectors were based on damped statistics and chronologically on the basis of decay function and cannot separately be handled for the evaluation and validation of classification while it is

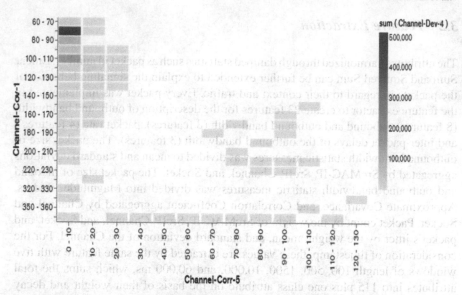

Fig. 15 Channel covariance versus channel correlation extracted from DBMango atlas

impossible to distinguish that last records are from 1 min or 10 min ago. In our
case, a prequential test-then-train evaluation was performed by Hoeffding Adaptive
Tree and Leveraging Bag separately. The evaluation process has taken 124.94 and
304.61 s respectively to complete the testing of every single point and train the
model while the majority class which process runtime on the data without evaluation
was 85.38 s. For both evaluations' frequency has been considered 10,000 packets
which are around 2 percent of the whole 764,137 packets. In leveraging Bag, five
Ensembles like bootstrap, Blast, and Stacking were chosen for the augmentation of
drifting with 3 leaves and 6 nodes from a single leaf root at the beginning of the Mirai
Bot activation. In Hoeffding Adaptive Tree, for the calculation of Information Gain
at the given confidence level, the tree grown from root to 4 leaves and 7 nodes, and
the leaf was again split to reciprocate the algorithm (Rehman et al., 2021; Suleman
et al., 2021).

4 Classification Results: Ensemble Learning

The concept drift can detect changes according to the damped statistic, in Fig. 16
two streaming behaviors of the dataset and their detected anomaly are distinguished.
It is clear that at the beginning of the stream dataset the number of false alarms is
increased by the application of EDDM and Page Hinckley—the input values and
detected changes are identified.

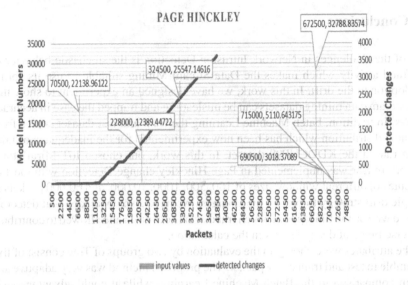

Fig. 16 Concept drift strategy for anomaly detection by EDDM performed on MOA

As it can be seen in Fig. 16, the data class attribute of normal type is less than attacked type and started on the 121 K records and is not happening. In Fig. 17, the accuracy of two tree classification is shown which can be compared with the majority class and can be best explained by temporal dependence of the Dataset.

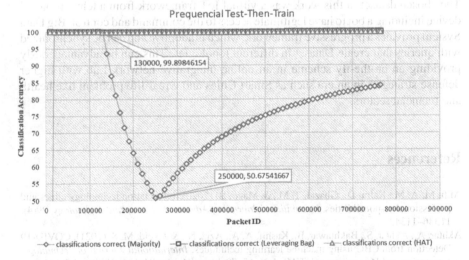

Fig. 17 Classification accuracy on majority class, leveraging bag and hoeffding adaptive tree

5 Conclusions

One of the challenges in Network Intrusion Detection is the stochastic behavior of the fraud activity which makes the Data Stream Mining suitable to use its ability of adoption to the drift. In this work, we have designed an experiment to show that even stream data mining datasets can be misleading and trapped the case in accuracy pitfalls. In tradition, batch Machine Learning the train and test dataset comes from the same distribution which has been now experimented for the inadequacy of false alarm rate in the KDDCUP99 dataset. In this work, our chosen IoT Mirai attack network mirror was implemented in Page Hinckley change Detection with the 115 attributes of different window sizes of 100, 500, 1000, 10,000, and 60,000 packets in a single drift strategy to predict the changes. We have also concluded that detectors worked well on the timespan in which fraud events occurred and allowed to contribute a diverse range of decay factors in the calculation.

The attributes were chosen in the evaluation by two groups of Tree consist of five Ensemble to test and train every packet. The proposed method was very adaptive and fast in comparison to the Batch Machine Learning, while it could advantageously get the frequency variation in comparison to the No-Change ADWIN which could process the more recent past.

6 Future Work

The chosen dataset in this work was a virtual IoT framework from a telnet protocol device including a bot to have legitimate access to the command and control. Big Data System provides a progressive framework, which is adoptive and can be synchronized with queries that create Dataset in different topologies. This can be advantageously providing an on-the-fly scheme in an online integrative network Lab with higher defense strategies in an era such as Smart Cities and a real-life problem like health and financial sectors.

References

Afifi, M. A. M., Kalra, D., Ghazal, T. M., & Mago, B. (2020). Information technology ethics and professional responsibilities. *International Journal of Advanced Science and Technology, 29*(4), 11336–11343.

Akhtar, A., Akhtar, S., Bakhtawar, B., Kashif, A. A., Aziz, N., & Javeid, M. S. (2021). COVID-19 Detection from CBC using machine learning techniques. *International Journal of Technology, Innovation and Management (IJTIM), 1*(2), 65–78. https://doi.org/10.54489/ijtim.v1i2.22

Al Ali, A. (2021). The impact of information sharing and quality assurance on customer service at UAE banking sector. *International Journal of Technology, Innovation and Management (IJTIM), 1*(1), 01–17. https://doi.org/10.54489/ijtim.v1i1.10

Al Batayneh, R. M., Taleb, N., Said, R. A., Alshurideh, M. T., Ghazal, T. M., & Alzoubi, H. M. (2021). IT governance framework and smart services integration for future development of Dubai infrastructure utilizing ai and big data, Its reflection on the citizens standard of living. *The International Conference on Artificial Intelligence and Computer Vision*, 235–247.

Al Shebli, K., Said, R. A., Taleb, N., Ghazal, T. M., Alshurideh, M. T., & Alzoubi, H. M. (2021). RTA's employees' perceptions toward the efficiency of artificial intelligence and big data utilization in providing smart services to the residents of Dubai. *The International Conference on Artificial Intelligence and Computer Vision*, 573–585.

AlHamad, A., Alshurideh, M., Alomari, K., Kurdi, B., Alzoubi, H., Hamouche, S., & Al-Hawary, S. (2022). The effect of electronic human resources management on organizational health of telecommuni-cations companies in Jordan. *International Journal of Data and Network Science*, 6(2), 429–438.

Alhamad, A. Q. M., Akour, I., Alshurideh, M., Al-Hamad, A. Q., Kurdi, B. A., & Alzoubi, H. (2021). Predicting the intention to use google glass: A comparative approach using machine learning models and PLS-SEM. *International Journal of Data and Network Science*, 5(3). https://doi.org/10.5267/j.ijdns.2021.6.002

Ali, N., Ahmed, A., Anum, L., Ghazal, T. M., Abbas, S., Khan, M. A., Alzoubi, H. M., & Ahmad, M. (2021). Modelling supply chain information collaboration empowered with machine learning technique. *Intelligent Automation and Soft Computing*, 30(1), 243–257. https://doi.org/10.32604/iasc.2021.018983

Ali, N., M. Ghazal, T., Ahmed, A., Abbas, S., A. Khan, M., Alzoubi, H., Farooq, U., Ahmad, M., & Adnan Khan, M. (2022). Fusion-Based supply chain collaboration using machine learning techniques. *Intelligent Automation & Soft Computing*, 31(3), 1671–1687. https://doi.org/10.32604/iasc.2022.019892

Alnazer, N. N., Alnuaimi, M. A., & Alzoubi, H. M. (2017). Analysing the appropriate cognitive styles and its effect on strategic innovation in Jordanian universities. *International Journal of Business Excellence*, 13(1), 127–140. https://doi.org/10.1504/IJBEX.2017.085799

Alnuaimi, M., Alzoubi, H. M., Ajelat, D., & Alzoubi, A. A. (2021). Towards intelligent organisations: An empirical investigation of learning orientation's role in technical innovation. *International Journal of Innovation and Learning*, 29(2), 207–221. https://doi.org/10.1504/IJIL.2021.112996

Alsharari, N. (2021). Integrating blockchain technology with internet of things to efficiency. *International Journal of Technology, Innovation and Management (IJTIM)*, 1(2), 1–13.

Alshurideh, M., Gasaymeh, A., Ahmed, G., Alzoubi, H., & Kurd, B. A. (2020). Loyalty program effectiveness: Theoretical reviews and practical proofs. *Uncertain Supply Chain Management*, 8(3). https://doi.org/10.5267/j.uscm.2020.2.003

Alshurideh, M. T., Al Kurdi, B., Alzoubi, H. M., Ghazal, T. M., Said, R. A., AlHamad, A. Q., Hamadneh, S., Sahawneh, N., & Al-kassem, A. H. (2022). Fuzzy assisted human resource management for supply chain management issues. *Annals of Operations Research*, 1–19.

Alzoubi, Ali. (2021a). The impact of process quality and quality control on organizational competitiveness at 5-star hotels in Dubai. *International Journal of Technology, Innovation and Management (IJTIM)*, 1(1), 54–68. https://doi.org/10.54489/ijtim.v1i1.14

Alzoubi, Asem. (2021b). Renewable green hydrogen energy impact on sustainability performance. *International Journal of Computations, Information and Manufacturing (IJCIM)*, 1(1), 94–110. https://doi.org/10.54489/ijcim.v1i1.46

Alzoubi, H. M., Alshurideh, M., & Ghazal, T. M. (2021a). Integrating BLE beacon technology with intelligent information systems IIS for Operations' performance: a managerial perspective. In: *The International Conference on Artificial Intelligence and Computer Vision*, 527–538.

Alzoubi, H. M., & Aziz, R. (2021). Does emotional intelligence contribute to quality of strategic decisions? the mediating role of open innovation. *Journal of Open Innovation: Technology, Market, and Complexity*, 7(2), 130. https://doi.org/10.3390/joitmc7020130

Alzoubi, H. M., Vij, M., Vij, A., & Hanaysha, J. R. (2021b). What leads guests to satisfaction and loyalty in UAE five-star hotels? AHP analysis to service quality dimensions. *Enlightening Tourism, 11*(1), 102–135. https://doi.org/10.33776/et.v11i1.5056

Alzoubi, H. M., & Yanamandra, R. (2020). Investigating the mediating role of information sharing strategy on agile supply chain. *Uncertain Supply Chain Management, 8*(2), 273–284. https://doi.org/10.5267/j.uscm.2019.12.004

Alzoubi, H., Ahmed, G., Al-Gasaymeh, A., & Kurdi, B. (2020a). Empirical study on sustainable supply chain strategies and its impact on competitive priorities: The mediating role of supply chain collaboration. *Management Science Letters, 10*(3), 703–708.

Alzoubi, H., Alshurideh, M., Kurdi, B., Akour, I., & Aziz, R. (2022). Does BLE technology contribute towards improving marketing strategies, customers' satisfaction and loyalty? The role of open innovation. *International Journal of Data and Network Science, 6*(2), 449–460.

Alzoubi, H., & Ahmed, G. (2019). Do TQM practices improve organisational success? A case study of electronics industry in the UAE. *International Journal of Economics and Business Research, 17*(4), 459–472. https://doi.org/10.1504/IJEBR.2019.099975

Alzoubi, H., Alshurideh, M., Kurdi, B. A., & Inairat, M. (2020b). Do perceived service value, quality, price fairness and service recovery shape customer satisfaction and delight? A practical study in the service telecommunication context. *Uncertain Supply Chain Management, 8*(3), 579–588. https://doi.org/10.5267/j.uscm.2020.2.005

Aziz, N., & Aftab, S. (2021). Data mining framework for nutrition ranking: methodology: SPSS modeller. *International Journal of Technology, Innovation and Management (IJTIM), 1*(1), 85–95.

Cruz, A. (2021). Convergence between blockchain and the internet of things. *International Journal of Technology, Innovation and Management (IJTIM), 1*(1), 35–56.

Eli, T. (2021). Students perspectives on the use of innovative and interactive teaching methods at the university of nouakchott al aasriya, mauritania: english department as a case study. *International Journal of Technology, Innovation and Management (IJTIM), 1*(2), 90–104.

Farouk, M. (2021). The universal artificial intelligence efforts to face coronavirus COVID-19. *International Journal of Computations, Information and Manufacturing (IJCIM), 1*(1), 77–93. https://doi.org/10.54489/ijcim.v1i1.47

Ghazal, T., Alshurideh, M., & Alzoubi, H. (2021a). Blockchain-enabled internet of things (IoT) platforms for pharmaceutical and biomedical research. *The International Conference on Artificial Intelligence and Computer Vision*, 589–600.

Ghazal, T. M. (2021). Positioning of UAV base stations using 5g and beyond networks for IoMT applications. *Arabian Journal for Science and Engineering*, 1–12.

Ghazal, T. M., Anam, M., Hasan, M. K., Hussain, M., Farooq, M. S., Ali, H. M., Ahmad, M., & Soomro, T. R. (2021b). Hep-Pred: Hepatitis C staging prediction using fine gaussian SVM. *Comput Mater Continua, 69*, 191–203.

Ghazal, T. M., Hasan, M. K., Alshurideh, M. T., Alzoubi, H. M., Ahmad, M., Akbar, S. S., Al Kurdi, B., & Akour, I. A. (2021c). IoT for smart cities: machine learning approaches in smart healthcare—a review. *Future Internet, 13*(8), 218. https://doi.org/10.3390/fi13080218

Ghazal, T. M., Hussain, M. Z., Said, R. A., Nadeem, A., Hasan, M. K., Ahmad, M., Khan, M. A., & Naseem, M. T. (2021d). *Performances of K-means clustering algorithm with different distance metrics.*

Ghazal, T. M., Said, R. A., & Taleb, N. (2021e). Internet of vehicles and autonomous systems with AI for medical things. *Soft Computing*, 1–13.

Ghazal, T., Soomro, T. R., & Shaalan, K. (2013). Integration of project management maturity (PMM) based on capability maturity model integration (CMMI). *European Journal of Scientific Research, 99*(3), 418–428.

Guergov, S., & Radwan, N. (2021). Blockchain convergence: analysis of issues affecting IoT, AI and Blockchain. *International Journal of Computations, Information and Manufacturing (IJCIM), 1*(1), 1–17. https://doi.org/10.54489/ijcim.v1i1.48

Hamadneh, S., Keskin, E., Alshurideh, M., Al-Masri, Y., & Al Kurdi, B. (2021a). The benefits and challenges of RFID technology implementation in supply chain: A case study from the Turkish construction sector. *Uncertain Supply Chain Management, 9*(4), 1071–1080.

Hamadneh, Samer, Pedersen, O., & Al Kurdi, B. (2021b). An investigation of the role of supply chain visibility into the scottish bood supply chain. *Journal of Legal, Ethical and Regulatory Issues, 24*(Special Issue 1), 1–12.

Hanaysha, J. R., Al-Shaikh, M. E., Joghee, S., & Alzoubi, H. (2021a). Impact of Innovation Capabilities on Business Sustainability in Small and Medium Enterprises. *FIIB Business Review*, 1–12. https://doi.org/10.1177/23197145211042232

Hanaysha, J. R., Al Shaikh, M. E., & Alzoubi, H. M. (2021b). Importance of marketing mix elements in determining consumer purchase decision in the retail market. *International Journal of Service Science, Management, Engineering, and Technology (IJSSMET), 12*(6), 56–72.

Joghee, S., Alzoubi, H. M., & Dubey, A. R. (2020). Decisions effectiveness of FDI investment biases at real estate industry: Empirical evidence from Dubai smart city projects. *International Journal of Scientific and Technology Research, 9*(3), 3499–3503.

Kalra, D., Ghazal, T. M., & Afifi, M. A. M. (2020). Integration of collaboration systems in hospitality management as a comprehensive solution. *International Journal of Advanced Science and Technology, 29*(8s), 3155–3173.

Kashif, A. A., Bakhtawar, B., Akhtar, A., Akhtar, S., Aziz, N., & Javeid, M. S. (2021). Treatment response prediction in hepatitis C patients using machine learning techniques. *International Journal of Technology, Innovation and Management (IJTIM), 1*(2), 79–89. https://doi.org/10.54489/ijtim.v1i2.24

Khan, M. A. (2021). Challenges facing the application of IoT in medicine and healthcare. *International Journal of Computations, Information and Manufacturing (IJCIM), 1*(1), 39–55. https://doi.org/10.54489/ijcim.v1i1.32

Khan, M. F., Ghazal, T. M., Said, R. A., Fatima, A., Abbas, S., Khan, M. A., Issa, G. F., Ahmad, M., & Khan, M. A. (2021a). An IoMT-Enabled smart healthcare model to monitor elderly people using machine learning technique. *Computational Intelligence and Neuroscience, 2021a.*

Khan, Q.-T.-A., Ghazal, T. M., Abbas, S., Khan, W. A., Khan, M. A., Said, R. A., Ahmad, M., & Asif, M. (2021b). *Modeling Habit Patterns Using Conditional Reflexes in Agency.*

Lee, C., & Ahmed, G. (2021). Improving IoT privacy, data protection and security concerns. *International Journal of Technology, Innovation and Management (IJTIM), 1*(1), 18–33. https://doi.org/10.54489/ijtim.v1i1.12

Lee, K., Azmi, N., Hanaysha, J., Alzoubi, H., & Alshurideh, M. (2022a). The effect of digital supply chain on organizational performance: An empirical study in Malaysia manufacturing industry. *Uncertain Supply Chain Management, 10*(2), 495–510.

Lee, K., Romzi, P., Hanaysha, J., Alzoubi, H., & Alshurideh, M. (2022b). Investigating the impact of benefits and challenges of IOT adoption on supply chain performance and organizational performance: An empirical study in Malaysia. *Uncertain Supply Chain Management, 10*(2), 537–550.

Lee, S.-W., Hussain, S., Issa, G. F., Abbas, S., Ghazal, T. M., Sohail, T., Ahmad, M., & Khan, M. A. (2021). Multi-Dimensional trust quantification by artificial agents through evidential fuzzy multi-criteria decision making. *IEEE Access, 9*, 159399–159412.

Matloob, F., Ghazal, T. M., Taleb, N., Aftab, S., Ahmad, M., Khan, M. A., Abbas, S., & Soomro, T. R. (2021). Software defect prediction using ensemble learning: A systematic literature review. *IEEE Access.*

Mehmood, T. (2021). Does information technology competencies and fleet management practices lead to effective service delivery? empirical evidence from e-commerce industry. *International Journal of Technology, Innovation and Management (IJTIM), 1*(2), 14–41.

Mehmood, T., Alzoubi, H. M., & Ahmed, G. (2019). Schumpeterian entrepreneurship theory: evolution and relevance. *Academy of Entrepreneurship Journal, 25*(4).

Miller, D. (2021). The best practice of teach computer science students to use paper prototyping. *International Journal of Technology, Innovation and Management (IJTIM)*, *1*(2), 42–63. https://doi.org/10.54489/ijtim.v1i2.17

Mondol, E. P. (2021). The impact of block chain and smart inventory system on supply chain performance at retail industry. *International Journal of Computations, Information and Manufacturing (IJCIM)*, *1*(1), 56–76. https://doi.org/10.54489/ijcim.v1i1.30

Naqvi, R., Soomro, T. R., Alzoubi, H. M., Ghazal, T. M., & Alshurideh, M. T. (2021). The nexus between big data and decision-making: a study of big data techniques and technologies. In: *The International Conference on Artificial Intelligence and Computer Vision*, 838–853.

Nuseir, M. T., Al Kurdi, B. H., Alshurideh, M. T., & Alzoubi, H. M. (2021). Gender discrimination at workplace: do artificial intelligence (AI) and machine learning (ML) have opinions about it. In: *The International Conference on Artificial Intelligence and Computer Vision*, 301–316.

Obaid, A. J. (2021). Assessment of smart home assistants as an IoT. *International Journal of Computations, Information and Manufacturing (IJCIM)*, *1*(1), 18–36. https://doi.org/10.54489/ijcim.v1i1.34.

Radwan, N., & Farouk, M. (2021). The growth of internet of things (IoT) in the management of healthcare issues and healthcare policy development. *International Journal of Technology, Innovation and Management (IJTIM)*, *1*(1), 69–84. https://doi.org/10.54489/ijtim.v1i1.8.

Rehman, E., Khan, M. A., Soomro, T. R., Taleb, N., Afifi, M. A., & Ghazal, T. M. (2021). Using blockchain to ensure trust between donor agencies and ngos in under-developed countries. *Computers, 10*(8), 98.

Salloum, S. A., Alshurideh, M., Elnagar, A., & Shaalan, K. (2020). Mining in educational data: review and future directions. In *Advances in Intelligent Systems and Computing: Vol. 1153 AISC*. https://doi.org/10.1007/978-3-030-44289-7_9.

Shamout, M., Ben-Abdallah, B., Alshurideh, M., Alzoubi, H., Al Kurdi, B., & Hamadneh, S. (2022). A conceptual model for the adoption of autonomous robots in supply chain and logistics industry. *Uncertain Supply Chain Management, 10*, 1–16.

Suleman, M., Soomro, T. R., Ghazal, T. M., & Alshurideh, M. (2021). Combating against potentially harmful mobile apps. In: *The International Conference on Artificial Intelligence and Computer Vision*, 154–173.

Machine Learning-Based Intrusion Detection Approaches for Secured Internet of Things

Taher M. Ghazal, Mohammad Kamrul Hasan,
Siti Norul Huda Sheikh Abdullah, Khairul Azmi Abu Bakar,
Nidal A. Al-Dmour, Raed A. Said, Tamer Mohamed Abdellatif,
Abdallah Moubayed, Haitham M. Alzoubi⊙, Muhammad Alshurideh⊙,
and Waleed Alomoush

Abstract Nowadays, protecting communication and information for Internet of Things (IOT) has emerged as a critical challenge. Existing systems use firewalls to ensure that they are safe from any unexpected occurrences that may disrupt the desired systems and applications. Intrusion detection systems (IDSs) are an acceptable second line of defence for IOT applications. IDS play a crucial role ensuring that

T. M. Ghazal · W. Alomoush
School of Information Technology, Skyline University College, Sharjah, UAE
e-mail: taher.ghazal@skylineuniversity.ac.ae

W. Alomoush
e-mail: waleed.alomoush@skylineuniversity.ac.ae

T. M. Ghazal · M. K. Hasan · S. N. H. S. Abdullah · K. A. A. Bakar
Faculty of Information Science and Technology, Center for Cyber Security, University Kebansaan Malaysia (UKM), Bangi, Malaysia
e-mail: mkhasan@ukm.edu.my

S. N. H. S. Abdullah
e-mail: Snhsabdullah@ukm.edu.my

K. A. A. Bakar
e-mail: khairul.azmi@ukm.edu.my

N. A. Al-Dmour
Department of Computer Engineering, College of Engineering, Mutah University, Mu'tah, Jordan

R. A. Said
Faculty of Management, Canadian University Dubai, Dubai, UAE

T. M. Abdellatif
Faculty of Engineering, Applied Science and Technology, Canadian University Dubai, Dubai, UAE
e-mail: tamer.mohamed@cud.ac.ae

A. Moubayed
Electrical & Computer Engineering Department, University of Western Ontario, Ontario, ON, Canada
e-mail: amoubaye@uwo.ca

M. Alshurideh et al. (eds.), *The Effect of Information Technology on Business and Marketing Intelligence Systems*, Studies in Computational Intelligence 1056,
https://doi.org/10.1007/978-3-031-12382-5_110

it enhances the IOT security level maintaining sophisticated framework. Attackers have continuously been attempting to determine novel ways to circumnavigate security frameworks that prevent the structures. This paper reviews the security advances, threats and countermeasures for the IOT applications. A state of art review has accomplished using the references from 2009 to 2020 to encompass the real demography of the IOT security research data. This work also highlights the deep learning-based intrusion detection approaches for Internet of Things (IOT) security. With the systematic literature review approach, the review suggests that implementing existing security measures, such as encryption, authentication, access control, network and application security for IoT systems and their intrinsic amenability is ineffective for the IOT systems.

Keywords Machine learning · Internet of Things security · Intrusion detection

1 Introduction

The extent of the use of the Internet of Things (IOT) has made it imperative to have several devices on a network. As such, most of these are computers, but there are sensors, digital tools, and vehicles (Ghazal et al., 2021a, 2021b, 2021c, 2021d, 2021e; Lee et al., 2022a, 2022b). The large size of the network of devices and anonymous or uncontrolled structure of the Internet is imperative to consider (Al Kurdi et al., 2021; Alshurideh et al., 2019, 2021). Protecting data and communication systems is imperative in IoT security. A plethora of frameworks used in firewalls in IoT are useful in protecting each of these systems. Intrusion Detection Systems (IDS) are useful in IoT security and ensure that these frameworks are safe all the time. In this regard, attackers have continuously been attempting to determine novel ways to bypass the systems that prevent the structures (Akhtar et al., 2021; Al Ali, 2021; AlHamad et al., 2022; Alhamad et al., 2021). The IOT system requires ensuring that they have kept malware out of their operations. It is of the essence to achieve this objective because it means that they focus on ensuring that they have kept their devices and systems safe. Spark-Chi-Support-Vector-Machine is an example of an IDS algorithm (Ali et al., 2022; Alnazer et al., 2017; Alnuaimi et al., 2021; Alsharari, 2021; Alshurideh et al., 2022a, 2022b). The model reduces training time, increases performance, and is ideal for big data. It undergoes loading a dataset and exporting it in the resilient distributed datasets and data frame that deploys apache spark.

H. M. Alzoubi (✉)
School of Business, Skyline University College, Sharjah, UAE
e-mail: haitham.alzubi@skylineuniversity.ac.ae

M. Alshurideh
Department of Marketing, School of Business, The University of Jordan, Amman 11942, Jordan
e-mail: m.alshurideh@ju.edu.jo; malshurideh@sharjah.ac.ae

Department of Management, College of Business, University of Sharjah, 27272 Sharjah, United Arab Emirates

The data then undergoes pre-processing before the selection of features, training, testing, and evaluating in another dataset. It is an incremental task to ensure that intrusion detection systems are of the highest capabilities from the perspective of IOT security. However, it does not imply that companies should avoid trying to understand the implications of failing to detect intrusions into the systems that they have set. If the algorithms that they are deploying at a specific time have been outdated, it becomes easy for an attacker to take advantage of the system by attacking it at any given time. As a matter of consequence, the use of machine learning techniques in intrusion detection systems should help IoT companies avoid unwanted issues that may occur over time (Alshurideh et al., 2020a, 2020b; Akour et al., 2021; Nuseir et al., 2021; Salloum et al., 2020) Consequently, the essence of this paper is to outline the usefulness of machine learning in intrusion detection systems. As such, it will focus on deep learning and how best it is useful in the context of intrusion detection systems for IOT systems to protect systems and valuable information (Alshurideh et al., 2020a, 2020b; Ali Alzoubi, 2021a, 2021b; Alzoubi et al., 2020a, 2020b, 2022).

Big data analytics has become imperative in the analysis of data generated from IoT. Connected devices help in the process of data generation, and the analysis is imperative in improving the decision-making process. The role of big data in IoT is to ensure that the data has undergone processing in real-time. The data then undergoes storage using a variety of technologies available in the current world. IoT has become a backbone of the big data trend with output information becoming the data or input for another system. The data is further useful in guiding specific processes. IDS in IoT can be useful in enforcing its security, as shown in Fig. 1—for example, it is helpful in detecting attacks once a hacker breaks the cryptography (Alzoubi & Ahmed, 2019; Alzoubi & Aziz, 2021; Alzoubi & Yanamandra, 2020; Alzoubi et al., 2020a, 2020b, 2021a).

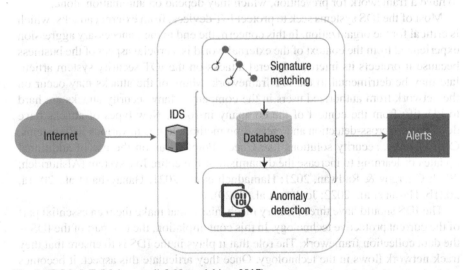

Fig. 1 IDS in IoT (Mohammadi & Namadchian, 2017)

The contribution of this paper is as follows:

1. An in-depth discussion of recent algorithm and techniques used in big data incentives.
2. Taxonomy of deep learning and its usefulness in intrusion detection systems.
3. Discuss the recent IDS threats; advance techniques and figuring out the research directions and its limitations in IOT & block chain applications.

2 Literature Review

Block-Chain-based security solutions for IoT systems help secure peer-to-peer networks. The blocks are composed of transaction records that users initiate. They help prevent unauthorised operations on IoT data by using blocks to register all transactions. The evolution of Intrusion Detection Systems hinges on the fact that there are specific elements that control them. The IOT network systems have increased in complexity to the point that they are prone to mistakes. In turn, these errors are easy to exploit from the context of attackers or intruders. The present IOT network systems have several critical deficiencies in their security networks, which makes it easy for attackers to target them for their gain (Aziz & Aftab, 2021; Cruz, 2021; Eli, 2021; Farouk, 2021; Ghazal et al., 2021a, 2021b, 2021c, 2021d, 2021e). Although there is a need for additional work or tools that attempt to find and fix deficiencies, it is an incremental task to close all existing loopholes. Despite that intrusion detection systems exist, absolute prevention in a specific system is an additional task. As a matter of consequence, intrusion detection systems have emerged as the best way to find and identify any intrusion in a specific system. Afterward, it becomes imperative to have a framework for prevention, which may depend on automation alone.

Most of the IDS systems seek to protect IoT devices, from external attacks, which is critical for the organization. In this concern, the end to the unnecessary aggression experienced from the context of the external world is a crucial aspect of the business because it protects its interests. Several attacks on the IOT security system articulate may be detrimental to the IoT framework. Many of the attacks may occur on the network from authorized users in the company. Many security attacks are hard to identify from the context of the company in focus. New types of attacks have developed in cross-detection and prevention methodologies in various IoT systems. Consequently, security solutions like Check Point focus on the use of additional updates or learning to increase the dynamism of the entire IoT system (Alshurideh, 2022; Guergov & Radwan, 2021; Hamadneh et al., 2021; Hanaysha et al., 2021a, 2021b; Hasan et al., 2022; Joghee et al., 2020).

The IDS should have three primary mechanisms that make them an essential part of the current protective technology. In this contemplation, the first part of the IDS is the data collection framework. The role that it plays in the IDS is to ensure that they track network flows in the technology. Once they articulate this aspect, it becomes critical to transmitting the data in the form of packets. The second part of the IDS is vectorization, where data is used for various features; as a matter of consequence,

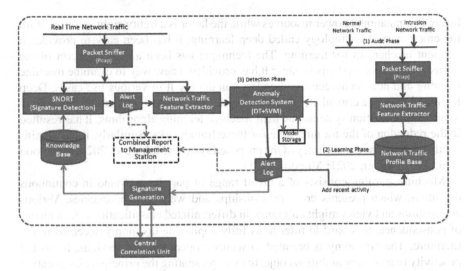

Fig. 2 IDS framework

they create feature vectors. The final aspect of IDS is the classification engine that gives classified packets to state whether the system is standard or there has been an intrusion. The use of vectors in the engine traces the flow of all packets within the system using previous knowledge to identify if there has been an intrusion or if the system is typical, as shown in IDS framework, Fig. 2.

The classification engine is an imperative aspect of the intrusion detection system. From a notable perspective, it decides if the converted features are part of an intrusion in the specific locality or if normalcy is present. the part is signature-based, which has the implications of classifications following anomalies or defined signatures based on the normal flow of packets in the systems (Kashif et al., 2021; Khan, 2021; Lee & Ahmed, 2021; Lee et al., 2022a, 2022b) the first one has excellent reliability and decision time, but it cannot detect new types of attacks; as a matter of consequence, it produces a low detection rate. Simultaneously, the latter has low acceptable scalability, robustness, and flexibility. As a matter of consequence, each of these systems needs to imminent the correct dynamism to ensure that they reach the highest level of performance. Majorly, they need to use the best techniques to ensure that they follow machine-learning mechanisms.

Machine learning-based IDSs for enhancing the IOT security have become critical challenge for the industries. One of the crucial aspects of such systems is the need to have a significant training time for processing a large dataset from the network's previous flow of data. Traditional machine learning approaches include decision trees, logistic regression, and support vector machine (SVM). SVM is an algorithm that finds optimized hyper-planes to classify between two classes. Building an excellent SVM model majorly depends on finding an excellent kernel. Random Forest is built on several decision trees that combine the prediction of a single tree. Xgboost and Naïve Bayes are based on machine learning whereby the former is useful for heavy computational power resources while the latter is a lightweight solution. With the emergence of technology called deep learning, it has been easy to provide an efficient mechanism for learning. The technique has been a critical aspect of the intrusion detection systems because it has provided a new way to examine machine learning and how technicians and coders can deploy it in various use cases. Deep learning has been a critical aspect of these systems because it is a parallel technology to intrusion detection systems and their machine learning algorithms. It has resulted in the reduction of the training time for these frameworks. Similarly, it has significantly helped raise the accuracy of the proposed system (Mehmood, 2021; Mehmood et al., 2019; Miller, 2021; Mondol, 2021).

Machine learning consists of a broad range of paradigms found in continuous evolution, which presents cross relationships and weakened boundaries. Various applications and views might culminate in differentiated classifications. As a matter of consequence, it is hard to refer to as fully applicable taxonomy accepted in the literature. The taxonomy is oriented to security operations professions. It has the proclivity to avoid any ambitious objectives of presenting the principal classification for all application use cases and artificial technology professionals. Machine learning is divided into shallow learning and deep learning alternatives. Shallow learning has supervised and unsupervised variants. The supervised version uses shallow neural networks, K-nearest neighbour, hidden Markov models, Random Forests, support vector machines, logistic regression, and Naïve Bayes. The unsupervised version has an association and clustering as examples. Deep learning also has supervised and unsupervised variants. Supervised deep learning has recurrent deep neural networks, feedforward, convolutional neural networks, and feedforward, fully connected neural networks. Unsupervised deep learning has deep belief networks and stacked Auto-Encoders. Unsupervised deep learning techniques do not need a representative and large dataset than that which human experts have previously classified. In contrast to that, supervised deep learning variants do not need a pre-labelled training dataset. Shallow neural networks are algorithms based on neural networks and have a pre-set of processing elements called neurons, which are organized into two or more communication layers. They include all neural networks with a limited number of layers and neurons. Despite their existence, they have been most useful as classification tools in cyber security. K-Nearest Neighbour is useful in various classification models and may be imperative in problems that relate to multiple classes. In contrast to that, a significant issue exists in the testing and training phases of the software. From a notable perspective, it is computationally demanding to classify all test samples and compare them against all the training samples used in the previous phase.

The Hidden Markov Model is an imperative model that entails a set of states to produce outputs by deploying a range of probabilities. The objective is to determine the states' sequence produced at observable outputs. The model is a useful way of understanding temporary behaviours and for quantifying the probabilities of specified event sequences. The latter use cases have been the most applicable in the context of cyber security despite that training of the model can occur on datasets with or without labels (Ahmad et al., 2021; Obaid, 2021; Radwan & Farouk, 2021; Shamout et al., 2022; Tariq et al., 2022).

Random Forest is a set of decision trees that contemplates the outputs of all tresses before giving a final response, which is typically unified. All decision trees are conditional classifiers, which mean that they are visited from the top, and at every node. The given conditions are checked against one or more aspects of the data that has been analysed. The methodology is useful for a large dataset and excels at several issues in the multiple class segments. In contrast to that, over fitting is a critical aspect of deeper trees and needs articulation to attain its goal. Support vector machines are classifiers with a non-probabilistic take on data. They are useful mapping samples of datasets in feature spaces with the objective of distance maximization between sample categories. There is no assumption on input features, but the performance is unfitting for multiple class classifications. The limitation in scalability might culminate in long processing time.

Logistic regression is a categorical classifier that has the objective of adopting a discriminative model (Afifi et al., 2020; Ali et al., 2021; Alzoubi et al., 2021b). It is similar to the Naïve Bayes algorithm in that it makes the prior independence assumption of features of inputs. The performance primarily depends on the training data's size (Al Batayneh et al., 2021; Ghazal, 2021; Ghazal et al., 2021a, 2021b, 2021c, 2021d, 2021e). In this consideration, the Naïve Bayes algorithm is a proba-bilistic classifier that makes prior assumptions that input features of a dataset have independence. In this concern, therefore, these are scalable and do not need a large training dataset to ensure that the results they produce are appreciable. The gap in the literature is a lack of succinct consideration of deep learning methodologies for intrusion detection systems.

3 Methodology

The methodology of choice for this paper was a review of the literature concerning the use of deep learning approach to implementing an intrusion detection system for information systems security. The objective was to find papers that discussed deep learning algorithms used in intrusion detection systems. The researcher was determined to find 56 updated articles from the year 2009 onward. Most of the deep learning algorithms have been present in the contemporary world from this time onwards. As a matter of consequence, it was imperative to include articles in English to ease readability and understanding. As a matter of impact, the researcher excluded articles written in a language rather than English. The database deployed for the research was Google Scholar, since it is a free database, and it offers peer-reviewed articles. These were critical in the study because they made sure of peer-reviewed information being used.

Research ethics is a critical aspect of a study that needs to be considered. The current research did not meet any ethics issues that were worthy of attention. The present paper did not need research ethics to be met at any point. There was no need for respondents to take part in the study because it would have mandated the use of a consent form to which they would agree to participate in the research or reject it outright based on personal preference. The choice not to use respondents meant that it was relatively easy to conduct the study from the viewpoint of secondary analysis of data. The information obtained from the investigation would be deployed in helping understand the impacts of deep learning in intrusion detection systems.

4 Analysis

Intrusion detection SYSTEMS are a plethora of examples, but the most common are fuzzy logic, rule-based frameworks, support vector machines, artificial neural networks, and statistical intrusion detection systems. Fuzzy logic relies based on human thinking intending to convert information obtained into mathematical equivalents. Persons who have a specialty in one specific area develop rule-based systems. Their role is to examine the traffic in particular systems. Then, they develop attack detention and specific rules that the system should follow. Data mining is a technique used for reaching out for information in a big data setting. The data extraction depends on a set of rules whereby the objective is to ensure that the correlation between users and data has been established (Alshurideh et al., 2022a, 2022b; Ghazal et al., 2013, 2021a, 2021b, 2021c, 2021d, 2021e). It uses the fuzzy set theory to determine the usefulness of fuzzy logic in data. Support Vector Machines are the most preferred methodology for intrusion detection systems. They can select a specific feature vector, and they can distinguish data from two class sets using a particular feature vector for each one of them. They are deployed in several classification systems like sound analysis and face recognition systems. Artificial neural networks are models that give graphs of neurons of data. They have a way to correlate specific vectors with algorithms that help create new data. The method is imperative in examining and learning data behaviours in a system. An enhanced form of artificial neural networks has been preferred in data mining contexts due to increased efficiency. Statistical intrusion detection systems are based on measurements in statistics. In this context, the examination of system or user behaviour is imperative because it helps create a specific statistical model. New intrusions correlate with a particular mathematical model. Some of the methodologies used in intrusion detection systems include Gaussian Mixture Distribution, Chi-square distribution, and Principal Component Analysis (Kalra et al., 2020; Khan et al., 2021a, 2021b).

An intrusion detection system that is based on a signature is useful in detecting attacks that have already been identified. The objective is to reduce the false rates while increasing the detection accuracy, which needs to be considerably high. In contrast to that, performance may suffer when the detection of a new or unknown attack occurs. The system typically has a limitation in the rules used and can be installed beforehand in the intrusion detection system (Lee et al., 2021; Matloob et al., 2021; Naqvi et al., 2021). From another perspective, an intrusion detection system based on an anomaly is best suited for a context, whereby the attack is new or unknown. Although such a type of an intrusion detection system is best suited for such settings and has a high false-positive capability, it has broad acceptance in the current research community. Two challenges arise when developing a flexible and effective intrusion detection system to get to understand new issues. The first of these issues is the proper feature selection from the dataset of network traffic to detect the anomaly. The second problem is the lack of availability of a labelled traffic dataset from a real network that should be useful in the development of the intrusion detection system. Deep learning is a technique in the class of machine learning methodologies; it deploys consecutive information layer and processing levels in the hierarchy. The objective is to help in the representation and learning of features as well as the classification of patterns. From a notable perspective, the essence of deep learning is to help the classification results of images to increase. Besides, technology is useful in signal processing, natural language processing, video, images, audio, and speech. It is also helpful in pattern recognition, graphic modelling, and language, which are useful for intrusion detection systems. It means that several deep learning methodologies include Deep Auto Encoder, Auto Encoder, Deep Neural Network, Deep Boltzmann Machine, Restricted Boltzmann Machine, and Deep Belief Network.

Several advancements exist for algorithms used in learning contexts to improve the capabilities within the intrusion detection system that help it reach a low false alarm rate and a high detection rate. From a notable perspective, the deep learning-based approach is useful in overcoming issues that hinder the development of an efficient intrusion detection system. The unbiased comparison of the efficiency of two deep learning algorithms expects that their training be on a similar dataset. Similarly, their testing needs to occur on the same dataset. Although several proposals in the context of cyber security are prevalent in the contemporary world, they rely on a few and old public datasets. While this may appear to be a mistake in the modern manufacture of the intrusion detection systems, the results are hard to compare due to a plethora of issues. First, they may deploy a different split between the dataset used for training and that deployed in the testing phase. Second, one or both of these algorithms might implement a pre-filtering operation that can alter the dataset used in the training phase. Finally, the algorithms used may have differentiated featured. For these reasons, a meaningful comparison between the detection performances of various intrusion detection systems is an incremental task. In this context, the literature may discuss machine-learning methods for two distinct cyber security issues. However, it may fail to consider the training and testing phases for the analysis deployed. As a matter of consequence, some of the solutions obtained may have a higher level of accuracy than others, which is highly problematic. It means that the results might change significantly if the training settings have been differentiated. Additionally, there is a lack of a guarantee that a specific method that works on a given dataset is the best when used in some other datasets. While it may work for one set of data, it might fail to depict the same performance when used on a different one. Security administration professionals have noted this problem, but some remain unaware of its existence. It is imperative to thoroughly query the methods used in the evaluation phase before accepting the performance results from various deep learning algorithms (Rehman et al., 2021; Suleman et al., 2021).

Misclassification in the cyber security context may have an implicit cost that might result in serious problems. A false positive in intrusion detection systems and malware classification may become the point of annoyance for security operators because it hinders the remediation in the contexts of actual infections. It is critical to get the correct information in all cases to ensure that the intrusion detection system achieves its objective within the shortest time possible. In this context, the phishing detection may cause legitimate and essential messages to fail to reach the specific individuals who need it. In this context, the failure to detect malware may have significant issues in the organization. Notably, phishing emails and network intrusions can be the point of compromise for entire organizations rather than just one aspect of the company in focus. Companies need to be ready to consider the performance of various deep learning algorithms and solutions in the context of organizations. It should be noted that they should then devote them to phishing detection and malware analysis in the context of the systems they have set. It should be noted that they need to perform experiments for intrusion detection systems with the orientation of detecting problems before they can implement them in specific use cases. In this context, the objective is to reduce problems that may make it hard for the organization to exist as a business entity.

Traditional intrusion detection systems have been a point of limitation for a plethora of companies. It should be noted that they are based on static detection rules, which are less than ideal for detecting malware and phishing attacks. There is a need for continuous and frequent updates in a specific system that is imperative for the detection of malware and phasing. For example, companies need daily updates of virus definitions in their antivirus libraries. If they fail to achieve this objective, it is highly likely that they cannot meet the ultimate goal of ensuring that they have the best performance. It should be noted that they could fail to get to know the specific malware they are facing, which may hinder their ability to detect specific instances of malware. When this happens, they compromise the entire system, and an intruder takes advantage of the whole company's security defences.

A similar problem influences advanced deep learning incentives that companies may want to ensure they keep malware out of their existing systems. Some companies may be relying on outdated datasets used for the training phase but think they will achieve the same goals they expect from the framework. It is a critical issue that affects companies because they need to supervise their learning approaches, which should help label training datasets in the best way possible. Creating a manual for similar datasets is expensive for most companies. The implication of cost lies in the fact that the dataset should be considerably large and comprehensive, which should allow the deep learning algorithm to know how to differentiate between various classes. Additionally, the operations are prone to errors and may result in incorrectness in the classification made. Lastly, most companies are unwilling to share data about internal networks because they believe it may give other firms an advantage over their existing data, which could compromise them. The scenario results in an overall scarcity of publicly available data that may be labelled for cyber security operations. The definitive implication is that the periodic retraining of a dataset to ensure organizations use deep learning incentives correctly might be impossible or extremely difficult.

5 Discussion and Open Issues

The IDSs are of various forms, which make them usable in the context of IoT frameworks. The usefulness of these systems depends on the specific use cases, where they are applied. One algorithm may fail to work in a context different from that for which it is designed. Thus, care should be taken to ensure that the correct methodology is deployed in a specific context. The failure to perform an analysis on the issues that may occur may result in unwanted complications. Table 1 is the summary of most related work from the literature along with techniques it also discusses the advantages and disadvantages of the machine learning algorithms.

6 Conclusion

IoT connects computers, but there are sensors, digital tools, and vehicles. IDS are an acceptable second line of defence for a plethora of companies. Big data analytics has become imperative in the analysis of data from IoT. IoT is the backbone of the big data trend with output information becoming the data or input for another system. The paper discusses recent algorithm and techniques used in big data incentives, the taxonomy of deep learning and its usefulness in IDS, and recent IDS threats. Examples of IDS are fuzzy logic, rule-based frameworks, support vector machines, artificial neural networks, and statistical intrusion detection systems. The basis of an IDS is on a signature is useful in detecting attacks that have already been identified. Developing excellent IDS expects understanding proper feature selection from the dataset of network traffic to detect the anomaly and the availability of a labelled traffic dataset from a real network that should be useful in the development of the intrusion detection system. Companies want to use these systems but are unwilling to share their data with competitors. Others fear the cost of installing IDS that will help keep their IoT devices safe. Unless the situation changes, it will be increasingly hard to implement IDS in IoT contexts.

Table 1 Summary of the most related works

Authors and year	Features	Advantages	Disadvantages
Munoz-Gonza´lez et al. (2017)	Convolutional neural network	High error detection rate	No way to be spatially invariant to the input data, cannot encode the orientation or position of the object, circumventing the problems needs adding more data
Tariq et al. (2022)	Convolutional neural network	Accurate, precise	No way to be spatially invariant to the input data, cannot encode the orientation or position of the object, circumventing the problems needs adding more data
Li et al. (2017)	Convolutional neural network	Accurate	No way to be spatially invariant to the input data, cannot encode the orientation or position of the object, circumventing the problems needs adding more data
Miller (2021)	Conditional deep belief networks	Accurate	Require a large volume of data, circumventing the problems needs adding more data
Mathew et al. (2017)	Conditional deep belief networks	Accurate	Require a large volume of data, circumventing the problems needs adding more data
Shamout et al. (2022)	Conditional deep belief networks	True positive detection	Require a large volume of data, circumventing the problems needs adding more data
Liu et al. (2017)	Conditional deep belief networks	Precise	Require a large volume of data, circumventing the problems needs adding more data
Mehmood et al. (2019)	Conditional deep belief networks	Accurate	Require a large volume of data, circumventing the problems needs adding more data

(continued)

Table 1 (continued)

Authors and year	Features	Advantages	Disadvantages
Wang et al. (2017)	Deep neural network	True positive detection	Limited capacity to transfer data, lack of transparency, learning requires a large data sample, circumventing the problem requires using a large training dataset and an expert
Rehman et al. (2021)	Deep neural network	Accurate	Limited capacity to transfer data, lack of transparency, learning requires a large data sample, circumventing the problem requires using a large training dataset and an expert
Saxe & Berlin (2015)	Deep neural network	Low training time and detection accuracy	Limited capacity to transfer data, lack of transparency, learning requires a large data sample, circumventing the problem requires using a large training dataset and an expert
Matloob et al. (2021)	Deep neural network	Classification accuracy	Limited capacity to transfer data, lack of transparency, learning requires a large data sample, circumventing the problem requires using a large training dataset and an expert
Wu et al. (2017)	Deep neural network	Accuracy	Limited capacity to transfer data, lack of transparency, learning requires a large data sample, circumventing the problem requires using a large training dataset and an expert
Obaid (2021)	Stacked auto-encoder	Accuracy	Limited capacity to transfer data, lack of transparency, learning requires a large data sample, circumventing the problem requires using a large training dataset and an expert

(continued)

Table 1 (continued)

Authors and year	Features	Advantages	Disadvantages
Wei & Mendis (2016)	Stacked auto-encoder	Detection accuracy	Limited capacity to transfer data, lack of transparency, learning requires a large data sample, circumventing the problem requires using a large training dataset and an expert
Lee et al. (2021)	Stacked auto-encoder	Detection accuracy	Limited capacity to transfer data, lack of transparency, learning requires a large data sample, circumventing the problem requires using a large training dataset and an expert
Zolotukhin et al. (2016)	Stacked auto-encoder	Classification accuracy	Limited capacity to transfer data, lack of transparency, learning requires a large data sample, circumventing the problem requires using a large training dataset and an expert
Naqvi et al. (2021)	Relevance deep learning	False positives and detection accuracy	Limited capacity to transfer data, lack of transparency, learning requires a large data sample, circumventing the problem requires using a large training dataset and an expert
Hou et al. (2016)	Deep belief network	Accuracy	Limited capacity to transfer data, lack of transparency, learning requires a large data sample, circumventing the problem requires using a large training dataset and an expert
Kashif et al. (2021)	Deep belief network	F1-score	Limited capacity to transfer data, lack of transparency, learning requires a large data sample, circumventing the problem requires using a large training dataset and an expert

(continued)

Table 1 (continued)

Authors and year	Features	Advantages	Disadvantages
Diro & Chilamkurti (2018)	Deep belief network	True positive, accuracy	Limited capacity to transfer data, lack of transparency, learning requires a large data sample, circumventing the problem requires using a large training dataset and an expert
Suleman et al. (2021)	Deep belief network	Accuracy, recall, precision	Limited capacity to transfer data, lack of transparency, learning requires a large data sample, circumventing the problem requires using a large training dataset and an expert
Aminanto et al. (2018)	Deep belief network	Accuracy	Limited capacity to transfer data, lack of transparency, learning requires a large data sample, circumventing the problem requires using a large training dataset and an expert
Kalra et al. (2020)	Deep belief network	False positive, detection accuracy	Limited capacity to transfer data, lack of transparency, learning requires a large data sample, circumventing the problem requires using a large training dataset and an expert
Thing (2017)	Deep belief network	Recall, detection ratio	Limited capacity to transfer data, lack of transparency, learning requires a large data sample, circumventing the problem requires using a large training dataset and an expert

(continued)

Table 1 (continued)

Authors and year	Features	Advantages	Disadvantages
Salloum et al. (2020)	Sparse auto-encoder	False positive	Limited capacity to transfer data, lack of transparency, learning requires a large data sample, circumventing the problem requires using a large training dataset and an expert
Yuan et al. (2017)	Sparse auto-encoder	Time for number of packets running	Limited capacity to transfer data, lack of transparency, learning requires a large data sample, circumventing the problem requires using a large training dataset and an expert
Hasan et al. (2022)	Sparse auto-encoder	Precision, recall, F-1 score, accuracy	Limited capacity to transfer data, lack of transparency, learning requires a large data sample, circumventing the problem requires using a large training dataset and an expert
Vinayakumar et al. (2017)	Sparse auto-encoder	Precision, recall, F-1 score, accuracy	Limited capacity to transfer data, lack of transparency, learning requires a large data sample, circumventing the problem requires using a large training dataset and an expert
Radwan & Farouk (2021)	Recurrent neural network	False positive	Limited capacity to transfer data, lack of transparency, learning requires a large data sample, circumventing the problem requires using a large training dataset and an expert

(continued)

Table 1 (continued)

Authors and year	Features	Advantages	Disadvantages
Ishitaki et al. (2017)	Recurrent neural network	Detection positive	Limited capacity to transfer data, lack of transparency, learning requires a large data sample, circumventing the problem requires using a large training dataset and an expert
Guergov & Radwan (2021)	Recurrent neural network	Error rate	Limited capacity to transfer data, lack of transparency, learning requires a large data sample, circumventing the problem requires using a large training dataset and an expert
Javaid et al. (2016)	Recurrent neural network	False positive	Limited capacity to transfer data, lack of transparency, learning requires a large data sample, circumventing the problem requires using a large training dataset and an expert
Nuseir et al. (2021)	Recurrent neural network	False positive	Limited capacity to transfer data, lack of transparency, learning requires a large data sample, circumventing the problem requires using a large training dataset and an expert
Fiore et al. (2019)	Support vector machine	Accuracy	Limited capacity to transfer data, lack of transparency, learning requires a large data sample, circumventing the problem requires using a large training dataset and an expert

(continued)

Table 1 (continued)

Authors and year	Features	Advantages	Disadvantages
Farouk (2021)	Stacked de-noising auto-encoders	Accuracy	Limited capacity to transfer data, lack of transparency, learning requires a large data sample, circumventing the problem requires using a large training dataset and an expert
Taormina & Galelli (2017)	Recursive neural tensor network	Accuracy	Limited capacity to transfer data, lack of transparency, learning requires a large data sample, circumventing the problem requires using a large training dataset and an expert
Mondol (2021)	Recursive neural tensor network	Precision recall F-1 score	Limited capacity to transfer data, lack of transparency, learning requires a large data sample, circumventing the problem requires using a large training dataset and an expert
Shibahara et al. (2016)	Auto-encoder	Classification accuracy	Limited capacity to transfer data, lack of transparency, learning requires a large data sample, circumventing the problem requires using a large training dataset and an expert
Cruz (2021)	Auto-encoder	Classification accuracy	Limited capacity to transfer data, lack of transparency, learning requires a large data sample, circumventing the problem requires using a large training dataset and an expert

(continued)

References

Afifi, M. A. M., Kalra, D., Ghazal, T. M., & Mago, B. (2020). Information technology ethics and professional responsibilities. *International Journal of Advanced Science and Technology, 29*(4), 11336–11343.

Ahmad, A., Alshurideh, M. T., Al Kurdi, B. H., & Alzoubi, H. M. (2021). Digital strategies: A systematic literature review. *The International Conference on Artificial Intelligence and Computer Vision*, 807–822.

Akhtar, A., Akhtar, S., Bakhtawar, B., Kashif, A. A., Aziz, N., & Javeid, M. S. (2021). COVID-19 detection from CBC using machine learning techniques. *International Journal of Technology, Innovation and Management (IJTIM), 1*(2), 65–78. https://doi.org/10.54489/ijtim.v1i2.22

Akour, I., Alshurideh, M., Al Kurdi, B., Al Ali, A., & Salloum, S. (2021). Using machine learning algorithms to predict people's intention to use mobile learning platforms during the COVID-19 pandemic: Machine learning approach. *JMIR Medical Education, 7*(1), 1–17. https://doi.org/10.2196/24032

Al Ali, A. (2021). The impact of information sharing and quality assurance on customer service at UAE banking sector. *International Journal of Technology, Innovation and Management (IJTIM), 1*(1), 01–17. https://doi.org/10.54489/ijtim.v1i1.10

Al Batayneh, R. M., Taleb, N., Said, R. A., Alshurideh, M. T., Ghazal, T. M., & Alzoubi, H. M. (2021). IT governance framework and smart services integration for future development of Dubai infrastructure utilizing AI and big data, its reflection on the citizens standard of living. *The International Conference on Artificial Intelligence and Computer Vision*, 235–247.

Al Kurdi, B., Alshurideh, M., Nuseir, M., Aburayya, A., & Salloum, S. A. (2021). The effects of subjective norm on the intention to use social media networks: An exploratory study using PLS-SEM and machine learning approach. *Advanced Machine Learning Technologies and Applications: Proceedings of AMLTA, 2021*, 581–592.

AlHamad, A., Alshurideh, M., Alomari, K., Kurdi, B., Alzoubi, H., Hamouche, S., & Al-Hawary, S. (2022). The effect of electronic human resources management on organizational health of telecommunications companies in Jordan. *International Journal of Data and Network Science, 6*(2), 429–438.

Alhamad, A. Q. M., Akour, I., Alshurideh, M., Al-Hamad, A. Q., Kurdi, B. A., & Alzoubi, H. (2021). Predicting the intention to use google glass: A comparative approach using machine learning models and PLS-SEM. *International Journal of Data and Network Science, 5*(3). https://doi.org/10.5267/j.ijdns.2021.6.002

Ali, N., Ahmed, A., Anum, L., Ghazal, T. M., Abbas, S., Khan, M. A., Alzoubi, H. M., & Ahmad, M. (2021). Modelling supply chain information collaboration empowered with machine learning technique. *Intelligent Automation and Soft Computing, 30*(1), 243–257. https://doi.org/10.32604/iasc.2021.018983

Ali, N., Ghazal, M. T., Ahmed, A., Abbas, S., A. Khan, M., Alzoubi, H., Farooq, U., Ahmad, M., & Adnan Khan, M. (2022). Fusion-based supply chain collaboration using machine learning techniques. *Intelligent Automation & Soft Computing, 31*(3), 1671–1687. https://doi.org/10.32604/iasc.2022.019892

Alnazer, N. N., Alnuaimi, M. A., & Alzoubi, H. M. (2017). Analysing the appropriate cognitive styles and its effect on strategic innovation in Jordanian universities. *International Journal of Business Excellence, 13*(1), 127–140. https://doi.org/10.1504/IJBEX.2017.085799

Alnuaimi, M., Alzoubi, H. M., Ajelat, D., & Alzoubi, A. A. (2021). Towards intelligent organisations: An empirical investigation of learning orientation's role in technical innovation. *International Journal of Innovation and Learning, 29*(2), 207–221. https://doi.org/10.1504/IJIL.2021.112996

Alsharari, N. (2021). Integrating blockchain technology with internet of things to efficiency. *International Journal of Technology, Innovation and Management (IJTIM), 1*(2), 1–13.

Alshurideh, M. (2022). Does electronic customer relationship management (E-CRM) affect service quality at private hospitals in Jordan? *Uncertain Supply Chain Management, 10*(2), 1–8.

Alshurideh, M., Al Kurdi, B., Alzoubi, H., Ghazal, T., Said, R., AlHamad, A., Hamadneh, S., Sahawneh, N., & Al-kassem, A. (2022b). Fuzzy assisted human resource management for supply chain management issues. *Annals of Operations Research*, 1–19.

Alshurideh, M., Al Kurdi, B., Salloum, S. A., Arpaci, I., & Al-Emran, M. (2020b). Predicting the actual use of m-learning systems: A comparative approach using PLS-SEM and machine learning algorithms. *Interactive Learning Environments*, 1–15.

Alshurideh, M., Gasaymeh, A., Ahmed, G., Alzoubi, H., & Kurd, B. A. (2020a). Loyalty program effectiveness: Theoretical reviews and practical proofs. *Uncertain Supply Chain Management*, *8*(3). https://doi.org/10.5267/j.uscm.2020.2.003

Alshurideh, M., Salloum, S. A., Al Kurdi, B., & Al-Emran, M. (2019). Factors affecting the social networks acceptance: An empirical study using PLS-SEM approach. In *PervasiveHealth: Pervasive Computing Technologies for Healthcare, Part F1479*. https://doi.org/10.1145/3316615.331 6720

Alshurideh, M. T., Al Kurdi, B., Alzoubi, H. M., Ghazal, T. M., Said, R. A., AlHamad, A. Q., Hamadneh, S., Sahawneh, N., & Al-kassem, A. H. (2022a). Fuzzy assisted human resource management for supply chain management issues. *Annals of Operations Research*, 1–19.

Alshurideh, M. T., Hassanien, A. E., & Masa'deh, R. (2021). *The effect of coronavirus disease (COVID-19) on business intelligence*. Springer.

Alzoubi, A. (2021a). The impact of process quality and quality control on organizational competitiveness at 5-star hotels in Dubai. *International Journal of Technology, Innovation and Management (IJTIM)*, *1*(1), 54–68. https://doi.org/10.54489/ijtim.v1i1.14

Alzoubi, A. (2021b). Renewable green hydrogen energy impact on sustainability performance. *International Journal of Computations, Information and Manufacturing (IJCIM)*, *1*(1), 94–110. https://doi.org/10.54489/ijcim.v1i1.46

Alzoubi, H. M., Alshurideh, M., & Ghazal, T. M. (2021a). Integrating BLE beacon technology with intelligent information systems IIS for operations' performance: A managerial perspective. *The International Conference on Artificial Intelligence and Computer Vision*, 527–538.

Alzoubi, H. M., & Aziz, R. (2021). Does emotional intelligence contribute to quality of strategic decisions? The mediating role of open innovation. *Journal of Open Innovation: Technology, Market, and Complexity*, *7*(2), 130. https://doi.org/10.3390/joitmc7020130

Alzoubi, H. M., Vij, M., Vij, A., & Hanaysha, J. R. (2021b). What leads guests to satisfaction and loyalty in UAE five-star hotels? AHP analysis to service quality dimensions. *Enlightening Tourism*, *11*(1), 102–135. https://doi.org/10.33776/et.v11i1.5056

Alzoubi, H. M., & Yanamandra, R. (2020). Investigating the mediating role of information sharing strategy on agile supply chain. *Uncertain Supply Chain Management*, *8*(2), 273–284. https://doi.org/10.5267/j.uscm.2019.12.004

Alzoubi, H., & Ahmed, G. (2019). Do TQM practices improve organisational success? A case study of electronics industry in the UAE. *International Journal of Economics and Business Research*, *17*(4), 459–472. https://doi.org/10.1504/IJEBR.2019.099975

Alzoubi, H., Ahmed, G., Al-Gasaymeh, A., & Kurdi, B. (2020a). Empirical study on sustainable supply chain strategies and its impact on competitive priorities: The mediating role of supply chain collaboration. *Management Science Letters*, *10*(3), 703–708.

Alzoubi, H., Alshurideh, M., Kurdi, B., Akour, I., & Aziz, R. (2022). Does BLE technology contribute towards improving marketing strategies, customers' satisfaction and loyalty? The role of open innovation. *International Journal of Data and Network Science*, *6*(2), 449–460.

Alzoubi, H., Alshurideh, M., Kurdi, B. A., & Inairat, M. (2020b). Do perceived service value, quality, price fairness and service recovery shape customer satisfaction and delight? A practical study in the service telecommunication context. *Uncertain Supply Chain Management*, *8*(3), 579–588. https://doi.org/10.5267/j.uscm.2020.2.005

Aminanto, M. E. et al. (2018). Deep abstraction and weighted feature selection for Wi-Fi impersonation detection. *IEEE Transactions on Information Forensics and Security, 13*(3), 621–636.

Aziz, N., & Aftab, S. (2021). Data mining framework for nutrition ranking: Methodology: SPSS modeller. *International Journal of Technology, Innovation and Management (IJTIM), 1*(1), 85–95.

Cruz, A. (2021). Convergence between blockchain and the internet of things. *International Journal of Technology, Innovation and Management (IJTIM), 1*(1), 35–56.

Diro, A. A., & Chilamkurti, N. (2018). Distributed attack detection scheme using deep learning approach for Internet of Things. *Future Generation Computer Systems, 82*:761–768.

Eli, T. (2021). Students perspectives on the use of innovative and interactive teaching methods at the University of Nouakchott Al Aasriya, Mauritania: English Department as a case study. *International Journal of Technology, Innovation and Management (IJTIM), 1*(2), 90–104.

Farouk, M. (2021). The universal artificial intelligence efforts to face coronavirus COVID-19. *International Journal of Computations, Information and Manufacturing (IJCIM), 1*(1), 77–93. https://doi.org/10.54489/ijcim.v1i1.47

Fiore, U. et al. (2019). Using generative adversarial networks for improving classification effectiveness in credit card fraud detection. *Information Sciences, 479*, 448–455.

Ghazal, T., Alshurideh, M., & Alzoubi, H. (2021a). Blockchain-enabled internet of things (IoT) platforms for pharmaceutical and biomedical research. *The International Conference on Artificial Intelligence and Computer Vision*, 589–600.

Ghazal, T., Soomro, T. R., & Shaalan, K. (2013). Integration of project management maturity (PMM) based on capability maturity model integration (CMMI). *European Journal of Scientific Research, 99*(3), 418–428.

Ghazal, T. M. (2021). Positioning of UAV base stations using 5G and beyond networks for IoMT applications. *Arabian Journal for Science and Engineering*, 1–12.

Ghazal, T. M., Anam, M., Hasan, M. K., Hussain, M., Farooq, M. S., Ali, H. M., Ahmad, M., & Soomro, T. R. (2021b). Hep-Pred: Hepatitis C staging prediction using fine Gaussian SVM. *Comput Mater Continua, 69*, 191–203.

Ghazal, T. M., Hasan, M. K., Alshurideh, M. T., Alzoubi, H. M., Ahmad, M., Akbar, S. S., Al Kurdi, B., & Akour, I. A. (2021c). IoT for smart cities: Machine learning approaches in smart healthcare—A review. *Future Internet, 13*(8), 218. https://doi.org/10.3390/fi13080218

Ghazal, T. M., Hussain, M. Z., Said, R. A., Nadeem, A., Hasan, M. K., Ahmad, M., Khan, M. A., & Naseem, M. T. (2021d). *Performances of K-means clustering algorithm with different distance metrics*.

Ghazal, T. M., Said, R. A., & Taleb, N. (2021e). Internet of vehicles and autonomous systems with AI for medical things. *Soft Computing*, 1–13.

Guergov, S., & Radwan, N. (2021). Blockchain convergence: Analysis of issues affecting IoT, AI and blockchain. *International Journal of Computations, Information and Manufacturing (IJCIM), 1*(1), 1–17. https://doi.org/10.54489/ijcim.v1i1.48

Hamadneh, S., Pedersen, O., & Al Kurdi, B. (2021). An investigation of the role of supply chain visibility into the Scottish bood supply chain. *Journal of Legal, Ethical and Regulatory Issues, 24*(Special Issue 1), 1–12.

Hanaysha, J. R., Al-Shaikh, M. E., Joghee, S., & Alzoubi, H. (2021a). Impact of innovation capabilities on business sustainability in small and medium enterprises. *FIIB Business Review, 1–12,*. https://doi.org/10.1177/23197145211042232

Hanaysha, J. R., Al Shaikh, M. E., & Alzoubi, H. M. (2021b). Importance of marketing mix elements in determining consumer purchase decision in the retail market. *International Journal of Service Science, Management, Engineering, and Technology (IJSSMET), 12*(6), 56–72.

Hasan, O., McColl, J., Pfefferkorn, T., Hamadneh, S., Alshurideh, M., & Kurdi, B. (2022). Consumer attitudes towards the use of autonomous vehicles: Evidence from United Kingdom taxi services. *International Journal of Data and Network Science, 6*(2), 537–550.

Hou, S. et al. (2016). Deep4maldroid: a deep learning framework for android malware detection based on Linux kernel system call graphs. In *2016 IEEE/WIC/ACM international conference on web intelligence workshops (WIW)*. IEEE.

Ishitaki, T. et al. (2017). Application of deep recurrent neural networks for prediction of user behavior in tor networks. In *2017 31st international conference on advanced information networking and applications workshops (WAINA)*. IEEE.

Javaid, A., et al. (2016). A deep learning approach for network intrusion detection system. In *Proceedings of the 9th EAI international conference on bio-inspired information and communications technologies (formerly BIONETICS)*. ICST (Institute for Computer Sciences, Social-Informatics and Telecommunications Engineering.

Joghee, S., Alzoubi, H. M., & Dubey, A. R. (2020). Decisions effectiveness of FDI investment biases at real estate industry: Empirical evidence from Dubai smart city projects. *International Journal of Scientific and Technology Research, 9*(3), 3499–3503.

Kalra, D., Ghazal, T. M., & Afifi, M. A. M. (2020). Integration of collaboration systems in hospitality management as a comprehensive solution. *International Journal of Advanced Science and Technology, 29*(8s), 3155–3173.

Kashif, A. A., Bakhtawar, B., Akhtar, A., Akhtar, S., Aziz, N., & Javeid, M. S. (2021). Treatment response prediction in hepatitis C patients using machine learning techniques. *International Journal of Technology, Innovation and Management (IJTIM), 1*(2), 79–89. https://doi.org/10.54489/ijtim.v1i2.24

Khan, M. A. (2021). Challenges facing the application of IoT in medicine and healthcare. *International Journal of Computations, Information and Manufacturing (IJCIM), 1*(1), 39–55. https://doi.org/10.54489/ijcim.v1i1.32

Khan, M. F., Ghazal, T. M., Said, R. A., Fatima, A., Abbas, S., Khan, M. A., Issa, G. F., Ahmad, M., & Khan, M. A. (2021a). An IoMT-enabled smart healthcare model to monitor elderly people using machine learning technique. *Computational Intelligence and Neuroscience, 2021a*.

Khan, Q.-T.-A., Ghazal, T., Abbas, S., Khan, W., Khan, A., Muhammad, A., Said, Raed, A., Ahmad, M., & Asif, M. (2021b). Modeling habit patterns using conditional reflexes in agency. *Intelligent Automation and Soft Computing, 30*(2), 539–552.

Lee, C., & Ahmed, G. (2021). Improving IoT privacy, data protection and security concerns. *International Journal of Technology, Innovation and Management (IJTIM), 1*(1), 18–33. https://doi.org/10.54489/ijtim.v1i1.12

Lee, K., Azmi, N., Hanaysha, J., Alzoubi, H., & Alshurideh, M. (2022a). The effect of digital supply chain on organizational performance: An empirical study in Malaysia manufacturing industry. *Uncertain Supply Chain Management, 10*(2), 495–510.

Lee, K. L., Romzi, P. N., Hanaysha, J. R., Alzoubi, H. M., & Alshurideh, M. (2022b). Investigating the impact of benefits and challenges of IOT adoption on supply chain performance and organizational performance: An empirical study in Malaysia. *Uncertain Supply Chain Management, 10*(2), 537–550. https://doi.org/10.5267/j.uscm.2021.11.009

Lee, S.-W., Hussain, S., Issa, G. F., Abbas, S., Ghazal, T. M., Sohail, T., Ahmad, M., & Khan, M. A. (2021). Multi-dimensional trust quantification by artificial agents through evidential fuzzy multi-criteria decision making. *IEEE Access, 9*, 159399–159412.

Li, Z. et al. (2017). Intrusion detection using convolutional neural networks for representation learning. In *International conference on neural information processing*.

Liu, Y. et al. (2017). Fault injection attack on deep neural network. In *Proceedings of the 36th international conference on computeraided design*. IEEE Press.

Matloob, F., Ghazal, T. M., Taleb, N., Aftab, S., Ahmad, M., Khan, M. A., Abbas, S., & Soomro, T. R. (2021). Software defect prediction using ensemble learning: A systematic literature review. *IEEE Access*.

Mathew, A. et al. (2017). An improved transfer learning approach for intrusion detection. *Procedia Computer Science, 115*, 251–257.

Mehmood, T. (2021). Does information technology competencies and fleet management practices lead to effective service delivery? Empirical evidence from E-commerce industry. *International Journal of Technology, Innovation and Management (IJTIM), 1*(2), 14–41.

Mehmood, T., Alzoubi, H. M., & Ahmed, G. (2019). Schumpeterian entrepreneurship theory: Evolution and relevance. *Academy of Entrepreneurship Journal, 25*(4).

Miller, D. (2021). The best practice of teach computer science students to use paper prototyping. *International Journal of Technology, Innovation and Management (IJTIM)*, *1*(2), 42–63. https://doi.org/10.54489/ijtim.v1i2.17

Mondol, E. P. (2021). The impact of block chain and smart inventory system on supply chain performance at retail industry. *International Journal of Computations, Information and Manufacturing (IJCIM)*, *1*(1), 56–76. https://doi.org/10.54489/ijcim.v1i1.30

Mohammadi, S., & Namadchian, A. (2017). A new deep learning approach for anomaly base IDS using memetic classifier. *International Journal of Computers Communications & Control*, *12*(5), 677–688.

Munoz-Gonza´lez, L. et al. (2017). Towards poisoning of deep learning algorithms with back-gradient optimization. In *Proceedings of the 10th ACM workshop on artificial intelligence and security*. ACM.

Naqvi, R., Soomro, T. R., Alzoubi, H. M., Ghazal, T. M., & Alshurideh, M. T. (2021). The nexus between big data and decision-making: A study of big data techniques and technologies. *The International Conference on Artificial Intelligence and Computer Vision*, 838–853.

Nuseir, M. T., Al Kurdi, B. H., Alshurideh, M. T., & Alzoubi, H. M. (2021). Gender discrimination at workplace: Do artificial intelligence (AI) and machine learning (ML) have opinions about it. *The International Conference on Artificial Intelligence and Computer Vision*, 301–316.

Obaid, A. J. (2021). Assessment of smart home assistants as an IoT. *International Journal of Computations, Information and Manufacturing (IJCIM)*, *1*(1), 18–36. https://doi.org/10.54489/ijcim.v1i1.34

Radwan, N., & Farouk, M. (2021). The growth of internet of things (IoT) in the management of healthcare issues and healthcare policy development. *International Journal of Technology, Innovation and Management (IJTIM)*, *1*(1), 69–84. https://doi.org/10.54489/ijtim.v1i1.8

Rehman, E., Khan, M. A., Soomro, T. R., Taleb, N., Afifi, M. A., & Ghazal, T. M. (2021). Using blockchain to ensure trust between donor agencies and NGOs in under-developed countries. *Computers, 10*(8), 98.

Salloum, S. A., Alshurideh, M., Elnagar, A., & Shaalan, K. (2020). Machine learning and deep learning techniques for cybersecurity: A review. In *Joint European-US Workshop on Applications of Invariance in Computer Vision* (pp. 50–57).

Saxe, J., & Berlin, K. (2015). Deep neural network-based malware detection using twodimensional binary program features. In *2015 10th international conference on malicious and unwanted software (MALWARE)*. IEEE.

Shamout, M., Ben-Abdallah, B., Alshurideh, M., Alzoubi, H., Al Kurdi, B., & Hamadneh, S. (2022). A conceptual model for the adoption of autonomous robots in supply chain and logistics industry. *Uncertain Supply Chain Management, 10*, 1–16.

Shibahara, T. et al. (2016). Efficient dynamic malware analysis based on network behavior using deep learning. In *2016 IEEE on global communications conference (GLOBECOM)*. IEEE.

Suleman, M., Soomro, T. R., Ghazal, T. M., & Alshurideh, M. (2021). Combating against potentially harmful mobile apps. *The International Conference on Artificial Intelligence and Computer Vision*, 154–173.

Taormina, R., & Galelli, S. (2017). Real-time detection of cyberphysical attacks on water distribution systems using deep learning. In *World environmental and water resources congress 2017*.

Tariq, E., Alshurideh, M., Akour, I., & Al-Hawary, S. (2022). The effect of digital marketing capabilities on organizational ambidexterity of the information technology sector. *International Journal of Data and Network Science, 6*(2), 401–408.

Thing, V. L. (2017). IEEE 802.11 network anomaly detection and attack classification: A deep learning approach. In *2017 IEEE on wireless communications and networking conference (WCNC)*. IEEE.

Vinayakumar, R., Soman, K., & Poornachandran, P. (2017). Evaluating effectiveness of shallow and deep networks to intrusion detection system. In *2017 international conference on advances in computing, communications and informatics (ICACCI)*. IEEE.

Wang, Q. et al. (2017). Adversary resistant deep neural networks with an application to malware detection. In *Proceedings of the 23rd ACM SIGKDD international conference on knowledge discovery and data mining*. ACM.

Wei, J., & Mendis, G. J. (2016). A deep learning-based cyber-physical strategy to mitigate false data injection attack in smart grids. In *Joint workshop on cyber-physical security and resilience in smart grids (CPSR-SG)*. IEEE.

Wu, T. et al. (2017). Twitter spam detection based on deep learning. In *Proceedings of the Australasian computer science week multiconference*. ACM.

Yuan, X., Li, C., & Li, X. (2017). Deep Defense: identifying DDoS attack via deep learning. In *2017 IEEE international conference on smart computing (SMARTCOMP)*. IEEE.

Zolotukhin, M. et al. (2016). Increasing web service availability by detecting application-layer DDoS attacks in encrypted traffic. In *2016 23rd international conference on telecommunications (ICT)*. IEEE.

An Integrated Cloud and Blockchain Enabled Platforms for Biomedical Research

Taher M. Ghazal, Mohammad Kamrul Hasan, Siti Norul Huda Sheikh Abdullah, Khairul Azmi Abu Bakar, Nasser Taleb, Nidal A. Al-Dmour, Eiad Yafi, Ritu Chauhan, Haitham M. Alzoubi⑩, and Muhammad Alshurideh⑩

Abstract In the current pandemic scenario, healthcare data tends to be an important asset among organizations. The major challenge is to handle the data effectively while maintaining the privacy and security of the data. In a real-world, context healthcare data proves to be heterogeneous. Hence, managing such significance to big data has ardently laid numerous challenges among researchers and scientists around the globe. Cloud environment and blockchain technology can be discussed as usable platforms

T. M. Ghazal
School of Information Technology, Skyline University College, Sharjah, UAE
e-mail: taher.ghazal@skylineuniversity.ac.ae

T. M. Ghazal · M. K. Hasan · S. N. H. S. Abdullah · K. A. A. Bakar
Faculty of Information Science and Technology, Center for Cyber Security, University Kebansaan Malaysia (UKM), Bangi, Malaysia
e-mail: mkhasan@ukm.edu.my

S. N. H. S. Abdullah
e-mail: Snhsabdullah@ukm.edu.my

K. A. A. Bakar
e-mail: khairul.azmi@ukm.edu.my

N. Taleb
Faculty of Management, Canadian University Dubai, Dubai, UAE
e-mail: naser.taleb@cud.ac.ae

N. A. Al-Dmour
Department of Computer Engineering, College of Engineering, Mutah University, Mu'tah, Jordan

E. Yafi
Institute of Business, Dili, Timor-Leste
e-mail: eiad.yafi@uts.edu.au

R. Chauhan
Center for Computational Biology and Bioinformatics, Amity University, Noida, India
e-mail: rchauhan@amity.edu

H. M. Alzoubi (✉)
School of Business, Skyline University College, Sharjah, UAE
e-mail: haitham.alzubi@skylineuniversity.ac.ae

© The Author(s), under exclusive license to Springer Nature Switzerland AG 2023   2037
M. Alshurideh et al. (eds.), *The Effect of Information Technology on Business and Marketing Intelligence Systems*, Studies in Computational Intelligence 1056,
https://doi.org/10.1007/978-3-031-12382-5_111

which can deliver a comprehensive centralized data privacy system. In the current approach study, we have integrated both technologies to provide usability in medical systems. Further, we have also proposed and implemented a blockchain application with an integrated cloud-based environment regarding heterogeneous medical databases. The study is proposed in 2 phases to maintain the privacy and the accessibility of the data. The double-spending problem is also presented, as mentioned above, using Blockchain's consensus process. Each network node independently verifies the validity of individual transactions and entire blocks. As a result, there is no need to put faith in a single entity or other nodes. As a result, third parties are no longer required for network actions or blockchain management.

Keywords Blockchain technology · Cloud environment · Medical system · Healthcare databases

1 Introduction

In the recent era of digitization, data has exponentially grown from megabytes to terabytes. The challenge is to gather information from such volumes of data and provide an eccentric environment to maintain privacy and security (Alnazer et al., 2017; Hamadneh et al., 2021a, 2021b; Obaid, 2021; Shamout et al., 2022). Hence, a traditional computation system doesn't offer a paving technology to handle such generosity among the data (Ahmad et al., 2021; Alshurideh et al., 2020a, 2020b; Kurdi et al., 2020; Mehrez et al., 2021). We can say that blockchains are an emerging domain that may provide a substitute for outdated databases (Akhtar et al., 2021; Cruz, 2021; Guergov & Radwan, 2021; Khan, 2021). We may consider the different scenario in which domain may be used as a candidate solution Blockchain technology is a domain that offers us a global solution to our problems, which will create trust and guarantee transparency among several application domains (Alzoubi et al., 2020a, 2020b; Aziz & Aftab, 2021; Joghee et al., 2020). Hence, Blockchain manages privacy by maintaining centrally defined criteria. All connected chain members can check the centralized information by providing the right paths of the chain. Blockchain systems may have the capacity (Alzoubi & Ahmed, 2019; Alzoubi & Aziz, 2021; Alzoubi et al., 2022). The blockchain system is dependent on temperature conditions. During the documented and the transport, it assists in the cold chain. In this way, patients benefit from accessible documentation and are tampered with tracing the drugs flow transportation (Ali et al., 2022; Alzoubi et al., 2020a, 2020b; Mehmood, 2021).

The introduction of the Blockchain in the healthcare sector is primarily about increasing security and improving data protection. In addition to the possibilities of

M. Alshurideh
Department of Marketing, School of Business, University of Jordan, Amman, Jordan
e-mail: m.alshurideh@ju.edu.jo; malshurideh@sharjah.ac.ae

Department of Management, College of Business Administration, University of Sharjah, Sharjah, UAE

the Blockchain for doctors, authorities, and patients, the interaction of the data flow between patients and hospitals or service providers in the healthcare sector can also be significantly improved (Alshurideh et al., 2022; Hanaysha et al., 2021a, 2021b). The Blockchain can be an ideal solution wherever information has to be stored and passed on. Even when companies exchange data from patients in the healthcare sector, the Blockchain improves data protection and transmission security (Alsharari, 2021; Alzoubi & Yanamandra, 2020; Lee & Ahmed, 2021; Lee et al., 2022a, 2022b).

Another feature of Blockchain which unprecedently benefits end-users is smart contracts. The concept of smart contracts was introduced by Nick Szabo back in 1997 and is defined as a computer-based transaction protocol that implements the terms of a contract. Due to its properties, the Blockchain offers a suitable medium for implementing such contracts for the first time. Smart contracts are to be understood as computer programs that can make decisions when certain conditions are met. For this purpose, external information can be used as input through the smart contract, which then induces a certain action via the specified rules of the contract. The corresponding scripts with the contract details will be too stored in a specific address on the Blockchain (Alhamad et al., 2021; Ali et al., 2021; Alshurideh et al., 2020a, 2020b; Ghazal et al., 2021). If the specified external event occurs, a transaction is sent to the address, after that the terms of the contract are executed accordingly. Smart Contracts are tools with which human interactions are automated by executing, enforcing, verifying, and inhibiting contracts through algorithms. All interactions with the blockchain system, such as making payments, exchanging the digital key, or accessing the patient data, can be carried out by the user and the healthcare professional using a smartphone (Alzoubi, 2021a; Eli, 2021; Lee et al., 2022a, 2022b; Mehmood et al., 2019; Miller, 2021).

Health data management is crucial in the healthcare system. Their design may aid patients in receiving the correct diagnosis and avoiding drug errors. Cybercriminals had also turned their attention to patient data. Smart contracts can be used to correctly and safely manage patient health data (Al Ali, 2021; AlHamad et al., 2022; Mondol, 2021; Radwan & Farouk, 2021). Smart contracts will be used to store this information on a distributed ledger. When a patient needs to be transferred from one hospital to another, it can be done swiftly and efficiently without completing additional documents. Blockchain technology enables the patient's healthcare professional to examine his health data and medical records easily. Many hospitals and healthcare providers still use traditional databases. If these health data are saved in smart contracts and disseminated via Blockchain, they will be accessible to hospitals and research institutes worldwide. Smart contracts can also operate as gatekeepers, limiting access to data to only those with a private key, making it more accessible and transparent for both patients and providers (Alnuaimi et al., 2021; Alzoubi, 2021b; Farouk, 2021).

In Parallel, Blockchain can be implemented in patients' health records, which could be encrypted and stored in the Blockchain with a private key that only allows certain users access. Hence, efforts are aligned to ensure that research is conducted by HIPAA laws (securely and confidentially). Further, the study evidence can be stored in a blockchain and automatically sent to the payers as evidence of performance

(AlShamsi et al., 2021; Alshurideh, 2022; Alzoubi et al., 2021; Kashif et al., 2021). A directory could also be used for general health management such as drug monitoring, compliance with regulations, test results, and health product management to monitor the entire process. The patient data could be mapped in a digital contract for doctors and insurance companies. Hence, all the parties in the proposed contracts will be able to visualize the data at a glance; also, patients can decide individually which information they want to disclose (Alwan & Alshurideh, 2022; Hamadneh et al., 2021a, 2021b; Tariq et al., 2022a, 2022b).

There are several studies in the past conducted dealing with blockchain technology with varied application domains. We can say that Blockchain has unprecedently advanced and benefited biomedical research. For example, If the entire drug production cycle is handled in the Blockchain, all processes can be monitored. This includes production, transport, storage, and dispensing at the pharmacy (Al-Hamadi et al., 2021; Al-Naymat et al., 2021; Zitar, 2021). Manufacturers could check that the cold chain was adhered to, if necessary, and assign a QR code to the drugs. Patients can use this to verify the authenticity of medication and at the same time check the transport. In drug research, the Blockchain helps to obtain declarations of consent from testers more quickly and reliably (Al-Hamadi et al., 2015; Al Neaimi et al., 2020; Bibi et al., 2021). Here, too, data can be sent more easily between the various parties involved.

The advert facts and literature review show that blockchain technology has opened vast opportunities in several application domains. But certainly, medical systems are the need of time due to the current pandemic scenario. To address the factors related to healthcare application, the current study of approach is focused on shedding some light on the recent developments regarding the use of Blockchain in the context of public healthcare and discussing the workings and security features of blockchain technology in moderate detail. Further, we have also proposed and implemented a blockchain application with an integrated cloud-based environment regarding heterogeneous medical databases. The study is presented in 2 phases to maintain the privacy and the accessibility of the data.

In general, the paper's outline is discussed as an extensive literature study is conducted to understand the role of Blockchain in varied application domains. Then, a comprehensive framework is proposed in context with the medical system. Further, the study discusses the diverse impacts of Blockchain in the medical system. In the end, the conclusion and discussion is widely elaborated.

2 How Blockchain Works

A "blockchain" is a data structure in which data design is stored in an unchangeable manner. This takes place in the form of chained data blocks (hence the term "block" "chain"). Newly added data are regularly merged into a new block, provided with a data "fingerprint" (hash) and a timestamp or an ascending ordinal number (Al Hamadi et al., 2017; Aslam et al., 2021; Ghazal, 2021; Maasmi et al., 2021). In some blockchains, executable program logic, i.e., Smart Contracts, can also be saved in a forgery-proof manner.

To protect against tampering, exact copies are automatically generated and distributed via many independent computers, ideally globally distributed and operated by a wide variety of participants and organizations. Since the stored data is usually a log of transactions, it also comes from a distributed ledger, cashbook, or register, i.e., "Distributed Ledger Technology" (DLT). The hash of every data block in a blockchain is unique. Since the data fingerprint of the previous block is also taken into account when calculating this value, the hash of the newest block is also the data fingerprint of the entire Blockchain. This means that even the slightest change in older data is immediately noticeable when compared with other computers. Manipulation security is achieved with the help of automatic consensus building in the network.

As a distributed ledger, the Blockchain manages without a central authority, so the consistency of the data has to be ensured in another way. A new block to be appended must be identical on all computers involved in the network. To ensure this, certain conditions are specified which a block must meet to be attached to a blockchain or accepted by all nodes (Siddiqui et al., 2021). In the event of discrepancies, it is automatically determined based on a predefined algorithm which copies are discarded and recognized by all computers. This process is also called consensus-building. Traditionally, transferring values (e.g., amounts of money, property titles, certificates, ID cards, rights, etc.) requires intermediaries (e.g., banks or authorities) whom both transaction partners trust and who handle the transaction. Users of a blockchain trust its cryptographic protocol and that the majority of those involved (peers) behave honestly. In the future, distributed ledgers will take over intermediary functions in many application areas and thus enable a wide variety of peer-to-peer transactions ("trust through transparency") (Al-Dmour & Teahan, 2005). The concept of public-key cryptography was introduced in 1976 by Diffie and Hellman. An algorithm is used to generate a mathematically linked vital pair consisting of a private and a public key. This key pair can be used to create a digital signature. To do this, the sender signs or combines a message with his private key, which only he knows, and sends the resulting signed message to the recipient. He can now check the signed message with the sender's public key and thus verify the message's authenticity (if the two keys correspond). Overall, three goals can be achieved with a digital signature. Since only the sender knows the private key, the message's authenticity can be proven. In addition, the sender cannot deny having signed the message. Furthermore, the message cannot be changed unnoticed by the asymmetrical encryption, which guarantees its content integrity.

With the appropriate query tools often available online, the entire blockchain database can be searched and readout. If visibility is to be prevented, the data can also be encrypted and stored in the Blockchain. This leads to further challenges, such as the question of where the key material should be stored. In the case of public-permission less Blockchain, new data is not appended directly but initially forwarded to all computers in the network. After a few minutes, a computer "packs" all the accumulated data into a new block and attaches it to the Blockchain. Hash-secured data can neither be deleted nor overwritten by individual participants. This is the most important unique selling point of blockchain technology, especially in public healthcare. Incorrect entries can always be corrected with a valid correction entry. Some blockchains only store simple information. The Bitcoin blockchain contains entries such as "Participant A transfers participant B two Bitcoins." Other blockchains offer more extensive functions: The Ethereum blockchain allows conditional transactions such as "After electronic confirmation of the delivery of the product P, A transfers the amount X to B." This enables business logic to be carried out automatically; this is known as a smart contract (Al-Dmour, 2016; Ali & Dmour, 2021; Zitar et al., 2021). With digital values stored on a DLT system, smart contracts are becoming a fundamental innovation for new business models. It should be noted that smart contracts are executed locally on each of the computers in the network. Compared to a central solution, smart contracts are very resource-intensive.

3 Integrated Network of Cloud and Blockchain for Medical System

In the past several studies are comprehended to understand the ease of technology in the real-world application domain. As we know, the current pandemic scenario has opened vast challenges among healthcare practitioners and scientists around the globe to determine the factors that can influence and improve healthcare costs. In comparison, the focus was to provide patients with utmost facilities with network capabilities. Hence, blockchain and cloud networks with healthcare databases can revolutionize administrators and healthcare professionals to work and enhance the captivity of their operations.

3.1 Development of Phase I Framework

The research framework was designed in 2 phases, Fig. 1 represents the first phase where the focus is to assimilate and regulate the factors responsible for sharing the information among healthcare professionals. Moreover, determining the benefits of technology can affluently impact the social and wellbeing of patients' healthcare.

3.1.1 Heterogeneous Medical Data

Patient data has immensely grown over the past decade, which has created massive challenges among researchers and scientists to design new adaptive technology. We may say that gathering the data from varied resources or heterogeneous has consolidated in the form of big data. This data has created a background to enhance our

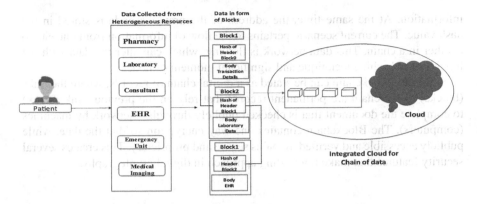

Fig. 1 The framework of phase I

technology to discover, analyze, and provide end-user services in a secure environment. Hence, Blockchain can be discussed as the technology which can cross many hindrances and give substantial privacy among the data shared. In Fig. 1, we have applied a coagulative blockchain and cloud computing approach in the healthcare system.

In the healthcare system, we can find data of several types such as patient symptoms, financial data, societal factors such as age, gender, origin, past abusive records, and others stored in the alphanumeric form. Moreover, audio, videos, imaging, including MRI, CT, Pet scan, and other aspects of data have different formats for storage. Then, we integrate the data from heterogeneous resources then we are submerged in the pool of data. Real-time accessibility of data in a private mode is the challenging factor; hence the Blockchain is the key technology that end-users can provide. I the current study of approach, we have designed a framework that can adhere and cope with the heterogeneous nature of data.

In the proposed approach, all the patient data was collected from varied heterogeneous resources such as Laboratory, Imaging data, Electronic Health Records (EHR) of the patients, Pharmacy data, and other clinical aspects of the patients. The first framework level extracts the data in different formats. It generates the data in blocks while uploading the same on the cloud-based architecture using recommender from the medical practitioners.

3.1.2 Blocks of Data

As we know, data has been gathered from disseminated resources, which has prudently influenced the medical system. Moreover, the data is stored in nodes where each node is subdivided into sub-headers to organize the data in the structured format eventually. The node consists of a header with a body; we can infer it to be a form of linked list where the body refers to having actual information or data of patient's information. At the same time, the address of the previous node is stored in the hash value. The current scenario pertains to the flow of information from one end to another in a chain. This data network is flowing, which can alter the data with the flexibility of hashing technique and significant moment of nodes.

Similarly, Blockchain can be stated as a digital chain of records, where the links (blocks) of the chain are permanent records that rely on the previous link (block) to complete the document that is checked publicly through a network by machines (computers). The Blockchain contains one-way encryption so that the data, while publicly accessible and verified, remains secure and proprietary. This creates several security features that make it a leading approach in digital cryptography.

3.1.3   Integrated Network of Cloud and Blockchain

In general, the availability of medical data at a critical time can be essential for patient wellbeing. In several conditions, the monitoring of the patients can occur in a heterogeneous environment where the team of doctors from various hospitals needs to investigate the patients on different parameters. So, the challenge is to create an environment where patient healthcare data and financial transactions are easily maintained from one end to another. Adopted blockchain technology can play a crucial role in enhancing interaction among healthcare practitioners and providing a collaborative environment for end-users to handle such limitations.

Thanks to Blockchain, patients have more control over their medical records thanks to Blockchain, and it could also be used to transmit real-time data from wearables and medical equipment with doctors. Using the technology's secure network, medical gadgets might be tracked correctly from manufacturing through installation during patient operations. Regulatory authorities, who would gain from mandating the integration of blockchain technology to simplify medical device approvals and post-market surveillance, are likely to push blockchain adoption by medical device manufacturers. Blockchain will make it easier to collect and review patient data and medical device performance data during clinical trials. Clinical trial data could also be manipulated or falsified with the use of technology.

Further, blockchains make a distinction between public and private variants depending on the visibility of the data. In the personal variant, visibility is limited to authorized participants. The right to append blocks to the chain can be controlled in both cases: one speaks here of the properties "permissioned" or "permission-less." Here, public permissioned Blockchain is used in the public healthcare scenario, while a more restricted version is employed in biomedical research. In all cases, data stored on a blockchain can be encrypted so that the participants do the transactions but cannot see the content transported in it. The Blockchain in a different form to access the data can be discussed as below:

1. **Public Permissionless Blockchain**: All entries are open, while all participants remain anonymous or pseudonymous. Examples: Bitcoin blockchain, Ethereum, EW blockchain.

2. **Public Permissioned Blockchain**: Access for participants is controlled by technology. With this type of Blockchain, the system's end users are subject to the specified framework conditions and rules.
3. **Private Permissioned Blockchain**: All participants are known. Since the participants often know each other and their number is usually lower, reaching a consensus is simplified so that the creation of blocks is speedy.
4. **Private Permissionless Blockchain**: This type of Blockchain is suspected of being out for the snowball effect and is more likely to pursue fraudulent purposes.

Moreover, cloud-based computing and Blockchain can quickly investigate, store, and provide a private environment for a healthcare-based system. The cloud-based system has attracted several users to store the data, process the data, and deliver the services to end-users as per the need of the time. Nowadays, it has attracted several small and large organizations where they pay for the data or services which are required at an essential time. Hence, forfeited benefiting many organizations in terms of cost while enhancing the flexibility of time and availability. In the current proposed study, we have integrated the cloud and blockchain technology to provide the healthcare system with the flexibility of data storage, analysis of the study, and maintaining the private environments for patients' data.

3.2 Framework for Phase II

In a previous phase I, data is generated from patients and further stored in nodes in a cloud-based interactive environment. However, the challenge is to access the data while maintaining privacy and availability to end-users. Hence, blockchain technology can be easily applied to medical systems as it can easily allow a transaction to work securely while maintaining privacy among the user's access. In Fig. 2, a representative model depicts the flow of information patients' data in the distributed environment where confidentiality and privacy are maintained. For example, if the insurance companies want to access the patient's records, the organization needs to send a request to a cloud-based system. Where the receiver will access the request and confirm the ownership of data while verifying through the e-stamp. The e-stamp is the value that is generated during the transaction.

Although data on a blockchain is viewable by everyone, it is stored encrypted (depending on the technology) or for data security purposes. Encryption is often accomplished through the use of tried-and-true technologies such as public/private key cryptography. Participants are responsible for their key and identity maintenance, whether physicians or patients; a blockchain does not set any requirements for this. In the event of a private-permissioned blockchain, the patient's data can also be packaged behind a firewall, further restricting access. It should be highlighted at this time that future breakthroughs in the realm of quantum computing might significantly call into question the security of encryption and distributed ledger technology. Because the volumes of data recorded on a blockchain are tiny, it is common for merely

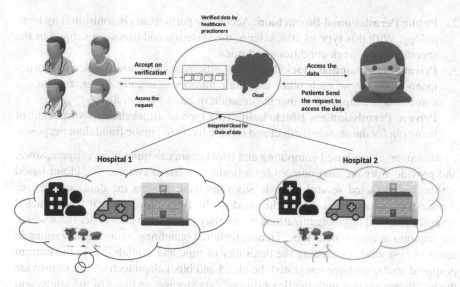

Fig. 2 The framework of phase II

a mathematical summary of the data to be placed there (usually as a hash). The chain member must then find out how to store the original data on safe servers but may subsequently establish its status at a particular time by making an entry in the Blockchain. All notarial applications are based on this.

Medical information is available with a high degree of certainty at all times due to the sheer technology, distributed, and fragmented storage, provided that a sufficient number of nodes keep the complete data set available so that partial failures of the underlying network can also be tolerated. Targeted attacks on a single central point are not feasible, while an attacker can restrict availability in private blockchains with considerable effort. A higher connection rate strengthens the network's resistance to failures caused by targeted attacks or technical faults.

To affect a blockchain of the public-permission kind with proof-of-work as a whole, an attacker must have control of the majority of all computers. It is problematic and quickly becomes apparent when the code is released as open-source and spreads across different nations, jurisdictions, and operators. The bulk of attacks on smaller systems have already succeeded. The great majority of attacks, on the other hand, are phishing efforts that steal access details. Mining malware, selfish mining, distributed denial of service, hacked crypto wallets, transfer Trojans, and other indirect attack scenarios are also feasible.

Hence, an identical architecture can be followed in the distributed environment of the medical system. Several hospitals can treat a patient while maintaining the financial ecosystem of the patients and providing flexibility among the healthcare practitioners to access the patients' data and provide comprehensive services to end-users.

Blockchain and IoT Enabled Healthcare

In the field of medical technology, various applications of Blockchain and smart contract ideas are conceivable. Although no concrete application scenarios have yet been created and published, research interest has already been aroused. In the age of digital medicine, more and more health-related data are available in different years. This leads to longitudinal sensor data such as radiological images, clinical information, outstanding patient files, and compassionate data, for example, genetic tests.

Modern preventive medicine requires evaluating such data across the population, for example, using current model-free statistics (deep learning). Traditional modeling in this environment is not used since data cannot be collected and made available in a centralized manner. There are no socially accepted methods of secure data storage and secure usage monitoring. One way out could be mechanisms that allow the owner of the data transactional, auditable control over its use. That is precisely the value proposition of blockchain technology. Not all blockchain characteristics are required; for example, it is unnecessary to decentralize the verification. According to the prevailing opinion, there are at least three significant technical challenges to be mastered:

1. **Complete data usage monitoring**. It must be demonstrably excluded that data usage bypasses the system. This can also make it necessary to certify secure hardware.
2. **Group and person-specific usage permits**. It must be possible to restrict the use of data to certain people or institutions or specific parts of the data.
3. **Complete decentralized data and logs**. Data must be transported across systems in a scenario where data is kept decentrally (in hospitals, insurance firms, mobile devices, etc.). Securing the communication from unauthorised usage is a significant problem.

In all three areas, some projects and approaches could be a basis for opening up a new type of access to the area of personalized, data-driven health management.

Data-centric medicine is the name for this type of tailored therapy. Today, data are abundant on a patient, and it continues to grow dramatically. Whereas a laboratory result and an X-ray image may have been the basis of a medical judgment in the past, many metrics will be used to produce an entire picture for medical recommendations in the future. In the context of patient management, this will also modify structures and processes. In the whole therapeutic chain, patient data takes the lead. The rising specialization of professions is also driven by data-based knowledge about a patient's health and the potential diagnoses, remedies, or preventive actions.

Traditional providers in the health sector and new suppliers such as companies in the food industry, sports, fitness, medical technology, or the IT sector will constitute the nodes of dynamic health networks that revolve around the individual. The rate at which people make decisions about preventative measures or therapies is increasing, as is the number of new products and health services available. Furthermore, this type of tailored medicine allows you to find new places for health and arrange health themes flexibly, such as at home, in the car, in a restaurant, or at work.

As a result, data protection in the future must imply that patients have control over their data and decide how it is handled. He must trust that his data is constantly current and accessible while also being safeguarded from illegal third-party access. Blockchain technology plays a critical role in this scenario. The digital currency Bitcoin is the most well-known and oldest blockchain application. It is no longer enough to attack a single server to steal or alter sensitive patient data; now, every single computer in the Blockchain must be hacked. As a result, the system is highly safe and appropriate for sending sensitive health or medical information.

Apart from personal healthcare, Blockchain applications have proven useful in clinical studies, especially those involving several centers, placing high logistics and documentation demands. By using a blockchain, the history of the events and the existence of the data can be proven beyond doubt. Subsequent changes are practically impossible. This helps in monitoring and controlling clinical trials. It then makes it easier for licensing authorities to check the integrity of the data and thus evaluate the study results.

A case that exemplifies the use of Blockchain is in the scope of the supply chain is the cold chain of drugs. To guarantee the effectiveness of certain medications, it is necessary to transport them under precise temperature conditions. The negative consequences of an ineffective drug can be financially far-reaching for companies and, most importantly, ultimate for patients. Guaranteeing a constant cold chain and documenting it remains a crucial challenge to this day. The complicated transport logistics, in which many different parties and systems are involved, is inefficient and prone to problems such as lack of transparency and traceability of the transport routes and temperature conditions. It can result in retailers receiving drugs that they cannot understand the quality of and patients ultimately taking ineffective drugs.

While the sensor technology for the control of drugs is already partially available and the interface to the Internet of Things can ensure that the corresponding analysis data is fed into digital systems, a piece of the puzzle is missing for a holistic solution to the problem. This Blockchain can create trust between the individual participants in the supply chain and the end consumer. Without a blockchain, the information about the medication is stored and passed on to the consumer. Central servers or even in paper form. This creates a dependency on each intermediary involved in the supply chain—government officials, customs, banks, transportation and storage companies, etc. Coordinating all of these parties is also highly time-consuming and costly.

Another critical point here is increasing the efficiency of the supply chain, especially in terms of administration and payment. Here, Smart Contracts can call necessary information, and payments are automatically processed between the respective parties. For example, a contaminated medication package discovered afterward can trigger a corresponding complaint payment directly. Salaries and fees to transport companies can also be paid directly upon successful delivery.

4 Conclusion

While Blockchain promises to improve interoperability, security, and data protection, its limits must be recognized. The Blockchain is not intended to replace existing databases. Furthermore, the solutions are not appropriate for large amounts of data that require exclusive access inside a single firm. Blockchain solutions are designed to capture specific transaction data and distribute it to many users in a network where transparency and data security are critical.

The combination of a decentralized peer-to-peer network and a consensus technique for determining the state of the Blockchain offers up a flood of new possibilities. Because all network users often replicate both the Blockchain and individual processes such as digital signature verification, there is no single point of failure inside the network. This demonstrates that both network failure security and data availability are high. The consensus process also avoids the previously noted double-spending concern. Each network node separately verifies individual transactions and whole blocks. As a result, there is no need for trust in a single entity or other nodes; as a result, third parties are obsolete for network actions and blockchain maintenance.

Blockchain can be a game-changer in the healthcare sector due to its versatility and security. Blockchain as a forerunner in safe digital infrastructure Blockchain technology has several applications in the healthcare field. However, it is not yet thoroughly developed, nor is it a panacea that can be used immediately. First and foremost, several technological, organizational, and behavioral economic difficulties must be overcome before healthcare providers can adopt a blockchain tailored to the healthcare sector.

References

Ahmad, A., Alshurideh, M. T., Al Kurdi, B. H., & Alzoubi, H. M. (2021). Digital strategies: A systematic literature review. *The International Conference on Artificial Intelligence and Computer Vision*, 807–822.

Akhtar, A., Akhtar, S., Bakhtawar, B., Kashif, A. A., Aziz, N., & Javeid, M. S. (2021). COVID-19 detection from CBC using machine learning techniques. *International Journal of Technology, Innovation and Management (IJTIM)*, *1*(2), 65–78. https://doi.org/10.54489/ijtim.v1i2.22

Al-Dmour, N. (2016). Using unstructured search algorithms for data collection in IoT-based WSN. *International Journal of Engineering Research and Technology. ISSN*, 974–3154.

Al-Dmour, N. A., & Teahan, W. J. (2005). Peer-to-peer protocols for resource discovery in the grid. *Parallel and Distributed Computing and Networks*, 319–324.

Al-Hamadi, H., Gawanmeh, A., & Al-Qutayri, M. (2015). An automatic ECG generator for testing and evaluating ECG sensor algorithms. In *2015 10th International Design & Test Symposium (IDT)* (pp. 78–83).

Al-Hamadi, H., Nasir, N., Yeun, C. Y., & Damiani, E. (2021). A verified protocol for secure autonomous and cooperative public transportation in smart cities. In *IEEE International Conference on Communications Workshops (ICC Workshops), 2021* (pp. 1–6).

Al-Naymat, G., Hussain, H., Al-Kasassbeh, M., & Al-Dmour, N. (2021). Accurate detection of network anomalies within SNMP-MIB data set using deep learning. *International Journal of Computer Applications in Technology, 66*(1), 74–85.

Al Ali, A. (2021). The impact of information sharing and quality assurance on customer service at UAE banking sector. *International Journal of Technology, Innovation and Management (IJTIM), 1*(1), 01–17. https://doi.org/10.54489/ijtim.v1i1.10

Al Hamadi, H., Gawanmeh, A., & Al-Qutayri, M. (2017). Guided test case generation for enhanced ECG bio-sensors functional verification. *International Journal of E-Health and Medical Communications (IJEHMC), 8*(4), 1–20.

Al Neaimi, M., Al Hamadi, H., Yeun, C. Y., & Zemerly, M. J. (2020). Digital forensic analysis of files using deep learning. In *2020 3rd International Conference on Signal Processing and Information Security (ICSPIS)* (pp. 1–4).

AlHamad, A., Alshurideh, M., Alomari, K., Kurdi, B., Alzoubi, H., Hamouche, S., & Al-Hawary, S. (2022). The effect of electronic human resources management on organizational health of telecommunications companies in Jordan. *International Journal of Data and Network Science, 6*(2), 429–438.

Alhamad, A. Q. M., Akour, I., Alshurideh, M., Al-Hamad, A. Q., Kurdi, B. A., & Alzoubi, H. (2021). Predicting the intention to use google glass: A comparative approach using machine learning models and PLS-SEM. *International Journal of Data and Network Science, 5*(3). https://doi.org/10.5267/j.ijdns.2021.6.002

Ali, L., & Dmour, N. (2021). The shift to online assessment due to COVID-19: An empirical study of university students, behaviour and performance, in the region of UAE. *International Journal of Information and Education Technology, 11*(5), 220–228.

Ali, N., Ahmed, A., Anum, L., Ghazal, T. M., Abbas, S., Khan, M. A., Alzoubi, H. M., & Ahmad, M. (2021). Modelling supply chain information collaboration empowered with machine learning technique. *Intelligent Automation and Soft Computing, 30*(1), 243–257. https://doi.org/10.32604/iasc.2021.018983

Ali, N., Ghazal, M. T., Ahmed, A., Abbas, S., Khan, A. M., Alzoubi, H., Farooq, U., Ahmad, M., & Adnan Khan, M. (2022). Fusion-based supply chain collaboration using machine learning techniques. *Intelligent Automation & Soft Computing, 31*(3), 1671–1687. https://doi.org/10.32604/iasc.2022.019892

Alnazer, N. N., Alnuaimi, M. A., & Alzoubi, H. M. (2017). Analysing the appropriate cognitive styles and its effect on strategic innovation in Jordanian universities. *International Journal of Business Excellence, 13*(1), 127–140. https://doi.org/10.1504/IJBEX.2017.085799

Alnuaimi, M., Alzoubi, H. M., Ajelat, D., & Alzoubi, A. A. (2021). Towards intelligent organisations: An empirical investigation of learning orientation's role in technical innovation. *International Journal of Innovation and Learning, 29*(2), 207–221. https://doi.org/10.1504/IJIL.2021.112996

AlShamsi, M., Salloum, S. A., Alshurideh, M., & Abdallah, S. (2021). Artificial intelligence and blockchain for transparency in governance. In *Artificial intelligence for sustainable development: Theory, practice and future applications* (pp. 219–230). Springer.

Alsharari, N. (2021). Integrating blockchain technology with internet of things to efficiency. *International Journal of Technology, Innovation and Management (IJTIM), 1*(2), 1–13.

Alshurideh, M. (2022). Does electronic customer relationship management (E-CRM) affect service quality at private hospitals in Jordan? *Uncertain Supply Chain Management, 10*(2), 1–8.

Alshurideh, M., Gasaymeh, A., Ahmed, G., Alzoubi, H., & Kurd, B. A. (2020a). Loyalty program effectiveness: Theoretical reviews and practical proofs. *Uncertain Supply Chain Management, 8*(3). https://doi.org/10.5267/j.uscm.2020.2.003

Alshurideh, M. T., Al Kurdi, B., Alzoubi, H. M., Ghazal, T. M., Said, R. A., AlHamad, A. Q., Hamadneh, S., Sahawneh, N., & Al-kassem, A. H. (2022). Fuzzy assisted human resource management for supply chain management issues. *Annals of Operations Research*, 1–19.

Alshurideh, M., Al Kurdi, B., Salloum, S. A., Arpaci, I., & Al-Emran, M. (2020b). Predicting the actual use of m-learning systems: A comparative approach using PLS-SEM and machine learning algorithms. *Interactive Learning Environments*, 1–15.

Alwan, M., & Alshurideh, M. T. (2022). *The effect of digital marketing on purchase intention: Moderating effect of brand equity. March.* https://doi.org/10.5267/j.ijdns.2022.2.012

Alzoubi, A. (2021a). The impact of process quality and quality control on organizational competitiveness at 5-star hotels in Dubai. *International Journal of Technology, Innovation and Management (IJTIM)*, *1*(1), 54–68. https://doi.org/10.54489/ijtim.v1i1.14

Alzoubi, A. (2021b). Renewable green hydrogen energy impact on sustainability performance. *International Journal of Computations, Information and Manufacturing (IJCIM)*, *1*(1), 94–110. https://doi.org/10.54489/ijcim.v1i1.46

Alzoubi, H. M., & Aziz, R. (2021). Does emotional intelligence contribute to quality of strategic decisions? The mediating role of open innovation. *Journal of Open Innovation: Technology, Market, and Complexity, 7*(2), 130. https://doi.org/10.3390/joitmc7020130

Alzoubi, H. M., Vij, M., Vij, A., & Hanaysha, J. R. (2021). What leads guests to satisfaction and loyalty in UAE five-star hotels? AHP analysis to service quality dimensions. *Enlightening Tourism, 11*(1), 102–135. https://doi.org/10.33776/et.v11i1.5056

Alzoubi, H. M., & Yanamandra, R. (2020). Investigating the mediating role of information sharing strategy on agile supply chain. *Uncertain Supply Chain Management, 8*(2), 273–284. https://doi.org/10.5267/j.uscm.2019.12.004

Alzoubi, H., & Ahmed, G. (2019). Do TQM practices improve organisational success? A case study of electronics industry in the UAE. *International Journal of Economics and Business Research, 17*(4), 459–472. https://doi.org/10.1504/IJEBR.2019.099975

Alzoubi, H., Ahmed, G., Al-Gasaymeh, A., & Kurdi, B. (2020a). Empirical study on sustainable supply chain strategies and its impact on competitive priorities: The mediating role of supply chain collaboration. *Management Science Letters, 10*(3), 703–708.

Alzoubi, H., Alshurideh, M., Kurdi, B., Akour, I., & Aziz, R. (2022). Does BLE technology contribute towards improving marketing strategies, customers' satisfaction and loyalty? The role of open innovation. *International Journal of Data and Network Science, 6*(2), 449–460.

Alzoubi, H., Alshurideh, M., Kurdi, B. A., & Inairat, M. (2020b). Do perceived service value, quality, price fairness and service recovery shape customer satisfaction and delight? A practical study in the service telecommunication context. *Uncertain Supply Chain Management, 8*(3), 579–588. https://doi.org/10.5267/j.uscm.2020.2.005

Aslam, M. S., Ghazal, T. M., Fatima, A., Said, R. A., Abbas, S., Khan, M. A., Siddiqui, S. Y., & Ahmad, M. (2021). *Energy-efficiency model for residential buildings using supervised machine learning algorithm.*

Aziz, N., & Aftab, S. (2021). Data mining framework for nutrition ranking: Methodology: SPSS modeller. *International Journal of Technology, Innovation and Management (IJTIM)*, *1*(1), 85–95.

Bibi, R., Saeed, Y., Zeb, A., Ghazal, T. M., Rahman, T., Said, R. A., Abbas, S., Ahmad, M., & Khan, M. A. (2021). Edge AI-based automated detection and classification of road anomalies in VANET using deep learning. *Computational Intelligence and Neuroscience, 2021.*

Cruz, A. (2021). Convergence between blockchain and the internet of things. *International Journal of Technology, Innovation and Management (IJTIM)*, *1*(1), 35–56.

Eli, T. (2021). Students perspectives on the use of innovative and interactive teaching methods at the University of Nouakchott Al Aasriya, Mauritania: English Department as a case study. *International Journal of Technology, Innovation and Management (IJTIM)*, *1*(2), 90–104.

Farouk, M. (2021). The universal artificial intelligence efforts to face coronavirus COVID-19. *International Journal of Computations, Information and Manufacturing (IJCIM)*, *1*(1), 77–93. https://doi.org/10.54489/ijcim.v1i1.47

Ghazal, T. M. (2021). Internet of things with artificial intelligence for health care security. *Arabian Journal for Science and Engineering*, 1–12.

Ghazal, T. M., Hasan, M. K., Alshurideh, M. T., Alzoubi, H. M., Ahmad, M., Akbar, S. S., Al Kurdi, B., & Akour, I. A. (2021). IoT for smart cities: Machine learning approaches in smart healthcare—A review. *Future Internet, 13*(8), 218. https://doi.org/10.3390/fi13080218

Guergov, S., & Radwan, N. (2021). Blockchain convergence: Analysis of issues affecting IoT, AI and blockchain. *International Journal of Computations, Information and Manufacturing (IJCIM), 1*(1), 1–17. https://doi.org/10.54489/ijcim.v1i1.48

Hamadneh, S., Keskin, E., Alshurideh, M., Al-Masri, Y., & Al Kurdi, B. (2021a). The benefits and challenges of RFID technology implementation in supply chain: A case study from the Turkish construction sector. *Uncertain Supply Chain Management, 9*(4), 1071–1080.

Hamadneh, S., Pedersen, O., & Al Kurdi, B. (2021b). An investigation of the role of supply chain visibility into the Scottish bood supply chain. *Journal of Legal, Ethical and Regulatory Issues, 24*(Special Issue 1), 1–12.

Hanaysha, J. R., Al-Shaikh, M. E., Joghee, S., & Alzoubi, H. (2021a). Impact of innovation capabilities on business sustainability in small and medium enterprises. *FIIB Business Review, 1–12.* https://doi.org/10.1177/23197145211042232

Hanaysha, J. R., Al Shaikh, M. E., & Alzoubi, H. M. (2021b). Importance of marketing mix elements in determining consumer purchase decision in the retail market. *International Journal of Service Science, Management, Engineering, and Technology (IJSSMET), 12*(6), 56–72.

Joghee, S., Alzoubi, H. M., & Dubey, A. R. (2020). Decisions effectiveness of FDI investment biases at real estate industry: Empirical evidence from Dubai smart city projects. *International Journal of Scientific and Technology Research, 9*(3), 3499–3503.

Kashif, A. A., Bakhtawar, B., Akhtar, A., Akhtar, S., Aziz, N., & Javeid, M. S. (2021). Treatment response prediction in hepatitis C patients using machine learning techniques. *International Journal of Technology, Innovation and Management (IJTIM), 1*(2), 79–89. https://doi.org/10.54489/ijtim.v1i2.24

Khan, M. A. (2021). Challenges facing the application of IoT in medicine and healthcare. *International Journal of Computations, Information and Manufacturing (IJCIM), 1*(1), 39–55. https://doi.org/10.54489/ijcim.v1i1.32

Kurdi, B. A., Alshurideh, M., & Salloum, S. A. (2020). Investigating a theoretical framework for e-learning technology acceptance. *International Journal of Electrical and Computer Engineering, 10*(6). https://doi.org/10.11591/IJECE.V10I6.PP6484-6496

Lee, C., & Ahmed, G. (2021). Improving IoT privacy, data protection and security concerns. *International Journal of Technology, Innovation and Management (IJTIM), 1*(1), 18–33. https://doi.org/10.54489/ijtim.v1i1.12

Lee, K., Azmi, N., Hanaysha, J., Alzoubi, H., & Alshurideh, M. (2022a). The effect of digital supply chain on organizational performance: An empirical study in Malaysia manufacturing industry. *Uncertain Supply Chain Management, 10*(2), 495–510.

Lee, K., Romzi, P., Hanaysha, J., Alzoubi, H., & Alshurideh, M. (2022b). Investigating the impact of benefits and challenges of IOT adoption on supply chain performance and organizational performance: An empirical study in Malaysia. *Uncertain Supply Chain Management, 10*(2), 537–550.

Maasmi, F., Morcos, M., Al Hamadi, H., & Damiani, E. (2021). Identifying applications' state via system calls activity: A pipeline approach. In *2021 28th IEEE International Conference on Electronics, Circuits, and Systems (ICECS)* (pp. 1–6).

Mehmood, T. (2021). Does information technology competencies and fleet management practices lead to effective service delivery? Empirical evidence from E-commerce industry. *International Journal of Technology, Innovation and Management (IJTIM), 1*(2), 14–41.

Mehmood, T., Alzoubi, H. M., & Ahmed, G. (2019). Schumpeterian entrepreneurship theory: Evolution and relevance. *Academy of Entrepreneurship Journal, 25*(4).

Mehrez, A. A. A., Alshurideh, M., Kurdi, B. A., & Salloum, S. A. (2021). Internal factors affect knowledge management and firm performance: A systematic review. In *Advances in intelligent systems and computing, AISC* (Vol. 1261). https://doi.org/10.1007/978-3-030-58669-0_57

Miller, D. (2021). The best practice of teach computer science students to use paper prototyping. *International Journal of Technology, Innovation and Management (IJTIM)*, *1*(2), 42–63. https://doi.org/10.54489/ijtim.v1i2.17

Mondol, E. P. (2021). The impact of block chain and smart inventory system on supply chain performance at retail industry. *International Journal of Computations, Information and Manufacturing (IJCIM)*, *1*(1), 56–76. https://doi.org/10.54489/ijcim.v1i1.30

Obaid, A. J. (2021). Assessment of smart home assistants as an IoT. *International Journal of Computations, Information and Manufacturing (IJCIM)*, *1*(1), 18–36. https://doi.org/10.54489/ijcim.v1i1.34

Radwan, N., & Farouk, M. (2021). The growth of internet of things (IoT) in the management of healthcare issues and healthcare policy development. *International Journal of Technology, Innovation and Management (IJTIM)*, *1*(1), 69–84. https://doi.org/10.54489/ijtim.v1i1.8

Shamout, M., Ben-Abdallah, B., Alshurideh, M., Alzoubi, H., Al Kurdi, B., & Hamadneh, S. (2022). A conceptual model for the adoption of autonomous robots in supply chain and logistics industry. *Uncertain Supply Chain Management, 10*, 1–16.

Siddiqui, S. Y., Haider, A., Ghazal, T. M., Khan, M. A., Naseer, I., Abbas, S., Rahman, M., Khan, J. A., Ahmad, M., & Hasan, M. K. (2021). IoMT cloud-based intelligent prediction of breast cancer stages empowered with deep learning. *IEEE Access, 9*, 146478–146491.

Tariq, E., Alshurideh, M., Akour, I., & Al-Hawary, S. (2022a). The effect of digital marketing capabilities on organizational ambidexterity of the information technology sector. *International Journal of Data and Network Science, 6*(2), 401–408.

Tariq, E., Alshurideh, M., Akour, I., Al-Hawary, S., & Al, B. (2022b). The role of digital marketing, CSR policy and green marketing in brand development. *International Journal of Data and Network Science, 6*(3), 1–10.

Zitar, R. A. (2021). A review for the genetic algorithm and the red deer algorithm applications. In *2021 14th International Congress on Image and Signal Processing, BioMedical Engineering and Informatics (CISP-BMEI)* (pp. 1–6).

Zitar, R. A., Abualigah, L., & Al-Dmour, N. A. (2021). Review and analysis for the red deer algorithm. *Journal of Ambient Intelligence and Humanized Computing*, 1–11.

Analysis of Issues Affecting IoT, AI, and Blockchain Convergence

Nasser Taleb, Nidal A. Al-Dmour, Ghassan F. Issa,
Tamer Mohamed Abdellatif, Haitham M. Alzoubi ⓘ,
Muhammad Alshurideh ⓘ, and Mohammed Salahat

Abstract The purpose of this project was to appraise the integration or convergence issues influencing the mutual functioning of blockchain, AI, and IoT. The study argued that the recent developments in the field of IoT and blockchain prediction have involved the integration of innumerable classification schemes to establish a hybrid model. The introduction of the hybrid technique relies on the prediction performance that strives to override the limitations of any available architectural

N. Taleb
Faculty of Management, Canadian University Dubai, Dubai, UAE
e-mail: nasser.taleb@cud.ac.ae

N. A. Al-Dmour
Department of Computer Engineering, College of Engineering, Mutah University, Mu'tah, Jordan

G. F. Issa
School of Information Technology, Skyline University College, Sharjah, UAE
e-mail: ghassan.issa@skylineuniversity.ac.ae

T. M. Abdellatif
Faculty of Engineering, Applied Sciences and Technology, Canadian University Dubai, Dubai, UAE
e-mail: tamer.mohamed@cud.ac.ae

H. M. Alzoubi (✉)
School of Business, Skyline University College, Sharjah, UAE
e-mail: haitham.alzubi@skylineuniversity.ac.ae

M. Alshurideh
Department of Marketing, School of Business, University of Jordan, Amman, Jordan
e-mail: m.alshurideh@ju.edu.jo; malshurideh@sharjah.ac.ae

Department of Management, College of Business Administration, University of Sharjah, Sharjah, UAE

M. Salahat
College of Engineering and Information Technology, University of Science and Technology of Fujairah, Fujairah, UAE
e-mail: m.salahat@ustf.ac.ae

© The Author(s), under exclusive license to Springer Nature Switzerland AG 2023   2055
M. Alshurideh et al. (eds.), *The Effect of Information Technology on Business
and Marketing Intelligence Systems*, Studies in Computational Intelligence 1056,
https://doi.org/10.1007/978-3-031-12382-5_112

scheme. This study offers a comprehensive exploratory appraisal of the issues influencing the successful integration of IoT and blockchain in regards to functionality and effectiveness of security, trust, and flawless communication issues. The exploratory research methodology was used in analyzing the issues affecting the integration of blockchain, artificial intelligence (AI), and the internet of things (IoT). The findings indicated that the integration challenges influencing the effective operations of blockchain, AI, and IoT as a single system involve security, scalability, accountability, and trust of communications. The study recommends that successful and effective integration will enhance the development of new business models as well as the digital transformation of market corporations. Accordingly, new approaches to convergence should ensure that executives address the new technology demands to obtain significant gains in efficiency.

Keywords Artificial intelligence · Internet of things · Blockchain convergence

1 Introduction

New developments in the integration frameworks between IoT and blockchain in different fields have sparked debates concerning the actual framework that should be adopted by firms desiring to benefit from the convergence of the two technologies (Akhtar et al., 2021; Al Ali, 2021; Alhamad et al., 2021; AlHamad et al., 2022). As one of the principal concepts guiding new prospects of industrial revolution, the Internet of Things (IoT) has achieved immeasurable milestones. Globally, the projected growth of IoT was about $170 billion in 2017 and is expected to be about $560 billion by end of the financial year 2022 (Ali et al., 2022; Alnazer et al., 2017; Alnuaimi et al., 2021; Alsharari, 2021; Alshurideh et al., 2022). Although countless professionals have indicated that IoT is the new industrial revolution, key challenges have affected the performance of IoT, starting from the ancient days in which there was the lack of a protected ecosystem covering all the construction blocks of IoT design as well as the scalability issues affecting the entire system. The number of devices operating in any IoT system has been one of the primary issues affecting the performance of IoT since its introduction.

It is commendable that the IoT has enhanced a common operating picture (COP) handling many applications and aspects of modern day life (Alshurideh et al., 2020; Alzoubi, 2021a, 2021b; Alzoubi et al., 2020a, 2020b, 2022). Blockchain, in this regard, has enhanced the effectiveness of COP because it has advanced the operations of wireless network and sensor devices that could not otherwise communicate via the conventional IoT network. Blockchain, artificial intelligence (AI), and IoT are the principal technologies that have driven the next phase of digital transformation. It is projected that these technologies will enable for the creation of new business models, including autonomous agents, digital version of IoT, receiving or sending money via blockchain technology, and autonomous decision-making as independent agents of economy.

The IoT has enabled a complex connection of things or objects, powered by sensing, communication units, and processing, to identify physical events, interact with their environments, and exchange data. The objective of such interactions is to monitor processes or make decisions concerning events requiring human interventions. Perhaps the most renowned inspirations related to the rise of IoT systems was the need to foster the real-time information collection as well as the need to offer remote and automatic control mechanisms that have replaced current conventional control and monitoring systems across industries (Alzoubi & Ahmed, 2019; Alzoubi & Aziz, 2021; Alzoubi & Yanamandra, 2020; Alzoubi et al., 2020a, 2020b, 2021a, 2021b). The integration between IoT, AI, and blockchain will introduce a new system architecture that will control and advance most of the ineffective procedures associated with the welfare of humankind.

2 Theoretical Framework

The theoretical framework adopted in this study considered the limitations of the existing shreds of research on the most effective convergence architecture for AI, blockchain, and IoT. The current research, in this regard, utilizes generalized philosophies and theories concerning the effectiveness of convergence between the three aspects of modern industrialization, including blockchain, AI, and IoT (Aziz & Aftab, 2021; Cruz, 2021; Eli, 2021; Farouk, 2021; Ghazal et al., 2021b). The theories considered in this depends on the fact that the distributive ledger aspect of blockchain is one of the solutions for the existing security and privacy challenges.

2.1 Operational Definitions

- *IoT*—refers to the Internet of Things
- *COP*—denotes common operating picture
- *AI*—refers to artificial intelligence
- *Blockchain*—defines the digital transaction ledger that is often distributed or duplicated across the whole network of devices and computers operating in the chain of blocks.

2.2 Industry Description

The general industry guiding the integration of IoT, AI, and blockchain has produced innumerable scales of information requiring power, network connectivity, storage, and processing (Alhashmi et al., 2020; Al Shebli et al., 2021; AlShamsi et al., 2021; Nuseir et al., 2021; Yousuf et al., 2021). The objective of convergence, in this regard, is to transform available data into meaningful services and information (Alshurideh, 2022; Alwan & Alshurideh, 2022; Tariq et al., 2022a, 2022b; Emad Tariq et al., 2022a, 2022b). Along with concerns such as network scalability and reliable connectivity, data privacy and cybersecurity are issues of critical importance regarding the networks serving IoT and related systems (Guergov & Radwan, 2021; Hamadneh et al., 2021; Hanaysha et al., 2021a, 2021b; Joghee et al., 2020). The current industry involves centralized designs that have widely been used to connect, authorize, and authenticate different IoT network nodes.

3   Literature Review

The integration and performance issues affecting convergence of blockchain with the Internet of Things (IoT) has been addressed by many scholars. One of the main concerns of the issues affecting the seamless integration involve the fact that integrating these technologies focused on the prediction performance by filling the gap of limited literatures on the previous classification or convergence techniques. In modern world, artificial intelligence, IoT, and blockchain technologies have been acknowledged as innovations that can promote the existing business processes, disrupt entire market economies, and establish new business models. For instance, blockchain can enhance business process efficiency, security, transparency, and trust because of its decentralized, distributed, and shared ledger (Kashif et al., 2021; Khan, 2021; Lee & Ahmed, 2021; Lee et al., 2022a, 2022b).

Issues affecting convergence or integration between AI blockchain, and IoT has often been neglected based on many factors. For example, these three technologies are often used separately and selectively based on the demands of a specific firm. These innovations, however, should be implemented collectively now and in the future. One potential integration platform between these technologies is the use of IoT to provide and solicit data, with blockchain offering the setup rules of engagement and infrastructure while the AI maximizing the rules and process optimization.

Information process integration for security, trust, and seamless flow is one of the principal objectives of connection IoT, AI, and blockchain. There are many solutions for addressing the vulnerabilities and threats affecting the operations of IoT. IoT has played a crucial role in the daily organization level by ensuring the ease of working in diverse enterprises. Accordingly, the threat level existing in IoT devices is relatively high, indicating that the assessment and integration demands of the different models must be guaranteed to ensure orientation (Mehmood, 2021; Mehmood et al., 2019; Miller, 2021; Mondol, 2021; Obaid, 2021). Because IoT security is critical commensurate with the activities of hackers and malicious users, it is arguable the current architectures are often prone to attacks.

Convergence between AI, blockchain, and IoT has also been discussed by several shreds of literature. It is outlined that the use of AI, along with the abilities of IoT has enhanced treatments targeting patients suffering from Hepatitis C, a blood-borne infection that is often asymptomatic in the initial phases. The progression of hepatitis

C throughout the final phases often complicates the treatment and diagnosis process. Accordingly, a system based on AI and machine learning algorithms can assist healthcare providers in offering effective diagnoses in the early stages (Radwan & Farouk, 2021; Shamout et al., 2022). Based on the effectiveness of the blockchain in ensuring seamless, secure, and confirmed flow of information, it is arguable that the convergence of AI, blockchain, and IoT can significantly enhance the services given to patients suffering from hepatitis C.

4 Problem Statement, Research Gap, Research Contribution

Many pieces of research have tackled the issues, capabilities, benefits, and challenges facing the integration of AI, blockchain, and IoT. The existing studies, however, have only focused on particular areas, including health, finance, and agriculture, with finance being at the forefront of most emerging studies (Afifi et al., 2020; Al Batayneh et al., 2021; Ali et al., 2021; Alzoubi et al., 2021a). Despite the evident success of the convergence of blockchain with IoT in some of these fields, the analysis of converging AI, blockchain, and IoT has not been addressed in most of the existing bodies of literature. Accordingly, this study strives to determine the opportunities and challenges of integrating these three technologies using the best architectural model to enhance the accountability and accuracy of offering a broad range of services.

4.1 Research Model and Hypotheses

The analytical approach was considered as the best research design for this research. An analytical mechanism denotes the application of appraisals to decipher an issue down to its specific elements appropriate for finding a solution. Generally, the analytical approach is also referred to as formal analysis (Ghazal, 2021; Ghazal et al., 2021a, 2021c, 2021d). The primary challenge linked with the analytic approach, however, is that the existing tools are limited to the specific problems they can identify and solve. The approach adopted in this study is founded on the fact that the issues influencing integration between IoT, blockchain, and AI include the common elements such as efficiency, trust, accuracy, and scalability. Commensurate, the research design or model hypothesizes that integrating the three technologies can produce significant outcomes for organizations over time.

4.2 Methodology and Research Design

The adopted methodology was the quantitative research design. The selected research method focused on the performance of organizations that have implemented the AI, blockchain, and IoT in delivering services to public. Quantitative research was important for this study because it is crucial to identifying trends and averages to predict or evaluate causal associations that generalizes outcomes to a broader population in the end (Ghazal et al., 2013; Kalra et al., 2020; Khan et al., 2021a, 2021b). This research method was adopted as an experimental and correlational research technique because it formally examines the predictions or hypotheses based on statistics.

Based on the required data to perform a comprehensive quantitative analysis of the data on convergence issues affecting the integration of blockchain, Ai, and IoT, trust issues were the major element involved in the process. Trust is one of the primary concerns that has affected the integration of blockchain with IoT and AI. The quantitative approach, in this regard, involved the different procedural steps appropriate to the existing study. Principally, the advantage of using a quantitative research method is that the design is necessary regarding the use of factual data needed to address research questions.

In the process of determining the most appropriate quantitative method for the study, the study relied on the proportion supporting the assisted perception, which concerned the distribution of respondents according to age, residence, household income, and marital status, among others (Lee et al., 2021; Matloob et al., 2021; Naqvi et al., 2021). Additionally, the study data focused on the number of companies that have performed based on issues such as level of education and understanding of key issues guiding data use and the need for historical evidence.

4.3 Population/Sample/Unit of Analysis

The sampling, population, and unit of analysis was regarded as one of the fundamental elements of the study. In this research, it was considered that the companies that have implemented some of the suggested frameworks have understood some of the most influential elements, including, among others, the challenges, benefits, and opportunities associated with any of the emerging issues. Regarding the study and associated measures, it is arguable that the study determined the efficiency of the suggested approach because of the urgent need to foster the transaction rate occurring through blockchain transactions.

According to the exploratory design of the suggested framework, this study considered the fact that the efficiency of any applicable approach should follow the contribution of the following specific elements: transaction efficiency (described as the study equation—Eq. 1), n (number of business transactions happening through the integrated AI, IoT, and blockchain platforms), communications trust via trust (t), and the nature of security associated with the transactions (s). It is noteworthy that this model relies on the findings and design of the study conducted by Ghazal et al., in which the analysis relied on the equation outlined below

$$\text{Effectivity of performance} = s + n + t\,(n) \tag{1}$$

The inclusion of n depends on the fact that the data used in the appraisal relied on the procedural developments of blockchain in the development of convergence metrics between AI, blockchain, and IoT. In one of the papers published by the individuals supporting the convergence of blockchain, AI, and IoT, Ghazal et al. argue that the computation of the immediate communication trust focuses on enhancing the trust of consumers concerning the most detailed model of the blockchain.

5 Analyzing Data

Equation 1 presented above outlines all the variables that have been used in presenting all the data aspects in this study. Accordingly, once this process improves, it is arguable that the integrated architecture covering convergence of IoT, AI, and blockchain will outline a comprehensive sequence of tests to determine the accuracy of handling the challenges influencing the existing block chain models. In this study, the immediate test concerned the computation of trust transaction that majorly focused on the capacity of companies to improve the level of trust in the suggested blockchain models. It is arguable that this aspect was obtained via seeking the attitudes of the different users concerning their ability and capacities to obtain the outlined requirements.

The data model, commensurate with the qualitative analysis study has many inferences. Firstly, the level of consumer/customer trust is determined by adding the scale of successful transactions (St) to the number of unsuccessful transactions (Zt). The outcome of this figure is divided by tt (the aggregate or total transactions). In this regard, Eq. 2 provides the actual number of transactions associated with any type of transaction.

$$\text{T or transactions trust} = (St + Zt)/tt \tag{2}$$

Based on the equation above, it is difficult to determine the security of the transactions that can happen without allowing or understanding of the detected threats. In this regard, it is arguable that the threat numbers is computed by dividing ts (solved threats) with the overall risks or threats identified in the entire convergence system.

The number of secure transactions following the convergence of AI, IoT, and blockchain (S) is denoted by Eq. 3:

$$S = (td - ts)/td \tag{3}$$

$S = (td - ts)/td$. Which reflects the number of transactions as

N = transaction days (dt)/average daily transaction (adt); that is:
N = 9dt/adt.

5.1 Discussion of the Results

It is arguable that one of the many challenges affecting effective integration of AI, IoT, and blockchain is the existence of heterogeneous alternative outlining the variety of IoT devices and applications that needs to integrate AI and blockchain with IoT tech founded on their requirements and demands (Rehman et al., 2021; Suleman et al., 2021). Generally, these alternatives are only founded on particular use cases that cannot suit a broad range of devices and applications in this specific sector.

As a result, new studies should advocate for the development of a set of standards and protocols that can support the essential and basic needs of all IoT devices and applications rather than introducing applications/devices that can only operate via IoT networks. The combined potential of AI, IoT, and blockchain is immeasurable. Based on existing kinds of studies, it is arguable that the amalgamation of AI, IoT, and blockchain technology can unlock several new business architectures for the accrual of funds from IoT devices and applications. The results presented in the formulas and considerations outlined above resonates with the fact that the security of transactions have the accuracy that reflect the percentages produced by the following formulas. When comparing the models, it is appropriate that the architecture controlling the design of the tables indicate centralization as about 100% of the model. These results indicate that either of the algorithms applicable in the development of a perfect architecture can help address all the concerns of blockchain, AI, and IoT convergence in both the short and long-terms.

6 Conclusions and Recommendations

The objective of this study was to assess the most recent and adopted architectures of blockchain. The analysis involved comparing the most popular, recent, and interesting consensus algorithms as well as evaluating the integration between IoT, AI, and blockchain via illustrating the existing field research. The paper also offered a comprehensive overview of the disruptive studies on the topic that current authors have continued to investigate. The findings indicated that the convergence between AI, blockchain, and IoT can enhance the adequacy of computational level as well as efficiency in optimizing the energy consumption of connected devices. AI, IoT, and blockchain are technologies that will continue to be integrated in myriad dimensions. This paper contends that the integration of these innovative models will occur because services, products, and business models will benefit from the diversity of these technologies. Generally, these business models can be widely adopted by any independent agent, including cameras, trucks, machines, cars, and numerous sensors.

References

Afifi, M. A. M., Kalra, D., Ghazal, T. M., & Mago, B. (2020). Information technology ethics and professional responsibilities. *International Journal of Advanced Science and Technology, 29*(4), 11336–11343.

Akhtar, A., Akhtar, S., Bakhtawar, B., Kashif, A. A., Aziz, N., & Javeid, M. S. (2021). COVID-19 detection from CBC using machine learning techniques. *International Journal of Technology, Innovation and Management (IJTIM), 1*(2), 65–78. https://doi.org/10.54489/ijtim.v1i2.22

Al Ali, A. (2021). The impact of information sharing and quality assurance on customer service at UAE banking sector. *International Journal of Technology, Innovation and Management (IJTIM), 1*(1), 01–17. https://doi.org/10.54489/ijtim.v1i1.10

Al Batayneh, R. M., Taleb, N., Said, R. A., Alshurideh, M. T., Ghazal, T. M., & Alzoubi, H. M. (2021). IT governance framework and smart services integration for future development of Dubai infrastructure utilizing AI and big data, its reflection on the citizens standard of living. *The International Conference on Artificial Intelligence and Computer Vision*, 235–247.

Al Shebli, K., Said, R. A., Taleb, N., Ghazal, T. M., Alshurideh, M. T., & Alzoubi, H. M. (2021). RTA's employees' perceptions toward the efficiency of artificial intelligence and big data utilization in providing smart services to the residents of Dubai. *The International Conference on Artificial Intelligence and Computer Vision*, 573–585.

AlHamad, A., Alshurideh, M., Alomari, K,, Kurdi, B., Alzoubi, H., Hamouche, S., & Al-Hawary, S. (2022). The effect of electronic human resources management on organizational health of telecommunications companies in Jordan. *International Journal of Data and Network Science, 6*(2), 429–438.

Alhamad, A. Q. M., Akour, I., Alshurideh, M., Al-Hamad, A. Q., Kurdi, B. A., & Alzoubi, H. (2021). Predicting the intention to use google glass: A comparative approach using machine learning models and PLS-SEM. *International Journal of Data and Network Science, 5*(3). https://doi.org/10.5267/j.ijdns.2021.6.002

Alhashmi, S. F. S., Alshurideh, M., Al Kurdi, B., & Salloum, S. A. (2020). A systematic review of the factors affecting the artificial intelligence implementation in the health care sector. In *Advances in intelligent systems and computing, AISC* (Vol. 1153). https://doi.org/10.1007/978-3-030-442 89-7_4

Ali, N., Ahmed, A., Anum, L., Ghazal, T. M., Abbas, S., Khan, M. A., Alzoubi, H. M., & Ahmad, M. (2021). Modelling supply chain information collaboration empowered with machine learning technique. *Intelligent Automation and Soft Computing, 30*(1), 243–257. https://doi.org/10.32604/iasc.2021.018983

Ali, N., Ghazal, M. T., Ahmed, A., Abbas, S., A. Khan, M., Alzoubi, H., Farooq, U., Ahmad, M., & Adnan Khan, M. (2022). Fusion-based supply chain collaboration using machine learning techniques. *Intelligent Automation & Soft Computing, 31*(3), 1671–1687. https://doi.org/10.32604/iasc.2022.019892

Alnazer, N. N., Alnuaimi, M. A., & Alzoubi, H. M. (2017). Analysing the appropriate cognitive styles and its effect on strategic innovation in Jordanian universities. *International Journal of Business Excellence, 13*(1), 127–140. https://doi.org/10.1504/IJBEX.2017.085799

Alnuaimi, M., Alzoubi, H. M., Ajelat, D., & Alzoubi, A. A. (2021). Towards intelligent organisations: An empirical investigation of learning orientation's role in technical innovation. *International Journal of Innovation and Learning, 29*(2), 207–221. https://doi.org/10.1504/IJIL.2021.112996

AlShamsi, M., Salloum, S. A., Alshurideh, M., & Abdallah, S. (2021). Artificial intelligence and blockchain for transparency in governance. In *Artificial intelligence for sustainable development: Theory, practice and future applications* (pp. 219–230). Springer.

Alsharari, N. (2021). Integrating blockchain technology with internet of things to efficiency. *International Journal of Technology, Innovation and Management (IJTIM), 1*(2), 1–13.

Alshurideh, M. (2022). Does electronic customer relationship management (E-CRM) affect service quality at private hospitals in Jordan? *Uncertain Supply Chain Management, 10*(2), 1–8.

Alshurideh, M., Gasaymeh, A., Ahmed, G., Alzoubi, H., & Kurd, B. A. (2020). Loyalty program effectiveness: Theoretical reviews and practical proofs. *Uncertain Supply Chain Management*, *8*(3). https://doi.org/10.5267/j.uscm.2020.2.003

Alshurideh, M. T., Al Kurdi, B., Alzoubi, H. M., Ghazal, T. M., Said, R. A., AlHamad, A. Q., Hamadneh, S., Sahawneh, N., & Al-kassem, A. H. (2022). Fuzzy assisted human resource management for supply chain management issues. *Annals of Operations Research*, 1–19.

Alwan, M., & Alshurideh, M. T. (2022). *The effect of digital marketing on purchase intention: Moderating effect of brand equity. March.* https://doi.org/10.5267/j.ijdns.2022.2.012

Alzoubi, A. (2021a). The impact of process quality and quality control on organizational competitiveness at 5-star hotels in Dubai. *International Journal of Technology, Innovation and Management (IJTIM)*, *1*(1), 54–68. https://doi.org/10.54489/ijtim.v1i1.14

Alzoubi, A. (2021b). Renewable green hydrogen energy impact on sustainability performance. *International Journal of Computations, Information and Manufacturing (IJCIM)*, *1*(1), 94–110. https://doi.org/10.54489/ijcim.v1i1.46

Alzoubi, H. M., Alshurideh, M., & Ghazal, T. M. (2021a). Integrating BLE beacon technology with intelligent information systems IIS for operations' performance: A managerial perspective. *The International Conference on Artificial Intelligence and Computer Vision*, 527–538.

Alzoubi, H. M., & Aziz, R. (2021). Does emotional intelligence contribute to quality of strategic decisions? The mediating role of open innovation. *Journal of Open Innovation: Technology, Market, and Complexity*, *7*(2), 130. https://doi.org/10.3390/joitmc7020130

Alzoubi, H. M., Vij, M., Vij, A., & Hanaysha, J. R. (2021b). What leads guests to satisfaction and loyalty in UAE five-star hotels? AHP analysis to service quality dimensions. *Enlightening Tourism*, *11*(1), 102–135. https://doi.org/10.33776/et.v11i1.5056

Alzoubi, H. M., & Yanamandra, R. (2020). Investigating the mediating role of information sharing strategy on agile supply chain. *Uncertain Supply Chain Management*, *8*(2), 273–284. https://doi.org/10.5267/j.uscm.2019.12.004

Alzoubi, H., & Ahmed, G. (2019). Do TQM practices improve organisational success? A case study of electronics industry in the UAE. *International Journal of Economics and Business Research*, *17*(4), 459–472. https://doi.org/10.1504/IJEBR.2019.099975

Alzoubi, H., Ahmed, G., Al-Gasaymeh, A., & Kurdi, B. (2020a). Empirical study on sustainable supply chain strategies and its impact on competitive priorities: The mediating role of supply chain collaboration. *Management Science Letters*, *10*(3), 703–708.

Alzoubi, H., Alshurideh, M., Kurdi, B., Akour, I., & Aziz, R. (2022). Does BLE technology contribute towards improving marketing strategies, customers' satisfaction and loyalty? The role of open innovation. *International Journal of Data and Network Science*, *6*(2), 449–460.

Alzoubi, H., Alshurideh, M., Kurdi, B. A., & Inairat, M. (2020b). Do perceived service value, quality, price fairness and service recovery shape customer satisfaction and delight? A practical study in the service telecommunication context. *Uncertain Supply Chain Management*, *8*(3), 579–588. https://doi.org/10.5267/j.uscm.2020.2.005

Aziz, N., & Aftab, S. (2021). Data mining framework for nutrition ranking: Methodology: SPSS modeller. *International Journal of Technology, Innovation and Management (IJTIM)*, *1*(1), 85–95.

Cruz, A. (2021). Convergence between blockchain and the internet of things. *International Journal of Technology, Innovation and Management (IJTIM)*, *1*(1), 35–56.

Eli, T. (2021). Students perspectives on the use of innovative and interactive teaching methods at the University of Nouakchott Al Aasriya, Mauritania: English Department as a case study. *International Journal of Technology, Innovation and Management (IJTIM)*, *1*(2), 90–104.

Farouk, M. (2021). The universal artificial intelligence efforts to face coronavirus COVID-19. *International Journal of Computations, Information and Manufacturing (IJCIM)*, *1*(1), 77–93. https://doi.org/10.54489/ijcim.v1i1.47

Ghazal, T. M. (2021). Positioning of UAV base stations using 5G and beyond networks for IoMT applications. *Arabian Journal for Science and Engineering*, 1–12.

Ghazal, T. M., Anam, M., Hasan, M. K., Hussain, M., Farooq, M. S., Ali, H. M., Ahmad, M., & Soomro, T. R. (2021a). Hep-Pred: Hepatitis C staging prediction using fine Gaussian SVM. *Comput Mater Continua, 69*, 191–203.

Ghazal, T. M., Hasan, M. K., Alshurideh, M. T., Alzoubi, H. M., Ahmad, M., Akbar, S. S., Al Kurdi, B., & Akour, I. A. (2021b). IoT for smart cities: Machine learning approaches in smart healthcare—A review. *Future Internet, 13*(8), 218. https://doi.org/10.3390/fi13080218

Ghazal, T. M., Hussain, M. Z., Said, R. A., Nadeem, A., Hasan, M. K., Ahmad, M., Khan, M. A., & Naseem, M. T. (2021c). *Performances of K-means clustering algorithm with different distance metrics.*

Ghazal, T. M., Said, R. A., & Taleb, N. (2021d). Internet of vehicles and autonomous systems with AI for medical things. *Soft Computing, 1–13.*

Ghazal, T., Soomro, T. R., & Shaalan, K. (2013). Integration of project management maturity (PMM) based on capability maturity model integration (CMMI). *European Journal of Scientific Research, 99*(3), 418–428.

Guergov, S., & Radwan, N. (2021). Blockchain convergence: Analysis of issues affecting IoT, AI and blockchain. *International Journal of Computations, Information and Manufacturing (IJCIM), 1*(1), 1–17. https://doi.org/10.54489/ijcim.v1i1.48

Hamadneh, S., Pedersen, O., & Al Kurdi, B. (2021). An investigation of the role of supply chain visibility into the Scottish bood supply chain. *Journal of Legal, Ethical and Regulatory Issues, 24*(Special Issue 1), 1–12.

Hanaysha, J. R., Al-Shaikh, M. E., Joghee, S., & Alzoubi, H. (2021a). Impact of innovation capabilities on business sustainability in small and medium enterprises. *FIIB Business Review, 1–12.* https://doi.org/10.1177/23197145211042232

Hanaysha, J. R., Al Shaikh, M. E., & Alzoubi, H. M. (2021b). Importance of marketing mix elements in determining consumer purchase decision in the retail market. *International Journal of Service Science, Management, Engineering, and Technology (IJSSMET), 12*(6), 56–72.

Joghee, S., Alzoubi, H. M., & Dubey, A. R. (2020). Decisions effectiveness of FDI investment biases at real estate industry: Empirical evidence from Dubai smart city projects. *International Journal of Scientific and Technology Research, 9*(3), 3499–3503.

Kalra, D., Ghazal, T. M., & Afifi, M. A. M. (2020). Integration of collaboration systems in hospitality management as a comprehensive solution. *International Journal of Advanced Science and Technology, 29*(8s), 3155–3173.

Kashif, A. A., Bakhtawar, B., Akhtar, A., Akhtar, S., Aziz, N., & Javeid, M. S. (2021). Treatment response prediction in hepatitis C patients using machine learning techniques. *International Journal of Technology, Innovation and Management (IJTIM), 1*(2), 79–89. https://doi.org/10.54489/ijtim.v1i2.24

Khan, M. A. (2021). Challenges facing the application of IoT in medicine and healthcare. *International Journal of Computations, Information and Manufacturing (IJCIM), 1*(1), 39–55. https://doi.org/10.54489/ijcim.v1i1.32

Khan, M. F., Ghazal, T. M., Said, R. A., Fatima, A., Abbas, S., Khan, M. A., Issa, G. F., Ahmad, M., & Khan, M. A. (2021a). An IoMT-enabled smart healthcare model to monitor elderly people using machine learning technique. *Computational Intelligence and Neuroscience, 2021a.*

Khan, Q.-T.-A., Ghazal, T., Abbas, S., Khan, W. A., Khan, M. A., Raed, A. S., Ahmad, M., & Asif, M. (2021b). Modeling habit patterns using conditional reflexes in agency. *Intelligent Automation and Soft Computing, 30*(2), 539–552.

Lee, C., & Ahmed, G. (2021). Improving IoT privacy, data protection and security concerns. *International Journal of Technology, Innovation and Management (IJTIM), 1*(1), 18–33. https://doi.org/10.54489/ijtim.v1i1.12

Lee, K., Azmi, N., Hanaysha, J., Alzoubi, H., & Alshurideh, M. (2022a). The effect of digital supply chain on organizational performance: An empirical study in Malaysia manufacturing industry. *Uncertain Supply Chain Management, 10*(2), 495–510.

Lee, K., Romzi, P., Hanaysha, J., Alzoubi, H., & Alshurideh, M. (2022b). Investigating the impact of benefits and challenges of IOT adoption on supply chain performance and organizational performance: An empirical study in Malaysia. *Uncertain Supply Chain Management, 10*(2), 537–550.

Lee, S.-W., Hussain, S., Issa, G. F., Abbas, S., Ghazal, T. M., Sohail, T., Ahmad, M., & Khan, M. A. (2021). Multi-dimensional trust quantification by artificial agents through evidential fuzzy multi-criteria decision making. *IEEE Access, 9*, 159399–159412.

Matloob, F., Ghazal, T. M., Taleb, N., Aftab, S., Ahmad, M., Khan, M. A., Abbas, S., & Soomro, T. R. (2021). Software defect prediction using ensemble learning: A systematic literature review. *IEEE Access.*

Mehmood, T. (2021). Does information technology competencies and fleet management practices lead to effective service delivery? Empirical evidence from E-commerce industry. *International Journal of Technology, Innovation and Management (IJTIM), 1*(2), 14–41.

Mehmood, T., Alzoubi, H. M., & Ahmed, G. (2019). Schumpeterian entrepreneurship theory: Evolution and relevance. *Academy of Entrepreneurship Journal, 25*(4).

Miller, D. (2021). The best practice of teach computer science students to use paper prototyping. *International Journal of Technology, Innovation and Management (IJTIM), 1*(2), 42–63. https://doi.org/10.54489/ijtim.v1i2.17

Mondol, E. P. (2021). The impact of block chain and smart inventory system on supply chain performance at retail industry. *International Journal of Computations, Information and Manufacturing (IJCIM), 1*(1), 56–76. https://doi.org/10.54489/ijcim.v1i1.30

Naqvi, R., Soomro, T. R., Alzoubi, H. M., Ghazal, T. M., & Alshurideh, M. T. (2021). The nexus between big data and decision-making: A study of big data techniques and technologies. *The International Conference on Artificial Intelligence and Computer Vision*, 838–853.

Nuseir, M. T., Al Kurdi, B. H., Alshurideh, M. T., & Alzoubi, H. M. (2021). Gender discrimination at workplace: Do artificial intelligence (AI) and machine learning (ML) have opinions about it. *The International Conference on Artificial Intelligence and Computer Vision*, 301–316.

Obaid, A. J. (2021). Assessment of smart home assistants as an IoT. *International Journal of Computations, Information and Manufacturing (IJCIM), 1*(1), 18–36. https://doi.org/10.54489/ijcim.v1i1.34

Radwan, N., & Farouk, M. (2021). The growth of internet of things (IoT) in the management of healthcare issues and healthcare policy development. *International Journal of Technology, Innovation and Management (IJTIM), 1*(1), 69–84. https://doi.org/10.54489/ijtim.v1i1.8

Rehman, E., Khan, M. A., Soomro, T. R., Taleb, N., Afifi, M. A., & Ghazal, T. M. (2021). Using blockchain to ensure trust between donor agencies and NGOs in under-developed countries. *Computers, 10*(8), 98.

Shamout, M., Ben-Abdallah, B., Alshurideh, M., Alzoubi, H., Al Kurdi, B., & Hamadneh, S. (2022). A conceptual model for the adoption of autonomous robots in supply chain and logistics industry. *Uncertain Supply Chain Management, 10*, 1–16.

Suleman, M., Soomro, T. R., Ghazal, T. M., & Alshurideh, M. (2021). Combating against potentially harmful mobile apps. *The International Conference on Artificial Intelligence and Computer Vision*, 154–173.

Tariq, E., Alshurideh, M., Akour, I., & Al-Hawary, S. (2022a). The effect of digital marketing capabilities on organizational ambidexterity of the information technology sector. *International Journal of Data and Network Science, 6*(2), 401–408.

Tariq, E., Alshurideh, M., Akour, I., Al-Hawary, S., & Al, B. (2022b). The role of digital marketing, CSR policy and green marketing in brand development. *International Journal of Data and Network Science, 6*(3), 1–10.

Yousuf, H., Zainal, A. Y., Alshurideh, M., & Salloum, S. A. (2021). Artificial intelligence models in power system analysis. In *Artificial intelligence for sustainable development: Theory, practice and future applications* (pp. 231–242). Springer.

Breast Cancer Prediction Using Machine Learning and Image Processing Optimization

Nidal A. Al-Dmour, Raed A. Said, Haitham M. Alzoubi⊙, Muhammad Alshurideh⊙, and Liaqat Ali

Abstract In this study, it is wanted to implement Naïve Bayes data mining techniques on a dataset gained from digitized Image of fine needle aspirate (or FNA) belonged to benign and malignant type of breast tumors. One of the most important factors causing cancers in the females, which can be measured from sample after maintaining in several hours' laboratory sample preparation at certain condition. As an alternative, Image processing can be substituted by measuring the geometrical features of FNA are well adopted economically and efficiently.

Keywords Data science · Female cancer tumor · Naïve Bayes · Clustering

N. A. Al-Dmour
Department of Computer Engineering, College of Engineering, Mutah University, Mu'tah, Jordan

R. A. Said
Faculty of Management, Canadian University Dubai, Dubai, UAE

H. M. Alzoubi (✉)
School of Business, Skyline University College, Sharjah, UAE
e-mail: haitham.alzubi@skylineuniversity.ac.ae

M. Alshurideh
Department of Marketing, School of Business, University of Jordan, Amman, Jordan
e-mail: m.alshurideh@ju.edu.jo; malshurideh@sharjah.ac.ae

Department of Management, College of Business Administration, University of Sharjah, Sharjah, UAE

L. Ali
College of Engineering and Information Technology, University of Science and Technology of Fujairah, Fujairah, UAE
e-mail: l.ali@ustf.ac.ae

© The Author(s), under exclusive license to Springer Nature Switzerland AG 2023
M. Alshurideh et al. (eds.), *The Effect of Information Technology on Business and Marketing Intelligence Systems*, Studies in Computational Intelligence 1056,
https://doi.org/10.1007/978-3-031-12382-5_113

1 Introduction

Despite of the fact that planar feature separation is the main point discussed on the literature, we would like all feature in a standalone model to use Naïve Base Prediction modelling along with the evaluation by a separate testing dataset (Akhtar et al., 2021; Al Ali, 2021; AlHamad et al., 2022). The objective is not only to evaluate the prediction on the cancer dataset but also to check if Bayes Naïve is a good substitute for feature extraction (Alnuaimi et al., 2021; Alsharari, 2021; Alshurideh et al., 2020a, 2020b, 2022; Alzoubi, 2021a, 2021b). The most important required libraries are discussed in the following section.

Naïve Bayes is the method used in Machine learning for the classification of two or more objects which are very similar to variance and regression analysis dependent variable is linearly relate to the other variables (Akour et al., 2021; Al Kurdi et al., 2021; Alhamad et al., 2021; Ali et al., 2021, 2022; Alnazer et al., 2017; Alshurideh et al., 2020a; Salloum et al., 2020). Technically this algorithm using Bayes conditional probability:

- C is the class label:

  - C ∈ {C1, C2, ... Cn}

- A is the observed object attributes

  - A = (a1, a2, ... am)
  - P(C | A) is the probability of C given A is observed Called the conditional probability.

For observed attributes A = (a1, a2, ... am), we want to compute and assign the classifier, Ci, with the largest P(Ci|A).

Two simplifications to the calculations.

Apply naïve assumption—each aj is conditionally independent of each other, then Denominator P(a1, a2, ... am) is a constant and can be ignored.

In this paper, the essential data preprocessing stages are described in Sect. 2 then in Sect. 3 the configuration of Python for the LDA analysis are explained. Modelling methods is the subject of Sect. 4 and their tuning parameters are described in Sect. 5. The results are shown for the classification accuracy are discussed and the confusion matrix is represented. At last, the results are summarized and the usage of the Machine Deep Learning in Tumor Dataset classification are discussed in the conclusion.

2 Dataset Description

A. Dataset Description: The dataset features are consisted of 10 * 3 features according to ten parameter's Mean, Standard Error and Worst Value as follow:

(a) radius (mean of distances from center to points on the perimeter)

(b) texture (standard deviation of gray-scale values)
(c) perimeter
(d) area
(e) smoothness (local variation in radius lengths)
(f) compactness (perimeter^2/area − 1.0)
(g) concavity (severity of concave portions of the contour)
(h) concave points (number of concave portions of the contour)
(i) symmetry
(j) fractal dimension ("coastline approximation" − 1).

These parameters are gained by Image Processing software. This method consisted of the following steps:

1. Kernel Boundary Estimation with image processing
2. Geometrical estimation of parameter statistical
3. Gradual Optimization by Kernel Estimation.

B. Identifying Python tools and libraries

For this module we will describe the required libraries and tools for python framework which can model the data as well as doing all data preparation. The code itself will be provided in html and pdf formats:

1. libraries for arraying algebra, Density and Probability scatters

    (a) pandas
    (b) numpy
    (c) plotly
    (d) os
    (e) seaborn
    (f) imblearn
    (g) plt
    (h) scatter_matrix
    (i) sklearn.

2. sklearn libraries for data preparation

    (a) train_test_split
    (b) preprocessing
    (c) LabelEncoder
    (d) RobustScaler.

3. sklearn libraries for data mining, validation and reporting

    (a) BernoulliNB
    (b) metrics
    (c) classification_report
    (d) confusion_matrix

(e)   accuracy_score.

In Tumor dataset 30 attributes are applied in the Gaussian Naïve Bayes to predict one target variable. At first data should imported in Jupiter notebook for further preparation by panda's library function read_csv.

All variables are real number other than diagnosis and ID which is integer which should be dropped from the variable raw_data and moved to train_raw. These variables' histograms are shown in Figs. 1 and 2.

3   Methodology

We have used python 3.6 to import the dataset into Jupyter Notebook. The details of results of coding within the description of all attributes are attached in the Appendix I; The distribution of diagnosis (target) is shown in Fig. 3 which should be drop to form to distinct variables X and then only diagnosis moved to variable y. Afterwards, the train_raw is splitted into train/test partitions with X_train, X_test, y_train and y_test. The train X_train has 398 with 30 columns while y_train would be 398with one column. On the other hand, the X_test has 171 records of data. The result of Gaussian fit function would be exported to pred variable.

After these preparation step test and train should be standardized to reduce the overfitting bias (Alzoubi & Aziz, 2021; Alzoubi & Yanamandra, 2020; Alzoubi et al., 2020a, 2020b, 2022; Nuseir et al., 2021). For this purpose, we have imported RobustScaler (Al-Hamadi et al., 2015, 2021; Ali & Dmour, 2021; Al-Naymat et al., 2021; Aslam et al., 2021; Bibi et al., 2021; Ghazal, 2021). The Robust scaler use a transform and fit functions which is robust to outlier and remove the median and scale all features (Alzoubi & Ahmed, 2019; Aziz & Aftab, 2021; Alzoubi et al., 2020a, 2020b, 2021; Cruz, 2021; Eli, 2021). The correlation map was also visualized in Fig. 4. The advantage is that it shows the range of variables with their color (Maasmi et al., 2021; Siddiqui et al., 2021; Zitar, 2021; Zitar et al., 2021).

4   Results

In the following a training table layout that allows visualization of the performance of Naïve Bayes model are shown.

Accuracy on train set 0.9422110552763819

        [[2427]
        [16133]]

precision recall   f1-score support

B 0.94 0.97 0.95 249
M 0.95 0.89 0.92 149

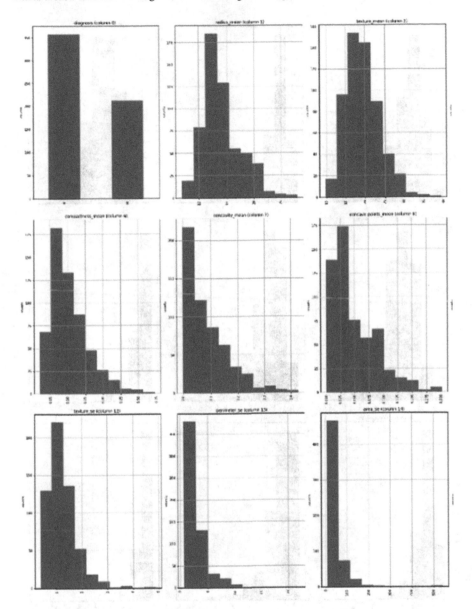

Fig. 1 Attribute histograms—first round

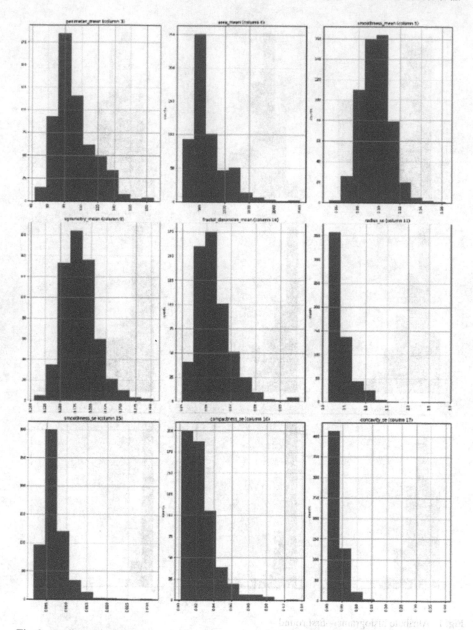

Fig. 2 Attribute histograms—second round

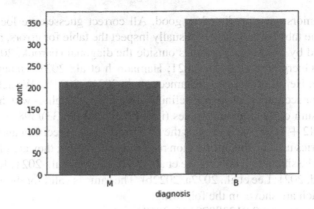

Fig. 3 Diagnosis class distribution

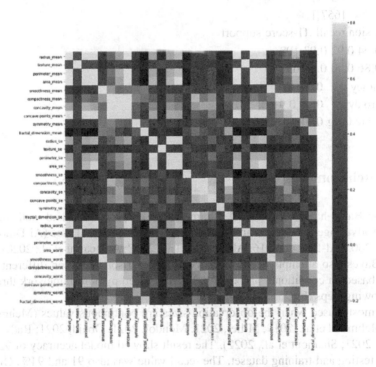

Fig. 4 Correlation map

accuracy   0.94 398
macro avg   0.94 0.93 0.94 398
weighted avg 0.94 0.94 0.94 398

Of 398 and 171 samples, the system predicted that there were M and B tumors, and of the 258 good tumor, the model predicted 16 as bad and similarly 7 of the actual bad tumors were predicted as good. All correct guesses are located in the diagonal of the table, so it's easy to visually inspect the table for errors, as they will be represented by any non-zero values outside the diagonal (Farouk, 2021; Ghazal et al., 2021; Guergov & Radwan, 2021; Hamadneh et al., 2021; Hanaysha et al., 2021a, 2021b; Hejazi et al., 2021; Khamees et al., 2021a, 2021b). We define overall success rate (or accuracy) as a metric defining—what we got right—which is the ratio between the sum of the diagonal values (i.e., TP = (242 + 133)/398 = 94.2% and TN = 242/(242 + 7) = 97.2%) versus the sum of the table. Precision and Recall are accuracy metrics used by the information retrieval community; they are often used to characterize classifiers as well (Joghee et al., 2020; Kashif et al., 2021; Khan, 2021; Lee & Ahmed, 2021; Lee et al., 2022a, 2022b). The same is valid for the 171 records of testing which are shown in the following:

Accuracy on test set 0.9122807017543859

      [[999]

      [657]]

precision recall f1-score support

B 0.94 0.92 0.93 108

M 0.86 0.90 0.88 63

accuracy   0.91 171

macro avg   0.90 0.91 0.91 171

weighted avg 0.91 0.91 0.91 171

5 Conclusion

W Naïve Bayes has a best optimal runtime based on probability and mathematics with the advantage of being computationally inexpensive for real time (Al-Dmour & Teahan, 2005; Al-Dmour, 2016; Al Hamadi et al., 2017; Al Neaimi et al., 2020). The Naïve Bayes also attempts to model pairwise comparisons between different data classes based on conditional probability. We have used python framework through the following steps (Fig. 5).

It is mostly used when the observation sizes are continuous values (Mehmood, 2021; Mehmood et al., 2019; Miller, 2021; Mondol, 2021; Obaid, 2021; Radwan & Farouk, 2021; Shamout et al., 2022). The result showed model accuracy of 92 and 94% in testing and training dataset. The recall value was also 91 and 94%. Of 398 and 171 samples, 375 and 166 records were correctly classified.

We have analyzed the features through different literature articles which was often modelled in planar projected dimension and our analysis showed that the result of our methodology can be simpler with less computation costs in Python framework.

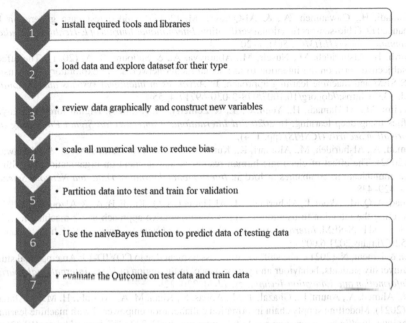

Fig. 5 Step by step python code

References

Akhtar, A., Akhtar, S., Bakhtawar, B., Kashif, A. A., Aziz, N., & Javeid, M. S. (2021). COVID-19 detection from CBC using machine learning techniques. *International Journal of Technology, Innovation and Management (IJTIM)*, *1*(2), 65–78. https://doi.org/10.54489/ijtim.v1i2.22

Akour, I., Alshurideh, M., Al Kurdi, B., Al Ali, A., & Salloum, S. (2021). Using machine learning algorithms to predict people's intention to use mobile learning platforms during the COVID-19 pandemic: Machine learning approach. *JMIR Medical Education, 7*(1), 1–17.

Al-Dmour, N. (2016). Using unstructured search algorithms for data collection in IoT-based WSN. *International Journal of Engineering Research and Technology. ISSN*, 974–3154.

Al-Dmour, N. A., & Teahan, W. J. (2005). Peer-to-peer protocols for resource discovery in the Grid. *Parallel and Distributed Computing and Networks*, 319–324.

Al-Hamadi, H., Gawanmeh, A., & Al-Qutayri, M. (2015). An automatic ECG generator for testing and evaluating ECG sensor algorithms. In *2015 10th International Design & Test Symposium (IDT)* (pp. 78–83).

Al-Hamadi, H., Nasir, N., Yeun, C. Y., & Damiani, E. (2021). A verified protocol for secure autonomous and cooperative public transportation in smart cities. *IEEE International Conference on Communications Workshops (ICC Workshops), 2021*, 1–6.

Al-Naymat, G., Hussain, H., Al-Kasassbeh, M., & Al-Dmour, N. (2021). Accurate detection of network anomalies within SNMP-MIB data set using deep learning. *International Journal of Computer Applications in Technology, 66*(1), 74–85.

Al Ali, A. (2021). The impact of information sharing and quality assurance on customer service at UAE banking sector. *International Journal of Technology, Innovation and Management (IJTIM)*, *1*(1), 01–17. https://doi.org/10.54489/ijtim.v1i1.10

Al Hamadi, H., Gawanmeh, A., & Al-Qutayri, M. (2017). Guided test case generation for enhanced ECG bio-sensors functional verification. *International Journal of E-Health and Medical Communications (IJEHMC), 8*(4), 1–20.

Al Kurdi, B., Alshurideh, M., Nuseir, M., Aburayya, A., & Salloum, S. A. (2021). The effects of subjective norm on the intention to use social media networks: An exploratory study using PLS-SEM and machine learning approach. In *Advances in intelligent systems and computing* (Vol. 1339). https://doi.org/10.1007/978-3-030-69717-4_55

Al Neaimi, M., Al Hamadi, H., Yeun, C. Y., & Zemerly, M. J. (2020). Digital forensic analysis of files using deep learning. In *2020 3rd International Conference on Signal Processing and Information Security (ICSPIS)* (pp. 1–4).

AlHamad, A., Alshurideh, M., Alomari, K., Kurdi, B., Alzoubi, H., Hamouche, S., & Al-Hawary, S. (2022). The effect of electronic human resources management on organizational health of telecommunications companies in Jordan. *International Journal of Data and Network Science, 6*(2), 429–438.

Alhamad, A. Q. M., Akour, I., Alshurideh, M., Al-Hamad, A. Q., Kurdi, B. A., & Alzoubi, H. (2021). Predicting the intention to use google glass: A comparative approach using machine learning models and PLS-SEM. *International Journal of Data and Network Science, 5*(3). https://doi.org/10.5267/j.ijdns.2021.6.002

Ali, L., & Dmour, N. (2021). The shift to online assessment due to COVID-19: An empirical study of university students, behaviour and performance, in the region of UAE. *International Journal of Information and Education Technology, 11*(5), 220–228.

Ali, N., Ahmed, A., Anum, L., Ghazal, T. M., Abbas, S., Khan, M. A., Alzoubi, H. M., & Ahmad, M. (2021). Modelling supply chain information collaboration empowered with machine learning technique. *Intelligent Automation and Soft Computing, 30*(1), 243–257. https://doi.org/10.32604/iasc.2021.018983

Ali, N., Ghazal, M. T., Ahmed, A., Abbas, S., A. Khan, M., Alzoubi, H., Farooq, U., Ahmad, M., & Adnan Khan, M. (2022). Fusion-based supply chain collaboration using machine learning techniques. *Intelligent Automation & Soft Computing, 31*(3), 1671–1687. https://doi.org/10.32604/iasc.2022.019892

Alnazer, N. N., Alnuaimi, M. A., & Alzoubi, H. M. (2017). Analysing the appropriate cognitive styles and its effect on strategic innovation in Jordanian universities. *International Journal of Business Excellence, 13*(1), 127–140. https://doi.org/10.1504/IJBEX.2017.085799

Alnuaimi, M., Alzoubi, H. M., Ajelat, D., & Alzoubi, A. A. (2021). Towards intelligent organisations: An empirical investigation of learning orientation's role in technical innovation. *International Journal of Innovation and Learning, 29*(2), 207–221. https://doi.org/10.1504/IJIL.2021.112996

Alsharari, N. (2021). Integrating blockchain technology with internet of things to efficiency. *International Journal of Technology, Innovation and Management (IJTIM), 1*(2), 1–13.

Alshurideh, M., Al Kurdi, B., Salloum, S. A., Arpaci, I., & Al-Emran, M. (2020a). Predicting the actual use of m-learning systems: A comparative approach using PLS-SEM and machine learning algorithms. *Interactive Learning Environments.* https://doi.org/10.1080/10494820.2020.1826982

Alshurideh, M., Gasaymeh, A., Ahmed, G., Alzoubi, H., & Kurd, B. A. (2020b). Loyalty program effectiveness: Theoretical reviews and practical proofs. *Uncertain Supply Chain Management, 8*(3). https://doi.org/10.5267/j.uscm.2020.2.003

Alshurideh, M. T., Al Kurdi, B., Alzoubi, H. M., Ghazal, T. M., Said, R. A., AlHamad, A. Q., Hamadneh, S., Sahawneh, N., & Al-kassem, A. H. (2022). Fuzzy assisted human resource management for supply chain management issues. *Annals of Operations Research*, 1–19.

Alzoubi, A. (2021a). The impact of process quality and quality control on organizational competitiveness at 5-star hotels in Dubai. *International Journal of Technology, Innovation and Management (IJTIM), 1*(1), 54–68. https://doi.org/10.54489/ijtim.v1i1.14

Alzoubi, A. (2021b). Renewable green hydrogen energy impact on sustainability performance. *International Journal of Computations, Information and Manufacturing (IJCIM), 1*(1), 94–110. https://doi.org/10.54489/ijcim.v1i1.46

Alzoubi, H. M., & Aziz, R. (2021). Does emotional intelligence contribute to quality of strategic decisions? The mediating role of open innovation. *Journal of Open Innovation: Technology, Market, and Complexity, 7*(2), 130. https://doi.org/10.3390/joitmc7020130

Alzoubi, H. M., Vij, M., Vij, A., & Hanaysha, J. R. (2021). What leads guests to satisfaction and loyalty in UAE five-star hotels? AHP analysis to service quality dimensions. *Enlightening Tourism, 11*(1), 102–135. https://doi.org/10.33776/et.v11i1.5056

Alzoubi, H. M., & Yanamandra, R. (2020). Investigating the mediating role of information sharing strategy on agile supply chain. *Uncertain Supply Chain Management, 8*(2), 273–284. https://doi.org/10.5267/j.uscm.2019.12.004

Alzoubi, H., & Ahmed, G. (2019). Do TQM practices improve organisational success? A case study of electronics industry in the UAE. *International Journal of Economics and Business Research, 17*(4), 459–472. https://doi.org/10.1504/IJEBR.2019.099975

Alzoubi, H., Ahmed, G., Al-Gasaymeh, A., & Kurdi, B. (2020a). Empirical study on sustainable supply chain strategies and its impact on competitive priorities: The mediating role of supply chain collaboration. *Management Science Letters, 10*(3), 703–708.

Alzoubi, H., Alshurideh, M., Kurdi, B., Akour, I., & Aziz, R. (2022). Does BLE technology contribute towards improving marketing strategies, customers' satisfaction and loyalty? The role of open innovation. *International Journal of Data and Network Science, 6*(2), 449–460.

Alzoubi, H., Alshurideh, M., Kurdi, B. A., & Inairat, M. (2020b). Do perceived service value, quality, price fairness and service recovery shape customer satisfaction and delight? A practical study in the service telecommunication context. *Uncertain Supply Chain Management, 8*(3), 579–588. https://doi.org/10.5267/j.uscm.2020.2.005

Aslam, M. S., Ghazal, T. M., Fatima, A., Said, R. A., Abbas, S., Khan, M. A., Siddiqui, S. Y., & Ahmad, M. (2021). *Energy-efficiency model for residential buildings using supervised machine learning algorithm.*

Aziz, N., & Aftab, S. (2021). Data mining framework for nutrition ranking: Methodology: SPSS modeller. *International Journal of Technology, Innovation and Management (IJTIM), 1*(1), 85–95.

Bibi, R., Saeed, Y., Zeb, A., Ghazal, T. M., Rahman, T., Said, R. A., Abbas, S., Ahmad, M., & Khan, M. A. (2021). Edge AI-based automated detection and classification of road anomalies in VANET using deep learning. *Computational Intelligence and Neuroscience, 2021.*

Cruz, A. (2021). Convergence between blockchain and the internet of things. *International Journal of Technology, Innovation and Management (IJTIM), 1*(1), 35–56.

Eli, T. (2021). Students perspectives on the use of innovative and interactive teaching methods at the University of Nouakchott Al Aasriya, Mauritania: English Department as a case study. *International Journal of Technology, Innovation and Management (IJTIM), 1*(2), 90–104.

Farouk, M. (2021). The universal artificial intelligence efforts to face coronavirus COVID-19. *International Journal of Computations, Information and Manufacturing (IJCIM), 1*(1), 77–93. https://doi.org/10.54489/ijcim.v1i1.47

Ghazal, T. M. (2021). Internet of things with artificial intelligence for health care security. *Arabian Journal for Science and Engineering*, 1–12.

Ghazal, T. M., Hasan, M. K., Alshurideh, M. T., Alzoubi, H. M., Ahmad, M., Akbar, S. S., Al Kurdi, B., & Akour, I. A. (2021). IoT for smart cities: Machine learning approaches in smart healthcare—A review. *Future Internet, 13*(8), 218. https://doi.org/10.3390/fi13080218

Guergov, S., & Radwan, N. (2021). Blockchain convergence: Analysis of issues affecting IoT, AI and blockchain. *International Journal of Computations, Information and Manufacturing (IJCIM), 1*(1), 1–17. https://doi.org/10.54489/ijcim.v1i1.48

Hamadneh, S., Pedersen, O., & Al Kurdi, B. (2021). An investigation of the role of supply chain visibility into the Scottish bood supply chain. *Journal of Legal, Ethical and Regulatory Issues, 24*(Special Issue 1), 1–12.

Hanaysha, J. R., Al-Shaikh, M. E., Joghee, S., & Alzoubi, H. (2021a). Impact of innovation capabilities on business sustainability in small and medium enterprises. *FIIB Business Review*, 1–12. https://doi.org/10.1177/23197145211042232

Hanaysha, J. R., Al Shaikh, M. E., & Alzoubi, H. M. (2021b). Importance of marketing mix elements in determining consumer purchase decision in the retail market. *International Journal of Service Science, Management, Engineering, and Technology (IJSSMET)*, 12(6), 56–72.

Hejazi, H. D., Khamees, A. A., Alshurideh, M., & Salloum, S. A. (2021). Arabic text generation: Deep learning for poetry synthesis. In *Advances in intelligent systems and computing* (Vol. 1339). https://doi.org/10.1007/978-3-030-69717-4_11

Joghee, S., Alzoubi, H. M., & Dubey, A. R. (2020). Decisions effectiveness of FDI investment biases at real estate industry: Empirical evidence from Dubai smart city projects. *International Journal of Scientific and Technology Research*, 9(3), 3499–3503.

Kashif, A. A., Bakhtawar, B., Akhtar, A., Akhtar, S., Aziz, N., & Javeid, M. S. (2021). Treatment response prediction in hepatitis C patients using machine learning techniques. *International Journal of Technology, Innovation and Management (IJTIM)*, 1(2), 79–89. https://doi.org/10.54489/ijtim.v1i2.24

Khamees, A. A., Hejazi, H. D., Alshurideh, M., & Salloum, S. A. (2021a). Classifying audio music genres using a multilayer sequential model. In *Advances in intelligent systems and computing* (Vol. 1339). https://doi.org/10.1007/978-3-030-69717-4_30

Khamees, A. A., Hejazi, H. D., Alshurideh, M., & Salloum, S. A. (2021b). Classifying audio music genres using CNN and RNN. In *Advances in intelligent systems and computing* (Vol. 1339). https://doi.org/10.1007/978-3-030-69717-4_31

Khan, M. A. (2021). Challenges facing the application of IoT in medicine and healthcare. *International Journal of Computations, Information and Manufacturing (IJCIM)*, 1(1), 39–55. https://doi.org/10.54489/ijcim.v1i1.32

Lee, C., & Ahmed, G. (2021). Improving IoT privacy, data protection and security concerns. *International Journal of Technology, Innovation and Management (IJTIM)*, 1(1), 18–33. https://doi.org/10.54489/ijtim.v1i1.12

Lee, K., Azmi, N., Hanaysha, J., Alzoubi, H., & Alshurideh, M. (2022a). The effect of digital supply chain on organizational performance: An empirical study in Malaysia manufacturing industry. *Uncertain Supply Chain Management*, 10(2), 495–510.

Lee, K., Romzi, P., Hanaysha, J., Alzoubi, H., & Alshurideh, M. (2022b). Investigating the impact of benefits and challenges of IOT adoption on supply chain performance and organizational performance: An empirical study in Malaysia. *Uncertain Supply Chain Management*, 10(2), 537–550.

Maasmi, F., Morcos, M., Al Hamadi, H., & Damiani, E. (2021). Identifying applications' state via system calls activity: A pipeline approach. In *2021 28th IEEE International Conference on Electronics, Circuits, and Systems (ICECS)* (pp. 1–6).

Mehmood, T. (2021). Does information technology competencies and fleet management practices lead to effective service delivery? Empirical evidence from E-commerce industry. *International Journal of Technology, Innovation and Management (IJTIM)*, 1(2), 14–41.

Mehmood, T., Alzoubi, H. M., & Ahmed, G. (2019). Schumpeterian entrepreneurship theory: Evolution and relevance. *Academy of Entrepreneurship Journal*, 25(4).

Miller, D. (2021). The best practice of teach computer science students to use paper prototyping. *International Journal of Technology, Innovation and Management (IJTIM)*, 1(2), 42–63. https://doi.org/10.54489/ijtim.v1i2.17

Mondol, E. P. (2021). The impact of block chain and smart inventory system on supply chain performance at retail industry. *International Journal of Computations, Information and Manufacturing (IJCIM)*, 1(1), 56–76. https://doi.org/10.54489/ijcim.v1i1.30

Nuseir, M. T., Al Kurdi, B. H., Alshurideh, M. T., & Alzoubi, H. M. (2021). Gender discrimination at workplace: Do artificial intelligence (AI) and machine learning (ML) have opinions about it. *The International Conference on Artificial Intelligence and Computer Vision*, 301–316.

Obaid, A. J. (2021). Assessment of smart home assistants as an IoT. *International Journal of Computations, Information and Manufacturing (IJCIM)*, 1(1), 18–36. https://doi.org/10.54489/ijcim.v1i1.34

Radwan, N., & Farouk, M. (2021). The growth of internet of things (IoT) in the management of healthcare issues and healthcare policy development. *International Journal of Technology, Innovation and Management (IJTIM)*, *1*(1), 69–84. https://doi.org/10.54489/ijtim.v1i1.8

Salloum, S. A., Alshurideh, M., Elnagar, A., & Shaalan, K. (2020). Machine learning and deep learning techniques for cybersecurity: A review. In *Joint European-US Workshop on Applications of Invariance in Computer Vision* (pp. 50–57).

Shamout, M., Ben-Abdallah, B., Alshurideh, M., Alzoubi, H., Al Kurdi, B., & Hamadneh, S. (2022). A conceptual model for the adoption of autonomous robots in supply chain and logistics industry. *Uncertain Supply Chain Management*, *10*, 1–16.

Siddiqui, S. Y., Haider, A., Ghazal, T. M., Khan, M. A., Naseer, I., Abbas, S., Rahman, M., Khan, J. A., Ahmad, M., & Hasan, M. K. (2021). IoMT cloud-based intelligent prediction of breast cancer stages empowered with deep learning. *IEEE Access, 9*, 146478–146491.

Zitar, R. A. (2021). A review for the genetic algorithm and the red deer algorithm applications. In *2021 14th International Congress on Image and Signal Processing, BioMedical Engineering and Informatics (CISP-BMEI)* (pp. 1–6)

Zitar, R. A., Abualigah, L., & Al-Dmour, N. A. (2021). Review and analysis for the red deer algorithm. *Journal of Ambient Intelligence and Humanized Computing*, 1–11.

Development of Data Mining Framework Cardiovascular Disease Prediction

Raed A. Said, Nidal A. Al-Dmour, Mohammed Salahat, Ghassan F. Issa, Haitham M. Alzoubi⬤, and Muhammad Alshurideh⬤

Abstract One of the highest shares of data-driven technology of health sector happens for private insurance stakeholders. It is therefore clear that private insurance companies can only survive being competitive in covering different medical stages such as surgery, intervention and other clinical trials in a high-risk environment. Estimation of expected costs and coverage is also important for both patient and insurer. In this case study we as a Data Mining and Artificial Business consultant want to explore different techniques of data mining to find out business risks for patients. We have asked the insurer to provide us a sizable medical history to watch those features. We would like to predict if given biographical profile of the patient along with exam results can predict CVD so he can cover his costs with this Insurer. On the other hand, in case of higher error of misclassified CVD what kind of decision should be taken by risk holder and insurer. Which one of these attributes

R. A. Said
Faculty of Management, Canadian University Dubai, Dubai, UAE

N. A. Al-Dmour
Department of Computer Engineering, College of Engineering, Mutah University, Mu'tah, Jordan

M. Salahat
College of Engineering and Information Technology, University of Science and Technology of Fujairah, Fujairah, UAE
e-mail: m.salahat@ustf.ac.ae

G. F. Issa
School of Information Technology, Skyline University College, Sharjah, UAE
e-mail: ghassan.issa@skylineuniversity.ac.ae

H. M. Alzoubi (✉)
School of Business, Skyline University College, Sharjah, UAE
e-mail: haitham.alzubi@skylineuniversity.ac.ae

M. Alshurideh
Department of Marketing, School of Business, University of Jordan, Amman, Jordan
e-mail: m.alshurideh@ju.edu.jo; malshurideh@sharjah.ac.ae

Department of Management, College of Business Administration, University of Sharjah, Sharjah, UAE

M. Alshurideh et al. (eds.), *The Effect of Information Technology on Business and Marketing Intelligence Systems*, Studies in Computational Intelligence 1056,
https://doi.org/10.1007/978-3-031-12382-5_114

2081

causing this cost and what other stakeholders like target group of patients can be suffered from the loss? The ultimate goal is to develop a model that can predict the gap between those patients' perception of their disease and their real disease. This can further help stakeholders to develop specific insurance policy.

Keywords Data science · CVD · Data mining · Disease prediction

1 Introduction

The smart healthcare is composed of multiple participants that include patients, doctors, research institutions, and hospitals (Akhtar et al., 2021; Al Ali, 2021; AlHamad et al., 2022; Alhamad et al., 2021; Ali et al., 2021, 2022). In essence, it is considered as an organic whole that involves various dimensions such as disease control and monitoring, diagnosis, treatment, health decision making, hospital management and medical research (Aburayya et al., 2020a, 2020b, 2020c; Alhashmi et al., 2020; Taryam et al., 2020). It is clear that Artificial Intelligence can be considered from Information marketplace perspective as well as IT governance among stakeholders. But Cardiovascular Disease (CVD) exceptionally has been paid more attention among others (Alnazer et al., 2017; Alnuaimi et al., 2021; Alsharari, 2021; Alshurideh et al., 2020, 2022; Alzoubi, 2021a). Linkage between CVD and other diseases are the subject of debate for instance look at how Cerebrovascular diseases being accompanied with mental impairment resulted in higher fatal rate. On the other hand, any patient suffering from CVD should experience extremely condition for his or her rest of life. What Data science as a framework can serve for this diagnosis of CVD is a kind of unstructured machine learning which can be best interpreted when is coupled with biography of the patient in features such as age, weight, physical activities etc. (Alzoubi, 2021b; Alzoubi & Ahmed, 2019; Alzoubi & Aziz, 2021; Alzoubi & Yanamandra, 2020; Alzoubi et al., 2020a, 2020b, 2021b, 2022) (Fig. 1).

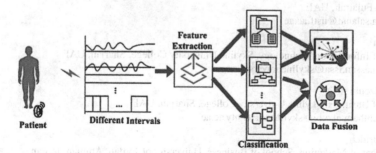

Fig. 1 Smart healthcare system

2 Dataset Description

A. Preprocessing

The raw version of dataset originally consists of 70,000 records in 12 features patient biographical, medical exams and some information given by the patient like the smoking attitude etc. needed to implement with various data mining techniques in the IBM Modeler software (Aziz & Aftab, 2021; Cruz, 2021; Eli, 2021; Farouk, 2021; Ghazal et al., 2021b). Among the 12 columns of data, six of them were categorical, five numerical. The two remaining namely ID and age in days decided to be dropped out from the analysis which was filtered in source node of IBM SPSS. All data roles were set to Input except than presence of CVD as target. There was no missing or duplicated column in data. The numerical fields was checked for the outlier by comparing the range in boxplots of 3σ (95%). Gender, smoking, alcohol and physical activities as binary categorical data (Guergov & Radwan, 2021; Hamadneh et al., 2021; Hanaysha et al., 2021a, 2021b; Joghee et al., 2020). Furthermore, the cardinality of the Cholesterol and Glucose categorical data was transferred from well above normal, above normal and normal to 0, 1, 2. Here the model prediction power can be enhanced according to the label's value which is more favorable. In Fig. 2, the name and type of all variables plus target role presense or absense of cardiovascular disease can be observed (following figure).

In Fig. 3, the distribution and statistical representation of all variables observed (following figure).

Measurement unit of height is centimeter and weight is kg. Systolic blood pressure is the exerted pressure when blood is flowed in arteries (Kashif et al., 2021; Khan, 2021; Lee & Ahmed, 2021; Lee et al., 2022a, 2022b; Mehmood, 2021). Its normal value is 120 mmhg or below. Diastolic blood pressure is the measured pressure of blood between arteries and heartbeat with the normal value of 80 mmhg or below. The Cholesterol is a kind of fat measure which is ranged between 200 and 239 mg/dl. Glucose level is ranged from 100 to 140 mg/dl 8 h mg/dl hours before and 2 h after eating. For people without Diabetes is around 70–80 mg/dl (Fig. 4).

B. Bivariate Analysis

By univariate analysis we have observed that people older than 55 years are more suffered from CVD. We have also implemented the other categorical variables grouped by age which are shown in Fig. 5.

3 Methodology

Machine Learning of the selected dataset could semi-structured in in methods nature which can not only be perceived as clustering problem but also can be seen as a classification problem (Mehmood et al., 2019; Miller, 2021; Mondol, 2021; Obaid, 2021; Radwan & Farouk, 2021; Shamout et al., 2022). For initiating the modelling,

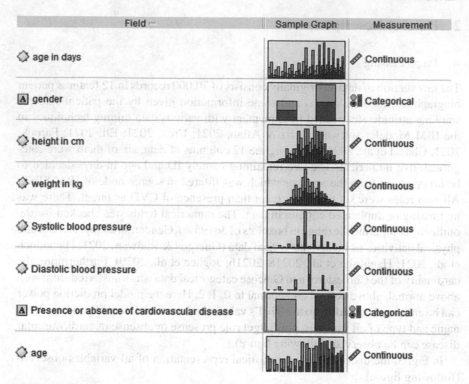

Fig. 2 Dataset fields

a feature selection was performed. The results of the feature selection are shown in Fig. 6.

4 Modelling

It is possible to derive more variable such BMI from our original dataset but because the coefficient of variation was far below threshold, we have selected only weight as the demonstration of the fat and glucose. By using the auto-classifier node we have set the baseline for model selection (Afifi et al., 2020; Al Batayneh et al., 2021; Ali et al., 2021; Alzoubi et al., 2021a; Ghazal, 2021; Ghazal et al., 2021a). The raw propensity with weighted voting Ensemble was selected between 9 different models. One advantage of this method was that exact feature could be defined by users. We have considered the selected features from the Fig. 6. On the other hand, the cut-off for calculation are outliers such as runtime, models accuracy and AUC configured to outrank when are below 50% (Ghazal et al., 2013; Ghazal et al., 2021c, 2021d; Kalra et al., 2020; Khan et al., 2021a, 2021b). This method is computationally cost effective ranking the models according to their number of features taken 3 min

Min	Max	Sum	Range	Mean	Mean Std. Err.	Std. Dev
14275	23713	1362850702	9438	19469.296	9.322	2466.411
--	--	--	--	--	--	--
140	188	11509355	48	164.419	0.029	7.718
32	117	5159050	85	73.701	0.050	13.214
76	177	8879851	101	126.855	0.063	16.581
51	111	5699522	60	81.422	0.035	9.328
--	--	--	--	--	--	--
39	64	3698930	25	52.842	0.026	6.764

Fig. 3 Statistical criteria

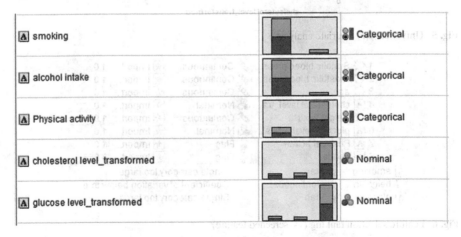

Fig. 4 Categorical variables

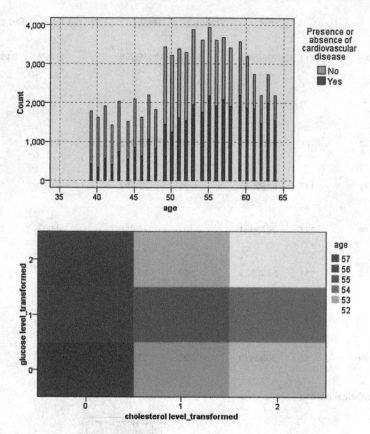

Fig. 5 Univariate and bivariate analysis

1 ◇ Systolic blood press...	✎ Continuous	★ Import...	1.0
2 ◇ Diastolic blood pres...	✎ Continuous	★ Import...	1.0
3 ◇ age	✎ Continuous	★ Import...	1.0
4 🅰 cholesterol level_tra...	🔸 Nominal	★ Import...	1.0
5 ◇ weight in kg	✎ Continuous	★ Import...	1.0
6 🅰 glucose level_transf...	🔸 Nominal	★ Import...	1.0
7 🅰 Physical activity	🔘 Flag	★ Import...	1.0
8 🅰 gender	🔘 Flag	★ Import...	0.968
🅰 smoking	🔘 Flag	Single category too large	
◇ height in ...	✎ Continuous	Coefficient of variation below thre...	
🅰 alcohol i...	🔘 Flag	Single category too large	

Fig. 6 Feature selection ranking (vs. screened feature)

and 22 s to finish calculation on 9 different models. The feature importance with modelling accuracies is summarized in Fig. 7.

The last two models namely Decision List and Discrimination Analysis was our choice for modelling. Linear Discrimination Analysis (LDA) is the method used in Machine learning for the recognition of two or more objects (Khan et al., 2021a, 2021b; Lee et al., 2021; Matloob et al., 2021). In LDA, which is very similar to variance and regression analysis dependent variable is linearly relate to the other variables. The difference between these methods is that in LDA, instead of distance relationship, dependent variable is solved by the ranking or ordinal outranking. The LDA is also closely related to Principal Component Analysis (PCA) core component analysis. Because both methods look for a linear combination of variables that best describe the data (Naqvi et al., 2021; Rehman et al., 2021; Suleman et al., 2021). The LDA also attempts to model differences between different data classes. The

Graph	Model	Build Time (mins)	Max Profit	Max Profit Occurs in	Lift[Top 30...	Overall Accuracy	No. Fields Used	Area Under Curve
	C5 1	1	83918.188	47	1.672	74.013	11	0.789
	Logistic regr...	1	80,850.0	48	1.649	72.974	11	0.793
	Bayesian N...	1	79,782.903	46	1.668	72.756	11	0.793
	Discriminan...	1	76,548.0	46	1.627	71.913	5	0.782
	Decision Lis...	1	73,789.935	34	1.654	71.183	3	0.718

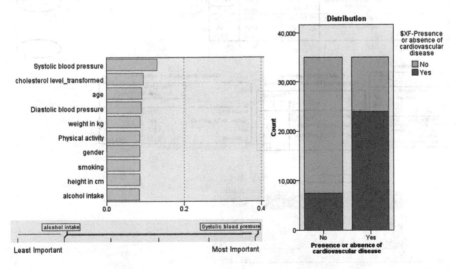

Fig. 7 Feature engineering

LDA is used when the observation sizes are continuous values. The other implemented algorithm was Decision List which is similar to decision tree. The method provides a set of classification rule that form an inherited order which make it easy for interpretation.

5 Results

The result of LDA was implemented for variable Systolic Blood Pressure fields. While no significant difference between Target grouping was observed in the covariance demonstrated for weight, the age parameter was significantly higher in No CVD patients (Fig. 8).

It should be noted that normal Value of Systolic Pressure is less than 120. This is related to age and weight according to the Discrimination function. The result of accuracy and confusion Matrix for the LDA analysis are shown in the following Fig. 9.

As can be seen in the above figure the more gain cannot be resulted by balancing the ratio of CVD to no CVD. The slight skewness between the cost of prediction FP and TP is on the basis of relationship between the Systolic Blood Pressure and

Covariance Matrices

Presence or absence of cardiovascular disease		weight in kg	Systolic blood pressure	age
Yes	weight in kg	180.624	41.658	-2.537
	Systolic blood pressure	41.658	287.982	3.988
	age	-2.537	3.988	40.378
No	weight in kg	157.068	40.617	5.813
	Systolic blood pressure	40.617	157.400	19.625
	age	5.813	19.625	45.957
Total	weight in kg	174.601	58.495	5.504
	Systolic blood pressure	58.495	274.921	23.446
	age	5.504	23.446	45.758

Structure Matrix

	Function
	1
Systolic blood pressure	.917
Diastolic blood pressure	.700
age	.463
weight in kg	.350

Pooled within-groups correlations between discriminating variables and standardized canonical discriminant functions
Variables ordered by absolute size of correlation within function.

Standardized Canonical Discriminant Function Coefficients

	Function
	1
weight in kg	.157
Systolic blood pressure	.758
Diastolic blood pressure	.119
age	.359

Classification Function Coefficients

	Presence or absence of cardiovascular disease	
	Yes	No
weight in kg	.322	.309
Systolic blood pressure	.483	.424
age	1.117	1.059
(Constant)	-75.772	-64.193

Fisher's linear discriminant functions

Predictor Importance
Target: Presence or absence of cardiovascular disease

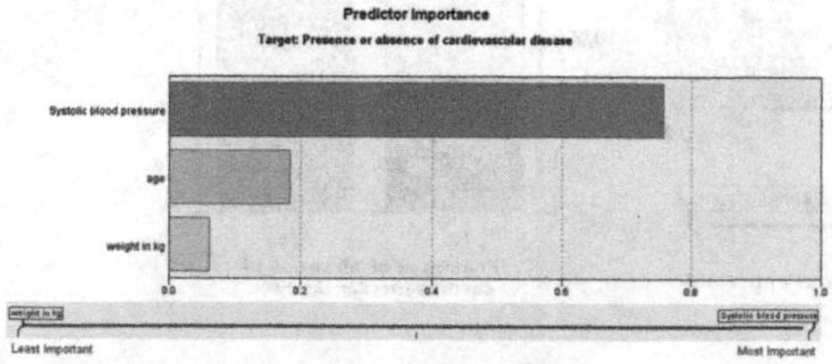

Fig. 8 Discrimination function centroids, coefficient and correlation

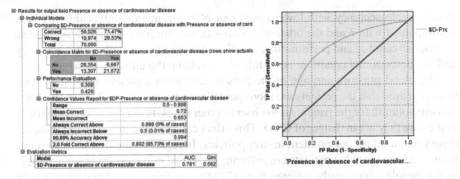

Fig. 9 Sensitivity versus specificity

id	Segment Rules	Score	Cover (n)	Frequency	Probability
	All segments including Remainder		70,000	34,979	49.97%
	⊟ Diastolic blood pressure, Systolic blood pressure				
1	Diastolic blood pressure > 80.000 and	Yes	4,452	3,857	86.64%
	Systolic blood pressure > 150.000				
	⊟ Diastolic blood pressure, Systolic blood pressure				
2	Diastolic blood pressure > 80.000 and	Yes	1,090	3,430	85.71%
	Systolic blood pressure > 140.000 and				
	Systolic blood pressure <= 150.000				
	⊟ Diastolic blood pressure, Systolic blood pressure				
3	Diastolic blood pressure > 80.000 and	Yes	7,117	5,807	81.59%
	Systolic blood pressure > 130.000 and				
	Systolic blood pressure <= 140.000				
	⊟ Diastolic blood pressure, Systolic blood pressure				
4	Diastolic blood pressure > 79.000 and	Yes	3,584	2,950	82.31%
	Diastolic blood pressure <= 80.000 and				
	Systolic blood pressure > 130.000				
	⊟ cholesterol level_transformed				
5	cholesterol level_transformed = "0"	Yes	4,492	3,183	70.86%
	Remainder		46,353	15,752	33.98%

Fig. 10 Decision list result analysis

Age. This correlation happened just between weight and Systolic blood pressure and can also approved or further analyze in the decision list. We can see that with hypertension stages below reach the second stage of hypertension value of more than 150 and if it goes beyond 180 it forms a crisis (Fig. 10).

6 Discussion

In this case study we have implemented data analysis, we have checked all the variables for outliers, sharp variances. We have transferred two categorical variables namely Cholesterol and Glucose from categorical into ordinal. At the preprocessing stage performed univariate analysis and compare all distribution versus target groups. We have seen that between the selected fields three variables are not allowed to contribute to the modelling and outranked them. Then we have used 9 different modeling algorithm and watch the number of fields or set of fields that can resulted in acceptable accuracy with lower costs. Out of two remaining algorithm one was accompanied with 3 and the other with two 4 fields. We have chosen those models into our stream file and perform the modeling. The result was with 71.4% accuracy and the best optimal condition is slightly skewed from the unbalanced target value.

The blood pressure and age of the persons can form a pattern for Insurer. Our results proved that the patients are more perceived with more than 25% of the cases and our solution for the patients is to observe other cases, testing or interventionist to make sure they are in danger of CVD. This effect can be compared with the Prospect theory and used for our problem insurer policies. In regard to the framing, reference point we have seen that patients are suffering from bad nutrition are in same danger as the people who already suffered from CVD when they are old and their weight increased.

References

Aburayya, A., Alshurideh, M., Al Marzouqi, A., Al Diabat, O., Alfarsi, A., Suson, R., Salloum, S. A., Alawadhi, D., & Alzarouni, A. (2020a). Critical success factors affecting the implementation of TQM in public hospitals: A case study in UAE hospitals. *Systematic Reviews in Pharmacy, 11*(10). https://doi.org/10.31838/srp.2020.10.39

Aburayya, A., Alshurideh, M., Albqaeen, A., Alawadhi, D., & Al A'yadeh, I. (2020b). An investigation of factors affecting patients waiting time in primary health care centers: An assessment study in Dubai. *Management Science Letters, 10*(6). https://doi.org/10.5267/j.msl.2019.11.031

Aburayya, A., Alshurideh, M., Alawadhi, D., Alfarsi, A., Taryam, M., & Mubarak, S. (2020c). An investigation of the effect of lean six sigma practices on healthcare service quality and patient satisfaction: Testing the mediating role of service quality in Dubai primary healthcare sector. *Journal of Advanced Research in Dynamical and Control Systems, 12*(8), 56–72.

Afifi, M. A. M., Kalra, D., Ghazal, T. M., & Mago, B. (2020). Information technology ethics and professional responsibilities. *International Journal of Advanced Science and Technology, 29*(4), 11336–11343.

Akhtar, A., Akhtar, S., Bakhtawar, B., Kashif, A. A., Aziz, N., & Javeid, M. S. (2021). COVID-19 detection from CBC using machine learning techniques. *International Journal of Technology, Innovation and Management (IJTIM), 1*(2), 65–78. https://doi.org/10.54489/ijtim.v1i2.22

Al Ali, A. (2021). The impact of information sharing and quality assurance on customer service at UAE banking sector. *International Journal of Technology, Innovation and Management (IJTIM), 1*(1), 01–17. https://doi.org/10.54489/ijtim.v1i1.10

Al Batayneh, R. M., Taleb, N., Said, R. A., Alshurideh, M. T., Ghazal, T. M., & Alzoubi, H. M. (2021). IT governance framework and smart services integration for future development of Dubai infrastructure utilizing AI and big data, its reflection on the citizens standard of living. *The International Conference on Artificial Intelligence and Computer Vision*, 235–247.

AlHamad, A., Alshurideh, M., Alomari, K., Kurdi, B., Alzoubi, H., Hamouche, S., & Al-Hawary, S. (2022). The effect of electronic human resources management on organizational health of telecommunications companies in Jordan. *International Journal of Data and Network Science, 6*(2), 429–438.

Alhamad, A. Q. M., Akour, I., Alshurideh, M., Al-Hamad, A. Q., Kurdi, B. A., & Alzoubi, H. (2021). Predicting the intention to use google glass: A comparative approach using machine learning models and PLS-SEM. *International Journal of Data and Network Science, 5*(3). https://doi.org/10.5267/j.ijdns.2021.6.002

Alhashmi, S. F. S., Alshurideh, M., Al Kurdi, B., & Salloum, S. A. (2020). A systematic review of the factors affecting the artificial intelligence implementation in the health care sector. In *Advances in intelligent systems and computing, AISC* (Vol. 1153). https://doi.org/10.1007/978-3-030-442 89-7_4

Ali, N., Ahmed, A., Anum, L., Ghazal, T. M., Abbas, S., Khan, M. A., Alzoubi, H. M., & Ahmad, M. (2021). Modelling supply chain information collaboration empowered with machine learning technique. *Intelligent Automation and Soft Computing, 30*(1), 243–257. https://doi.org/10.32604/iasc.2021.018983

Ali, N., M. Ghazal, T., Ahmed, A., Abbas, S., A. Khan, M., Alzoubi, H., Farooq, U., Ahmad, M., & Adnan Khan, M. (2022). Fusion-based supply chain collaboration using machine learning techniques. *Intelligent Automation & Soft Computing, 31*(3), 1671–1687. https://doi.org/10.32604/iasc.2022.019892

Alnazer, N. N., Alnuaimi, M. A., & Alzoubi, H. M. (2017). Analysing the appropriate cognitive styles and its effect on strategic innovation in Jordanian universities. *International Journal of Business Excellence, 13*(1), 127–140. https://doi.org/10.1504/IJBEX.2017.085799

Alnuaimi, M., Alzoubi, H. M., Ajelat, D., & Alzoubi, A. A. (2021). Towards intelligent organisations: An empirical investigation of learning orientation's role in technical innovation. *International Journal of Innovation and Learning, 29*(2), 207–221. https://doi.org/10.1504/IJIL.2021.112996

Alsharari, N. (2021). Integrating blockchain technology with internet of things to efficiency. *International Journal of Technology, Innovation and Management (IJTIM), 1*(2), 1–13.

Alshurideh, M., Gasaymeh, A., Ahmed, G., Alzoubi, H., & Kurd, B. A. (2020). Loyalty program effectiveness: Theoretical reviews and practical proofs. *Uncertain Supply Chain Management, 8*(3). https://doi.org/10.5267/j.uscm.2020.2.003

Alshurideh, M. T., Al Kurdi, B., Alzoubi, H. M., Ghazal, T. M., Said, R. A., AlHamad, A. Q., Hamadneh, S., Sahawneh, N., & Al-kassem, A. H. (2022). Fuzzy assisted human resource management for supply chain management issues. *Annals of Operations Research*, 1–19.

Alzoubi, A. (2021a). The impact of process quality and quality control on organizational competitiveness at 5-star hotels in Dubai. *International Journal of Technology, Innovation and Management (IJTIM), 1*(1), 54–68. https://doi.org/10.54489/ijtim.v1i1.14

Alzoubi, A. (2021b). Renewable green hydrogen energy impact on sustainability performance. *International Journal of Computations, Information and Manufacturing (IJCIM), 1*(1), 94–110. https://doi.org/10.54489/ijcim.v1i1.46

Alzoubi, H. M., Alshurideh, M., & Ghazal, T. M. (2021a). Integrating BLE beacon technology with intelligent information systems IIS for operations' performance: A managerial perspective. *The International Conference on Artificial Intelligence and Computer Vision*, 527–538.

Alzoubi, H. M., & Aziz, R. (2021). Does emotional intelligence contribute to quality of strategic decisions? *The Mediating Role of Open Innovation*. https://doi.org/10.3390/joitmc7020130

Alzoubi, H. M., Vij, M., Vij, A., & Hanaysha, J. R. (2021b). What leads guests to satisfaction and loyalty in UAE five-star hotels? AHP analysis to service quality dimensions. *Enlightening Tourism, 11*(1), 102–135. https://doi.org/10.33776/et.v11i1.5056

Alzoubi, H. M., & Yanamandra, R. (2020). Investigating the mediating role of information sharing strategy on agile supply chain. *Uncertain Supply Chain Management, 8*(2), 273–284. https://doi.org/10.5267/j.uscm.2019.12.004

Alzoubi, H., Ahmed, G., Al-Gasaymeh, A., & Kurdi, B. (2020a). Empirical study on sustainable supply chain strategies and its impact on competitive priorities: The mediating role of supply chain collaboration. *Management Science Letters, 10*(3), 703–708.

Alzoubi, H., Alshurideh, M., Kurdi, B., Akour, I., & Aziz, R. (2022). Does BLE technology contribute towards improving marketing strategies, customers' satisfaction and loyalty? The role of open innovation. *International Journal of Data and Network Science, 6*(2), 449–460.

Alzoubi, H., & Ahmed, G. (2019). Do TQM practices improve organisational success? A case study of electronics industry in the UAE. *International Journal of Economics and Business Research, 17*(4), 459–472. https://doi.org/10.1504/IJEBR.2019.099975

Alzoubi, H., Alshurideh, M., Kurdi, B. A., & Inairat, M. (2020b). Do perceived service value, quality, price fairness and service recovery shape customer satisfaction and delight? A practical study in the service telecommunication context. *Uncertain Supply Chain Management, 8*(3), 579–588. https://doi.org/10.5267/j.uscm.2020.2.005

Aziz, N., & Aftab, S. (2021). Data mining framework for nutrition ranking: Methodology: SPSS modeller. *International Journal of Technology, Innovation and Management (IJTIM), 1*(1), 85–95.

Cruz, A. (2021). Convergence between blockchain and the internet of things. *International Journal of Technology, Innovation and Management (IJTIM), 1*(1), 35–56.

Eli, T. (2021). Students perspectives on the use of innovative and interactive teaching methods at the University of Nouakchott Al Aasriya, Mauritania: English Department as a case study. *International Journal of Technology, Innovation and Management (IJTIM), 1*(2), 90–104.

Farouk, M. (2021). The universal artificial intelligence efforts to face coronavirus COVID-19. *International Journal of Computations, Information and Manufacturing (IJCIM), 1*(1), 77–93. https://doi.org/10.54489/ijcim.v1i1.47

Ghazal, T. M. (2021). Positioning of UAV base stations using 5G and beyond networks for IoMT applications. *Arabian Journal for Science and Engineering*, 1–12.

Ghazal, T. M., Anam, M., Hasan, M. K., Hussain, M., Farooq, M. S., Ali, H. M., Ahmad, M., & Soomro, T. R. (2021a). Hep-Pred: Hepatitis C staging prediction using fine Gaussian SVM. *Comput Mater Continua, 69*, 191–203.

Ghazal, T. M., Hasan, M. K., Alshurideh, M. T., Alzoubi, H. M., Ahmad, M., Akbar, S. S., Al Kurdi, B., & Akour, I. A. (2021b). IoT for smart cities: Machine learning approaches in smart healthcare—A review. *Future Internet, 13*(8), 218. https://doi.org/10.3390/fi13080218

Ghazal, T. M., Hussain, M. Z., Said, R. A., Nadeem, A., Hasan, M. K., Ahmad, M., Khan, M. A., & Naseem, M. T. (2021c). *Performances of K-means clustering algorithm with different distance metrics.*

Ghazal, T. M., Said, R. A., & Taleb, N. (2021d). Internet of vehicles and autonomous systems with AI for medical things. *Soft Computing*, 1–13.

Ghazal, T., Soomro, T. R., & Shaalan, K. (2013). Integration of project management maturity (PMM) based on capability maturity model integration (CMMI). *European Journal of Scientific Research, 99*(3), 418–428.

Guergov, S., & Radwan, N. (2021). Blockchain convergence: Analysis of issues affecting IoT, AI and blockchain. *International Journal of Computations, Information and Manufacturing (IJCIM), 1*(1), 1–17. https://doi.org/10.54489/ijcim.v1i1.48

Hamadneh, S., Pedersen, O., & Al Kurdi, B. (2021). An investigation of the role of supply chain visibility into the Scottish bood supply chain. *Journal of Legal, Ethical and Regulatory Issues, 24*(Special Issue 1), 1–12.

Hanaysha, J. R., Al-Shaikh, M. E., Joghee, S., & Alzoubi, H. (2021a). Impact of innovation capabilities on business sustainability in small and medium enterprises. *FIIB Business Review*, 1–12. https://doi.org/10.1177/23197145211042232

Hanaysha, J. R., Al Shaikh, M. E., & Alzoubi, H. M. (2021b). Importance of marketing mix elements in determining consumer purchase decision in the retail market. *International Journal of Service Science, Management, Engineering, and Technology (IJSSMET), 12*(6), 56–72.

Joghee, S., Alzoubi, H. M., & Dubey, A. R. (2020). Decisions effectiveness of FDI investment biases at real estate industry: Empirical evidence from Dubai smart city projects. *International Journal of Scientific and Technology Research, 9*(3), 3499–3503.

Kalra, D., Ghazal, T. M., & Afifi, M. A. M. (2020). Integration of collaboration systems in hospitality management as a comprehensive solution. *International Journal of Advanced Science and Technology, 29*(8s), 3155–3173.

Kashif, A. A., Bakhtawar, B., Akhtar, A., Akhtar, S., Aziz, N., & Javeid, M. S. (2021). Treatment response prediction in hepatitis C patients using machine learning techniques. *International Journal of Technology, Innovation and Management (IJTIM), 1*(2), 79–89. https://doi.org/10.54489/ijtim.v1i2.24

Khan, M. A. (2021). Challenges facing the application of IoT in medicine and healthcare. *International Journal of Computations, Information and Manufacturing (IJCIM)*, *1*(1), 39–55. https://doi.org/10.54489/ijcim.v1i1.32

Khan, M. F., Ghazal, T. M., Said, R. A., Fatima, A., Abbas, S., Khan, M. A., Issa, G. F., Ahmad, M., & Khan, M. A. (2021a). An IoMT-enabled smart healthcare model to monitor elderly people using machine learning technique. *Computational Intelligence and Neuroscience, 2021*.

Khan, Q.-T.-A., Ghazal, T., Abbas, S., Khan, W. A., Khan, M. A., Raed, A. S., Ahmad, M., & Asif, M. (2021b). Modeling habit patterns using conditional reflexes in agency. *Intelligent Automation and Soft Computing, 30*(2), 539–552.

Lee, C., & Ahmed, G. (2021). Improving IoT privacy, data protection and security concerns. *International Journal of Technology, Innovation and Management (IJTIM)*, *1*(1), 18–33. https://doi.org/10.54489/ijtim.v1i1.12

Lee, K., Azmi, N., Hanaysha, J., Alzoubi, H., & Alshurideh, M. (2022a). The effect of digital supply chain on organizational performance: An empirical study in Malaysia manufacturing industry. *Uncertain Supply Chain Management, 10*(2), 495–510.

Lee, K., Romzi, P., Hanaysha, J., Alzoubi, H., & Alshurideh, M. (2022b). Investigating the impact of benefits and challenges of IOT adoption on supply chain performance and organizational performance: An empirical study in Malaysia. *Uncertain Supply Chain Management, 10*(2), 537–550.

Lee, S.-W., Hussain, S., Issa, G. F., Abbas, S., Ghazal, T. M., Sohail, T., Ahmad, M., & Khan, M. A. (2021). Multi-dimensional trust quantification by artificial agents through evidential fuzzy multi-criteria decision making. *IEEE Access, 9*, 159399–159412.

Matloob, F., Ghazal, T. M., Taleb, N., Aftab, S., Ahmad, M., Khan, M. A., Abbas, S., & Soomro, T. R. (2021). Software defect prediction using ensemble learning: A systematic literature review. *IEEE Access*.

Mehmood, T. (2021). Does information technology competencies and fleet management practices lead to effective service delivery? Empirical evidence from E-commerce industry. *International Journal of Technology, Innovation and Management (IJTIM)*, *1*(2), 14–41.

Mehmood, T., Alzoubi, H. M., & Ahmed, G. (2019). Schumpeterian entrepreneurship theory: Evolution and relevance. *Academy of Entrepreneurship Journal, 25*(4).

Miller, D. (2021). The best practice of teach computer science students to use paper prototyping. *International Journal of Technology, Innovation and Management (IJTIM)*, *1*(2), 42–63. https://doi.org/10.54489/ijtim.v1i2.17

Mondol, E. P. (2021). The impact of block chain and smart inventory system on supply chain performance at retail industry. *International Journal of Computations, Information and Manufacturing (IJCIM)*, *1*(1), 56–76. https://doi.org/10.54489/ijcim.v1i1.30

Naqvi, R., Soomro, T. R., Alzoubi, H. M., Ghazal, T. M., & Alshurideh, M. T. (2021). The nexus between big data and decision-making: A study of big data techniques and technologies. *The International Conference on Artificial Intelligence and Computer Vision*, 838–853.

Obaid, A. J. (2021). Assessment of smart home assistants as an IoT. *International Journal of Computations, Information and Manufacturing (IJCIM)*, *1*(1), 18–36. https://doi.org/10.54489/ijcim.v1i1.34

Radwan, N., & Farouk, M. (2021). The growth of internet of things (IoT) in the management of healthcare issues and healthcare policy development. *International Journal of Technology, Innovation and Management (IJTIM)*, *1*(1), 69–84. https://doi.org/10.54489/ijtim.v1i1.8

Rehman, E., Khan, M. A., Soomro, T. R., Taleb, N., Afifi, M. A., & Ghazal, T. M. (2021). Using blockchain to ensure trust between donor agencies and NGOs in under-developed countries. *Computers, 10*(8), 98.

Shamout, M., Ben-Abdallah, B., Alshurideh, M., Alzoubi, H., Al Kurdi, B., & Hamadneh, S. (2022). A conceptual model for the adoption of autonomous robots in supply chain and logistics industry. *Uncertain Supply Chain Management, 10*, 1–16.

Suleman, M., Soomro, T. R., Ghazal, T. M., & Alshurideh, M. (2021). Combating against potentially harmful mobile apps. *The International Conference on Artificial Intelligence and Computer Vision*, 154–173.

Taryam, M., Alawadhi, D., Aburayya, A., Albaqa'een, A., Alfarsi, A., Makki, I., Rahmani, N., Alshurideh, M., & Salloum, S. A. (2020). Effectiveness of not quarantining passengers after having a negative COVID-19 PCR test at arrival to dubai airports. *Systematic Reviews in Pharmacy*, *11*(11). https://doi.org/10.31838/srp.2020.11.197

Unknown-Unknown Risk Mitigation Through AI: Case of Covid-19

Mounir El Khatib, Amna Obaid, Fatima Al Mehyas, Fatma Ali Al Ali, Jawahir Abughazyain, Kayriya Alshehhi, Haitham M. Alzoubi(ID), and Muhammad Alshurideh(ID)

Abstract The purpose of this study is to investigate how artificial intelligence can mitigate unknown-unknown risks, taking COVID-19 as a case. The utilization of this technology has helped with the risk management and mitigations process by providing efficient solutions. Today, the artificial intelligence of applications plays a critical role in containing the fast spread of COVID-19 virus and reducing the number of infected individuals. This research will explore how countries around the world have benefited from contact tracing applications. The research discussed the UAE Alhosn application compared to the applications developed by China, Australia, Thailand, France, South Korea to identify the best features and better utilization of AI technology in the applications. In addition, a mixed-method approach used in this research, and a combination of a qualitative and a quantitative approach to make a clear understanding of the topic and find out accurate results. For qualitative research, interviews were conducted with five people from different companies in the UAE. For quantitative research, ten questions survey was sent to a sample of 50 random respondents who are the users of Alhosn application. Domain analysis was used to organize the ideas and to generate other ideas. The research will also gather secondary data based on reviews of literature related to the research topic. Based on the gathered information and research results recommendations were given.

Keywords AI · Covid-19 · Unknown risks · Risk mitigation · Project risk management

M. El Khatib · A. Obaid · F. Al Mehyas · F. A. Al Ali · J. Abughazyain · K. Alshehhi
Hamdan Bin Mohamad Smart University, Dubai, UAE

H. M. Alzoubi (✉)
School of Business, Skyline University College, Sharjah, UAE
e-mail: haitham.alzubi@skylineuniversity.ac.ae

M. Alshurideh
Department of Marketing, School of Business, University of Jordan, Amman, Jordan
e-mail: malshurideh@sharjah.ac.ae; m.alshurideh@ju.edu.jo

Department of Management, College of Business Administration, University of Sharjah, Sharjah, UAE

© The Author(s), under exclusive license to Springer Nature Switzerland AG 2023
M. Alshurideh et al. (eds.), *The Effect of Information Technology on Business and Marketing Intelligence Systems*, Studies in Computational Intelligence 1056,
https://doi.org/10.1007/978-3-031-12382-5_115

1 Introduction

With the fast spread of COVID-19 around the globe, many sectors got affected, such as economy, hospitalities, and tourism (Aljumah et al., 2021; Alshurideh et al., 2021). One of the main sectors is the health care; projects in this sector are focusing on minimizing chronic diseases and early prediction of dangerous conditions affecting people (Alshurideh et al., 2020, 2021a, 2021b; Lee et al., 2022a, 2022b). Artificial intelligence (AI) has played a significant role in performing business and governmental tasks. Many governments are leading in the utilization of applications to connect with the citizens (Alhashmi et al., 2020; AlShamsi et al., 2021; Alshurideh et al., 2022; Mehmood et al., 2019). Therefore, they have developed applications that use artificial intelligence to mitigate the unknown-unknown risk of the crises. Which will help to contain the fast spread of the virus and reduce the number of infected individuals (Alzoubi et al., 2021; Hanaysha et al., 2021a, 2021b).

Similarly, the UAE has launched Alhosn Application with a purpose to contain the spread factor. Those applications have emerged as great examples of the utilization of artificial intelligence for better health and life saving issues (Aburayya et al., 2020a, 2020b; Ali et al., 2022; Ghazal et al., 2021). For the reason of the high spread of the COVID-19, more than 500,000 people worldwide lost their lives. Governments around the world started to use artificial intelligence in order to mitigate the unknown-unknown risk of the crises (Ali et al., 2021; Alnazer et al., 2017; Obaid, 2021). So, they started to create applications for the reason of reducing the chances of people getting the virus (Alzoubi & Yanamandra, 2020; Alzoubi et al., 2020a, 2020b; Lee et al., 2022a, 2022b; Nuseir et al., 2021).

This research will study the impact of using artificial intelligence of applications in different countries, illustrating their features used to mitigate unknown-unknown risk of COVID-19. Emphasizing on Alhosn Application and provide recommendations for improvements.

2 Research Question and Hypotheses

Research Question

- How can we alleviate the known and unknown risks of COVID-19 by using the AI-based applications?

Hypotheses

- Artificial intelligence can mitigate unknown-unknown risks. COVID-19 as a case.

3   Literature Review

AI has been used by many organizations in different sectors to mitigate unknown-unknown risk (Alzoubi, 2021a; Guergov & Radwan, 2021; Khan, 2021). The purpose of this literature review is to evaluate and discuss how AI can help the health sector with mitigating the unknown-unknown risk through contact tracing applications. Many research papers and articles globally were reviewed to conduct the research.

Artificial intelligence is the science that allows machines to perform tasks that require human intelligence such as problem-solving, analyzing, speech recognition, etc. (Farouk, 2021; Mondol, 2021). Currently, AI is being used in different fields such as Human Resources, customer services, and banks for fraud detection (Alhamad et al., 2021; AlHamad et al., 2022; Alzoubi et al., 2022). However, organizations should understand that it's a double-edged weapon that has advantages and disadvantages (Alnuaimi et al., 2021; Alzoubi & Ahmed, 2019). One of the disadvantages is that it can be misused by attackers or hackers to expose the organization's data (Durbin, 2019).

There are many usages for the AI, one of them is that it is built in the applications as analytics tools (Hanaysha et al., 2021a, 2021b; Joghee et al., 2020). In the current case of COIVD-19, the AI was used in the applications to keep track of the infected people and analyze their input to mitigate the unknown-unknown risk of the rising numbers of the COVID-19 cases (Durbin, 2019).

Apple and Google announced a collaboration to slow the spread of the virus and accelerate going back to normal life (Akhtar et al., 2021; Cruz, 2021). Google stated that their initiative was to provide Application Programming Interfaces (APIs) to public health authorities around the world, which will allow them to trace the infected individuals and notify them to isolate themselves and take the necessary actions.

The applications are using Bluetooth and AI technology to operate (Aziz & Aftab, 2021; Mehmood, 2021). After downloading the application people are allowed to self-diagnose their health by feeding the application with their data. There are different uses for the AI in these applications; it has been used to analyze the data, and to protect the privacy of the users.

3.1   China

One of the successful implementations using the tracing Application was in China. Many tech companies developed different tracing applications to mitigate the unknown-unknown risk of the pandemic COVID-19. Jin et al. (2020) mentioned that 'the close contact detector' was added to the Chinese payment and social media applications in order to reach the largest number of people. These applications have a huge amount of user data that's been analyzed by AI technology (Alsharari, 2021; Eli, 2021; Lee & Ahmed, 2021). After downloading the applications, users have to add their details and scan a QR code to be able to enter or leave public places and transportations. Based on the color code system, the application notifies all those who've been in contact with the infected individual in the last 14 days to quarantine themselves and to get tested (Jin et al., 2020).

3.2 Australia

Australia CovidSafe is another application that uses the contact tracing technology to contain the spread of COVID-19 (Al Ali, 2021; Miller, 2021; Radwan & Farouk, 2021). After downloading the application, the users have to add their personal information (name, age range, mobile number, postcode) in order to be identified by the health authorities. Australia health authorities stated that the application uses Bluetooth to send an encrypted code to all users. If an individual was tested positive, the infected person has to update the status in the application and give permission to the health authorities to alert those who have been in contact with him within 5 feet for more than 15 min during the past 21 days (COVID Safe application, 2020).

3.3 Thailand

To control the cases of COVID-19 in Thailand, an application called Thai Chana (Thai Victory) was developed by Thailand's Center for COVID-19 Situation Administration (CCSA) which was introduced to the public on 17/May/2020. The application has many features. Dr. Taweesin Visanuyothin, the spokesperson in CCSA mentioned that all businesses including street food vendors should be registered for the application online to get a QR code, these codes will be posted in front of their shops so that any customer will scan it before entering. The purpose of scanning the code is that people can be tracked and will be receiving warning messages if infected people had visited the same place. By this, the government will be able to track infected people with COVID-19 or the people who got close to them. Another feature of the application is that people can check wither the place that they are willing to visit crowded or not and wither they follow the control measures of COVID-19. Dr. Taweesin has also added that all users' privacy is protected and won't be shared with third-party, the only department that can access the information is the Disease Control Department (Huaxia, 2020).

3.4 France

StopCovid is the application that France established and design in order to trace and reduce the invasion of COVID-19 pandemic. The application designed on the contact tracing function which recognizes any person may contact an infected person, this technology uses Bluetooth in the phones to trace any infected area or people that the user will contact. People who were diagnoses with coronavirus will have a special QR code to add it in the application, then the application stores this code in other people's devices to alert them if they are within one meter of the infected person (Wikipedia, 2022). The application was established on the 8th of April, as per TechCrunch, electronic news web site, their government hosted a press conference once the appl has been release to share the statistical analyses of the app. After three weeks of the release, 1.9 million users have downloaded the application from both stores and most of them activated it. However, most of the French population uninstalled it as they feel it won't help them, for many reasons. The majority of them see that the government will not have any record about the people as the application does not ask you to register your name, another reason that the application does not cover all European countries. France's digital minister Cédric O said. "If I have one regret it's obviously that it isn't compatible with other European countries." Moreover, the QR code should be entered manually in the application, which affected the accuracy of the application, this was a reason for them not to trust the application, however, Mr. Cédric O stated that the application will keep improving for better features and functions (Dillet, 2020).

3.5   South Korea

South Korea is considered to be a model since it did not take its citizens through a lockdown as in most of the countries. Moreover, there are fewer new cases in South Korea, with 244 reported deaths in three months since a sudden occurrence of a virus was announced in the country. This was achieved due to the good underlying basis of the country's coronavirus response with thorough testing mechanisms, which is backed by the countrywide contact tracers' network who dialogue with the infected people, followed by tracing their contacts. The technique involves the use of mobile phone alerts, though the Bluetooth applications are not used at the moment (Whitelaw et al., 2020). Thus, it is essential to be noted that the application "Corona 100 m" is based on the degree of surveillance that citizens of most of the countries do not find it easy to agree with.

Through this app, in case, an individual test positive for coronavirus, a text alert is sent to all the neighbors of the infected person who came in contact with him within 100 m, which is followed by contact tracing to curb the spread of the virus (Whitelaw et al., 2020). The alert text includes a link having many of the movements log of the infected people, and the neighbors are notified after every one minute. Moreover, these movements are traced from public data like television cameras' closed circuits. The results of the contact tracing are enhanced by the fact that the government of South Korean is permitted to access private records of people, such as the transactions of the credit-card (Whitelaw et al., 2020). Moreover, the most important segment of this application is that the movements of the people are published and anonymized online to ensure that the government's coronavirus guidelines are followed through supervision. The developer says the application is very successful reaching 20,000 downloads per hour. Therefore, the use of this application has led to quick and accurate contact tracing in South Korea leading to a low number of new cases and deaths from the virus on the global map (Alzoubi, 2021b; Hamadneh et al., 2021; Kashif et al., 2021).

3.6 Alhosn UAE

During this challenging time, UAE launched an application named Alhosn UAE to contain the spread of the virus. The application was developed by the National Emergency Crisis and Disaster Management Authority in collaboration with the Ministry of Health and Prevention and local Health Authorities (Alzoubi & Aziz, 2021; Alzoubi et al., 2020a, 2020b). The application depends on Bluetooth technology, AI and QR codes in order to operate. The two main functions of the applications are; to provide quick access to test results, track people movement and inform those who contacted an infected case about their current health states (Emirates News Agency-WAM, 2020b).

The users can register in the Application using their Emirates ID and phone number; each user will have a unique QR code that follows a color-coding system that indicates users' health states. Those codes might be used in the future to grant access to public places (Alhosan UAE App, 2022).

Abdul Rahman Al Owais Minister of Health and Prevention stresses on the importance of collective efforts all segments of society citizens and residents from all nationalities in the process of containing the rapid spread of the virus. The government encouraged everyone to download the application by making it available in three languages at both Apple Store and Google Play for free to stand united against the pandemic (Emirates News Agency-WAM, 2020a).

Everyone in the country should have an account; thus, the developers took into consideration the individuals who are not using phones such as children, elderly, and People of Determination. They can be added to one of the family member's account or the care provider's account (Alhosan UAE App, 2022).

4 Research Methods

To evaluate how artificial intelligence can mitigate the unknown-unknown risk-taking COVID-19 as a case, a study was applied to several risk executives and IT employees across the UAE. A combination of both qualitative and quantitative approaches was used to conduct the study since each one of them has its weakness and strength. The mixture between the methods allows us to get accurate results from the quantitative questionnaire and unpredictable answers from the qualitative interviews.

Five interviews were conducted from 5 different organizations around the UAE, interviewing risk management experts and IT executives. Questions were in a qualitative approach, which gave us the chance to interact with the interviewees directly. It also allowed us to understand their perceptions and opinions in a clear way.

To get more accurate analysis and results about the artificial intelligence and especially about the Alhosn application, a quantitative survey of 10 questions was distributed in a random sample among the seven Emirates on 50 people who are using Al Hosn app. This method helped us to get a fast analysis and data collection. The quantitative method of search is the best way to get exact statistics that can be compared easily.

To analyze the results of the questionnaire, the research used the domain analysis which is the best way to come with new and different ideas and then organize them. It also categorizes each idea to the one related to it. However, a narrative analysis was used to conduct the interviews which allowed us to analyze the results and observe them.

5 Results

5.1 Interviews

Through the interviews, it was clear that risks are different from one organization to another, in our research we had the chance to interview five employees from different organizations: Ministry of Community Development (MOCD), Khalifa Fund for Enterprise Development (KFED), Sharjah Documentation and Archives Authority (SDAA), Abu Dhabi Islamic Bank (ADIB), and Sharjah Girl Guides (SGG).

Our interviewer Mr. Saeed from MOCD said that "risk is divided into many categories starting from enterprise risks and breaking into IT, cybersecurity, operations, financial risks, legal liabilities, accidents, and natural disasters." Moreover, information lost and damage of documents duo to viruses or natural disasters is the main risk in SDAA as per Mr. Taryam.

Moreover, Mr. Loay Mahmoud from KFED stated that "the risk in the organization is low in severity, as the health risk and the financial risk is very minimum". Also, the interviewee Mr. Francies, IT manager, said: "There are some potential risks that could occur in SGG and have a negative impact and influence, such as operational risks, safety-related risks, work environment risks, natural disasters, financial risks, staff, and members related risks." The last interview was with Mr. Jaber Ali, Branch manager in ADIB stated that as a financial institution they have a variety of risks such as embezzlement, information leaking, and scammers.

All organizations have strategies and methodologies in dealing with risk, therefore all of our interviewees agreed that they deal with the unknown risk by planning, setting policies, and procedures. For example, ADIB has a risk department that includes employees who are experts in risk management, they hold a monthly meeting to discuss all the risks in the bank and might have emergency meetings in case of any risky situation.

Some of the interviewees mentioned that they are using and following defense-in-depth, risk-based, and business continuity policy approaches since they standardize the needed actions, resources, and crisis management communications during the unknown risk period.

Moreover, Mr. Francis said that their methodology is to find the root cause of the problems, look at the impact of each risk and the probability. Then, use that information to prioritize risks and come up with solutions as a response.

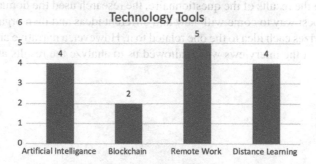

The organizations use several technology tools in the risk process, based on our interviews, all of them agreed on remote working, 4 out of 5 of the chosen organization uses AI and Distance learning, and 2 uses the blockchain.

ADIB uses AI to protect its customer's accounts form the risk of scammers as Mr. Jaber stated in his interview. They adopted the one-time password in any financial transaction that happens through their digital channels. Moreover, Mr. Saeed said that they use AI to implement risk responses, monitor and identify risk, and qualitative and quantitative analysis. Also, his organization uses AI to identify malicious codes and doing incident response in MOCD.

Remote working is another tool that been used by all mentioned organizations to mitigate the risk of COVID-19. Mr. Taryam illustrated that his organization can't stop working during the lockdown and their experience was successful. He added that the authority provided all the supporting applications and programs that mitigated the unknown-unknown risk, giving an example of that; they replaced the internal SharePoint by external SharePoint to ensure the employee's access to the needed information from home.

Similarly, ADIB head quarter replaced the employee's PCs with laptops with the required systems and applications ensuring the business continuity with a high level of security when working remotely to prevent the leaking of financial information.

5.2 Interview—AlHosn Application

Our primary data shows that it was not mandatory to download Alhosn application in 80% of the organizations. However, ADIB management stressed on the importance of downloading the application. On the other hand, Mr. Loay and Mr. Francis stated that all their employees must download the application as it Identifies any COVID-19 positive cases and regularly checks the employees' health status.

Most of the candidates suggested new features to be added to the application. Tracking other infected people without leaving the application open all the time which affects the battery life and the data as per Mr. Taryam. Mr. Loay and Mr. Jaber suggested keeping a record of recommended safe areas such as shopping malls and groceries as well as a history of the visited places. Also, Mr. Francis suggested linking the application to Human Resources in the organization to alert them for any infected cases.

Three of our interviewees think that China developed the most powerful and effective application, as it's mandatory to download the application that helped to decrease the spared of the virus. A security feature in the application needs to be improved as it keeps a record of personal information and the visited places. Mr. Taryam stated that a third party might use this information for advertising purposes.

6 Survey

The survey targeted 50 individuals based on their age and location. The majority of them were aged between 20 and 40, 50% of them were between 20 and 30 years old, and 28% were between 31 and 40. Among the 50 candidates, 12 of them are located in Sharjah, 10 in Dubai, 4 in Abu Dhabi, and the rest are from the other Emirates.

Questions 3 and 4 were based on the scaling from 1 to 5 as of 1 is least, 5 is most. 70% of them trust that the application can mitigate the risk of COVID-19 as they chose 5 from the scale. Moreover, 30–40% will likely recommend downloading the application to other people. The survey questions were followed by yes and no question, 70% of the candidates was not obligatory in their work or organization to download the application. Also, about the personal information in Alhosn's application, 80% of them find that their information is not safe and might be exposed. Moreover, the features in the application were satisfying for 60% of the users. Besides, all of the 50 candidates think that the application needs to be improved to mitigate the risk of COVID-19.

The candidate used Alhosn application in many places such as supermarkets, work, and closed shopping malls. Moreover, most of them voted for other countries' applications for the best application in their opinion, Thailand got 15 votes, China 12 votes, KSA 10 votes, Alhosn 3 votes, Qatar 2 vote and google 10 votes.

7 Discussion

It is evident that, the type and intensity of risks vary from one organization to another. Unknown -Unknown risks are the most critical of them all as they are usually unexpected an unplanned for. COVID-19 has changed all previous assumptions, the pandemic has proved the importance of having a Risk Unit at every organization, and it has also revealed which organizations have strong Risk Assessment and Business Continuity plans compared to those who failed to catch up with the others.

As we are in 2020, it is important to highlight the methods used to address risks have improved a lot. AI is a major tool used in many organizations to address risks such as COVID-19. As our research has proved the use of AI mainly through advanced tracking mobile applications by the governments of the countries illustrated in the literature review that helped control the spread of COVID-19. There is an evident link between the efficiency of the AI tools used and the number of cases and deaths in the country compared to the overall calculation. On the other hand, one example in our research, France, had a less impactful AI application which the public didn't trust, didn't download and the impact is disastrous as shown in the cases/deaths in the diagram below (Wikipedia, 2022).

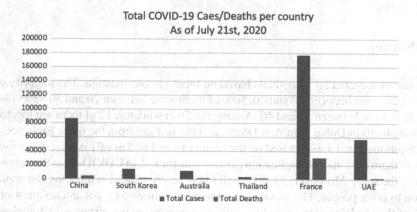

Taking UAE as a case, our interview and survey results show unlike the countries that made it mandatory, it is still not mandatory to download Alhosn Application at many organizations (80% of the interviewees and 70% of the individuals surveyed). Also, the interviewees stated that even if it is downloaded, unfortunately, many of them don't activate it as it consumes their battery-life. On the other hand, most of them agree on the effectiveness of this application in controlling the pandemic in the country and worldwide.

8 Conclusion

In conclusion, risk assessment is a critical exercise that is conducted in most orga-nizations. COVID-19 is one of the most critical risks in this century, if not the most critical of them all in both severity and coverage. The pandemic affected every country worldwide in all sectors mainly health and economy (Alameeri et al., 2021; Amarneh et al., 2021; Harahsheh et al., 2021; Leo et al., 2021; Shah et al., 2021). As our literature review showed the success of governments in controlling COVID-19 relies on the use of AI methods. Countries that have successfully launched advanced AI methods through mobile apps, access to analytical data, and linking it to the use of physical detectors have proved to control the pandemic with fewer cases of disease and fewer deaths. Our recommendation is to make Al Hosn Application in the UAE mandatory for citizens and residents to access public places. This will give the government a lot of insights on where the infected individuals have been and who they have encountered. Another recommendation is to increase the efficiency of the application by minimizing its battery usage. Finally, the Application can share the list of recommended or safe areas that implement all the precautionary measures to the users.

Appendix 1

1. Was it mandatory to download Alhosn App to resume the office work in your organization?
   ....
   ....

2. If yes, how did Alhosn App helped with mitigating the risk in your organization?
   ....
   ....

3. What feature can be added to the app in order to help the organization to mitigate the unknown un known risk?
   ....
   ....

4. In your point of view, which country developed the most powerful artificial intelligence app to face covid-19? Why?
   ....
   ....

5. In your opinion, dose this App have a threat on people privacy?
   ....
   ....

Appendix 2

Questionnaire to Evaluate the effectiveness of Alhosn app

This questionnaire is made for the purpose of evaluating Alhosn app which is an Artificial intelligence tool which is an app to track infected people with Covis-19 and how its impact on people in different areas in the UAE

Candidate Information

Name:

1- **Age**

   20 to 30 years   ☐

   31 to 40 years   ☐

   41 to 50 years   ☐

   Over 51   ☐

2- **Residential Area**

   Abu Dhabi   ☐

   Dubai   ☐

   Sharjah   ☐

   Ajman   ☐

   Um Al Quwain   ☐

   Al Fujairah   ☐

   Ras Al Khaimah   ☐

Al Hosn App Inquirations

In the scale of 5 choose the most appropriate answer [1 is least, 5 is most]

3- **Trust that app can mitigate the risk of Covid-19**

   1   2   3   4   5
   ☐   ☐   ☐   ☐   ☐

4- **How likely do you recommend others to download the app**

   1   2   3   4   5
   ☐   ☐   ☐   ☐   ☐

In your point of view answer by yes/no

5- Was it a mandatory to download the app in your institution or work?
Yes ☐ No ☐

6- Personal information in Al hosn app can be exposed by third party
Yes ☐ No ☐

7- Are the app feature satisfying?
Yes ☐ No ☐

8- Do you think that the app can be improved which will have more impact in mitigating the risk of covid 19?
Yes ☐ No ☐

9- Where do you use Al Hosn app in order to protect you from being infected from the covid-19? (you can choose more than one choice)

Work ☐
Closed shopping malls ☐
Open areas of shopping ☐
Gym ☐
Beauty Spa ☐
Supermarkets ☐
Other ☐
[Specify] _____

10- Which from the following apps you think has the best features?

Thailand app ☐
China app ☐
UAE Al hosn app ☐
Qatar app ☐
KSA app ☐
Australia app ☐
Other ☐
[Specify] _____

Thank you for your feedback!

References

Aburayya, A., Alshurideh, M., Alawadhi, D., Alfarsi, A., Taryam, M., & Mubarak, S. (2020a). An investigation of the effect of lean six sigma practices on healthcare service quality and patient satisfaction: Testing the mediating role of service quality in dubai primary healthcare sector. *Journal of Advanced Research in Dynamical and Control Systems, 12*(8), 56–72.

Aburayya, A., Alshurideh, M., Al Marzouqi, A., Al Diabat, O., Alfarsi, A., Suson, R., Bash, M., & Salloum, S. A. (2020b). An empirical examination of the effect of TQM practices on hospital service quality: An assessment study in uae hospitals. *Systematic Reviews in Pharmacy, 11*(9). https://doi.org/10.31838/srp.2020a.9.51

Akhtar, A., Akhtar, S., Bakhtawar, B., Kashif, A. A., Aziz, N., & Javeid, M. S. (2021). COVID-19 detection from CBC using machine learning techniques. *International Journal of Technology, Innovation and Management (IJTIM)*, *1*(2), 65–78. https://doi.org/10.54489/ijtim.v1i2.22

Al Ali, A. (2021). The impact of information sharing and quality assurance on customer service at UAE banking sector. *International Journal of Technology, Innovation and Management (IJTIM)*, *1*(1), 01–17. https://doi.org/10.54489/ijtim.v1i1.10

Alameeri, K. A., Alshurideh, M. T., & Al Kurdi, B. (2021). The effect of Covid-19 pandemic on business systems' innovation and entrepreneurship and how to cope with it: A theatrical view. In *Studies in systems, decision and control* (vol. 334). https://doi.org/10.1007/978-3-030-67151-8_16

Alhamad, A. Q. M., Akour, I., Alshurideh, M., Al-Hamad, A. Q., Kurdi, B. A., & Alzoubi, H. (2021). Predicting the intention to use google glass: A comparative approach using machine learning models and PLS-SEM. *International Journal of Data and Network Science*, *5*(3). https://doi.org/10.5267/j.ijdns.2021.6.002

AlHamad, A., Alshurideh, M., Alomari, K., Kurdi, B., Alzoubi, H., Hamouche, S., & Al-Hawary, S. (2022). The effect of electronic human resources management on organizational health of telecommuni-cations companies in Jordan. *International Journal of Data and Network Science*, *6*(2), 429–438

Alhashmi, S. F. S., Alshurideh, M., Al Kurdi, B., & Salloum, S. A. (2020). A systematic review of the factors affecting the artificial intelligence implementation in the health care sector. In *Advances in intelligent systems and computing* (vol. 1153). AISC. https://doi.org/10.1007/978-3-030-442 89-7_4

Alhosan UAE App. (2022). *National emergency crisis and disaster management*.

Ali, N., Ahmed, A., Anum, L., Ghazal, T. M., Abbas, S., Khan, M. A., Alzoubi, H. M., & Ahmad, M. (2021). Modelling supply chain information collaboration empowered with machine learning technique. *Intelligent Automation and Soft Computing*, *30*(1), 243–257. https://doi.org/10.32604/iasc.2021.018983

Ali, N., M. Ghazal, T., Ahmed, A., Abbas, S., A. Khan, M., Alzoubi, H., Farooq, U., Ahmad, M., & Adnan Khan, M. (2022). Fusion-based supply chain collaboration using machine learning techniques. *Intelligent Automation & Soft Computing*, *31*(3), 1671–1687. https://doi.org/10.32604/iasc.2022.019892

Aljumah, A., Nuseir, M. T., & Alshurideh, M. T. (2021). The impact of social media marketing communications on consumer response during the COVID-19: Does the brand equity of a university matter. *The Effect of Coronavirus Disease (COVID-19) on Business Intelligence*, *334*, 384–367.

Alnazer, N. N., Alnuaimi, M. A., & Alzoubi, H. M. (2017). Analysing the appropriate cognitive styles and its effect on strategic innovation in Jordanian universities. *International Journal of Business Excellence*, *13*(1), 127–140. https://doi.org/10.1504/IJBEX.2017.085799

Alnuaimi, M., Alzoubi, H. M., Ajelat, D., & Alzoubi, A. A. (2021). Towards intelligent organisations: An empirical investigation of learning orientation's role in technical innovation. *International Journal of Innovation and Learning*, *29*(2), 207–221. https://doi.org/10.1504/IJIL.2021.112996

AlShamsi, M., Salloum, S. A., Alshurideh, M., & Abdallah, S. (2021). Artificial intelligence and blockchain for transparency in governance. In *Studies in computational intelligence* (vol. 912). https://doi.org/10.1007/978-3-030-51920-9_11

Alsharari, N. (2021). Integrating blockchain technology with internet of things to efficiency. *International Journal of Technology, Innovation and Management (IJTIM)*, *1*(2), 1–13.

Alshurideh, M., Gasaymeh, A., Ahmed, G., Alzoubi, H., & Kurd, B. A. (2020). Loyalty program effectiveness: Theoretical reviews and practical proofs. *Uncertain Supply Chain Management*, *8*(3). https://doi.org/10.5267/j.uscm.2020.2.003

Alshurideh, M. T., Al Kurdi, B., AlHamad, A. Q., Salloum, S. A., Alkurdi, S., Dehghan, A., Abuhashesh, M., & Masa'deh, R. (2021a). Factors affecting the use of smart mobile examination platforms by universities' postgraduate students during the COVID-19 pandemic: An empirical study. *Informatics, 8*(2). https://doi.org/10.3390/informatics8020032

Alshurideh, M. T., Hassanien, A. E., & Masa'deh, R. (2021b). *The effect of coronavirus disease (COVID-19) on business intelligence.* Springer.

Alshurideh, Muhammad Turki, Al Kurdi, B., Alzoubi, H. M., Ghazal, T. M., Said, R. A., AlHamad, A. Q., Hamadneh, S., Sahawneh, N., & Al-kassem, A. H. (2022). Fuzzy assisted human resource management for supply chain management issues. *Annals of Operations Research*, 1–19.

Alzoubi, H., & Ahmed, G. (2019). Do TQM practices improve organisational success? A case study of electronics industry in the UAE. *International Journal of Economics and Business Research, 17*(4), 459–472. https://doi.org/10.1504/IJEBR.2019.099975

Alzoubi, H. M., & Aziz, R. (2021). Does emotional intelligence contribute to quality of strategic decisions? The mediating role of open innovation. *Journal of Open Innovation: Technology, Market, and Complexity, 7*(2), 130. https://doi.org/10.3390/joitmc7020130

Alzoubi, H. M., & Yanamandra, R. (2020). Investigating the mediating role of information sharing strategy on agile supply chain. *Uncertain Supply Chain Management, 8*(2), 273–284. https://doi.org/10.5267/j.uscm.2019.12.004

Alzoubi, H., Alshurideh, M., Kurdi, B. A., & Inairat, M. (2020b). Do perceived service value, quality, price fairness and service recovery shape customer satisfaction and delight? A practical study in the service telecommunication context. *Uncertain Supply Chain Management, 8*(3), 579–588. https://doi.org/10.5267/j.uscm.2020.2.005

Alzoubi, H., Alshurideh, M., Kurdi, B., Akour, I., & Aziz, R. (2022). Does BLE technology contribute towards improving marketing strategies, customers' satisfaction and loyalty? The role of open innovation. *International Journal of Data and Network Science, 6*(2), 449–460.

Alzoubi, H, Ahmed, G., Al-Gasaymeh, A., & Kurdi, B. (2020a). Empirical study on sustainable supply chain strategies and its impact on competitive priorities: The mediating role of supply chain collaboration. *Management Science Letters, 10*(3), 703–708.

Alzoubi, H. M., Vij, M., Vij, A., & Hanaysha, J. R. (2021). What leads guests to satisfaction and loyalty in UAE five-star hotels? AHP analysis to service quality dimensions. *Enlightening Tourism, 11*(1), 102–135. https://doi.org/10.33776/et.v11i1.5056

Alzoubi, A. (2021a). The impact of process quality and quality control on organizational competitiveness at 5-star hotels in Dubai. *International Journal of Technology, Innovation and Management (IJTIM), 1*(1), 54–68. https://doi.org/10.54489/ijtim.v1i1.14

Alzoubi, A. (2021b). Renewable Green hydrogen energy impact on sustainability performance. *International Journal of Computations, Information and Manufacturing (IJCIM), 1*(1), 94–110. https://doi.org/10.54489/ijcim.v1i1.46

Amarneh, B. M., Alshurideh, M. T., Al Kurdi, B. H., & Obeidat, Z. (2021). The impact of COVID-19 on E-learning: Advantages and challenges. *The International Conference on Artificial Intelligence and Computer Vision*, 75–89.

Aziz, N., & Aftab, S. (2021). Data mining framework for nutrition ranking: Methodology: SPSS modeller. *International Journal of Technology, Innovation and Management (IJTIM), 1*(1), 85–95.

Cruz, A. (2021). Convergence between blockchain and the internet of things. *International Journal of Technology, Innovation and Management (IJTIM), 1*(1), 35–56.

Dillet, R. (2020). *French contact-tracing app StopCovid has been activated 1.8 million times but only sent 14 notifications.*

Durbin, S. (2019). *Reducing the risks posed by artificial intelligence.*

Eli, T. (2021). Students perspectives on the use of innovative and interactive teaching methods at the University of Nouakchott Al Aasriya, Mauritania: English department as a case study. *International Journal of Technology, Innovation and Management (IJTIM), 1*(2), 90–104.

Emirates News Agency-WAM. (2020a). *Health sector launches new app "ALHOSN UAE" as part of efforts to contain COVID-19.*

Emirates News Agency-WAM. (2020b). *UAE public urged to join COVID-19 contact tracing app Alhosn to protect themselves, communities.*

Farouk, M. (2021). The universal artificial intelligence efforts to face coronavirus COVID-19. *International Journal of Computations, Information and Manufacturing (IJCIM)*, *1*(1), 77–93. https://doi.org/10.54489/ijcim.v1i1.47

Ghazal, T. M., Hasan, M. K., Alshurideh, M. T., Alzoubi, H. M., Ahmad, M., Akbar, S. S., Al Kurdi, B., & Akour, I. A. (2021). IoT for smart cities: Machine learning approaches in smart healthcare—a review. *Future Internet, 13*(8), 218. https://doi.org/10.3390/fi13080218

Guergov, S., & Radwan, N. (2021). Blockchain convergence: Analysis of issues affecting IoT, AI and blockchain. *International Journal of Computations, Information and Manufacturing (IJCIM)*, *1*(1), 1–17. https://doi.org/10.54489/ijcim.v1i1.48

Hamadneh, S., Pedersen, O., & Al Kurdi, B. (2021). An investigation of the role of supply chain visibility into the scottish bood supply chain. *Journal of Legal, Ethical and Regulatory Issues, 24*(Special Issue 1), 1–12.

Hanaysha, J. R., Al-Shaikh, M. E., Joghee, S., & Alzoubi, H. (2021a). Impact of innovation capabilities on business sustainability in small and medium enterprises. *FIIB Business Review*. https://doi.org/10.1177/23197145211042232

Hanaysha, J. R., Al Shaikh, M. E., & Alzoubi, H. M. (2021b). Importance of marketing mix elements in determining consumer purchase decision in the retail market. *International Journal of Service Science, Management, Engineering, and Technology (IJSSMET)*, *12*(6), 56–72.

Harahsheh, A. A., Houssien, A. M. A., & Alshurideh, M. T. (2021). The effect of transformational leadership on achieving effective decisions in the presence of psychological capital as an intermediate variable in private Jordanian. In *The effect of coronavirus disease (COVID-19) on business intelligence* (pp. 243–221). Springer Nature.

Huaxia. (2020). *Thailand to launch COVID-19 tracing application to facilitate disease-control tracking of customers as shops reopen.*

Jin, Y.-H., Cai, L., Cheng, Z.-S., Cheng, H., Deng, T., Fan, Y.-P., Fang, C., Huang, D., Huang, L.-Q., & Huang, Q. (2020). A rapid advice guideline for the diagnosis and treatment of 2019 novel coronavirus (2019-nCoV) infected pneumonia (standard version). *Military Medical Research, 7*(1), 1–23.

Joghee, S., Alzoubi, H. M., & Dubey, A. R. (2020). Decisions effectiveness of FDI investment biases at real estate industry: Empirical evidence from Dubai smart city projects. *International Journal of Scientific and Technology Research, 9*(3), 3499–3503.

Kashif, A. A., Bakhtawar, B., Akhtar, A., Akhtar, S., Aziz, N., & Javeid, M. S. (2021). Treatment response prediction in hepatitis C patients using machine learning techniques. *International Journal of Technology, Innovation and Management (IJTIM)*, *1*(2), 79–89. https://doi.org/10.54489/ijtim.v1i2.24

Khan, M. A. (2021). Challenges facing the application of IoT in medicine and healthcare. *International Journal of Computations, Information and Manufacturing (IJCIM)*, *1*(1), 39–55. https://doi.org/10.54489/ijcim.v1i1.32

Lee, K., Azmi, N., Hanaysha, J., Alzoubi, H., & Alshurideh, M. (2022a). The effect of digital supply chain on organizational performance: An empirical study in Malaysia manufacturing industry. *Uncertain Supply Chain Management, 10*(2), 495–510.

Lee, K., Romzi, P., Hanaysha, J., Alzoubi, H., & Alshurideh, M. (2022b). Investigating the impact of benefits and challenges of IOT adoption on supply chain performance and organizational performance: An empirical study in Malaysia. *Uncertain Supply Chain Management, 10*(2), 537–550.

Lee, C., & Ahmed, G. (2021). Improving IoT privacy, data protection and security concerns. *International Journal of Technology, Innovation and Management (IJTIM)*, *1*(1), 18–33. https://doi.org/10.54489/ijtim.v1i1.12

Leo, S., Alsharari, N. M., Abbas, J., & Alshurideh, M. T. (2021). From offline to online learning: A qualitative study of challenges and opportunities as a response to the COVID-19 pandemic in the UAE higher education context. *The Effect of Coronavirus Disease (COVID-19) on Business Intelligence, 334*, 203–217.

Mehmood, T., Alzoubi, H. M., & Ahmed, G. (2019). Schumpeterian entrepreneurship theory: Evolution and relevance. *Academy of Entrepreneurship Journal, 25*(4).

Mehmood, T. (2021). Does information technology competencies and fleet management practices lead to effective service delivery? Empirical evidence from E-commerce industry. *International Journal of Technology, Innovation and Management (IJTIM), 1*(2), 14–41.

Miller, D. (2021). The best practice of teach computer science students to use paper prototyping. *International Journal of Technology, Innovation and Management (IJTIM), 1*(2), 42–63. https://doi.org/10.54489/ijtim.v1i2.17

Mondol, E. P. (2021). The impact of block chain and smart inventory system on supply chain performance at retail industry. *International Journal of Computations, Information and Manufacturing (IJCIM), 1*(1), 56–76. https://doi.org/10.54489/ijcim.v1i1.30

Nuseir, M. T., Aljumah, A., & Alshurideh, M. T. (2021). How the business intelligence in the new startup performance in UAE during COVID-19: The mediating role of innovativeness. In *Studies in Systems, Decision and Control* (vol. 334). https://doi.org/10.1007/978-3-030-67151-8_4

Obaid, A. J. (2021). Assessment of smart home assistants as an IoT. *International Journal of Computations, Information and Manufacturing (IJCIM), 1*(1), 18–36. https://doi.org/10.54489/ijcim.v1i1.34

Radwan, N., & Farouk, M. (2021). The growth of Internet of Things (IoT) in the management of healthcare issues and healthcare policy development. *International Journal of Technology, Innovation and Management (IJTIM), 1*(1), 69–84. https://doi.org/10.54489/ijtim.v1i1.8

Shah, S. F., Alshurideh, M. T., Al-Dmour, A., & Al-Dmour, R. (2021). Understanding the influences of cognitive biases on financial decision making during normal and COVID-19 pandemic situation in the United Arab Emirates. In *Studies in Systems, Decision and Control* (vol. 334). https://doi.org/10.1007/978-3-030-67151-8_15

Whitelaw, S., Mamas, M. A., Topol, E., & Van Spall, H. G. C. (2020). Applications of digital technology in COVID-19 pandemic planning and response. *The Lancet Digital Health, 2*(8), e435–e440.

Wikipedia. (2022). *Template:COVID-19 pandemic data.*

Covid19 Unknown Risks—Using AI for Disaster Recovery

Mounir El khatib, Fatma Beshwari, Maryam Beshwari, Ayesha Beshwari, Haitham M. Alzoubi⊕, and Muhammad Alshurideh⊕

Abstract Artificial intelligence (AI) is a potentially powerful tool to mitigate the unknown unknown risk. Unknown-Unknown risks are unknown risks and accordingly unknown consequences. This research's objective is investigating the impact of AI on mitigating unknown unknown risk. For all six cases analyzed, it was possible to extract that unknown risk is hard to be reported in advance. Therefore, since the outbreak of the Coronavirus COVID-19, organizations have been eager to use AI in response to the consequences caused by this pandemic. As a result, AI played a significant role in rapid response, decision-making, early prediction, automated processes, and detection and tracking. This means AI has the potential to be used as a solution to mitigate the post event impacts and study the patterns for future prediction.

Keywords Artificial Intelligence · Machine Learning · Unknown risks · Uncertainty · Coronavirus · COVID-19

1 Introduction

Risk management is the guide to protect against any uncertainty and risk. The concept of 'unknown-unknown risk', which is the lack of awareness of the existence of a risk

M. El khatib · F. Beshwari · M. Beshwari · A. Beshwari
Hamdan Bin Mohamad Smart University, Dubai, UAE

H. M. Alzoubi (✉)
School of Business, Skyline University College, Sharjah, UAE
e-mail: haitham.alzubi@skylineuniversity.ac.ae

M. Alshurideh
Department of Marketing, School of Business, University of Jordan, Amman, Jordan
e-mail: malshurideh@sharjah.ac.ae; m.alshurideh@ju.edu.jo

Department of Management, College of Business Administration, University of Sharjah, Sharjah, UAE

that in itself is unfamiliar, has gained increasing attention in recent years, especially with the emerging crisis such as the Fukushima nuclear disaster. In many cases, the unknown unknown risk disturbed businesses, people, and caused huge losses.

Over last few months, as coronavirus spread across the world, countries are trying to learn about the COVID-19 to figure out how to slowdown the spread of the virus and defeat it (Akour et al., 2021; Alshurideh et al., 2021a, 2021b). It has been very difficult for business and health care systems to plan for the impact of COVID-19 due to the lack of historical comparisons of such outbreaks of this level and severity (Alshurideh et al., 2021a, 2021b; Hamadneh et al., 2021; Harahsheh et al., 2021). Countries struggled to act against this event due to the unknown nature of it and its consequence, which led many to follow the containment method to limit the spread of this virus and be able to focus on finding better solutions to react (Guergov & Radwan, 2021; Nuseir et al., 2021; Obaid, 2021; Taryam et al., 2020). Many new strategies were followed to decrease the impact of the unknown risk of this event such as school closure, traveling limitations, work from home, etc. (Ahmad et al., 2021a, 2021b; Aljumah et al., 2021). However, some of these strategies resulted in long-term damage to the economy due to the uncertainty of the shape of this virus during each stage and how to deal with it, which put a lot of stress on the healthcare system (Aburayya et al., 2020a, 2020b, 2020c; Cruz, 2021; Khan, 2021). In response to this risk, science and technologies have been working together to contain the negative consequences and reduce the losses for businesses and economies.

Allowing the potential of AI application in reacting against unknown unknown risk can be an opportunity to mitigate the impact of such risk (Al AlShamsi et al., 2021; Shebli et al., 2021; Yousuf et al., 2021). What makes it a good solution is the ability to increase in computing processing power, the rise of big data, and the adoption of cloud computing (Ali et al., 2022; Alzoubi et al., 2022; Mehmood, 2021). In addition, AI can adapt ambiguous or contradictory messages through visual perception, decision-making, and speech recognition as well as translating between languages (Hejazi et al., 2021; Khamees et al., 2021; Naqvi et al., 2021). AI is coming across as one of the main technologies that are accelerating the decision-making process based on real time insights.

2 Research Problem

Many papers were published regarding unknown unknown risk and how to deal and overcome the challenges posed by this type of risk. Unknown unknown risk is a type of risk that can be considered as an extremely dangerous risk because its possibility and consequences are undetermined. The danger of unknown unknown risk is that since it is unpredictable and no one is aware of it, it cannot be managed at its initial appearance that in some cases leads to disaster or organization bankruptcy and massive losses (Alshurideh et al., 2022; Shah et al., 2020, 2021). Therefore, this research will aim to study the ambiguity of managing and acting towards the unknown risk. Specifically, it will examine the contribution of AI in mitigating the unknown risk and the management of change when this risk occurs. The capability of this technology can play a significant role in lowering the chances of negative consequences. This study will be based on the COVID-19 pandemic. It has been chosen because it illustrated the importance of technology in allowing life and businesses to continue as usual under such circumstances.

This paper will be divided into five sections. Firstly, the introduction, followed by a literature review addressing the unknown unknown risk, and Artificial Intelligence. The third section will gather data from secondary sources about different uses of AI in combating unknown unknown risk and COVID-19 will be a case study. The fourth section will perform a qualitative and quantitative analysis for these case studies, and based on the outcome, it will answer the hypothesis of this paper and present the effectiveness of this technology in mitigating unknown unknown risk.

3 Literature Review

3.1 The Need for Technology-Based Solution

Typically, once a disease or biological disaster becomes a reality, humans and human contact will aid its transmission and become the main element for spreading and aggravating infections and deaths caused by the disease or disaster (Hanaysha et al., 2021a, 2021b). A typical example is the case of the ongoing COVID-19 pandemic. COVID-19 is a looming but avoidable catastrophe. During the COVID-19 pandemic, the potential to control and contain the pandemic and its consequences using human elements have been critically undermined and also led to infected health care professionals (Blumenthal et al., 2020; Cabarkapa et al., 2020; Pietsch, 2020). Considering the potential human and economic consequences disaster could have, immediate actions must be taken to effectively monitor, manage, and recover disasters. Instead of relying solely on overstretched healthcare professionals and government officials, it is important to seek a better and risk-free solution to prevent and control disaster (Alshurideh, 2014; Aburayya et al., 2020a, 2020b, 2020c). This, therefore, necessitated the need for a technology-based solution to address these crises.

One justification for the technology-based solution is that it is not highly reliant on physical interactions (Hanaysha et al., 2021a, 2021b; Lee & Ahmed, 2021). That is, it requires limited human interaction and resources as it is able to operate with little or no human input, it is immune to infectious diseases, and it can function effectively in infected environments (Alsharari, 2021). More so, a technology-based solution is less vulnerable to human-related issues like stress, mental health burdens, physical fatigue, and other health challenges that frontline health workers may encounter during unknown risks and disaster recovery.

3.2 Unknown Unknown Risks

An unknown unknown can be defined as the state of being unaware that a particular risk exists. The repetitive unknown unknown is not a word mistake; it refers to unknown likelihood and unknown impact. Another research defined it as events considered to be impossible to find or imagine in advance (Kim, 2012). Many researchers have explored how to understand the unknown unknown, but the challenge exists in the nature of the event. In some research, they related the unknown unknown risk to the black swan theory, which is an unpredictable event that is beyond what is normally expected of a situation and has potentially severe consequences (Nafday, 2009). This term came from an ancient saying that presumed black swans did not exist and it was developed by Nassim Nicholas (Nafday, 2009). This is because both share the condition of having a major impact, extreme infrequency, and the widespread insistence (Alzoubi, 2021a; Lee et al., 2022a, 2022b). It is described by some as an outlier because it lies outside the realm of regular expectation, since nothing in the past can predict its possibility (Hajikazemi et al., 2016).

Sometimes, problems exist in people rather than the event itself and the major obstacle to address this type of event is the lack of capability to accept it, and people who cannot cope with unknown unknown will sometimes actively ignore it (Alles, 2009). Another challenge is to turn as many unknown unknowns to known known and being proactive in being exploring organizational environment (Eli, 2021; Miller, 2021). Not all possibilities can be thought of and planned, but there are ways to minimize delays and negative impact caused by unknown unknowns. The biggest takeaway is to become better equipped at identifying, predicting, and managing the risks related to these occurrences (AlHamad et al., 2022; Lee et al., 2022a, 2022b).

Despite that risk management performs as forward looking, it is not possible to identify all risks in advance (Hillson, 2016). Unknown Risk cannot be managed proactively and usually in unknown risk events, a workaround approach is used to respond to unidentified risk. The approach is based on planning once the unknown risk occurs and managing it. It is hard to predict it, but possible to manage and minimize its negative impact once it has happened.

3.3 The Need for Advanced Technology

In a bid to effectively predict, control, manage, and contain biological disasters like COVID-19, which are occurrences that are difficult to detect preemptively but have a high negative impact, advanced technology is needed. Although several emerging technologies have been revealed to show potency for disaster recovery (Jia, 2019), an advanced technology-based solution that is more promising is AI (Artificial Intelligence) technology.

3.4 Artificial Intelligence and Machine Learning

AI, as we know it, is the short form of Artificial Intelligence. This is where the science and technology come to play, with multiple approaches, on where the advancement of science creates a paradigm of virtue for every sector in any industry to succeed (Zhang et al., 2019). The famous mathematician Alan Turing, changed the world forever by asking a simple question and that is, 'Can Machines Think'? With in-depth research, he was able to prove the main and fundamental goal of artificial intelligence. AI was created for a reason and that is basically to replicate the human intelligence in a machine (Graham-Cumming, 2012). In simple words, AI can be defined as the simulation of human intelligence in machines that are programmed to think like humans and mimic their activities (Al Ali, 2021; Mondol, 2021; Radwan & Farouk, 2021). This term can also be used for any machine that demonstrates characters associated with a human mind such as learning and problem solving.

Moreover, Machine Learning (ML) is a subdivision of AI and refers to the study of computer algorithms that allow computer programs to automatically improve through experience (Lord et al., 2009). It is one of the ways to achieve AI and functions on small to large datasets by inspecting and correlating between the data to find common pattern. It is very important to understand how to choose the correct algorithm type and its logistic progression in order to get a correct output and avoid any bias affecting the application. Several researchers found that using correct algorithms and tools can help in dealing with uncertainty projects and complex environments better. In addition, they believed that AI as a tool is better and more accurate than traditional tools (Martínez & Fernández-Rodríguez, 2015).

AI was created for the entire basis and the future of all complex decision making and it has been extremely efficient with its evolution of machine learning. There are several types of AI where machines are being taught and knowledge is learned (Alzoubi, 2021b). Machines are given the opportunities to analyze and learn data just like the human brain (Motoda & Yoshida, 1998). With technology as advanced as this, there has been significant progress in manufacturing, finance, healthcare and even retail, all with the help of artificial intelligence (Zhang et al., 2019).

3.5 Why AI is Trending Now?

Over the past decades, the field of AI has made great progress toward computerizing human reasoning and its breakthrough in a wide diversity of sectors from healthcare to transportation. AI techniques are used by many industries through exploiting data and optimizing business and production processes. It processes the data in ways that are more dynamic and create value with faster and more accurate decision that result in reduced operational costs and personalized customer experience (Ferma, 2019). In addition, ML is assisting financial services companies with advanced fraud detection, and computer vision is optimizing offline retail. AI can assist with an unbroken supply chain and boost productivity. Artificial intelligence is not only known to improve lives but has managed to save lives as well. AI has assisted with creating personalized drug protocols for a better diagnosis, with robots even assisting in surgeries (He, 2019). AI is helping the humankind to save the world by creating smart agriculture with efforts to solve global challenges (Motoda & Yoshida, 1998).

3.6 AI in Risk

AI is recognized for its potential in many fields including risk management. We can see that ML and AI are playing an important role in enhancing the risk management processes. Companies like oracle uses AI to conduct enterprise risk assessment through identifying risk material in the organization by calculating inherent risk using likelihood, impact, and analysis model (Folsom, 2020). Microsoft is another example that utilize AI to identify the root cause of risk occurrence and perform qualify analysis in earlier phases of risk management process (Microsoft, 2020). However, the level of implementation varies between the organizations, but it is clear that companies are in the beginning stages of understanding the full benefits of AI. The current usage in this field is only visible when losses occur or for known risks (Farouk, 2021).

3.7 Adopting AI Capabilities for Disaster Recovery

According to Turing et al., AI (Artificial Intelligence) is simply considered as a "thinking machine" (Turing & Braithwaite, 2004). It was described by Shukla and Jaiswal as a technique that can facilitate "a computer to do things which, when done by people, are said to involve intelligence" (Shukla Shubhendu & Vijay, 2013). With AI technology, machines can detect patterns that are too complex for humans to quickly detect and process. AI has been widely adopted in areas like machine vision, speech recognition, natural language processing, targeted marketing, and even in the healthcare management system, which includes using AI to combat COVID-19 (Davenport et al., 2019; Laghi, 2020). While other technologies like robotics, drones, smart sensors, virtual reality, and so on have all played positive roles in supporting health care professionals in managing and coping with the COVID-19 pandemic (Tavakoli et al., 2020; Zeng et al., 2020), AI will arguably be more instrumental in tackling some of the most striking and prevalent issues that both health care professionals and government officials are faced with. These issues include pandemic surveillance, drug and vaccine development for the coronavirus pandemic, and disaster recovery (Keshavarzi Arshadi et al., 2020; Laghi, 2020; Lin & Hou, 2020).

AI technology is very useful in searching for specific patterns or identifying trends and patterns across a large amount of data quickly and cheaply. For instance, AI can extract data retrospectively from medical records or prospectively in real-time and process this data statistically for insights that can supplement any existing structured data to develop actionable information (Sharma et al., 2019). During the outbreak of the coronavirus, AI has been used to gather and analyze publicly available data, such as geolocation data, social media content, and content timestamps to identify and record potential cases of coronavirus in a cost-effective way without expending money on testing devices or employing clinical resources that involve health care professional (Klein et al., 2021).

As seen above, AI is irreplaceable in terms of handling complex tasks, which include extracting useful information from large datasets. However, considering the increasing advancement in technology and its applications, AI can be used as the main component in other emerging technologies (Jia & Yang, 2020), which can be used for disaster management and recovery. All these insights drawn from the COVID-19 pandemic (Keshavarzi Arshadi et al., 2020; Lin & Hou, 2020) suggest that AI has a great potential in disaster monitoring and managing.

AI technology, alone or when coupled with other advanced technologies, can be very effective in elevating disaster monitoring, control, and management (Kashif et al., 2021). That is, AI technology can be adopted to handle issues, such as early detection and identification of disaster, treatment design and development, public health interventions, and disaster recovery (through sentiment analysis using social media content. This study specifically focuses on adopting AI for disaster recovery taking COVID-19 unknown risk as a case study.

3.8   The Role of AI in COVID-19

COVID-19 is a global pandemic that was first identified in Wuhan city of China and exponentially started to spread across the world. This pandemic has changed the world significantly, not only for health care systems, but also for politics, education, economics, transportation, etc. Consider what makes COVID-19, as a crisis is its unknown nature as a virus and its consequences, as well as the speed at which it is spreading across the world. Although World Health Organization (WHO) and Centers for Disease Control and Prevention (CDC) have published instructions and regulations to reduce the risk of COVID-19 from spreading, it is still breaking out. Many countries considered this epidemic as a normal flu and ignored all advice released by WHO and followed the approach of herd immunity, but unfortunately this method made it worsen. This approach results in exploding the number of infected cases that eventually led those countries to issue a 24 h lockdown while preventing any type of gathering as well as closing their borders (Mayo Clinic, 2019). In response to this unknown risk, AI played an important role and it has been applied successfully in managing this outbreak.

Countries like Japan, Hong Kong, and South Korea combined lessons learned from H1N1 and SARS with the AI powered capabilities to process the unknown risk in real time which helped them to build defense and response systems to mitigate losses associated with the COVID-19 epidemic (Rogers, n.d.). This proves that AI can analyze the anatomy of past crisis and apply that intelligence to the precautions that countries develop today as they prepare for the challenges of tomorrow. The truth is countries and organizations cannot predict the onset of the global pandemic (Akhtar et al., 2021; Aziz & Aftab, 2021). However, they can identify the telltale indications. AI can empower organizations to proactively spot growing unknown risk early and take decisive measures before massive loss occurs. Fortunately, AI and unsupervised machine learning are capable of analyzing, processing, and deriving actionable insight from all this data and can do so in real time.

3.9 Conceptual Framework and Hypothesized Relationships

Organizations are recognizing AI effects on the market and they became more interested in investigating AI as a new technology that can transform their management strategy. Unknown unknown risk is a big concern when it hurts the business without any contingency plan. Therefore, the ability of AI to predict and deal with the risk that might appear in the future is catching business attraction. Using AI to mitigate the upcoming unknown unknown risk in an organization can increase the chance of avoiding any loss. In addition, it will help the production to create new plans to reduce the risk. Moreover, they can manipulate it to their advantage, create an opportunity in having a change that exceeds that of opponents, and be the leader in the market. The objective of this research is to study AI capability towards unknown unknown risks negative consequences and the target audience for this research are organizations, project managers, and policy makers. Figure 1 represents the proposed hypothesis of this research.

   H1: AI technology has a positive effect on mitigating unknown unknown risk.

4 Methodology

After a complete review of the literature, a qualitative method has been decided to be used to meet the objective and answer the question of this study, which is how AI can mitigate the impact of unknown unknown risks. A combination of secondary data and interview methods will be utilized. As secondary data, six case studies of different utilizations of AI in combating COVID-19 will be chosen to discuss the

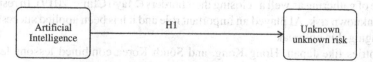

Fig. 1 Research conceptual framework

role AI played in mitigating the unknown unknown risk. In addition, a structured interview will be used to gather detailed information and the questions will be based on an open-ended approach. This data will help to better understand in which stage risk management processes can involve AI technology and the potential level of its capabilities in unknown events cases.

5 Data Collection

5.1 Case A: Canada BlueDot

The Canadian based AI model BlueDot that functions based on machine learning, predicts humans by spotting infectious disease outbreaks. It predicted the outbreak of the infection at the end of 2019 issuing a warning to its clients on 31 December 2019 before the WHO did so on 9 January 2020 (Kreuzhuber, 2020). A group of researchers who are working with the BlueDot listed the top 20-destination cities where passengers from Wuhan would arrive in the wake of the outbreak. They warned that these cities could be at the forefront of the global spread of the disease.

BlueDot aims not only to identify the threat of infectious diseases, but also to understand how diseases could spread to different parts of the world, then to determine the possible consequences of their spread. BlueDot was able not only to send out a warning, but also to correctly identify the cities that will be highly infected through using global air ticket data. It helped to predict where the infected people could potentially travel to. Eleven of the top listed cities were also the first areas affected by COVID-19 cases. For WHO, which is dependent on official statistics from Chinese authorities it was difficult to access such information and make forecasts early on (Pham et al., 2021).

5.2 Case B: China Detection System

CCTV based tracking was used in Asian countries to identify the potential infected people. China used infrared cameras to scan crowds for high temperature. Their AI powered system used facial recognition to identify individuals with high temperature and whether he/she is wearing mask. It is currently being used at Beijing's Qinghai Railway Station and it has the capability to inspect up to 200 people in one minute without interrupting passenger flow (Allam et al., 2020). This system also used to ensure citizens comply with self-quarantine instructions. According to reports, individuals who ignored the instruction and left home would get a call from the authorities after being tracked by the system (Preethika et al., 2020).

5.3 Case C: Dubai Police Movement Permission

The Dubai government announced a 24-h sterilization program, which aims to reduce the physical contact between people and speed up the sterilization program. Dubai Police used AI to find if the trip was for the reason claimed in the permit request. AI was used along with speed radars to identify which vehicle has a movement permit or belongs to people working for important sectors. In addition, if a vehicle did not get a permit and went out for necessary purposes, AI can recognize the route and check if this trip was in fact for an important matter. Plus, the system has the capability to identify the path a person takes when they work in vital sectors and it will know the route from home to work.

5.4 Case D: IBM Watson Assistant

Watson Assistant is a conversational AI platform that provides an engaging experience for users to get even their most complex issues solved by using supported chat and speech interfaces. It was preprogrammed with COVID-19 questions via the IBM cloud. Organizations in 12 countries from the city of Austin to the Polish Ministry of Health have been able to get up and running in as little as 24 h (Hanford, 2005). The objective of this type of technologies is to help reduce the waiting time for calls those users make to the telephone lines of state services relating to the novel coronavirus. In addition, it provides data driven insights to help employers make informed decisions on workplace re-entry, facilities management, space allocation, and other covered related priorities.

5.5 Case E: Microsoft Azure AI—ImmuneCODE

Biotechnology and Microsoft collaborated in launch an immuneCODE that is an open database to share an immune response toward COVID-19 blood samples from patients around the world. The database contains information details on "T" cells, which specifically recognize features of the virus with unprecedented scale and speed. This cell has trove of information that can provide one consistent trackable measure of immune response. In addition, it can help in diagnosing and managing COVID-19 and provide accurate assessment of immunity. As a result, data from the immuneCODE will speed up vaccination development, provide better diagnose, and reduce the spread of the virus (Microsoft, 2020). According to Kevin Scott (2020), Microsoft is using similar supercomputing capacity that have been used before for natural language process to run molecular simulation to identify potential therapy for COVID-19.

5.6 Case F: Facebook Fighting Misinformation

Facebook uses AI model to detect and track false news written in text and images effectively. The software is trained to find duplicated or slightly modified version of that content as it spread across the platform in different languages (Statt, 2020). It can recognize and track both the copy and the original image with perhaps one changed word in the post. It can recognize whether the images contain false claims or misleading information about COVID-19 and apply warning label automatically (Statt, 2020). The system is applied to every image that has been uploaded on each platform, Instagram, and Facebook, which means billions of images are being checked per day in their database.

6 Analysis

6.1 Case A

BlueDot was able to identify the crisis and the cities that will be affected by this disease that gives it the advantage of early prediction of such pandemic. It alerted the Canadian entity about this pandemic 8 days before the WHO announcement, which presents its effectiveness in early prediction (BlueDot, n.d.). It also assisted in understanding how disease might scatter to different parts of the world and then determined the potential impacts. It pointed out which cities had the risk of transmitted COVID-19 despite having no official cases (BlueDot, n.d.).

It required human interpretation and providing context to recognize the threat. In fact, even in the case of BlueDot, human remains central in evaluating and interpreting its output. Therefore, it is correct to stress that human input across disciplines is needed for the optimal application of AI.

6.2 Case B

Case B used IOT devices as CCTV based on AI and machine learning algorithm. It was able to detect infected people. In addition, it was used to track those who are not following the guidelines and report them. The data that was collected was interpreted and transferred to the concerned person in the organization for interpreting them, conducting results, and creating an action plan. Their main goal was to identify areas of vulnerability in early stages to avoid surprises by managing potential aspects, which can lead to occurrence of risk overtime. The shown AI system takes a whole CCTV as an input and runs a quick diagnosis to be able to provide the abnormal cases as an output.

6.3 Case C

In Case C, they used AI to track people who got the permission in advance along with the ability of taking a decision regarding the fines, whether it is eligible or should be removed. The new understanding of the technology can be result in evaluating the current situation and making sense of a complex pattern of different aspects. Their aim was to produce a fast-reactive mechanism to various changes and pro-act to them. The effectiveness of their method was depending on the speed of analyzing the vehicle data. This system assisted in automating the validation processes, which in usual cases take days to be verified and can sometimes include human error (UAE Government, n.d.).

6.4 Case D

Case D used machine learning to enable chatbots to interact with people and take decisions in responding to different inquiries from them later on. In addition, it automates the call center work to speed up the efficiency of this service and effectively carry out the process. The success of this mechanism depends on the level of maturity of the organization in feeding up the basic knowledge and questions to the AI. As well as, expanding the sharing data for more than 12 countries for better proactive responding. Effectiveness and coverage are the two measures that provide the overall performance of Watson. With COVID-19, it succeeded because of its capability in responding to most inquiries well (Benvie et al., 2020).

6.5 Case E

Azure ML empowers researchers and medical organizations in responding to COVID-19 and it accelerated solutions to react against the global pandemic. The accuracy of the immune response will be continuously enhanced and updated online in real time as extra samples are sequenced by using MS Azure cloud and ML capabilities. The objective of this collaboration is to speed up the vaccination research and identify the patterns of this virus in timely manner.

6.6 Case F

Facebook AI system's goal is to reduce the spread of misleading information across the program. In addition, it helps them to demonstrate their commitment to keep

Table 1 Summary of AI contribution

Case	Decision Making	Early prediction	Tracking and detection	Automation processes	Cost efficiency
Case A	•	•	•	•	•
Case B	•	•	•	•	•
Case C	•		•	•	•
Case D	•			•	•
Case E	•	•	•	•	•
Case F	•	•	•	•	•

Facebook and Instagram safe and genuine with the content uploaded in these platforms. It uses the three-pronged strategy of informing, reducing, and removing. It starts with informing users of recognizing false news. Then, it reduced the distribution and appearance of false news that has been labeled. Finally, it removes the fake posts and disable the accounts. According to Facebook Community Standards Enforcement Report, the fake accounts represent approximately 5% of their users and were around 1.5 billion accounts in the second quarter of 2020. In addition, 99.6% accounts were flagged with through their system before the other users reported them (Facebook, n.d.).

After analyzing, we came up with five main tangible roles that AI played in COVID-19. Table 1 illustrates the identified influences in each case.

Another analysis was done to demonstrate the AI contribution in each risk management process based on analyzing the six case studies. The results were based on correlating between the functionality AI provided in each case with the elements of each risk management processes as shown in Table 2. It is clearly shown that AI can help in developing a consistent, and integrated risk management methodology to help determine how best to identify, manage, and mitigate significant risk through the organizations.

6.7 Effectiveness of AI

Table 3 addresses the six cases with using and without using AI for measuring the effectiveness of utilizing AI in COVID-19 and any future unknown events.

In addition, we did a quantitative measurement of using AI technology in respect to the following metrics: speed, adoption, response, availability, and cost reduction. The scale of assessment will be from 1 to 5 where 5 presents excellent and 1 is low. The scores are based on the researcher's point of views from each case.

Table 4 illustrates that the five measurements reveal the effectiveness of utilizing AI in mitigating unknown risk. As shown in Table 4 most of the scores were between 3 and 5 which indicate that AI has a strong potential in mitigating unknown unknown risks. AI based systems are always available and were not impacted by working

Table 2 AI in risk management processes

Sr	Case	Planning risk management	Risk identification	Qualitative analysis	Quantitative analysis	Planning risk response	Implement risk response	Monitor risk
1	Case A	•	•	•	•	•		•
2	Case B	•		•	•		•	•
3	Case C	•		•	•		•	
4	Case D	•		•	•	•	•	
5	Case E		•	•			•	•
6	Case F		•	•	•	•	•	

from a distance and being on lockdown. AI quickly adapted the change with the pandemic and assisted in reducing the cost of operations that required human involvement such as responding to customer questions, monitoring people's movement, and checking the logs of violators. Moreover, it contributed in automation of many repetitive and manual processes that have been impacted by implementing working from a distance. In addition, AI participated in the better understanding of this unknown risk through analyzing the patterns and turning it to become known as well as provided recommendations on how to react in different stages of this disease.

7 Discussion of Results

After analyzing the six case studies, the results that were obtained within the previously described method for each case of using AI technology can be summarized in the main roles of rapid response, namely, decision making, early prediction, automate processes, and detection and tracking. We can consider AI as a game changer for accelerating cognitive abilities and economic benefits. AI and machine learning can help proactively prepare for unknown unknown risk. In the COVID-19 case, we found that most companies have focused so far on reactive measures, but now it is the perfect time to take transformative action and be proactive. Companies that have already introduced AI should press ahead immediately in order to achieve maximum advantage in the near term. They should not hesitate to scale up because AI will be a significant level that can help them manage this crisis and future unknown unknown risk. The following are the main AI roles found in mitigating unknown unknown risk:

Table 3 Measuring AI effectiveness—qualitative

Case	Without AI	With AI
Case A	Human needs time and effort to analyze and find the pattern to predict the pandemic as well as he will be observing and waiting for other countries to report the infected cases. WHO took 9 days after BlueDot results to announce the pandemic	Predicted the pandemic before the WHO announcement and reported the top 20 cities that will have the virus thorough correlating between the airlines' tickets and infected cases in city Wuhan. 11 cities were correct from the list
Case B	CCTV will not be able to detect and take actions against any infected cases and it will require human intervention to verify the videos and report the violators	AI directly detected the infected cases once they passed through the CCTV as well as identified the violators who are not wearing masks
Case C	Requires human review that will take more than a week to verify each video recorded and map it to the permissions	AI could identify the violators within seconds and take the decision to issue fines
Case D	The customer service agent will take time to ask for the answer and respond to the phone call inquiries and sometimes the responsible person is not available to answer and that can take more than one day	A direct response is available to any inquiries because of the data that has been fed to the machine learning and any unknown question will be reported to a higher level as a quick reaction
Case E	The process of analyzing the T cell will be very slow and it will require time and efforts to study all the samples and find the common pattern which in some cases requires months. Even sharing data will require many approvals and permissions. This will lead to put huge pressure on medical organizations which are already overwhelmed with infected cases	AI speed up the analyzing process because of the open public database to find the common attributes of the virus as well as it provided better diagnose. It helped to drive model in a day instead of months which is normally required using older technique
Case F	The process of tracking the source of fake news requires times as well as it is very hard to identify the news if its true or fake	AI speed up the process of verification and assisted in tracking the sources of fake news in seconds instead of weeks. In addition to its capability to automatically takes the decision to remove the news

Table 4 Measuring AI effectiveness—quantitative

Case	Metric				
	Speed	Adoption	Response	Availability	Reduce cost
Case A	5	4	4	5	4
Case B	4	4	4	5	4
Case C	4	4	5	5	5
Case D	4	4	3	5	4
Case E	4	3	4	5	4
Case F	4	4	5	5	3

7.1   Rapid Response

During any pandemic, the role of technology rises and shines. For example, in COVID-19, AI and big data showed an extraordinary role in solving problems and increasing the potential for response to any future circumstances. It played a significant role in diagnosing coronavirus patients. Thus, AI enabled rapid and ongoing adaption and recognition to the new abnormal and can be open to change (Alnazer et al., 2017; Haitham Alzoubi & Ahmed, 2019). It improves the adaptability required to address changing conditions at every step of the unknown unknown risks. It is able to process the combination of structured and unstructured data in a massive amount, which shows the patterns and results in decision-making.

7.2   Decision Making

AI is a decision support system that allows domain experts to generate diverse alternative scenarios of the future and imagine different possible outcomes including unlikely, but potentially impactful future (FERMA, 2019). Any unknown risk has an impact on any business and the effect can be long-term, so utilizing technology that can think and learn is a great opportunity. The top management can benefit from the technology through better decision-making, providing great productive insights and visibility of risk that can be delivered by AI. The organizations that benefited from the AI vital role in fighting against COVID-19 will assist the future developments in their decision-making strategy (Alzoubi et al., 2020a, 2020b). AI technology has the ability to use secondary data to provide reference suggestions when people are unsure or when there are multiple decisions (Jin et al., 2020).

7.3   Improve Cost Efficiency

AI helps organizations to improve their work efficiency through reducing cost of the processes by automating day-to-day assistance and guidance in the risk management process (FERMA, 2019). For a small organization with few resources, AI can propose a method to achieve rapid and accurate risk assessment, which result in proper allocation of the limited resources. This means that AI led to substantial savings in term of human resources and improved emergency repair, efficiency, and decreased the risk of infection for individuals within the organization (Alzoubi & Yanamandra, 2020; Mehmood et al., 2019). The ability of adapting to different requirements can be considered as a benefit in the resources.

7.4 Early Prediction

Unknown risk has warning signs and if we can identify them and act against minor events, then perhaps we can avoid losses. It is important to obtain an early indication of negative risk development that will become noticeable in the future (Hajikazemi et al., 2016). AI can be used to identify abnormalities in certain areas of data storage. With the huge amount of data, you would not see them as a person, whereas the machine now is enabling us to notice them. It has the potential to exceed humans, not only through speed, but also by dictating patterns, which humans have overlooked. These patterns will be taken under consideration and will generate many different scenarios exploring a variety of possible outcomes (Alzoubi & Yanamandra, 2020; Joghee et al., 2020). The top management roles come to prepare the organization to deal with realistic circumstances for both expected and unexpected futures.

7.5 Automate Processes

Automate repetitive tasks that a human does every day, but in doing so we are going to create work that will allow people to do more complex and judgment-based pieces of work that will enhance their ability to engage in the job. It conducts enterprise risk assessment through identifying risk materials in the organization, by calculating inherent risks using likelihood, impact, and analysis model. Second, it evaluates and redeems the risks by using context models to determine an action, such as accepting, monitoring, or treating. Then it performs residual risk analysis after controls are applied. Finally, it schedules risk assessment by running periodic risk assessment for analysis and evaluation. Not to mention, it has a self-learning feature which means the embedded self-learning feature of AI will keep learning for accurate results in the future. AI technology has shown good performance during the pandemic and can be used with different diagnosis levels, according to different policies that humans teach it. The system will keep learning and reach to experienced levels, which later on will automatically produce the output of the matter (Alshurideh et al., 2020; Alzoubi et al., 2020a, 2020b; Jin et al., 2020).

7.6 Detection and Tracking

Organizations need to analyze their risk profile through the competence of their risk management framework. The insufficient knowledge and the complete lack of awareness about their existence can make it hard on organizations to detect. This may lead to losses due to inadequate experience and interaction with the available risk management system (Emerging). Detecting and tracking symptoms with AI can lower the possibility of losses and will lead to the creation of a relevant credible management model. This creates the opportunity to plan different scenarios and strategic foresight, which must go beyond the conventional approaches due to the emerging risk occurring from the complex and interrelated event. For example, in the COVID-19 pandemic, AI technology was able to track the suspicious cases and forecast the nature of the virus from the feed up data about the risk infection and the likelihood of spreading (Alnuaimi et al., 2021). The monitoring feature resulted in predicting the possible positive cases and the regions in which they are likely to be present. Thus, it provided daily updated information, which aided in giving the possible measurements to prevent the virus from spreading accordingly.

8 Limitation

Many papers admitted that AI in some cases struggled due to lack of data. In fact, the successful implementation of artificial intelligence is dependent upon the data set and creativity of the humans who are deploying it to mitigate the unknown unknown risk. The value of AI and its application in exceptional times like the coronavirus pandemic stems from the human input that goes into its design process (Alzoubi & Aziz, 2021; Alzoubi et al., 2021). To realize the full potential of AI, it is necessary to marry innovation and human action together, so that we can win the war against any type of unknown risk with reduced losses.

9 Recommendations

The following are recommendations to enhance the role of AI in risk mitigation.

- Governments must put in place the necessary supporting technology, regulations, and governance to enable large-scale adoption in response to unknown unknown risk.
- Promote co-operation, and data exchange both on a national scale and globally by the AI community, and policy makers to contribute in preparing for the problem, recognizing important data and launching datasets for unknown unknown risk.
- Dedicate research on AI technologies that can understand from limited data.
- Build AI-powered monitoring tools that empower research without affecting privacy.
- Initiating emerging risk governance framework that contains new approaches to the risk management strategy, operational processes, and measurement of exposure to them.
- Integration AI with other technologies such as blockchain, big data, and cloud to achieve better efficiency in reacting to pre and post unknown unknown risk.

10 Conclusion

In conclusion, it is clearly seen that AI can be a game changer for mitigating unknown unknown risk. The capability of this technology, which was proven during the COVID-19 pandemic, is that it could be a valuable source to combat black swan events. In this research, we have presented a state by utilizing AI technology to address numerous challenges posed by the coronavirus pandemic. Six main applications of AI potentials for coping with COVID-19 crisis have been analyzed all through (Ali et al., 2021; Ghazal et al., 2021a). As a result, this research acknowledged that AI and machine-learning technology are enabling the shift by providing the tools to support remote communication, enable telemedicine, and protect food security. In addition to its capability of early prediction, tracking and detecting, decision-making, and process automation, we believe AI-driven algorithms could deliver valuable predictions and readings in the future. Companies do not require completely new processes for dealing with AI, but they will need to improve existing ones to consider AI and fill the necessary gaps (Alhamad et al., 2021; Kurdi et al., 2021). For best utilization, it is recommended to increase data sharing, in addition to having access to different historical data with the ability of pulling data from the right sources. However, the biggest challenge is to create a convincing and reliable global data set of different risks identified among the nations with possible different metrics. Adoption of AI requires from companies to go through a learning journey and to work in a more scientific mind-set.

References

Aburayya, A., Alshurideh, M., Alawadhi, D., Alfarsi, A., Taryam, M., & Mubarak, S. (2020a). An investigation of the effect of lean six sigma practices on healthcare service quality and patient satisfaction: Testing the mediating role of service quality in Dubai primary healthcare sector. *Journal of Advanced Research in Dynamical and Control Systems, 12*(8), 56–72.

Aburayya, A., Alshurideh, M., Al Marzouqi, A., Al Diabat, O., Alfarsi, A., Suson, R., Bash, M., & Salloum, S. A. (2020b). An empirical examination of the effect of TQM practices on hospital service quality: An assessment study in uae hospitals. *Systematic Reviews in Pharmacy, 11*(9). https://doi.org/10.31838/srp.2020a.9.51

Aburayya, A., Alshurideh, M., Albqaeen, A., Alawadhi, D., & Al A'yadeh, I. (2020c). An investigation of factors affecting patients waiting time in primary health care centers: An assessment study in Dubai. *Management Science Letters, 10*(6). https://doi.org/10.5267/j.msl.2019.11.031

Ahmad, A., Alshurideh, M. T., Al Kurdi, B. H., & Salloum, S. A. (2021). Factors impacts organization digital transformation and organization decision making during Covid19 pandemic. In *Studies in systems, decision and control* (vol. 334). https://doi.org/10.1007/978-3-030-67151-8_6

Akhtar, A., Akhtar, S., Bakhtawar, B., Kashif, A. A., Aziz, N., & Javeid, M. S. (2021). COVID-19 detection from CBC using machine learning techniques. *International Journal of Technology, Innovation and Management (IJTIM), 1*(2), 65–78. https://doi.org/10.54489/ijtim.v1i2.22

Akour, I., Alshurideh, M., Al Kurdi, B., Al Ali, A., & Salloum, S. (2021). Using machine learning algorithms to predict people's intention to use mobile learning platforms during the COVID-19 pandemic: Machine learning approach. *JMIR Medical Education, 7*(1), 1–17.

Al Ali, A. (2021). The impact of information sharing and quality assurance on customer service at UAE banking sector. *International Journal of Technology, Innovation and Management (IJTIM)*, *1*(1), 01–17. https://doi.org/10.54489/ijtim.v1i1.10

Al Shebli, K., Said, R. A., Taleb, N., Ghazal, T. M., Alshurideh, M. T., & Alzoubi, H. M. (2021). RTA's employees' perceptions toward the efficiency of artificial intelligence and big data utilization in providing smart services to the residents of Dubai. *The International Conference on Artificial Intelligence and Computer Vision*, 573–585.

Alhamad, A. Q. M., Akour, I., Alshurideh, M., Al-Hamad, A. Q., Kurdi, B. A., & Alzoubi, H. (2021). Predicting the intention to use google glass: A comparative approach using machine learning models and PLS-SEM. *International Journal of Data and Network Science, 5*(3), 311–320. https://doi.org/10.5267/j.ijdns.2021.6.002

AlHamad, A., Alshurideh, M., Alomari, K., Kurdi, B., Alzoubi, H., Hamouche, S., & Al-Hawary, S. (2022). The effect of electronic human resources management on organizational health of telecommuni-cations companies in Jordan. *International Journal of Data and Network Science, 6*(2), 429–438

Alhashmi, S. F. S., Alshurideh, M., Al Kurdi, B., & Salloum, S. A. (2020). A systematic review of the factors affecting the artificial intelligence implementation in the health care sector. In *Advances in Intelligent Systems and Computing* (Vol. 1153). AISC. https://doi.org/10.1007/978-3-030-44289-7_4

Ali, N., Ahmed, A., Anum, L., Ghazal, T. M., Abbas, S., Khan, M. A., Alzoubi, H. M., & Ahmad, M. (2021). Modelling supply chain information collaboration empowered with machine learning technique. *Intelligent Automation and Soft Computing, 30*(1), 243–257. https://doi.org/10.32604/iasc.2021.018983

Ali, N., M. Ghazal, T., Ahmed, A., Abbas, S., A. Khan, M., Alzoubi, H., Farooq, U., Ahmad, M., & Adnan Khan, M. (2022). Fusion-based supply chain collaboration using machine learning techniques. *Intelligent Automation & Soft Computing, 31*(3), 1671–1687. https://doi.org/10.32604/iasc.2022.019892

Aljumah, A., Nuseir, M. T., & Alshurideh, M. T. (2021). The impact of social media marketing communications on consumer response during the COVID-19: Does the brand equity of a university matter? In *Studies in systems, decision and control* (Vol. 334). https://doi.org/10.1007/978-3-030-67151-8_21

Allam, Z., Dey, G., & Jones, D. S. (2020). Artificial Intelligence (AI) provided early detection of the coronavirus (COVID-19) in China and will influence future urban health policy internationally. In *AI 2020*, (Vol. 1, pp. 156–165). https://doi.org/10.3390/AI1020009

Alles, M. (2009). Governance in the age of unknown unknowns. *International Journal of Disclosure and Governance, 6*(2), 85–88. https://doi.org/10.1057/JDG.2009.2

Alnazer, N. N., Alnuaimi, M. A., & Alzoubi, H. M. (2017). Analysing the appropriate cognitive styles and its effect on strategic innovation in Jordanian universities. *International Journal of Business Excellence, 13*(1), 127–140. https://doi.org/10.1504/IJBEX.2017.085799

Alnuaimi, M., Alzoubi, H. M., Ajelat, D., & Alzoubi, A. A. (2021). Towards intelligent organisations: An empirical investigation of learning orientation's role in technical innovation. *International Journal of Innovation and Learning, 29*(2), 207–221. https://doi.org/10.1504/IJIL.2021.112996

AlShamsi, M., Salloum, S. A., Alshurideh, M., & Abdallah, S. (2021). Artificial intelligence and blockchain for transparency in governance. In *Artificial intelligence for sustainable development: Theory, practice and future applications* (pp. 219–230). Springer.

Alsharari, N. (2021). Integrating blockchain technology with internet of things to efficiency. *International Journal of Technology, Innovation and Management (IJTIM), 1*(2), 1–13.

Alshurideh, M. (2014). The factors predicting students' satisfaction with universities' healthcare clinics' services. *Dirasat. Administrative Sciences, 41*(2), 451–464.

Alshurideh, M., Gasaymeh, A., Ahmed, G., Alzoubi, H., & Kurd, B. A. (2020). Loyalty program effectiveness: Theoretical reviews and practical proofs. *Uncertain Supply Chain Management, 8*(3). https://doi.org/10.5267/j.uscm.2020.2.003

Alshurideh, M.T., Al Kurdi, B., AlHamad, A. Q., Salloum, S. A., Alkurdi, S., Dehghan, A., Abuhashesh, M., & Masa'deh, R. (2021a). Factors affecting the use of smart mobile examination platforms by universities' postgraduate students during the COVID-19 pandemic: An empirical study. *Informatics, 8*(2). https://doi.org/10.3390/informatics8020032

Alshurideh, M. T., Hassanien, A. E., & Masa'deh, R. (2021b). *The effect of coronavirus disease (COVID-19) on business intelligence.* Springer.

Alshurideh, M. T., Al Kurdi, B., Alzoubi, H. M., Ghazal, T. M., Said, R. A., AlHamad, A. Q., Hamadneh, S., Sahawneh, N., & Al-kassem, A. H. (2022). Fuzzy assisted human resource management for supply chain management issues. *Annals of Operations Research,* 1–19.

Alzoubi, H., & Ahmed, G. (2019). Do TQM practices improve organisational success? A case study of electronics industry in the UAE. *International Journal of Economics and Business Research, 17*(4), 459–472. https://doi.org/10.1504/IJEBR.2019.099975

Alzoubi, H. M., & Aziz, R. (2021). Does emotional intelligence contribute to quality of strategic decisions? The mediating role of open innovation. *Journal of Open Innovation: Technology, Market, and Complexity, 7*(2), 130. https://doi.org/10.3390/joitmc7020130

Alzoubi, H. M., & Yanamandra, R. (2020). Investigating the mediating role of information sharing strategy on agile supply chain. *Uncertain Supply Chain Management, 8*(2), 273–284. https://doi.org/10.5267/j.uscm.2019.12.004

Alzoubi, H., Alshurideh, M., Kurdi, B., Akour, I., & Aziz, R. (2022). Does BLE technology contribute towards improving marketing strategies, customers' satisfaction and loyalty? The role of open innovation. *International Journal of Data and Network Science, 6*(2), 449–460.

Alzoubi, H. M., Ahmed, G., Al-Gasaymeh, A., & Al Kurdi, B. (2020a). Empirical study on sustainable supply chain strategies and its impact on competitive priorities: The mediating role of supply chain collaboration. *Management Science Letters, 10*(3), 703–708. https://doi.org/10.5267/j.msl.2019.9.008

Alzoubi, H., Alshurideh, M., Kurdi, B. A., & Inairat, M. (2020b). Do perceived service value, quality, price fairness and service recovery shape customer satisfaction and delight? A practical study in the service telecommunication context. *Uncertain Supply Chain Management, 8*(3), 579–588. https://doi.org/10.5267/j.uscm.2020.2.005

Alzoubi, H. M., Vij, M., Vij, A., & Hanaysha, J. R. (2021). What leads guests to satisfaction and loyalty in UAE five-star hotels? AHP analysis to service quality dimensions. *Enlightening Tourism, 11*(1), 102–135. https://doi.org/10.33776/et.v11i1.5056

Alzoubi, A. (2021a). The impact of process quality and quality control on organizational competitiveness at 5-star hotels in Dubai. *International Journal of Technology, Innovation and Management (IJTIM), 1*(1), 54–68. https://doi.org/10.54489/ijtim.v1i1.14

Alzoubi, A. (2021b). Renewable Green hydrogen energy impact on sustainability performance. *International Journal of Computations, Information and Manufacturing (IJCIM), 1*(1), 94–110. https://doi.org/10.54489/ijcim.v1i1.46

Aziz, N., & Aftab, S. (2021). Data mining framework for nutrition ranking: Methodology: SPSS modeller. *International Journal of Technology, Innovation and Management (IJTIM), 1*(1), 85–95.

Benvie, A., Wayne, E., & Arnold, M. (2020). *Watson assistant continuous improvement best practices.*

BlueDot. (n.d.). *BlueDot: Outbreak intelligence platform.* https://bluedot.global/

Blumenthal, D., Fowler, E. J., Abrams, M., & Collins, S. R. (2020). Covid-19—implications for the health care system. *New England Journal of Medicine, 383*(15), 1483–1488. https://doi.org/10.1056/nejmsb2021088

Cabarkapa, S., Nadjidai, S. E., Murgier, J., & Ng, C. H. (2020). The psychological impact of COVID-19 and other viral epidemics on frontline healthcare workers and ways to address it: A rapid systematic review. *Brain, Behavior, & Immunity—Health, 8*, 100144. https://doi.org/10.1016/j.bbih.2020.100144

Cruz, A. (2021). Convergence between blockchain and the internet of things. *International Journal of Technology, Innovation and Management (IJTIM), 1*(1), 35–56.

Davenport, T., Guha, A., Grewal, D., & Bressgott, T. (2019). How artificial intelligence will change the future of marketing. *Journal of the Academy of Marketing Science, 48*(1), 24–42. https://doi.org/10.1007/S11747-019-00696-0

Eli, T. (2021). Students perspectives on the use of innovative and interactive teaching methods at the University of Nouakchott Al Aasriya, Mauritania: English department as a case study. *International Journal of Technology, Innovation and Management (IJTIM), 1*(2), 90–104.

Facebook. (n.d.). *Community standards enforcement \ transparency center*. Facebook Transparency Report.

Farouk, M. (2021). The universal artificial intelligence efforts to face coronavirus COVID-19. *International Journal of Computations, Information and Manufacturing (IJCIM), 1*(1), 77–93. https://doi.org/10.54489/ijcim.v1i1.47

FERMA. (2019). *Artificial intelligence applied to risk management*.

Folsom, J. (2020). *Distance learning during the COVID-19 pandemic*.

Ghazal, T. M., Hasan, M. K., Alshurideh, M. T., Alzoubi, H. M., Ahmad, M., Akbar, S. S., Al Kurdi, B., & Akour, I. A. (2021). IoT for smart cities: Machine learning approaches in smart healthcare—a review. *Future Internet, 13*(8), 218. https://doi.org/10.3390/fi13080218

Graham-Cumming, J. (2012). *Alan turing: Intelligence and life \ New Scientist*.

Guergov, S., & Radwan, N. (2021). Blockchain convergence: Analysis of issues affecting IoT, AI and blockchain. *International Journal of Computations, Information and Manufacturing (IJCIM), 1*(1), 1–17. https://doi.org/10.54489/ijcim.v1i1.48

Hajikazemi, S., Ekambaram, A., Andersen, B., & Zidane, Y.J.-T. (2016). The black swan—knowing the unknown in projects. *Procedia - Social and Behavioral Sciences, 226*, 184–192. https://doi.org/10.1016/J.SBSPRO.2016.06.178

Hamadneh, S., Pedersen, O., & Al Kurdi, B. (2021). An investigation of the role of supply chain visibility into the Scottish bood supply chain. *Journal of Legal, Ethical and Regulatory Issues, 24*(Special Issue 1), 1–12.

Hanaysha, J. R., Al-Shaikh, M. E., Joghee, S., & Alzoubi, H. (2021a). Impact of innovation capabilities on business sustainability in small and medium enterprises. *FIIB Business Review*. https://doi.org/10.1177/23197145211042232

Hanaysha, J. R., Al Shaikh, M. E., & Alzoubi, H. M. (2021b). Importance of marketing mix elements in determining consumer purchase decision in the retail market. *International Journal of Service Science, Management, Engineering, and Technology (IJSSMET), 12*(6), 56–72.

Hanford, M. (2005). Defining program governance and structure. *Developer Works, IBM, April,* 1–12.

Harahsheh, A. A., Houssien, A. M. A., & Alshurideh, M. T. (2021). The effect of transformational leadership on achieving effective decisions in the presence of psychological capital as an intermediate variable in private Jordanian. In *The effect of coronavirus disease (COVID-19) on business intelligence* (pp. 243–221). Springer Nature.

He, Y. (2019). The importance of artificial intelligence to economic growth. *Korean Artificial Intelligence, 7*(1), 17–22. https://doi.org/10.24225/kjai.2019.7.1.17

Hejazi, H. D., Khamees, A. A., Alshurideh, M., & Salloum, S. A. (2021). Arabic text generation: deep learning for poetry synthesis. In *Advances in intelligent systems and computing* (Vol. 1339). https://doi.org/10.1007/978-3-030-69717-4_11

Hillson, D. (2016). *Project Management.com—Why some risks turn into surprises*. Risk Insights from The Risk Doctor Blog.

Jia, P. (2019). Spatial lifecourse epidemiology. *The Lancet Planetary Health, 3*(2), e57–e59. https://doi.org/10.1016/S2542-5196(18)30245-6

Jia, P., & Yang, S. (2020). China needs a national intelligent syndromic surveillance system. *Nature Medicine, 26*(7), 990–990. https://doi.org/10.1038/s41591-020-0921-5

Jin, C., Chen, W., Cao, Y., Xu, Z., Tan, Z., Zhang, X., Deng, L., Zheng, C., Zhou, J., Shi, H., & Feng, J. (2020). Development and evaluation of an artificial intelligence system for COVID-19 diagnosis. *Nature Communications, 11*(1), 1–14. https://doi.org/10.1038/s41467-020-18685-1

Joghee, S., Alzoubi, H. M., & Dubey, A. R. (2020). Decisions effectiveness of FDI investment biases at real estate industry: Empirical evidence from Dubai smart city projects. *International Journal of Scientific and Technology Research, 9*(3), 3499–3503.

Kashif, A. A., Bakhtawar, B., Akhtar, A., Akhtar, S., Aziz, N., & Javeid, M. S. (2021). Treatment response prediction in hepatitis C patients using machine learning techniques. *International Journal of Technology, Innovation and Management (IJTIM), 1*(2), 79–89. https://doi.org/10.54489/ijtim.v1i2.24

Keshavarzi Arshadi, A., Webb, J., Salem, M., Cruz, E., Calad-Thomson, S., Ghadirian, N., Collins, J., Diez-Cecilia, E., Kelly, B., Goodarzi, H., & Yuan, J. S. (2020). Artificial intelligence for COVID-19 drug discovery and vaccine development. *Frontiers in Artificial Intelligence, 3*, 65. https://doi.org/10.3389/FRAI.2020.00065

Khamees, A. A., Hejazi, H. D., Alshurideh, M., & Salloum, S. A. (2021). Classifying audio music genres using CNN and RNN. In *Advances in intelligent systems and computing* (Vol. 1339). https://doi.org/10.1007/978-3-030-69717-4_31

Khan, M. A. (2021). Challenges facing the application of IoT in medicine and healthcare. *International Journal of Computations, Information and Manufacturing (IJCIM), 1*(1), 39–55. https://doi.org/10.54489/ijcim.v1i1.32

Kim, S. D. (2012). *Characterizing unknown unknowns.*

Klein, A. Z., Magge, A., O'Connor, K., Amaro, J. I. F., Weissenbacher, D., & Hernandez, G. G. (2021). Toward using Twitter for tracking COVID-19: A natural language processing pipeline and exploratory data set. *Journal of Medical Internet Research, 23*(1), e25314.

Kreuzhuber, K. (2020). *How AI, big data and machine learning can be used against the Corona virus–ars electronica blog.* ARS Electronica Blog.

Kurdi, B., Al Elrehail, H., Alzoubi, H. M., Alshurideh, M., & Al-adaileh, R. (2021). The interplay among HRM practices, job satisfaction and intention to leave: an empirical investigation. *Journal of Legal, Ethical and Regulatory Issues, 24*(1), 1–14.

Laghi, A. (2020). Cautions about radiologic diagnosis of COVID-19 infection driven by artificial intelligence. *The Lancet Digital Health, 2*(5), e225. https://doi.org/10.1016/S2589-7500(20)30079-0

Lee, K., Azmi, N., Hanaysha, J., Alzoubi, H., & Alshurideh, M. (2022a). The effect of digital supply chain on organizational performance: An empirical study in Malaysia manufacturing industry. *Uncertain Supply Chain Management, 10*(2), 495–510.

Lee, K., Romzi, P., Hanaysha, J., Alzoubi, H., & Alshurideh, M. (2022b). Investigating the impact of benefits and challenges of IOT adoption on supply chain performance and organizational performance: An empirical study in Malaysia. *Uncertain Supply Chain Management, 10*(2), 537–550.

Lee, C., & Ahmed, G. (2021). Improving IoT privacy, data protection and security concerns. *International Journal of Technology, Innovation and Management (IJTIM), 1*(1), 18–33. https://doi.org/10.54489/ijtim.v1i1.12

Lin, L., & Hou, Z. (2020). Combat COVID-19 with artificial intelligence and big data. *Journal of Travel Medicine, 27*(5), 1–8.

Lord, P., Martin, K., Atkinson, M., & Mitchell, H. (2009). *Narrowing the gap in outcomes: What is the relationship between leadership and governance?* 8.

Martínez, D. M., & Fernández-Rodríguez, J. C. (2015). Artificial intelligence applied to project success: A literature review. *IJIMAI, 3*(5), 77–84.

Mayo Clinic. (2019). *Herd immunity and COVID-19 (coronavirus): What you need to know—Mayo Clinic.*

Mehmood, T., Alzoubi, H. M., Alshurideh, M., Al-Gasaymeh, A., & Ahmed, G. (2019). Schumpeterian entrepreneurship theory: Evolution and relevance. *Academy of Entrepreneurship Journal, 25*(4), 1–10

Mehmood, T. (2021). Does information technology competencies and fleet management practices lead to effective service delivery? Empirical evidence from E-commerce industry. *International Journal of Technology, Innovation and Management (IJTIM), 1*(2), 14–41.

Microsoft. (2020). *Adaptive biotechnologies and microsoft launch groundbreaking immune CODE database to share populationwide immune response to the COVID-19 virus—stories*. Microsoft News Center.

Miller, D. (2021). The best practice of teach computer science students to use paper prototyping. *International Journal of Technology, Innovation and Management (IJTIM)*, *1*(2), 42–63. https://doi.org/10.54489/ijtim.v1i2.17

Mondol, E. P. (2021). The impact of block chain and smart inventory system on supply chain performance at retail industry. *International Journal of Computations, Information and Manufacturing (IJCIM)*, *1*(1), 56–76. https://doi.org/10.54489/ijcim.v1i1.30

Motoda, H., & Yoshida, K. (1998). Machine learning techniques to make computers easier to use. *Artificial Intelligence, 103*(1–2), 295–321.

Nafday, A. M. (2009). Strategies for managing the consequences of black swan events. *Leadership and Management in Engineering, 9*(4), 191–197. https://doi.org/10.1061/(ASCE)LM.1943-5630.0000036

Naqvi, R., Soomro, T. R., Alzoubi, H. M., Ghazal, T. M., & Alshurideh, M. T. (2021). The nexus between big data and decision-making: A study of big data techniques and technologies. In *The international conference on artificial intelligence and computer vision* (pp. 838–853).

Nuseir, M. T., Aljumah, A., & Alshurideh, M. T. (2021). How the business inelligence in the new startup performance in UAE during COVID-19: The mediating role of innovativeness. In *Studies in systems, decision and control* (Vol. 334). https://doi.org/10.1007/978-3-030-67151-8_4

Obaid, A. J. (2021). Assessment of smart home assistants as an IoT. *International Journal of Computations, Information and Manufacturing (IJCIM)*, *1*(1), 18–36. https://doi.org/10.54489/ijcim.v1i1.34

Pham, Q.-V., Nguyen, D. C., Huynh-The, T., Hwang, W.-J., & Pathirana, P. N. (2021). Artificial intelligence (AI) and big data for coronavirus (COVID-19) pandemic: A survey on the state-of-the-arts. *IEEE Access, 8*, 130820–130839. https://doi.org/10.1109/access.2020.3009328

Pietsch, B. (2020). *Central and Southern California have 0 percent I.C.U. capacity—the New York times*. https://www.nytimes.com/2020/12/26/world/central-and-southern-california-icu-capacity.html

Preethika, T., Vaishnavi, P., Agnishwar, J., Padmanathan, K., Umashankar, S., Annapoorani, S., Subash, M., & Aruloli, K. (2020). *Artificial intelligence and drones to combat COVID-19*. https://doi.org/10.20944/PREPRINTS202006.0027.V1

Radwan, N., & Farouk, M. (2021). The growth of internet of things (IoT) in the management of healthcare issues and healthcare policy development. *International Journal of Technology, Innovation and Management (IJTIM)*, *1*(1), 69–84. https://doi.org/10.54489/ijtim.v1i1.8

Rogers, A. (n.d.). *Singapore was ready for Covid-19—other countries, Take Note | WIRED*.

Shah, S. F., Alshurideh, M., Al Kurdi, B., & Salloum, S. A. (2020). The impact of the behavioral factors on investment decision-making: A systemic review on financial institutions. In *International conference on advanced intelligent systems and informatics* (pp. 100–112).

Shah, S. F., Alshurideh, M. T., Al-Dmour, A., & Al-Dmour, R. (2021). Understanding the influences of cognitive biases on financial decision making during normal and COVID-19 pandemic situation in the United Arab Emirates. In *Studies in systems, decision and control* (Vol. 334). https://doi.org/10.1007/978-3-030-67151-8_15

Sharma, A., Gupta, G., Ahmad, T., Krishan, K., & Kaur, B. (2019). Next generation agents (synthetic agents): Emerging threats and challenges in detection, protection, and decontamination. In *Handbook on biological warfare preparedness* (pp. 217–256). https://doi.org/10.1016/B978-0-12-812026-2.00012-8

Shukla Shubhendu, S., & Vijay, J. (2013). Applicability of artificial intelligence in different fields of life. *International Journal of Scientific Engineering and Research (IJSER)*, *1*(1), 28–35.

Statt, N. (2020). *How facebook is using AI to combat COVID-19 misinformation and detect 'hateful memes'—the verge*. THE VERGE.

Taryam, M., Alawadhi, D., Aburayya, A., Albaqa'een, A., Alfarsi, A., Makki, I., Rahmani, N., Alshurideh, M., & Salloum, S. A. (2020). Effectiveness of not quarantining passengers after having a negative COVID-19 PCR test at arrival to dubai airports. *Systematic Reviews in Pharmacy, 11*(11). https://doi.org/10.31838/srp.2020.11.197

Tavakoli, M., Carriere, J., & Torabi, A. (2020). Robotics, smart wearable technologies, and autonomous intelligent systems for healthcare during the COVID-19 pandemic: An analysis of the state of the art and future vision. *Advanced Intelligent Systems, 2*(7), 2000071. https://doi.org/10.1002/aisy.202000071

Turing, A., & Braithwaite, R. (2004). Can automatic calculating machines be said to think? *The Essential Turing*. https://doi.org/10.1093/OSO/9780198250791.003.0020

UAE Government. (n.d.). *Movement permits during sanitisation drive—the official portal of the UAE government*.

Yousuf, H., Zainal, A. Y., Alshurideh, M., & Salloum, S. A. (2021). Artificial intelligence models in power system analysis. In *Artificial intelligence for sustainable development: Theory, practice and future applications* (pp. 231–242). Springer.

Zeng, Z., Chen, P. J., & Lew, A. A. (2020). From high-touch to high-tech: COVID-19 drives robotics adoption. *Tourism Geographies, 22*(3), 724–734. https://doi.org/10.1080/14616688.2020.1762118

Zhang, D., Peng, G., & Yao, Y. (2019). Artificial intelligence or intelligence augmentation? Unravelling the Debate through an Industry-level analysis. In *Proceedings of the 23rd pacific asia conference on information systems: Secure ICT platform for the 4th industrial revolution, PACIS 2019*. https://doi.org/10.2139/SSRN.3315946

The Role and Impact of Big Data in Organizational Risk Management

Mounir El Khatib, Ahmad Ankit, Ishaq Al Ameeri, Hamad Al Zaabi, Rehab Al Marqab, Haitham M. Alzoubi⊙, and Muhammad Alshurideh⊙

Abstract Technological advancement has exposed companies to various risks. With the adoption of technological infrastructures in many companies, various processes have been rendered vulnerable to different forms of threats. Evidence from the current empirical studies on organizational management indicates effective risk assessment is a crucial aspect in any organization. Global technological advancement has equally redefined risk assessment and management strategies. Global leading technology companies such as Apple Inc., Amazon Inc., and Google are leveraging modern technology to unlock hidden data to aid in risk assessment and management. As evidenced in this research report, the use of big data has revolutionized risk assessment in many companies across the world. Big data technology has enabled organizations to collect, store, and assess huge information to aid in risk assessment and management. The processes of risk identification, assessment, mitigation, monitoring, and reporting have been redefined due to the adoption of big data analytical technology. As a result, this research reviews the specific roles played by big data in organizational risks management. The research carries out a comparative case study analysis among three companies that utilize big data in risk management. Specifically, the roles of big data technology in risk management at Apple Inc., Amazon Inc., and at Google are identified. Results from this comparative analysis are used in formulating recommendations for various organizations that desire to adopt big data analytical technology in risk management.

M. El Khatib · A. Ankit · I. Al Ameeri · H. Al Zaabi · R. Al Marqab
Hamdan Bin Mohamad Smart University, Dubai, UAE

H. M. Alzoubi (✉)
School of Business, Skyline University College, Sharjah, UAE
e-mail: haitham.alzoubi@skylineuniversity.ac.ae

M. Alshurideh
Department of Marketing, School of Business, University of Jordan, Amman, Jordan
e-mail: malshurideh@sharjah.ac.ae; m.alshurideh@ju.edu.jo

Department of Management, College of Business Administration, University of Sharjah, Sharjah, UAE

© The Author(s), under exclusive license to Springer Nature Switzerland AG 2023   2139
M. Alshurideh et al. (eds.), *The Effect of Information Technology on Business and Marketing Intelligence Systems*, Studies in Computational Intelligence 1056,
https://doi.org/10.1007/978-3-031-12382-5_117

Keywords Big data · Big data analytics · Organizational risk management ·
Project risk management

1 Introduction

Over the years, organizations have been implementing various risk assessment strategies to identify and prevent threats that hinder competitiveness. Risk assessment has been a central aspect in many organizations based on its ability to identify factors undermining performance. The proliferation of digital technology has redefined the need for risk assessment in many organizations. The incorporation of technological infrastructures has influenced many organizations to prioritize risk assessment due to various threats (Alshurideh et al., 2019; Agha et al., 2021; Kurdi et al., 2020). The advent of digital technology has exposed many organizations to several risks. Arguably, unlike in the analogue era, organizations are currently utilizing modern technology to formulate effective risk assessment tools. Presently, technological risk analytic and data mining tools such as big data have been adopted to help organizations identity, assess, mitigate, monitor, and report the progress of various threats. Global-leading technology companies such as Apple Inc., Amazon Inc., and Google rely on data mining tools such as big data in assessing and managing risks (Al Shebli et al., 2021; Lee et al., 2022a, 2022b; Naqvi et al., 2021). While several data mining techniques are used by organizations in risk assessment, the use of big data has proved to be effective because it unlocks hidden information enabling an organization to make accurate and data-driven risk management decisions (Al Alwan & Alshurideh, 2022; Batayneh et al., 2021; Tariq et al., 2022a, 2022b).

2 Background

Organizations in the current world have adopted digital technology in almost all of their processes. Apart from leading to efficient, reliable, and accurate product and service delivery, technological adoption has led to the collection of information (Ahmed et al., 2021; Almaazmi et al., 2021; Dicuonzo et al., 2019). Also known as big data, the collected information has been valuable in enabling organizations to formulate customer-centric business models (Ahmad et al., 2021; Alshurideh et al., 2022). Guided by the importance of data in the contemporary world, Almeida (2017) argues that risks assessment can effectively utilize big data technology.

Big data has been a crucial aspect of risk management because it has helped organizations to collect, store, evaluate, and apply data originating from different sources. The approach adopted in big data analytical strategies has enabled many organizations to detect, assess, mitigate, monitor, and report all forms of risks that might hinder continuous operations (Alshurideh, 2022; Shabbir & Waheed, 2020).

Defined as a structured and unstructured software that is used in collecting information during the daily operations of an organization, big data has proved to be a vital risk assessment tool due to its ability to unlock some hidden information (Ahmad et al., 2021a, 2021b, 2021c; Ali et al., 2022).

According to Tang and Karim Khondkar (2017), the effectiveness of risk analytics is determined by the ability of an organization to carry out a holistic assessment. This process should assess and determine some of the hidden risks that might affect seamless operations. Notably, adopting data-driving analytical strategies enables organizations to unlock some hidden and crucial information that leads to the formulation of an effective risks assessment strategy (Alshraideh et al., 2017; Ben-Abdallah et al., 2022; Hanaysha et al., 2021a, 2021b). Guided by the vital roles played by big data in risk management, it is essential to carry out research that will provide a practical application of this data-driven technology in preventing various organizational threats.

2.1 Research Problem

Technological advancement has exposed several organizations to numerous risks. Apart from fostering efficient, reliable, and convenient products and services provision, the digital technology being used by organizations has exposed them to several risks. Unlike in the analogue era, the increased risks facing organizations have led to the demand for the formulation of a more accurate and data-driven strategy that can effectively assess and manage various risks. An increase in technological risks faced by organizations has led to the demand for a more effective management strategy (Tang & Karim Khondkar, 2017). As a result, this research focused on solving the current challenge of increasing cases of technological risks faced by organizations by analyzing the roles played by big data. Specifically, this research reviews the use of big data analytical technology in solving various risks faced by organizations (Alhamad et al., 2021; Shamout et al., 2022a, 2022b). With reference to the application of big data technology in leading companies, this research formulates effect strategies that should be utilized by organizations to identify, assess, mitigate, monitor, and report various risks (Alshurideh et al., 2021; Amarneh et al., 2021).

2.2 Research Objectives

The primary objective of this research was "To determine the roles played by big data in organizational risks management". Other objectives of the research were "To determine the practical applicability of risks assessment in leading organizations" and "To determine how big data is used in risk identification, assessment, mitigation, monitoring, and reporting". Achieving these objectives was effective in solving the research problem and enabling the researchers to identify the critical roles played by big data in organizational risk management.

2.3 Research Questions

The whole of the research process aimed at answering the following four questions.

i. What is the role of big data in organizational risk management?
ii. How is big data applied in the management of risks at Apple Inc.?
iii. How does Amazon Inc. utilize data mining technologies such as big data to assess and mitigate various risks?
iv. How does Google utilize big data in risk management?

Notably, answering the above four research questions was vital in enabling the researchers to formulate effective risks management strategies that organizations can utilize to overcome various risks. Specifically, the questions were significant in analyzing the practical application of big data technology in organizational risk management in the contemporary world.

2.4 Significance of the Research

Despite the existence of literature on the use of big data in achieving effective customer management, there is a dearth of data detailing the role of big data in organizational risk management. In effect, this research was vital in filling the existing gap by analyzing how big data is utilized in assessing and managing organizational risks (Dicuonzo et al., 2019). literature from this research will be vital in filling the existing research gap. Researchers will build on the findings of this study to carry out further analysis on the application of data mining technologies in organizational risks management.

Moreover, this research is very important because it formulates strategies that can be applied by organizational management and the risks management departments in the course of assessing and managing various threats (Hanaysha et al., 2021a, 2021b). Guided by the advent of digitalization in many organizations, the findings of this research will be used as a guide that will influence the adoption of big data analytical technology in risk assessment and management. Notably, carrying out this study is vital in identifying the roles and benefits of using big data mining technology in organizational risks management.

Lastly, the comparative case study analysis approach adopted in this research was vital in enabling the researcher to solve the current problem of technological risks being witnessed in various organizations. The adoption of digital technology in many organizations has exposed many processes to various risks (Hamadneh et al., 2021; Rui et al., 2022). These risks can only be solved by adopting a data-driven risk management strategy (Shabbir & Waheed, 2020). Adopting big data analytics fosters the aspects of accuracy during risk identification, assessment, mitigation, monitoring, and reporting. As a result, the comparative case study analysis carried out in this research was vital in creating practical applicability of big data in organizational risk management.

3 Literature Review

Technological advancement in the contemporary world has influenced scholars to carry out studies to identify the roles played by big data in organizational risks assessment and management (Lee et al., 2022a, 2022b). Evidence from the studies by Kumari (2018) indicates data mining technologies such as big data analytics have revolutionized organizational risk management. On the other hand, Shabbir and Waheed (2020) postulate that the application of big data analytics in risks management has enabling organizations to uncover critical data that has led to the formulation of a more accurate and applicable risk management strategy. The studies by Tang and Karim Khondkar (2017) justify the use of big data technologies in organizational risk management by holding that this approach enables managers to make accurate and data-driven decisions (AlHamad et al., 2022). Literature from the current studies identifies various roles played by big data analytics in enabling organizations to assess and manage different risks (Ali et al., 2021). Some of these roles are discussed in the subsequent sections.

3.1 Big Data Analytics is Used in Fraud Prevention

Studies by Shabbir and Waheed (2020) analyzed the role of big data analytics in preventing various forms of frauds experienced in organizations. Evidence from this study indicated that big data analytics have the capability of detecting different forms of fraud. Organizations in the contemporary world experience various forms of financial frauds caused by employees and other stakeholders. However, Jin and Liad (2021) suggest that big data analytics enables an organization to collect big volumes of data that detect and prevent any form of fraud in an organization (Shamout et al., 2022a, 2022b). On this regard, organizations should consider adopting big data analytics to prevent the rising cases of fraud.

3.2 Big Data Analytic Help Organizations to Manage Various Risks Associated with Third Parties

Most organization outsource different services from third party companies. According to Choi et al. (2018), some of these third parties end up causing significant risks to organizations. For instance, an organization outsourcing technological services such as cloud computing from third-party providers' risk facing challenges related to cybersecurity (Choi et al., 2018). On the other hand, an organization that deals with different vendors is always prone to several risks originating from these third party suppliers (Kumari, 2018). As explained by Kumari (2018), risks originating from third parties can easily lead to the collapse of an organization (Alnuaimi et al., 2021). Big data analytics has been crucial in enabling organizations to collect a huge volume of data that can easily detect any form of external threat (Alzoubi et al., 2022). The collected data has been utilized in assessing threats from third party companies and formulating effective mitigation strategies (Kumari, 2018). Results from the studies by Jin and Liad (2021) indicates that organizations that depend on big data analytics have well-established risk management approaches that prevent any form of external threat.

3.3 Big Data Analytics Have Been Vital in Reducing High Customer Churn Rates and Preventing High Employee Attrition

Customers and employees are two vital stakeholders in any organizations that determine its growth and development. According to Dicuonzo et al. (2019), organizations that achieve customer loyalty and retention register continuous growth and development due to the increased sales (Joghee et al., 2020). On the other hand, Cole et al. (2015) hold that employees are a driving force that ensures continuous operations in any organization. In this regard, increased customer churn rates pose critical risks that might result in the collapse of an organization. As explained by Cole et al. (2015), the increased customer churn rate reduces the total market share of an organization rendering it uncompetitive. Similarly, high employee attrition leads to low-quality service delivering hindering an organization from meeting its objectives. Consequently, the research by Almeida (2017) concludes that the adoption of big data analytics prevents the risks associated with increased customer churn rates and employee attrition. By applying big data analytics, organizations collect numerous information based on different customer segments (Alnazer et al., 2017). The company counters the risks of high customer churn rates by identifying various factors that hindering customer satisfaction (De Mauro et al., 2018). Additionally, high attrition rates among employees can be easily detected, analyzed, and manipulated by using big data analytics (Alzoubi & Aziz, 2021). The adopted big data analytics identify the specific needs of employees and formulates ways of fostering retention.

3.4 Big Data Analytics Foster Operational Assessment and Enables Organizations to Manage Various Forms of Digital Risks

Big data technology enables organizations to collect, store, analyze, and apply big chunks of data in the course of assessing and managing different forms of risks. Operational risks are some of the common threats facing organizations in the contemporary world. Companies experience different risks originating from different organizational processes (De Mauro et al., 2018). Ineffective management of these risks threatens the stability of any organization. In this regard, the adoption of big data analytics has proved to be effective in handling various operational management threats. Similarly, research by Vassakis et al. (2018) revealed that adopting digital technology has led to the existence of several risks in the contemporary world. However, Jin and Liad (2021) established that relying on big data analytics has proved to be effective in dealing with various threats associated with digitalization (Ghazal et al., 2021). Notably, organizations should consider adopting big data technologies in achieving accurate, data-driven, and appropriate risk assessment and management.

4 Methodological Approach

4.1 The Research Design

The methodological approaches adopted in any research determine the effectiveness and viability of the research findings, conclusions and recommendations. As explained by Schoonenboom and Burke (2017), the methodological approaches adopted in any research are equally important in determining the accuracy of the collected data. This research adopted a hybrid research strategy that collected both qualitative and quantitative data. The hybrid or mixed-method research design adopted a case study analysis. The comparative case studies aimed at determining the specific roles played by big data in organizational risk management (Alzoubi et al., 2021). As evidenced below, qualitative and qualitative secondary data was collected from the case studies on how Apple Inc., Amazon Inc., and Google utilize the data mining technology of big data analytics in managing various forms of risks.

4.2   Case Study Analysis

Guided by the developed research design, researchers carried out a comparative analysis that reviewed the application of big data analytics in risk management in Apple Inc. Amazon Inc., and at Google. The comparative analysis aimed at establishing the specific roles played by big data analytics in assessing and managing various risks.

4.2.1   Case Study 1: The Role of Big Data in Risk Management at Apple Inc.

The utilization of big data has proved to effective in managing various forms of risks at Apple Inc., as explained by PathakJan (2021). Apple Inc. adopts a Siri voice recognition big data technology that collects, stores, analysis, and applies the collected data. According to PathakJan (2021). Apple Inc. has been relying on its big data collection and analytical strategy to improve counter various risks. Just like other leading technology companies, Apple Inc. utilizes the big collected data to identify, assess, mitigate, monitor, and report various risks (Alzoubi et al., 2020a, 2020b). For instance, internal risks associated with financial transactions, employee attrition, high customer churn rates, and threats posed by third-party organizations (AppleInc, 2021). As of the leading technology companies, Apple Inc. has shifted to the use of data mining analytics to accurately assess and manage various risks (Alshurideh et al., 2020). The *Siri* voice recognition technology is utilized by this company in collecting big data has been essential in detecting different forms of risks and formulating mitigating strategies before occurrence. As explained by Sun et al. (2020), apart from enhancing customer relationships management, the big data analytics used by Apple Inc. have been vital in formulating recovery strategies that foster continuous growth and development. Similar to the risks management strategies used by both Amazon Inc. and Google, Apple Inc. has been utilizing this big data technology in unlocking some of the hidden information that threatens seamless operations in the company.

4.2.2   Case Study 2: The Role of Big Data in Risk Management at Amazon Inc.

The business model adopted by Amazon Inc. has exposed it to numerous risks. As a company that relies mostly on third-party vendors, Amazon Inc. is always vulnerable to various risks (Alzoubi & Ahmed, 2019). However, the company has been utilizing effective data mining technologies to detect, assess, mitigate, monitor, and report various forms of internal and external risks (Amazon, 2021). Ranked among the global leading technology companies, Amazon Inc. collects, stores, analyses and applies big data in all of its business processes (Mehmood et al., 2019). The cloud computing technology adopted by Amazon Inc. utilizes the *AWS* big data software to assess and manage different forms of risks, essential in collecting, storing, processing, and applying big data (Amazon, 2021). According to Galea-Pace (2021), the application of big data at Amazon Inc. has been vital in mitigating risks associated with business strategy, compliance and regulatory risks, financial risks, and other operational risks. As the leading e-commerce platform in the world, the big data collected by Amazon Inc. has been vital in enabling it to make accurate and data-driven risks management decisions. Just like Apple Inc. Amazon Inc. has been utilizing big data analytics to prevent various forms of internal and external fraud, lower retention and churn rates from employees and customers, respectively. Moreover, big data analytics used at Amazon Inc. have been vital in managing third-party risks and mitigating threats associated with operational processes.

4.2.3 Case Study 3: The Role of Big Data in Risk Management at Google

Google is the global-leading technology company that offers different products and services. As a global leading technology company, Google faced both internal and external risks that require effective management approaches (Aziz & Aftab, 2021). According to Sun et al. (2020), technological advancement has led to the adoption of big data analytics at Google. Apart from enabling the company to offer customer-centric products and services, the big data analytics adopted by Google Inc. has been essential in assessing and managing various forms of risks. Google collects big data using different software (Alzoubi et al., 2020a, 2020b). The collected data is then stored, analyzed, and applied in managing different forms of risks (Google, 2021). Google has prioritized the use of big data in risk assessment by identifying risks at an early stage and crafting effective mitigation strategies (Radwan & Farouk, 2021). According to Sun et al. (2020), the use of big data at Google has enabled this company to prevent risks associated with fraud and those originating from third party companies. Just like Apple Inc. and Amazon Inc., the use of big data at Google has been vital in mitigating operational risks that threaten the competitiveness of this company (Cruz, 2021). Notably, big data has proved to be effective because Google is among companies that detect and manage different forms of risks at an early stage.

5 Results and Discussions

It is evidenced from the comparative case study analysis that big data plays vital roles in organizational risk management (Eli, 2021). Guided by the data collected in this research, big data analytics is an essential data mining technology that is used in leading organizations such as Apple Inc., Amazon Inc., and Google to detect, assess, mitigate, monitor, and report various risks (Lee & Ahmed, 2021). As evidenced in the concepts developed in the literature review, big data analytics plays the following risk assessment and management roles.

Firstly, big data analytics collects large sums of data that is used in detecting and preventing any form of internal and external fraud (Kashif et al., 2021). As evidenced in a study by Almeida (2017), big data technologies have been essential in detecting any form of internal and external fraud in organizations (Akhtar et al., 2021). The detected threat is quickly analyzed, and accurate data-driven mitigation strategies are implemented. Secondly, evidence from this research showed that organizational risks associated with third-party service providers or vendors could effectively be management by big data technologies (Alzoubi & Yanamandra, 2020). According to Dicuonzo et al. (2019), technological adoption in many organizations has led to increased outsourcing. Leading companies like Amazon Inc. rely heavily on external vendors and other third-party product and service providers. Such companies are always exposed to several risks that can be effectively mitigated by big data analytics (Alsharari, 2021). The data mining technology of big data collects large sums of data that can easily detect and manage risks associated with third-party companies and customers (Mehmood, 2021).

Thirdly, evidence from this research shows that big data analytics has been vital in reducing employee attrition rates and countering the increasing customer churn rates. As stated by Shabbir and Waheed (2020), employees and customers are vital stakeholders in any organization. Any risks that prompt either employees or customers to lose interest or loyal in an organization causes detrimental effects (Tang & Karim Khondkar, 2017). In this regard, the adoption of big data analytics enables organizations to assess various factors that lead to increased rates of employee attrition and craft effective mitigation strategies (Miller, 2021). Similarly, big data technologies are used to collect data on different customer segments (Khan, 2021). The collected data is analyzed to determine factors leading to the increased rates of customer churns. Effective strategies are equally established to mitigate these challenges.

Lastly, results from this research showed that adopting big data analytics is crucial in assessing and managing risks associated with organizational processes and other digital threats (Mondol, 2021). As stated above, technological advancement has led to increased demand for digital technology. Most of the organizational processes are currently executed using digital technology (Cole et al., 2015). Despite being effective in fostering the aspects of reliability, convenience and efficiency, digitalization has caused numerous risks in many organizations (De Mauro et al., 2018). Many of the organizational processes have been exposed to both internal and external risks caused by digital technology (Alzoubi, 2021a). However, big data analytics has been crucial in enabling companies to counter risks associated with organizational processes and other digital infrastructures (Guergov & Radwan, 2021). As evidenced in this research, big data analytics detect, analyze, mitigate, monitor, and report different form of risks associated with organizational processes and digital technology.

6 Conclusion and Recommendations

It is evidenced from this research that organizations have been implementing various risk assessment strategies to identify and prevent threats that hinder competitiveness. The effectiveness of these strategies has been based on their ability to timely identify, analyze, mitigate, monitor, and report various risks facing organizations (Al Ali, 2021). With the growth of digital technology, the demand for data mining software such as big data has been on the rise (Farouk, 2021). As shown in this research, big data has been vital in enabling organizations to craft customer-centric business models. Apart from helping companies customize their products and services in various markets, big data has been vital in organizational risk management. It is shown that big data analytics enables organizations to identify and mitigate various forms of risks (Obaid, 2021). As a result, while several data mining techniques are used by organizations in risk assessment, the use of big data has proved to be effective because it unlocks hidden information enabling an organization to make accurate and data-driven risk management decisions.

Guided by the findings of this research, it is vital for companies to consider utilizing big data analytics in solving various risks associated with organizational management. Implementing big data risks management strategies will enable organizations to make data-driven decisions in the course of solving various threats (Alzoubi, 2021b). Additionally, big data technologies will be essential in enabling organizational managers and risk management officers to detect risks at the early stages and utilize accurate mitigation strategies.

References

Agha, K., Alzoubi, H. M., & Alshurideh, M. T. (2021). Measuring reliability and validity instruments of technologically driven cognitive intrusion towards work-life balance. In *The international conference on artificial intelligence and computer vision* (pp. 601–614).

Ahmad, A., Alshurideh, M., Al Kurdi, B., Aburayya, A., & Hamadneh, S. (2021a). Digital transformation metrics: A conceptual view. *Journal of Management Information and Decision Sciences, 24*(7), 1–18.

Ahmad, A., Alshurideh, M. T., Al Kurdi, B. H., & Salloum, S. A. (2021b). Factors impacts organization digital transformation and organization decision making during Covid19 Pandemic. In *Studies in systems, decision and control* (Vol. 334). https://doi.org/10.1007/978-3-030-67151-8_6

Ahmed, A., Alshurideh, M., Al Kurdi, B., & Salloum, S. A. (2021c). Digital transformation and organizational operational decision making: A systematic review. In *Advances in intelligent systems and computing* (Vol. 1261). AISC. https://doi.org/10.1007/978-3-030-58669-0_63

Akhtar, A., Akhtar, S., Bakhtawar, B., Kashif, A. A., Aziz, N., & Javeid, M. S. (2021). COVID-19 detection from CBC using machine learning techniques. *International Journal of Technology, Innovation and Management (IJTIM), 1*(2), 65–78. https://doi.org/10.54489/ijtim.v1i2.22

Al Ali, A. (2021). The impact of information sharing and quality assurance on customer service at UAE banking sector. *International Journal of Technology, Innovation and Management (IJTIM), 1*(1), 01–17. https://doi.org/10.54489/ijtim.v1i1.10

Al Batayneh, R. M., Taleb, N., Said, R. A., Alshurideh, M. T., Ghazal, T. M., & Alzoubi, H. M. (2021). IT governance framework and smart services integration for future development of Dubai infrastructure utilizing AI and big data, its reflection on the citizens standard of living. In *The international conference on artificial intelligence and computer vision* (pp. 235–247).

Al Shebli, K., Said, R. A., Taleb, N., Ghazal, T. M., Alshurideh, M. T., & Alzoubi, H. M. (2021). RTA's Employees' perceptions toward the efficiency of artificial intelligence and big data utilization in providing smart services to the residents of Dubai. In *The international conference on artificial intelligence and computer vision* (pp. 573–585).

Alhamad, A. Q. M., Akour, I., Alshurideh, M., Al-Hamad, A. Q., Kurdi, B. A., & Alzoubi, H. (2021). Predicting the intention to use google glass: A comparative approach using machine learning models and PLS-SEM. *International Journal of Data and Network Science, 5*(3), 311–320. https://doi.org/10.5267/j.ijdns.2021.6.002

AlHamad, A., Alshurideh, M., Alomari, K., Kurdi, B. Al, Alzoubi, H., Hamouche, S., & Al-Hawary, S. (2022). The effect of electronic human resources management on organizational health of telecommuni-cations companies in Jordan. *International Journal of Data and Network Science, 6*(2), 429–438. https://doi.org/10.5267/j.ijdns.2021.12.011

Ali, N., Ahmed, A., Anum, L., Ghazal, T. M., Abbas, S., Khan, M. A., Alzoubi, H. M., & Ahmad, M. (2021). Modelling supply chain information collaboration empowered with machine learning technique. *Intelligent Automation and Soft Computing, 30*(1), 243–257. https://doi.org/10.32604/iasc.2021.018983

Ali, N., M. Ghazal, T., Ahmed, A., Abbas, S., A. Khan, M., Alzoubi, H., Farooq, U., Ahmad, M., & Adnan Khan, M. (2022). Fusion-based supply chain collaboration using machine learning techniques. *Intelligent Automation & Soft Computing, 31*(3), 1671–1687. https://doi.org/10.32604/iasc.2022.019892

Almaazmi, J., Alshurideh, M., Al Kurdi, B., & Salloum, S. A. (2021). The effect of digital transformation on product innovation: A critical review. In *Advances in intelligent systems and computing* (Vol. 1261). AISC. https://doi.org/10.1007/978-3-030-58669-0_65

Almeida, F. L. F. (2017). Benefits, challenges and tools of big data management. *Journal of Systems Integration, 8*(4), 12–20.

Alnazer, N. N., Alnuaimi, M. A., & Alzoubi, H. M. (2017). Analysing the appropriate cognitive styles and its effect on strategic innovation in Jordanian universities. *International Journal of Business Excellence, 13*(1), 127–140. https://doi.org/10.1504/IJBEX.2017.085799

Alnuaimi, M., Alzoubi, H. M., Ajelat, D., & Alzoubi, A. A. (2021). Towards intelligent organisations: An empirical investigation of learning orientation's role in technical innovation. *International Journal of Innovation and Learning, 29*(2), 207–221. https://doi.org/10.1504/IJIL.2021.112996

Alsharari, N. (2021). Integrating blockchain technology with internet of things to efficiency. *International Journal of Technology, Innovation and Management (IJTIM), 1*(2), 01–13.

Alshraideh, A., Al-Lozi, M., & Alshurideh, M. (2017). The impact of training strategy on organizational loyalty via the mediating variables of organizational satisfaction and organizational performance: An empirical study on jordanian agricultural credit corporation staff. *Journal of Social Sciences (COES&RJ-JSS), 6*, 383–394.

Alshurideh, M. (2022). Does electronic customer relationship management (E-CRM) affect service quality at private hospitals in Jordan? *Uncertain Supply Chain Management, 10*(2), 1–8.

Alshurideh et al. (2019). Understanding the quality determinants that influence the intention to use the mobile learning platforms: A practical study. *International Journal of Interactive Mobile Technologies (IJIM), 13*(11), 157–183.

Alshurideh, M., Gasaymeh, A., Ahmed, G., Alzoubi, H., & Kurd, B. A. (2020). Loyalty program effectiveness: Theoretical reviews and practical proofs. *Uncertain Supply Chain Management, 8*(3), 599–612. https://doi.org/10.5267/j.uscm.2020.2.003

Alshurideh, M. T., Hassanien, A. E., & Masa'deh, R. (2021). *The effect of coronavirus disease (COVID-19) on business intelligence.* Springer.

Alshurideh, M. T., Al Kurdi, B., Alzoubi, H. M., Ghazal, T. M., Said, R. A., AlHamad, A. Q., Hamadneh, S., Sahawneh, N., & Al-kassem, A. H. (2022). Fuzzy assisted human resource management for supply chain management issues. *Annals of Operations Research*, 1–19.

Alwan, M., & Alshurideh, M. T. (2022). *The effect of digital marketing on purchase intention : Moderating effect of brand equity. March.* https://doi.org/10.5267/j.ijdns.2022.2.012

Alzoubi, H., & Ahmed, G. (2019). Do TQM practices improve organisational success? A case study of electronics industry in the UAE. *International Journal of Economics and Business Research, 17*(4), 459–472. https://doi.org/10.1504/IJEBR.2019.099975

Alzoubi, H. M., & Aziz, R. (2021). Does emotional intelligence contribute to quality of strategic decisions? The mediating role of open innovation. *Journal of Open Innovation: Technology, Market, and Complexity, 7*(2), 130. https://doi.org/10.3390/joitmc7020130

Alzoubi, H. M., & Yanamandra, R. (2020). Investigating the mediating role of information sharing strategy on agile supply chain. *Uncertain Supply Chain Management, 8*(2), 273–284. https://doi.org/10.5267/j.uscm.2019.12.004

Alzoubi, H., Alshurideh, M., Kurdi, B. A., & Inairat, M. (2020b). Do perceived service value, quality, price fairness and service recovery shape customer satisfaction and delight? A practical study in the service telecommunication context. *Uncertain Supply Chain Management, 8*(3), 579–588. https://doi.org/10.5267/j.uscm.2020.2.005

Alzoubi, H., Alshurideh, M., Kurdi, B., Akour, I., & Aziz, R. (2022). Does BLE technology contribute towards improving marketing strategies, customers' satisfaction and loyalty? The role of open innovation. *International Journal of Data and Network Science, 6*(2), 449–460.

Alzoubi, H. M., Ahmed, G., Al-Gasaymeh, A., & Al Kurdi, B. (2020a). Empirical study on sustainable supply chain strategies and its impact on competitive priorities: The mediating role of supply chain collaboration. *Management Science Letters, 10*(3), 703–708. https://doi.org/10.5267/j.msl.2019.9.008

Alzoubi, H. M., Vij, M., Vij, A., & Hanaysha, J. R. (2021). What leads guests to satisfaction and loyalty in UAE five-star hotels? AHP analysis to service quality dimensions. *Enlightening Tourism, 11*(1), 102–135. https://doi.org/10.33776/et.v11i1.5056

Alzoubi, A. (2021a). The impact of process quality and quality control on organizational competitiveness at 5-star hotels in Dubai. *International Journal of Technology, Innovation and Management (IJTIM), 1*(1), 54–68. https://doi.org/10.54489/ijtim.v1i1.14

Alzoubi, A. (2021b). Renewable Green hydrogen energy impact on sustainability performance. *International Journal of Computations, Information and Manufacturing (IJCIM), 1*(1), 94–110. https://doi.org/10.54489/ijcim.v1i1.46

Amarneh, B. M., Alshurideh, M. T., Al Kurdi, B. H., & Obeidat, Z. (2021). The impact of COVID-19 on E-learning: Advantages and Challenges. In *The international conference on artificial intelligence and computer vision* (pp. 75–89).

Amazon, I. (2021). *About Amazon Inc.* [Accessed 6th May 2021].

AppleInc, I. (2021). Apple Inc. *About Apple Inc, 6.*

Aziz, N., & Aftab, S. (2021). Data mining framework for nutrition ranking: methodology: SPSS modeller. *International Journal of Technology, Innovation and Management (IJTIM), 1*(1), 85–95.

Ben-Abdallah, R., Shamout, M., & Alshurideh, M. (2022). Business development strategy model using EFE, IFE and IE analysis in a high-tech company: An empirical study. *Academy of Strategic Management Journal, 21*(Special Issue 2), 1–9.

Choi, T., Wallace Stein, W., & Yulan, W. (2018). Big data analytics in operations management. *Production and Operations Management, 27*(10), 1868–1883.

Cole, D., Jasmine, N., & Brian, M. (2015). Benefits and risks of big data. In *Association for information systems AIS electronic library (AISeL) SAIS* (pp. 1–6).

Cruz, A. (2021). Convergence between blockchain and the internet of things. *International Journal of Technology, Innovation and Management (IJTIM), 1*(1), 35–56.

De Mauro, A., Greco, M., Grimaldi, M., & Ritala, P. (2018). Human resources for Big Data professions: A systematic classification of job roles and required skill sets. *Information Processing & Management, 54*(5), 807–817.

Dicuonzo, G., Galeone, G., Zappimbulso, E., & Dell'Atti, V. (2019). Risk management 4.0: the role of big data analytics in the bank sector. *International Journal of Economics and Financial Issues, 9*, 6.

Eli, T. (2021). Students perspectives on the use of innovative and interactive teaching methods at the University of Nouakchott Al Aasriya, Mauritania: English department as a case study. *International Journal of Technology, Innovation and Management (IJTIM)*, *1*(2), 90–104.

Farouk, M. (2021). The Universal artificial intelligence efforts to face coronavirus COVID-19. *International Journal of Computations, Information and Manufacturing (IJCIM)*, *1*(1), 77–93. https://doi.org/10.54489/ijcim.v1i1.47

Galea-Pace, S. (2021). How Amazon uses Big Data to transform operations. *SupplyChain*.

Ghazal, T. M., Hasan, M. K., Alshurideh, M. T., Alzoubi, H. M., Ahmad, M., Akbar, S. S., Al Kurdi, B., & Akour, I. A. (2021). IoT for smart cities: Machine learning approaches in smart healthcare—a review. *Future Internet, 13*(8), 218. https://doi.org/10.3390/fi13080218

Google, I. (2021). No title. *Google. About Google*.

Guergov, S., & Radwan, N. (2021). Blockchain convergence: Analysis of issues affecting IoT, AI and blockchain. *International Journal of Computations, Information and Manufacturing (IJCIM)*, *1*(1), 1–17. https://doi.org/10.54489/ijcim.v1i1.48

Hamadneh, S., Pedersen, O., Alshurideh, M., Kurdi, B. Al, & Alzoubi, H. (2021). An investigation of the role of supply chain visibility into the Scottish blood supply chain. *Journal of Legal, Ethical and Regulatory Issues, 24*(Special Issue 1), 1–12.

Hanaysha, J. R., Al-Shaikh, M. E., Joghee, S., & Alzoubi, H. (2021a). Impact of innovation capabilities on business sustainability in small and medium enterprises. *FIIB Business Review*. https://doi.org/10.1177/23197145211042232

Hanaysha, J. R., Al Shaikh, M. E., & Alzoubi, H. M. (2021b). Importance of marketing mix elements in determining consumer purchase decision in the retail market. *International Journal of Service Science, Management, Engineering, and Technology (IJSSMET)*, *12*(6), 56–72.

Jin, G. Z., & Liad, W. (2021). Big data at the crossroads of antitrust and consumer protection. *Information Economics and Policy, 54*, 1–17.

Joghee, S., Alzoubi, H. M., & Dubey, A. R. (2020). Decisions effectiveness of FDI investment biases at real estate industry: Empirical evidence from Dubai smart city projects. *International Journal of Scientific and Technology Research, 9*(3), 3499–3503.

Kashif, A. A., Bakhtawar, B., Akhtar, A., Akhtar, S., Aziz, N., & Javeid, M. S. (2021). Treatment response prediction in hepatitis C patients using machine learning techniques. *International Journal of Technology, Innovation and Management (IJTIM)*, *1*(2), 79–89. https://doi.org/10.54489/ijtim.v1i2.24

Khan, M. A. (2021). Challenges facing the application of IoT in medicine and healthcare. *International Journal of Computations, Information and Manufacturing (IJCIM)*, *1*(1), 39–55. https://doi.org/10.54489/ijcim.v1i1.32

Kumari, A. (2018). Verification and validation techniques for streaming big data analytics in the internet of things environment. *IET Networks, 8*(3), 155–163.

Kurdi, B. A., Alshurideh, M., Salloum, S. A., Obeidat, Z. M., & Al-dweeri, R. M. (2020). An empirical investigation into examination of factors influencing university students' behavior towards elearning acceptance using SEM approach. *International Journal of Interactive Mobile Technologies, 14*(2). https://doi.org/10.3991/ijim.v14i02.11115

Lee, K., Azmi, N., Hanaysha, J., Alzoubi, H., & Alshurideh, M. (2022a). The effect of digital supply chain on organizational performance: An empirical study in Malaysia manufacturing industry. *Uncertain Supply Chain Management, 10*(2), 495–510.

Lee, K. L., Romzi, P. N., Hanaysha, J. R., Alzoubi, H. M., & Alshurideh, M. (2022b). Investigating the impact of benefits and challenges of IOT adoption on supply chain performance and organizational performance: An empirical study in Malaysia. *Uncertain Supply Chain Management, 10*(2), 537–550. https://doi.org/10.5267/j.uscm.2021.11.009

Lee, C., & Ahmed, G. (2021). Improving IoT privacy, data protection and security concerns. *International Journal of Technology, Innovation and Management (IJTIM)*, *1*(1), 18–33. https://doi.org/10.54489/ijtim.v1i1.12

Mehmood, T., Alzoubi, H. M., Alshurideh, M., Al-Gasaymeh, A., & Ahmed, G. (2019). Schumpeterian entrepreneurship theory: Evolution and relevance. *Academy of Entrepreneurship Journal, 25*(4), 1–10.

Mehmood, T. (2021). Does information technology competencies and fleet management practices lead to effective service delivery? Empirical evidence from E-commerce industry. *International Journal of Technology, Innovation and Management (IJTIM), 1*(2), 14–41.

Miller, D. (2021). The best practice of teach computer science students to use paper prototyping. *International Journal of Technology, Innovation and Management (IJTIM), 1*(2), 42–63.

Mondol, E. P. (2021). The impact of block chain and smart inventory system on supply chain performance at retail industry. *International Journal of Computations, Information and Manufacturing (IJCIM), 1*(1), 56–76. https://doi.org/10.54489/ijcim.v1i1.30

Naqvi, R., Soomro, T. R., Alzoubi, H. M., Ghazal, T. M., & Alshurideh, M. T. (2021). The nexus between big data and decision-making: A study of big data techniques and technologies. In *The international conference on artificial intelligence and computer vision* (pp. 838–853).

Obaid, A. J. (2021). Assessment of smart home assistants as an IoT. *International Journal of Computations, Information and Manufacturing (IJCIM), 1*(1), 18–36. https://doi.org/10.54489/ijcim.v1i1.34

PathakJan, R. (2021). How apple uses AI and big data. *Big Data*.

Radwan, N., & Farouk, M. (2021). The growth of Internet of Things (IoT) in the management of healthcare issues and healthcare policy development. *International Journal of Technology, Innovation and Management (IJTIM), 1*(1), 69–84. https://doi.org/10.54489/ijtim.v1i1.8

Rui, L. S., Khai, L. L., & Alzoubi, H. M. (2022). Determinants of emerging technology adoption for safety among construction businesses. *Academy of Strategic Management Journal, 21*(Special Issue 4), 1–20.

Schoonenboom, J., & Burke, J. R. (2017). How to construct a mixed methods research design. *KZfSS Kölner Zeitschrift Für Soziologie Und Sozialpsychologie, 69*(2), 107–131.

Shabbir, M. Q., & Waheed, G. S. B. (2020). Application of big data analytics and organizational performance: The mediating role of knowledge management practices. *Journal of Big Data, 7*(1), 1–17.

Shamout, M., Ben-Abdallah, B., Alshurideh, M., Alzoubi, H., Al Kurdi, B., & Hamadneh, S. (2022a). A conceptual model for the adoption of autonomous robots in supply chain and logistics industry. *Uncertain Supply Chain Management, 10*, 1–16.

Shamout, M., Elayan, M., Rawashdeh, A., Kurdi, B., & Alshurideh, M. (2022b). E-HRM practices and sustainable competitive advantage from HR practitioner's perspective: A mediated moderation analysis. *International Journal of Data and Network Science, 6*(1), 165–178.

Sun, H., Rabbani, M. R., Sial, M. S., Yu, S., Filipe, J. A., & Cherian, J. (2020). Identifying big data's opportunities, challenges, and implications in finance. *Mathematics, 8*, 10.

Tang, J. J., & Karim Khondkar, E. (2017). Big data in business analytics: Implications for the audit profession. *The CPA Journal, 87*(6), 34–39.

Tariq, E, Alshurideh, M., Akour, I., & Al-Hawary, S. (2022a). The effect of digital marketing capabilities on organizational ambidexterity of the information technology sector. *International Journal of Data and Network Science, 6*(2), 401–408.

Tariq, E., Alshurideh, M., Akour, I., Al-Hawary, S., & Al, B. (2022b). The role of digital marketing, CSR policy and green marketing in brand development. *International Journal of Data and Network Science, 6*(3), 1–10.

Vassakis, K., Petrakis, E., & Kopanakis, I. (2018). Big data analytics: Applications, prospects and challenges. In *Mobile big data* (pp. 3–20). Cham.

Marketing Mix, Services and Branding

Customer Awareness Towards Green Marketing Mix in 5-Star Hotels in Jordan

Anber Abraheem Shlash Mohammad, Faraj Mazyed Faraj Aldaihani,
Sara M. Alrikabi, Muhammad Turki Alshurideh ⓘD,
Riad Ahmad Mohammed Abazeed, Doa'a Ahmad Odeh Al-Husban,
Ayat Mohammad, Sulieman Ibraheem Shelash Al-Hawary,
and Barween H. Al Kurdi ⓘD

Abstract This descriptive study determines the level of customers' and visitors' awareness regarding the green marketing mix activities and applications in green five-stars hotels in Jordan. The study population consisted of the visitors of 16 green

A. A. S. Mohammad · S. M. Alrikabi
Marketing Department, Faculty of Administrative and Financial Sciences, Petra University,
D.O.Dox 961343, Amman 11196, Jordan

F. M. F. Aldaihani
Kuwait Civil Aviation, Ishbiliyah Bloch 1, Street 122, Home 1, Kuwait City, Kuwait

M. T. Alshurideh
Department of Marketing, School of Business, The University of Jordan, Amman 11942, Jordan
e-mail: m.alshurideh@ju.edu.jo; malshurideh@sharjah.ac.ae

Department of Management, College of Business, University of Sharjah, Sharjah 27272, UAE

R. A. M. Abazeed
Business Management and Public Administration, Department of Business Administration,
Faculty of Finance and Business Administration, Al Al-Bayt University, P.O.BOX 130040,
Mafraq 25113, Jordan

D. A. O. Al-Husban
Department of Human Fundamental Sciences, Faculty of Alia College, Al- Balqa Applied
University, Amman, Jordan
e-mail: D_husban@bau.edu.jo

A. Mohammad
Business and Finance Faculty, The World Islamic Science and Education University (WISE),
11947, P.O Box 1101, Amman, Jordan

S. I. S. Al-Hawary (✉)
Department of Business Administration, Faculty of Economics and Administrative Sciences, Al
Al-Bayt University, P.O.BOX 130040, Mafraq 25113, Jordan
e-mail: dr_sliman73@aabu.edu.jo; dr_sliman@yahoo.com

B. H. Al Kurdi
Department of Marketing, Faculty of Economics and Administrative Sciences, The Hashemite
University, Zarqa, Jordan
e-mail: barween@hu.edu.jo

© The Author(s), under exclusive license to Springer Nature Switzerland AG 2023   2157
M. Alshurideh et al. (eds.), *The Effect of Information Technology on Business
and Marketing Intelligence Systems*, Studies in Computational Intelligence 1056,
https://doi.org/10.1007/978-3-031-12382-5_118

labeled five-stars hotels in Jordanian cities and tourism locations (Amman, Dead Sea, Petra and Aqaba). Non-random (purposive) sample was used to collect data and a structured questionnaire was distributed to (460) hotel customers, (412) of which were retrieved and (367) were valid for analysis with a response rate of (79.78%). A set of statistical methods were used through the (SPSS v23) program and results showed only two of the dimensions of green marketing mix had a high level of customer awareness. These dimensions were green product and green price. The overall level of customers' environmental concern was moderate along with the rest of the green marketing dimensions (promotion, place, process, people, and physical evidence). Based on results, it is recommended that hotel managers and marketers carefully design their business and marketing plans to truly develop credible green hotel image, avoid misrepresentation and green washing, and encourage customers to participate in protecting the environment.

Keywords Customer awareness · Green marketing mix · 5-star hotels · Jordan

1 Introduction

The vast development in the hospitality industry in recent years had a huge contribution to the environmental damage and global warming due to the massive amounts of water supply, energy, and natural resources consumed (Alshurideh et al., 2019; Leonidou et al., 2013). During this time, and specifically in the early 90's, an emerging movement for protecting the environment has developed. This movement forced the hotels managements shift their direction to this cause as an attempt to attract the novel segment of green customers and urge them into adopting a variety of green marketing activities (Joghee et al., 2021; Shishan et al., 2022; Tariq et al., 2022). The green marketing mix in hotels range from designing products that reduce emissions and hazards, altering the pricing strategies in exchange of the environmental commitment, developing online third-party partnerships to reduce transport pollution, expressing the hotel orientation through related marketing messages and promotions, providing appropriate training for employees, altering and redesigning the internal processes, and validating customer's expectations through physical evidence (Al-Hawary & Mohammad, 2011; Metabis & Al-Hawary, 2013).

The outcome of customers' awareness regarding a hotel's green marketing mix include influencing the buying behavior (Altarifi et al., 2015; Al-Hawary & Alhajri, 2020; Al-Nady et al., 2016), portraying a source of competitive advantage (Al-Hawary & Al-Hamwan, 2017; Al-Hawary & Ismael, 2010; Al-Hawary et al., 2011; Nadanyiova & Gajanova, 2018), affect the customers' willing to pay premium (Dimara et al., 2015) and increase customer loyalty (Al-Hawary & Al-Fassed, 2021; Al-Hawary & Al-Menhaly, 2016; Al-Hawary & Al-Smeran, 2017; Al-Hawary & Harahsheh, 2014; Al-Hawary & Hussien, 2017; Al-Hawary, 2013a, 2013b; Martinez, 2015).

This study discusses one of the most important issues at present. It fills a gap in giving precise measurements of using green marketing mix by hotels and the degree of customers' awareness of these practices. The study will provide guidance to hotels that need to develop new marketing mix practices for satisfying current customers and attracting new ones. It also explores the customers' needs and desires when they choose a hotel. And importantly, this research will educate hotels in deciding on whether this new approach (green marketing) is really fulfilling its purpose or not.

Given the importance of environmental protection, Green Hotels Association (GHA) made tremendous efforts to transform hotels into green entities. Not only providing extra positive publicity, but green marketing also helped increasing the competitive advantage, brand loyalty, and sustainable development. 5-star hotels begun leaning towards the green and sustainable trends and are most committed to the governmental regulations concerning energy and water conservation and proper waste management (Alketbi et al., 2020; Al-Quran et al., 2020).

This study identifies to hotels' managers and marketers the extent of customer awareness and knowledge of their hotels regarding the green hotel initiatives and consequently encourages marketers to develop successful green marketing strategies matching customers' requirements. This study examines the degree of customers' awareness of green marketing mix at the hotels in Jordan.

2 Theoretical Framework

2.1 Green Marketing Concept

When it comes to Green Marketing, the majority associate it with merely promoting or advertising the environmentally friendly products. The green marketing trend first began in the late '70 s when the Marketing Association of America conducted the first ecological-issues workshop. This workshop resulted in a book called "Ecological Marketing" by Heniun and Keny in 1979. These results weren't taken seriously until the late '80 s due to the growing movements to increase public awareness and urging customers to purchase green products. Later, the green marketing phenomena have gained distinctive importance in the modern market and have emerged in both developed and developing countries while being seen as a fundamental approach to aid sustainable development (Al-Hawary & Abu-Laimon, 2013; Al-Hawary & Aldaihani, 2016; Al-Hawary & Al-Syasneh, 2020; Al-Hawary, 2013a, 2013b; Alolayyan et al., 2018; Alshurideh et al., 2017a, 2017b; Lekhanya, 2014).

The relevant literature review showed a lack of sufficient research and the absence of a unique and defined definition. The American Marketing Association defined ecological marketing as studying both the positive and negative effects of pollution by carbon depletion and resources degradation by various marketing practices. Herbig and Butler (1993) defined green marketing as all products and associated packaging manufactured with criteria that are less toxic, durable, reusable and recyclable.

Then, green marketing was broadly defined by (Polonsky, 1994) who described green marketing as the practices intended to generate and facilitate human needs with a minimal harm to the environment. Later, Peattie (1995) defined green marketing as the holistic managerial process that distinguishes, predicts, and fulfils customer sustainable requirements. While the techniques used to promote products in terms of green marketing were described by Prakash (2002) as the use of ecological statements, either on the characteristics and features of the products or about the processes, policies, and systems of the companies that manufacture or sell these products. Jain and Kaur (2004) defined green marketing is all those marketing activities that companies undertake to create a positive impact or lessen the negative impact of their products or services on the environment. Chen and Chai (2010) stretched the definition of green marketing to include practices by companies that are apprehensive about environmental or green issues and are therefore selling products or services that are environmentally responsible to satisfy customers and society in general. Table 1 summarize the research papers used green marketing mix dimensions through their research.

Green products are percieved as valuable marketing tools, with many hoteliers attempting to offer green products that draw customer interest (Jones et al., 2014). Herberger first defined green products as products that have an ecological

Table 1 Green marketing mix dimensions (7Ps)

Authors	Year	Study title
Lapian & Tumbel	2018	Analysis of Green Marketing Mix Factors on Hotel Industry (Study on Sintesa Peninsula Hotel Manado; Discovery Kartika Plaza Denpasar; Hyatt Regency Yogyakarta; Grand Melia Jakarta)
Pomering,	2017	Marketing for sustainability: Extending the conceptualisation of the marketing mix to drive value for individuals and society at large
Eneizan & Wahab	2016	Effects of green marketing strategy on the financial and non-financial performance of firms: A conceptual paper
Kirimi Gitonga	2014	The influence of green marketing mix strategies on performance of fast-moving customer goods companies in Nairobi County. University of Nairobi
Kotler, Keller, Brady, Goodman & Hansen	2012	Marketing Management (2nd ed.)
Larashati	2012	7Ps of Green Marketing as Factors Influencing Willingness to Buy Towards Environmentally Friendly Beauty Products
Ivy	2008	A new higher education marketing mix: The 7Ps for MBA marketing

orientation, where its environmental viability is perceived, recognized, and under-stood. This definition was followed by many, but it rarely is a clear definition. Perhaps it was considered as a common term that is already well-known. For example, (Chen & Chai, 2010) simply defined green products as ecological product or environmental friendly product. A wildly mentioned definition is of who defined the green product development as greening the entire production process starting with the manufacturing phase and resulting in disposal. In their research, De Medeiros and Ribeiro (2017) proposed a different approach in defining green products. They suggested that sustainable products are those capable of reducing customer stress related to the environmental responsibility without affecting the overall product quality. In a green hotel context, we find customers also identify features, design, brand name, service quality, corporate image, and credibility before selecting a final hotel destination (Jones et al., 2014).

Green Price: The debate of whether green hotels charge premium or not is never ending due to its direct relation to whether customers are willing to pay extra or not. According to D'Souza et al. (2006) eco-friendly products should be expected to require premium charges in exchange. However, Yazdanifard and Mercy (2011) argue that premium green prices need to be reasonable and justifiable for customers. However, Roos and Nyrud (2008) contradict this argument as they mentioned that the green customers are less price sensitive that traditional customers. On the contrary, research indicates a link between willing to pay premium and the level of environment concern. Price was briefly defined by Engel et al. (2008) as the value offered in exchange of a product. It is considered the most critical dimension as most customers are only willing to pay additional value in exchange of additional product value and attributes. Such attributes might include improved performance, design, and visual appeal. Prakash (2002) stated that customers are not always ready to pay extra for green products. The author also mentioned that during inconvenience, customers perceive price as a huge barrier.

Green place plays a significant role in deciding the eco-friendly place channels for any green product. Rivera-Camino (2007) defined green place as the system that adheres to the green attributes of green products and ensures the environmental characteristics.

According to Eneizan et al. (2016) green place implies the method of selecting the place routes that result in a less environmental damage, since the majority of harm occurs during transportation. Green place highlights the importance of reducing the carbon footprint by rearranging and managing logistics and transportation emissions (Shil, 2012).

In order for green hoteliers to optimize their marketing efforts effectively, they need to decide where and how to position their place routs. It is therefore essential to place their place channels in the right place and time for the right target, which is the green customers (Kotler et al., 2012). It is no doubt that the internet is a powerful communication medium which is often utilized to educate the target market about the hotels' commitment to the cause and the operational sustainable practices. Websites are one wildly used media where a hotel can display all their credentials, awards, environmental programs, and policies. This allows hotelier to be directly in touch with their customers to actively decrease the number of customers visiting the hotel or any related partner for more information.

Green promotion: Kamath et al. (2016) stated that green promotion refers to a type of advertising that promotes the organization's sustainable policies, their eco-friendly operations, and the environmental measures adopted. This involves developing the necessary promotional tools, such as advertising, marketing materials, signage, web sites, videos and presentations while remaining environmentally friendly (Al Kurdi & Alshurideh, 2021; Al-Dmour & Al-Shraideh, 2008; Alshurideh et al., 2016, 2018). According to Polonsky and Mintu (1997) green promotion helps customers to overcome the lack of environmental information. While Zint & Frederick suggested that customers are not to be assumed knowledgeable regarding the environmental damage and hazards. However, green promotion as explained by Ankit and Mayur (2013) and Alshurideh et al., (2017a, 2017b), might increase and satisfy the demand of green customers with environmental concerns.

Green process was defined as the methodology the service will be provided with minimum harm to the environment (Larashati, 2012). The author suggested greening the entire procedure line, if possible, to establish an overall green plan. Eneizan et al. (2016) argued that some changes to internal processes are expected by businesses who plan to implement green marketing as a business strategy. The company's processes might require some fundamental changes which can be achieved by using less energy, removing damaged products, and less destruction to the natural resources (Alsalaymeh, 2013). This can reflect sequential measures taken by a variety of different workers when seeking to fulfil a job which is incredibly important for the company objectives and procedures to comply. According to research, green process is determined based on possessing little to no potential in harming the environment starting with production, usage, and disposal. Green process should also contribute to saving non-renewable resources and energy. Lapian and Tumbel (2018) clarified that energy-efficient equipment involved in a hotel's operational procedures are of great necessity.

Green people: According to Larashati (2012), people are usually the employees of the company who contribute to both providing and marketing green services and products. It is also essential to possess a green mindset to reflect credibility in their duties. Sangeeta explains that when employees and management are aware of the probable damage to the environment and their role in minimizing this damage, the pro-environmental results and image are more visible. Involving the employees is more efficient and reduces the costs and expenses of the marketing efforts. In their research, Welmilla, and Ranasinghe (2020) explained that employees should be well educated on ecological issues, highly motivated and with a positive attitude, as the credibility of the company lies in their hands. **Green Physical Evidence**: According to (Pomering, 2017) green physical evidence is the environment where the service provided takes place and the customer-employee interaction occurs. Ivy (2008) explained that a potential overlapping between the physical evidence variable in the marketing mix and processes is possible. This strengthens the efforts of green hotels to convey their beliefs and persuades customers at the same time. The Chartered Institute of Marketing stated that the physical evidence provided by the company will validate the customer's expectations. For example, the physical setting of the reception will display the hotel's green intentions in a desirable fashion. The space should be designed with elegant and innovative furniture, recycled materials, or reused products. The lobby should be equipped with digital reading materials instead of paper format to help hotel's customers read these digital materials via their computers or smartphones. Satisfaction and loyalty for a hotel could be highly improved by incorporating green physical evidence.

2.2 *Customer Awareness*

Despite its importance to governments, organizations, and scholars, customer awareness seems to be poorly discussed in a proper manner in literature. The term itself is usually misleadingly used in articles with vague discussions and definitions (Rhein & Schmid, 2020). The available literature typically associate customer awareness with either environmental concerns or customers purchase decisions amongst alternatives (Alshurideh, 2022; Alshurideh et al., 2020; Alzoubi et al., 2022). According to Hammami et al. (2017) these studies barely measure the environmental concerns and fail to capture the entire complete dimensions of the customer awareness. Regarding this study context, the last three decades witnessed a rise in customer awareness of the environment, as it has evolved from a secondary concern to a primary one. This has been the result of a variety of causes, including increasing media attention, growing knowledge of environmental issues, the rise of advocacy campaigns and strict regulation (national and international) and the influence of industrial incidents on public opinion (Abuhashesh et al., 2021; Alkitbi et al., 2020; Alshurideh, 2019).

The green customer is known to be someone whose actions are motivated by environmental issues. Peattie (2001) describes a green customer as anyone who willingly participates in environmentally sustainable practices approved by researchers. Critical customers have started to emerge as a new driver in green customers, requiring corporate social responsibility (Maheshwari, 2014). Gradually, the emergence of sustainable customer society has contributed to an ever-wider definition of consumption called ethical customers which refers to customer behaviour that expresses concerns about problems arising from illegal and discriminatory global trade (Hamelin et al., 2013). These behaviours might include abusing human rights, trafficking with animals, child labor, trade union repression, pollution, low-paid labour, and inequality in trade relations. When the mid 90's witnessed the advent of both the green customers and ethical customers, customers demanded an opinion in the manufacturing processes starting from production, place and even the resources allocation. Charter and Polonsky (2017) stated the anticipation of this ongoing disruption of customers, by advocating for sustainable marketing by academics. They also defined sustainability marketing as creation and preservation of a healthy relationship between customers and the environment (Charter & Polonsky, 2017). On the other hand, Rousseau and Venter (1995) defined the customer awareness as the customers' levels of alertness regarding their rights and their obligations in the marketplace.

This definition was further explained by five distinct forces as follows: bargain hunting, general customer knowledge, product knowledge, information search and price consciousness (Rousseau & Venter, 1996).

3 The Study Methodology

3.1 Study Population and Sample

The study population consisted of customers of 5-star hotels in Jordan to recognize their awareness toward green marketing mix in these hotels, where the number of hotels of this category reached 16 hotels in Jordan, most of them in Amman. Due to the large number of hotel customers, it was not possible to conduct a complete census of all customers, so the non-random (purposive) sample was used to collect data related to the study. The study instrument was distributed to (460) hotel customers to achieve the appropriate response rate. A retrieved questionnaire (412) where it was found that (45) were not suitable for statistical analysis due to its incompleteness or the similarity of the answers. Consequently, the analyzed questionnaires were (367) which constitutes a response percentage (79.78%) of the distributed questionnaires.

3.2 Study Instrument

The questionnaire was used as a basic instrument to collect data from the target study sample, where this questionnaire consisted of two parts, the first part contains an introductory introduction about the nature of the study as well as questions related to demographic variables (gender, age, educational level, and familiar with the study variables). The second part, it consists of (50) items related to the variable represented by the customers' awareness of green marketing mix, which was developed based on appendix (3). This variable is formed by eight dimensions that were: general environmental concern which was measured by items (Q1–Q9), green product which was measured by items (Q10–Q16), green prices which was measured by items (Q17–Q22), green promotion which was measured by items (Q23–Q29), green place which was measured by items (Q30–Q33), green process which was measured by items (Q34–Q39), green people which was measured by items (Q40–Q44), green physical evidence which was measured by items (Q45–Q50).

3.3 Validity and Reliability

To verify of face validity of the study instrument and its suitability for conducting the study, the tool was presented to a number of referees from the academic doctors in order to identify the belonging of the items to their latent variables and their good linguistic formulation. In addition, the exploratory factor analysis (EFA) was used as a statistical method used to identify the validity and reliability of the instrument through the values of the average variance extracted (AVE) to test the convergent validity, and compared the values of the average variance extracted (AVE) with the maximum shared variance (MSV) to test the discriminate validity. Moreover, the reliability of the instrument was tested using McDonald's Omega coefficient (C.R) to evaluate composite reliability of the study instrument. The results of these tests are presented in Table 2.

The results in Table 2 showed that the values of items loadings on their latent variables ranged between (0.681–0.769) which was greater than the lowest acceptable value of 0.50 (Al-Hawary et al., 2018; Haig, 2018), and the values of the average variance extracted (AVE) within the ambit (0.507–0.558) which were higher than the minimum threshold for judging convergent validity (Howard, 2018). In addition, the obtained average variance extracted (AVE) were greater than the values of maximum shared variance (MSV), which indicates the achievement of the discriminate validity. In terms of reliability, the results listed in the table (1) showed that the values of McDonald's Omega coefficients (C.R) ranged between (0.813–0.905) that were higher than 0.70 which is the minimum acceptable value for the composite stability (Heale & Twycross, 2015).

4 Data Analysis

4.1 Correlation

Table 3 shows the results of Pearson's correlation matrix to exhibit the relationship between the study variables and ensuring that the data free of a multicollinearity problem.

Through the results shown in Table 3 that that the study variables are related to each other, where the values of the correlation coefficients within the range (0.126–0.650) and they were all statistically significant at a significance level less than 0.05. Furthermore, the results show that the data is free from the problem of multicollinearity where the correlation coefficients values were less than 0.80 (Senaviratna & Cooray, 2019).

Table 2 Results of a measurement model validity and reliability

Variables	Items	Loadings	AVE	MSV	C.R
General environmental concern	Q1	0.708	0.514	0.502	0.905
	Q2	0.725			
	Q3	0.716			
	Q4	0.707			
	Q5	0.748			
	Q6	0.706			
	Q7	0.693			
	Q8	0.689			
	Q9	0.757			
Green product	Q10	0.751	0.542	0.470	0.881
	Q11	0.698			
	Q12	0.681			
	Q13	0.731			
	Q14	0.725			
	Q15	0.686			
	Q16	0.743			
Green prices	Q17	0.734	0.521	0.492	0.867
	Q18	0.702			
	Q19	0.721			
	Q20	0.725			
	Q21	0.687			
	Q22	0.760			
Green promotion	Q23	0.708	0.511	0.180	0.880
	Q24	0.703			
	Q25	0.694			
	Q26	0.697			
	Q27	0.755			
	Q28	0.724			
	Q29	0.723			
Green place	Q30	0.700	0.521	0.154	0.813
	Q31	0.696			
	Q32	0.733			
	Q33	0.755			
Green process	Q34	0.769	0.558	0.302	0.883
	Q35	0.748			
	Q36	0.721			

(continued)

Table 2 (continued)

Variables	Items	Loadings	AVE	MSV	C.R
	Q37	0.735			
	Q38	0.754			
	Q39	0.758			
Green people	Q40	0.725	0.531	0.163	0.850
	Q41	0.700			
	Q42	0.724			
	Q43	0.733			
	Q44	0.760			
Green physical evidence	Q45	0.721	0.507	0.110	0.860
	Q46	0.690			
	Q47	0.756			
	Q48	0.708			
	Q49	0.686			
	Q50	0.707			

Table 3 Results of Pearson's correlation matrix

Variables	1	2	3	4	5	6	7	8
1. General environmental concern	1							
2. Green product	0.608^{**}	1						
3. Green prices	0.650^{**}	0.365^{**}	1					
4. Green promotion	0.378^{**}	0.219^{**}	0.262^{**}	1				
5. Green place	0.332^{**}	0.221^{**}	0.301^{**}	0.169^{*}	1			
6. Green process	0.490^{**}	0.376^{**}	0.379^{**}	0.222^{**}	0.230^{**}	1		
7. Green people	0.352^{**}	0.226^{**}	0.336^{**}	0.141^{**}	0.126^{*}	0.190^{**}	1	
8. Green physical evidence	0.288^{**}	0.266^{**}	0.190^{**}	0.145^{**}	0.156^{*}	0.179^{*}	0.216^{**}	1

Note *Correlation is significant at the 0.05 level, **Correlation is significant at the 0.01 level

4.2 Descriptive Statistics

This section is providing a detailed presentation of data analysis that was collected through the study instrument, which was distributed to the study sample, as well as it also answers to the study questions and achieving its objectives. Table 4 displays the results of means and standard deviations used to answer the major question that was "What is the level of customers' awareness toward green marketing mix in the five stars hotels in Jordan?"

Table 4 Results of statistics related to green marketing mix

Dimensions	Means	SD	Rank	Level
Green prices	3.74	0.589	1	High
Green product	3.68	0.565	2	High
Green people	3.65	0.615	3	Moderate
General environmental concern	3.57	0.552	4	Moderate
Green physical evidence	3.56	0.574	5	Moderate
Green promotion	3.56	0.566	6	Moderate
Green process	3.55	0.629	7	Moderate
Green place	3.54	0.621	8	Moderate
Green marketing mix	3.61	0.352	–	Moderate

The results in Table 4 indicate that means of the responses of the study sample ranged between (3.54–3.74) with standard deviations between (0.552–0.629). These results referred that the relative importance of green marketing mix dimensions ranged between high and moderate level, where the dimension of green prices came in first rank with a mean (3.74) with a standard deviation (0.589), where it indicates to a high level, followed by the dimension of green product with a mean (3.68) with a standard deviation (0.565) that was refereed also to a high level, while the dimension of green place ranked in the last place with a mean (3.54) with a standard deviation (0.621) which was at a moderate level. Moreover, the results demonstrated that customers' awareness toward green marketing mix in the five stars hotels in Jordan was at a moderate level, where mean was (3.61) with a standard deviation (0.352). Besides, one sample t-test was conducted on the study instrument related to customers' awareness toward green marketing mix in the five stars hotels in Jordan by assuming the hypothetical mean is 3, where the table 5 shows the results obtained.

The results in Table 5 indicated that the calculated t value was (32.930) at a significance level (0.000) which was less than 0.05. Therefore, customers' awareness toward green marketing mix in the five stars hotels in Jordan was at a moderate level and has a statistically significant. One sample t-test was conducted on the study instrument related to customers' awareness toward each dimension of green marketing mix in the five stars hotels in Jordan by assuming the hypothetical mean is 3, where the Table 6 shows the results obtained.

Based on the calculated t value and the significance level which were for all less than 0.05. Therefore, the customers' awareness toward green product, Green Promotion, Green Place, Green Process, Green People, general environmental concern and

Table 5 Results of one-sample t test for green marketing mix

Variable	Mean	SD	t-value	df	P-value*
Green marketing mix	3.61	0.352	32.930	366	0.000

Note *Refers to a significance level (P ≤ 0.05)

Table 6 Results of one-sample t test for green product

Variable	Mean	SD	t-value	df	P-value*
Green product	3.68	0.565	22.977	366	0.000
Green prices	3.74	0.589	24.042	366	0.000
Green promotion	3.55	0.566	18.905	366	0.000
Green place	3.54	0.621	16.571	366	0.000
Green process	3.55	0.629	16.809	366	0.000
Green people	3.65	0.615	20.350	366	0.000
Green physical evidence	3.56	0.574	18.767	366	0.000
General environmental concern	3.57	0.552	19.652	366	0.000

Note *Refers to a significance level ($P \leq 0.05$)

Green Physical Evidence in the five stars hotels in Jordan was at a high level and has a statistically significant, while customers' awareness toward green prices in the five stars hotels in Jordan was at a high level and has a statistically significant.

5 Discussion, Conclusion and Recommendations

5.1 Results Discussion

Based on the statistical results, it was found that the Jordanian green hotels' customers have mixed beliefs when it comes to green hotels' practices as only (56.40%) of the study sample stated that they were familiar with these practices. When compared to the general environmental concern part of the questionnaire, obtained data showed moderate levels of awareness. According to literature, these two factors are positively engaged and are highly related to each other, as the higher the environmental concern the higher the level of familiarity of green practices. In order for hotel managers and marketers to benefit from the positive publicity, competitive advantage and brand loyalty, they need to identify the level of customers awareness not only regarding the environmental conservation but also the green hotel marketing mix. This helps them to better elevate or promote their efforts to their guests and also measure the effectiveness of these efforts.

According to this study, we have been able to answer this question by identifying the level of green marketing mix awareness and found it was moderate. This would consequently encourage marketers to develop successful green marketing strategies to match their customers' requirements and needs. This result was based on measuring the sub questions of green marketing mix dimensions which include (product, price, place, promotion, process, people, and physical evidence).

Findings related to these questions clearly show that the level of customers' awareness regarding prices was the highest. Respondents stated that they were willing to pay extra to green hotels especially if it was helping reducing pollution and being invested in eco-services. Price is also incredibly important to hoteliers as it provides insights regarding the competitive advantage and sufficient revenue. Green hotel managers will need to coordinate the prices they charge their customers with the additional costs associated to green technologies, processes, and promotion.

The second green marketing dimension to rank high in this study was green products. Respondents showed high interest in using eco-friendly products, expressing their understanding of the environmental phrases, symbols, and green claims. Respondents also stated that they enjoy environmentally friendly products and healthy amenities provided by green hotels. As the second most important dimension to customers, hoteliers need to constantly provide authentic green products along with environmentally friendly facilities to please customers and emphasize differentiation.

The remaining green marketing mix dimensions showed similar moderate awareness level results. According to questionnaire respondents, hotels usually fail to make their green promotions noticeable by clients and they expect to be informed regarding the hotel's green activities. The respondents also stated that when they are educated regarding the hotel's green policies, they will advise their friends and families to visit such hotels. As customers are probably suspicious regarding hotels honest eco-claims. Promotion plays a huge role in translating the hoteliers' efforts in conserving the environment. Promotion further aids customers in easily finding green hotels, their products, and facilities without wasting money, time and efforts.

Like promotion, green place dimension also plays a role when it comes to convenience. According to respondents, the hotel location needs to be in alignment to the green claims. Online booking services are usually preferred by customers as it provides all the information required regarding staying at a green hotel. Partners and vendors are required to have sufficient information regarding the greenness of a hotel as it helps maintain the hotel brand image.

As stated above, the green process showed moderate awareness levels amongst respondents. They were able to acknowledge that their hotels implement e-commerce to preserve the environment. They expressed that paperless interactions and water preservation are visible and effective approaches. Literature states that green processes can highly improve the financial performance. Eco-equipment and technology -although somewhat expensive- ultimately prove efficient in reducing cost as it reduces the overall natural resources consumption and effectively reduces the associated waste.

Green hotels' employees are a major asset to the hotel management as they are the frontline with customers. According to hotels' customers, having well trained employees who are ready to answer inquiries regarding green products, services, facilities, process or even green activities is of importance. Approaching these customers with positive and motivational green attitude is also recommended.

And finally, the green physical evidence translates all the above without human–human interaction. Green buildings are not only an architectural trend, but it also helps expressing the level of hotel greenness. Placing digital or recyclable green materials is preferable by customers. The general atmosphere, furniture and accessories should also reflect the nature and the eco-friendly efforts by hoteliers.

5.2 Conclusion

Looking back to literature, we can notice the majority highlighting the importance of customer concern towards the environment and the impact of customer awareness of the hotel green marketing mix on several behaviours ranging from attitude, purchase intention and word of mouth.

Based on the above discussion, we can confidently claim that the possibility of having more environmentally friendly hotels with better service quality and benefits as the customer general environment concern and awareness grow. Even though the respondents were not able to prove they fully understand the green marketing mix implications, but the overall idea of the concept shows a certain level of awareness. Naturally, we can't emphasise more the importance of reducing customer doubt and the potential mistrust of the green hotels as it increases the susceptibility of damaging the brand image by constantly promoting real actions and strategies while providing the sufficient training for the staff.

In conclusion, many hotels have begun to transform their objectives into greener ones to attract new customer segments and generate revenue. It is also essential to correctly promote these objectives and practices to increase the customer awareness and motivate purchase decisions. Hoteliers are also advised not to take advantage of the customers' confusion regarding the hotel's environmental claims and to righteously explain these claims to eliminate green washing. Developing guides signage to self-explain the ecological terms and claims might prove beneficial. This might be implemented after examining the hotel's surrounding environment (nature), the workflow processes and policies, the technology and innovation in accordance with natural resources, renewable energy and waste management.

As a growing concept in Jordan, green marketing is gaining a shy popularity in the hospitality industry as several hotels are adopting the concept with minimal efforts of promotions and advertisements as a source of gaining the knowledge and increasing the customers awareness. Making the hotels practices more visible to the public benefits the hotels, customers and finally the environment.

5.3 Marketing Practical Implications and Recommendations

Based on all of the above, we can positively state that the interest in green hotels will grow in the near future. In this study, specific recommendations and measures are suggested for an efficient and effective green policy implementation in the hotel sector in Jordan:

1. Hotel managers and marketers are suggested to carefully design their business and marketing plans to truly develop credible green hotel image and avoid misrepresentation. In addition, acquiring international green awards and accreditation enhances customers' trust and confidence regarding the hotels' green products and services, increase loyalty, and differentiates hotels by achieving a competitive advantage.
2. The utilization of a holistic sustainable framework in hotels results in higher efficiency in operations. For instance, improved efficiency in the use of energy and natural resources results in substantial cost savings, and a better waste management system will not only reduce the amount of waste but will also ensure its safe disposal.
3. The main purpose of green products is to provide benefit and value without harming the environment. Green hotels are suggested to alter their hotel equipment in rooms and hotel amenities into more efficiently energy conservative equipment and minimize waste. They can also provide organic food at restaurants and source from local markets, present ecological cleaning products with less plastic and distribute garbage separating bins.
4. Hotels can implement a new promotional and branding strategy while protecting the environment by providing branded reusable water bottles to their guests and placing water-filling stations throughout the hotel. They can also involve the local community and surrounding environment in the promotional ad campaigns, events and celebrations to encourage customers to make local purchases from vendors, promote tourism and benefit the community.
5. Digital marketing plays a major role nowadays. Incorporating social media, websites and online booking reduces the amount of paper consumed by both the hotel staff and the hotel customers and helps customers be closer to the information they require prior their arrival which eventually helps reduce the carbon footprint.
6. For customers to pay premium prices at green hotels they need to perceive great value in exchange. Hotels are advised to keep hotel guests happy, satisfied and actively engaged in the sustainable practices provided. Conducting interactive workshops, activities and minigames attracts more customers and satisfies the existing ones.
7. Hotel staff play a major role in providing necessary information regarding environment preservation and hotel procedures to their guests. Frequently providing the required relevant training is highly recommended as it reflects in their day-to-day work activities. Sufficient training also gives the staff confidence in responding to customers regarding the green hotel practices without the need to consult with the management.
8. The physical environment also plays a major role in translating sustainability. Wood furniture, greenery sights, natural aromas are some of the first interactions with hotel customers. Consulting with architects and interior designers to reflect the sustainability in hotels will prove beneficial on the long run.

References

Abuhashesh, M. Y., Alshurideh, M. T., & Sumadi, M. (2021). The effect of culture on customers' attitudes toward Facebook advertising: the moderating role of gender. Review of international business and strategy. *Review of International Business and Strategy, 31*(3), 416–437.

Al Kurdi, B. H., & Alshurideh, M. T. (2021). Facebook advertising as a marketing tool: Examining the influence on female cosmetic purchasing behaviour. *International Journal of Online Marketing (IJOM), 11*(2), 52–74.

Al- Quran, A. Z., Alhalalmeh, M. I., Eldahamsheh, M. M., Mohammad, A. A., Hijjawi, G. S., Almomani, H. M., & Al-Hawary, S. I. (2020). Determinants of the green purchase intention in Jordan: The moderating effect of environmental concern. *International Journal of Supply Chain Management, 9*(5), 366–371.

Al-Dmour, H., & Al-Shraideh, M. T. (2008). The influence of the promotional mix elements on Jordanian consumer's decisions in cell phone service usage: An analytical study. *Jordan Journal of Business Administration, 4*(4), 375–392.

Al-Hawary, S. I., & Al-Fassed, K. J. (2021). *The impact of social media marketing on building brand loyalty through customer engagement in Jordan*. In Press.

Al-Hawary, S. I. (2013a). The role of perceived quality and satisfaction in explaining customer brand loyalty: Mobile phone service in Jordan. *International Journal of Business Innovation and Research, 7*(4), 393–413.

Al-Hawary, S. I. (2013b). The roles of perceived quality, trust, and satisfaction in predicting brand loyalty: The empirical research on automobile brands in Jordan market. *International Journal of Business Excellence, 6*(6), 656–686.

Al-Hawary, S. I. S., & Alhajri, T. M. S. (2020). Effect of electronic customer relationship management on customers' electronic satisfaction of communication companies in Kuwait. *Calitatea, 21*(175), 97–102.

Al-Hawary, S. I. S., Abdul Aziz Allahow, T. J., & Aldaihani, F. M. F. (2018). Information technology and administrative innovation of the central agency for information technology in Kuwait. *Global Journal of Management and Business, 18*(11-A), 1–16.

Al-Hawary, S. I., & Abu-Laimon, A. A. (2013). The impact of TQM practices on service quality in cellular communication companies in Jordan. *International Journal of Productivity and Quality Management, 11*(4), 446–474.

Al-Hawary, S. I., & Aldaihani, F. M. (2016). Customer relationship management and innovation capabilities of Kuwait airways. *International Journal of Academic Research in Economics and Management Sciences, 5*(4), 201–226.

Al-Hawary, S. I., & Al-Hamwan, A. (2017). Environmental analysis and its impact on the competitive capabilities of the commercial banks operating in Jordan. *International Journal of Academic Research in Accounting, Finance and Management Sciences, 7*(1), 277–290.

Al-Hawary, S. I., & Al-Menhaly, S. (2016). The quality of E-government services and its role on achieving beneficiaries satisfaction. *Global Journal of Management and Business Research: A Administration and Management, 16*(11), 1–11.

Al-Hawary, S. I., & Al-Smeran, W. (2017). Impact of electronic service quality on customers satisfaction of Islamic banks in Jordan. *International Journal of Academic Research in Accounting, Finance and Management Sciences, 7*(1), 170–188.

Al-Hawary, S. I., & Al-Syasneh, M. S. (2020). Impact of dynamic strategic capabilities on strategic entrepreneurship in presence of outsourcing of five stars hotels in Jordan. *Business: Theory and Practice, 21*(2), 578–587.

Al-Hawary, S. I., & Harahsheh, S. (2014). Factors affecting Jordanian consumer loyalty toward cellular phone brand. *International Journal of Economics and Business Research, 7*(3), 349–375.

Al-Hawary, S. I., & Hussien, A. J. (2017). The Impact of electronic banking services on the customers loyalty of commercial banks in Jordan. *International Journal of Academic Research in Accounting, Finance and Management Sciences, 7*(1), 50–63.

Al-Hawary, S. I., & Ismael, M. (2010). The Effect of using information technology in achieving competitive advantage strategies: A field study on the jordanian pharmaceutical companies. *Al Manara for Research and Studies, 16*(4), 196–203.

Al-Hawary, S. I., & Mohammad, A. A. (2011). The role of the internet in marketing the services of travel and tourism agencies in Jordan. *Abhath Al-Yarmouk, 27*(2B), 1339–1359.

Al-Hawary, S. I., Mohammad, A. A., & Al-Shoura, M. (2011). The impact of E-marketing on achieving competitive advantage by the Jordanian pharmaceutical firms. *DIRASAT, 38*(1), 143–160.

Alketbi, S., Alshurideh, M., & Al Kurdi, B. (2020). The Influence of service quality on customers' retention and loyalty in the UAE hotel sector with respect to the impact of customer' satisfaction, trust, and commitment: A qualitative study. *International Journal of Innovation, Creativity and Change, 14*(7), 734–754.

Al-Nady, B. A., Al-Hawary, S. I., & Alolayyan, M. (2016). The role of time, communication, and cost management on project management success: An empirical study on sample of construction projects customers in Makkah City, Kingdom of Saudi Arabia. *International Journal of Services and Operations Management, 23*(1), 76–112.

Alolayyan, M., Al-Hawary, S. I., Mohammad, A. A., & Al-Nady, B. A. (2018). Banking service quality provided by commercial banks and customer satisfaction. A structural equation modelling approaches. *International Journal of Productivity and Quality Management, 24*(4), 543–565.

Al-Salaymeh, M. (2013). The application of the concept of green marketing in the productive companies from the perspective of workers. *Interdisciplinary Journal of Contemporary Research Business, 4*(12), 634–641.

Alshurideh, D. M. (2019). Do electronic loyalty programs still drive customer choice and repeat purchase behaviour? *International Journal of Electronic Customer Relationship Management, 12*(1), 40–57.

Alshurideh, M. (2022). Does electronic customer relationship management (E-CRM) affect service quality at private hospitals in Jordan? *Uncertain Supply Chain Management, 10*(2), 1–8.

Alshurideh, M., Al Kurdi, B. H., Vij, A., Obiedat, Z., & Naser, A. (2016). Marketing ethics and relationship marketing-An empirical study that measure the effect of ethics practices application on maintaining relationships with customers. *International Business Research, 9*(9), 78–90.

Alshurideh, M., Al Kurdi, B., Abu Hussien, A., & Alshaar, H. (2017a). Determining the main factors affecting consumers' acceptance of ethical advertising: A review of the Jordanian market. *Journal of Marketing Communications, 23*(5), 513–532.

Alshurideh, M., Al Kurdi, B., Abumari, A., & Salloum, S. (2018). Pharmaceutical promotion tools effect on physician's adoption of medicine prescribing: Evidence from Jordan. *Modern Applied Science, 12*(11), 210–222.

Alshurideh, M., Al-Hawary, S. I., Batayneh, A. M., Mohammad, A., & Al-Kurdi, B. (2017b). The impact of Islamic banks' service quality perception on Jordanian customers loyalty. *Journal of Management Research, 9*(2), 139–159.

Alshurideh, M., Gasaymeh, A., Ahmed, G., Alzoubi, H., & Kurd, B. (2020). Loyalty program effectiveness: Theoretical reviews and practical proofs. *Uncertain Supply Chain Management, 8*(3), 599–612.

Alshurideh, M., Kurdi, B. A., Shaltoni, A. M., & Ghuff, S. S. (2019). Determinants of pro-environmental behaviour in the context of emerging economies. *International Journal of Sustainable Society, 11*(4), 257–277.

Altarifi, S., Al-Hawary, S. I. S., & Al Sakkal, M. E. E. (2015). Determinants of E-shopping and its effect on consumer purchasing decision in Jordan. *International Journal of Business and Social Science, 6*(1), 81–92.

Alzoubi, H., Alshurideh, M., Kurdi, B., Akour, I., & Aziz, R. (2022). Does BLE technology contribute towards improving marketing strategies, customers' satisfaction and loyalty? The role of open innovation. *International Journal of Data and Network Science, 6*(2), 449–460.

Ankit, G., & Mayur, R. (2013). Green marketing: Impact of green advertising on consumer purchase intention. *Advances In Management, Advances in Management, 6*(9).

Charter, M., & Polonsky, M. J. (Eds.). (2017). *Greener marketing: A global perspective on greening marketing practice* (2nd ed.). Routledge.

Chen, T. B., & Chai, L. T. (2010). Attitude towards the environment and green products: Customers' perspective. *Management Science and Engineering, 4*(2), 27–39.

D'Souza, C., Taghian, M., Lamb, P., & Peretiatkos, R. (2006). Green products and corporate strategy: An empirical investigation. *Society and Business Review, 1*(2), 144–157.

De Medeiros, J. F., & Ribeiro, J. L. D. (2017). Environmentally sustainable innovation: Expected attributes in the purchase of green products. *Journal of Cleaner Production, 142*(1), 240–248.

Dimara E., Manganari, E., & Skuras, D. (2015). Consumers' willingness to pay premium for green hotels: Fact or Fad? In *Proceedings international marketing trends conference.*

Eneizan, B. M., Abd-Wahab, K., & Sharif., Z. M. (2016). Effects of green marketing strategy on the financial and non-financial performance of firms: A conceptual paper. *Oman Chapter of Arabian Journal of Business and Management Review, 5*(12), 14–27.

Engel, S., Pagiola, S., & Wunder, S. (2008). Designing payments for environmental services in theory and practice: An overview of the issues. *Ecological Economics: THe Journal of the International Society for Ecological Economics, 65*(4), 663–674.

Haig, B. D. (2018). Exploratory factor analysis, theory generation, and scientific method. In B. D. Haig (Ed.), *Method matters in psychology* (Vol. 45, pp. 65–88). Springer International Publishing.

Hamelin, N., Harcar, T., & Benhari, Y. (2013). Ethical customerism: A view from the food industry in Morocco. *Journal of Food Products Marketing, 19*(5), 343–362.

Hammami, M. B., Mohammed, E., Haghom, A., Al-Khafaji, M., Alqahtani, F., Alzaabi, S., & Dash, N. (2017). Survey on awareness and attitudes of secondary school students regarding plastic pollution: Implications for environmental education and public health in Sharjah city, UAE. *Environmental Science and Pollution Research International.* https://doi.org/10.1007/s11356-017-9625-x

Heale, R., & Twycross, A. (2015). Validity and reliability in quantitative studies. *Evidence Based Nursing, 18*(3), 66–67.

Herbig, P. A., & Butler, D. D. (1993). The greening of international marketing. *Journal of Teaching in International Business, 5*(1/2), 63–76.

Howard, M. C. (2018). The convergent validity and nomological net of two methods to measure retroactive influences. *Psychology of Consciousness: Theory, Research, and Practice, 5*(3), 324–337.

Ivy, J. (2008). A new higher education marketing mix: The 7Ps for MBA marketing. *International Journal of Educational Management, 22,* 288–299.

Jain, S. K., & Kaur, G. (2004). Green marketing: An attitudinal and behavioural analysis of Indian customers. *Global Business Review, 5*(2), 187–205.

Joghee, S., Alzoubi, H. M., Alshurideh, M., & Al Kurdi, B. (2021). The role of business intelligence systems on green supply chain management: Empirical analysis of FMCG in the UAE. In *The international conference on artificial intelligence and computer vision* (pp. 539–552). Springer.

Jones, P., Hillier, D., & Comfort, D. (2014). Sustainability in the global hotel industry. *International Journal of Contemporary Hospitality Management, 26*(1), 5–17.

Kamath, N. (Ed.). (2016). *Handbook of research on strategic supply chain management in the retail industry.* Idea Group.

Kotler, P., Keller, K. L., Brady, M., Goodman, M., & Hansen, T. (2012). *Marketing management* (2nd ed.). Pearson Education.

Lapian & Tumbel, (2018). Analysis of green marketing mix factors on hotel industry (Study on Sintesa Peninsula Hotel Manado; Discovery Kartika Plaza Denpasar; Hyatt Regency Yogyakarta; Grand Melia Jakarta). *International Review of Management and Marketing, 8*(6), 112–121.

Larashati, H. (2012). *7Ps of green marketing as factors influencing willingness to buy towards environmentally friendly beauty products.*

Lekhanya, L. M. (2014). The impact of viral marketing on corporate brand reputation. *International Business & Economics Research Journal (IBER), 13*(2), 213.

Leonidou, L. C., Leonidou, C. N., Fotiadis, T. A., & Zeriti, A. (2013). Resources and capabilities as drivers of hotel environmental marketing strategy: Implications for competitive advantage and performance. *Tourism Management, 35*, 94–110.

Maheshwari, P. S. (2014). Awareness of green marketing and its influence on buying behavior of customers: Special reference to Madhya Pradesh, India. *AIMA Journal of Management & Research, 8*(1/4).

Martínez, P. (2015). Customer loyalty: Exploring its antecedents from a green marketing perspective. *International Journal of Contemporary Hospitality Management, 27*(5), 896–917.

Metabis, A., & Al-Hawary, S. I. (2013). The impact of internal marketing practices on services quality of commercial banks in Jordan. *International Journal of Services and Operations Management, 15*(3), 313–337.

Nadanyiová, M., & Gajanová, L., (2018). Customers' perception of green marketing as a source of competitive advantage in the hotel industry.

Peattie, K. (1995). *Environmental marketing management: Meeting the green challenge.* Financial Times Prentice Hall.

Peattie, K. (2001). Towards sustainability: The third age of green marketing. *The Marketing Review, 2*(2), 129–146.

Polonsky, M., & Mintu, A. (1997). The future of Environmental marketing: Food for thought. *Environmental marketing strategies, practice, theory and research Haworth* 389–391.

Polonsky M. J. (1994). An introduction to Green marketing. *Electronic Green Journal, 1*(2).

Pomering, A. (2017). Marketing for sustainability: Extending the conceptualisation of the marketing mix to drive value for individuals and society at large. *Australasian Marketing Journal (AMJ), 25*(2), 157–165.

Prakash, A. (2002). Green marketing, public policy and managerial strategies. *Business Strategy and the Environment, 11*(5), 285–297.

Rhein, S., & Schmid, M. (2020). Customers' awareness of plastic packaging: More than just environmental concerns. *Resources, Conservation, and Recycling, 162*(105063), 105063.

Rivera-Camino, J. (2007). Re-evaluating green marketing strategy: A stakeholder perspective. *European Journal of Marketing, 41*(11/12), 1328–1358.

Roos, A., & Nyrud, A. Q. (2008). Description of green versus environmentally indifferent consumers of wood products in Scandinavia: Flooring and decking. *Journal of Wood Science, 54*(5), 402–407.

Rousseau, G. G., & Venter, D. J. L. (1995). Measuring customer awareness in Zimbabwe. *Journal of Industrial Psychology, 21*(1), 18–24.

Rousseau, G. G., & Venter, D. J. L. (1996). A comparative analysis of customer awareness in South Africa during 1992 and 1994: Implications for the reconstruction and development programme (RDP). *Journal of Industrial Psychology, 22*(2), 26–31.

Senaviratna, N. A. M. R., & Cooray, T. M. J. A. (2019). Diagnosing multicollinearity of logistic regression model. *Asian Journal of Probability and Statistics, 5*(2), 1–9.

Shil, P. (2012). Evolution and future of environmental marketing. *Asia Pacific Journal of Marketing and Management Review, 1*(3), 74–81.

Shishan, F., Mahshi, R., Al Kurdi, B., Alotoum, F. J., & Alshurideh, M. T. (2022). Does the past affect the future? An analysis of consumers' dining intentions towards green restaurants in the UK. *Sustainability, 14*(1), 1–14.

Tariq, E., Alshurideh, M., Akour, E., Al-Hawaryd, S., & Al Kurdi, B. (2022). The role of digital marketing, CSR policy and green marketing in brand development at UK. *International Journal of Data and Network Science, 6*(3), 1–10.

Welmilla, I., & Ranasinghe, V. (2020). Green employee engagement.

Yazdanifard, R., & Mercy, I. E. (2011). The impact of green marketing on customer satisfaction and environmental safety.

Customers' Perception of the Social Responsibility in the Private Hospitals in Greater Amman

Faraj Mazyed Faraj Aldaihani, Anber Abraheem Shlash Mohammad, Hanan AlChahadat, Sulieman Ibraheem Shelash Al-Hawary, Mohammad Fathi Almaaitah, Nida'a Al-Husban, Abdullah Ibrahim Mohammad, Muhammad Turki Alshurideh, and Ayat Mohammad

Abstract The study determines the level of patients' and visitors' perception of social responsibility in private hospitals in Greater Amman. The population of the study consisted of all patients and visitors of private hospitals in Greater Amman.

F. M. F. Aldaihani
Kuwait Civil Aviation, Ishbiliyah Bloch 1, Street 122, Home 1, Kuwait City, Kuwait

A. A. S. Mohammad · H. AlChahadat
Marketing Department, Faculty of Administrative and Financial Sciences, Petra University,
P.O.Box 961343, Amman 11196, Jordan

S. I. S. Al-Hawary (✉) · N. Al-Husban
Department of Business Administration, Faculty of Economics and Administrative Sciences, Al
Al-Bayt University, P.O.BOX 130040, Mafraq 25113, Jordan
e-mail: dr_sliman73@aabu.edu.jo; dr_sliman@yahoo.com

M. F. Almaaitah
Faculty of Economic and Administration Sciences, Department of Business Administration &
Public Administration, Al Al-Bayt University Jordan, P.O. Box, 130040, Mafraq 25113, Jordan
e-mail: m.maaitah@aabu.edu.jo

A. I. Mohammad
Department of Basic Scientific Sciences, Al-Huson University College, Al-Balqa Applied
University, Irbid, Jordan

M. T. Alshurideh
Department of Marketing, School of Business, The University of Jordan, Amman 11942, Jordan
e-mail: malshurideh@sharjah.ac.ae

Department of Management, College of Business, University of Sharjah, 27272 Sharjah, United
Arab Emirates

A. Mohammad
Business and Finance Faculty, The World Islamic Science and Education University (WISE),
11947, P.O Box 1101, Amman, Jordan
e-mail: dr_ayatt@yahoo.com

M. Alshurideh et al. (eds.), *The Effect of Information Technology on Business
and Marketing Intelligence Systems*, Studies in Computational Intelligence 1056,
https://doi.org/10.1007/978-3-031-12382-5_119

The sample of the study consisted of patients and visitors of three private hospitals (Jordan Hospital, Istishari Hospital, AlKhalidi Hospital),100 patients and visitors from each hospital, the total number of patients and visitors are three hundred. Those patients and visitors are randomly selected, three hundred questionnaires were distributed by hand, out of them 210 questionnaires are returned complete with a response rate of 70%. This study identifies to managers and decision-makers in private hospitals that to which extent the patients and visitors of private hospitals perceive their social responsibility programs and consequently encourages them to develop successful social responsibility programs which compatible with customer's perceptions and preferences. The results show that all dimensions of social responsibility are moderately perceived by patients and visitors of private hospitals in Greater Amman.

Keywords Social responsibility · Private hospital · Customers' perception · Amman

1 Introduction

The healthcare sector has a variety of challenges such as stringent regulatory compliance, intense labor shortages in nursing, increased and costly technological advancements, implementation of international quality standards, and substantial community dependence (Al-Hawary, 2012; Al-Hawary et al., 2011; Abu Qaaud et al., 2011; Al-Hawary et al., 2011; Al-Hawary et al., 2017; Al-Hawary & Al-Namlan, 2018; Mohammad et al., 2020). Corporate social responsibility could play a major role in this context that's why hospitals with the mission of providing medical services, should engage in social responsibility programs effectively (Al-Hawary & Al-Khazaleh, 2020; Alhalalmeh et al., 2020; Abu Zayyad et al., 2021).

Social responsibility programs in hospitals range from voluntary programs and partnerships to mitigate the environmental impact of industrial plants and production methods to the development of sourcing and marketing initiatives that protect social welfare and commit to environmental benefits (Alshurideh et al. 2016, 2019a, b). In turn, the outcomes of social responsibility programs include achieving competitive advantage (Sen et al., 2006), gauging corporate reputation (Ellen et al., 2006), positive corporate financial outcomes and increased market value (Peloza & Shang, 2011), and increased willing to pay premium (Parsa et al., 2015; Alshurideh et al., 2021).

This study discusses one of the most important issues at present. The importance raised by many factors which are: evolved new concerns and expectations of citizens, customers, public authorities, and investors about corporate social responsibility, increased concerns about the harm caused by economic practices to the environment, and transparency of business activities brought by the media and modern information and communication technologies (Almazrouei et al., 2020; Aljumah et al., 2021; Al-Maroof et al., 2021; Khasawneh et al., 2021a, 2021b; Kurdi et al., 2021; Shamout et al., 2022; Vallaster et al., 2012). With a view to those factors, service providers are allocating significant resources to manifest their commitment, as well as their ethical vision and socially responsible behavior. The first main objective is determining the patients' and visitors' perception of social responsibility dimensions in private hospitals in Greater Amman.

2 Theoretical Framework

2.1 Social Responsibility

There have been different definitions of social responsibility offered in the literature which represents different conceptualizations (Maignan & Ferrell, 2004). The social responsibility has emerged as a key concept in the management, ethical, marketing, and communication literatures (AlShehhi et al., 2020; Alshurideh et al., 2017; Alyammahi et al., 2020; Zadek, 2006). Bolton and Mattila (2015) defined corporate social responsibility as "a company's commitment to minimizing or eliminating any harmful effects and maximizing its long-run beneficial impact on society". Chen (2016) defined corporate social responsibility as "the extent to which the social responsibility actions contribute to customers' perceptions on corporate social responsibility commitment, values-driven motives, customer orientation, and trustworthiness".

Jin et al. (2017) defined social responsibility as "the responsibility of an organization for the impacts of its decisions and activities on society and the environment, through transparent and ethical behavior that contributes to sustainable development including the health and the welfare of society, takes into account the expectations of stakeholders, is in compliance with applicable law and consistent with international norms of behavior, and is integrated throughout the organization and practiced in its relationships". Finally, Schwartz and Carroll (2018) defined social responsibility by three domain approach which presents in three core domains of social responsibility: economic, legal, and ethical responsibilities.

In literatures social responsibility examined as multi-dimensional, where its initiatives vary from voluntary programs and partnerships to mitigate the environmental impact of industrial plants and production methods to the development of initiatives which protect social welfare and commit to environmental benefits. This study will discuss six dimensions of social responsibility as follows:

- **Economic Responsibility**: Bello et al. (2016) defines economic responsibility as corporate responsibility to make decisions that will enhance the shareholders' wealth and found jobs with fair wages, while at the same time, producing quality products that are needed and desired by customers and also sell those products at a reasonable price. Moreover, Schwartz and Carroll (2018) in three domain approach embraces two types of economic activities direct and indirect. Direct economic activities include actions intended to increase sales or avoid litigation. Indirect economic activities include activities that are designed to improve employees' morals or the company's public image (Al-Jarrah et al., 2012; Al-Gasaymeh et al., 2015, 2020; Assad and Alshurideh, 2020a, 2020b).

- **Legal Responsibility**: Bello et al. (2016) defined legal responsibility as the responsibility of the company to operate within the laws, rules and regulations set down by recognized institutions in the society. Schwartz and Carroll (2018) in three domain approach explain that "the legal category of corporate social responsibility relates to the corporate response to legal anticipations imposed and expected by society as emerged in case law".

- **Ethical Responsibility**: The ethical dimension of social responsibility encompasses those activities that are based on their adherence to a set of ethical or moral standards or principles. Interestingly, Nussbaum (2009) mentioned ethical corporate social responsibility activities as follow: publish a detailed (annual report) on the CSR activities of its members, partnerships between locals' investments in awareness campaign, proactively go beyond what is only the legal thing to do, sticking to the core business of life-threatening diseases (support patient associations), and not giving up business ethics for commercial purposes.

- **Philanthropic Responsibility**: Corporate philanthropy is no longer a simple matter of donating money to a deserving cause. It is increasingly expected to serve long term business interests (Alshurideh et al., 2014; Rampal & Bawa, 2008). Accordingly, Peloza and Shang (2011) mentioned different types of philanthropic activities as follow cause-related marketing (where a charity donation is attached to a commercial exchange), donations of cash (different from cause-related marketing because donations are not attached to a sale), statements of support for charities without stating explicitly how that support is given, promotion of a social issue, donations of products, licensing, event sponsorship, customer donations and non-specific support for charities and employee volunteerism.

- **Environmental Contribution**: The environmental activities or green activities focus on firm activities that do not corrode natural resources through corporate environmental responsibility. Parsa et al. (2015) identified different types of environmental practices as follows: incorporating product features that are environmentally friendly (green products), minimizing waste (recycling program), energy conservation (power and water conservation), reduction of pollutants (reduction of fluorocarbons and Styrofoam), and good-neighbor activities such as (neighborhood clean-up).

- **Customer Protection**: Customers are noted to be the main stakeholder group whose continuous protection is essential for the continuance of business. The customers also have notable interactions with corporations on a consistent basis and this leads to raising the possibilities for potential conflicts between the two parties, especially on the issues relating to non-observance of customers' rights by the corporation's (Al-Hawary & Hussien, 2017; Al-Hawary & Obiadat, 2021; Alolayyan et al., 2018). Thus, issues relating to corporate social responsibility and customer protection have remained contemporary topics in the marketing literature (Bello et al., 2016). Sun and price (2016) mentioned stakeholder theory which argues that corporate social responsibility contributes to customer satisfaction by establishing customer relationships and meeting customer needs.

2.2 Customer Perception

It is important to differentiate between customers' expectations of social responsibility and customers' perceptions of social responsibility. Golob et al. (2008) found that customers have high expectations of corporate social responsibility, especially in the legal and ethical-philanthropic domains. Moreover, expectations for the ethical-philanthropic dimension of CSR are higher amongst customers holding high self-transcendent values and practicing high involvement. Gleim (2011) represented that the environmental sustainability is becoming increasingly important to customers, it is critical that firms understand the role that it can have with regards to perceptions of the firm.

JIN et al. (2017) examined the effect of corporate social responsibility on customer behavior and found significant differences in brand attitude and perception of credibility when companies did not engage in corporate social responsibility activities. Su et al. (2017) found that corporate social responsibility positively affects perceived corporate reputation and customer satisfaction, which in turn, significantly affects customer commitment and behavioral responses (i.e., loyalty intentions and word-of-mouth) (Alzoubi et al., 2020; Sweiss et al., 2021).

3 Study Methodology

3.1 Population and Sample of the Study

The population of the study consisted of all patients and visitors of private hospitals in Greater Amman. The sample of the study was selected from patients and visitors of three private hospitals (Jordan Hospital, Istishari Hospital, AlKhalidi Hospital), 100 patients and visitors from each hospital. the total number of study participants are 300 individuals. Those participants are randomly selected.

4 Data Analysis

4.1 Exploratory Factor Analysis (EFA)

To verify the psychometric properties of the questionnaire, an exploratory factor analysis (EFA) was conducted. Table 1 shows the rotation matrix for the items of the questionnaire.

The results in Table 1 show that the factor loadings of all items are ranged from (0.734) to (0.941). Factor loadings greater than (0.4) were identified as acceptable. Factorability of questionnaire items was identified based on KMO (KMO = 0.686,

Table 1 Results of exploratory factor analysis

Items	Factors					
	1	2	3	4	5	6
1	0.852					
2	0.845					
3	0.888					
4	0.842					
5	0.91					
6	0.734					
7		0.865				
8		0.852				
9		0.921				
10		0.742				
11			0.924			
12			0.854			
13			0.889			
14			0.778			
15				0.931		
16				0.852		
17				0.863		
18				0.859		
19				0.738		
20					0.941	
21					0.854	
22					0.821	
23					0.886	
24					0.773	
25						0.854
26						0.894
27						0.932
28						0.784
29						0.852
30						0.861
31						0.841
32						0.863
33						0.766

Determinant = 0.005, KMO = 0.686, Bartlett's Test significant at $\alpha \leq 0.05$

greater than 0.05) and Bartlett's Test (Significant at $\alpha \leq 0.05$). The table illustrates that the items of the questionnaire were loaded on six factors: Factor 1 (6 items), factor 2 (4 items), factor 3 (4 items), factor 4 (5 items), factor 5 (5 items) and factor 6 (9 items).

4.2 Confirmatory Factor Analysis (CFA)

In order to examine the goodness-of-fit of the measurement model, confirmatory factor analysis (CFA) was used. The results in Fig. 1 show that the goodness-of-fit of the measurement model is adequate (Chi-square $= 1.778 < 2.0$, GFI $= 0.933 > 0.90$, CFI $= 0.917 > 0.90$ and RMSEA $= 0.068 < 0.08$).

4.3 Testing Validity and Reliability

In order to validate the research instrument, the researchers have presented the questionnaire to panel composed of marketing professors. Each of them on read and / or contributed to the instrument to ensure the questions were representative of the purpose of the study. Based on the feedback provided, the researcher made the necessary modifications to the newly developed instrument, including editing the content and order of the survey instrument. After they have confirmed the instruments in Arabic language, it was distributed to a sample of (10) patients and visitors to ensure it is understandable and easy to read.

Reliability is defined as "the degree to which majors are free from error and therefore yield consistent roles". The researcher used Petrick test, which is an indicator of consistency that is believed to determine the correlations between the items of the scales from which the questionnaire has been composed. The scale is considered reliable when it has Petrick test (Composite Reliability) of (≥ 0.70). For the purpose of this study, researcher used SPSS package to determine Petrick test (Composite Reliability) for the scale questions. The results in Table 2 confirm that the study tool is reliable to be used for data collection.

As the data presented in Table 2, the scale is reliable and set of items are closely related to a group.

5 Data Analysis

To test this hypothesis, descriptive statistics means and standard deviations were used in measuring the customers' perception of social responsibility and its dimensions.

Table 3 shows the means scores for the total perceived social responsibility and its dimensions.

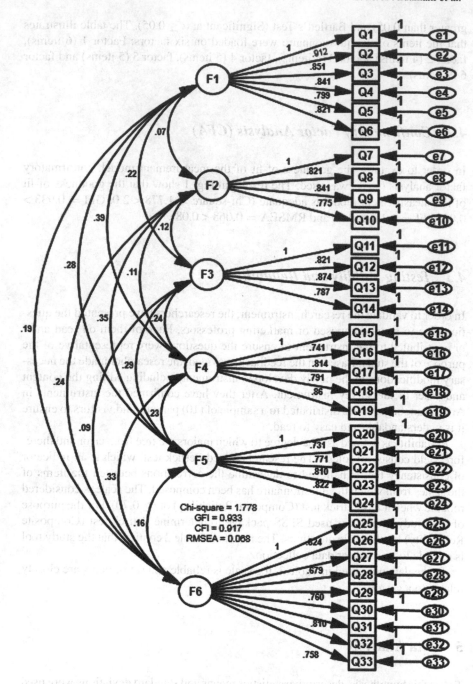

Fig. 1 Results of CFA of social responsibility items

Table 2 Results of validity and reliability test (Petrick test)

Dimensions	Reliability
	Petrick test (composite reliability)
Philanthropic responsibility	0.938
Ethical responsibility	0.910
Legal responsibility	0.921
Economic responsibility	0.929
Environmental contribution	0.932
Customer protection	0.959

Table 3 Mean and standard deviations of social responsibility dimensions

Dimensions	Mean	Std. deviation	t-value	P-value
Philanthropic responsibility	2.79	0.74	14.03	0.000
Ethical responsibility	3.53	0.61	12.42	0.000
Legal responsibility	3.59	0.65	13.12	0.000
Economic responsibility	3.58	0.56	14.89	0.000
Environmental contribution	3.39	0.56	10.15	0.000
Customer protection	3.54	0.59	13.29	0.000
Grand mean	3.40	Moderate		

In order to measure the customers' perception of social responsibility in the private hospitals in Greater Amman, the grand mean (3.40) was matched with the statistical rational formula and it was found that it is in the second category (2.34—less than 3.68), which means that the level of customers' perception of social responsibility in the private hospitals in Greater Amman is moderate.

6 Conclusion and Discussion

The private sector is gaining a much bigger role and responsibility for economic development globally. This responsibility is not limited to economic issues but must also include social and environmental contribution. Hospitals that fail to recognize this responsibility are at the risk of being denied the social acceptance that comes from the community. Without this acceptance, hospitals can never function in a profitable and sustainable manner. Utilization of a holistic corporate social responsibility framework in hospitals can result in higher efficiency in operations, for instance, improved efficiency in the use of energy and natural resources can result in substantial cost savings. A better waste management system will not only reduce the amount of waste but will also ensure its safe disposal.

Reputational risk is considered as one of the most crucial threats facing an organization and is even more critical for a hospital. This includes risks not only to loss of patients, but goes beyond to hospital itself, and may even impact the whole industry. In case of a reputational crisis involving the hospital, the consequences could be huge in terms of lost trust, legal costs and patient loyalty. A damaged reputation might require years to rebuild and cost a large sum of money. A hospital which has a sound corporate social responsibility mechanism and a history of exceptional service to society and environment often does not suffer as much as a hospital with no corporate social responsibility plans, in the incidence of a reputation crisis.

Patients need to be able to trust a hospital in order to recommend it to someone. Trust is probably the most valuable currency in the healthcare industry and it doesn't come easy. In order for hospitals to earn patient trust and loyalty, they need to go beyond healthcare services and create an emotional bond with the patient through ethical business practices. Patient loyalty goes a long way in contributing towards sustainable business growth of a hospital. When competitors adopt less costly but not socially responsible and ethically sound healthcare solutions, a hospital should take advantage of this challenge and explore new innovative and green solutions. This raises the barriers to entry and will make CSR as the industry norm with your company being the pioneer.

Investors no more only rely on financial data but also look at how a company deals with the relevant social and environmental issues. If a company is not prudent enough to pay any attention to these issues, with time it will lose credibility and no investor wants to invest in a company that has no credibility or stakeholder trust. In order for a healthcare service provider to attract investors who can fund their expansion, they need to focus on social, environmental and economic performance in addition to financial performance.

Appendix 1

See Table 4.

Table 4 Social responsibility dimensions

Author/year	Social responsibility dimensions
Carroll (1991)	(a) Economic—making a profit, (b) Legal—obeying the law and following government regulations, (c) Ethical—doing what is right and avoiding harm, and (d) Philanthropic—contributing resources to the community to improve the quality of life
Chattananon et al. (2007)	Social responsibility Programs: program symbolism (visual symbolism and other tangibles), behavior (management behavior, employees' behavior and other intangibles), program communications (primary, secondary and tertiary communications)
Golob et al. (2008)	Economic, Legal, Ethical, and Philanthropic
Swaen and Chumpitaz (2008)	Philanthropic activities, Respect for the environment, Respect for customers, Respect for workers
Peloza and Shang (2011)	Philanthropy, Business practices, Product related
Chabowski et al. (2011)	Environmental dimension, Social dimension, Economic dimension
Smith and Langford (2011)	Ethical CSR, Economic CSR, Philanthropic CSR, Legal CSR, Social responsibility, and Environmental responsibility
Vallaster et al. (2012)	CSR entrepreneurs, CSR performers, and vocal CSR converts or quietly conscientious
Blagov and Savchenko (2012)	Philanthropic activities: corporate donations in money; corporate donations in kind; rendering services; employees' donations; and corporate sponsorship
Lombart and Louis (2014)	Respect for the environment, Respect for customers, and Philanthropic activities put in place
Garcia and Greenwood (2015)	Environmental sustainability, Quality products and Services, Financial sustainability, Ethics, Philanthropy, Employee welfare, and Legal responsibility
Parsa et al. (2015)	Nutritional concerns, Environmental concerns, Social concerns

(continued)

Table 4 (continued)

Author/year	Social responsibility dimensions
Bolton and Mattila (2015)	Corporate Philanthropy, and Sustainability
Garcia and Greenwood (2015)	Environmental sustainability
Santhosh and Baral (2015)	Social Responsibility Towards Employees: Sound policies and practices, employee participation in decision making, equity in employee remuneration, good working conditions, elimination of child labor, avoidance of gender bias Social Responsibility Towards Customers: Practicing ethical advertising, adhering to product standards, prioritizing customer health and safety, and providing products at right price Social Responsibility Towards Community: Building schools, hospitals, temples, etc Social Responsibility Towards Environment: Adhere to environmental standards
Chen (2016)	Cause donation, Cause sponsorship, Social alliance, and Value-chain social responsibility
Bello et al. (2016)	Economic, Legal, Ethical, and Philanthropic
Su et al. (2017)	Conserve the natural and cultural resources, and Boost the local economy by hiring the local residents and bringing jobs to the region
Jin et al. (2017)	Donations, Employee volunteering, and Nonperformance of CSR
Schwartz and Carroll (2018)	Legal domain, Ethical Domain, Economic domain

References

Abu Qaaud, F., Al-Shoura, M., & Al-Hawary, S, I. (2011). The impact of the service marketing mix in the service quality of health services from the viewpoint of patients in government hospitals in Amman "A Field study." *Abhath Al-Yarmouk, 27*(1B), 417–441.

Abu Zayyad, H. M., Obeidat, Z. M., Alshurideh, M. T., Abuhashesh, M., Maqableh, M., & Masa'deh, R. E. (2021). Corporate social responsibility and patronage intentions: the mediating effect of brand credibility. *Journal of Marketing Communications, 27*(5), 533–510.

Al-Gasaymeh, A., Almahadin, A., Alshurideh, M., Al-Zoubid, N., & Alzoubi, H. (2020). The role of economic freedom in economic growth: Evidence from the MENA region. *International Journal of Innovation, Creativity and Change, 13*(10), 759–774.

Al-Gasaymeh, A., Kasem, J., & Alshurideh, M. (2015). Real exchange rate and purchasing power parity hypothesis: Evidence from Adf unit root test. *International Research Journal of Finance and Economics, 14*, 450–2887.

Alhalalmeh, M. I., Almomani, H. M., Altarifi, S., Al-Quran, A. Z., Mohammad, A. A., & Al-Hawary, S. I. (2020). The nexus between corporate social responsibility and organizational performance in Jordan: The mediating role of organizational commitment and organizational citizenship behavior. *Test Engineering and Management, 83*, 6391–6410.

Al-Hawary, S. I. (2012). Health care services quality at private hospitals, from patient's perspective: A comparative study between Jordan and Saudi Arabia. *African Journal of Business Management, 6*(22), 6516–6529.

Al-Hawary, S. I. & Al-Khazaleh A, M. (2020). The mediating role of corporate image on the relationship between corporate social responsibility and customer retention. *Test Engineering and Management, 83*(516), 29976–29993.

Al-Hawary, S. I., & Al-Namlan, A. (2018). Impact of electronic human resources management on the organizational learning at the private hospitals in the State of Qatar. *Global Journal of Management and Business Research: A Administration and Management, 18*(7), 1–11.

Al-Hawary, S. I., & Hussien, A. J. (2017). The impact of electronic banking services on the customers loyalty of commercial banks in Jordan. *International Journal of Academic Research in Accounting, Finance and Management Sciences, 7*(1), 50–63.

Al-Hawary, S. I., Alghanim, S., & Mohammad, A. (2011). Quality level of health care service provided by king Abdullah educational hospital from patient's viewpoint. *Interdisciplinary Journal of Contemporary Research in Business, 2*(11), 552–572.

Al-Hawary, S. I., AL-Zeaud, H., & Batayneh, A. M. (2011). The relationship between transformational leadership and employee's satisfaction at Jordanian private hospitals. *Business and Economic Horizons, 5*(2), 35–46.

Al-Hawary, S. I., Batayneh, A. M., Mohammad, A. A., & Alsarahni, A. H. (2017). Supply chain flexibility aspects and their impact on customers satisfaction of pharmaceutical industry in Jordan. *International Journal of Business Performance and Supply Chain Modelling, 9*(4), 326–343.

Al-Hawary, S. I. S., & Obiadat, A. A. (2021). Does mobile marketing affect customer loyalty in Jordan? *International Journal of Business Excellence, 23*(2), 226–250.

Al-Jarrah, I., Al-Zu'bi, M. F., Jaara, O., & Alshurideh, M. (2012). Evaluating the impact of financial development on economic growth in Jordan. *International Research Journal of Finance and Economics, 94*, 123–139.

Aljumah, A., Nuseir, M. T., & Alshurideh, M. T. (2021). The impact of social media marketing communications on consumer response during the COVID-19: Does the brand equity of a university matter. In *The effect of coronavirus disease (COVID-19) on business intelligence* (pp. 367–384).

Al-Maroof, R., Ayoubi, K., Alhumaid, K., Aburayya, A., Alshurideh, M., Alfaisal, R., & Salloum, S. (2021). The acceptance of social media video for knowledge acquisition, sharing and application: A comparative study among YouYube users and TikTok users' for medical purposes. *International Journal of Data and Network Science, 5*(3), 197–214.

Almazrouei, F. A., Alshurideh, M., Kurdi, B. A., & Salloum, S. A. (2020). Social media impact on business: A systematic review. In *International conference on advanced intelligent systems and informatics* (pp. 697–707). Springer.

Alolayyan, M., Al-Hawary, S. I., Mohammad, A. A., & Al-Nady, B. A. (2018). Banking service quality provided by commercial banks and customer satisfaction. A structural equation modelling approaches. *International Journal of Productivity and Quality Management, 24*(4), 543–565.

AlShehhi, H., Alshurideh, M., Al Kurdi, B., & Salloum, S. A. (2020). The impact of ethical leadership on employees performance: A systematic review. In *International conference on advanced intelligent systems and informatics* (pp. 417–426). Springer.

Alshurideh, M. T., & Shaltoni, A. M. (2014). Marketing communications role in shaping consumer awareness of cause-related marketing campaigns. *International Journal of Marketing Studies, 6*(2), 163–168.

Alshurideh, M. T., Al Kurdi, B., & Salloum, S. A. (2021). The moderation effect of gender on accepting electronic payment technology: a study on United Arab Emirates consumers. Review of international business and strategy. *Review of International Business and Strategy, 31*(3), 375–396.

Alshurideh, M., Al Kurdi, B. H., Vij, A., Obiedat, Z., & Naser, A. (2016). Marketing ethics and relationship marketing-An empirical study that measure the effect of ethics practices application on maintaining relationships with customers. *International Business Research, 9*(9), 78–90.

Alshurideh, M., Al Kurdi, B., Abu Hussien, A., & Alshaar, H. (2017). Determining the main factors affecting consumers' acceptance of ethical advertising: A review of the Jordanian market. *Journal of Marketing Communications, 23*(5), 513–532.

Alshurideh, M., Kurdi, B. A., Shaltoni, A. M., & Ghuff, S. S. (2019a). Determinants of pro-environmental behaviour in the context of emerging economies. *International Journal of Sustainable Society, 11*(4), 257–277.

Alshurideh, M., Salloum, S. A., Al Kurdi, B., & Al-Emran, M. (2019b). Factors affecting the social networks acceptance: an empirical study using PLS-SEM approach. In *Proceedings of the 2019b 8th international conference on software and computer applications* (pp. 414–418).

Alyammahi, A., Alshurideh, M., Al Kurdi, B., & Salloum, S. A. (2020). The impacts of communication ethics on workplace decision making and productivity. In *International conference on advanced intelligent systems and informatics* (pp. 488–500). Springer.

Alzoubi, H. M., Alshurideh, M., Al Kurdi, B., & Inairat, M. (2020). Do perceived service value, quality, price fairness and service recovery shape customer satisfaction and delight? A practical study in the service telecommunication context. *Uncertain Supply Chain Management, 8*(3), 579–588.

Assad, N. F., & Alshurideh, M. T. (2020a). Financial reporting quality, audit quality, and investment efficiency: Evidence from GCC economies. *WAFFEN-UND Kostumkd. J, 11*(3), 194–208.

Assad, N. F., & Alshurideh, M. T. (2020b). Investment in context of financial reporting quality: A systematic review. *WAFFEN-UND Kostumkd. J, 11*(3), 255–286.

Bello, K., Jusoh, A., & Nor, K. (2016). Corporate social responsibility and consumer rights awareness: A research agenda. *Indian Journal of Science and Technology, 9*(46), 3–9.

Bolton, L., & Mattila, B. (2015). How does corporate social responsibility affect consumer response to service failure in buyer-seller relationships? *Journal of Retailing, 91*(1), 140–153.

Chen, X., (2016). Bridging corporate social responsibility and consumers' corporate brand evaluations—understanding consumers' psychological processes. *Journal of research of consumer, 31.*

Ellen, P., Webb, D., & Mohr, L. (2006). Building corporate associations: Consumer attributions for corporate socially responsible programs. *Journal of the Academy of Marketing Science, 34*(2), 147–157.

Gleim, M., (2011). From green to gold: three essays on corporate social responsibility as a marketing strategy. 3–15.

Golob, U., Lah, M., & Jancic, Z. (2008). Value orientations and consumer expectations of corporate social responsibility. *Journal of Marketing Communications., 14*(2), 83–96.

Jin, Y., Park, S., & Yoo, J. (2017). Effects of corporate social responsibility on consumer credibility perception and attitude toward luxury brands. *Social Behavior and Personality, 45*(5), 795–808.

Khasawneh, M. A., Abuhashesh, M., Ahmad, A., Masa'deh, R., & Alshurideh, M. T. (2021a). Customers online engagement with social media influencers' content related to COVID 19. In *The effect of coronavirus disease (COVID-19) on business intelligence* (pp. 385–404). Springer.

Khasawneh, M. A., Abuhashesh, M., Ahmad, A., Alshurideh, M. T., & Masa'deh, R. (2021b). Determinants of e-word of mouth on social media during COVID-19 outbreaks: An empirical study. In *The effect of coronavirus disease (COVID-19) on business intelligence* (pp. 347–366). Springer.

Kurdi, B. A., Alshurideh, M., Nuseir, M., Aburayya, A., & Salloum, S. A. (2021). The effects of subjective norm on the intention to use social media networks: An exploratory study using PLS-SEM and machine learning approach. In *International conference on advanced machine learning technologies and applications* (pp. 581–592). Springer.

Maignan, I., & Ferrell, O. (2004). Corporate social responsibility and marketing: An integrative framework. *Journal of the Academy of Marketing Science, 32*(1), 3–30.

Mohammad, A. A., Alshura, M. S., Al-Hawary, S. I. S., Al-Syasneh, M. S., & Alhajri, T. M. (2020). The influence of Internal Marketing Practices on the employees' intention to leave: A study of the private hospitals in Jordan. *International Journal of Advanced Science and Technology, 29*(5), 1174–1189.

Nussbaum, A. (2009). Ethical corporate social responsibility (CSR) and the pharmaceutical industry: A happy couple? *Journal of Medical Marketing, 9*(1), 67–76.

Parsa, H., Lord, K., Putrevu, S., & Kreeger, J. (2015). Corporate social and environmental responsibility in services: Will consumers pay for it? *Journal of Retailing and Consumer Services, 22,* 250–260.

Peloza, J., & Shang, J. (2011). How can corporate social responsibility activities create value for stakeholders? A systematic review. *Academy of Marketing Science, 39,* 117–135.

Rampal, M., & Bawa, A. (2008). Corporate philanthropy: A study of consumer perceptions. *The Journal of Business Perspective, 12*(2), 24–33.

Schwartz, M. S., & Carroll, A. B. (2018). Corporate social responsibility: A three-domain approach. *Business Ethics Quarterly, 13*(4), 503–530.

Sen, S., Bhattacharya, C. B., & Korschun, D. (2006). The role of corporate social responsibility in strengthening multiple stakeholder relationships: A field experiment. *Journal of the Academy of Marketing Science., 34*(2), 158–166.

Shamout, M., Elayan, M., Rawashdeh, A., Kurdi, B., & Alshurideh, M. (2022). E-HRM practices and sustainable competitive advantage from HR practitioner's perspective: A mediated moderation analysis. *International Journal of Data and Network Science, 6*(1), 165–178.

Su, L., Pan, Y., & Chen, X. (2017). Corporate social responsibility: Findings from the Chinese hospitality industry. *Journal of Retailing and Consumer Services, 34,* 240–247.

Sun, W., & Price, J. (2016). The impact of environmental uncertainty on increasing customer satisfaction through corporate social responsibility. *European Journal of Marketing, 50*(8), 1209–1238.

Sweiss, N., Obeidat, Z. M., Al-Dweeri, R. M., Mohammad Khalaf Ahmad, A., M. Obeidat, A., & Alshurideh, M. (2021). The moderating role of perceived company effort in mitigating customer misconduct within Online Brand Communities (OBC). *Journal of Marketing Communications,* 1–24.

Vallaster, C. H., Lindgreen, A., & Maon, F. (2012). Strategically leveraging corporate social responsibility: A corporate branding perspective. *California Management Review, 54*(3), 34–60.

Zadek, S. (2006). Responsible competitiveness: Reshaping global markets through responsible business practices. *Corporate Governance: The International Journal of Business in Society, 6*(4), 334–348.

The Impact of Brand Loyalty Determinants on the Tourists' Choice of Five Stars Hotels in Jordan

Reem Abu Qurah, Nida'a Al-Husban, Anber Abraheem Shlash Mohammad, Faraj Mazyed Faraj Aldaihani, Sulieman Ibraheem Shelash Al-Hawary, Riad Ahmad Mohammed Abazeed, Ibrahim Rashed Soliaman AlTaweel, Muhammad Turki Alshurideh ⓘ, and Barween Al Kurdi ⓘ

Abstract This Study aims to investigate the determinants of brand loyalty (Customer satisfaction, Price sensitivity and word of mouth) affecting tourists' choice of five stars hotels on Jordan. The population of the study consisted of people who frequent select these hotels to stay. The researcher study data were collected using the non-random sampling method, where (400) questionnaires were distributed to

R. A. Qurah · A. A. S. Mohammad
Marketing Department, Faculty of Administrative and Financial Sciences, Petra University, Jordan, P.O. Box 961343, Amman 11196, Jordan

N. Al-Husban · S. I. S. Al-Hawary (✉)
Department of Business Administration, Faculty of Economics and Administrative Sciences, Al Al-Bayt University, P.O. BOX 130040, Mafraq 25113, Jordan
e-mail: dr_sliman73@aabu.edu.jo; dr_sliman@yahoo.com

F. M. F. Aldaihani
Kuwait Civil Aviation, Ishbiliyah Block 1, Street 122, Home 1, Kuwait, Kuwait

R. A. M. Abazeed
Business Management and Public Administration, Department of Business Administration, Faculty of Finance and Business Administration, Al Al-Bayt University, P.O.BOX 130040, Mafraq 25113, Jordan

I. R. S. AlTaweel
Department of Business Administration, Faculty of Business School, Qussim University, P.O.BOX 6502, Al Russ City 51452, Saudi Arabia
e-mail: toiel@qu.edu.sa

M. T. Alshurideh
Department of Marketing, School of Business, The University of Jordan, Amman 11942, Jordan
e-mail: m.alshurideh@ju.edu.jo; malshurideh@sharjah.ac.ae

Department of Management, College of Business, University of Sharjah, 27272 Sharjah, United Arab Emirates

B. Al Kurdi
Department of Marketing, Faculty of Economics and Administrative Sciences, The Hashemite University, Zarqa, Jordan
e-mail: barween@hu.edu.jo

© The Author(s), under exclusive license to Springer Nature Switzerland AG 2023   2193
M. Alshurideh et al. (eds.), *The Effect of Information Technology on Business and Marketing Intelligence Systems*, Studies in Computational Intelligence 1056, https://doi.org/10.1007/978-3-031-12382-5_120

people who visited five-star hotels in Amman, Aqaba and the Dead Sea. To achieve the objective of the study, and to test hypotheses, the researcher used statistical program of social Sciences (SPSS) and (AMOS). The study results revealed that there is a statistical significant impact of brand loyalty determinants on tourists' choice in five stars hotel in Jordan, based on the study results the researchers recommends managers and decision makers should provide a superior service quality also training the staff and every person who would contact the tourists, finally they should undertake promotional and informational marketing activities. Thus, the guest has wonderful experience which will ultimately lead customer to be loyalty to the brand and repeat their visits to the same hotel.

Keywords Brand loyalty (BL) · Customer satisfaction (SC) · Price sensitivity (PS) · Word of mouth (WOM) · Tourists choice · Five stars hotels · Amman · Aqaba and Dead Sea—Jordan

1 Introduction

The hotel industry is growing very fast in the whole world. Competition is increasing among the hotel operators. Therefore, managers are concerned about increasing the profitability of hotels. Earlier, hotel operators were only interested in attracting new customers but now the trend is changing. Now, retaining an existing customer is more important than attracting new customers (Alshurideh et al., 2020; Alzoubi et al., 2022). Therefore, managers are concerned with making existing customers loyal to the hotel and they want to create repurchase intent in the customers.

The loyalty and the impacts on the customer choice have attracted interest in many empirical as well as exploratory investigations (Alshurideh, 2019, 2022). Researchers have created several constructs to refer to as well as look into the part of impact as well as loyalty incidents working with an assortment of conceptual frameworks (Alshurideh et al., 2012, 2017a, 2017b). Marketing research as well as marketing methods has compensated growing focus on procedures related to creating a good connection between consumer and brand (Aljumah et al., 2021; Hamadneh et al., 2021; Sweiss et al., 2021). It's frequently believed the brand would be the best asset of every business (Abu Zayyad et al., 2021; Tariq et al., 2022; Vukasovic, 2015). Therefore, this study is dedicated to investigate the impact of brand loyalty determinants on the tourists' choice of five stars hotels in Jordan.

2   Theoretical Framework and Hypotheses Development

2.1   The Brand Loyalty Concept

Brand is a name, sign, term, symbol, design or signaling combination that is intended to identify the products and services (Kotler & Armstrong, 2010). American Marketing Association (AMA, 2011) defined brand as "A brand is a "name, sign, term, symbol, or design, or a combination of them, intended to identify the products and services of one seller or group of sellers and to differentiate them from their competitor"." According to Tumewu et al. (2017) identified the brand as a factor for helping the customer by making data about the quality of the product; a customer who always purchases a product with specific brands knows well that these products will have, advantages, characteristics and qualities. Reviewing the literature, Kotler and Keller (2015), defined that a brand is "anything that identifies a seller's goods or services and distinguishes them from others". However, brand recognized as the apparent quality about the brand's segment, the express or the deceptive (Al-Hawary & Harahsheh, 2014, Alwan & Alshurideh, 2022).

Generally, the brand is a gathering of behaviors, the brand's customers and individuals from the association's channels which have the brand, empower the firm to has unendingly and competitive advantage, gets the brand's an incentive from its origins which comprises of the brand's notoriety, its apparent quality, picture, and its guess to the purchasers (Al-Hawary, 2013a, 2013b; Alshurideh et al., 2015).

American Marketing Association (AMA, 2011) defined brand loyalty as "the situation in which a consumer generally buys the same manufacturer-originated product or service repeatedly over time rather than buying from multiple suppliers within the category". Brand loyalty is also defined as the degree of consumer's attachment to a specific brand (Lee & Kang, 2012). However, Ali and Muqadas (2015) defined brand loyalty as a concept related to re-purchase intention and psychological commitment while Dehdashti et al. (2012) defined brand loyalty as a long-term relationship grounded on a customer's re-purchase intention and continuous commitment towards a brand. In this regard Wulandari (2016) adopted Oliver's (2010) definition of brand loyalty in which the concept identified as a variable with two sub-factors: a customer commitment to re-purchase and to re-patronize a good or a service. Moreover, Bozok et al. (2015) defined brand loyalty dependent on the behavior methodology accentuate the real addiction coming from the buy inclinations of customers, the definitions based on the attitudinal viewpoint center around the loyalty of consumer requests to the brand.

A review of the literature brand loyalty, as one of brand value basic dimensions, is defined as "the degree to which consumers buy a brand of product." In other words, it is the shoppers' inclination to continually pick a particular brand among restricting brands and dismissing the others. Most of studies that focused on brand loyalty, which is an important element of brand equation, take some concepts into consideration such as the brand customer value, the eagerness of consumers to purchase the brand, the price, and the attractiveness of the brand in the consumers eyes (Kautish, 2010).

Moreover, Rather (2018) reported that brand loyalty represents customer's positive attitude towards a brand or offering, in addition to repeat buying behaviour. Liu et al. (2012) find out that affective loyalty is positively related to attitudes toward co-branded products.

Previous research suggests that there are two dimensions of brand loyalty: attitude and behaviour (Chiu et al., 2013). Attitudinal loyalty is defined as the internal and psychological feeling of a consumer, such as liking, emotion, and obligation to a specific brand in the hope to engage continued buying without showing the actual purchase action (Yasmin & Raju, 2020). The attitude-based method is interested in discovering the influencing factors on purchase behavior, preference, commitment, word of mouth recommendation, and customer willingness to pay a higher price are how attitudinal loyalty is commonly measured (Shahzad et al., 2020).

Moreover, attitudinal loyalty is also seen as the extent of the client's psychological attachments, and the ability to help the business (Forgas et al., 2010). Accordingly, loyalty to attitudes requires strong word of mouth, the ability to refer to others and inspire them to use the products and services of an organization (Becker et al., 2016). Tomalieh (2015) pointed out that to increase sales and revenue, a company has to take behavioural loyalty into account since it is one of the core outcomes of the customer-company relationship. It is important to know this impact because attitudinal and behavioural loyalty has different impacts on brand performance or the firm performance (Farrukh et al., 2020).

2.2 The Tourists' Choice

It is a crucial of understanding why consumers will in general choose a specific destination and what kind of components impacts those (Wu & Lee, 2011). According to Kumar and Singh (2014) as contrast with the more youthful more established consumers are more loyal the brand where they use to stay while making a trip to a particular destination, as they are more conscious about the services of these hotel brands and it is very hard to change this discernment.

Moreover, Mueller and Szolnoki (2010), Almli et al. (2011) and Rahman reported that consumers are shaping quality desires based on quality signals during the buy choice cycle, henceforth extraneous attributes (for brand name, instance price, country of origin, etc.) are liable for buy choice, while natural attributes will influence preferring an item or service and influence brand loyalty. A generous assortment of consumer research affirms that extraneous item signals, for example, branding and packaging impact the buyers' item assessment on explicit item bunch like food, articles of clothing, etc.

Furthermore, Baruca and Čivre (2012) defined selection between alternatives important to consumers because they have to choose between two or more alternatives; otherwise, they are forced to buy a particular product or service. Consumers seek to obtain information about products and services and use them as part of logical problem solving in the decision-making process. Traditionally, these purchases required brochures and contacts with a travel agent, however, today all the required information can be found on the Internet and consumers can compare the information relatively easily. Although before the final purchase decision, consumers evaluate various alternatives based on information received. The evaluation process becomes particularly difficult when the product is a hotel residence, because the hospitality service is intangible in nature. If a hotel is chosen, this information will depend on the characteristics of the product or service, quality, price, facilities, location, reputation, etc. When making a real buying decision, consumers weigh which of these traits gives them greater relevance and concern related to their personal values, needs and preferences. Stylos et al. (2016) destination choice includes choosing a particular place by assessing guests' emotions about another alternatives.

Kozak and Baloglu (2010) Consumer discernments may impact the choice of a location, the utilization of products and services while in holiday, and the choice to return. Since consistently, guests are offered a more noteworthy assortment of destinations, more choice of convenience, a more extensive scope of activities, and visits which are intended for explicit interests, it has now gotten genuinely hard for a person to choose where and how to go and where to remain, accordingly, the utilization of choice sets has gotten fundamental in buyer dynamic about the travel industry, friendliness, and recreation administrations.

Kim et al. (2012) explained the consideration of competitiveness of destination in various choice sets could also be significant in understanding the brand value of destination, including brand awareness, brand image, brand identity, brand loyalty.

2.3 Determinants of Brand Loyalty that Affect the Tourist's Choice

The researchers examined the factors affecting brand loyalty from different perspectives, shown in the following table.

Based on the above-reported determinants, three determinants were chosen for the purpose of the current study. These determinants are customer satisfaction, price sensitivity and word of mouth. The following paragraphs of the current section are dedicated to clarify these determinants.

Customer Satisfaction: According to Jana and Chandra (2016), customer satisfaction has become an important point in the hotel industry, unlike other industries; the hotel industry thrives due to customer retention. A hotel can only retain its customers through customer satisfaction. Customer loyalty depends mainly on the quality of services they receive at the hotel. Moreover, the hospitality industry has experienced the entry of many investors wanting to meet their customers' needs. Therefore, guests will look for better services if they were not satisfied with the services they receive from the hotel they lodge in (Jana & Chandra, 2016). Moreover, satisfaction has been defined as "satisfaction is a consumer's fulfilment response. It is a judgment that a product/service feature, or the product or service itself, provided a pleasurable level of consumption related loyalty, including level of under or over loyalty" (Oliver, 2010).

Moreover, Bianchi et al (2014), stated that higher satisfaction would result from trust on the brand. Therefore, customer satisfaction with said brand is a primary driver of loyalty. According to Han et al. (2011), customer satisfaction depends on the feeling about product compares with expectations of customers. If satisfaction is lower than expectation, the buyer is not satisfied; if expectation is appropriate, customer will be satisfaction. To measure customer satisfaction with the product/service, it is based on a comparison between the products/services feeling and service expectations.

Subagio and Dimyati stated the customer satisfaction is linked with customers' expectations concerning products or services perceived performance. Gaskari and Naeimavi examine customer satisfaction between the perceptual experiences of the overall performance of a product with all the expectations. Adhitama et al. (2017) stated that customer satisfaction of a product might be described through looking at the expectations of service with his knowledge of the services provided. If provided services exceed customer expectations, they're known as service that is outstanding. Based on above literature the hypotheses can be as:

H1: Customer Satisfaction has a significant impact on tourists' choice of five stars hotel in Jordan

Price Sensitivity: Kotler and Armstrong (2010) define price as "the amount of money that is collected for a product or service". Levy and Weitz (2012) defined price as the money that customers exchange in terms of service or product, or the value they receive (Kotler & Armstrong, 2010). It is important for the company to retain loyal customers who are willing to pay higher prices for their favourite brand and not buy on a low price basis (Alzoubi et al., 2020; Alshurideh et al., 2021). Brand loyalty customers are eager to pay more money and are less price sensitive.

According to Malc et al. (2016) examined the effects of a consumer's perception of price fairness on his buying decision and referred to it as an appropriate indicator of a consumer's buying decision. Also, Khuraim stated that price has a significant effect on brand loyalty which in turn affects the buying intent of the customers. Consequently, and according to the discussions. Moreover, Beristain and Zorrilla (2011) proof affirms that price is a significant determinants of customer loyalty to a specific brand and the customer loyalty to a brand is significantly affected by the way that shoppers see the purchase of these brands as value for the money, and the higher price sensitivity affectability of customer intensity of brand loyalty. Based on above literature the hypotheses can be as:

H2: Price Sensitivity has a significant impact on tourists' choice of five stars hotel in Jordan

Word of Mouth: Customer WOM is arguably the most important outcome of customer–firm relationships as a relational outcome in a business context. Customers consider verbal information reliable because the company cannot tamper with it, so it attracts the interest of potential clients, as a reliable source of information, due to its personalized transmission, content and context that reflects personal experiences. Scholars citing the importance of word of mouth acknowledge that it serves as peers' advice, which can influence consumer choice, as well as product evaluations and purchase decisions (Zhang et al., 2010).

Whether intentional or not, if WOM receives positive, communication between customer can go beyond sharing direct experiences, to become storytelling and folklore, then narrative discourse about the company shapes the "truth" about the company in customers' minds. WOM communication appears reliable and impartial because the sender of the information usually does not earn it if the recipient decides to make a purchase (Cheung and Thadani, 2012). Furthermore, customer to customer interactions is increasingly important, which has considerable implications for firms; the exchange of information about customers' experiences influences each party's attitudes and behaviours (Blazevic et al., 2013). Through their extensive meta-analytic review of antecedents of WOM communication illustrate the important roles of satisfaction, loyalty, quality, commitment, trust and perceived value. WOM connections are also co-produced in consumer networks. However, customers who interact with a brand and/or service provider are expected to build positive attitudes more immediately than customers who do not engage with the brand or provider (Harrigan et al., 2016; Kurdi et al., 2020; So et al., 2014). Such attitudes are expected to be favorable which may lead to increased loyalty and/or favouritism intent (Al Kurdi et al., 2020; Hollebeek et al., 2016).

Kher et al., WOM provided by customers, online or offline, positive or negative, is closely reviewed by potential clients, making it a powerful marketing tool, such that trust in customer relationships and engagement with the company is a necessary strategic focus to companies. Loyal customer often act as trusted advocates through social media (e-WOM), informally linking networks of friends, relatives, and other potential clients with the organization (Khasawneh et al., 2021). Based on above literature the hypotheses can be as:

H3: Word of Mouth has a significant impact on tourists' choice of five stars hotel in Jordan

3 Study Model

Figure 1 shows the conceptual model of the study, in which three independent variables' (i.e., Brand Loyalty determinants: Customer Satisfaction, Price Sensitivity and Word of Mouth) were assumed to have significant impact on tourists choice. Studies that used to develop the model are shown in Table 1.

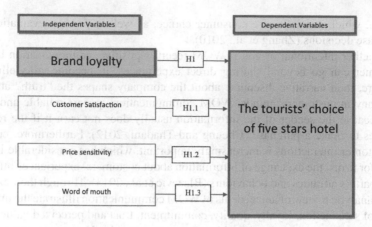

Fig. 1 Study conceptual model. *Source* This model has been prepared by the researcher based on these sources

Table 1 Examples of brand loyalty determinants on the choice

Factors	Authors
Service quality, perceived value, customer satisfaction, and brand image	Suhartanto (2011)
Service quality, perceived value, and customer satisfaction	Al-Hawary (2013a, 2013b)
Perceived quality, satisfaction and trust	Al-Hawary (2013a, 2013b)
Customer satisfaction, perceived quality, price, perceived value	Al-Hawary and Harahsheh (2014)
Brand name, brand price and brand quality	Devi (2014)
Price, perceived quality, perceived value, satisfaction	Al-Hawary and Harahsheh (2014)
Trust, price, customer satisfaction and perceived value	Fazal and Kanwal
Price and quality perception	Alić et al. (2019)
Brand equity, brand reputation, brand trust and word of mouth	Kumar et al. (2020)

4 Study Methodology

4.1 Study Population

The current study focuses on an important sector in the Jordanian economy, which is the luxury hotel industry in Amman, Aqaba, and the Dead Sea. The study population consists of people over 20 years old who frequent select these hotels to stay and eat, whether they are families, business owners or employees that require their job is to visit hotels to perform their duties. Since the large size of the study population and the inability to conduct a complete census, the study data were collected using the non-random sampling method, where (400) questionnaires were distributed to people who visited five-star hotels in Amman, Aqaba and the Dead Sea. The questionnaires were distributed in both languages' Arabic and English due to some of the hotel's visitors are not realizing Arabic. Moreover, the questionnaires were distributed on two ways (150) paper questionnaires and (250) electronic questionnaires through Google forms. The retrieved questionnaires were (367) questionnaires, while it was found that (15) questionnaires were incomplete and not suitable for statistical analysis. Therefore, (352) questionnaires were used for statistical analysis, which represents a response rate (88%).

4.2 Study Instrument

The questionnaire was designed as a basic instrument to accomplish the study's purposes for identifying the impact of brand loyalty determinants on tourists' choice of five stars hotels in Jordan. This questionnaire consisted of three parts: the first part was composed of the introductory and questions related to demographic variables represented by (gender, age, educational level, and nationality). The second part was devoted to measuring the independent variable represented by brand loyalty determinants, which was developed by (Al-Hawary & Al-Khazaleh, 2020; Al-Hawary & Al-Menhaly, 2016; Al-Hawary & Al-Smeran, 2017; Al-Hawary & Alhajri, 2020; Al-Hawary & Hussien, 2017; Al-Hawary & Metabis, 2012; Al-Hawary & Obiadat, 2021; Al-Hawary et al., 2011, 2017; Alić et al., 2019; Alolayyan et al., 2018; Alshurideh et al., 2017a, 2017b; Cetin & Dincer, 2013; Metabis & Al-Hawary, 2013), where this variable divided into customer satisfaction measured through items (Q1–Q11), price sensitivity which was measured by items (Q12–Q16), and word of mouth which was measured by items (Q17–Q21). While the third part used to measure the dependent variable, which is the tourists' choice, where this variable's items were developed by (Unurlu & Uca, 2017), which was measured by items (Q22–Q25) used for this purpose.

4.3 Validity and Reliability

For identifying the validity and reliability of the study instrument, a set of methods were used that began to identify the face validity by presenting the questionnaire to a number of academic doctoral of various ranks and those with specialization in the field of study for determining the degree of phrases suitability to measure the study variables and linguistic validation in order to be able respondents easily understood it, where their names, academic ranks, and their work place listed in appendix (1). In addition, exploratory factor analysis (EFA) was used to test the convergent validity of the study instrument through the values of the average variance extracted (AVE), as well as the discriminant validity test by comparing the values of the average variance extracted (AVE) with both maximum shared variance (MSV) and square root of average variance extracted ($\sqrt{}$AVE). The reliability of the study instrument was also confirmed by using Cronbach Alpha coefficient (α) to ensure internal consistency and McDonald's Omega coefficient (ω) to ensure the composite reliability. The results of these tests demonstrated in Table 2.

The results listed in Table 2 showed that the values of the items' loadings on the variables associated with them ranged between (0.704–0.814) which were higher than the minimum threshold for retaining the item represented by the value of 0.50 (Haig, 2018). The results also showed that the values of the average variance extracted (AVE) were higher than the minimum value for accepted of 0.50, where it was within the range (0.548–0.613) that is an indication of the convergent validity (Antunes et al., 2017). Moreover, the values of average variance extracted (AVE) of the latent

Table 2 Exploratory factor analysis results

Variables	Items	Loadings	AVE	MSV	$\sqrt{}$AVE	ω	α
Customer satisfaction	Q1	0.814	0.613	0.530	0.746	0.934	0.945
	Q2	0.734					
	Q3	0.782					
	Q4	0.775					
	Q5	0.786					
	Q6	0.779					
	Q7	0.776					
	Q8	0.784					
	Q9	0.787					
	Q10	0.791					
	Q11	0.789					
Price sensitivity	Q12	0.774	0.548	0.365	0.755	0.858	0.859
	Q13	0.729					
	Q14	0.704					
	Q15	0.759					
	Q16	0.733					
Word of mouth	Q17	0.767	0.574	0.327	0.776	0.876	0.871
	Q18	0.764					
	Q19	0.751					
	Q20	0721					
	Q21	0.785					
Tourists' choice	Q22	0.745	0.586	0.530	0.738	0.850	0.848
	Q23	0.786					
	Q24	0.738					
	Q25	0.791					

variables were greater than the maximum shared variance (MSV), as well as the values of square root average variance extracted ($\sqrt{AVE}$) were higher than the values of the correlation with the rest of the latent variables, where these are indicators of achieving discriminant validity (Crego & Widiger, 2016).

Regarding to reliability of the study instrument, the results presented in Table 2 indicate that the values of Cronbach Alpha coefficient (α) were within the range (0.848–0.945) which were higher than 0.70, which is the minimum value for realizing the internal consistency of the study instrument (Heale & Twycross, 2015). Further, the results stated that the values of McDonald's Omega coefficient (ω) that were ranged (0.850–0.934), thus it is greater than 0.80 the minimum limit for considering the instrument having a composite reliability (Zhang & Yuan, 2016).

Together with exploratory factor analysis (EFA), confirmatory factor analysis (CFA) was used to test the structural validity of the study model used through a bundle of indicators represented by the chi-square ratio (CMIN/DF), goodness of fit index (GFI), comparative fit index (CFI), Tucker-Lewis coefficient (TLI), and root mean square error of approximation (RMSEA). Figure 2 shows the results obtained by confirmatory factor analysis.

The results in Fig. 1 showed that the chi-square ratio (CMIN/DF) was (1,574) that was greater than 3 that expresses the high acceptable value for this indicator, and that the value of each of goodness of fit index (GFI), comparative fit index (CFI), and Tucker-Lewis coefficient (TLI) was higher than the acceptable minimum limit of 0.90. Moreover, the results illustrated that the value of root mean square error of approximation (RMSEA) reached (0.040) which is smaller than the acceptable upper threshold of 0.050, thus the study model has a structural validity (Gu et al., 2017).

5 Testing Hypotheses

Structural equation modelling is one of the statistical methods used to analyze the structural relationship between measured variables and latent structures, as well as to testing studies hypotheses. This statistical method was used for testing the current study hypotheses, where the Fig. 3 illustrates the structural model used to test the impact of brand loyalty determinants on tourists' choice of five stars hotels in Jordan.

Moreover, Table 3 shows the results of the main study hypothesis test that was "There is no statistically significant impact of brand loyalty on tourists' choice of five stars hotel in Jordan".

The results in Table 3 indicate that the value of the correlation coefficient (R) between brand loyalty determinants and the tourists' choice (0.653), and the value of the determination coefficient (R^2) that was (0.427), this value indicates that (42.7%) of the variance in the tourists' choice was due to the change in brand loyalty determinants of five stars hotel in Jordan. Also, the probability value (P) was less than 0.05, thus the null hypothesis was rejected and the alternative hypothesis that referred to "There is a statistically significant impact of brand loyalty determinants on tourists'

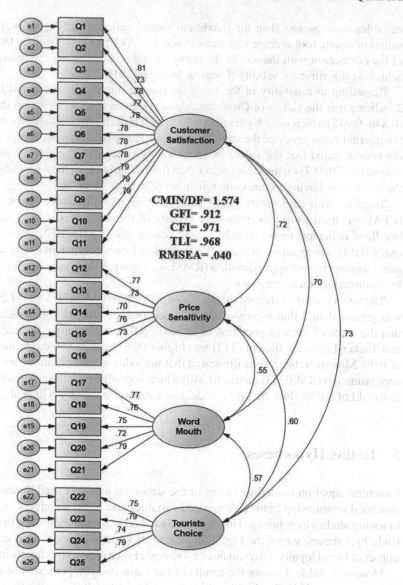

Fig. 2 Result of confirmatory factor analysis for the measurement model

choice of five stars hotel in Jordan" was accepted, and the standard impact value (β) was (0.653).

In addition to testing the main hypothesis, the sub-hypotheses of the study were tested that were referring to "There is no statistically significant impact of brand loyalty determinants dimensions on tourists' choice of five stars hotel in Jordan" (Fig. 4). Figure 3 shows the structural model used for that.

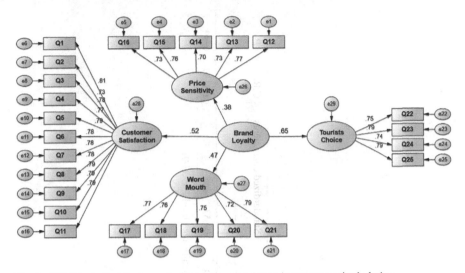

Fig. 3 SEM for testing the impact of brand loyalty determinants on tourists' choice

Table 4 shows the results obtained using the structural equation modelling (SEM) method to test the impact of customer satisfaction, price sensitivity, and word of mouth on the tourists' choice of five-star hotels in Jordan.

The results in Table 4 indicate that the value of the correlation coefficient (R) between brand loyalty determinants dimensions and the tourists' choice (0.672), and the value of the determination coefficient (R^2) that was (0.452), this value indicates that (45.2%) of the variance in the tourists' choice was due to the change in the dimensions of brand loyalty determinants of five stars hotel in Jordan. Moreover, the probability values (P) of all dimensions of brand loyalty determinants were less than 0.05, thus the null sub-hypotheses were rejected and the alternative hypotheses that referred to "There is a statistically significant impact of brand loyalty determinants dimensions on tourists' choice of five stars hotel in Jordan" was accepted, where the first rank was for the dimension of customer satisfaction with standard impact value (β) which was (0.486), followed by the dimension of price sensitivity with standard impact value (β) that was (0.147), while the last rank for the dimension of word of mouth with standard impact value (β) which was (0.118).

6 Discussion and Conclusion

The aim of this study was to investigate the impact of brand loyalty determinants on the tourists' choice of five stars hotel in Jordan. The independent variable (brand loyalty) was measured by using three determinants variables, i.e., customer satisfaction, price sensitivity and word of mouth. It was hypothesized that there is a significant positive impact of brand loyalty on the tourists' choice. More specifically, the three

Table 3 Result of testing the impact of brand loyalty determinants on tourists' choice

Path			R	R²	Unstandardized coefficients		Standardized coefficients	C.R.	P
					Estimate	S.E.	β		
Brand loyalty determinants	→	Tourists' choice	0.653	0.427	0.812	0.050	0.653	16.137	***

Note * Refers to significance level $P < 0.05$, ** Refers to significance level $P < 0.01$, *** Refers to significance level $P < 0.001$

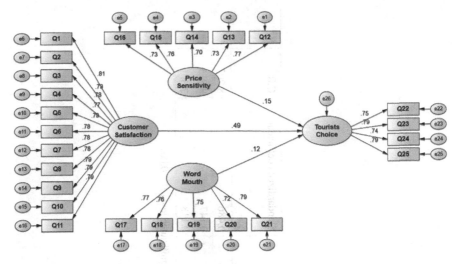

Fig. 4 SEM used to test the impact of brand loyalty determinants dimensions on tourists' choice

determinants of the brand loyalty were assumed to have significant positive impacts on the tourists' choice. On the basis of the results reached by the current study, it was concluded that brand loyalty as measured by the determinants of brand loyalty: customer satisfaction, price sensitivity and word of mouth has a significant impact on tourists' choice. Price sensitivity is highest determinant in the tourist choice process. When it comes to the quality of the service the hotel provide which should satisfy the guest of the hotel and meet the guest expectations that will drive the guest to spread word of mouth positively in front of others.

However, numerous studies have been accepted the hypothesis that there is a significant and positive impact of brand loyalty on tourists' choice. Bozok et al. (2015) examined the dimension of brand equity in choosing a hotel and found an impact of brand loyalty on customer choice. Alhedhaif et al. indicate that Saudi consumers of cosmetics product have significant brand loyalty towards their choice. Malik et al. (2012) analysed that brand loyalty have positive association with Intention to buy.

The results of the study showed that the customer satisfaction has a significant and positive impact on tourists' choice. The participants assessed customer satisfaction at a medium level. This means the services that the hotel provides moderately meets customer expectation. This moderate degree of customer satisfaction revealed in the current study can be explained by looking at the measure customer satisfaction with the service, it is based on a comparison between the services feeling and service expectations. In other words, the quality that the customer gets from the service. Suhartanto (2011) indicates that, satisfied guests will not automatically become behaviourally loyal (i.e., resettle in the future) unless they realize the hotel performs better on key features than other hotels. Tourist satisfaction positively affects the intention to return to place, local authorities may implement policies

Table 4 Result of testing the impact of brand loyalty determinants dimensions on tourists' choice

Path			R	R^2	Unstandardized coefficients		Standardized coefficients	C.R.	P
					Estimate	S.E.	β		
Customer Satisfaction	→	Tourists' Choice	0.672	0.452	0.507	0.062	0.486	8.207	***
Price Sensitivity	→	Tourists' Choice			0.159	0.057	0.147	2.815	**
Word of Mouth	→	Tourists' Choice			0.123	0.053	0.118	2.301	**

Note * Refers to significance level $P < 0.05$, ** Refers to significance level $P < 0.01$, *** Refers to significance level $P < 0.001$

to control the perceived quality of the tourism experience, improve services and security, and compare destinations in terms of performance.

In a study conducted by Souri to determine the factors that involved in the customer satisfaction that leads to competitive advantages and finally customer loyalty and repeat purchase. According to Liata et al. Customers who are satisfied with the services provided by the hotel are very likely to be satisfied with the services experienced, which in turn will create loyalty among them.

On the basis of the data used in the current study, it was acknowledged that there was a significant positive impact of price sensitivity on tourists' choice. Price sensitivity studied in this study was evaluated by respondents was fairly evaluated. This may be due to the fact that in a consumer's mind, a high price is often associated with high quality. Price and quality is a reason to prefer a certain provider.

Concerning the significant impact of price sensitivity on tourists' choice, numerous previous studies confirmed this result. In a study conducted by Madupu and Cooley brand loyalty related to a consumer's brand attitude, as it reflects a consumer's commitment to the brand and their likelihood to repeatedly purchase and pay high prices to stay with the brand. In the same context, consumers' loyalty like intention to revisit and willingness to pay high prices are developed through their long-term interactions with the brand.

The results of the study showed that word of mouth has a significant positive impact on tourists' choice. Word of mouth as a determinant of brand loyalty was moderately rated by respondents. Brand loyalty described by the consumer behaviour of spreading positive word-of-mouth. A consumer's intention to advocate for the brand and spread positive word-of-mouth was found to sustain the consumer's brand loyalty and potentially to help the brand expand its market share Madupu and Cooley. Word of mouth it serves as peers' advice, which can influence consumer decision-making, as well as product evaluations and purchase decisions Zhang et al. (2010). Moreover, word of mouth is a type of direct reference for consumers, and its effect on brand attributes, on consumers' perceptions and attitudes when consumers select tourism destinations.

7 Recommendations

In light of its findings, on the basis of the results reached by the current study, it was concluded that brand loyalty as measured by satisfaction, price and word of mouth has a significant impact on tourists' choice; the researcher recommends five-star hotel managers to Provide a superior service quality with emphasizing on the staff ability to increase guest confidence, serve at the right time, and understanding the individual needs of the hotel guests. Undertake promotional and informational marketing activities via advertising tools and integrated marketing communication efforts. And Train the staff and every person who would contact the tourists, so the guest has wonderful experience in culture tourism which may turn into positive word of mouth.

References

Abu Zayyad, H. M., Obeidat, Z. M., Alshurideh, M. T., Abuhashesh, M., Maqableh, M., & Masa'deh, R. E. (2021). Corporate social responsibility and patronage intentions: The mediating effect of brand credibility. *Journal of Marketing Communications, 27*(5), 510–533.

Adhitama, R., Kusumawati, A., & Abdillah, Y. (2017). The influence of service quality on brand image and customer satisfaction in airline service. (Survey of the Garuda Indonesia passenger domiciled in Malang, East Java). *Jurnal Administrasi Bisnis (JAB), 23*(1), 122–130.

Al Kurdi, B., Alshurideh, M., & Al afaishat, T. (2020). Employee retention and organizational performance: Evidence from banking industry. *Management Science Letters, 10*(16), 3981–3990.

Al-Hawary, S. I. S. (2013a). The role of perceived quality and satisfaction in explaining customer brand loyalty: Mobile phone service in Jordan. *International Journal of Business Innovation and Research, 7*(4), 393–413.

Al-Hawary, S. I. S. (2013b). The roles of perceived quality, trust, and satisfaction in predicting brand loyalty: The empirical research on automobile brands in Jordan market. *International Journal of Business Excellence (IJBEX), 6*(6), 656–686.

Al-Hawary, S. I. S., & Alhajri, T. M. S. (2020). Effect of electronic customer relationship management on customers' electronic satisfaction of communication companies in Kuwait. *Calitatea, 21*(175), 97–102.

Al-Hawary, S. I., & Al-, S. (2016). The quality of E-government services and its role on achieving beneficiaries satisfaction. *Global Journal of Management and Business Research: A Administration and Management, 16*(11), 1–11.

Al-Hawary, S. I., & Al-, W. (2017). Impact of electronic service quality on customers satisfaction of Islamic banks in Jordan. *International Journal of Academic Research in Accounting, Finance and Management Sciences, 7*(1), 170–188.

Al-Hawary, S. I. S., & Harahsheh, S. A. (2014). Factors affecting Jordanian consumer loyalty toward cellular phone brand. *International Journal of Economics and Business Research (IJEBR), 7*(3), 349–375.

Al-Hawary, S. I., & Hussien, A. J. (2017). The impact of electronic banking services on the customers loyalty of commercial banks in Jordan. *International Journal of Academic Research in Accounting, Finance and Management Sciences, 7*(1), 50–63.

Al-Hawary, S. I., & Metabis, A. (2012). Service quality at Jordanian commercial banks: What do their customers say? *International Journal of Productivity and Quality Management, 10*(3), 307–334.

Al-Hawary, S. I. S., & Obiadat, A. A. (2021). Does mobile marketing affect customer loyalty in Jordan? *International Journal of Business Excellence, 23*(2), 226–250.

Al-Hawary, S. I. S., Alhamali, R. M., & Alghanim, S. A. (2011). Banking service quality provided by commercial banks and customer satisfaction. *American Journal of Scientific Research, 27*, 68–83.

Al-Hawary, S. I., Batayneh, A. M., Mohammad, A. A., & Alsarahni, A. H. (2017). Supply chain flexibility aspects and their impact on customers satisfaction of pharmaceutical industry in Jordan. *International Journal of Business Performance and Supply Chain Modelling, 9*(4), 326–343.

Al-Hawary, S. I., & Al-Khazaleh, A. M. (2020). The mediating role of corporate image on the relationship between corporate social responsibility and customer retention. *Test Engineering and Management, 83*(516), 29976–29993.

Ali, F., & Muqadas, S. (2015). The impact of brand equity on brand loyalty: The mediating role of customer satisfaction. *Pakistan Journal of Commerce and Social Sciences, 9*(3), 890–915.

Alić, A., Peštek, A., & Merdić, E. (2019). Determinants influencing consumer's loyalty towards a private brand. *Poslovna Izvrsnost Zagreb, 13*(2), 31–51.

Aljumah, A., Nuseir, M. T., & Alshurideh, M. T. (2021). The impact of social media marketing communications on consumer response during the COVID-19: Does the brand equity of a university matter. In *The effect of coronavirus disease (COVID-19) on business intelligence* (pp. 367–384).

Almli, V. L., Verbeke, W., Vanhonacker, F., Naes, T., & Hersleth, M. (2011). General image and attribute perceptions of traditional food in six European Countries. *Food Quality and Preference, 22*(1), 129–138.

Alolayyan, M., Al-Hawary, S. I., Mohammad, A. A., & Al-Nady, B. A. (2018). Banking service quality provided by commercial banks and customer satisfaction. A structural equation modelling approaches. *International Journal of Productivity and Quality Management, 24*(4), 543–565.

Alshurideh, D. M. (2019). Do electronic loyalty programs still drive customer choice and repeat purchase behaviour? *International Journal of Electronic Customer Relationship Management, 12*(1), 40–57.

Alshurideh, M. (2022). Does electronic customer relationship management (E-CRM) affect service quality at private hospitals in Jordan? *Uncertain Supply Chain Management, 10*(2), 1–8.

Alshurideh, M., Nicholson, M., & Xiao, S. (2012). The effect of previous experience on mobile subscribers' repeat purchase behaviour. *European Journal of Social Sciences, 30*(3), 366–376.

Alshurideh, M., Bataineh, A., Alkurdi, B., & Alasmr, N. (2015). Factors affect mobile phone brand choices—Studying the case of Jordan universities students. *International Business Research, 8*(3), 141–155.

Alshraideh, A. T. R., Al-Lozi, M., & Alshurideh, M. T. (2017a). The impact of training strategy on organizational loyalty via the mediating variables of organizational satisfaction and organizational performance: An empirical study on Jordanian agricultural credit corporation staff. *Journal of Social Sciences (COES&RJ-JSS), 6*(2), 383–394.

Alshurideh, M., Al-Hawary, S. I., Batayneh, A. M., Mohammad, A., & Al-Kurdi, B. (2017b). The impact of Islamic banks' service quality perception on Jordanian customers loyalty. *Journal of Management Research, 9*(2), 139–159.

Alshurideh, M., Gasaymeh, A., Ahmed, G., Alzoubi, H., & Kurd, B. (2020). Loyalty program effectiveness: Theoretical reviews and practical proofs. *Uncertain Supply Chain Management, 8*(3), 599–612.

Alshurideh, M. T., Al Kurdi, B., & Salloum, S. A. (2021). The moderation effect of gender on accepting electronic payment technology: A study on United Arab Emirates consumers. *Review of International Business and Strategy, 31*(3), 375–396.

Alwan, M., & Alshurideh, M. (2022). The effect of digital marketing on purchase intention: Moderating effect of brand equity. *International Journal of Data and Network Science, 10*(3), 1–12.

Alzoubi, H. M., Alshurideh, M., Al Kurdi, B., & Inairat, M. (2020). Do perceived service value, quality, price fairness and service recovery shape customer satisfaction and delight? A practical study in the service telecommunication context. *Uncertain Supply Chain Management, 8*(3), 579–588.

Alzoubi, H., Alshurideh, M., Kurdi, B., Akour, I., & Aziz, R. (2022). Does BLE technology contribute towards improving marketing strategies, customers' satisfaction and loyalty? The role of open innovation. *International Journal of Data and Network Science, 6*(2), 449–460.

American Marketing Association. (2011). *AMA Marketing power Dictionary.* Retrieved November 5, 2011, from http://www.marketingpower.com/_layouts/Dictionary.aspx

Antunes, A. C., Caetano, A., & Pina e Cunha, M. (2017). Reliability and construct validity of the Portuguese version of the psychological capital questionnaire. *Psychological Reports, 120*(3), 520–536.

Baruca, P. Z., & Čivre, Z. (2012). How do guests choose a hotel? *Academica Turistica, Year, 5*(1), 75–84.

Becker, S. J., Midoun, M. M., Zeithaml, V. A., Clark, M. A., & Spirito, A. (2016). Dimensions of treatment quality most valued by adolescent substance users and their caregivers. *Professional Psychology: Research and Practice, 47*(2), 120–129.

Beristain, J. J., & Zorrilla, P. (2011). The relationship between store image and store brand equity: A conceptual framework and evidence from hypermarkets. *Journal of Retailing and Consumer Services, 18*(6), 562–574.

Bianchi, C., Drennan, J., & Proud, W. (2014). Antecedents of consumer brand loyalty in the Australian wine industry. *Journal of Wine Research, 25*(2), 91–104.

Blazevic, V., Hammedi, W., Garnefeld, I., Rust, R. T., Keiningham, T., Andreassen, T. W., Donthu, N., & Carl, W. (2013). Beyond traditional word-of-mouth: An expanded model of customer-driven influence. *Journal of Service Management, 24*(3), 294–313.

Bozok, D., Gul, K., Gul, M., & Saatcı, G. (2015). Measurement of brand equity at the facility level in accommodation sector and an application. *3*(1), 1–10.

Cetin, G., & Dincer, F. I. (2013). Influence of customer experience on loyalty and word-of-mouth in hospitality operations. *An International Journal of Tourism and Hospitality Research, 25*(2), 181–194.

Cheung, C., & Thadani, D. (2012). The impact of electronic word-of-mouth communication: A literature analysis and integrative model. *Decision Support Systems, 54*(1), 461–470.

Chiu, C.-M., Fang, Y.-H., Cheng, H.-L., & Yen, C. (2013). On online repurchase intentions: Antecedents and the moderating role of switching cost. *Human Systems Management, 32*(4), 283–296.

Crego, C., & Widiger, T. A. (2016). Convergent and discriminant validity of alternative measures of maladaptive personality traits. *Psychological Assessment, 28*(12), 1561–1575.

Dehdashti, Z., Kenari, M., & Bakhshizadeh, A. (2012). The impact of social identity of brand on brand loyalty development. *Management Science Letters, 2*(4), 1425–1434.

Devi, S. R. (2014). Determinants of brand loyalty of skin care products in rural areas. *International Journal of Exclusive Global Research, 1*(6), 1–11.

Farrukh, M., Meng, F., Sajid, M., & Shahzad, I. (2020). Does strategic fit matter in measuring organizational performance? An empirical analysis. *Corporate Social Responsibility and Environmental Management, 27*(4), 1800–1808.

Forgas, S., Moliner, M., Sánchez-García, J., & Palau-Saumell, R. (2010). Antecedents of airline passenger loyalty: Low-cost versus traditional airlines. *Journal of Air Transport Management, 16*(4), 229–233.

Gu, H., Wen, Z., & Fan, X. (2017). Structural validity of the Machiavellian personality scale: A bifactor exploratory structural equation modeling approach. *Personality and Individual Differences, 105*, 116–123.

Haig, B. D. (2018). Exploratory factor analysis, theory generation, and scientific method. In *Method matters in psychology. Studies in applied philosophy, epistemology and rational ethics* (Vol. 45, pp. 65–88).

Hamadneh, S., Hassan, J., Alshurideh, M., Al Kurdi, B., & Aburayya, A. (2021). The effect of brand personality on consumer self-identity: The moderation effect of cultural orientations among British and Chinese consumers. *Journal of Legal, Ethical and Regulatory Issues, 24*, 1–14.

Han, H., Kim, W., & Hyun, S. S. (2011). Switching intention model development: Role of service performances, customer satisfaction, and switching barriers in the hotel industry. *International Journal of Hospitality Management, 30*(3), 619–629.

Harrigan, P., Evers, U., Miles, M., & Daly, T. (2016). Customer engagement with tourism social media brands. *Tourism Management, 59*, 597–609.

Heale, R., & Twycross, A. (2015). Validity and reliability in quantitative studies. *Evidence Based Nursing, 18*(3), 66–67.

Hollebeek, L. D., Srivastava, R. K., & Chen, T. (2016). SD logic–informed customer brand engagement: Integrative framework, revised fundamental propositions, and application to CRM. *Journal of the Academy of Marketing Science, 47*, 161–185.

Jana, A., & Chandra, B. (2016). Mediating role of customer satisfaction in the mid-market hotels: An empirical analysis. *Indian Journal of Science and Technology, 9*(1), 1–16.

Kautish, P. (2010). Empirical study on influence of extraversion on consumer passion and brand evangelism with word-of-mouth communication. *Review of Economic and Business Studies, 3*(6), 187–198.

Khasawneh, M. A., Abuhashesh, M., Ahmad, A., Alshurideh, M. T., & Masa'deh, R. (2021). Determinants of e-word of mouth on social media during COVID-19 outbreaks: An empirical study. In

The effect of coronavirus disease (COVID-19) on business intelligence (pp. 347–366). Springer, Cham.

Kim, W., Ok, C., & Canter, D. D. (2012). Moderating role of a priori customer-firm relationship in service recovery situations. *The Service Industries Journal, 32*(1), 59–82.

Kotler, P., & Armstrong, G. (2010). *Principles of marketing* (13th ed.). Pearson Education.

Kotler, P., & Keller, K. L. (2015). *Marketing management* (15th ed.). Pearson Education.

Kozak, M., Baloglu, S., & Bahar, O. (2010). Measuring destination competitiveness: Multiple destinations versus multiple nationalities. *Journal of Hospitality Marketing & Management, 19*(1), 56–71.

Kumar, C. P., Senthil, Priya, M. K., Balaji, P. & Rameshkumar, P. M. (2020). Determinants of brand loyalty among female millennial cyber shoppers—An empirical investigation. *International Journal of Advanced Science and Technology, 29*(7), 8861–8871.

Kumar, S., & Singh, D. (2014). Exploring hotel selection motives among Indian tourists. *Indian Journal of Research in Management, Business and Social Sciences IJRMBSS, 2*(1), 82–87.

Kurdi, B., Alshurideh, M., & Alnaser, A. (2020). The impact of employee satisfaction on customer satisfaction: Theoretical and empirical underpinning. *Management Science Letters, 10*(15), 3561–3570.

Lee, H. J., & Kang, M. S. (2012). The effect of brand experience on brand relationship quality. *Academy of Marketing Studies Journal, 16*(1), 87–98.

Levy, M., & Weitz, B. (2012). *Retailing management* (8th ed.). McGraw-Hill.

Liu, F., Li, J., Mizerki, D., & Soh, H. (2012). Self-congruity, brand attitude, and brand loyalty: A study on luxury brands. *European Journal of Marketing, 46*(7/8), 922–937.

Malc, D., Mumel, D., & Pisnik, A. (2016). Exploring price fairness perceptions and their influence on consumer behavior. *Journal of Business Research, 69*(9), 3693–3697.

Malik, P. M., Ghafoor, M. M., & Iqbal, H. K. (2012). Impact of brand image, service quality and price on customer satisfaction in Pakistan telecommunication sector. *International Journal of Business and Social Science, 3*(23), 123–129.

Metabis, A., & Al-Hawary, S. I. (2013). The impact of internal marketing practices on services quality of commercial banks in Jordan. *International Journal of Services and Operations Management, 15*(3), 313–337.

Mueller Loose, S., & Szolnoki, G. (2010). The relative influence of packaging, labelling, branding and sensory attributes on liking and purchase intent: Consumers differ in their responsiveness. *Food Quality and Preference, 21*(7), 774–783.

Oliver, R. L. (2010). *Satisfaction: A behavioral perspective on the consumer* (2nd ed.). Armonk.

Rather, R. A. (2018). Consequences of consumer engagement in service marketing: An empirical exploration. *Journal of Global Marketing, 32*(2), 116–135.

Shahzad, I. A., Farrukh, M., & Yasmin, N. (2020). Career growth opportunities as non-financial compensation—A new induction: Reciprocation of performance by combining social exchange theory & organizational support theory. *TEST Engineering & Management, 83*, 16905–16920.

So, K. K. F., King, C., Sparks, B. A., & Wang, Y. (2014). The role of customer brand engagement in building consumer loyalty to tourism brands. *Journal of Travel Research, 55*(1), 64–78.

Stylos, N., Vassiliadis, C., Bellou, V., & Andronikidis, A. (2016). Destination images, holistic images and personal normative beliefs: Predictors of intention to revisit a destination. *Tourism Management, 53*, 40–60.

Suhartanto, D. (2011). An examination of the structure and determinants of brand loyalty across hotel brand origin. *ASEAN Journal on Hospitality and Tourism, 10*(2), 146–161.

Sweiss, N., Obeidat, Z. M., Al-Dweeri, R. M., Mohammad Khalaf Ahmad, A., Obeidat, A. M., & Alshurideh, M. (2021). The moderating role of perceived company effort in mitigating customer misconduct within Online Brand Communities (OBC). *Journal of Marketing Communications*, 1–24.

Tariq, E., Alshurideh, M., Akour, E., Al-Hawaryd, S., & Al Kurdi, B. (2022). The role of digital marketing, CSR policy and green marketing in brand development at UK. *International Journal of Data and Network Science, 6*(3), 1–10.

Tomalieh, E. (2015). The impact of customer loyalty programs on customer retention. *International Journal of Business and Social Science, 6*(8), 78–93.

Tumewu, A. M., Pangemanan, S., & Tumewu, F. (2017). The effect of customer trust, brand brand image, and service quality on customer loyalty of airline e tickets (A Study on Traveloka) (Studiterhadap Traveloka). *Journal EMBA, 5*(2), 552–563.

Unurlu, C., & Uca, S. (2017). The effect of culture on brand loyalty through brand performance and brand personality. *International Journal of Tourism Research, 19*(6), 672–681.

Vukasovic, T. (2015). Managing consumer-based brand equity in higher education. *Managing Global Transitions, 13*(1), 75–90.

Wu, M.-Y., & Lee, Y.-R. (2011). The effects of internal marketing, job satisfaction and service attitude on job performance among high-tech firm. *African Journal of Business Management, 5*(32), 12551–12562.

Wulandari, N. (2016). Brand experience in banking industry: Direct and indirect relationship to loyalty. *Expert Journal of Marketing, 4*(1), 1–9.

Yasmin, N., & Raju, V. (2020). A review on the antecedents of brand loyalty among Bangladeshi customers of ready-made garments. © *IJCIRAS. International Journal of Creative and Innovative Research in All Studies, 3*(3), 22–29.

Zhang, Z., & Yuan, K.-H. (2016). Robust coefficients alpha and omega and confidence intervals with outlying observations and missing data: Methods and software. *Educational and Psychological Measurement, 76*(3), 387–411.

Zhang, J., Craciun, G., & Shin, D. (2010). When does electronic word-of-mouth matter? A study of consumer product reviews. *Journal of Business Research, 63*(12), 1336–1341.

The Impact of Branded Mobile Applications on Customer Loyalty

MoayyadMohammed Shaqrah, Anber Abraheem Shlash Mohammad,
Faraj Mazyed Faraj Aldaihani, Sulieman Ibraheem Shelash Al-Hawary,
Muhammad Turki Alshurideh⬭, Ibrahim Rashed Soliaman AlTaweel,
Riad Ahmad Mohammed Abazeed, Ayat Mohammad,
and D. Barween Al Kurdi⬭

Abstract This Study aims to investigate the impact of using branded mobile applications on customer loyalty. The population of the study consisted of all users of Pharmacy One's mobile application in Jordan. The researcher used the purposive

M. Shaqrah
Researcher, Marketing Department, Faculty of Administrative and Financial Sciences, Petra
University, P.O. Box: 961343, Amman 11196, Jordan

A. A. S. Mohammad
Marketing, Marketing Department, Faculty of Administrative and Financial Sciences, Petra
University, P.O. Box: 961343, Amman 11196, Jordan

F. M. F. Aldaihani
Kuwait Civil Aviation, Ishbiliyah Bloch 1, Street 122, Home 1, Kuwait, Kuwait

S. I. S. Al-Hawary (⬭)
Department of Business Administration, Faculty of Economics and Administrative Sciences, Al
Al-Bayt University, P.O.BOX 130040, Mafraq 25113, Jordan
e-mail: dr_sliman73@aabu.edu.jo

M. T. Alshurideh
Department of Marketing, School of Business, The University of Jordan, Amman 11942, Jordan
e-mail: m.alshurideh@ju.edu.jo; malshurideh@sharjah.ac.ae

Department of Management, College of Business, University of Sharjah, 27272 Sharjah, United
Arab Emirates

I. R. S. AlTaweel
Department of Business Administration, Faculty of Business School, Qussim University, Al
Russ51452, P.O. BOX 6502, Al Russ City, Saudi Arabia
e-mail: toiel@qu.edu.sa

R. A. M. Abazeed
Business Management and Public Administration, Department of Business Administration,
Faculty of Finance and Business Administration, Al Al-Bayt University, P.O. BOX 130040,
Mafraq 25113, Jordan

A. Mohammad
Business and Finance Faculty, the World Islamic Science and Education University (WISE),
Postal Code 11947, P.O. Box 1101, Amman, Jordan

© The Author(s), under exclusive license to Springer Nature Switzerland AG 2023   2215
M. Alshurideh et al. (eds.), *The Effect of Information Technology on Business
and Marketing Intelligence Systems*, Studies in Computational Intelligence 1056,
https://doi.org/10.1007/978-3-031-12382-5_121

sampling by taking 2035 active users of the on-line applications from 70,251 down-loader; 500 questionnaires were distributed randomly. To achieve the objective of the study, and to test hypotheses, the researcher used statistical program of social studies (SPSS) & (AMOS). The study results revealed that there is a statistically effect of using branded mobile applications on customer loyalty, based on the study results the researcher recommends managers and decision makers to improving the system of customer services in order to improve their loyalty, by reviewing the system after receiving feedback from customers through mobile applications. Also enhancing the quality of system information by ensuring that the content is relevant to customer requirements. and improving the perceived quality of the customer by identifying the expectations of the customers and seeking to achieve these expectations, so that the gap between what they expect and what they actually get.

Keywords Branded mobile applications (BMA) · Customer loyalty (CL) · Healthcare technologies · Amman-Jordan

1 Introduction

After the booming of smart phones and tablets that helping people not only to contacting each other but doing their jobs and tasks through it, the on-line smart phones' applications take place as a service for the community starting from using applications for work and ending with gaming and socialization applications at home, in another word it's became a life style. This wide spread of the mobile devices is due to the extensive usage of the mobile phone's applications (Al Dmour et al., 2014; Alshurideh et al., 2019a, 2019b). Mobile applications are currently used by compa-nies to help and serve their customer and clients without being on the spot. This action will save time and decrease the level of human contact within the services given (Akour et al., 2021; Alshurideh et al., 2021).

Medical sector and pharmacies are investing with the mobile applications to have a direct contact with patients and increase their compliance to medications. Also, is used as an informative reference for clients about health tips and drug-drug inter-actions and other related services. Chain of Pharmacies today are launching their own branded mobile applications to have a loyal customer and increase their sales through it.

Consumers' behavior has been observed during their usage of different mobile devices, results showed that consumers had positive emotion toward the responses they receive when visiting the mobile site which is considered as basic application of the pharmacies systems (Alshurideh et al., 2012a, 2012b; Al-Hamad et al., 2021).

D. B. Al Kurdi
Department of Marketing, Faculty of Economics and Administrative Sciences, The Hashemite University, Zarqa, Jordan
e-mail: barween@hu.edu.jo

In pharmaceutical sector, there are many applications of mobile devices, there usage has led to useful changes in pharmacy practices, this refers to the fact that many of the pharmacy daily activities can be achieved by using the mobile devices, according to the vision statement issued by the American Association of the pharmacists of the Health system regarding the proposed role of information technology in practicing pharmacy. There are many features through which technology can participate in improving the pharmacy general performance (Aburayya et al., 2020a, 2020b; Taryam et al., 2020; Ghazal et al., 2021; Svoboda et al., 2021).

But there are advantages and disadvantages of using the mobile devices, applications with positive and tangible impact on the pharmaceutical products should be studied, this increasing trend towards using the mobile devices as instruments to manage the health information in a better way requires more education of the consumers (users), and in some cases might need placing some laws. The used mobile devices applications have been developed remarkably in the medical clinical practice with a number of applications for the clinical practitioners. Determining and selecting the reliable application from thousands of medical applications available on the internet or the applications stores a basic factor in assuring the effectiveness of these devices.

In summary, this research concerned itself with the impact of using branded mobile applications on customer loyalty in chain of pharmacies (a case study on Pharmacy One) in Jordan, Understanding customers responses to each of branded mobile applications sub variable (Service information quality, Service system quality, Awareness about mobile application and Perceived quality about mobile phone application) in order to help firm to measure the impact of this application and whether its' helpful and useful for their clients, and how important it is to invest in this online technology to formulate a new marketing plan and strategy to get the customer loyalty.

2 Theoretical Frame and Hypotheses Development

2.1 The Brand

The trade mark known to the customers makes them feel secure toward the use of the product, in a severe competing business environment, which contributes to high share to the brand, and placing high value to the product in the customer's mind (Hamadneh et al., 2021; Sweiss et al., 2021). The consumer distrust the brand if it quality changes, since the perceived quality is not the real product quality, it is the customer personal evaluation which shows the product's real quality, since quality represents the images established in the customer's mind, since he does not have enough information to evaluate the quality, the price is placed as a positive indicator with the perceived quality. Brand offers multiple benefits to the organization; these benefits include increased the likely hood to select the brand which leads to customers loyalty (Alwan & Alshurideh, 2022; Tariq et al., 2022). Also, brand expands to products from other categories, increases the product/ service market value because customers are ready to pay high prices (Abu Zayyad et al., 2021; Aljumah et al., 2021).

2.2 Branded Mobile Applications

In 2008, Apple opened its iOS App Store that was the primary application distribution service. This set the quality for applications distribution services for alternative mobile firms. This led to shopper's exploitation their mobile devices as the way of connecting to the online. Consequently, mobile phones began to be referred to as smartphone because of their various capabilities. Mobile and software package developers accomplished that they might take mobile web site capabilities up with new applications named mobile applications. The applications would have identical functions and capabilities as websites designed for a desktop or laptop computer. The websites were scaled down so they'd work mobile devices. This led to anover plus of mobile applications being developed and enforced. In 2010, the word "app" was listed because the "word of the year" by the yank idiom Society (2011). Early applications for mobile apps were for email, calendars, contacts, the securities market, and weather data. High demand from users led to mobile applications being employed in mobile games, factory automation, banking, order-tracking, and price ticket purchases.

The mobile business has continuing to grow quickly, with a complete of 3.6 billion unique mobile subscribers at the top of 2014 associated an addition of one billion subscribers predicted by 2020 globally. Branded Mobile Applications may be defined as the software that is downloadable and available on online stores for mobile phones and tablets, carrying the name or the brand of an existing firm to deliver products and services through it. In the theoretical frame, the study has addressed and demonstrated both the independent and dependent variable, the independent variable (branded mobile application) with its dimension (service information quality, service system quality, perceived quality of the trade mark and awareness about the mobile phone trade mark). The study has benefited from some these previous studies mainly that have addressed these dimensions, like (Monica et al., 2017); (Francisquinho, 2013).

Service System Quality: The researcher defines quality of the system as a general performance statement of the system, which can be measured by the customer's point of view and the degree of ease of use in marketing and satisfaction with the system. Quality of service, customer satisfaction and customer loyalty are three key elements for business organizations to strive to focus on (Al-Dmour et al., 2021; Alshurideh, 2022). An organization that does not focus on quality is not successful in its business, so, quality is an important tool for strong competitive advantage market. (Tehrani & Jamshidi, 2015; Alzoubi et al., 2020).

Service Information Quality: If the provided information by the mobile phone applications are inaccurate, insufficient and without visibility, or uncompleted, this will contribute to reduction in customers satisfaction. The researcher sees that the availability of high quality information brings and attract more customers because of their loyalty to the quality of the information provided by the different applications for the pharmaceutical sector. Service Information Quality may be defined as the quality level of processes and procedures provided through the service whether it is available at the time needed for and work smoothly or not.

Awareness about the Mobile Phone Application: Awareness about the brand is identified as "the prospected buyer's ability to remember the brand in a product type". (Chen & Ching, 2007).Also identified as "the customer's ability to know the brand in the different conditions" while Ross identified it as "strength of the brand's presence in the customers mind".

The researcher sees that the consumer's ability to remember and distinguish some components of the brand within different conditions and provisions is considered basic and important condition to create the awareness about the brand. Awareness about the Mobile Phone Application may be defined as the level of consumer concessions about the brand of the mobile phone application and to which extent they can recall or recognize it.

The Perceived Quality about the Mobile Phone Application: Perceived quality is identified as "the consumer perception of the total quality or superiority of the product or the service". Swinker & Hines classify the perceived quality to four categories; essential, external, appearance and performance. Perceived quality and brand's image are determinants of loyalty to the brand. Many studies have indicated at the positive influence of the perceived quality on buying intention. The Perceived Quality about the Mobile Phone Application may be defined as the customer's judgment and perception about the total quality of the mobile phone application and its superiority to another.

2.3 Brand Loyalty

Loyalty to the brand is considered one of the important factors in the success of the commercial businesses since studying the client's loyalty through his behavior and the attitudes are basic methods in building the client's loyalty. Attitudes about loyalty were identified as the degree of the behavioral commitment regarding the unique correlations regarding their kind with the brand's value (Suhartanto & Noor, 2013: 65). Several researches believe that loyalty in the attitudes affecting the clients purchase repetition of the brand. Brand Loyalty may be defined as Deep commitment of customer to a product or service despite situational influences and marketing efforts having the potential to cause switching to another product or service.

Attitudinal Brand Loyalty: Attitudinal method issued in order to measure the clients' loyalty to the brand and increasing the number of the new clients. The clients' perspective method about the client's loyalty and the intention to re-buy the product or recommending it (Kandampully & Hu, 2007). So, the researcher sees that many studies use this kind of instruments to perform the customer's loyalty to the brand but measuring the loyalty to the brand with one instrument is an effective measure to loyalty compared to customer satisfaction on level of retaining them. Also, attitudinal method issued in order to measure the clients' loyalty to the brand and increasing the number of the new clients. The clients' perspective method about the client's loyalty and the intention to re-buy the product or recommending it (Kandampully & Hu, 2007). So, the researcher sees that many studies use this kind of instruments to perform the customers loyalty to the brand but measuring the loyalty to the brand with one instrument is an effective measure to loyalty compared to customer satisfaction on level of retaining them.

Behavioral Brand Loyalty: From an analytical study conducted by (Glasman & Albarracin, 2006), the study indicated that the attitude greatly predicts the future behavior and understanding loyalty by the client is basic factor for the businesses because it helps marketing directors in designing programs to modify the clients' behavior in the future, especially changing the behavior from one specific brand to another. To find the motive for buying behavior, directors need understanding the clients' cognitive mechanisms which are considered important part of these attitude (Back, 2005).

Rewards programs issued by the companies to promote the client's loyalty to its brands increase the client's loyalty to that brand, because the client might depend on his loyalty to the brand on specific and desired behaviors and provide him with comfort. It the customer attitude to the brand is positive, the percentage of his loyalty to the brand will be higher because his attitude will contribute to be as an advertisement and promotion to the brand, also, those attitudes in loyalty will contribute to creating long-term loyalty relationship between the brand and the client (Meyer-Waarden, 2015). The researcher sees that, in order to achieve specific behavior from the client, there should be process to control the clients' behaviors in the relationship between the brand and the client, by including the client's psychological factor and his attitude. So, the researcher sees that, one group of behavioral intentions can be design as positive behavioral intentions. one of this is loyalty, and certain behaviors signal that customer are forging bond with a company, and when customer praise the firm, express preference for the company over other, recommend the company of service to other.

2.4 Branded Mobile Applications and Loyalty

Bellman et al. (2011) have investigated the effect of branded mobile applications, including information and experimental applications related to attitudes toward the brand, and buying intention, results from their study showed that brand applications had positive impact of the attitude towards the brand, but less impact on the buying intention, also, found that the branded applications are new form of interaction information means, such as the websites, but differ from the advertisements thorough the internet because of the applications that characterizes by high level of the customers participation. Yu has investigated the influences of the mobile advertisement messages in building positive attitudes toward the band and the customer's buying intentions. Results of his study showed that the mobile phone advertisements with interactive features have led to positive attitudes toward the brand and the buying intentions. Hoogendoorn study has proved the extent of the impact of the mobile applications with brand on the brand's cognitive and emotional shares and found that the interaction with branded applications has resulted in an increase in the positive aspects of the brand. Also, the results of that study showed that the interaction in the mobile applications with brand improved the brand's cognitive share.

Xie et al. study found that design of the mobile devices such as the screen's size, advertisement size and ease of uses influence the customers' emotional response to the mobile advertisements that might generate the positive emotions and increasing the customers buying intention. Kumar and Mukherjee have explained the attitudes of the mobile phone users in the shopping and purchase intention through the devices and found that the mobile phone does not always push the users to buy. Based on the above literature the study hypotheses can be formulated as:

H01: There is a significant impact of branded mobile applications) on customer loyalty of Pharmacy One

3 Research Model

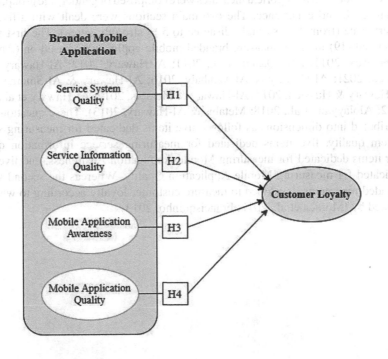

4 Methodology

4.1 Population and Sample Selection

The sample of the study is defined as part of the sample that contains appropriate information about the study population. The study population was 70,251 downloaders of the mobile application. In order to obtain the appropriate information owned by some members of the population, the best technique to select the sample is the purposive sampling. According to Sekaran and Bougie (2016), a purposive sample is the one that contains individuals who are able to provide the desired information. Since the current study seeks to collect data from online users of the application, a purposive sample consisted of 2035 real active users of the mobile application, 500 users were randomly selected. Out of them, 423 users were accepted since they download and activate the intended application and answers the questionnaire form link.

4.2 Measurement Instrument

A self-reported questionnaire that consists of two main sections along with a section regarding control variables was used as the measurement instrument. Control variables considered as categorical measures were composed of gender, age group, educational level, and experience. The two main sections were dealt with a five-point Likert scale (from 1 = strongly disagree to 5 = strongly agree). The first section contained (19) items to measure branded mobile applications based on (Abbad & Al-Hawary, 2011; Abu Qaaud et al., 2011; Al-Hawary, 2012; Al-Hawary & Al-Fassed, 2021; Al-Hawary & Al-Menhaly, 2016; Al-Hawary & Al-Smeran, 2017; Al-Hawary & Hussien, 2017; Al-Hawary & Metabis, 2012; Al-Hawary et al., 2011, 2012; Alolayyan et al., 2018; Metabis & Al-Hawary, 2013). These questions were distributed into dimensions as follows: five items dedicated for measuring service system quality, five items dedicated for measuring service information quality, four items dedicated for measuring Mobile Application Awareness, and five items dedicated for measuring Mobile Application Quality. Whereas the second section included nine items developed to measure customer loyalty according to what was pointed by (Monica et al., 2017; Francisquinho, 2013).

5 Findings

5.1 Measurement Model Evaluation

This study was conducted structural equation modeling (SEM) to test hypotheses, which represents a contemporary statistical technique for testing and estimating the relationship between factors and variables (Wang & Rhemtulla, 2021). Accordingly, the reliability and validity of the constructs were tested using confirmatory factor analysis (CFA) through the statistical program AMOSv24. Table 1 summarizes the results of convergent and discriminant validity, as well the indicators of reliability.

Table 1 shows that the standard loading values for the individual items were within the domain (0.624–0.872), these values greater than the minimum retention of the elements based on their standard loads (Al-Lozi et al., 2018; Sung et al., 2019). Average variance extracted (AVE) is a summary indicator of the convergent validity of constructs that must be above 0.50 (Howard, 2018). The results indicate that the AVE values were greater than 0.50 for all constructs, thus the used measurement model has an appropriate convergent validity. Rimkeviciene et al. (2017) suggested the comparison approach as a way to deal with discriminant validity assessment in covariance-based SEM. This approachis based on comparing the values of maximum shared variance (MSV) with the values of AVE, as well as comparing the values of

Table 1 Results of validity and reliability tests

Constructs	1	2	3	4	5
Service system quality	**0.755**				
Service information quality	0.564	**0.760**			
Mobile application Awareness	0.612	0.538	**0.762**		
Mobile application quality	0.488	0.542	0.493	**0.722**	
Customer loyalty	0.692	0.689	0.665	0.677	**0.748**
VIF	5	5	4	5	9
Loadings range	0.622–0.822	0.638–0.854	0.624–0.866	0.682–0.766	0.631–0.872
AVE	0.570	0.577	0.580	0.521	0.559
MSV	0.511	0.502	0.488	0.469	0.509
Internal consistency	0.866	0.868	0.841	0.840	0.915
Composite reliability	0.868	0.871	0.845	0.844	0.918

Note VIF: variance inflation factor, AVE: average variance extracted, MSV: maximum shared variance, Bold numbers refer to $\sqrt{AVE}$

square root of AVE ($\sqrt{AVE}$) with the correlation between the rest of the structures. The results show that the values of MSV were smaller than the values of AVE, and that the values of $\sqrt{AVE}$ were higher than the correlation values among the rest of the constructs. Therefore, the measurement model used is characterized by discriminative validity. The internal consistency measured through Cronbach's Alpha coefficient (α) and compound reliability by McDonald's Omega coefficient (ω) was conducted as indicators to evaluate measurement model. The results listed in Table 1 demonstrated that both values of Cronbach's Alpha coefficient and McDonald's Omega coefficient were greater than 0.70, which is the lowest limit for judging on measurement reliability (De Leeuw et al., 2019).

5.2 Structural Model

The structural model illustrated no multicollinearity issue among predictor constructs because variance inflation factor (VIF) values are below the threshold of 5, as shown in Table 1 (Hair et al., 2017). This result is supported by the values of model fit indices shown in Fig. 1.

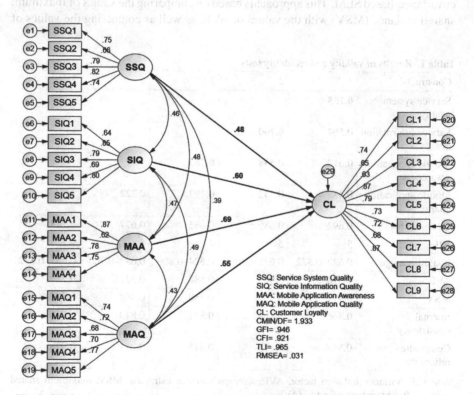

Fig. 1 SEM results of the branded mobile applications effect on customer loyalty

Table 2 Hypothesis testing

Hyp	Relation			Standard beta	t value	p value
H1	Service system quality	→	Customer Loyalty	0.483	20.04*	0.03
H2	Service information quality	→	Customer Loyalty	0.609	27.83**	0.002
H3	Mobile application awareness	→	Customer Loyalty	0.692	30.72***	0.000
H4	Mobile application quality	→	Customer Loyalty	0.550	25.37**	0.005

Note $* p < 0.05$, $** p < 0.01$, $*** p < 0.001$

The results in Fig. 1 indicated that the chi-square to degrees of freedom (CMIN/DF) was 1.933, which is less than 3 the upper limit of this indicator. The values of the goodness of fit index (GFI), the comparative fit index (CFI), and the Tucker-Lewis index (TLI) were upper than the minimum accepted threshold of 0.90. Moreover, the result of root mean square error of approximation (RMSEA) indicated to value 0.031, this value is a reasonable error of approximation because it is less than the higher limit of 0.08. Consequently, the structural model used in this study was recognized as a fit model for predicting the DEP and generalization of its result (Ahmad et al., 2016; Shi et al., 2019). To verify the results of testing the study hypotheses, structural equation modeling (SEM) was used, the results of which are listed in Table 2.

The results demonstrated in Table 2 show that mobile application awareness has the greatest positive impact on customer loyalty ($\beta = 0.692$, $t = 30.72$, $p = 0.000$), followed by service information quality ($\beta = 0.609$, $t = 27.83$, $p = 0.002$), then mobile application quality ($\beta = 0.550$, $t = 25.37$, $p = 0.005$), and finally the lowest impact was for service system quality ($\beta = 0.483$, $t = 20.04$, $p = 0.03$). Thus, all the minor hypotheses of the study were supported based on these results.

6 Discussion

The main hypothesis presumed that there is a significant and positive effect of the branded mobile applications on customer loyalty. This hypothesis was tested based on the data collected from the respondents. The branded mobile application was measured using the items (1–19) of the questionnaire, while customer loyalty was measured by the items (20–28) of the questionnaire. Since the independent variable, the branded mobile applications consisted of four dimensions in the current study, which were the dimensions used in the sub-hypotheses, the discussion of the main hypothesis was combined with the discussion of the sub-hypotheses.

However, numerous studies have been accepted the hypothesis that there is a significant and positive impact of branded mobile applications on customer; loyalty.

Lin and Wang examined the determinants of customer loyalty in mobile commerce contexts and found an impact of perceived value of mobile applications on customer loyalty. Aydin and Özer analyzed the antecedents of customer loyalty in mobile telecommunication in Turkey and revealed a significant effect of perceived service quality on customer loyalty. One of the most pivotal intangible advantages of mobile applications is the improved customer service.

7 Recommendations

In light of its findings, the researchers recommend managers and decision makers to improve the system of customer services in order to improve their loyalty, by reviewing the system after receiving feedback from customers, and enhancing the quality of system information by ensuring that the content is relevant to customer requirements, the timing of the information and its practical usefulness to customers. Also Raising customer awareness by focusing on advertising, promoting and benefiting from the applications provided and encouraging customers to download and use them. The service provider can use the rewards to achieve this goal. Finally Improving the perceived quality of the customer by identifying the expectations of the customers and seeking to achieve these expectations, so that the gap between what they expect and what they actually get.

References

Abbad, J., & Al-Hawary, S. I. (2011). Measuring banking service quality in Jordan: A case study of Arab bank. *Abhath Al-Yarmouk, 27*(3), 2179–2196.

Abu Qaaud, F., Al-Shoura, M., & Al-Hawary, S. I. (2011). The impact of the service marketing mix in the service quality of health services from the viewpoint of patients in Government Hospitals in Amman "A Field study." *Abhath Al-Yarmouk, 27*(1B), 417–441.

Abu Zayyad, H. M., Obeidat, Z. M., Alshurideh, M. T., Abuhashesh, M., Maqableh, M., & Masa'deh. (2021). Corporate social responsibility and patronage intentions: The mediating effect of brand credibility. *Journal of Marketing Communications, 27*(5), 510–533.

Aburayya, A., Alshurideh, M., Al Marzouqi, A., Al Diabat, O., Alfarsi, A., Suson, R., & Salloum, S. A. (2020a). An empirical examination of the effect of TQM practices on hospital service quality: An assessment study in UAE hospitals. *Systematic Reviews in Pharmacy, 11*(9), 347–362.

Aburayya, A., Alshurideh, M., Al Marzouqi, A., Al Diabat, O., Alfarsi, A., Suson, R., & Alzarouni, A. (2020b). Critical success factors affecting the implementation of tqm in public hospitals: A case study in UAE Hospitals. *Systematic Reviews in Pharmacy, 11*(10), 230–242.

Ahmad, S., Zulkurnain, N., & Khairushalimi, F. (2016). Assessing the validity and reliability of a measurement model in structural equation modeling (SEM). *British Journal of Mathematics & Computer Science, 15*(3), 1–8. https://doi.org/10.9734/BJMCS/2016/25183

Akour, I., Alshurideh, M., Al Kurdi, B., Al Ali, A., & Salloum, S. (2021). Using machine learning algorithms to predict people's intention to use mobile learning platforms during the COVID-19 pandemic: Machine learning approach. *JMIR Medical Education, 7*(1), 1–17.

Al Dmour, H., Alshurideh, M., & Shishan, F. (2014). The influence of mobile application quality and attributes on the continuance intention of mobile shopping. *Life Science Journal, 11*(10), 172–181.

Al-Dmour, R., AlShaar, F., Al-Dmour, H., Masa'deh, R., & Alshurideh, M. T. (2021). The effect of service recovery justices strategies on online customer engagement via the role of "customer satisfaction" during the Covid-19 pandemic: An empirical study. *The Effect of Coronavirus Disease (COVID-19) on Business Intelligence, 334*, 325–346.

Al-Hamad, M., Mbaidin, H., AlHamad, A., Alshurideh, M., Kurdi, B., & Al-Hamad, N. (2021). Investigating students' behavioral intention to use mobile learning in higher education in UAE during Coronavirus-19 pandemic. *International Journal of Data and Network Science, 5*(3), 321–330.

Al-Hawary, S. I. (2012). Health care services quality at private hospitals, from patient's perspective: A comparative study between Jordan and Saudi Arabia. *African Journal of Business Management, 6*(22), 6516–6529.

Al-Hawary, S. I., & Al-Menhaly, S. (2016). The quality of E-government services and its role on achieving beneficiaries satisfaction. *Global Journal of Management and Business Research: A Administration and Management, 16*(11), 1–11.

Al-Hawary, S. I., & Al-Smeran, W. (2017). Impact of electronic service quality on customers satisfaction of Islamic banks in Jordan. *International Journal of Academic Research in Accounting, Finance and Management Sciences, 7*(1), 170–188.

Al-Hawary, S. I. S,. & Harahsheh, S. A. (2014). Factors affecting Jordanian consumer loyalty toward cellular phone brand. *International Journal of Economics and Business Research (IJEBR), 7*(3), 349–375.

Al-Hawary, S. I., & Hussien, A. J. (2017). The impact of electronic banking services on the customers loyalty of commercial banks in Jordan. *International Journal of Academic Research in Accounting, Finance and Management Sciences, 7*(1), 50–63.

Al-Hawary, S. I., & Metabis, A. (2012). Service quality at Jordanian commercial banks: What do their customers say? *International Journal of Productivity and Quality Management, 10*(3), 307–334.

Al-Hawary, S. I., Alghanim, S., & Mohammad, A. (2011). Quality level of health care service provided by King Abdullah educational hospital from patient's viewpoint. *Interdisciplinary Journal of Contemporary Research in Business, 2*(11), 552–572.

Al-Hawary, S. I., & Al-Fassed, K. J. (2021). The impact of social media marketing on building brand loyalty through customer engagement in Jordan. *International Journal of Business Innovation and Research*, In Press.

Al-Hawary, S. I., AL-Zeaud, H., & Matabes, A. (2012). Measuring the quality of educational services offered to postgraduate students at the faculty of business and finance: A field study on the Universities of the North Region. *Al Manara for Research and Studies, 18*(1), 241–278.

Aljumah, A., Nuseir, M. T., & Alshurideh, M. T. (2021). The impact of social media marketing communications on consumer response during the COVID-19: Does the brand equity of a University matter. *The Effect of Coronavirus Disease (COVID-19) on Business Intelligence*, 367–384.

Al-Lozi, M. S., Almomani, R. Z. Q., & Al-Hawary, S. I. S. (2018). Talent management strategies as a critical success factor for effectiveness of human resources information systems in commercial banks working in Jordan. *Global Journal of Management and Business Research: A Administration and Management, 18*(1), 30–43.

Alolayyan, M., Al-Hawary, S. I., Mohammad, A. A., & Al-Nady, B. A. (2018). Banking service quality provided by commercial banks and customer satisfaction. A structural equation modelling approaches. *International Journal of Productivity and Quality Management, 24*(4), 543–565.

Alshurideh, M. (2022). Does electronic customer relationship management (E-CRM) affect service quality at private hospitals in Jordan? *Uncertain Supply Chain Management, 10*(2), 1–8.

Alshurideh, M., Masa'deh, R. M. D. T., & Alkurdi, B. (2012a). The effect of customer satisfaction upon customer retention in the Jordanian mobile market: An empirical investigation. *European Journal of Economics, Finance and Administrative Sciences, 47*(12), 69–78.

Alshurideh, M., Nicholson, M., & Xiao, S. (2012b). The effect of previous experience on mobile subscribers' repeat purchase behaviour. *European Journal of Social Sciences, 30*(3), 366–376.

Alshurideh, M., Al Kurdi, B., & Salloum, S. A. (2019a). Examining the main mobile learning system drivers' effects: A mix empirical examination of both the Expectation-Confirmation Model (ECM) and the Technology Acceptance Model (TAM). In *International Conference on Advanced Intelligent Systems and Informatics* (pp. 406–417). Cham: Springer.

Alshurideh, M., Salloum, S. A., Al Kurdi, B., Monem, A. A., & Shaalan, K. (2019b). Understanding the quality determinants that influence the intention to use the mobile learning platforms: A practical study. *International Journal of Interactive Mobile Technologies, 13*(11), 157–183.

Alshurideh, M. T., Kurdi, B. A., AlHamad, A. Q., Salloum, S. A., Alkurdi, S., Dehghan, A., & Masa'deh, R. E. (2021). Factors affecting the use of smart mobile examination platforms by universities' postgraduate students during the COVID 19 pandemic: an empirical study. In *Informatics* (vol. 8 2, pp 1–21). Multidisciplinary Digital Publishing Institute.

Alwan, M., & Alshurideh, M. (2022). The effect of digital marketing on purchase intention: Moderating effect of brand equity. *International Journal of Data and Network Science, 10*(3), 1–12.

Alzoubi, H. M., Alshurideh, M., Al Kurdi, B., & Inairat, M. (2020). Do perceived service value, quality, price fairness and service recovery shape customer satisfaction and delight? A practical study in the service telecommunication context. *Uncertain Supply Chain Management, 8*(3), 579–588.

Back, K. J. (2005). The effects of image congruence on customers' brand loyalty in the upper middle-class hotel industry. *Journal of Hospitality & Tourism Research, 29*(4), 448–467.

Bellman, S., Potter, R., Treleaven-Hassard, S., Robinson, J., & Varan, D. (2011). The effectiveness of branded mobile phone apps. *Journal of Interactive Marketing, 25*(4), 191–200.

Chen, J. S., & Ching, K. H. (2007). The effects of mobile customer relationship management on customer loyalty: Brand image does matter. In *Proceedings of the 40th Hawaii International Conference on System Sciences* (pp. 1–10).

De Leeuw, E., Hox, J., Silber, H., Struminskaya, B., & Vis, C. (2019). Development of an international survey attitude scale: Measurement equivalence, reliability, and predictive validity. *Measurement Instruments for the Social Sciences, 1*(1), 9. https://doi.org/10.1186/s42409-019-0012-x

Francisquinho, A. N. (2013). Consumer attitudes and perceptions towards medicine types: Brand medicines versus generic medicines. Project submitted as partial requirement for the conferral of Master in Marketing, Lisbon University Institute.

Ghazal, T. M., Alshurideh, M. T., & Alzoubi, H. M. (2021). Blockchain-enabled internet of things (IoT) platforms for pharmaceutical and biomedical research. In *The International Conference on Artificial Intelligence and Computer Vision* (pp. 589–600). Cham: Springer.

Glasman, L. R., & Albarracín, D. (2006). Forming attitudes that predict future behavior: A meta-analysis of the attitude behavior relation. *Psychological Bulletin, 132*(5), 778–822.

Hair, J. F., Babin, B. J., & Krey, N. (2017). Covariance-based structural equation modeling in the journal of advertising: Review and recommendations. *Journal of Advertising, 46*(1), 163–177. https://doi.org/10.1080/00913367.2017.1281777

Hamadneh, S., Hassan, J., Alshurideh, M., Al Kurdi, B., & Aburayya, A. (2021). The effect of brand personality on consumer self-identity: The moderation effect of cultural orientations among British and Chinese consumers. *Journal of Legal, Ethical and Regulatory Issues, 24*, 1–14.

Howard, M. C. (2018). The convergent validity and nomological net of two methods to measure retroactive influences. *Psychology of Consciousness: Theory, Research, and Practice, 5*(3), 324–337. https://doi.org/10.1037/cns0000149

Kandampully, J., & Hu, H. (2007). Do hoteliers need to manage image to retain loyal customers? *International Journal of Contemporary Hospitality Management, 19*(6), 435–443.

Metabis, A., & Al-Hawary, S. I. (2013). The impact of internal marketing practices on services quality of commercial banks in Jordan. *International Journal of Services and Operations Management, 15*(3), 313–337.

Meyer-Waarden, L. (2015). Effects of loyalty program rewards on store loyalty. *Journal of Retailing and Consumer Services, 24*, 22–32.

Monica, E., Dharmmesta, B. S., & Syahlani, S. P. (2017). Correlation analysis between the service quality, customer satisfaction, and customer loyalty of Viva Generik Pharmacy in Semarang analysis Journal FarmasiaSanisDan. *Komuntitas, 14*(2), 86–92.

Rimkeviciene, J., Hawgood, J., O'Gorman, J., & De Leo, D. (2017). Construct validity of the acquired capability for suicide scale: Factor structure, convergent and discriminant validity. *Journal of Psychopathology and Behavioral Assessment, 39*(2), 291–302. https://doi.org/10.1007/s10862-016-9576-4

Sekaran, U., & Bougie, R. (2016). *Research methods for business: A skill-building approach* (Seventh edition). Wiley.

Shi, D., Lee, T., & Maydeu-Olivares, A. (2019). Understanding the model size effect on SEM fit indices. *Educational and Psychological Measurement, 79*(2), 310–334. https://doi.org/10.1177/0013164418783530

Suhartanto, D., & Noor, A. (2013). Attitudinal loyalty in the budget hotel industry: What are the important factors? *Journal of Tourism, Hospitality & Culinary Arts, 5*(2), 64–74.

Sung, K.-S., Yi, Y. G., & Shin, H.-I. (2019). Reliability and validity of knee extensor strength measurements using a portable dynamometer anchoring system in a supine position. *BMC Musculoskeletal Disorders, 20*(1), 1–8. https://doi.org/10.1186/s12891-019-2703-0

Svoboda, P., Ghazal, T. M., Afifi, M. A., Kalra, D., Alshurideh, M. T., & Alzoubi, H. M. (2021). Information systems integration to enhance operational customer relationship management in the pharmaceutical industry. In *The International Conference on Artificial Intelligence and Computer Vision* (pp. 553–572). Cham: Springer.

Sweiss, N., Obeidat, Z. M., Al-Dweeri, R. M., Mohammad Khalaf Ahmad, A. M., Obeidat, A., & Alshurideh, M. (2021). The moderating role of perceived company effort in mitigating customer misconduct within Online Brand Communities (OBC). *Journal of Marketing Communications*, 1–24.

Tariq, E., Alshurideh, M., Akour, E., Al-Hawaryd, S., & Al Kurdi, B. (2022). The role of digital marketing, CSR policy and green marketing in brand development at UK. *International Journal of Data and Network Science, 6*(3), 1–10.

Taryam, M., Alawadhi, D., Aburayya, A., Albaqa'een, A., Alfarsi, A., Makki, I., & Salloum, S. A. (2020). Effectiveness of not quarantining passengers after having a negative COVID-19 PCR test at arrival to dubai airports. *Systematic Reviews in Pharmacy, 11*(11), 1384–1395.

Tehrani, R., & Jamshidi, H. (2015). Analysis of the impact factors information quality, system quality, interface design quality on customer loyalty system websites according to the role of satisfaction and trust. *Iranian Journal of Information Processing Management, 30*(4), 1085–1106.

Wang, Y. A., & Rhemtulla, M. (2021). Power analysis for parameter estimation in structural equation modeling: A discussion and tutorial. *Advances in Methods and Practices in Psychological Science, 4*(1), 1–17. https://doi.org/10.1177/2515245920918253

Impact of Internal Marketing Practices on Intention to Stay in Commercial Banks in Jordan

Mohammad Issa Ghafel Alkhawaldeh, Faraj Mazyed Faraj Aldaihani, Bahaa Addin Ali Al-Zyoud, Sulieman Ibraheem Shelash Al-Hawary, Nancy Abdullah Shamaileh, Anber Abraheem Shlash Mohammad, Muhammad Turki Alshurideh ⓘ, and Omar Atallah Ali Al-Adamat

Abstract The study aimed at identifying the impact of internal marketing practices on the intention to stay in commercial banks in Jordan. The internal marketing practices represented by (empowerment, training programs, incentives and rewards, internal communication and administrative support). The study population consists of employees of Commercial Banks in the northern region of Jordan; a sample was taken from the study population. The questionnaire was distributed to 550 of study samples, 521 were valid for statistical analysis. To achieve the objectives of the study and test hypotheses statistical analysis programs (IBM SPSS 24) and (SMARTPSLS, v. 3.2.6) were used. The study found that there is a statistically significant effect of internal marketing practices on Intention to stay in commercial banks in Jordan. Based on the study results, the researcher recommended the managers and decision makers in the commercial banks in Jordan to emphasize the importance of job

M. I. G. Alkhawaldeh
Ministry of Local Administration, Amman, Jordan

F. M. F. Aldaihani
Kuwait Civil Aviation, Ishbiliyah Bloch 1, Street 122, Home 1, Al Farwaniyah, Kuwait

B. A. A. Al-Zyoud · S. I. S. Al-Hawary (✉) · N. A. Shamaileh
Department of Business Administration, School of Business, Al al-Bayt University, P.O. Box 130040, Mafraq 25113, Jordan
e-mail: dr_sliman73@aabu.edu.jo

A. A. S. Mohammad
Marketing Department, Faculty of Administrative and Financial Sciences, Petra University, P.O. Box 961343, Amman 11196, Jordan

M. T. Alshurideh
Department of Marketing, School of Business, The University of Jordan, Amman 11942, Jordan
e-mail: m.alshurideh@ju.edu.jo; malshurideh@sharjah.ac.ae

Department of Management, College of Business, University of Sharjah, 27272 Sharjah, United Arab Emirates

O. A. A. Al-Adamat
Ministry of Education, Amman, Jordan

commitment, which generates the desire of employees to maintain their membership in the banks and their willingness to pursue organizational goals and accept their values and objectives.

Keywords Internal marketing practices · Intention to stay · Commercial banks · Jordan

1 Introduction

Internal marketing practices concept emerged to allow organizations to prove themselves and achieve success, and since organizations are the main pillar in the service sector depends mainly on human resources, therefore, employees and trying to retain them are considered the organization success cornerstone. Service providers are more interactive with the customer, with employees bearing the brunt of success. Accordingly, marketing directed towards employees within organizations is considered one of the essentials for the organizations work success (Al-Hawary & Obiadat, 2021; Al-Hawary & Alhajri, 2020; Al-Hawary, 2013; Al-Hawary & Batayneh, 2010; Al-Hawary et al., 2013a, b, 2017; Altarifi et al., 2015; Al-Hawary & Al-Menhaly, 2016; Al-Hawary & Al-Smeran, 2017; Al-Hawary & Harahsheh, 2014; Al-Hawary & Hussien, 2017; Alolayyan et al., 2018; Alshurideh et al., 2017). And since each organization sets a set of goals and strives to achieve them through the application of activities carried out by employees who have specific qualifications and capabilities, and make their efforts to achieve the goals of the bank and in return they get many benefits from it that satisfy their needs, so it can be said that the relationship between employees and the bank is an integrated relationship (Al-Hawary & Al-Syasneh, 2020; AlTaweel & Al-Hawary, 2021; Al-Hawary et al., 2020; Alhalalmeh et al., 2020; Al-Nady et al., 2016; Al-Hawary & Nusair, 2017; Al-Lozi, et al., 2018; Al-Hawary & Al-Namlan, 2018; Al-Hawary & Mohammed, 2017; Al-Lozi, et al., 2017; Al-Hawary et al., 2012; Al-Hawary & Alajmi, 2017). For the organization to achieve goals, it must retain competent employees and set up appropriate compensation programs in order to ensure their commitment and survival (Alshurideh et al., 2020; Kurdi et al., 2020). Successful organizations are those who can harness their human resources to develop their products and their customers desires, and work to provide all material and moral capabilities that help these resources to develop and be creative in carrying out the required tasks, and always strive to preserve these resources due to their cost and high value in helping the organization achieve its goals, through research briefing, its found that commercial banks strive to secure an encouraging and attractive work environment, which motivates employees to stay and continue working for these banks (Al Kurdi et al., 2020; Alshamsi et al., 2020; Alshurideh, 2022). Therefore, this study came to determine the extent of the impact of internal marketing practices on the intention to stay in commercial banks in Jordan.

2 Literature Review and Hypotheses Development

2.1 Internal Marketing

A new concept of marketing appeared in the early eighties of the last century, called internal marketing, where the interest in the human resource increased significantly, in order to reach the quality level required of customers (Al-Hawary & Abu-Laimon, 2013; Al-Hawary et al., 2013a, b; Metabis & Al-Hawary, 2013; Mohammad et al., 2020). Participation processes between different departments had a prominent role in building and strengthening the relationship between employees through successful internal marketing practices application, and the internal marketing concept confirms that collective work among employees in all departments is an important development factor, and a method to maintain the continued organization success (ELSamen & Alshurideh, 2012; Alkitbi et al., 2020). It also indicates that the relationship between the organization and its employees must be a positive relationship, which in turn can lead to the development of understanding among employees and enhance their sense of belonging, as well as providing products and services to employees and building a long-term relationship with them (Alameeri et al., 2020; Khasawneh, 2016).

Internal marketing stresses the importance of the marketing within the company by focusing on the employee and providing a source of competitive advantage (AlShehhi et al., 2020; Alshurideh et al., 2015). Through the internal marketing strategy, employees are treated as internal customers who must be convinced of the company's vision and value just as strongly as external customers (Alsuwaidi et al., 2020; Al-bawaia et al., 2022). Thus, satisfied employees are those who have a high commitment to roles within the organization in terms of achieving organizational goals, and the implementation of internal marketing may be some kind of difficulty if internal customers are not satisfied, and therefore there will be difficulty in providing good service to external customers from internal customers (Aburayya et al., 2020; Alzoubi et al., 2022). Those who are not satisfied, and this will affect the organizations productivity (Al-Khayyal et al., 2020, 2021; Solomon, 2017).

It is necessary to establish a long-term relationship with employees, and consider them internal customers, models have been developed based on the idea of making the worker's job attractive; because employee satisfaction will lead to customer satisfaction, and make employees more loyal to their organizations, and this works to attract better employees, and increases their motivation (Alshurideh, 2019; Al-Dhuhouri et al., 2020). Thus, the increased recognition of the employee's importance role in the service industry prompted organizations to adopt the internal marketing concept, and thus treat them as internal customers. Meeting the internal customer needs makes the business in a better position; to provide the quality required to satisfy external customers (Alshurideh et al., 2016; Svoboda et al., 2021). Establishing internal marketing involves developing and retaining successful employees and demonstrating a positive relationship between organizations and their employees (Aljumah et al., 2021; Hayajneh et al., 2021). Internal marketing is a complement to human resource management; good internal programs create employee satisfaction, and this inevitably leads to external customer satisfaction (Shehada & Omar, 2015; Khasawneh et al., 2021a, b).

It should be noted the advantages that internal marketing gives organizations: encourage employees to perform better, empower the employee so that they are able to accept powers and assume responsibilities, achieve a greater understanding of the organization's goals, policies, procedures, and directions, and encourage the employee to provide distinguished service to customers, by commending their contribution to the organization successes, and achieving a better level of coordination and cooperation between the various organization departments and divisions (Abu Hamra, 2017; Lee et al., 2022a, b).

Many studies came to show the internal marketing dimensions, and these dimensions differed according to the different researchers viewpoints. Empowerment is a transfer of responsibility and authority, and an invitation to organization employees to participate in the information and knowledge provided by the organization through the database and decision-making, and therefore in the decision-making power becomes responsible for the quality of what he decides or what he performs, which leads to the transfer of powers from the president to employees (Alzoubi et al., 2022; Metabis & Al-Hawary, 2013; Tariq et al., 2022). Training programs are the systematic development of employees' knowledge, skills and attitudes required by employees; to perform sufficiently in a particular task or function. (Ramos, 2015; Alshraideh et al., 2017) and organizations resort to training their employee; In order to qualify him for the work that he will perform within the organization, or in order to raise his production capacity, or work to provide him with the necessary skills that appear due to environmental and technical changes that require new skills, Rewards and incentives are some driving force within the employees through which they try to achieve some goals in order to satisfy some needs, it is a feeling or desire to do or not to do something to satisfy a certain need (Abidin et al., 2017; Alshurideh et al., 2022). Internal communication is the work of internal communication that facilitates transactions between individuals and groups within the organization, and obtaining correct, reliable and timely information for the employee, in order to enhance knowledge, in addition to providing a healthy work environment that promotes motivation, creativity, effectiveness, and cooperation through the culture components, values and organizational assumptions, that develops a sense of belonging and loyalty among all employees (Alshura et al., 2016; Alyammahi et al., 2020; Sweiss et al., 2021).

2.2 Intention to Stay

Many researchers have noticed the lack of specific and clear definitions of intentions to stay at work in much of the literature that talks about the intention to stay at work or even leave it, and there are some researchers who have explained this; Since they considered intentions to remain in work a self-explanatory factor without the need to set definitions, Many researchers consider the intentions to stay at work or leave it as the last step in the decision-making process towards staying at work within the organization.

Leaving work is one of the tendencies that lead to complete separation from belonging to a particular social system at the exclusive and personal initiative of the employee (Mohammad et al., 2020; AlHamad et al., 2022). Halawa (2015) adds that in contrast to the intention to stay at work, there is a feeling that some employees may have job instability; If the employee feels dissatisfaction, his future career is threatened in the organization in which he works, and the unknown makes him doubt whether he will stay within the organization or leave it, and this feeling will always provoke the employee to a continuous review his survival in his works, and trying to search for a better job to improve his conditions, and provide him with work stability and tranquility (Al-Dmour et al., 2021; Allozi et al., 2022; Shamout et al., 2022).

Sadiq (2018) adds that the intention to stay or leave work is affected by organizational commitment, as it plays a key role in the employee's continuation to work within the organization, as organizational commitment works to reduce the levels of some negative phenomena in organizations such as absence, leaving work, and evading performance of duties. The researcher believes that the role of internal marketing shines through at this stage; As the internal marketing and the dimensions that it contains, such as employees training, the system of rewards and incentives, administrative support, and internal communications between the manager and employees and among the employees themselves, play an important role in influencing the employees psychology, which inevitably leads them to organizational commitment, improving the their work quality and their performance, Thus, this improves the organization reputation in which they work, and increases the customer satisfaction outside the organization, and at this moment, the employee can only expect everything that is wonderful, so the employee has that wonderful feeling that accompanies him all the time that he does not want to leave this organization, which provided all the material and moral support his needs; thus puts all his energy into trying to stay in his job.

2.3 Relationship Between Internal Marketing and Intent to Stay

Internal marketing practices lead to employees' adherence and increased desire to stay in the workplace, and to achieve organizational commitment that generates a sense of belonging to the organization, which in turn is considered the motivation for innovation and maximum energies. Internal marketing is also considered as a strategic direction concerned with providing a suitable work environment for employee. The success in applying internal marketing practices is reflected positively on external marketing practices, as training and motivating employees and securing a suitable work environment for them makes them capable of their best to serve customers (Alshura et al., 2016; Hasan et al., 2022).

Also, internal marketing has a significant and clear impact on organizational processes, an important factor that led to employee commitment is behavior with employees as internal organizational clients, in order to enhance and improve employee commitment and engagement. Al-Majali et al. (2016) add, that internal marketing is a social process through the organization's application to manage the communication process, and the interactions that take place between the organization and its employees, friendship, the needs of respect between the manager and the employees, in addition to the needs of self-realization such as promotion to a higher position, all of which can only be achieved through the application of internal marketing; Internal marketing is an integrated administrative process among all functions in the organization, which mainly focuses on that all employees possess the experience and skill, to implement operations to satisfy the external customers needs and desires, and this is only done by preparing and motivating them to perform their work effectively and efficiently.

Al-Omari and Al-Yafi (2017) add that there are some things that affect the individual's feeling of wanting to stay in his work or leave; The work environment has a great impact on the employee's desire to stay in his work within the organization, and many organizations have sought to make basic changes in the work environment to make it more comfortable for employees; to keep the inner feeling that he is seriously willing to remain in his organization, The work environment includes several aspects, including what is material, psychological, and social. During the development of industrial psychology, the physical environment for work was given high importance that it has a significant and important impact on employee productivity. The work environment does not consist only of the physical components, but also includes the psychological and social aspect. Accordingly, the research hypothesis can be formulated as follows.

There is a statistically significant impact of internal marketing practices on the intention to stay in the commercial banks in Jordan.

3 Study Model

See Fig. 1.

4 Methodology

4.1 Study Population

The study population represented all employees of commercial banks in the northern region of Jordan in the governorates of (Irbid, Mafraq, Jerash and Ajloun). The researcher used a Convenience Sampling represented by workers in commercial

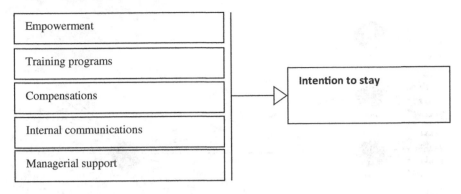

Fig. 1 Theoretical model

banks in the northern region who cooperated with the researcher in distributing the questionnaire. To achieve some criteria such as ease of access to the study sample of workers, and coverage of the northern region in Jordan. The study tool was distributed to the sample with 550 questionnaires, (521) questionnaires capable of statistical analysis were retrieved.

4.2 Study Tool

The study tool included four sections: the first section was for the purpose of collecting data on the demographic factors of the study sample represented by (gender, age, educational qualification, years of experience), while the second section related to the internal marketing practices with its dimensions (empowerment, training programs, and compensations, internal communication, and managerial support). In order to develop this section, the researcher used studies (Khasawneh, 2016; Toa & Martin, 2015; Wilmark, 2018). The third section is represented by the intention to stay based on studies (Hyun-Woom & Yuan, 2015).

4.3 Validity and Reliability

The researcher used the Cronbach's alpha coefficient test to verify the internal consistency of the study tool, in order to verify the reliability of the tool. Convergent Validity and Discriminant Validity were also used to verify the validity of the tool.

Factor loading: To ensure the reliability of the questionnaire statements by examining the path loadings values in order to ensure the existence of a relationship between indicators with latent variables or the absence of (latent variables) for each of them, the researcher also used factor analysis to assess the ability of the factors model to express data. The accepted value is the one with loading values (0.55) or

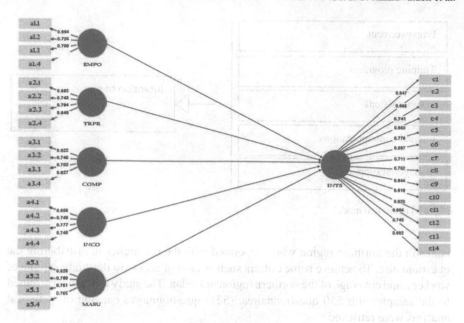

Fig. 2 Factor loading: the proposed model after deleting the statements

higher (de Leeuw et al., 2019). All of the questionnaire statements were accepted except for two statements for the variable (intent to stay) (C13, C14), which got less than (0.55), while the rest of the statements got values higher than (0.55 ≤ Path Loading-Factors Analysis Result-), which are accepted as greater values from (0.55), shown in Fig. 2.

Discriminant Validity: The discriminant validity test measures the degree of divergence of the variables of the study from each other, and based on that judgment is that there is no similarity between the variables and that each variable represents itself, by verifying that there is no level of linear correlation with a high degree (Multicollinearity) between the domains of the independent variable and the domains of the dependent using the criterion Fornell–Larcker Criterion) and using (Squared Average Variance Extracted) and as shown in Table 1.

The square root of the values of the test coefficient of confidence (AVE): Average Variance Extracted must be higher than the values of the correlation of the corresponding row and column variables up to half of the study tool with its differential validity (Hair et al., 2017), and this shows that the square roots of (AVE) values It is higher than the values of the correlation coefficient between the row dimensions and the corresponding columns, which reflects that the tool is characterized by its discriminant validity. Therefore, the results of the tests (convergent validity and discriminant validity) showed the validity and reliability of all measures of the study tool. The results, as they are in Table 1, indicate that the degree of correlation

Table 1 The (Fornell–Larcker criterion for multicollinearity) between the dimensions of the independent variable

	Empowerment	Training programs	Compensations	Internal communications	Managerial support
Empowerment	0.880				
Training programs	0.650	0.870			
Compensations	0.660	0.550	0.854		
Internal communications	0.590	0.731	0.831	0.863	
Managerial support	0.570	0.717	0.849	0.666	1

is medium between the independent study variables, and the values of the correlation coefficients ranged between the independent variables (0.55–0.88) and the significance level ($\alpha \leq 0.05$) for all correlation coefficients.

Composite Reliability (CR): The analysis was used to ensure that all the variables of the proposed model are reliable. This test is considered to be better compared to Cronbach's alpha test because it takes all path loading values into account, unlike Cronbach's alpha test, which considers all path loading values to be equal. And constant for all indicators (Rimkeviciene et al., 2017), the values of the study variables showed acceptance (CR $\geq$ 0.70), because all values (CR) greater than (0.70), and this is an indication of the stability of the test results. The application of this tool results in the ability to generalize the results of the study in light of its practical limitations.

Average Variance Extracted (AVE): This test is used in order to verify that all study variables in the model are reliable, and (AVE) values are acceptable if the values are greater than (0.50) (Rimkeviciene et al., 2017), and all (AVE) values are greater than (0.50). This is an indication of the acceptance of the reliability of the variables of the proposed model. Table 2 shows the summary of the values of (AVE) which is greater than (0.5) and also the (CR) values of the model variables, which are greater than the values (0.70) and indicate the acceptance of the reliability of the model variables and reflects the convergent validity of the study tool.

Table 2 Results of (AVE), (CR) and (Cronbach's alpha) test

	(AVE)	CR	Cronbach's alpha
Empowerment	0.57	0.853	0.837
Training programs	0.600	0.90	0.881
Compensations	0.741	0.891	0.874
Internal communications	0.630	0.840	0.892
Managerial support	0.54	0.799	0.822
Empowerment	0.540	0.900	0.902

Table 3 Results of the variance inflation factor (VIF)

	VIF
Empowerment	2.331
Training programs	1.893
Compensations	3.422
Internal communications	2.589
Managerial support	2.173

Table 4 Quality indicators of conformity to the study data model (Model Fit)

Indicators	Acceptance indicator	Result
Standardized root mean square residual (SRMR)	Less than 0.08	Accepted
(NFI)	More than 0.90	Accepted
(RMS_theta)	Less than 0.12	Accepted

Autocorrelation test: The researcher used the Variance Inflation Factor (VIF) test to identify the degree of Autocorrelation between the independent variables. The results show as they are in Table 3 that the independent study variables do not suffer from the problem of Autocorrelation, as the values of the variance inflation factor (VIF) are less from (5).

5 Model Fit

Table 4 shows that all the values of indicators used in evaluating the validity of the model were acceptable, as the SRMR ratio reached (0.035) and appears to be less than (0.08), and the NFI value was (0.91) which is greater than the value The minimum permissible value is (0.90), and the value of the square root of the approach error (RMS_theta) is (0.06), which is also less than (0.12).

6 Testing Hypotheses

Table 5 shows that internal marketing practices affect, with statistical significance, the intention to stay in commercial banks in Jordan, with a value of F (30,881). It is a statistically significant value ($\alpha \leq 0.05$), and the R^2 value of (788.) indicates that internal marketing practices explain (78.8%) of the changes in survival intention, and the correlation coefficient reached $R = (62.2\%)$, and this indicates the existence of a relationship Between internal marketing practices and the intention to stay, and the results of the partial analysis of this hypothesis showed that

Table 5 Hypothesis testing

Relation	Standard beta	t value	p value	Result
Empowerment → intention to stay	0.105	1.200	0.363	Not supported
Training programs → intention to stay	−0.224	−2.511	0.033	Supported
Compensations → intention to stay	0.490	5.587	0.000	Supported
Internal communications → intention to stay	0.268	2.793	0.004	Supported
Managerial support → intention to stay	0.210	2.151	0.059	Not supported

Note $* p < 0.05, ** p < 0.01, *** p < 0.001$

internal marketing practices (internal communication, managerial support, training programs) contribute to the direct impact on the intention to stay, and this is shown through the (β) values of (0.105), (0.224), (490.), (268.) respectively at the level of ($0.05 \geq \alpha$), As for each of the practices (empowerment, incentives, and rewards), their effect was not statistically significant, as the significance level reached (0.363) (0.059), respectively, and explained above. The significance ($\alpha \leq 0.05$) of internal marketing practices (empowerment, training programs, managerial support, internal communication, incentives and rewards) in the intention to stay in commercial banks in Jordan.

7 Discussions

Study results showed that Jordanian commercial banks encourage communication, interaction and participation between departments as a major key to achieving and strengthening the relationship between employees through the application of internal marketing concepts, as internal marketing emphasizes teamwork among employees, which is an important factor in their development and maintaining successful business in them. study results also showed that the intention to stay works to reduce many negative phenomena such as the leaving work, absenteeism, and evasion from performing work, increases the level of belonging, leads to increased productivity, and reduces the cost resulting from employee absenteeism, as it creates a feeling that this bank provides, a better career path, better workers effort, plus workers usually believe that there is no discrimination in reward or distribution of other fringe benefits. The study results indicate that there is an impact of internal marketing practices on the intention to stay at work, that the process of empowering the employee and involving him in decision-making has an impact on the intention to stay at work by activating his role as a member of the decision-making process and this enhances the employee's desire to maintain his membership within the organization, When transferring responsibility and authority to the employee, transferring and inviting employees to share information and knowledge, the employee generates a kind of job stability. Because it helps with continuous communication between employees and to ensure that individuals are committed to staying for a longer period of time, and that they have the desire to work, and internal communication works to identify employees who seek to achieve the work system goals. Administrative support also reflects the individual's sense of commitment to work, as he feels the need to continue working, and usually the bank seeks to make basic changes in the work environment to make it more comfortable for employees, and the employee remains with an internal feeling that he seriously wants to stay. The results showed that the training programs carried out by the bank have an impact on the intention to stay in the work by helping the employee, developing skills, improving abilities to perform the tasks assigned to him in a better way, helping to know and taking a lot of new information, and providing him with serious methods for better performance, this generates his ability to perform the duties assigned to him and thus remains attached to the organization and does not want to leave.

8 Recommendations

Based on the results, the study recommends that managers and decision-makers in commercial banks in Jordan invite employees in banks to share the information and knowledge provided by banks through the database and take the necessary decisions. Working on employees' systematic development knowledge, skills and attitudes required by the management of the study sample; In order to qualify them for the work, and in order to raise their production capabilities, or to work on providing them with the necessary skills that appear due to environmental and technical changes that require new skills. The study also recommends giving great attention to the internal communication of the study sample, which can lead to increased employee dedication, exchange and sharing of knowledge and skill, improve overall performance, and communication with the aim of providing a greater information understanding. And working to arouse employee's desire and intent with complete determination without any feelings of doubt to remain within the banks, which generates the desire to stay at work, a positive and comfortable feeling that the employee has towards a particular job or a certain environment.

References

Abidin, H. A. Z., Roslin, R. M., & Kamaluddin, N. (2017). Internal marketing and employees' performance: Relating marketing strategies in human resource efforts. In *Proceedings of the 2nd Advances in Business Research International Conference* (pp. 1–9).

Abu Hamra, S. S. (2017). *Internal marketing in Palestinian Universities and its relationship to achieving competitive advantage.* Unpublished Master's thesis, The Islamic University, Gaza.

Aburayya, A., Alshurideh, M., Alawadhi, D., Alfarsi, A., Taryam, M., & Mubarak, S. (2020). An investigation of the effect of lean six sigma practices on healthcare service quality and patient satisfaction: Testing the mediating role of service quality in Dubai primary healthcare sector. *Journal of Advanced Research in Dynamical and Control Systems, 12*(8), 56–72.

Al Kurdi, B., Alshurideh, M., & Al Afaishat, T. (2020). Employee retention and organizational performance: Evidence from banking industry. *Management Science Letters, 10*(16), 3981–3990.

Alameeri, K., Alshurideh, M., Al Kurdi, B., & Salloum, S. A. (2020, October). The effect of work environment happiness on employee leadership. In *International Conference on Advanced Intelligent Systems and Informatics* (pp. 668–680). Cham: Springer.

Al-bawaia, E., Alshurideh, M., Obeidat, B., & Masa'deh, R. (2022). The impact of corporate culture and employee motivation on organization effectiveness in Jordanian banking sector. *Academy of Strategic Management Journal, 21*(Special Issue 2), 1–18.

Al-Dhuhouri, F. S., Alshurideh, M., Al Kurdi, B., & Salloum, S. A. (2020, October). Enhancing our understanding of the relationship between leadership, team characteristics, emotional intelligence and their effect on team performance: A critical review. In *International Conference on Advanced Intelligent Systems and Informatics* (pp. 644–655). Cham: Springer.

Al-Dmour, A., Al-Dmour, H., Al-Barghuthi, R., Al-Dmour, R., & Alshurideh, M. T. (2021). Factors influencing the adoption of e-payment during pandemic outbreak (COVID-19): Empirical evidence. In *The Effect of Coronavirus Disease (COVID-19) on Business Intelligence* (Vol. 334), pp. 133–154.

Alhalalmeh, M. I., Almomani, H. M., Altarifi, S., Al-Quran, A. Z., Mohammad, A. A., & Al-Hawary, S. I. (2020). The nexus between corporate social responsibility and organizational performance in Jordan: The mediating role of organizational commitment and organizational citizenship behavior. *Test Engineering and Management, 83*, 6391–6410.

AlHamad, A., Alshurideh, M., Alomari, K., Kurdi, B., Alzoubi, H., Hamouche, S., & Al-Hawary, S. (2022). The effect of electronic human resources management on organizational health of telecommunications companies in Jordan. *International Journal of Data and Network Science, 6*(2), 429–438.

Al-Hawary, S. I. S., & Mohammed, A. K. (2017). Impact of team work traits on organizational citizenship behavior from the viewpoint of the employees in the education directorates in north region of Jordan. *Global Journal of Management and Business, 17*(2-A), 23–40.

Al-Hawary, S. I., & Al-Syasneh, M. S. (2020). Impact of dynamic strategic capabilities on strategic entrepreneurship in presence of outsourcing of five stars hotels in Jordan. *Business: Theory and Practice, 21*(2), 578–587.

Al-Hawary, S. I. (2013). The role of perceived quality and satisfaction in explaining customer brand loyalty: Mobile phone service in Jordan. *International Journal of Business Innovation and Research, 7*(4), 393–413.

Al-Hawary, S. I., & Abu-Laimon, A. A. (2013). The impact of TQM practices on service quality in cellular communication companies in Jordan. *International Journal of Productivity and Quality Management, 11*(4), 446–474.

Al-Hawary, S. I., & Alajmi, H. M. (2017). Organizational commitment of the employees of the ports security affairs of the state of Kuwait: The impact of human recourses management practices. *International Journal of Academic Research in Economics and Management Sciences, 6*(1), 52–78.

Al-Hawary, S. I., Al-Awawdeh, W., & Abden, M. A. (2012). The impact of the leadership style on organizational commitment: A field study on Kuwaiti telecommunications companies. *ALEDARI, 130*, 53–102.

Al-Hawary, S. I. S., & Alhajri, T. M. S. (2020). Effect of electronic customer relationship management on customers' electronic satisfaction of communication companies in Kuwait. *Calitatea, 21*(175), 97–102.

Al-Hawary, S. I., Al-Hawajreh, K., Al-Zeaud, H., & Mohammad, A. (2013b). The impact of market orientation strategy on performance of commercial banks in Jordan. *International Journal of Business Information Systems, 14*(3), 261–279.

Al-Hawary, S. I., & Al-Menhaly, S. (2016). The quality of e-government services and its role on achieving beneficiaries satisfaction. *Global Journal of Management and Business Research: A Administration and Management, 16*(11), 1–11.

Al-Hawary, S. I., & Al-Namlan, A. (2018). Impact of electronic human resources management on the organizational learning at the private hospitals in the state of Qatar. *Global Journal of Management and Business Research: A Administration and Management, 18*(7), 1–11.

Al-Hawary, S. I., Al-Qudah, K., Abutayeh, P., Abutayeh, S., & Al-Zyadat, D. (2013a). The impact of internal marketing on employee's job satisfaction of commercial banks in Jordan. *Interdisciplinary Journal of Contemporary Research in Business, 4*(9), 811–826.

Al-Hawary, S. I., & Al-Smeran, W. (2017). Impact of electronic service quality on customers satisfaction of Islamic Banks in Jordan. *International Journal of Academic Research in Accounting, Finance and Management Sciences, 7*(1), 170–188.

Al-Hawary, S. I., & Batayneh, A. M. (2010). The effect of marketing communication tools on non-Jordanian students' choice of Jordanian Public Universities: A field study. *International Management Review, 6*(2), 90–99.

Al-Hawary, S. I., Batayneh, A. M., Mohammad, A. A., & Alsarahni, A. H. (2017). Supply chain flexibility aspects and their impact on customers satisfaction of pharmaceutical industry in Jordan. *International Journal of Business Performance and Supply Chain Modelling, 9*(4), 326–343.

Al-Hawary, S. I., & Harahsheh, S. (2014). Factors affecting Jordanian consumer loyalty toward cellular phone brand. *International Journal of Economics and Business Research, 7*(3), 349–375.

Al-Hawary, S. I., & Hussien, A. J. (2017). The impact of electronic banking services on the customers loyalty of commercial banks in Jordan. *International Journal of Academic Research in Accounting, Finance and Management Sciences, 7*(1), 50–63.

Al-Hawary, S. I. S., Mohammad, A. S., Al-Syasneh, M. S., Qandah, M. S. F., & Alhajri, T. M. S. (2020). Organisational learning capabilities of the commercial banks in Jordan: Do electronic human resources management practices matter? *International Journal of Learning and Intellectual Capital, 17*(3), 242–266.

Al-Hawary, S. I., & Nusair, W. (2017). Impact of human resource strategies on perceived organizational support at Jordanian Public Universities. *Global Journal of Management and Business Research: A Administration and Management, 17*(1), 68–82.

Al-Hawary, S. I. S., & Obiadat, A. A. (2021). Does mobile marketing affect customer loyalty in Jordan? *International Journal of Business Excellence, 23*(2), 226–250.

Aljumah, A., Nuseir, M. T., & Alshurideh, M. T. (2021). The impact of social media marketing communications on consumer response during the COVID-19: Does the brand equity of a university matter. In *The Effect of Coronavirus Disease (COVID-19) on Business Intelligence* (pp. 367–384).

Al-Khayyal, A., Alshurideh, M., Al Kurdi, B., & Salloum, S. A. (2021). Factors influencing electronic service quality on electronic loyalty in online shopping context: Data analysis approach. In *Enabling AI Applications in Data Science* (pp. 367–378). Cham: Springer.

Al-Khayyal, A., Alshurideh, M., Al Kurdi, B., & Aburayya, A. (2020). The impact of electronic service quality dimensions on customers' e-shopping and e-loyalty via the impact of e-satisfaction and e-trust: A qualitative approach. *International Journal of Innovation, Creativity and Change, 14*(9), 257–281.

Alkitbi, S. S., Alshurideh, M., Al Kurdi, B., & Salloum, S. A. (2020, October). Factors affect customer retention: A systematic review. In *International Conference on Advanced Intelligent Systems and Informatics* (pp. 656–667). Cham: Springer.

Allozi, A., Alshurideh, M., AlHamad, A., & Al Kurdi, B. (2022). Impact of transformational leadership on the job satisfaction with the moderating role of organizational commitment: Case of UAE and Jordan manufacturing companies. *Academy of Strategic Management Journal, 21*, 1–13.

Al-Lozi, M., Almomani, R. Z., & Al-Hawary, S. I. (2017). Impact of talent management on achieving organizational excellence in Arab Potash Company in Jordan. *Global Journal of Management and Business Research: A Administration and Management, 17*(7), 15–25.

Al-Lozi, M., Almomani, R. Z., & Al-Hawary, S. I. (2018). Talent management strategies as a critical success factor for effectiveness of human resources information systems in commercial banks working in Jordan. *Global Journal of Management and Business Research: A Administration and Management, 18*(1), 30–43.

Al-Nady, B. A., Al-Hawary, S. I., & Alolayyan, M. (2016). The role of time, communication, and cost management on project management success: An empirical study on sample of construction projects customers in Makkah City, Kingdom of Saudi Arabia. *International Journal of Services and Operations Management, 23*(1), 76–112.

Alolayyan, M., Al-Hawary, S. I., Mohammad, A. A., & Al-Nady, B. A. (2018). Banking service quality provided by commercial banks and customer satisfaction. A structural equation modelling approaches. *International Journal of Productivity and Quality Management, 24*(4), 543–565.

Al-Omari, Mb. S., & Al-Yafi, R. S. (2017). The impact of the elements of quality of work life on the performance of the general employee: An applied study on civil servants in the Kingdom of Saudi Arabia. *The Jordanian Journal of Business Administration, 13*(1), 65–94.

Alshamsi, A., Alshurideh, M., Al Kurdi, B., & Salloum, S. A. (2020, October). The influence of service quality on customer retention: A systematic review in the higher education. In *International Conference on Advanced Intelligent Systems and Informatics* (pp. 404–416). Cham: Springer.

AlShehhi, H., Alshurideh, M., Al Kurdi, B., & Salloum, S. A. (2020, October). The impact of ethical leadership on employees performance: A systematic review. In *International Conference on Advanced Intelligent Systems and Informatics* (pp. 417–426). Cham: Springer.

Alshraideh, A. T. R., Al-Lozi, M., & Alshurideh, M. T. (2017). The impact of training strategy on organizational loyalty via the mediating variables of organizational satisfaction and organizational performance: An empirical study on Jordanian agricultural credit corporation staff. *Journal of Social Sciences (COES&RJ-JSS), 6*(2), 383–394.

Alshura, M., Nusair, W., & Aldaihani, F. (2016). Impact of internal marketing practices on the organizational commitment of the employees of the insurance companies in Jordan. *International Journal of Academic Research in Economics and Management Sciences, 5*(4), 168–187.

Alshurideh, M. T., Al Kurdi, B., Alzoubi, H. M., Ghazal, T. M., Said, R. A., AlHamad, A. Q., et al. (2022). Fuzzy assisted human resource management for supply chain management issues. *Annals of Operations Research*, 1–19.

Alshurideh, D. M. (2019). Do electronic loyalty programs still drive customer choice and repeat purchase behaviour? *International Journal of Electronic Customer Relationship Management, 12*(1), 40–57.

Alshurideh, M. (2022). Does electronic customer relationship management (E-CRM) affect service quality at private hospitals in Jordan? *Uncertain Supply Chain Management, 10*(2), 1–8.

Alshurideh, M., Al Kurdi, B. H., Vij, A., Obiedat, Z., & Naser, A. (2016). Marketing ethics and relationship marketing—An empirical study that measure the effect of ethics practices application on maintaining relationships with customers. *International Business Research, 9*(9), 78–90.

Alshurideh, M., Alhadid, A. Y., & Barween, A. (2015). The effect of internal marketing on organizational citizenship behavior an applicable study on the University of Jordan employees. *International Journal of Marketing Studies, 7*(1), 138–145.

Alshurideh, M., Al-Hawary, S. I., Batayneh, A. M., Mohammad, A., & Al-Kurdi, B. (2017). The impact of Islamic banks' service quality perception on Jordanian customers loyalty. *Journal of Management Research, 9*(2), 139–159.

Alshurideh, M., Gasaymeh, A., Ahmed, G., Alzoubi, H., & Kurd, B. (2020). Loyalty program effectiveness: Theoretical reviews and practical proofs. *Uncertain Supply Chain Management, 8*(3), 599–612.

Alsuwaidi, M., Alshurideh, M., Al Kurdi, B., & Salloum, S. A. (2020, October). Performance appraisal on employees' motivation: a comprehensive analysis. In *International Conference on Advanced Intelligent Systems and Informatics* (pp. 681–693). Cham: Springer.

Altarifi, S., Al-Hawary, S. I. S., & Al Sakkal, M. E. E. (2015). Determinants of e-shopping and its effect on consumer purchasing decision in Jordan. *International Journal of Business and Social Science, 6*(1), 81–92.

AlTaweel, I. R., & Al-Hawary, S. I. (2021). The mediating role of innovation capability on the relationship between strategic agility and organizational performance. *Sustainability, 13*(14), 7564.

Alyammahi, A., Alshurideh, M., Al Kurdi, B., & Salloum, S. A. (2020, October). The impacts of communication ethics on workplace decision making and productivity. In *International Conference on Advanced Intelligent Systems and Informatics* (pp. 488–500). Cham: Springer.

Alzoubi, H., Alshurideh, M., Kurdi, B., Akour, I., & Aziz, R. (2022). Does BLE technology contribute towards improving marketing strategies, customers' satisfaction and loyalty? The role of open innovation. *International Journal of Data and Network Science, 6*(2), 449–460.

de Leeuw, E., Hox, J., Silber, H., Struminskaya, B., & Vis, C. (2019). Development of an international survey attitude scale: Measurement equivalence, reliability, and predictive validity. *Measurement Instruments for the Social Sciences, 1*(1), 9. https://doi.org/10.1186/s42409-019-0012-x

ELSamen, A. A., & Alshurideh, M. (2012). The impact of internal marketing on internal service quality: A case study in a Jordanian pharmaceutical company. *International Journal of Business and Management, 7*(19), 84–95.

Hair, J. F., Babin, B. J., & Krey, N. (2017). Covariance-based structural equation modeling in the journal of advertising: Review and recommendations. *Journal of Advertising, 46*(1), 163–177. https://doi.org/10.1080/00913367.2017.1281777

Halawa, I. M. A. (2015). *The ambiguity of career life and its impact on job performance.* Unpublished Master's thesis, The Islamic University, Gaza.

Hasan, O., McColl, J., Pfefferkorn, T., Hamadneh, S., Alshurideh, M., & Kurdi, B. (2022). Consumer attitudes towards the use of autonomous vehicles: Evidence from United Kingdom taxi services. *International Journal of Data and Network Science, 6*(2), 537–550.

Hayajneh, N., Suifan, T., Obeidat, B., Abuhashesh, M., Alshurideh, M., & Masa'deh, R. (2021). The relationship between organizational changes and job satisfaction through the mediating role of job stress in the Jordanian telecommunication sector. *Management Science Letters, 11*(1), 315–326.

Hyun-Woo, L., & Yuan, J. (2015). Investigating relationships between internal marketing practices and employee organizational commitment in the foodservice industry. *International Journal of Contemporary Hospitality Management, 27*(7), 1618–1640.

Khasawneh, M. A., Abuhashesh, M., Ahmad, A., Masa'deh, R., & Alshurideh, M. T. (2021b). Customers online engagement with social media influencers' content related to COVID 19. In *The Effect of Coronavirus Disease (COVID-19) on Business Intelligence* (pp. 385–404). Cham: Springer.

Khasawneh, M. A., Abuhashesh, M., Ahmad, A., Alshurideh, M. T., & Masa'deh, R. (2021a). Determinants of e-word of mouth on social media during COVID-19 outbreaks: An empirical study. In *The Effect of Coronavirus Disease (COVID-19) on Business Intelligence* (pp. 347–366). Cham: Springer.

Khasawneh, Y. (2016). Impact of application of the internal marketing on job satisfaction for the employees in the Islamic banks a case study of Jordan. *International Journal of Managerial Studies and Research, 4*(6), 33–45.

Kurdi, B., Alshurideh, M., & Alnaser, A. (2020). The impact of employee satisfaction on customer satisfaction: Theoretical and empirical underpinning. *Management Science Letters, 10*(15), 3561–3570.

Lee, K., Azmi, N., Hanaysha, J., Alshurideh, M., & Alzoubi, H. (2022a). The effect of digital supply chain on organizational performance: An empirical study in Malaysia manufacturing industry. *Uncertain Supply Chain Management, 10*(2), 1–16.

Lee, K., Ramiz, P., Hanaysha, J., Alzoubi, H., & Alshurideh, M. (2022b). Investigating the impact of benefits and challenges of IOT adoption on supply chain performance and organizational performance: An empirical study in Malaysia. *Uncertain Supply Chain Management, 10*(2), 1–14.

Majali, M. M., Al-Bashabsha, A. A., & Al-Majali, H. M. (2016). The effect of internal marketing dimensions on employee satisfaction and its impact on their performance in commercial banks operating in Karak Governorate. *12*(3), 579–626.

Metabis, A., & Al-Hawary, S. I. (2013). The impact of internal marketing practices on services quality of commercial banks in Jordan. *International Journal of Services and Operations Management, 15*(3), 313–337.

Mohammad, A. A., Alshura, M. S., Al-Hawary, S. I. S., Al-Syasneh, M. S., & Alhajri, T. M. (2020). The influence of internal marketing practices on the employees' intention to leave: A study of the private hospitals in Jordan. *International Journal of Advanced Science and Technology, 29*(5), 1174–1189.

Ramos, W. (2015). Internal marketing (IM) dimensions and organizational commitment (OC) of universal banks' employees. In *Proceedings of the Second Asia-Pacific Conference on Global Business, Economics, Finance and Social Sciences (AP15Vietnam Conference).* ISBN: 978-1-63415-833-6.

Rimkeviciene, J., Hawgood, J., O'Gorman, J., & De Leo, D. (2017). Construct validity of the acquired capability for suicide scale: Factor structure, convergent and discriminant validity. *Journal of Psychopathology and Behavioral Assessment, 39*(2), 291–302. https://doi.org/10.1007/s10862-016-9576-4

Sadiq, R. M. (2018). The role of organizational climate dimensions in enhancing organizational commitment. *Journal of Kirkuk University for Administrative and Economic Sciences, 8*(2), 141–169.

Shamout, M., Elayan, M., Rawashdeh, A., Kurdi, B., & Alshurideh, M. (2022). E-HRM practices and sustainable competitive advantage from HR practitioner's perspective: A mediated moderation analysis. *International Journal of Data and Network Science, 6*(1), 165–178.

Shehada, M., & Omar, K. (2015). The impact of internal marketing on retaining employees at private hospitals in Amman, Jordan. *International Journal of Management and Commerce Innovations, 2*(2), 606–619.

Solomon, M. (2017). *The effect of internal marketing on customer satisfaction: The case of commercial bank of Ethiopia (CBE).* Master thesis, Addis Ababa University.

Svoboda, P., Ghazal, T. M., Afifi, M. A., Kalra, D., Alshurideh, M. T., & Alzoubi, H. M. (2021, June). Information systems integration to enhance operational customer relationship management in the pharmaceutical industry. In *The International Conference on Artificial Intelligence and Computer Vision* (pp. 553–572). Cham: Springer.

Sweiss, N., Obeidat, Z. M., Al-Dweeri, R. M., Mohammad Khalaf Ahmad, A., Obeidat, A. M., & Alshurideh, M. (2021). The moderating role of perceived company effort in mitigating customer misconduct within Online Brand Communities (OBC). *Journal of Marketing Communications,* 1–24.

Tariq, E., Alshurideh, M., Akour, I., & Al-Hawary, S. (2022). The effect of digital marketing capabilities on organizational ambidexterity of the information technology sector. *International Journal of Data and Network Science, 6*(2), 401–408.

Toa & Martin. (2015). Effect of management commitment to internal marketing on employee work attitude. *International Journal of Hospitality Management, 45*, 14–21.

Wilmark, J. R. (2018). Internal marketing dimensions and organizational commitment of universal banks' employees. *Review of Integrative Business and Economics Research, 7*(2), 39–51.

The Effect of Electronic Marketing on Customer Satisfaction of the Insurance Companies in Jordan

Abdullah Matar Al-Adamat, Mohammad Issa Ghafel Alkhawaldeh, Sabah Sameer Mansour, Mohammad Mousa Eldahamsheh, Anber Abraheem Shlash Mohammad, Mohammed Saleem Khlif Alshura, Muhammad Turki Alshurideh ⓘ, Sulieman Ibraheem Shelash Al-Hawary, and D. Barween Al Kurdi ⓘ

Abstract This study aimed to identify the impact of electronic marketing on customer satisfaction of the insurance companies in Jordan, the researchers relied on the quantitative approach by the descriptive analytical method, A questionnaire was

A. M. Al-Adamat · S. S. Mansour · S. I. S. Al-Hawary (✉)
Department of Business Administration & Public Administration, School of Business, Al Al-Bayt University Jordan, P.O. Box 130040, Mafraq 25113, Jordan
e-mail: dr_sliman@yahoo.com; dr_sliman73@aabu.edu.jo

A. M. Al-Adamat
e-mail: aaladamat@aabu.edu.jo

S. S. Mansour
e-mail: aaladamat@aabu.edu.jo

M. I. G. Alkhawaldeh
Directorates of Building and Land Tax, Ministry of Local Administration, Zarqa Municipality, Jordan

M. M. Eldahamsheh
Strategic Management, Amman, Jordan

A. A. S. Mohammad
Marketing Department, Faculty of Administrative and Financial Sciences, Petra University, P.O. Box: 961343, Amman 11196, Jordan

M. S. K. Alshura
Management Department, Faculty of Money and Management, The World Islamic Science University, P.O. Box 1101, Amman 11947, Jordan

M. T. Alshurideh
Department of Marketing, School of Business, The University of Jordan, Amman 11942, Jordan
e-mail: m.alshurideh@ju.edu.jo; malshurideh@sharjah.ac.ae

Department of Management, College of Business, University of Sharjah, 27272 Sharjah, United Arab Emirates

D. B. Al Kurdi
Department of Marketing, Faculty of Economics and Administrative Sciences, The Hashemite University, Zarqa, Jordan

© The Author(s), under exclusive license to Springer Nature Switzerland AG 2023
M. Alshurideh et al. (eds.), *The Effect of Information Technology on Business and Marketing Intelligence Systems*, Studies in Computational Intelligence 1056,
https://doi.org/10.1007/978-3-031-12382-5_123

developed and distributed to (440) clients of insurance companies affiliated with the Jordanian Federation of Insurance Companies, which numbered (22) companies in Jordan. The study reached several results, the most important of which are: There is an impact of e-marketing on customer satisfaction, and the study concluded a number of recommendations, including: A focus on their website to be designed in an easy-to-use way and to provide all the required information about the company that pertain to the customer and the requirements of his transactions, leading to its progression more and more instead of regressing.

Keywords E-marketing · Customer satisfaction · Insurance companies · Jordan

1 Introduction

This era is witnessing a lot of developments in many fields; Perhaps the most important of them is technological development, as this development has led to differences in human needs on the one hand, and the growth of companies and their business management on the other hand, in addition to their reliance on electronic marketing, which has become widely used instead of traditional methods of marketing, which provides many services to customers and companies on both (Al-Hawary & Mohammad, 2011; Al-Hawary & Obiadat, 2021; Al-Hawary et al., 2011a, 2011b).

The insurance sector is one of the sectors concerned with customer satisfaction and loyalty; hence, it is necessary to use the best methods to reach them and provide the best services, especially insurance companies in Jordan, as they have tended to adopt electronic marketing to keep pace with the development of customers' needs and meet their desires in a more effective manner (Aburayya et al., 2020; Al-Dmour et al., 2021; Alshurideh, 2022; Alshurideh et al., 2012; Alzoubi et al., 2022; Kurdi et al., 2020). Insurance companies are concerned with the rights, duties, penalties and transactions of individuals. Individuals prefer to resort to insurance companies out of a sense of safety from any expected or actual damage; As a result of vehicle accidents, fire, theft, and others, which are circumstances beyond their control, insurance comes with the least losses to the customer. The insurance also preserves the relations between members of society from conflicts and disagreements as much as possible.

Insurance companies seek to know the extent of customer satisfaction with their services provided to him, especially with the increase in competition, so they try to adopt everything new to improve customer satisfaction (Allozi et al., 2022; Alzoubi et al., 2020; Sultan et al., 2021). E-marketing may play an important role in gaining customer satisfaction, which is one of the important indicators of the success of that organization and enhancing its market share (Al-Khayyal et al., 2021; Alshurideh, 2019; Alshurideh et al., 2021). Through the review of the previous studies, it became

e-mail: barween@hu.edu.jo

clear that there is a dearth of studies that dealt with the issue of the impact of electronic marketing on achieving customer satisfaction in the insurance sector in developing countries, including Jordan. The study also attempts to verify the extent to which electronic marketing is employed in the insurance sector in Jordan, and to reach customer satisfaction in order to meet their needs and satisfy their desires in conducting their transactions and the procedures followed. It will also help decision makers in the insurance sector to take the best means and channels used in e-marketing. Therefore, this study attempts to study the impact of electronic marketing on customer satisfaction for insurance companies in Jordan.

2 Theoretical Framework and Hypothesis Building

2.1 Electronic Marketing

Electronic marketing has the largest share in the field of marketing compared to traditional marketing, and with the passage of time and the communications revolution and the tremendous development in technology, electronic marketing has become of great importance in the business sectors as a whole, and it is the cornerstone in order to increase its market share (Al-Hawary & Alhajri, 2020; Al-Hawary & Hussien, 2017; AlHamad et al., 2022). E-marketing is the latest way to promote goods and services, and companies have to find a direction in order to follow up on the buyer's requirements and know their opinions (Alwan & Alshurideh, 2022; Tariq et al., 2022a, 2022b). However, this creates competitive capabilities for them, enabling them to achieve a competitive advantage that would increase their market share and thus provide the best products of goods and services at competitive prices and fast services, and quality (Al Kurdi & Alshurideh, 2021; Al-Hawary et al., 2011a, 2011b; Aljumah et al., 2021).

There are many concepts of e-marketing by different researchers, this is due to the multiplicity of electronic technologies that can be used in building electronic marketing and the methods and forms that it may take, but they are similar in their content. Electronic marketing has been defined as: the use of a market without being governed by local or temporal behaviors, so that the Internet is used as a platform for institutions to adapt to the needs of customer while minimizing costs (Al-Hawary & Al-Menhaly, 2016; Al-Hawary & Al-Smeran, 2017; Watson et al., 2002). Tawfiq and Omar (2020), see it as a technology of change because it witnessed transformations in the philosophy and applications of marketing, as it is an activity that uses media technology, especially the Internet, to market products of goods and services without direct contact. Accordingly, e-marketing can be considered as a group of activities managed through electronic tools and technologies provided by the Internet at a low cost, less effort and a shorter time than in traditional marketing, with the aim of achieving common interests between organizations and individuals within well-studied strategies to reach the desired goals and required visions (Al-Hawary et al., 2013; Al-Nady et al., 2016; Alshurideh et al., 2016).

The importance of e-marketing lies in the fact that it is a double relationship between the seller and the buyer, as e-marketing addresses many of the problems of traditional marketing, especially with regard to crowded markets and the remoteness of the site. It has become possible to access goods and services without having to leave the house, with the opportunity to get to know what the customer wants before buying and what is being traded in the market for goods and services on the one hand, On the other hand, organizations and stores work to deliver products of goods and services to the consumer wherever he is through networks of advanced delivery systems and through systems supporting e-marketing, and e-marketing helps to support any new product or commodity in its various stages, e-marketing is one of the electronic works within a group of the businesses that constitute an integrated approach, such as production, finance, human resources, research and development, and information technology, which led to the establishment of the so-called electronic economy (Al-Hawary & Abu-Laimon, 2013; Al-Lozi et al., 2017, 2018; Alkalha et al., 2012; Alshurideh et al., 2022; Shamout et al., 2022).

One of the main goals of organizations is to reach customers in easy ways, communicate with them and know their needs and desires constantly, and provide them with the best products and services. This is not done in electronic marketing except through tools that provide many advantages; Such as low cost, and reduces the time to complete operations (Al-Hawary, 2013, 2015; Al-Hawary & Al-Syasneh, 2020; Al-Hawary & Alwan, 2016). Therefore, organizations resort to taking benefit of the advantages of electronic marketing tools and its role in developing the relationship with customers, communicating with them and maintaining them in the long term, by achieving value for them. This study also shows the most important of these tools as follows:

E-mail: E-mail is one of the tools that has kept pace with change well. E-mail marketing has become one of the largest platforms for successfully and profitably marketing products and services. One of the most important companies that use e-commerce to increase purchases of their products is Amazon (Al-Masaeed, 2021).

Website design: The first step that organizations which rely on the Internet is to design a website, with amazing technical methods and characteristics; So that it is a site that displays the strengths of the organization in its products and services, and the way of displaying the information in easy way to access for customers, companies offer through the site many benefits to customers such as pricing and means of payment, so organizations must ensure that it is of real benefit and not just a site that carries only name only.

Social networks: It is a set of Internet-based applications that rely on the ideological and technological foundations of the Web, which allow the creation and exchange of content as an important form of new media, social media has been dedicated to social interactions as well (Alshurideh et al., 2019; Khasawneh et al., 2021a, 2021b; Kurdi et al., 2021 Shen & Bissell, 2013).

Mobile applications: Mobile applications are meant as different mobile platforms running different operating systems in smart mobile devices, multimedia and communication functions as well as access to phones that support those systems (Sansour, 2016). Applications were the way to conduct the necessary transactions and procedures in insurance companies, especially during the period of the Corona pandemic, as the Jordanian Insurance Federation launched several programs, including "Safe in Your Car" and it was a program that prevents customers from mixing and crowding during the pandemic (JIF, 2021).

Internet advertisements: These advertisements are interactive between what the organization offers and customer reactions, with the least effort, cost and time period (Draz and Muslim, 2015). The interest in e-marketing has started in recent years remarkably, especially during the Corona pandemic, as insurance companies affiliated with the Jordan Insurance Federation adopted many tools to market their services electronically, as they launched the (Safe in Your Car) program, which was announced on their website and Facebook page. This program made the task easier for customers, protecting them from Corona at this time, so e-marketing was more successful in reaching the largest segment of customers and clarifying the program for all of them at the same time.

2.2 Customers' Satisfaction

Customer satisfaction is known through his reaction to the product. In marketing, customer satisfaction is the difference between the quality level of the product and his expectations. The organization must give it priority, since the customer and his satisfaction have a positive impact on profit and non-profit organizations, focusing on that, which was Satisfied the customer in the past, does not satisfy him today; This is due to the fluctuations of the market and its change, and the change of customers by what affects those around them (AlTaweel & Al-Hawary, 2021). Companies that follow electronic marketing in their transactions through Internet advertisements via websites must be characterized by credibility in the content of the advertisement first, and then complete an approach that suits the customer by providing information and supplying them with it to discover the characteristics of the product, whatever it is a good or service. Advertisement on unconventional rigid texts creates a type of entertainment that contributes to attracting the consumer, all of which is sufficient to gain customer satisfaction, and increases his purchase intention (Al-Hawary et al., 2011a, 2011b; Google, 2015; Shishan et al., 2022).

2.3 Electronic Marketing and Customer Satisfaction

The researchers dealt with the relationship between e-marketing and customer satisfaction in important economic, service and industrial sectors, where Mahmood &

Mahmood (2021) found a positive impact of e-marketing dimensions in achieving customer satisfaction of the Iraqi Oil Distribution Company. As confirmed by Bizhanova et al. (2019) On the impact of electronic marketing on entrepreneurship, the researcher also found that the increasing digitization leads to serious problems for executives in the field of marketing, as they face growing, complex and rapidly changing markets that are out of their control, as a result, companies have realized these changes and how to deal with them, as Electronic marketing is used for the regular review, analysis, interpretation and promotion process, It helps create strong consumer and brand relationships through search engine optimization, promotion, email, mailing list, contextual advertising and advertising in social networks, all of which increase sales while facilitating the purchase of goods and services. The findings of the Nakra and Pandey study (2019) underscore the importance of assisting hospital administration, governments, and other organizations in formulating appropriate policy for the use of smartphones by healthcare professionals. To find out the impact of social media marketing activities, Khan (2019) conducted a study that confirmed that social media marketing activities significantly affect brand loyalty, purchase intentions, value awareness and brand awareness. Salehi (2018) also examined the impact of smartphone use on health care, and the investigation of the applicability of routine clinical smartphones, and it was found that (85%) of the participants, their smartphones contribute to improving the quality of health care communications and help clinicians to prioritize tasks among them. Al-Sharif and Nassraldeen (2017) explored the influence of branding on the relationship between social media marketing on customer satisfaction and emphasized the importance of social media dimensions on customer satisfaction. The study of Bekar (2016) in Turkey also added an emphasis on the impact of brand social media marketing on the intent to purchase fashion and emphasized the need for social media to convince customers of product excellence and thus influence their purchase intent. Based on the above, the study hypothesis can be formulated as follows:

There is a statistically significant effect of e-marketing on the satisfaction of insurance companies' customers in Jordan.

3   Study Model

See Fig. 1.

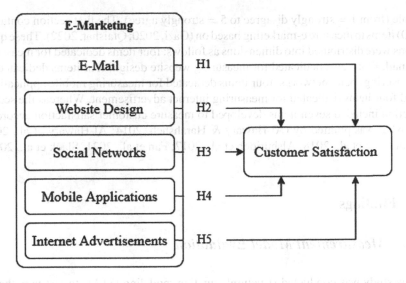

Fig. 1 Research model

4 Methodology

4.1 Population and Sample Selection

A qualitative method based on a questionnaire was used in this study for data collection and sample selection. The major aim of the study was to examine the impact of e-marketing on customer satisfaction. Therefore, it focused on Jordan-based companies operating in the insurance sector. Data were primarily gathered through self-reported questionnaires created by Google Forms which were distributed to a non-probability sample of companies' customers via email. In total, (463) responses were received including (23) invalid to statistical analysis due to uncompleted or inaccurate. Hence, the final sample contained (440) responses suitable to analysis requirements that were formed a response rate of (95%), where it proved to be sufficient to the extent that was predictable and allowed for a presumption of data saturation (Sekaran & Bougie, 2016a, 2016b).

4.2 Measurement Instrument

A self-reported questionnaire that consists of two main sections along with a section regarding control variables was used as the measurement instrument. Control variables considered as categorical measures were composed of gender, age group, educational level, and experience. The two main sections were dealt with a five-point Likert scale (from 1 = strongly disagree to 5 = strongly agree). The first section contained (20) items to measure e-marketing based on (Gazi, 2020; Qtaishat, 2022). These questions were distributed into dimensions as follows: four items dedicated for measuring e-mail, four items dedicated for measuring website design, four items dedicated for measuring social networks, four items dedicated for measuring mobile applications, and four items dedicated for measuring internet advertisement. Whereas the second section included seven items developed to measure customer satisfaction according to what was pointed by (Al-Hawary & Harahsheh, 2014; Al-Hawary et al., 2017; Alolayyan et al., 2018; Alshurideh et al., 2017; Fan et al., 2021; Slack et al., 2020).

5 Findings

5.1 Measurement Model Evaluation

This study was conducted structural equation modeling (SEM) to test hypotheses, which represents a contemporary statistical technique for testing and estimating the relationship between factors and variables (Al-Adamat et al., 2020; Al-Gasawneh and Al-Adamat, 2020; Wang & Rhemtulla, 2021). Accordingly, the reliability and validity of the constructs were tested using confirmatory factor analysis (CFA) through the statistical program AMOSv24. Table 1 summarizes the results of convergent and discriminant validity, as well the indicators of reliability.

Table 1 shows that the standard loading values for the individual items were within the domain (0.674–0.873), these values greater than the minimum retention of the elements based on their standard loads (Al-Lozi et al., 2018; Sung et al., 2019). Average variance extracted (AVE) is a summary indicator of the convergent validity of constructs that must be above 0.50 (Howard, 2018). The results indicate that the AVE values were greater than 0.50 for all constructs, thus the used measurement model has an appropriate convergent validity. Rimkeviciene et al. (2017) suggested the comparison approach as a way to deal with discriminant validity assessment in covariance-based SEM. This approach is based on comparing the values of maximum shared variance (MSV) with the values of AVE, as well as comparing the values of square root of AVE ($\sqrt{AVE}$) with the correlation between the rest of the structures. The results show that the values of MSV were smaller than the values of AVE, and that the values of $\sqrt{AVE}$ were higher than the correlation values among the rest of the constructs. Therefore, the measurement model used is characterized by discriminative validity. The internal consistency measured through Cronbach's Alpha coefficient (α) and compound reliability by McDonald's Omega coefficient (ω) was conducted as indicators to evaluate measurement model. The results listed in Table 1 demonstrated that both values of Cronbach's Alpha coefficient and McDonald's Omega coefficient were greater than 0.70, which is the lowest limit for judging on measurement reliability (De Leeuw et al., 2019).

Table 1 Results of validity and reliability tests

Constructs	1	2	3	4	5	6
1. E-mail	**0.762**					
2. Website design	0.524	**0.781**				
3. Social networks	0.481	0.405	**0.779**			
4. Mobile applications	0.449	0.556	0.492	**0.790**		
5. Internet advertisements	0.537	0.571	0.538	0.553	**0.768**	
6. Customer satisfaction	0.637	0.684	0.624	0.607	0.638	**0.757**
VIF	1.964	2.058	2.248	2.391	2.178	–
Loadings range	0.692–0.814	0.703–0.834	0.683–0.873	0.704–0.836	0.691–0.834	0.674–0.846
AVE	0.581	0.610	0.606	0.625	0.590	0.573
MSV	0.415	0.503	0.495	0.511	0.497	0.514
Internal consistency	0.845	0.859	0.855	0.866	0.848	0.899
Composite reliability	0.847	0.862	0.859	0.869	0.851	0.903

Note Bold fonts refers to root square of average variance extracted

5.2 Structural Model

The structural model illustrated no multicollinearity issue among predictor constructs because variance inflation factor (VIF) values are below the threshold of 5, as shown in Table 1 (Hair et al., 2017). This result is supported by the values of model fit indices shown in Fig. 1.

The results in Fig. 2 indicated that the chi-square to degrees of freedom (CMIN/DF) was 2.466, which is less than 3 the upper limit of this indicator. The values of the goodness of fit index (GFI), the comparative fit index (CFI), and the Tucker-Lewis index (TLI) were upper than the minimum accepted threshold of 0.90. Moreover, the result of root mean square error of approximation (RMSEA) indicated to value 0.051, this value is a reasonable error of approximation because it is less than the higher limit of 0.08. Consequently, the structural model used in this study was recognized as a fit model for predicting the DEP and generalization of its result (Ahmad et al., 2016; Shi et al., 2019). To verify the results of testing the study hypotheses, structural equation modeling (SEM) was used, the results of which are listed in Table 2.

The results demonstrated in Table 2 show that all e-marketing dimensions had a positive impact relationship on customer satisfaction except website design ($\beta = -0.029, t = -0.0568, p = 0.571$). On another hand, the results indicated that the highest

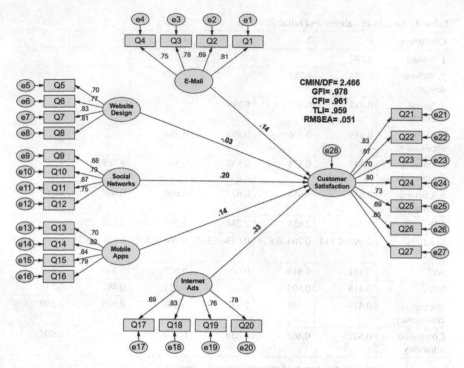

Fig. 2 SEM results of the e-marketing effect on customer satisfaction

Table 2 Hypothesis testing

Hypothesis	Relation	Standard beta	t value	p value
H1	e-mail → customer satisfaction	0.143	2.880	0.004
H2	website design → customer satisfaction	−0.029	−0.568	0.571
H3	social networks → customer satisfaction	0.203	3.826	0.000
H4	mobile applications → customer satisfaction	0.141	2.602	0.010
H5	internet advertisements → customer satisfaction	0.332	6.373	0.000

*Note * p < 0.05, ** p < 0.01, *** p < 0.001*

impact was for internet advertisements ($\beta = 0.332$, $t = 6.373$, $p = 0.000$), followed by social networks ($\beta = 0.203$, $t = 3.826$, $p = 0.000$), then e-mail ($\beta = 0.143$, $t = 2.880$, $p = 0.004$), and finally the lowest impact was for mobile applications ($\beta = 0.141$, $t = 2.602$, $p = 0.010$). Thus, most the minor hypotheses of the study were supported based on these results.

6 Discussion

The results of the analysis revealed that there is an impact of electronic marketing on customer satisfaction of the insurance companies in Jordan, and this result may be due to the fact that electronic marketing help insurance companies to manage their marketing programs and ways of presenting their advertisements to customers to display their services in line with customers' needs and desires and to deal with them easily. And this is what most companies aspire to marketing their products through the Internet, which is now occupying the world with its electronic services, more than traditional marketing. It was also found that there is an impact of e-mail on customer satisfaction of insurance companies in Jordan. The results of the analysis indicated that e-mail is a safe tool for collecting information related to insurance company customer, who preserves the privacy of customers compared to other marketing tools, and e-mail allows more interaction between the insurance company and its customer with recent electronic publications quickly. It was also found that there is no statistically significant effect of website design on customer satisfaction of the insurance companies in Jordan. This result is due to the fact that the website is designed in a somewhat complex manner and is not easy to use by customers, and this makes insurance companies reconsider their website and design method; In order to make it serve customers better, and perform the goal for which it was designed more, the site must be easier and more attractive to customers, and this result agrees with the study of Al-Sabbagh.

The results of the study showed that there is an impact of social networks on customer satisfaction of the insurance companies in Jordan, and this result may be due to the fact that social networks are characterized by always sticking to customers; Therefore, insurance companies have found that these networks are an excellent platform to reach customers easily to market their services and manage their dealings with customers, and insurance companies publish on social networks as a media about their activities and events for their customers, and this result is consistent with Al-Sharif and Nassraldeen (2017). The results of the data analysis added that there is an impact of mobile applications on the satisfaction of insurance companies' customers in Jordan, and this result is due to the fact that the use of mobile applications saves time and effort, especially if they are effectively managed as insurance companies do, as they rely on social networks as a marketing tool to market its services to customers, and this is what the customer prefers, especially if those applications are easy to use, and this result agrees with Salehi (2018) and Nakra and Pandey (2019). The results of the study add that there is an impact of Internet advertisements on customer satisfaction of the insurance companies in Jordan, and this result may be due to the fact that Internet advertisements via social networks help in choosing the appropriate service for customers to conduct insurance in the appropriate manner, and whenever the advertisement online is continuously updated and in a timely manner, whenever it is a tool that targets the desired customers, and directs their behavior towards the concerned institution well, and the result agrees with Al-Sharif and Nassraldeen (2017).

7 Recommendations

In light of the findings of the study, the study recommends that insurance companies need to constantly develop their tools used in electronic marketing, keep pace with contemporary updates for each of them, and link all insurance companies affiliated with the Jordanian Federation on one website, with a focus on being designed in an easy-to-use way and providing all information Required about the companies that pertain to the customer and the requirements of his transactions, the study also recommends employing mobile applications to send notifications of its activities to customers constantly and at the appropriate times in order to achieve the desired goal, and not be satisfied with the programs used, but rather convert them into applications used through smart phones. Finally, insurance companies should intensify their advertisements on various social media sites, promote them to reach the largest segment of customers, and work to intensify control on the insurance companies' website, to check the information presented about the insurance company and the service provided, and to ensure that it matches reality.

References

Aburayya, A., Alshurideh, M., Alawadhi, D., Alfarsi, A., Taryam, M., & Mubarak, S. (2020). An investigation of the effect of lean six sigma practices on healthcare service quality and patient satisfaction: Testing the mediating role of service quality in Dubai primary healthcare sector. *Journal of Advanced Research in Dynamical and Control Systems, 12*(8), 56–72.

Ahmad, S., Zulkurnain, N., & Khairushalimi, F. (2016). Assessing the validity and reliability of a measurement model in structural equation modeling (SEM). *British Journal of Mathematics & Computer Science, 15*(3), 1–8. https://doi.org/10.9734/BJMCS/2016/25183

Al Kurdi, B. H., & Alshurideh, M. T. (2021). Facebook advertising as a marketing tool: Examining the influence on female cosmetic purchasing behaviour. *International Journal of Online Marketing (IJOM), 11*(2), 52–74.

Al-Adamat, A., Al-Gasawneh, J., & Al-Adamat, O. (2020). The impact of moral intelligence on green purchase intention. *Management Science Letters, 10*(9), 2063–2070.

Al-Dmour, R., AlShaar, F., Al-Dmour, H., Masa'deh, R., & Alshurideh, M. T. (2021). The effect of service recovery justices strategies on online customer engagement via the role of "Customer Satisfaction" during the Covid-19 pandemic: An empirical study. In *the effect of coronavirus disease (COVID-19) on business intelligence* (Vol. 334, pp. 325–346).

Al-Gasawneh, J. A., & Al-Adamat, A. M. (2020). The relationship between perceived destination image, social media interaction and travel intentions relating to Neom city. *Academy of Strategic Management Journal, 19*(2), 1–12.

AlHamad, A., Alshurideh, M., Alomari, K., Kurdi, B., Alzoubi, H., Hamouche, S., & Al-Hawary, S. (2022). The effect of electronic human resources management on organizational health of telecommunications companies in Jordan. *International Journal of Data and Network Science, 6*(2), 429–438.

Al-Hawary, S. I. (2013). The role of perceived quality and satisfaction in explaining customer brand loyalty: Mobile phone service in Jordan. *International Journal of Business Innovation and Research, 7*(4), 393–413.

Al-Hawary, S. I. (2015). Human resource management practices as a success factor of knowledge management implementation at health care sector in Jordan. *International Journal of Business and Social Science, 6*(11/1), 83–98.

Al-Hawary, S. I., & Abu-Laimon, A. A. (2013). The impact of TQM practices on service quality in cellular communication companies in Jordan. *International Journal of Productivity and Quality Management, 11*(4), 446–474.

Al-Hawary, S. I. S., & Alhajri, T. M. S. (2020). Effect of electronic customer relationship management on customers' electronic satisfaction of communication companies in Kuwait. *Calitatea, 21*(175), 97–102.

Al-Hawary, S. I., & Al-Menhaly, S. (2016). The Quality of E-government services and its role on achieving beneficiaries satisfaction. *Global Journal of Management and Business Research: A Administration and Management, 16*(11), 1–11.

Al-Hawary, S. I., & Al-Smeran, W. (2017). Impact of electronic service quality on customers satisfaction of Islamic Banks in Jordan. *International Journal of Academic Research in Accounting, Finance and Management Sciences, 7*(1), 170–188.

Al-Hawary, S. I. S., & Alwan, A. M. (2016). Knowledge management and its effect on strategic decisions of Jordanian public universities. *Journal of Accounting-Business & Management, 23*(2), 24–44.

Al-Hawary, S. I., & Harahsheh, S. (2014). Factors affecting Jordanian consumer loyalty toward cellular phone brand. *International Journal of Economics and Business Research, 7*(3), 349–375.

Al-Hawary, S. I., & Hussien, A. J. (2017). The impact of electronic banking services on the customers loyalty of commercial banks in Jordan. *International Journal of Academic Research in Accounting, Finance and Management Sciences, 7*(1), 50–63.

Al-Hawary, S. I., & Mohammad, A. A. (2011). The role of the internet in marketing the services of travel and tourism agencies in Jordan. *Abhath Al-Yarmouk, 27*(2B), 1339–1359.

Al-Hawary, S. I. S., & Obiadat, A. A. (2021). Does mobile marketing affect customer loyalty in Jordan? *International Journal of Business Excellence, 23*(2), 226–250.

Al-Hawary, S. I. S., Alhamali, R. M., & Alghanim, S. A. (2011a). Banking service quality provided by commercial banks and customer satisfaction. *American Journal of Scientific Research, 27*, 68–83.

Al-Hawary, S. I., Mohammad, A. A., & Al-Shoura, M. (2011b). The impact of E-marketing on achieving competitive advantage by the Jordanian pharmaceutical firms. *DIRASAT, 38*(1), 143–160.

Al-Hawary, S. I., Batayneh, A. M., Mohammad, A. A., & Alsarahni, A. H. (2017). Supply chain flexibility aspects and their impact on customers satisfaction of pharmaceutical industry in Jordan. *International Journal of Business Performance and Supply Chain Modelling, 9*(4), 326–343.

Al-Hawary, S. I., & Al-Syasneh, M. S. (2020). Impact of dynamic strategic capabilities on strategic entrepreneurship in presence of outsourcing of five stars hotels in Jordan. *Business: Theory and Practice, 21*(2), 578–587.

Al-Hawary, S. I., Al-Hawajreh, K., AL-Zeaud, H., & Mohammad, A. (2013). The impact of market orientation strategy on performance of commercial banks in Jordan. *International Journal of Business Information Systems, 14*(3), 261–279.

Aljumah, A., Nuseir, M. T., & Alshurideh, M. T. (2021). The impact of social media marketing communications on consumer response during the COVID-19: Does the brand equity of a university matter. In *The effect of coronavirus disease (COVID-19) on business intelligence* (pp. 367–384).

Alkalha et al., 2012Alkalha, Z., Al-Zu'bi, Z., Al-Dmour, H., Alshurideh, M., & Masa'deh, R. (2012). Investigating the effects of human resource policies on organizational performance: An empirical study on commercial banks operating in Jordan. *European Journal of Economics, Finance and Administrative Sciences, 51*(1), 44–64.

Al-Khayyal, A., Alshurideh, M., Al Kurdi, B., & Salloum, S. A. (2021). Factors influencing electronic service quality on electronic loyalty in online shopping context: Data analysis approach. In *Enabling AI applications in data science* (pp. 367–378). Springer, Cham.

Al-Lozi, M., Almomani, R. Z., & Al-Hawary, S. I. (2017). Impact of talent management on achieving organizational excellence in Arab Potash Company in Jordan. *Global Journal of Management and Business Research: A Administration and Management, 17*(7), 15–25.

Al-Lozi, M. S., Almomani, R. Z. Q., & Al-Hawary, S. I. S. (2018). Talent Management strategies as a critical success factor for effectiveness of human resources information systems in commercial banks working in Jordan. *Global Journal of Management and Business Research: A Administration and Management, 18*(1), 30–43.

Allozi, A., Alshurideh, M., AlHamad, A., & Al Kurdi, B. (2022). Impact of transformational leadership on the job satisfaction with the moderating role of organizational commitment: Case of UAE and Jordan manufacturing companies. *Academy of Strategic Management Journal, 21*, 1–13.

Al-Masaeed, M. (2021). The impact of electronic marketing tools on customer interaction in Jordanian telecom companies. MBA Thesis, unpublished, Al al-Bayt University, Jordan.

Al-Nady, B. A., Al-Hawary, S. I., & Alolayyan, M. (2016). The role of time, communication, and cost management on project management success: An empirical study on sample of construction projects customers in Makkah City, Kingdom of Saudi Arabia. *International Journal of Services and Operations Management, 23*(1), 76–112.

Alolayyan, M., Al-Hawary, S. I., Mohammad, A. A., & Al-Nady, B. A. (2018). Banking service quality provided by commercial banks and customer satisfaction. A structural equation modelling approaches. *International Journal of Productivity and Quality Management, 24*(4), 543–565.

Al-Sharif, A. &Nasseraldeen, H. (2017). *The impact of social media marketing on customer satisfaction through brand image (Field Study based on Customers of Jordan Telecommunication Companies, Applied on the Students of the Private Universities that Located in Amman-Jordan).* Middle East University.

Alshurideh, D. M. (2019). Do electronic loyalty programs still drive customer choice and repeat purchase behaviour? *International Journal of Electronic Customer Relationship Management, 12*(1), 40–57.

Alshurideh, M. (2022). Does electronic customer relationship management (E-CRM) affect service quality at private hospitals in Jordan? *Uncertain Supply Chain Management, 10*(2), 325–332.

Alshurideh, M., Masa'deh, R. M. D. T., & Alkurdi, B. (2012). The effect of customer satisfaction upon customer retention in the Jordanian mobile market: An empirical investigation. *European Journal of Economics, Finance and Administrative Sciences, 47*(12), 69–78.

Alshurideh, M., Al Kurdi, B. H., Vij, A., Obiedat, Z., & Naser, A. (2016). Marketing ethics and relationship marketing—An empirical study that measure the effect of ethics practices application on maintaining relationships with customers. *International Business Research, 9*(9), 78–90.

Alshurideh, M., Al-Hawary, S. I., Batayneh, A. M., Mohammad, A., & Al-Kurdi, B. (2017). The impact of Islamic Banks' service quality perception on Jordanian Customers loyalty. *Journal of Management Research, 9*(2), 139–159.

Alshurideh, M., Salloum, S. A., Al Kurdi, B., & Al-Emran, M. (2019). Factors affecting the social networks acceptance: An empirical study using PLS-SEM approach. In *Proceedings of the 2019 8th International Conference on Software and Computer Applications* (pp. 414–418).

Alshurideh, M. T., Al Kurdi, B., & Salloum, S. A. (2021). The moderation effect of gender on accepting electronic payment technology: A study on United Arab Emirates consumers. *Review of International Business and Strategy, 31*(3), 375–396.

Alshurideh, M. T., Al Kurdi, B., Alzoubi, H. M., Ghazal, T. M., Said, R. A., AlHamad, A. Q., et al. (2022). Fuzzy assisted human resource management for supply chain management issues. *Annals of Operations Research*, 1–19.

AlTaweel, I. R., & Al-Hawary, S. I. (2021). The mediating role of innovation capability on the relationship between strategic agility and organizational performance. *Sustainability, 13*(14), 7564.

Alwan, M., & Alshurideh, M. (2022). The effect of digital marketing on purchase intention: Moderating effect of brand equity. *International Journal of Data and Network Science, 10*(3), 1–12.

Alzoubi, H. M., Alshurideh, M., Al Kurdi, B., & Inairat, M. (2020). Do perceived service value, quality, price fairness and service recovery shape customer satisfaction and delight? A practical study in the service telecommunication context. *Uncertain Supply Chain Management, 8*(3), 579–588.

Alzoubi, H., Alshurideh, M., Kurdi, B., Akour, I., & Aziz, R. (2022). Does BLE technology contribute towards improving marketing strategies, customers' satisfaction and loyalty? The role of open innovation. *International Journal of Data and Network Science, 6*(2), 449–460.

Bekar, G. (2016). *Impacts of luxury fashion brand's social media marketing on purchase intention in Turkey: A comparative study on Louis Vuitton and Chanel.* Eastern Mediterranean University.

Bizhanova, K., Mamyrbekov, A., Umarov, I., Orazymbetova, A., & Khairullaeva, A. (2019). *Impact of digital marketing development on entrepreneurship* (p. 135). Al-Farabi Kazakh National University.

De Leeuw, E., Hox, J., Silber, H., Struminskaya, B., & Vis, C. (2019). Development of an international survey attitude scale: Measurement equivalence, reliability, and predictive validity. *Measurement Instruments for the Social Sciences, 1*(1), 9. https://doi.org/10.1186/s42409-019-0012-x

Draz, I., & Muslim, M. (2015). Awareness of the role of Internet advertisements and its relationship to the purchasing decision to furnish a home for those who are about to get married. *Journal of Specific Education Research, Egypt,* (40).

Fan, D., Xiao, C., Zhang, X., & Guo, Y. (2021). Gaining customer satisfaction through sustainable supplier development: The role of firm reputation and marketing communication. *Logistics and Transportation Review, 154*, 102453. https://doi.org/10.1016/j.tre.2021.102453

Gazi, M. (2020). E-marketing practice in Bangladesh: An empirical study on trend of use and expansion in business. *Canadian Journal of Business and Information Studies, 2*(1), 12–23.

Google, S. (2015). The impact of mobile advertising on the purchasing behavior of consumers: A field study on customers of telephone companies in Syria. Master's thesis, Damascus University, Faculty of Economics, Department of Business Administration, Marketing Specialization, Syria.

Hair, J. F., Babin, B. J., & Krey, N. (2017). Covariance-based structural equation modeling in the journal of advertising: Review and recommendations. *Journal of Advertising, 46*(1), 163–177. https://doi.org/10.1080/00913367.2017.1281777

Howard, M. C. (2018). The convergent validity and nomological net of two methods to measure retroactive influences. *Psychology of Consciousness: Theory, Research, and Practice, 5*(3), 324–337. https://doi.org/10.1037/cns0000149

Khan, M. (2019). The impact of perceived social media marketing activities: An empirical study in Saudi. King Saud University, Saudi Arabia. *International Journal of Marketing Studies, 11*(1).

Khasawneh, M. A., Abuhashesh, M., Ahmad, A., Alshurideh, M. T., & Masa'deh, R. (2021a). Determinants of e-word of mouth on social media during COVID-19 outbreaks: An empirical study. In *The effect of coronavirus disease (COVID-19) on business intelligence* (pp. 347–366). Springer, Cham.

Khasawneh, M. A., Abuhashesh, M., Ahmad, A., Masa'deh, R., & Alshurideh, M. T. (2021b). Customers online engagement with social media influencers' content related to COVID 19. In *The effect of coronavirus disease (COVID-19) on business intelligence* (pp. 385–404). Springer, Cham.

Kurdi, B., Alshurideh, M., & Alnaser, A. (2020). The impact of employee satisfaction on customer satisfaction: Theoretical and empirical underpinning. *Management Science Letters, 10*(15), 3561–3570.

Kurdi, B. A., Alshurideh, M., Nuseir, M., Aburayya, A., & Salloum, S. A. (2021). The effects of subjective norm on the intention to use social media networks: an exploratory study using PLS-SEM and machine learning approach. In *International conference on advanced machine learning technologies and applications* (pp. 581–592). Springer, Cham.

Mahmood, I., & Mahmood, A. (2021). The impact of E-marketing in achieving the client satisfaction: An exploratory study in Iraqi oil products distribution company. *Palarch's Journal of Archaeology of Egypt/Egyptology, 18*(3).

Nakra, N., & Pandey, M. (2019). Smartphone as an intervention to intention-behavior of patient care. *Health Policy and Technology, 8*(2).

Qtaishat, N. (2022). The impact of E-marketing on consumer purchase decision in Jordan. *WSEAS Transactions on Business and Economics, 19*, 161–168.

Rimkeviciene, J., Hawgood, J., O'Gorman, J., & De Leo, D. (2017). construct validity of the acquired capability for suicide scale: Factor structure, convergent and discriminant validity. *Journal of Psychopathology and Behavioral Assessment, 39*(2), 291–302. https://doi.org/10.1007/s10862-016-9576-4

Salehi, H. (2018). Smartphone for healthcare communication. University of Iowa, Iowa, USA. *Journal of Hospital Administration, 7*(5).

Sansour, R. (2016). Towards a modified scrum for multimedia mobile software development. Master thesis in computer sciences, AL-Quds University, Jerusalem-Palestine.

Sekaran, U., & Bougie, R. (2016a). *Research method for business: A skill building approach*. Wiley.

Sekaran, U., & Bougie, R. (2016b). *Research methods for business: A skill-building approach* (7th ed.). Wiley.

Shamout, M., Elayan, M., Rawashdeh, A., Kurdi, B., & Alshurideh, M. (2022). E-HRM practices and sustainable competitive advantage from HR practitioner's perspective: A mediated moderation analysis. *International Journal of Data and Network Science, 6*(1), 165–178.

Shen, B., & Bissell, K. (2013). *Social media, social me: A content analysis of beauty companies' use of Facebook in marketing and branding*. University of Alabama.

Shi, D., Lee, T., & Maydeu-Olivares, A. (2019). Understanding the model size effect on SEM fit indices. *Educational and Psychological Measurement, 79*(2), 310–334. https://doi.org/10.1177/0013164418783530

Shishan, F., Mahshi, R., Al Kurdi, B., Alotoum, F. J., & Alshurideh, M. T. (2022). Does the past affect the future? An analysis of consumers' dining intentions towards green restaurants in the UK. *Sustainability, 14*(1), 1–14.

Slack, N., Singh, G., & Sharma, S. (2020). The effect of supermarket service quality dimensions and customer satisfaction on customer loyalty and disloyalty dimensions. *International Journal of Quality and Service Sciences, 12*(3), 297–318. https://doi.org/10.1108/IJQSS-10-2019-0114

Sultan, R. A., Alqallaf, A. K., Alzarooni, S. A., Alrahma, N. H., AlAli, M. A., & Alshurideh, M. T. (2021). How students influence faculty satisfaction with online courses and do the age of faculty matter. In *The international conference on artificial intelligence and computer vision* (pp. 823–837). Springer, Cham.

Sung, K.-S., Yi, Y. G., & Shin, H.-I. (2019). Reliability and validity of knee extensor strength measurements using a portable dynamometer anchoring system in a supine position. *BMC Musculoskeletal Disorders, 20*(1), 1–8. https://doi.org/10.1186/s12891-019-2703-0

Tariq, E., Alshurideh, M., Akour, E., Al-Hawaryd, S., & Al Kurdi, B. (2022a). The role of digital marketing, CSR policy and green marketing in brand development at UK. *International Journal of Data and Network Science, 6*(3), 1–10.

Tariq, E., Alshurideh, M., Akour, I., & Al-Hawary, S. (2022b). The effect of digital marketing capabilities on organizational ambidexterity of the information technology sector. *International Journal of Data and Network Science, 6*(2), 401–408.

Tawfik, B., & Omar, H. (2020). *The role of electronic marketing in improving the financial performance of the institution: A field study of the Economic Corporation of Algeria Telecom*. University of Colonel Ahmed Deraya.

Wang, Y. A., & Rhemtulla, M. (2021). Power analysis for parameter estimation in structural equation modeling: A discussion and tutorial. *Advances in Methods and Practices in Psychological Science, 4*(1), 1–17. https://doi.org/10.1177/2515245920918253

Watson, R., Pitt, L., Berthon, P., & Zinkhan, G. (2002). U-commerce: Expanding the universe of marketing. *Journal of the Academy of Marketing Science, 30*(4). http://www.joif.org/

Share Your Beautiful Journey: Investigating User Generated Content (UGC) and Webrooming Among Malaysian Online Shoppers

Wan Nadiah Mohd Nadzri, Azreen Jihan Che Hashim, Muhammad Majid, Nur Aina Abdul Jalil, Haitham M. Alzoubi[ID], and Muhammad T. Alshurideh[ID]

Abstract The electric commerce (e-commerce) platform in Malaysia has been growing exponentially since the spread of Coronavirus Disease 2019 (COVID-19) pandemic. Many local businesses were badly affected when the first phase of Movement Control Order (MCO) was implemented for several weeks. The increasing number of social media platform users stems from wide internet accessibility and smartphone ownership. As such, this study investigated online purchasing behaviour in Malaysia by linking webrooming and user generated content (UGC) with brand experience and brand engagement. Essentially, this study unravelled consumer purchasing behaviour during the COVID-19 pandemic. This study explored the effects of UGC and webrooming on brand experience and brand engagement among

W. N. M. Nadzri (✉) · A. J. C. Hashim
Faculty of Economics and Muamalat, Universiti Sains Islam Malaysia, Nilai, Malaysia
e-mail: wan.nadiah.nadzri@gmail.com

A. J. C. Hashim
e-mail: azreenjihan@usim.edu.my

M. Majid
Faculty of Business and Management, Universiti Teknologi MARA (UiTM) Cawangan Johor, Johor, Malaysia
e-mail: muhdmajid@uitm.edu.my

N. A. A. Jalil
Faculty of Business Management and Professional Studies Management and Science University (MSU), Shah Alam, Selangor, Malaysia
e-mail: nur_aina@msu.edu.my

H. M. Alzoubi
School of Business, Skyline University College, University City of Sharjah, Sharjah, United Arab Emirates
e-mail: haitham.alzubi@skylineuniversity.ac.ae

M. T. Alshurideh
Department of Management, College of Business Administration, University of Sharjah, Sharjah, United Arab Emirates
e-mail: malshurideh@sharjah.ac.ae

Malaysian online consumers who purchased beauty product. As a result, UGC and webrooming displayed positive correlation with brand experience. However, brand experience was insignificant for brand engagement. Additionally, brand experience was unsuitable to function as a mediator. Thus, the impact of social influencers and shoppertainment on brand engagement should be assessed. Future studies may look into webrooming from the light of Stimulus-Organism-Response (SOR) Theory. Lastly, fake reviews, fake accounts, and bots that influence online purchasing behaviour among Malaysians should be investigated as well.

Keywords E-commerce · Online shopping · User generated content · Webrooming · Brand experience · Brand engagement

1 Introduction

With higher digital service awareness and easier access to internet connectivity, online shopping becomes a norm and disregards the traditional brick-and-mortar retail outlets that demands engagement with customers. Customers have taken a liking and have been adapting rapidly the online shopping model due to its ease of use, flexibility, controllable, and reference to rich information prior to making a decision to purchase tangible or intangible service or goods. Online shopping has been progressing since the past two decades, while simultaneously improving automated processing, ordering system, and integrated logistics companies that offer customers seamless purchase experience at a minimum cost.

The Malaysian government is a supporting pillar that drives the growth of electric commerce (e-commerce) business, whereby customers have been actively purchasing online for various products and services over the years. The Malaysian Department of Statistics reported higher adoption rate of e-commerce in 2019 than the previous years with double-digit growth. As stated in the Usage of Information and Communication Technology (ICT) in 2020 report, higher percentages of business-to-business (B2B) rates were recorded at 13, 53, and 22.6% from 2017 to 2019 (Department of Statistics Malaysia, 2021). This was due to the surge in ICT consumption and availability of internet access, which encouraged the growth of e-commerce amidst local and global players. A statistical study conducted by Hootsuite in 2019 identified some top-rated products purchased online, including fashion and beauty items, electronics, and sports-related products. The growing preference to shop online was motivated by consumers' perception that vast information is available, which aids or facilitates the decision-making process.

The COVID-19 pandemic has immensely affected everyone worldwide, particularly the retail industry ecosystem. A study forecasted the closure of 20,000 high street retail outlets in the UK due to the COVID-19 lockdown. During the troubled period of COVID-19 pandemic from 2020 to 2021, consumer behaviour was driven more towards e-commerce or online purchase, thus exerting drastic changes to and impacts on businesses across the globe. Businesses with e-commerce facility may reap benefits, while those without any online facility are bound to suffer from loss unless they urgently adapt to change. With higher online shopping adaption, more Malaysians have begun adopting the spending habit during the lockdown period. The surge in Malaysian online spending is due to work at home and more consumption of time online with better internet accessibility.

The Malaysian government has extensively facilitated local businesses to venture into e-commerce. Interestingly, many local businesses are positively forced to change their nature of selling into e-commerce and social media. A recent study showed that 76% of Malaysians paid additional 5% premium on the product for value, convenience, and benefits. Online shopping helps them to purchase products without the hassle of going out and getting infected by COVID-19. Therefore, it is imminent to investigate the factors that influence Malaysian consumers to purchase online and to understand how consumers engaged with the community via social media platform.

2 Study Importance and Objectives

The e-commerce platform in Malaysia has grown rapidly due to the COVID-19 pandemic. In the beginning, the local businesses had problems in their operation during the first phase of Movement Control Order (MCO) implemented for several weeks. Such scenario had forced the local businesses to invest in e-commerce, thus driving consumers to purchase online. Statista in 2021 reported that the most clicked e-commerce platform in Malaysia 2020 was Shopee, followed by Lazada, PG Mall, Zalora, eBay, GoShop, Lelong, Sephora, Wowshop, and Applecrumby. Furthermore, 91% of Malaysians appeared to be avid users of the internet with high rate of smartphone consumption. The e-commerce platform is favourable among Gen Y and Gen Z (Muda et al., 2016) mainly because online purchase has become easier, faster, and more transparent. These younger generations are adopting shopping as a lifestyle and as an escapism mechanism from their daily routine. They would spend longer time and money to shop online. Besides, Malaysians working at home are expected to spend more as stated by Pricewaterhouse Coopers (PWC) in 2021. A survey by Rakuten Insight in 2020 reported that Malaysia imported beauty and skincare products from France ($67.6 million), followed by the US and South Korea. Hence, it is crucial to decipher the behaviour of Malaysian consumers who purchase beauty products via online platform.

The social media has grown into a platform to gather information about goods and services (Department of Statistics Malaysia, 2021). Internet consumption has changed as more consumers are spending more time using video, messaging applications, and social media platform. This paradigm shift has continued to increase as more Malaysians practice social distancing stated by PWC in 2020 and work from home. More people are exploring the social media due to the COVID-19 pandemic. Malaysians use social media to conduct product research based on reviews, feedback, and comments as stated by Malaysian Communication and Multimedia Commission (MCMC) in 2020 Mostly, individuals use the social media to interact, engage, and share brand experiences about a particular product. At present, people are getting more creative in developing their own contents and sharing them widely on the social media platform. The increasing number of users using the social media platform is a result of better internet accessibility and smartphone ownership (Arif et al., 2020). Therefore, more people rely on user generated content (UGC) as reference before purchasing as they feel that the information shared on social media is trustworthy and credible. They believe more in the experience of other consumers, instead of the e-commerce platform due to lack of trust, security, and confidence. As suggested by Herrero et al. (2015), more studies should assess the synergy between credibility of a source and worthiness of information on consumer behaviour. Additionally, it is important to decipher the videos and contents generated by users that influence purchasing decision.

In reference to consumer research online prior to purchase, webrooming has an important role in the decision-making process. About 70% of consumers world-wide had conducted product research online before purchasing as stated by PWC in 2015. With the expansion of smartphones and internet, webrooming is becoming favourable among consumers. While many consumers visit the e-commerce platform (e.g., websites & social media) prior to purchase decision making, studies on this research area are in scarcity (Arora & Sahney, 2019; Flavián et al., 2016). Hence, it is crucial to investigate the importance of UGC and webrooming variables in the context of purchasing online beauty products by Malaysians.

This study assessed the online purchasing behaviour in the context of Malaysians by linking webrooming and UGC with brand experience and brand engagement during this COVID-19 pandemic. The objectives of this study are listed in the following:

1. To investigate factors that influence brand experience that lead to brand engagement through SOR Theory
2. To analyse the mediating effect of brand experience and brand engagement
3. To explore the direct relationship between user generated content (UGC) and brand engagement
4. To examine the direct effect between webrooming and brand engagement.

3 Literature Review

3.1 SOR Theory and Brand Experience

Brand experience refers to one's experience interacting with a product or service through brand-related stimuli that evokes emotions, such as pleasure, arousal, dominance, and intelligence. In return, such emotions alleviate the brand value. Notably, brand experiences are "sensations, feelings, cognitions, and behavioural responses evoked by brand-related stimuli that are part of a brand's design and identity, packaging, communications, and environments" (Brakus et al., 2009).

The SOR Theory introduced by Mehrabian and Russell (1974) enables the investigation of cognitive and effective impacts on consumer behaviour. The underlying dimensions of SOR are stimuli, organism, and response. Stimuli denote the external factors that stimulate one's evaluation. Organism reflects the internal emotions derived from environment stimuli. Response is the final consequence if one should approach or avoid certain behaviour. Turning to this present study, the SOR Theory was deployed in the Malaysian online shopping scenario due to the limited studies in doing so.

The fundamental of the SOR Theory is to identify consumer emotions that influence consumer response (Ha & Lennon, 2010) such as purchase intention (Giovanis et al., 2014). Consumers who experience positive emotional state have the tendency or willingness to purchase. However, more researchers need to investigate the environmental cues that may influence brand experience quality, especially in e-commerce landscape. With the context of experience, it is essential to measure the emotional scale in response to environmental stimuli. Based on the literature gaps, more studies are required to examine environmental cues that evoke brand experience quality. Therefore, this study assessed the effect of environmental stimuli (UGC and webrooming) on brand experience using the SOR Theory. In reference to this present study, brand experience reflects a customer's experience influenced by virtual experience, which in turn, affects their online engagement towards online communities, such as Facebook and Instagram. Brand experience elements have three basic dimensions of emotions, namely pleasure, dominance, and arousal, in responses to environmental stimuli (Mehrabian & Russell, 1974). It is crucial to decipher the effect of webrooming and UGC on brand experience towards brand engagement.

3.2 Webrooming

Webrooming is a process of product information seeking behaviour by consumers via online channels, continued by verification of information, and finally, the actual purchase at physical stores (Aw et al., 2021). Webrooming denotes consumers gathering information about a product from online platform, but the real purchase happens in a physical store (Pimentel et al., 2019).

Many studies have assessed the role of webrooming and other factors in strengthening certain behaviour (Herrero-Crespo et al., 2021). The four drivers linked with webrooming are psychographic, shopping motivation, channel-related, and product-related variables (Kleinlercher et al., 2020). The literature depicts a broad range of reasons on why consumers webroom. Webrooming helps consumers minimise the risks of buying products online (Chiu et al., 2011).

Consumers webroom to gain information from the online platform by reading online reviews in order to make clearer and better choices. This decrease worries that one would feel in process of online purchase (Flavián et al., 2016). Webroom boosts selection trust and confidence towards purchasing products from physical stores as a result of gathering product-knowledge via reviews accessible online. Some consumers webroom to limit their selection before purchasing products from physical stores (Wolny & Charoensuksai, 2014). The webrooming behaviour is conceptualised only lately (Arora & Sahney, 2019; Aw, 2020) and the academic research on this concept is still in its infancy stage.

Flavián et al. (2019) emphasised on the importance of expanding approaches to enhance consumer confidence especially when it comes to purchase process. Studies have highlighted those marketers should seek to improve consumers' participation and engagements towards brand, product, and proses situation. The proses situation is associated with emotional attachment to the representational meaning of purchase (Balasubramanian et al., 2005) that leads to consumer confidence in purchase selection. Based on data retrieved from 302 customers in Hong Kong, (Cheung et al., 2020) found that ongoing information search behaviour by consumers is the key driver of brand engagement. They verified the role of enduring involvement to improve continuous search behaviour and brand engagement. Consumer brand engagement and the importance of customer involvement is based on the argument that consumers are more likely to be engaged, to expend more cognitive effort, and to participate more resources in ongoing information search about their product selection.

According to Flavián et al. (2016), consumers webroom to acquire information online, while the contrary was reported by Phan et al. The latter found that information readiness was insignificant for brand engagement. Consumers actively participate in multiple brand-related activities to satisfy their individual needs. Consumers' utilitarian benefits are related to information availability. According to Chahal and Rani (2017), customers are engaged with a brand more likely to satisfy their demands. Rohm et al. (2013) stated that product information expands customers' connection, such as social media and engagement with brands (brand engagement).

According to Aw (2020), Shankar and Jain (2021), studies on webrooming behaviour trend is in its initial stages. Thus, this present study assessed the links among webrooming, UGC, brand experience, and brand engagement in the Malaysian online shopping context.

3.3 User Generated Content (UGC)

Social media platform, such as Facebook, Instagram, and Snapchat, offer an outlet for companies to expand their successful brands. The UGC has gained popularity among online users as companies recognise the importance of UGC to engage with costumers. This development has urged many studies to reveal the more significant relevance of digital platform (e.g., Instagram & Facebook) when compared to conventional marketing (Oliveira & Casais, 2019). The present trend shows that consumers spend more time on UGC and create contents, such as videos, reviews, and attractive images, to be shared online, as well as gain more followers and popularity. Some online consumers explore the social media platform to become social media influencers.

As marketers acknowledge the importance of UGC in social media, they provide limitless opportunity for social media users to share videos, post photos, and leave comments about their experiences consuming a particular product (Raji Ridwan et al., 2017). Interestingly, the e-commerce platform offers UGC features, such as review, rate, and comment sections, for users to share their brand experience. This valuable tool influences consumers in future purchasing decisions as they gather content at the platform. Lazada Malaysia, for instance, provides these features that allow consumers to share UGC with the community. This helps the e-commerce platform to gain confidence from consumers, which in turn, generates higher sales.

The advent of social media has changed the dynamics of disseminating information. The UGC was designed as a new type of brand communication for companies. In order to maintain the UGC (content) quality, marketers constantly monitor their content from malicious comments and negative content by unanimous users (Kim et al., 2012). The UGC platform serves as a marketing tool managed by influencers or marketing team. Marketers and influencers use their social media platform to target a specific target market, such as Tiktok among Gen Z and Gen Alpha. Thus, it is imminent to investigate the behaviour of young generations towards UGC usage as their avenue. Essentially, it is timely to unravel the influence of UGC on brand experience and brand engagement among young savvy Malaysians who indulged in online purchasing activities.

3.4 Brand Engagement

A brand is no longer what customers are told, but it is what the customers are telling themselves. The use of a combination of social media for marketing has never been more important than the present time. Generally, branding consists of several perception concepts, namely perceived knowledge and attitude, where assessment of consumers' perceptions is integral for they shape their brand attitudes (Hollebeek, 2011). Consumer brand experience is a critical factor for a business to succeed in the social media setting. The branding process has been the subject of a great deal of scientific debate since the past few decades (Kramoliš & Kopečková, 2013). As products and services have their lifecycles; consistent quality, credibility, and experience are vital.

Brand is a value and many businesses automatically add brand to their balance sheets due to its positive association. Creating a brand as a 'trademark' is not just applying a logo or graphic, mainly because brand blankets the overall customer experience. It is inevitable to track and talk about brand experience, as a good and strong brand does not flow from each other (Wroblowská, 2016). Brand creation is important in the online environment (Hollebeek et al., 2014) because the social media make brand more visible and available. Strategic branding work ensures customers have positive feeling towards a brand in their experience.

Past studies have established that brand leads to purchase intention (Chou et al., 2016). The famous the brand; the higher the tendency for customers to purchase the brand due to the trust they have built towards the brand (Chou et al., 2016). As such, the brand becomes the first choice during purchase process. Brand has been extensively linked with loyalty. Loyalty has turned into the goal for each brand in the market because retaining the existing customers is easier than seeking new. According to Hollebeek (2011), customers who are loyal to a brand will influence others to try out the offer and share similar feeling and experience.

During this period of transition and adjustment to the digital economy, a new marketing approach is sought to help marketers anticipate and capitalise on disruptive technology platforms to increase brand engagement of skincare product. Does brand continue to play a critical role in engaging with customers? According to Kotler et al. (2017), globalisation breaks down the boundaries between large and small businesses. To date, new, smaller, younger, and local businesses can compete with the biggest, the oldest, and global companies with strong brand.

Instead of being highly dominating, a company is more competitive if it can connect with client and partner communities for co-creation, as well as with competitors for co-operatives. In the past, companies believed that innovation should spark from brand support, while focusing on research and market development (Kotler et al., 2017). Consequently, they realised that in-house innovation rate was insufficient to compete in a constantly changing market. Procter & Gamble (P&G), for example, found out in early 2000 when its sales of new products flattened out. Next, the company turned its R&D model into a connection-development model. This also happened to Unilever, which evolved in the same direction where innovation is now horizontal, where the market generates ideas and the companies commercialise idea. Such new internal strategy builds positive or negative brand experience, thus leading to brand engagement. This study established the much-sought links of UGC, webrooming, and brand experience with brand engagement on social media in order to identify the path to engage customers with skincare product brands.

4 Theoretical Framework

The Stimulus-Organism-Response (SOR) Theory was implemented in this study as it assessed organisational characteristics (e.g., UGC & webrooming) as stimuli and responses (e.g., brand experience & brand engagement). The SOR Theory initiated by Woodworth in 1929 incorporated stimuli (S)—response (R) characters. (Mehrabian & Russell, 1974) expanded the theory by inculcating organism (O) human cognitive and affective elements that influence the stimuli response ecosystem. Scholars, including (Musa et al., 2020; Wang et al., 2020), who adopted the SOR Theory had explored the social media setting. Nonetheless, only a handful of studies have assessed webrooming (Arora & Sahney, 2017; Flavián et al., 2016). Very few UGC studies (Arif et al., 2020; Mohammad et al., 2020) had assessed emotional and functional elements in social media during COVID-19 pandemic.

Accordingly, this study investigated the SOR Theory within the online beauty industry in the context of Malaysia. Figure 1 illustrates the conceptual framework

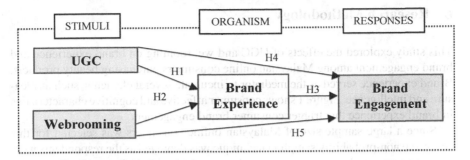

Fig. 1 Conceptual framework

adapted from Mehrabian and Russell (1974), Wan Nadiah Mohd Nadzri (2017). This present study has incorporated new variables to suit the COVID-19 pandemic context. Despite the past studies on consumer online purchase behaviour, more studies have recommended to investigate the SOR Theory within the beauty industry.

5 Formation of Hypotheses and Constructs

The growing demand for shopping online by Malaysians is driven by accessible internet, high smartphone usage, and boundless information that aid consumers to make decision prior to online purchase. Malaysians love to indulge in online purchasing as they spend longer time browsing through the vast social media platform during the pandemic. The pandemic has changed the consumers' purchase pattern as they continue to work remotely from home. Consumers have more time to conduct online research by referring to reviews, videos, and stories posted across the social media platform. Influential information can entice consumers to purchase a particular product. Thus, it is relevant to unravel elements such as UGC (H1) and webrooming (H2). A well-researched consumer will influence brand experience (H3) that leads to brand engagement. Consumers willingly share authentic recommendations to online communities (H4) in order to build brand engagement. Website browsing, such as webrooming (H5), may influence consumers to share their experiences by engaging with communities. Hence, the following hypotheses are proposed:

H1: User generated content is positively related to brand experience.

H2: Webrooming is positively related to brand experience.

H3: Brand experience is positively related to brand engagement.

H4: User generated content is positively related to brand engagement.

H5: Webrooming is positively related to brand engagement.

6 Research Methodology

This study explored the effects of UGC and webrooming on brand experience and brand engagement among Malaysian online consumers purchasing beauty products. Brand experience served as the mediator to inculcate several elements, such as pleasure and intelligence. Figure 1 shows consumer affective and cognitive characteristics in brand experience that trigger consumer brand engagement.

Since a large sample size of Malaysian online consumers was selected for this study, the non-probability sampling technique was deployed as the respondents were not given an equal opportunity to participate (Etikan & Bala, 2017). Data were collected from Malaysians experienced in using the social media platform or e-commerce to purchase beauty products. Malaysians are a technology savvy society and are highly influenced by social media, particularly online shopping (Mohd Fudzi et al., 2021). The respondents must be users of smartphone and had internet access. Hence, the purposive sampling method was employed. Due to MCO restriction and social distancing, the questionnaire was distributed via online using Google Form in English version. Although the suitable sample size for this study is 250, 276 usable questionnaires were used based on the study by Mohd Fudzi et al. (2021) that involved 267 Malaysians who purchased online.

The questionnaire has two sections: Section A emphasises on items that measure UGC, webrooming, brand experience, and brand engagement. Section B focuses on demographic information, such as gender, age, and education level. The respondents were required to rate their disagreement or agreement on a seven-point Likert scale ranging from $1 =$ Very strongly disagree to $7 =$ Very strongly agree; adopted from Mohd Nadzri (2017). The UGC was measured based on Kim et al. (2012) webrooming was obtained from Vasudevan and Arokiasamy (2021), while brand experience and brand engagement were adopted from Bansal et al. (2005). The next step was data collection and analysis using software. The outcomes are presented in the next section.

7 Research Findings

This section explains the three stages of findings derived from descriptive, measurement model, and structural model analyses. In light of descriptive analysis, most of the respondents were females (87%) at the age of 18–30 years old (74.6%). They were mostly single (78.3%) and possessed a Bachelor's Degree (55.1%). As for beauty products, the respondents mostly purchased toner, serums, and moisturisers (66.7%) via online platform. Interestingly, the respondents referred to Instagram (70.3%) to gather information prior to purchase. The details are tabulated in Table 1.

The data were analysed using IBM SPSS and Smart PLS version 3. The results showed that the data did not have multivariate normality based on Mardia's multivariate skewness ($\beta = 2.667, p < 0.01$) and Mardia's multivariate kurtosis ($\beta = 30.86$,

$p < 0.01$), as extracted from IBM SPSS. The results of multivariate skewness and kurtosis were assessed as prescribed by Nadzri and Musa (2017), Hair et al. (2017). The link is given in the following:

- https://webpower.psychstat.org/models/kurtosis/results.php?url=fb9771ad6508 7c96bdc6a313929fa338

Smart PLS is a non-parametric software analysis and is beneficial for studies that have relatively large or small sample size (Hair et al., 2011). Besides, Smart PLS is suitable for studies with limited theory available in the literature, as the case of this present study. In order to analyse the research model, two steps of analytical procedure were executed (Anderson & Gerbing, 1988). In Step 1, the measurement model was assessed to investigate convergent validity and discriminant validity. Next, Step 2 assessed the structural model that tested the formulated hypotheses. In order to identify the significance of path coefficient of each construct and its loadings, the bootstrapping method was performed with 5000 resamples as suggested by Hair et al. (2017).

7.1 Measurement Model Analysis

The measurement model was analysed for its convergent validity and discriminant validity. As depicted by Nadzri and Musa (2017), convergent validity examines average variance extracted (AVE), loadings, and composite reliability. Based on Table 2, both loadings and composite reliability exceeded 0.708 (Nadzri & Musa, 2017), except for several loadings. The items were not deleted as they affected both AVE and composite reliability of the variables. The AVE for all constructs exceeded 0.5, as prescribed by Nadzri and Musa (2017), Hair et al. (2017).

Table 3 tabulates the discriminant validity for this study based on Fornell and Larcker (1981) although several studies Henseler et al. (2016) criticised its reliability in analysing discriminant validity while prescribing the Monte Carlo simulation. Table 4 shows the HTMT values below HTMT0.85 value of 0.85 (Kline, 2011) and HTMT0.90 value of 0.90 (Ramayah et al., 2017). Thus, there was no discriminant validity issue using (Fornell & Larcker, 1981) in this present study.

7.2 Structural Model Analysis

The last step was analysing the structural model. To test the model fitness, Normed Fit Index (NFI) and Standardised Root Mean Square Residual (SRMR) were deployed. The model displayed SRMR = 0.064 that indicated good fit as the value is below 0.08 (Hair et al., 2011). The value of NFI that exceeds 0.9 represents acceptable fit.

In assessing the structural model, a bootstrapping procedure with 5000 resamples (Hair et al., 2017) was conducted to investigate R^2, beta (β), and t-value (see results in Table 5 and Fig. 2). Table 5 presents the results of UGC ($\beta = 0.259$, $t = 3.334$, $p < 0.01$), webrooming ($\beta = 0.619$, $t = 9.220$, $p < 0.01$), and brand experience ($\beta = 0.069$, $t = 0.522$, $p > 0.01$). Referring to Table 6, brand experience was unsuitable to serve as a mediator for the following links: UGC → Brand Experience → Brand Engagement, and Webrooming → Brand Experience → Brand Engagement. However, no direct effect was noted for Hypothesis H4 (UGC → Brand Engagement), while Hypothesis H5 (Webrooming → Brand Engagement) exhibited a direct relationship.

8 Discussion and Conclusion

In reference to descriptive analysis, most of the respondents were 18–30 years old Malaysians literate in technology and social media. A similar age pattern was also reported by Arora and Sahney (2018). Generation Alpha (born 2010–2025) and Gen Z (18–25 years old) are inclined to be proficient in technology (Karasek & Hysa, 2020). Interestingly, these generations cannot function well daily without digital technology or social media platform. Despite their efficiency in researching information via online, these young generations face problem focusing on a specific activity (Grenčíková & Vojtovič, 2017). The behaviour of Gen Z and Generation Alpha promotes online shopping and e-commerce as they serve as a crucial marketing segmentation strategy.

This study assessed the impact of webrooming and UGC on brand experience towards brand engagement. It also investigated the direct relationship between UGC and webrooming towards brand engagement, as reflected in Hypotheses H4 and H5. As a result, UGC displayed a positive relationship with brand experience. Similarly, (Arif et al., 2020; Cheung et al., 2020) reported that UGC influenced emotion response towards consumers' decision-making process. Emotional responses, such as pleasure and intelligence, triggered Malaysians to share their experiences with their family and friends. However, UGC exerted a negative impact on brand engagement in this present study. This contradicts with that reported by Yasin et al. (2020), in which brand engagement positively impacted UGC. The result retrieved in this present study revealed that the customers were overwhelmed by the massive information retrieved from reviews, testimonials, tutorials, and videos. Additionally, fake news or irrelevant information that circulated across the social media platform affected the consumers' decision-making process. However, irrelevant or fake information was seldom shared due to fear of spreading unauthentic facts that may get them in hot water by the Malaysian Communications and Multimedia Commission (MCMC). The existence of fake reviews, fake accounts, bots, and fake content can adversely affect company revenue (Katona, 2018).

In this study, webrooming exhibited a positive impact on brand experience. Similarly, Aw (2020), Hollebeek et al. (2014), Shankar and Jain (2021) highlighted that webrooming behaviour can lead to purchase intention. However, only a handful of studies have assessed webrooming and brand experience. Turning to this present study, most consumers preferred gathering information directly from website as

65.9% of the respondents relied on e-commerce (official website). The advantages of researching and browsing e-commerce websites are that they have more experiential tools, such as live-chat customer service support, friendly user navigation tools, specific information on product offer, current promotions, and authentic information. When consumers face issues about a particular product or inquiry, they can chat with the customer service representative to gain prompt reply. Essentially, the product and customer service team eases the purchase decision-making process. Consumers would feel more engaged with the company website as they feel that the e-commerce platform is indeed trustworthy and credible to be shared within their circle of network.

Brand experience appeared to be unfit to function as a mediating variable in this study. Brand experience had no significant relationship with brand engagement, which contradicted to that reported by Nadzri and Musa (2017), Katona (2018). This showed that consumers relied on social media influencers with higher followers and popularity. Marketers tend to hire social media influencers to increase their company revenue (Katona, 2018). Younger generations, such as Gen Z and Generation Alpha, are more exposed to influencers with the most followers. The more the followers, the greater is the impact of a brand on consumers. Hence, consumers highly depend on influencers when compared to brand name. The digital marketing landscape that incorporates social media influencers has higher impact on brand image, in comparison to brand experience.

9 Theoretical and Practical Implications

The framework for this study was adopted and adapted from SOR Theory and brand experience. Since brand experience had no significant effect as a mediator, it did not have an important role in the society. However, social media influencers have the power to increase brand awareness and conversion rate (Keller & Fay, 2016). Most studies have acknowledged the importance of social media influencers to have a greater impact as they are followed by millions of followers. Marketers have begun showering social media influencers with product endorsements so that the latter can create product content and post the content in the social media platform. Besides, consumers' preference has changed as they rely more on the experiences of influential people as their guide in purchase decision-making process.

Apart from theoretical contributions, this study unravels the importance of webrooming towards consumer experience and engagement. This suggests that marketers should focus on the timeliness aspect, while keeping their website content current. Transparency and credibility enhance both user experience and brand engagement among consumers. The quality of e-commerce platform can be enhanced by displaying quality seal badge, friendly user navigation tool, as well as display review quantity and quality. It is crucial to invest in Google Page rank that measures the quality and the quantity of the website content. Marketers can enhance brand engagement for online consumers by developing a mobile friendly user navigation tool as many consumers prefer web-browsing using their smartphones. Eye catching images, interactive polls, and rich content enable better accessibility with smartphone.

Marketers should also invest in stimulating the senses of consumers to elevate consumer brand experience. The colour of the website layout, interactive videos (GIF or flash videos), and readable font sizes help to evoke the sight senses. Marketers should monitor the information circulating across the social media platform to avoid fake news that might tarnish the brand image of the company. Leveraging quality content and authentic information via social media platform or websites may aid companies to increase their growth and help consumers to feel confident when purchasing online.

10 Limitations and Future Research

Moving forward, future studies should investigate the effect of social influencers and shoppertainment upon brand engagement. The shoppertainment phenomenon draws more consumers as it is fun and addictively engaging with consumers via e-commerce platform or social media. The e-commerce platform, such as Lazada and Shoppee, has invested a lot in their live-streaming channel, thus shaping the future of shopping globally. With the expansion of online shopping, assessing the purchasing behavioural of Generation Alpha and their shopping motivation is imminent. This is relevant as the younger generations embrace social media and digital technology more than the older generations.

Future studies should also evaluate webrooming that inculcates the SOR Theory. It will be interesting to improve the webrooming construct by studying the credibility of online reviews and interactive navigation tools. This can help companies to improve their websites in portraying a reputable image to the consumers. A longitudinal study on this topic would be interesting as it will look into diversified demographic and geographical area to better understand other cultural differences. Finally, future researchers are invited to investigate how fake reviews, fake accounts, and bots influence the online purchasing behaviour among Malaysians.

Appendices

See Tables 1, 2, 3, 4, 5, 6 and Fig. 2.

Table 1 Results from descriptive analysis

Demographic			
	Details	Frequency (n: 276)	Percentage (%)
Age	**18–30 years old**	**206**	**74.6**
	31–40 years old	50	18.1
	41–50 years old	16	5.8
	51–60 years old	4	1.4
Gender	Female	240	87.0
	Male	36	13.0
Marital status	**Single**	**216**	**78.3**
	Married with children	41	14.9
	Married without child	11	4.0
	Single parent with children	6	2.2
	Widowed	2	0.7
Education	SPM	19	6.9
	Diploma	56	20.3
	Bachelor's Degree	**152**	**55.1**
	Postgraduate	44	15.9
	Professional Cert	2	0.7
	STAM	1	0.4
	Foundation	1	0.4
	STPM	1	0.4
Skin care products	Body Lotion or Moisturiser	171	62.2
	Facial Wash or Cleanser	192	70.3
	Daily Sunscreen	137	50.2
	Toner, Serums and skin treatment	**182**	**66.7**
Social media preference	**Instagram**	**194**	**70.3**
	Facebook	72	26.1
	Product website	182	65.9
	Tiktok	88	31.9
	Pinterest	8	2.9
	Search Engine (Google, Yahoo)	168	60.9
	YouTube	128	46.4
	Twitter	18	6.52
	Shopee	6	2.17

(continued)

Table 1 (continued)

Demographic

	Details	Frequency (n: 276)	Percentage (%)
	Online Review from reliable sites	3	1.08
	Skincarisma	1	0.4
	Self-care website	1	0.4

Table 2 Convergent validity

Convergent validity

Construct	Items with codes	Factor loadings	AVE	Cronbach's α	Composite reliability
UGC	A1	0.737	0.577	0.803	0.804
	A2	0.754			
	A3	0.788			
Webrooming	A4	0.688	0.602	0.883	0.881
	A5	0.693			
	A6	0.769			
	A7	0.816			
	A8	0.946			
Brand Expe	A12	0.706	0.552	0.827	0.83
	A13	0.818			
	A14	0.769			
	A15	0.669			
Brand Engage	A9	0.963	0.585	0.805	0.802
	A10	0.689			
	A11	0.693			

Table 3 Discriminant validity

Discriminant validity-Fornell Larker

	BEngage	BExpe	UGC	Webrooming
BEngage	0.765			
BExpe	0.600	0.743		
UGC	0.531	0.639	0.760	
Webrooming	0.709	0.777	0.615	0.776

Table 4 HTMT criterion

HTMT criterion				
	BEngage	BExpe	UGC	Webrooming
BEngage				
BExpe	0.598			
UGC	0.521	0.642		
Webrooming	0.677	0.78	0.618	

Table 5 Hypotheses testing

Hypothesis testing						
	Relationship	Standard beta	Standard error	t-value	P value	Results
H1	UGC → BExpe	0.259	0.078	3.334	0.001	**Accepted**
H2	Webrooming → BExpe	0.619	0.067	9.220	0.000	**Accepted**
H3	BExpe → BEngagement	0.069	0.133	0.522	0.602	Rejected
H4	UGC → BEngagement	0.135	0.094	1.435	0.151	Rejected
H5	Webrooming → BEngage	0.57	0.116	4.906	0.000	**Accepted**

Table 6 Indirect effect testing

Indirect effect					
Relationship	Standard beta	Standard error	t-value	P value	Results
UGC → BExpe → BEngage	0.018	0.039	0.458	0.647	Rejected
Webrooming → BExpe → BEngage	0.043	0.082	0.522	0.602	Rejected

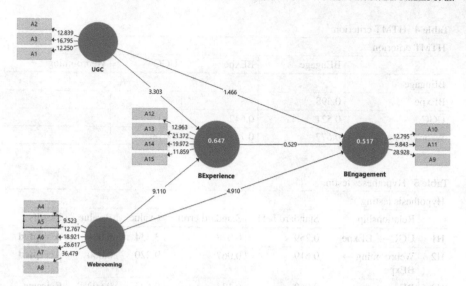

Fig. 2 Bootstrapping results

References

Anderson, J. C., & Gerbing, D. W. (1988). Structural equation modeling in practice: A review and recommended two-step approach. *Psychological Bulletin, 103*(3), 411.

Arif, I., Aslam, W., & Siddiqui, H. (2020). Influence of brand related user-generated content through Facebook on consumer behaviour: A stimulus-organism-response framework. *International Journal of Electronic Business, 15*(2), 109–132.

Arora, S., & Sahney, S. (2017). Webrooming behaviour: A conceptual framework. *International Journal of Retail & Distribution Management*.

Arora, S., & Sahney, S. (2018). Consumer's webrooming conduct: An explanation using the theory of planned behavior. *Asia Pacific Journal of Marketing and Logistics*.

Arora, S., & Sahney, S. (2019). Examining consumers' webrooming behavior: An integrated approach. *Marketing Intelligence & Planning*.

Aw, E. C. X. (2020). Understanding consumers' paths to webrooming: A complexity approach. *Journal of Retailing and Consumer Services, 53*, 101991.

Aw, E. C. X., Basha, N. K., Ng, S. I., & Ho, J. A. (2021). Searching online and buying offline: Understanding the role of channel-, consumer-, and product-related factors in determining webrooming intention. *Journal of Retailing and Consumer Services, 58*, 102328.

Balasubramanian, S., Raghunathan, R., & Mahajan, V. (2005). Consumers in a multichannel environment: Product utility, process utility, and channel choice. *Journal of Interactive Marketing, 19*(2), 12–30.

Bansal, H. S., Taylor, S. F., & James, Y. S. (2005). "Migrating" to new service providers: Toward a unifying framework of consumers' switching behaviors. *Journal of the Academy of Marketing Science, 33*(1), 96–115.

Brakus, J. J., Schmitt, B. H., & Zarantonello, L. (2009). Brand experience: What is it? How is it measured? Does it affect loyalty? *Journal of Marketing, 73*(3), 52–68.

Chahal, H., & Rani, A. (2017). How trust moderates social media engagement and brand equity. *Journal of Research in Interactive Marketing, 11*(3), 312–335. https://doi.org/10.1108/JRIM10-2016-0104

Cheung, M. L., Pires, G. D., & Rosenberger III, P. J. (2020). Exploring consumer–brand engagement: A holistic framework. *European Business Review*.

Chiu, H. C., Hsieh, Y. C., Roan, J., Tseng, K. J., & Hsieh, J. K. (2011). The challenge for multichannel services: Cross-channel free-riding behavior. *Electronic Commerce Research and Applications, 10*(2), 268–277.

Chou, S. Y., Shen, G. C., Chiu, H. C., & Chou, Y. T. (2016). Multichannel service providers' strategy: Understanding customers' switching and free-riding behavior. *Journal of Business Research, 69*(6), 2226–2232.

Department of Statistics Malaysia. (2021). ICT use and access by individuals and household survey report 2020. Press Release.

Etikan, I., & Bala, K. (2017). Sampling and sampling methods. *Biometrics & Biostatistics International Journal, 5*(6), 00149.

Flavián, C., Gurrea, R., & Orús, C. (2016). Choice confidence in the webrooming purchase process: The impact of online positive reviews and the motivation to touch. *Journal of Consumer Behaviour, 15*(5), 459–476.

Flavián, C., Gurrea, R., & Orús, C. (2019). Feeling confident and smart with webrooming: Understanding the consumer's path to satisfaction. *Journal of Interactive Marketing, 47*, 1–15.

Fornell, C., & Larcker, D. F. (1981). Evaluating structural equation models with unobservable variables and measurement error. *Journal of Marketing Research, 18*(1), 39–50.

Giovanis, A. N., Zondiros, D., & Tomaras, P. (2014). The antecedents of customer loyalty for broadband services: The role of service quality, emotional satisfaction and corporate image. *Procedia-Social and Behavioral Sciences, 148*, 236–244.

Grenčíková, A., & Vojtovič, S. (2017). Relationship of generations X, Y, Z with new communication technology. *Problems and Perspectives in Management, 15*(2), 558–564, 27. https://doi.org/10.21511/ppm.15(si).2017.09

Ha, Y., & Lennon, S. J. (2010). Online visual merchandising (VMD) cues and consumer pleasure and arousal: Purchasing versus browsing situation. *Psychology & Marketing, 27*(2), 141–165.

Hair, J. F., Ringle, C. M., & Sarstedt, M. (2011). PLS-SEM: Indeed a silver bullet. *Journal of Marketing Theory and Practice, 19*(2), 139–152.

Hair, J. F., Jr., Sarstedt, M., Ringle, C. M., & Gudergan, S. P. (2017). *Advanced issues in partial least squares structural equation modeling*. Sage publications.

Henseler, J., Hubona, G., & Ray, P. A. (2016). Using PLS path modeling in new technology research: updated guidelines. Industrial management & data systems.

Herrero, Á., San Martín, H., & Hernández, J. M. (2015). How online search behavior is influenced by user-generated content on review websites and hotel interactive websites. *International Journal of Contemporary Hospitality Management, 27*(7), 1573–1597.

Herrero-Crespo, A., Viejo-Fernández, N., Collado-Agudo, J., & Pérez, M. J. S. (2021). Webrooming or showrooming, that is the question: explaining omnichannel behavioural intention through the technology acceptance model and exploratory behaviour. *Journal of Fashion Marketing and Management: An International Journal*.

Hollebeek, L. (2011). Exploring customer brand engagement: Definition and themes. *Journal of Strategic Marketing, 19*(7), 555–573.

Hollebeek, L. D., Glynn, M. S., & Brodie, R. J. (2014). Consumer brand engagement in social media: Conceptualization, scale development and validation. *Journal of Interactive Marketing, 28*(2), 149–165.

Ioanid, A., Militaru, G., & Mihai, P. (2015). Social media strategies for organizations using influencers' power. *European Scientific Journal, 11*(10), 139–143.

Karasek, A., & Hysa, B. (2020). Social media and generation Y, Z–a challenge for employers.

Katona, Z. (2018). Competing for influencers in a social network. Available at SSRN 2335679.

Keller, E., & Fay, B. (2016). How to use influencers to drive a word-of-mouth strategy. *Warc Best Practice, 1*, 2–8.

Kim, C., Jin, M. H., Kim, J., & Shin, N. (2012). User perception of the quality, value, and utility of user-generated content. *Journal of Electronic Commerce Research, 13*(4), 305.

Kleinlercher, K., Linzmajer, M., Verhoef, P. C., & Rudolph, T. (2020). Antecedents of webrooming in omnichannel retailing. *Frontiers in Psychology, 11*, 3342.

Kline, R. B. (2011). Convergence of structural equation modelling and multilevel modelling.

Kotler, P., Kartajaya, H., & Hooi, D. H. (2017). Marketing for competitiveness: Asia to the world! in the age of digital consumers.

Kramoliš, J., & Kopečková, M. (2013). Product placement: A smart marketing tool shifting a company to the next competitive level. *Journal of Competitiveness*.

Mehrabian, A., & Russell, J. A. (1974). *An approach to environmental psychology*. The MIT Press.

Mohamed, R. N., Musa, R., Krishnan, R., & Ismail, S. (2013). The modern retail customer's experience on customer engagement: Evidence from health and personal care stores in Malaysia using structural equation modelling approach (SEM). *Global Journal of Business and Social Science Review, 1*(4), 28–34.

Mohammad, J., Quoquab, F., Thurasamy, R., & Alolayyan, M. N. (2020). The effect of user-generated content quality on brand engagement: The mediating role of functional and emotional values. *Journal of Electronic Commerce Research, 21*(1), 39–55.

Mohd Fudzi, W. F. H., Ismail, S., & Syed, S. F. (2021). Online shopping behaviour attributes during Covid-19 in Malaysia.

Mohd Nadzri, W. N. (2017). Modelling factors influencing brand experience and its consequences on Malaysia's national automotive brands.

Muda, M., Mohd, R., & Hassan, S. (2016). Online purchase behavior of generation Y in Malaysia. *Procedia Economics and Finance, 37*, 292–298.

Musa, R., Ahmad, N. S., & Janiffa, S. (2020). Hypothesized model of determinants and consequences of social media content marketing experience quality (Smcmeq) in muslimah fashion advertisement. *Malaysian Journal of Consumer and Family Economics, 24*(1), 247–261.

Nadzri, W. N. M., & Musa, R. (2017). Investigating psychometric elements measuring brand experience scale. *Advanced Science Letters, 23*(8), 7521–7524.

Oliveira, B., & Casais, B. (2019). The importance of user-generated photos in restaurant selection. *Journal of Hospitality and Tourism Technology*.

Pimentel, L. M., Dias, H. B. A., & de Mendonça, D. Jr. (2019). How to avoid webrooming behavior improving consumer experience using their online touchpoints with brand.

Raji Ridwan, A., Mohd Rashid, S., & Ishak, M. S. (2017). User-generated contents in Facebook, functional and hedonic brand image and purchase intention. In *SHS Web of Conferences* (Vol. 33, pp. 1–6).

Ramayah, T., Yeap, J. A., Ahmad, N. H., Halim, H. A., & Rahman, S. A. (2017). Testing a confirmatory model of Facebook usage in SmartPLS using consistent PLS. *International Journal of Business and Innovation, 3*(2), 1–14.

Rohm, A., Kaltcheva, V. D., & Milne, G. R. (2013). A mixed-method approach to examining brand-consumer interactions driven by social media. *Journal of Research in Interactive Marketing, 7*(4), 295–311. https://doi.org/10.1108/JRIM-01-2013-0009

Shankar, A., & Jain, S. (2021). Factors affecting luxury consumers' webrooming intention: A moderated-mediation approach. *Journal of Retailing and Consumer Services, 58*, 102306.

Vasudevan, P., & Arokiasamy, L. (2021). Online shopping among young generation in Malaysia. *Electronic Journal of Business and Management, 6*(1), 31–38.

Wang, Z., Kortana, T., & Kuang, W. (2020). Improving brand loyalty through social media marketing: Is it possible? An empirical study of SOR paradigm. In *E3S Web of Conferences* (Vol. 214, p. 01039). EDP Sciences.

Wolny, J., & Charoensuksai, N. (2014). Mapping customer journeys in multichannel decision-making. *Journal of Direct, Data and Digital Marketing Practice, 15*(4), 317–326.

Wroblowská, Z. (2016). Brand manager as a knowledge worker. *Trends of Management in the Contemporary Society*, 251.

Yasin, M., Porcu, L., Abusharbeh, T. M., & Liébana-Cabanillas, F. (2020). The impact of customer personality and online brand community engagement on intention to forward company and users generated content: Palestinian banking industry a case. *Economic Research-Ekonomska istraživanja, 33*(1), 1985–2006.

Zuzana, W. (2016). Requirements for brand managers and product managers responsible for competitiveness of product and brands. *Journal of Competitiveness, 8*(3), 5–21.

Sustaining Competitive Advantage During COVID-19 Pandemic: A Study of Aerospace Manufacturing Industry in Malaysia

Masnita Abdul Ghani, Nurhanan Syafiah Abdul Razak, Putri Rozita Tahir, and Muhammad Alshurideh[ID]

Abstract This study is one of the very few studies investigating the aerospace manufacturing industries hit severely by the coronavirus (COVID-19) pandemic, focusing on one of Malaysia's aerospace players. In addition, this study emphasises how an action is engaged to minimise or overcome the impact and effectiveness of strategic moves. PLS-SEM method is used to analyse data gathered from a survey. 30 leadership and executives have been chosen in this survey. Through a comprehensive literature study and survey, this paper presents a guide that illustrates general concepts and processes execution, allowing strategic management personnel to comprehend their market position and identify their long-term objectives. It is an approach to sustained competitive advantages by deliberate internal and external influences towards the change that scribe the marketplace competitive advantage landscape.

Keywords Competitive advantages · Strategic management · Internal influence · External influence · Aerospace manufacturer

M. A. Ghani (✉) · N. S. A. Razak · P. R. Tahir
Faculty of Business and Management, DRB, HICOM University of Automotive Malaysia (DHUAM), Pahang, Malaysia
e-mail: masnita.ghani@gmail.com

N. S. A. Razak
e-mail: nurhanan@meritus.edu.my; hana.syafiah@gmail.com

P. R. Tahir
e-mail: putri@dhu.edu.my

N. S. A. Razak
MERITUS University, Kuala Lumpur, Malaysia

M. Alshurideh
Department of Management, College of Business Administration, University of Sharjah, Sharjah, United Arab Emirates
e-mail: malshurideh@sharjah.ac.ae; m.alshurideh@ju.edu.jo

Marketing Department, School of Business, The University of Jordan, Amman, Jordan

© The Author(s), under exclusive license to Springer Nature Switzerland AG 2023   2287
M. Alshurideh et al. (eds.), *The Effect of Information Technology on Business and Marketing Intelligence Systems*, Studies in Computational Intelligence 1056,
https://doi.org/10.1007/978-3-031-12382-5_125

1 Introduction

A company's strategic purpose and a deep understanding of its principal strengths and assets reflect successful strategies. Successful businesses understand their customers' needs and want, their rivals' strengths and weaknesses, and how they can generate value for their stakeholders (Alketbi et al., 2020; Al-Khayyala et al., 2020; Alshamsi et al., 2020). The approach intends to position a company for competitive advantage (Altamony et al., 2012; Kluyver & Pearce, 2015; Obeidat et al., 2021). It results from executive decisions about what to provide, where to participate, and how to win to optimise long-term value (Alshurideh, 2019; Alzoubi et al., 2020; Joghee et al., 2021a).

The set of basic marketing principles that guide businesses in competitive markets defines marketing's role in leading and shaping strategic management (Alshurideh et al., 2020; Hooley et al., 2018; Kurdi et al., 2020). It is creating a marketing plan implemented, including defining where the company will engage and how it will compete.

Aside from the requirement for strategy as a single, interdependent process, the fourth industrial revolution necessitates agility and coordination of internal and external environment changes (Wiraeus & Creelman, 2019). While keeping a strong emphasis on strategy and the client, an agile business can evolve, driving transformation change, and be versatile (Ahmad et al., 2021a; Alaali, et al., 2021; Al-Dmour et al., 2021a; Alshraideh et al., 2017; Altamony et al., 2012; Ben-Abdallah et al., 2021; Shakhour et al., 2021).

1.1 Overview of Selected Company

The selected company CTRM Aero Composited Sdn Bhd is a manufacturer that fabricates composite components, named CTRM AC, along with this article. CTRM AC is a composite component manufacturer specialising in aircraft applications, mainly on nacelle and wing components. The CTRM AC's ultimate goal and commitment are to foster people, capabilities, and markets in Malaysia's advanced composites industry (CTRM AC, 2016).

Due to airline cancellations and reduced global production, this national composite vendor have adjusted their manufacturing rates. CTRM AC supplies wing and engine cover panels to Airbus SE and Boeing Co. As a result, delayed or cancelled aircraft production affects CTRM AC. Boeing announced 60 cancellations of 737 Max orders in June, on top of the previous 47. At the same time, customers cancelled 355 737 Max orders in the first half of 2020. Airbus plans to eliminate 15,000 jobs over two years and reduce production by 40% (Shan, 2020). It is also hard to see the direct impact, like order cancellations. Pandemic COVID-19 had significant effects on the firm revenue generation and forecast (Ahmad et al., 2021b; Akour et al., 2021; Alameeri et al., 2021; Al-Dmour et al., 2021b; Alshurideh et al., 2021; Amarneh et al., 2021; Leo et al., 2021; Nuseir et al., 2021a; Shah et al., 2021; Taryam et al., 2020). Before the pandemic happened, CTRM AC was forecasted to gain around 919 Million for Financial Year 2020. However, the post-pandemic impact brings down the demand; thus, the adjusted revenue forecasted for Financial Year 2020 will only be around 531.4 Million (CTRM AC, 2020). Overall revenue forecast declined by 42.2%, from 919 Million to 531.4 Million.

1.2 Overview of Covid-19 Lingerie Impact Towards Aerospace Manufacturer

The COVID-19 evidence predicts that the crisis will lead to market restructuring and a significant reduction in interest in air travel. Airflight is vulnerable to oil short-ages, force majeure, war, terror acts, global depressions, and outbreaks. In addition, demand severely weakened due to lower disposable income and behavioural changes (Suau-sanchez et al., 2020). Previous outbreaks apex around one to three months and back to the earlier situation in the next seven months. There are two possible outcomes expected from this situation. In 2020, the 'Limited Spread' scenario predicted a $63 billion loss in passenger revenues (11%), while the 'Extensive Spread' scenario predicted a $113 billion loss in passenger revenues worldwide (19%) (Pearce, 2020). In addition, stock markets are now expecting a substantial drop in airline profits internationally, far beyond the influence of SARS (Fig. 1).

The International Civil Aviation Organisation (ICAO) reported a 60% reduction in airline passengers (both regional and international) in 2020 compared to 2019, with a total decrease of 1,376 million passengers, or 74%. Loss of $250 billion in

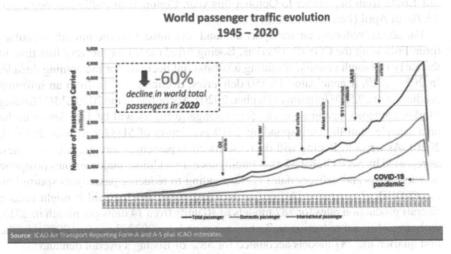

Fig. 1 World passenger traffic (*Source* ICAO)

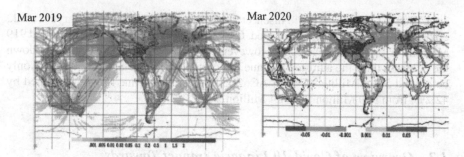

Fig. 2 March 2019 versus 2020 on Global ABO coverage observation (*Source* James et al., 2020)

gross operating income for airlines due to lower passenger numbers (Air Transport Bureau, 2020). According to the Airports Council International (ACI), 40% of traffic passengers and 50% of the terminal income by 2020 could wipe out due to COVID-19 (Airports Council International, 2020).

The plane observation range is densest over Europe and, Northern America subsequently followed by East Asia and Australia (James et al., 2020). 665 airports had at minimum one daily climb, 400 from European or Northern American terminals in late November 2019. However, flight cancellations due to the pandemic reduced aircraft observations by 80% by late April 2020.

Global COVID-19 flight reductions are shown in Fig. 2. Information coverage amid March 2019 and 2020 resulted in reduced air traffic worldwide. However, several flight routes and terminals have had coverage increased (James et al., 2020). The most affected industry is related to the airline industry (Haydon & Kumar, 2020). Commercial traffic fell 42.4% in October 2020, and traffic was 43% below 2019 levels in September and August, an insignificant gain (−45.2%). In contrast, traffic fell 1.67% from September to October this year. Commercial traffic was depressed 73.7% in April (Petchenik, 2020).

Far ahead, reducing air transport demand and impact on the aircraft manufacturer. Following the COVID-19 crisis, Boeing lifted the lid for the very first time in the early twentieth century, revealing a substantial decrease for the coming decade. In 2020–2029, it anticipated 18,350 deliveries, declining 10.7% from an informal prediction of 20,550 (Johnson & Hepher, 2020). According to the 2020 BMO (Boeing Market Outlook), revenue for 18,350 passenger airliners will be 11% lesser in the next decade than the corresponding 2019 prediction of $US2.9 trillion (Ratings, 2020). As a result, Boeing will decrease its 2020 passenger aircraft sales volumes, as reported by S&P Global Market Intelligence. In addition, major airlines postpone new jet orders and delivery dates corresponding to reducing passengers spurred by the COVID-19 pandemic. According to a statement, Boeing said it might reduce aircraft production rates for 787 models to 10 units from 14 units per month in 2020, with an additional reduction to 7 units per month by 2022 (Lazzaro, 2020). In the first quarter, the 787 models accounted for 58% of Boeing's overall demand.

Based on (Oestergaard, 2020), Airbus has slashed output on several programs and aims to keep the primary plane yield at 40% below levels before the pandemic for the next two years. Boeing reduced the production capacity of 787 jet models from 14 to 10 per month in 2020 and a further drop to 7 aircraft in 2022 (Lazzaro, 2020). As a result of the COVID-19, another aircraft OEM, Airbus, will reduce output across its entire manufacturing line. According to Garcia (2020), Airbus is cutting main aircraft models by a third to combat the COVID-19. 40 A320s will be built per month, down from 6. A330s will be produced at a rate of two per month instead of six. Likewise, A350 production will drop to six per month from ten aircraft sets.

COVID-19 spread has substantial waves on global production and supply chain networks. Manufacturing facilities have closed or reduced capacity, and the raw and finished product supply chains are disrupted by trading and transportation issues. This pandemic has a major influence on reintegrating industrial workers. It reduced human resources, lowering manufacturing process efficiency. Proper safety measures are required to consider social distance in the workplace and implement frequent health monitoring for employees to enhance the human aspect of production. Supply chain disruptions included (a) decreased demand for goods (automobiles, public transportation, and textiles), (b) decreased demand for services (temperature scanners, ventilation systems, face masks, disinfectants, protective clothing, and essential food products), and (c) distribution failures and unpredictability in raw materials, (d) Reducing the ability to deliver and obtain goods on time inadequate supply and transportation constraints; and (e) Assure employees capability to construct and transport goods (Kumar et al., 2020). The pandemic virus affects all industries, particularly aviation, passenger and cargo airlines, airline food caterers, airport authorities, aircraft parts manufacturers, travel agents, and MROs (Rahman et al., 2020). The virus's spread has varying effects on the aviation industry. The virus reduced airline capacity by nearly 50%. It is happening in nearly every country.

Despite the background and list of critical factors that could influence aerospace manufacturer demand reduction, it also represents a general insight into national aerospace composite manufacturers currently struggling due to demand reduction from the COVID-19 outbreak impacts. The delivery demand is pushed out to the right and with an increase of concern on ordering cancellation due to airline bankruptcy (AlShurideh et al., 2019; Hamadneh et al., 2021; Joghee et al., 2021b; Lazzaro, 2020; Oestergaard, 2020). This outbreak badly affects aircraft manufacturer's revenue and faces high inventory levels. Nevertheless, the manufacturer obliges to manage in-transit material, ordering material with a long lead-time, and several government commendations and requirements that affect different global manufacturing production flow to remain resilient in market share.

1.2.1 Impact of COVID-19 on the Aviation Players

Rahman et al. (2020) investigated the impact of COVID-19 on aviation such as airlines, airport authorities, air cargo providers, in-flight food caterers, aircraft manufacturers, maintenance, repair, and overhaul organisation. Their study recommended empirical research on COVID-19 and its impact on the aviation business. Due to COVID-19, research can focus on business improvement initiatives and current strategies to improve the performance of aviation organisations. Each aeronautical business can explore micro, macro, and meso factors to define key measurements. They also recommend future research on the relationship between foreign and local regulators to identify the best business practices for aviation players.

While (Thorbecke, 2020) used sectoral stock price responses to study the COVID-19 crisis' impact on the US economy, affected industries include funerals and airlines. Massive areas of the economy whose recovery is dependent on pandemic control rather than macroeconomic conditions. It will be interesting to see how other countries' stocks have responded and how industrial structures, development levels, and macroeconomic policy responses differ across countries.

(Mhalla, 2020) examined the COVID-19 consequences on the worldwide economy, focusing on oil and air travel. The long-term effects of the coronavirus are still unknown, but demand is expected to remain weak as the epidemic spreads to other major regional economies. In addition, due to the new coronavirus epidemic, many European airlines are reducing flight schedules and cutting costs. As a result, the major airlines will have to decide whether to ground planes or redeploy them.

An airline, airport, aircraft manufacturer, and other industry-specific issues are discussed in the Liptakova article. (Liptakova et al., 2020) believed that recovering from the current aviation crisis will require economic recovery and regaining public confidence. Nobody knows how long this will go on. Travel volumes will not resume for 3 to 5 years, according to IATA. However, the present predicament may purge the manufacturing in multiple approaches, such as eliminating weak airlines, as recent airline bankruptcies have shown.

1.2.2 Responsive Towards Firm Resilient

Xu et al. (2020) investigate the global supply chain's efficiency and reaction and propose a set of managerial strategies to reduce threats and improve agility in different manufacturing segments. An article discusses the implications of COVID-19's lengthier period, marketplace, and scope. The COVID-19 pandemic has reportedly disrupted major global supply chains in pharmaceuticals, food, electronics, and the automotive industries. Temporary closures of industries and shops, travel restrictions, and compulsory quarantine of citizens were used to stop the spread of COVID-19. Shortages of human resources, active ingredients, and raw materials resulted. Tightening controls exacerbated the issue. Delays and inventory shortages result in global supply chains, making two long-term competitiveness enablers recommendations. Understanding critical vulnerabilities and monitoring disruptive risks in real-time is the first. The subsequent endorsement collaborates with internal stakeholders and strategic and critical suppliers to develop risk management strategies (Ahmad et al., 2021c; Alshurideh, 2017; Alshurideh et al., 2019).

Aviation is currently focused on short endurance. An expectation of this "new normal" in this way must happen now, to take into account change over the next six to two years. The aerospace industry is seeing widespread requests for crisis state assistance and, more importantly, legislative intervention. Deregulation in aviation must be adequately managed to avoid twisted rivalry (Singh, 2020).

According to Hader (2020), firms and governments must thoroughly ensure the critical industrial skills are occupied to avoid driving the entire sector at risk. As a result, the industry must quickly agree on the "new normal" manufacturing rates, develop a mutual strategy for transforming the sector to the "new normal", recognise and support vulnerable components in the transition process.

Various industries have evolved immensely. The need to propagate massive R&D investments across many markets drove the globalisation of the aircraft manufacturing industry. Monitoring globalisation is an essential strategic management undertaking. Even more important is knowing how to use that data to gain an advantage (David et al., 2017). The ability to gain and hold a competitive advantage is critical to long-term success. A company's competitive advantage typically lasts only a short time due to competitor copying and weakening.

Typically, COVID-19 focused on the perspective of tourism, medical science, and airlines services as mentioned by Suau-sanchez et al. (2020); Rahman et al., 2020; Mhalla, 2020; Sun et al., 2020; Iacus et al., 2020; Madi Odeh et al., 2021). In contrast, a clear industrial organisation perspective of COVID-19 is inattentive. Only some researchers, which is (Singh, 2020), emphasised COVID-19 impact on aerospace, which conducts investigations to determine various methodologies embraced to determine the market impact and showcase the aerospace industry's development rate and downturn patterns because of the COVID-19 pandemic. So far, limited study has worked on the implementation to overcome the lingering COVID-19 effects, specifically on aerospace manufacturers in Malaysia and how the related firm reacts to sustain its competitive advantages and remain resilient.

2 Literature Review

2.1 Competitive Advantages

Corporate competition is inevitable. To compete, one must constantly distinguish, understand the market, consumer needs, and their variations. It is essential to consider and control a wide range of resources (Kuncoro & Suriani, 2018). Competitive advantages occur when companies outperform rivals economically (Allozi et al., 2021; Kabrilyants et al., 2021; Shamout et al., 2021). Financial value is the difference between a consumer's perceived benefits and the total cost of the goods or services (Al-Jarrah et al., 2012; Assad & Alshurideh, 2020a, 2020b; Barney & Hesterly, 2018; Shah et al., 2020). The study of heterogeneous firm features necessitated various methods to identify, measure, and comprehend firm resources and capabilities. In the resource-based approach, supporters argue that each firm has unique resources and skills to maintain competitive advantages (Dagnino & Cinici, 2016). Each theory relies on observations and predictions about how competition in this industry will evolve to gain a competitive advantage. The more accurately these theories and statements reflect current competitiveness in this sector, the more likely it will achieve a strategic edge by applying its strategies. If these conclusions and theories are incorrect, a firm's tactics are unlikely to provide a competitive advantage. Difficulty in predicting market competition, determining a company's strategy is challenging.

The five reasons that propel strong growth companies are human capital, human resource management, strategy, skills, and creativity (Al Kurdi et al., 2021; Demir et al., 2017; Harahsheh et al., 2021; Nuseir et al., 2021b). Competitiveness measure by past success or future competitiveness. Examples are financial performance (profit, earnings growth, returns on investment), non-financial performance (customer retention, employee growth), benchmarking balanced scorecard, metrics like earnings, pricing, productivity, and market share were used (Sachitra, 2016). A competitive advantage allows a company to outperform its competitors. Thus, profitability is a key indicator for assessing competition, and turnover is a key profit margin. However, efficiency, market share, and profitability are insufficient to determine firm-level competitive advantage.

Competitive advantage is assessed by comparing the results of strategic attempts to firm profitability or market share stabilisation. Although competition appears to revert corporate performance to the mean, long-term competitive advantages ensure great returns (Maury, 2016). A company's financial stability allows it to make prudent financial decisions and maintain or reduce market competition. According to Kliestik et al. (2020), interim financial stability reduces future financial threats. A company's overall financial strength is determined by the ratios listed above and other factors such as its size, industry, and location. Profits and losses are the most critical indicators of a firm's performance, especially compared to other manufacturing representatives. The results show that the liquidity ratios that measure a firm's ability to repay short-term debt obligations are crucial. The current ratio measures a company's ability to pay short-term obligations using current assets. A higher current ratio is better than a lower current ratio because it shows the firm pays a current debt. In the following period, short-term liabilities are outstanding. The liabilities-to-total-assets ratio measures the number of assets hidden by liabilities. The ratio of liabilities to total assets is high, indicating low stockholder equity and potential creditworthiness issues. Liabilities-to-total-assets ratios are high in distressed firms. Then there are the total sales to total assets ratio, which measures corporate efficiency by comparing assets to revenue. The higher the ratio, the healthier the firm. Finally, the equity-to-asset ratio regulates a firm's financial leverage. A higher equity-to-asset ratio is required to convince creditors that a firm is sustainable and low-risk.

Long-term success requires gaining and maintaining a competitive advantage. A company's competitive advantage typically lasts only a short time due to competitor copying and weakening. So gaining a competitive advantage is not enough. Assemble, implement, and evaluate plans that capitalise on external and internal variables to achieve long-term competitive advantage. Different industries go global for various reasons. The need to propagate R&D investments across many markets drove the globalisation of the aircraft manufacturing industry (David et al.,

2017). Monitoring industry globalisation is a strategic management task. Even more crucial knows how to use that data to acquire or optimise advantages.

A wide range of methods for assessing an enterprise's competitiveness and market position are being investigated, such as PEST analysis, McKinsey, Model Boston Consultative Group, GAP analysis, Michael Porter, situational analysis (SWOT analysis), expert assessment method, financial and economic method, and others (Khanenko, 2019).

Companies that can maintain a competitive advantage provide sustained benefits. They help a company's reputation by associating it with a long-lasting, appealing offer. A company can create a competitive advantage by combining a temporary competitive advantage (TCA) and a sustainable competitive advantage (SCA). This benefit adapts to changes in the company's background. Another management implication is that combining SCA and TCA does not guarantee a competitive advantage. By combining its strengths, the company can seize new opportunities while maintaining its core business. This method necessitates the establishment of fundamental assumptions (Sołoducho-Pelc & Sulich, 2020). Following these principles increases the likelihood of achieving outstanding short- and long-term results. Having an SCA and TCA focused not only on economic values but also on social and environmental issues may be reasonable.

Nobody has unlimited resources, and they must assess the company's value of various options. An organisation's long-term strategy decisions bind it to specific goods and markets. To win in the long run, strategists can help decision-makers predict significant events that could impact the company. To make assumptions, one needs to know what is coming up in the company's business. Strategists must adjust if actual events differ significantly from forecasts. No policy can be made without rational expectations. Accurate assumptions are a competitive advantage for well-informed firms (David et al., 2017). Only after competitors attempted to copy a firm's strategy have stalled or failed can a firm claim one or more critical strategic benefits. Also, no competitive advantage is permanent. How a competitor's responsiveness to obtain the abilities to imitate a company's value-creating strategy defines the duration a competitive advantage can last (Hitt et al., 2017).

According to Barney and Hesterly (2018), the best way for a firm to choose its strategy is to follow strategic management principles. The strategic management framework is a set of evaluations and decisions that will help the firm find a competitive advantage. The strategic management method of a series collection of analyses and choices will improve the probability of a company selecting a successful approach that produces competitive advantages, as shown in Fig. 3 below.

(Gurel & Tat, 2017) and (Zu'bi et al., 2012) suggested that the external and internal forces' assessment is one core element in strategic management. An external review should help the company identify key competitive risks and opportunities. It also points out that the world of competition is changing, posing new challenges and opportunities for businesses. An internal review identifies a company's strengths and weaknesses. It also shows a company's strengths and resources as sources of competitive advantage rather than gains.

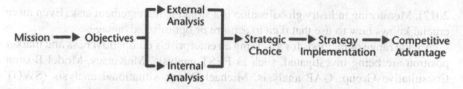

Fig. 3 Strategic management process (*Source* Hesterly & Barney, 2008)

2.2 Scanning Internal and External Influence

Many firms' success or failure is determined by the factors that influence their operations. Scanning the internal and external environments is critical for business planning and organisational analysis (Amuna et al., 2017). It is vital to both for-profit and charitable companies' business practices. However, it is challenging to plan excellent strategy management without considering environmental influences.

According to Rahman et al. (2020), to remain competitive, the company should carefully evaluate its external and internal factors. The approach is linking a company to its environment, both internal and external. These findings were reflected in the SWOT analysis (Guerras-Martín et al., 2014). Complex political, social, economic, technological, and cultural changes are now the norm. In a complex environment, an entity that is not responsive cannot survive. An environmental scan is a method of analysing the climate that is dependent on several variables. External factors include political, economic, scientific, technological factors. The internal environment includes resources, capability, culture, management style, and structure. This study's Independent variable is a latent variable, environment factor that influences business landscape. A competitive market will derive from PEST (Political, Economic, Social, and Technology) and SWOT analysis to assess CTRM AC key external and internal factors (Fig. 4).

2.2.1 Relationship Between Internal Influence and Competitive Advantages

Internal analysis is a third aspect of the strategic planning process that describes the successes besides limitations. Recognising the magnitude and nature of a business's tools and expertise and structure, particular talents, and company-specific or distinctive capabilities are addressed when investigating the origins of successful adviewing (Petchenik, 2020). Build and maintain a competitive advantage entails a business attaining superior performance, consistency, creativity, and responsiveness for its consumers. The company's strength contributes to outstanding performance, while its limitations interpret as unsatisfactory performance. The internal analysis supports a firm in identifying its strengths and weaknesses (Haydon & Kumar, 2020). It also assistances in understanding that its resources and capabilities are likely to be competitively advantageous and less likely to be the source of such advantages.

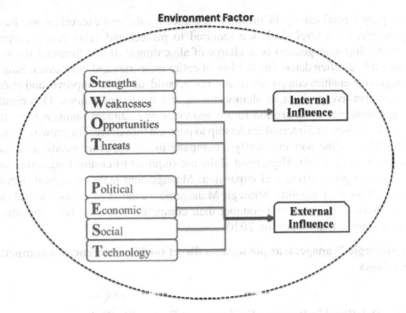

Fig. 4 Correlation between SWOT and PEST towards internal and external influence

H₁: Internal factors possess a direct positive influence on competitive advantages

2.2.2 Relationship Between External Influence and Competitive Advantages

External analysis is a second aspect of the strategic management phase in studying the firms' exterior operational climate. External evaluation's primary goal is to recognise strategic prospects and risks in the firms' operating atmosphere that could influence how it pursues its task. The firm recognises critical threats and opportunities in its competitive environment by conducting an external analysis (Haydon & Kumar, 2020). It scrutinises how the rivalry is likely to develop in this environment and its consequences for a firm's threats and opportunities.

H₂: External factors possess a direct positive influence on competitive advantages

2.2.3 Relationship Between Strategic Management and Competitive Advantages

The strategic management method is a sequential collection of analyses and choices which improve the probability of a company selecting a successful approach to produces competitive advantages (Barney & Hesterly, 2018). The approaches define a company's productivity in meeting customer needs, and executives are heavily responsible. High-level skills are required to protect and grow future corporate strengths. Top management is in charge of allocating staff and financial resources. These findings often determine the fate of entire industries and businesses. Strategic Management enables companies to add value, build, discover, improve, and conquer their competitive position by demonstrating the necessary actions. Organisations can use strategies to determine future addresses or treatment plans, actions to be taken, timetables, and internal leadership to position themselves for growth (Ratings, 2020). They define how efficiently a company meets customer needs, and executives are heavily liable. High-level skills are required to ensure long-term corporate strength preservation and expansion. Management makes essential decisions about staffing and budgeting. Strategic Management enables businesses to add value, build, discover, improve, and conquer their competitive position by demonstrating the necessary actions (Lazzaro, 2020).

H$_3$: Strategic Management possesses a direct positive influence on competitive advantages

2.2.4 Relationship Between Environment Factor, Strategic Management and Competitive Advantages

According to Oestergaard (2020), strategic management is a cognitive science that intersects with sociology, economics, psychology, finance, and marketing. The current state of the field can present a complex picture of numerous approaches and paradigms. As a result, there is no conclusive general theory that encompasses strategic management. In strategic management, decisions are made. Strategic decisions are those that determine a company's strategic direction. Their mission is to ensure the company's long-term success, which includes: (a) recognising and understanding path dependencies, (b) responding to current change dynamics, and (c) implementing strategic programs on the previous two points' foundations. Each resolution's action will define future growth. It includes (a) external and internal climate appraisal, (b) strategy formulation (focus on the long-term planning), and (c) strategy implementation (Strategic management focuses on managing and accessing potential threats and opportunities while taking into account a company's strengths and weaknesses). Companies must constantly update their organisation and search for new strategic goals and execution approaches to keep up with the changing climate (Garcia, 2020). The global market's ability to sustain competitive resources may be another competitive factor. Global affiliates or strategic alliances should learn these techniques. Concepts and ideas, especially international coalition methodologies, continue to shape market competition.

H$_4$: Strategic Management mediates the relationship between environmental factors and competitive advantages

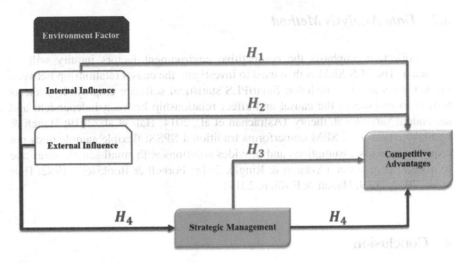

Fig. 5 Research framework

2.3 Research Framework

See Fig. 5.

3 Methodology

3.1 Research Method

The study utilises a quantitative approach by using the survey and literature method to discuss the research topic to determine the mediation effect of strategic management on the relationship between environmental influence and competitive advantages.

(a) Survey:
   The questionnaires were adapted and modified from the previous study by Borges and Gaia (2010); Sigalas et al., 2013; Bismark et al., 2018). Five Likert-point scales use to measure and evaluate response. The environment influences are the exogenous latent variable, while competitive advantages represent the endogenous latent variable and strategic management is the mediating latent variable in this study. The survey will be distributed online amongst 30 personnel directly involved with strategic management activities.

(b) Literature:
   Comprehend articles (from media and relevant websites) and company reports, archives, and presentations.

3.2 Data Analysis Method

Data collection emphasis the competitive environment factors inquiry will be collected. The PLS-SEM is then used to investigate the causal relationship between the variables and the mediator. SmartPLS statistical software employs hypothesis testing to investigate the causal and effect relationship between independent and dependent variables in theory. (Astrachan et al., 2014; Hair et al., 2019; Hanafi & Fadilah, 2017). PLS-SEM outperforms traditional SPSS; flexible sample does not require normality assumptions and provides solutions with small sample sizes due to a small population. (Avkiran & Ringle, 2018; Fornell & Bookstein, 1982; Hair et al., 2017, 2019; Hanafi & Fadilah, 2017).

4 Conclusion

The study's scope focuses on the environment's influence and lingering effects of the uncertainties from the COVID-19 pandemic to market share, disregarding other associate factors such as bailout, policy changes, and government austerity drive.

This study is one of the few that focuses on the aerospace manufacturing industries that have been significantly impacted by the coronavirus (COVID-19) outbreak by focusing on one of Malaysia's aerospace players. Furthermore, this research reflects how action could reduce or eliminate the impact and effectiveness of the strategic decision.

Competitiveness has evolved into the economic criterion for orienting and evaluating the company's performance from the inside out. The organisation must understand and determine which approaches are most likely to provide a competitive advantage over its rivals based on internal and external influences. Through a comprehensive literature study, this paper presents a guide that illustrates how general concepts and processes can be done, allowing strategic management personnel to understand their market position and, from there, identify where they hunger to reach in the future. Strategic Management enables businesses to provide value, build, discover, strengthen, and maintain competitive advantages by specifying what measures to achieve this position. It enables firms to stand out the addresses or course of action in the future, identifying the action rules, tracking behaviour in time, and outlining the company's internal leadership to set the organisation in the best competitive environment to achieve success. Business success necessitates a company's constant adaptation to its surroundings. Although it is not a guarantee of success, it enables organisations to make long-term decisions, put them into practice effectively, and initiate corrective steps as appropriate. Integration of intuition and analysis is essential for realistic strategic judgement.

References

Ahmad, A., Alshurideh, M. T., Al Kurdi, B. H., & Alzoubi, H. M. (2021a). Digital strategies: A systematic literature review. In *The International Conference on Artificial Intelligence and Computer Vision* (pp. 807–822).

Ahmad, A., Alshurideh, M. T., Al Kurdi, B. H., & Salloum, S. A. (2021b). Factors impacts organization digital transformation and organization decision making during covid19 pandemic. *The Effect of Coronavirus Disease (COVID-19) on Business Intelligence*, 95.

Ahmad, A., Alshurideh, M., Al Kurdi, B., Aburayya, A., & Hamadneh, S. (2021c). Digital transformation metrics: a conceptual view. *Journal of Management Information and Decision Sciences*, 24(7), 1–18.

Air Transport Bureau. (2020). Effects of novel coronavirus (COVID-19) on civil aviation: Economic impact analysis.

Airports Council International. (2020). The impact of COVID-19 on the airport business: Two-fifths of passenger traffic and almost half of revenues wiped out in 2020. *ACI Adviser Bullet*.

Akour, I., Alshurideh, M., Al Kurdi, B., Al Ali, A., & Salloum, S. (2021). Using machine learning algorithms to predict people's intention to use mobile learning platforms during the COVID-19 pandemic: Machine learning approach. *JMIR Medical Education*, 7(1), 1–17.

Al Kurdi, B., Elrehail, H., Alzoubi, H., Alshurideh, M., & Al-Adaileh, R. (2021). The interplay among HRM practices, job satisfaction and intention to leave: An empirical investigation. *Journal of Legal, Ethical and Regulatory Issues*, 24(1), 1–14.

Alaali, N., et al. (2021). The impact of adopting corporate governance strategic performance in the tourism sector: A case study in the Kingdom of Bahrain. *Journal of Legal, Ethical and Regulatory Issues*, 24(1), 1–18.

Alameeri, K. A., Alshurideh, M. T., & Al Kurdi, B. (2021). The effect of covid-19 pandemic on business systems' innovation and entrepreneurship and how to cope with it: A theatrical view. *The Effect of Coronavirus Disease (COVID-19) on Business Intelligence*, 334, 275–288.

Al-Dmour, A., Al-Dmour, H., Al-Barghuthi, R., Al-Dmour, R., & Alshurideh, M. T. (2021b). Factors influencing the adoption of e-payment during pandemic outbreak (COVID-19): Empirical evidence. *The Effect of Coronavirus Disease (COVID-19) on Business Intelligence*, 334, 154–133.

Al-Dmour, R., AlShaar, F., Al-Dmour, H., Masa'deh, R., & Alshurideh, M. T. (2021a). The effect of service recovery justices strategies on online customer engagement via the role of 'customer satisfaction' during the Covid-19 pandemic: An empirical study. *The Effect of Coronavirus Disease (COVID-19) on Business Intelligence*, 334, 346–325.

Al-Jarrah, I., Al-Zu'bi, M. F., Jaara, O., & Alshurideh, M. (2012). Evaluating the impact of financial development on economic growth in Jordan. *International Research Journal of Finance and Economics*, 94, 123–139.

Alketbi, S., Alshurideh, M., & Al Kurdi, B. (2020). The influence of service quality on customers' retention and loyalty in the UAE hotel sector with respect to the impact of customer' satisfaction, trust, and commitment: A qualitative study. *PalArch's J. Archaeol. Egypt/Egyptology*, 17(4), 541–561.

Al-Khayyala, A., Alshuridehb, M., Al Kurdic, B., & Aburayyad, A. (2020). The impact of electronic service quality dimensions on customers' E-shopping and E-loyalty via the impact of E-satisfaction and E-trust: A qualitative approach. *International Journal of Innovation, Creativity and Change*, 14(9), 257–281.

Allozi, B., Alshurideh, A., AlHamad, M., & Al Kurdi. (2021). Impact of transformational leadership on the job satisfaction with the moderating role of organizational commitment: case of UAE and Jordan manufacturing companies. *Academy of Strategic Management Journal*, 20(1), 1–13.

Alshamsi, A., Alshurideh, M., Al Kurdi, B., & Salloum, S. A. (2020). The influence of service quality on customer retention: A systematic review in the higher education. In *International Conference on Advanced Intelligent Systems and Informatics*, pp. 404–416.

Alshraideh, A., Al-Lozi, M., & Alshurideh, M. (2017). The impact of training strategy on organizational loyalty via the mediating variables of organizational satisfaction and organizational performance: An empirical study on jordanian agricultural credit corporation staff. *Journal of Social Sciences, 6*, 383–394.

Alshurideh. (2017). A theoretical perspective of contract and contractual customer-supplier relationship in the mobile phone service sector. *International Journal of Business and Management, 12*(7), 201–210.

Alshurideh, et al. (2019). Determinants of pro-environmental behaviour in the context of emerging economies. *International Journal of Sustainable Society, 11*(4), 257–277.

Alshurideh, M. T., et al. (2021). Factors affecting the use of smart mobile examination platforms by universities' postgraduate students during the COVID 19 pandemic: An empirical study. *Informatics, 8*(2), 32.

Alshurideh. (2019). Do electronic loyalty programs still drive customer choice and repeat purchase behaviour? *International Journal of Electronic Customer Relationship Management, 12*(1), 40–57.

AlShurideh, M., Alsharari, N. M., & Al Kurdi, B. (2019). Supply chain integration and customer relationship management in the airline logistics. *Theoretical Economics Letters, 9*(2), 392–414.

Alshurideh, M., Gasaymeh, A., Ahmed, G., Alzoubi, H., & Al Kurd, B. (2020). Loyalty program effectiveness: Theoretical reviews and practical proofs. *Uncertain Supply Chain Management, 8*(3), 599–612.

Altamony, H., Masa'deh, R. M. T., Alshurideh, M., & Obeidat, B. Y. (2012). Information systems for competitive advantage: Implementation of an organisational strategic management process. In *Innovation and Sustainable Competitive Advantage: From Regional Development to World Economies—Proceedings of the 18th International Business Information Management Association Conference* (vol. 1, pp. 583–592).

Alzoubi, H., Alshurideh, M., Al Kurdi, B., & Inairat, M. (2020). Do perceived service value, quality, price fairness and service recovery shape customer satisfaction and delight? A practical study in the service telecommunication context. *Uncertain Supply Chain Management, 8*(3), 579–588.

Amarneh, B. M., Alshurideh, M. T., Al Kurdi, B. H., & Obeidat, Z. (2021). The impact of COVID-19 on e-learning: advantages and challenges. In *The International Conference on Artificial Intelligence and Computer Vision* (pp. 75–89).

Amuna, Y. M. A., Al Shobaki, M. J., & Naser, S. S. A. (2017). Strategic environmental scanning: An approach for crises management. *International Journal of Information Technology and Electrical Engineering, 6*(3), 28–34.

Assad, N. F., & Alshurideh, M. T. (2020a). Investment in context of financial reporting quality: A systematic review. *WAFFEN-UND Kostumkunde Journal, 11*(3), 255–286.

Assad, N. F., & Alshurideh, M. T. (2020b). Financial reporting quality, audit quality, and investment efficiency: Evidence from GCC economies. *WAFFEN-UND Kostumkunde Journal, 11*(3), 194–208.

Astrachan, C. B., Patel, V. K., & Wanzenried, G. (2014). A comparative study of CB-SEM and PLS-SEM for theory development in family firm research. *Journal of Family Business Strategy, 5*(1), 116–128.

Avkiran, N. K., & Ringle, C. M. (2018). *Partial least squares structural equation modeling: Recent advances in banking and finance: international series in operations research & management science.* Springer International Publishing.

Barney, J. B., & Hesterly, W. S. (2018). *Strategic management and competitive advantage concepts* (6th ed.). Pearson Education.

Ben-Abdallah, M., Shamout, R., & Alshurideh, M. (2021). Business development strategy model using EFE, IFE and IE analysis in a high-tech company: An empirical study. *Academy of Strategic Management Journal, 21*(1), 1–8.

Bismark, O., Kofi, O. A., Frank, A.-G., & Eric, H. (2018). Utilizing Mckinsey 7s model, SWOT analysis, PESTLE and balance scorecard to foster efficient implementation of organizational strategy. Evidence from the community hospital group- Ghana limited. *International Journal of Business, Economics & Management, 2*(3), 94–113.

Borges, R., & Gaia, S. (2010). Evaluation of a strategic management process in a small business. *International Conference on Industrial Engineering and Operations Management, 12*, 1–13.

Chitra, M., CTRM AC. (2016). Corporate CTRM AC—About Us.

Chitra, M., CTRM AC. (2020). AMP FY2020 slide presentation (finance-FACT).

Dagnino, G. B., & Cinici, M. C. (2016). *Research methods for strategic management* (1st ed.). Routledge.

David, F. R., & David, F. R. (2017). *Strategic managment. A competitive advantage approach concepts and cases* (16th ed.). Pearson Education.

De Kluyver, C. A., & Pearce, J. A., II. (2015). *Strategic management: An executive perspective, first.* Business Expert Press.

Demir, R., Wennberg, K., & McKelvie, A. (2017). The strategic management of high-growth firms: A review and theoretical conceptualization. *Long Range Planning, 50*(4), 431–456.

Fornell, C., & Bookstein, F. L. (1982). Two structural equation models: LISREL and PLS applied to consumer exit-voice theory. *Journal of Marketing Research, 19*(4), 440–452.

Garcia, M. (2020). Airbus cuts production by a third, adjusting to reduced demand as coronavirus decimates air travel. *Forbes.*

Guerras-Martín, L. Á., Madhok, A., & Montoro-Sánchez, Á. (2014). The evolution of strategic management research: Recent trends and current directions. *BRQ Business Research Quarterly, 17*(2), 69–76.

Gurel, E., & Tat, M. (2017). SWOT analysis: A theoretical review. *Journal of International Social Research, 10*(51), 994–1006.

Hader, M. (2020). Plunge in air traffic will deeply impact demand for new aircraft. *Roland Berger*

Hair, J. F., Jr., Matthews, L. M., Matthews, R. L., & Sarstedt, M. (2017). PLS-SEM or CB-SEM: Updated guidelines on which method to use. *International Journal of Multivariate Data Analysis, 1*(2), 107–123.

Hair, J. F., Risher, J. J., Sarstedt, M., & Ringle, C. M. (2019). When to use and how to report the results of PLS-SEM. *European Business Review, 31*(1), 2–24.

Hamadneh, S., Pedersen, O., & Al Kurdi, B. (2021). An investigation of the role of supply chain visibility into the Scottish Blood supply chain. *Journal of Legal, Ethical and Regulatory Issues, 24*(1), 1–12.

Hanafi, M. A. O., & Fadilah, P. (2017). Quantitative data analysis: Choosing between SPSS, PLS and AMOS in social science research. *International Interdisciplinary Journal of Scientific Research, 3*(1), 14–25.

Harahsheh, A. A., Houssien, A. M. A., & Alshurideh, M. T. (2021). The effect of transformational leadership on achieving effective decisions in the presence of psychological capital as an intermediate variable in private Jordanian. In *The Effect of Coronavirus Disease (COVID-19) on Business Intelligence*, Springer Nature, pp. 243–221.

Haydon, D., & Kumar, N. (2020). industries most and least impacted by COVID-19 from a probability of default perspective—September 2020 update I S&P Global Market Intelligence. *S&P Global*, 2020.

Hesterly, W., & Barney, J. (2008). Strategic management and competitive advantage. Pearson Education Limited. www. pearson.com/us.

Hitt, M. A., Ireland, R. D., & Hoskisson, R. E. (2017). *Strategic management : Competitiveness & globalization : Concepts* (12th ed.). South-Western Cengage Learning.

Hooley, G., Piercy, N. F., Nicoulaud, B., & Rudd, J. M. (2018). *Marketing Strategy and Competitive Positioning*, Sixth. Pearson.

Iacus, S. M., Natale, F., Santamaria, C., Spyratos, S., & Vespe, M. (2020). Estimating and projecting air passenger traffic during the COVID-19 coronavirus outbreak and its socio-economic impact. *Safety Science, 129*, 104791.

James, E. P., Benjamin, S. G., & Jamison, B. D. (2020). Commercial-aircraft-based observations for NWP: Global coverage, data impacts, and COVID-19. *Journal of Applied Meteorology and Climatology, 59*(11), 1809–1825.

Joghee, S. H. S., Al Kurdi, B., Alshurideh, M., Alzoubi, H., Vij, A., Muthusamy, M. (2021a). Expats impulse buying behaviour in UAE: A customer perspective. *Journal of Management Information and Decision Sciences, 24*(1), 1–24, 2021a.

Joghee, S., Alzoubi, H. M., Alshurideh, M., & Al Kurdi, B. (2021b). The role of business intelligence systems on green supply chain management: Empirical analysis of FMCG in the UAE. In *The International Conference on Artificial Intelligence and Computer Vision* (pp. 539–552).

Johnson, Y E. M., & Hepher, T. (2020). Boeing cuts jet demand forecast on pandemic crisis I Reuters. *Reuters.*

Kabrilyants, R., Obeidat, B. Y., Alshurideh, M., & Masa'deh, R. (2021). The role of organizational capabilities on e-business successful implementation. *International Journal of Data and Network Science, 5*(3), 417–432.

Khanenko, A. V. (2019). The assessment of competitiveness and the formation of an economic entity. *2*(10), 103–111.

Kliestik, T., Valaskova, K., Lazaroiu, G., Kovacova, M., & Vrbka, J. (2020). Remaining financially healthy and competitive: The role of financial predictors. *Journal of Competitiveness, 12*(1), 74–92.

Kumar, A., Luthra, S., Mangla, S. K., & Kazançoğlu, Y. (2020). COVID-19 impact on sustainable production and operations management. *Sustainable Operations and Computers, 1*, 1–7.

Kuncoro, W., & Suriani, W. O. (2018). Achieving sustainable competitive advantage through product innovation and market driving. *Asia Pacific Management Review, 23*(3), 186–192.

Kurdi, B., Alshurideh, M., & Alnaser, A. (2020). The impact of employee satisfaction on customer satisfaction: Theoretical and empirical underpinning. *Management Science Letters, 10*(15), 3561–3570.

Lazzaro, N. (2020). Boeing to cut 2020 aircraft output rates on weak demand from pandemic: company I S&P Global Platts. *S&P Global Market Intelligence.*

Leo, S., Alsharari, N. M., Abbas, J., & Alshurideh, M. T. (2021). From offline to online learning: A qualitative study of challenges and opportunities as a response to the COVID-19 pandemic in the UAE higher education context. *The Effect of Coronavirus Disease (COVID-19) on Business Intelligence, 334*, 203–217.

Liptakova, D., Kolesar, J., & Keselova, M. (2020). Challenges to the global aerospace industry due to the pandemic epidemy of COVID-19. *NTinAD 2020—New Trends in Aviation Development 2020—15th International Science Conference Proceedings* (pp. 155–158).

Madi Odeh, R. B. S., Obeidat, B. Y., Jaradat, M. O., Masa'deh, R., & Alshurideh, M. T. (2021). The transformational leadership role in achieving organizational resilience through adaptive cultures: the case of Dubai service sector. *International Journal of Productivity and Performance Management.*

Maury, B. (2018). Sustainable competitive advantage and profitability persistence: Sources versus outcomes for assessing advantage. *Journal of Business Research, 84*, 100–113.

Mhalla, M. (2020). The impact of novel coronavirus (COVID-19) on the global oil and aviation markets. *Asian Journal of Scientific Research, 10*(2), 96–104.

Nuseir, M. T., Aljumah, A., & Alshurideh, M. T. (2021a). How the business intelligence in the new startup performance in UAE during COVID-19: The mediating role of innovativeness. *The Effect of Coronavirus Disease (COVID-19) on Business Intelligence*, 63–79.

Nuseir, M. T., Al Kurdi, B. H., Alshurideh, M. T., & Alzoubi, H. M. (2021b). Gender discrimination at workplace: Do artificial intelligence (AI) and machine learning (ML) have opinions about it. In *The International Conference on Artificial Intelligence and Computer Vision* (pp. 301–316).

Obeidat, U., Obeidat, B., Alrowwad, A., Alshurideh, M., Masadeh, R., & Abuhashesh, M. (2021). The effect of intellectual capital on competitive advantage: The mediating role of innovation. *Management Science Letters, 11*(4), 1331–1344.

Oestergaard, J. K. (2020). Airbus and boeing report September 2020 commercial aircraft orders and deliveries. *Defense Security Monitor.*

Pearce, B. (2020). COVID-19 updated impact * assessment of the novel coronavirus. *International Air Transport Association.*

Petchenik, I. (2020). Bumping along: charting October's flight activity | Flightradar24 Blog. *flightradar24.*

Rahman, N. A. A., Rahim, S. A., Ahmad, M. F., & Hafizuddin-Syah, B. A. M. (2020). Exploring COVID-19 pandemic: Its impact to global aviation industry and the key strategy. *International Journal of Advanced Science and Technology, 29*(6), 1829–1836.

Airline Ratings. (2020). Boeing cuts forecast aircraft demand by 11% over the next decade. *Airline Ratings.*

Sachitra, V. (2016). Review of competitive advantage measurements: Reference on agribusiness sector. *Journal of Scientific Research and Reports, 12*(6), 1–11.

Shah, S. F., Alshurideh, M. T., Al-Dmour, A., & Al-Dmour, R. (2021). Understanding the influences of cognitive biases on financial decision making during normal and COVID-19 pandemic situation in the United Arab Emirates. *The Effect of Coronavirus Disease (COVID-19) on Business Intelligence, 334*, 274–257.

Shah, S. F., Alshurideh, M., Al Kurdi, B., & Salloum, S. A. (2020). The impact of the behavioral factors on investment decision-making: a systemic review on financial institutions. In *International Conference on Advanced Intelligent Systems and Informatics* (pp. 100–112).

Shah, S. A. (2020). Covid-19 curtails CTRM AC's production. *The Malaysian Reserve*

Shakhour, M. R., Obeidat, N., Jaradat, B., Alshurideh, M. (2021). Agile-minded organizational excellence: empirical investigation. *Academy of Strategic Management Journal, 20*(6), 1–25.

Shamout, M., Elayan, M., Rawashdeh, M., Al Kurdi, A., & Alshurideh, B. (2021). E-HRM practices and sustainable competitive advantage from HR practitioner's perspective: A mediated moderation analysis. *International Journal of Data and Network Science, 4*(4), 1–14, 2021.

Sigalas, C., Pekka Economou, V., & Georgopoulos, N. B. (2013). Developing a measure of competitive advantage. *Journal of Strategy and Management, 6*(4), 320–342.

Singh, B. (2020). Downtown in aerospace industry and managing recovery plan during and post COVID 19. *International Journal of Research and Analytical Reviews, 7*(4), 670–673.

Sołoducho-Pelc, L., & Sulich, A. (2020). Between sustainable and temporary competitive advantages in the unstable business environment. *Sustainability, 12*(21), 1–16.

Suau-sanchez, P., Voltes-dorta, A., & Cugueró-escofet, N. (2020). An early assessment of the impact of COVID-19 on air transport : Just another crisis or the end of aviation as we know it ? *Journal of Transport Geography, 86*(May), 39–43.

Sun, X., Wandelt, S., & Zhang, A. (2020). How did COVID-19 impact air transportation? A first peek through the lens of complex networks. *Journal of Air Transport Management, 89*, 101928.

Taryam, M., Alawadhi, D., & Aburayya, A. (2020). Effectiveness of not quarantining passengers after having a negative COVID-19 PCR test at arrival to dubai airports. *Systematic Reviews in Pharmacy, 11*(11), 1384–1395.

Thorbecke, W. (2020). The impact of the COVID-19 pandemic on the U.S. economy: Evidence from the stock market. *Journal of Risk and Financial Management, 13*(10), 233.

Wiraeus, D., & Creelman, J. (2019). *Agile strategy management in the digital age.* Palgrave Macmillan.

Xu, Z., Elomri, A., Kerbache, L., & El Omri, A. (2020). Impacts of COVID-19 on global supply chains: Facts and perspectives. *IEEE Engineering Management Review, 48*(3), 153–166.

Zu'bi, Z., Al-Lozi, M., Dahiyat, S., Alshurideh, M., & Al Majali, A. (2012). Examining the effects of quality management practices on product variety. *European Journal of Economics, Finance and Administrative Sciences, 51*(1), 123–139.

Barriers to Reverse Logistic on Implementation of Reverse Logistic: A Case of Malaysian Small and Medium Enterprise

Mohamad Arif Izuddin Rahmat, Nurhanan Syafiah Abdul Razak, Suriati Deraman, and Muhammad Alshurideh ⓘ

Abstract Globally, reverse logistics is gaining traction due to increased awareness and as a result of resource depletion and climate deterioration. Firms have obstacles implementing reverse logistics from a variety of stakeholders, both internal and external. This study covers a quantitative research method. The target respondents of this study are small and medium enterprises facing barriers to reverse logistic implementation. In this research, the number of respondents is 113 out of a sample size of 81. The collection method is carried out through an email questionnaire. Considering the nature of the study, SPSS version 23 is used to analyse the quantitative data. As a reminder to researchers, future research initiatives may expand this model to include different types of businesses and industries, as well as a broader demographic, which will have an effect on reverse logistics implementation.

Keywords Reverse logistic · Green supply chain · Barriers · Sustainability · Green operation management · Waste management

M. A. I. Rahmat · N. S. A. Razak · S. Deraman
Faculty of Business and Management, DRB-HICOM University of Automotive Malaysia, 26607 Pahang, Malaysia
e-mail: arif_izzuddin94@yahoo.com

N. S. A. Razak
e-mail: hana.syafiah@gmail.com; nurhanan@meritus.edu.my

S. Deraman
e-mail: suriati.deraman@dhu.edu.my

N. S. A. Razak
Azman Hashim International Business School, MERITUS University, Kuala Lumpur, Malaysia

M. Alshurideh (✉)
Department of Management, College of Business Administration, University of Sharjah, Sharjah, United Arab Emirates
e-mail: m.alshurideh@ju.edu.jo; malshurideh@sharjah.ac.ae

Marketing Department, School of Business, The University of Jordan, Amman, Jordan

© The Author(s), under exclusive license to Springer Nature Switzerland AG 2023 2307
M. Alshurideh et al. (eds.), *The Effect of Information Technology on Business and Marketing Intelligence Systems*, Studies in Computational Intelligence 1056,
https://doi.org/10.1007/978-3-031-12382-5_126

1 Introduction

The rapid population growth worldwide and the booming technological progress have increased the manufacture and consumption of short-lived goods. As a result, more raw materials have been consumed by rapid production, thus contributing to the growth and sustainment of sites. Therefore, companies in their supply chain management need to integrate strategies that deal effectively and efficiently with sustainability issues.

A lack of top-level management interest, insufficient time commitment, shifting priorities within and across companies, and an inability to establish a corporate supply chain planning target for reverse logistics are among the reasons firms believe reverse logistics is undervalued in supply chain management general (AlShurideh et al., 2019; Hamadneh et al., 2021; Joghee et al., 2021; Vieira et al., 2020).

In practice, it has been demonstrated that the most effective reverse logistics operations result in increased sales income and decreased operational costs (Pacheco et al., 2018). Additionally, researchers have identified several potential benefits of reverse logistics, including improved resource utilisation and environmental preservation (Abbas, 2018). That is why this paper aims to investigate the most dominant factor of barriers in the implementation of reverse logistics within the small and medium enterprises in Pahang, Malaysia.

1.1 Background of the Study

This study presents an empirical analysis of the obstacles to reverse logistics by small and medium enterprises. These barriers are the main obstacle in the way the barriers function. The Reverse Logistics implementation has various obstacles. The Four barriers and twenty-nine sub-barriers are essential for implementing Reverse Logistics by small and medium enterprises.

According to Waqas et al. (2018), Kaviani et al., (2020a), the Four general categories of main barriers that block businesses from following reverse logistical practice are financial, knowledge, law and regulation, and management. Some of the barriers used by companies to enforce reverse logistics have been clarified. According to these companies, reverse logistics activities were implemented either to gain the economic advantages included in the competitive advantage or compelled by government law and public environmental knowledge. These four variables, 'traditional economic theory,' are necessary to understand, and their sole goal is to produce as much income as possible so that their shareholders are pleased (Rubio et al., 2019).

1.2   Research Discussion and Research Gap

This study presents an empirical analysis of the obstacles to reverse logistics by small and medium enterprises. These barriers are the main obstacle in the way the barriers function. The Reverse Logistics implementation has various obstacles. There are four barriers and twenty-nine sub-barriers necessary for implementing the Reverse Logistics by small and medium enterprise that has been compiled from multiple studies in the past.

The study outlines barriers to reverse logistics adoption in the world's leading sector. The purpose of this study is to look at the variables that prohibit businesses from adopting reverse logistics and to identify the primary hurdles that affect these businesses. According to Ritzén and Sandström (2017), the three general categories of main barriers that block businesses to follow reverse logistical practice are economic, legal and corporate citizenship. Some of the barriers used by companies to enforce reverse logistics have been clarified. According to these companies, reverse logistics activities were implemented either to gain the economic advantages included in the competitive advantage or compelled by government law and public environmental knowledge. These three variables, 'traditional economic theory,' are necessary to understand, and their sole goal is to produce as much income as possible so that their shareholders are pleased (Frei et al., 2016; Ritzén & Sandström, 2017; Yin, 2017). Even the company's role in the management of social responsibility diverges from its main aim to make money.

Prior research has examined the constraints to reverse logistics solely to ascertain its sustainability (Vieira et al., 2020). However, as environmental awareness rises, corporations face significant pressure from stakeholders, including government and customers, to mitigate their negative ecological consequences (Yin, 2017). Indeed, the private sector must consider integrating market practices with sustainability and manufacturing to maintain competitiveness (Frei et al., 2016; Ritzén & Sandström, 2017). In addition, global warming, climate change, trash, and air pollution have increased in recent decades, prompting specialists worldwide to be more ecologically friendly and seek the most acceptable feasible solution (Škapa, 2011).

In addition, several studies (Badenhorst, 2016; Kaviani et al., 2020b; Meyer et al., 2017; Sari et al., 2018; Škapa, 2011) have demonstrated the importance of reverse procedures in improving the performance of a firm's supply chain. Small and medium enterprises in Pahang, Malaysia, focused on this study paper's analysis of reverse logistics practices. A small and medium enterprise is a very unexplored and complex topic because of the requirements enterprises must meet. Using proven quantitative methods, the study has attempted to analyse reverse logistics activities in small and medium-sized enterprises towards the implementation.

2 Literature Review

2.1 Reverse Logistics

Reverse logistics consists of practices including material recycling, reuse, reduction, and replacement (Badenhorst, 2016). Furthermore, according to Sari et al. (2018), In light of all the new laws and regulations that have emerged in recent years, a method called reverse logistics has been making its mark among numerous industries. This approach helps these organisations achieve their profitability and growth share strategic goals by reducing their potential impacts and increasing their environmental sustainability. Therefore, knowledge and learning management practices are vital to the concept of informational boundaries for an organisation (Yin, 2017).

Besides that, most top-performing businesses understand the competitive advantage comes from effective distribution and supply chain operations. According to Meyer et al. (2017), the display of an efficient supply chain comes from an information system that predetermines the customer's satisfaction without obstacles during the initial process. According to Waqas et al. (2018), During transport, reusable containers can minimise waste products and product damage, helping to reduce ergonomic and safety issues further. The product development process will aid the 3R Recycling, Reuse, and Restoration principle. To minimise unnecessary packaging, 3R should be considered in product design.

2.2 Reverse Logistics Barriers

Over the past decade, environmental problems have increased, to ecological issues have been noted. They also thought that reverse logistics could contribute to industrial growth in conjunction with supply chain management (Chileshe et al., 2015). Therefore, green supply chains need to sustain global market competitiveness (Meyer et al., 2017).

Greening the supply chain can be inhibited by numerous police and government sectors around the world. Several environmental problems in India are driven by the rapidly increasing population and economic growth because of the growing and massive intensification of farming, the unregulated development of urbanisation, industrialisation, and forest destruction (Ravi & Shankar, 2015). (Kirchherr, et al., 2017) Have identified vital barriers encountered in reverse logistics: higher costs, poorly trusted policies, lack of knowledge, resources and supplier expertise, and lack of policy by the government.

According to Aryal (2020), literature reviews show that many researchers found barriers to effective implementation of reverse logistics systems that need to be established and explored. A comprehensive analysis has been done to identify the barriers that are currently impeding a total return to logistics system implementation and determine the fundamental barriers that lack management understanding and support. They underlined the fiscal, technical, and management systems constraints that prevent the effective implementation of sustainable processes.

2.3 Relationship Between Financial Related Barriers and Reverse Logistics Implementation

First of all, (Younas & Bajwa, 2019) analysed the 26 obstacles to reverse logistics theories implementation. Barriers are described as one of the most impacting walls, namely high investment and lower return on investment.

Direct and transaction costs include environmental management costs (Sari et al., 2018). These two prices serve as an essential barrier to sustainable practice implementation. Adoption of state-of-the-art technology, construction of IT infrastructure, and recruitment of highly skilled workers all go toward providing knowledge and training to sustainability workers. The key barriers are high investment start-ups, slow return on investment, and low-profit margins (Shao et al., 2019).

There are no consultants, green architects, green developers, contractors, and specialists among the corporate departments. Thus, this is the "lack of consultants, green architects, green developers, contractors, and specialists". A high-quality implementation offers new concepts for the organisation's overall growth, learns innovative technologies, and efficiently communicates knowledge, making it easier to incorporate sustainable practices. However, attracting high-quality staff is a barrier to financial restrictions. It also lacks IT infrastructure networks, a lack of waste management technology, reuse, and recycling. On the other hand, the utilisation of IT infrastructure decreases the substantial use of paper and time consumption (Shohan et al., 2020).

If the implementation of sustainable practices fails, the organisation will be deemed worthless. In addition, the brand value and competitiveness of the business may be affected (Singh et al., 2016). The goal of reverse logistics practices is to strengthen their environmental and financial efficiency, IR, eco-design, or ecological practices design. The emphasis on business organisation's sustainability is that today the company's environment is being transformed (Sivakumar et al., 2018).

Heavy investment is needed in reverse logistics programs concerning infrastructure, technology, and necessary software that is a significant impediment in small and medium enterprises to reverse logistics programs (Škapa, 2011). For example, it is estimated that relocating a product from consumer to the manufacturer was nine times the cost of relocating a product from supplier to customer. Financial constraints thus serve as a significant obstacle to reverse logistics.

According to Starostka-Patyk et al. (2014), several technological changes that demand high investment may be needed to introduce the reverse logistics industry successfully. However, investors should note the returns and the recovery of the investment. It has been a considerable challenge for technology to introduce new technologies and save investment costs professionally. Often because of economic constraints, the implementation of reverse logistics was met with obstacles. For example, one of the biggest problems in the chemical industry is toxic goods. Therefore, the introduction of reverse logistics becomes troublesome because the cost increases and the reverse knowledge is missing.

Successful reverse logistics requires staff training (Tesfaye & Kitaw, 2020). However, this training calls for financial resources, which can or may not be available for the organisation. This makes it very difficult for the company to handle reverse logistical activities. Therefore, training and education are the key demands for an organisation's success; the more significant the reverse logistics operation's productivity, the more skilled personnel available.

H1: There is a significant relationship between financial related barriers towards the implementation of reverse logistics.

2.4 Relationship Between Knowledge Related Barriers and Reverse Logistics Implementation

Apart from the financial factors, personnel and instruments are critical for executing related reverse logistics activities. Often essential are knowledge-based tools such as expertise and related knowledge relating to management capabilities and technology (Sivakumar et al., 2018). Therefore, the appropriate reverse logistics capital commitment will improve the logistics competence of an organisation.

Not only that, according to Bianchini et al. (2019), an organisation must need to have Data, method, commodity, or organisational support technologies deemed essential to the management of a green supply chain. Proper preparation and efficient human capital development are critical parts of the green supply chain. Small and medium enterprises encountered difficulties achieving adequate efficiency in their initiatives because of a shortage of human capital and internal innovation (Ormazabal et al., 2018). They also noted that many small and medium enterprises face new criteria, mainly because they lack information, skills, and financial and human capital.

Next, (Reddy & Audo, 2016) The absence of expertise in reverse logistics is a hindrance to proper implementation. In addition, the results demonstrate that reverse logistics is connected to knowledge and experience. In addition, it finds that reverse logistics has an enormous impact on a firm. Thus companies should also have a more robust understanding of reverse logistics' value.

(Moktadir et al., 2020) also suggest that reverse logistics awareness can carry economic advantages by taking back the commodity returned to use. Many of those who know the benefits of reverse logistics and what potentials or goods can be used in their companies are therefore essential to analyse the degree of awareness and perception. Besides that, according to Rubio et al. (2019), in the execution of related reverse logistics activities, workforce and capital are essential in addition to financial considerations. As a result, the hiring process must include knowledge-based instruments, such as talents and critical skills that are applicable to operating resources and technology. Improving a company's logistics expertise will necessitate an appropriate role for staff involved in reverse logistics.

Also, (Sharma et al., 2011) suggest that consumers are unaware of eco-friendly goods and their advantages. The most basic form of external pressure for any company is customer requirements. When a customer requests an eco-friendly product, the company must modify or update the current infrastructure to meet the same criteria. Sustainable practices are tough to follow for companies because of the increasing operational costs. According to Nikmehr et al. (2017), the lack of client awareness contributes to the low acceptance of recycled items. As a consequence, they get lower returns for further investment.

Lastly, according to Dieckmann et al. (2020), most logistical information systems do not support the reverse flow of materials. The lack of sound information systems is one of the most critical issues facing organisations when adopting the reverse logistics. Good support for knowledge is a prerequisite in reverse logistics for successful decision-making. Many companies have sophisticated information and technology systems for forwarding logistics, but additional work is needed to support reverse logistics. Manual data recording is also required because errors can be introduced because of duplication of operations, equipment absence, and inadequate IT tools (Thiyagarajan & Ali, 2016).

H2: There is a significant relationship between knowledge related barriers towards the implementation of reverse logistics.

2.5 Relationship Between Law and Regulation Related Barriers and Reverse Logistics Implementation

First of all, according to Yin (2017), e-waste is illegal in developed countries because of the absence of appropriate legal measures. Therefore, all reverse logistics stakeholders are urgently required to establish an efficient long-term strategy for e-waste disposal and manage informal recycling. Regarded the lack of state assistance as the key obstacle to reverse logistics implementation.

To implement reverse logistics, governments must provide financial support and implement tax regulations. Governments will diminish an organisation's motivation to pursue reverse logistics if they fail to adopt environmental rules that can be implemented (Pacheco et al., 2018). In addition, fiscal policies and government financial assistance are also important. Propose that the government can implement reverse logistics strategies or incentives (Abbas & Farooquie, 2018). We must expect that subsidies from the government will encourage the implementation of reverse logistics. Legislation is a logistic element that goes in the other direction.

The method of analytical hierarchy is used to determine the barriers to implementing sustainable plastic purchasing and supply initiatives in India (Abbas, 2013). The most important of the barriers was considered a barrier, that is to say, government funding and policies. Using the ISM methodology, the main obstacle factors of sustainability in Indian petroleum and gas industries have been identified and modelled (Raut et al., 2018). The findings showed that the lack of global warming pressure and ecological scarcity is the most critical barriers to other barriers.

According to Lototsky et al. (2019), the ten barriers to reverse logistics in the foundry industry have already been discussed. The most critical factors were present: barriers to progress and advancement were a lack of government control and legislation. There are established impediments and roadblocks for using reverse logistics strategies in Nigerian building enterprises (Dieckmann et al., 2020). The study showed that the lack of legal compliance and government constituted the primary barriers to reverse logistics practices.

It means that either the government does not follow industrial-friendly sustainability policies or offer organisation's unique advantages and incentives for applying them. The laws and regulations of the government are the required instruments for proper corporate governance (Shohan et al., 2020). The industry will inspire top manager companies to introduce reverse logistics in every industry sector by giving strict guidelines on reversing logistics practices and awarding the best reverse logistics guidelines (Kaviani et al., 2020a). The lack of supportive regulations and standards has also been appraised as a critical hurdle to reverse logistics in developing countries. When governments change, so do the guidelines, as witnessed in India. Specific norms and regulations are required to collaborate domestically and create a basic safety standard in the business.

Government laws and financial support can play an important role in reverse logistics practices within an organisation. For example, a lack of enforceable government environmental policy decreases motivation and encourages organisation's to reverse logistics procedures (Badenhorst, 2016). Furthermore, the lack of government fiscal policies and financial support to reverse logistics has disincentivised organisations. The goods returned to the reverse chain contribute to a high degree of tax complexity and exposure. Consequently, the most external effect on reverse logistics activities of a company is government policy.

H3: There is a significant relationship between law and regulation related barriers towards the implementation of reverse logistics.

2.6 Relationship Between Management Related Barriers and Reverse Logistics Implementation

Firstly, according to Starostka-Patyk et al. (2014), in management-related barriers, less cooperative partners within the supply chain are significant challenges in reverse logistics. Therefore, the supply chain could be revamped into the closed or a separate reverse supply chain. It indicated that enormous resources and finance require reverse logistical practices to be introduced.

Next, according to Reddy and Audo (2016), the analysed developing countries could not attain a reasonable recovery rate due to the reduced return volume. Noted that there are barriers to the technology of reverse logistics practices in return tracking, collaborative with 3PL providers. In knowledge exchange activities such as seminars and conferences, collaborative supply chains are implemented to support practitioner training and growth efforts in environmental matters (Pacheco et al., 2018). The use can treat the waste of different management, such as landfill or waste incineration, and return valuable materials to the supply chain.

By partnering with manufacturers, shippers, retailers and consumers to coordinate green supply chain operations, companies may make significant changes in business efficiency (Ormazabal et al., 2018). They find that environmentally friendly logistics can be used to minimise costs and improve overall manufacturing. Effective reverse logistics networks have shown that good economic benefits can be achieved and corporate competitiveness increased.

As a result, the management's awareness as a strategic decision in developing countries is vital in promoting reverse logistics implementation (Waqas et al., 2018). Although policymakers are crucial for implementing the reverse logistics process, they are concerned about how an inverse logistics system can be initiated and enforced.

Above all, there are numerous impediments to implementing reverse logistics systems, classified as management difficulties in various studies (Aryal, 2020). He identified two significant limitations and drivers of reverse logistics in his research: a lack of understanding of reverse logistics activities and a lack of senior management engagement. Thus, critical impediments to reverse logistics have been discovered. The reasons for these difficulties include a lack of senior management involvement and a lack of concern for green supply chain management in the Indian manufacturing industry (Frei et al., 2016).

Furthermore (Sari et al., 2018), short-term and long-term strategic priorities and high waste disposal costs have also been misaligned because of the lack of a high-level management commitment to logistical sustainability and a top barrier to implementation. Depending on whether these barriers can be eliminated, some businesses can introduce reverse logistics using their self-support scheme, while some organisation's externalise their clients and recycle services to third-party logistics suppliers (Chileshe et al., 2015). A self-support system is suitable for manufacturing high-end electronic goods since it allows a business to gather practical product knowledge for continuous improvement (Ormazabal et al., 2018). It requires, however, considerable expenditure in money.

Top management support and dedication are essential to embrace, incorporate and excel in sustainable practices in every enterprise. Therefore, top management support is vital. In addition, senior management will support a new integrated sustainable structure through employee empowerment, benefits and promotions, training and communication networks through different departments and promoting teamwork (Sivakumar et al., 2018).

H4: There is a significant relationship between management related barriers towards the implementation of reverse logistics.

2.7 A Theoretical Model for Reverse Logistics Implementation

Over several types of research, numerous obstacles to the implementation of reverse logistics have been identified. Even though global awareness, policies, resources, and government backing for reverse logistics is there, most companies saw little to no advantage in applying reverse logistics strategies. While it is generally recognised that reverse logistics is not a necessary process in the majority of the globe, we identify many significant barriers to implementation in the European setting, including a lack of management attention and a lack of acknowledgement of reverse logistics as a competitive element. (Abbas & Farooquie, 2007; Vieira et al., 2020; Campos et al., 2020; Younas & Bajwa, 2019; Kumar, 2017; Lakshmi, 2017; Sivakumar et al., 2018). The above issues can be grouped as financial, knowledge, law and regulation, and finally management.

In Malaysia, the situation is much different. Customers place a higher priority on garbage disposal by contractors, with approximately 80 percent of rubbish generated being sold to individual private collectors and recycled informally (Nik Abdullah et al., 2011). Over 90% of Malaysians are reluctant to pay for waste recycling.

The two main waste collection routes include formal methods utilised by public buildings and more informal options, which involve individual components and raw material sellers. Since the start of informal recycling in numerous locations throughout the world, studies have shown environmental and human health benefits (Abbas, 2013; Vieira et al., 2020; Badenhorst, 2016; Younas & Bajwa, 2019; Raut et al., 2018). However, a high percentage of material in the informal sector never makes it official or formal recycling and disposal sites, limiting the economies of scale in manufacturers' use of reverse logistics procedures (Meyer et al., 2017).

We classify the barriers to reverse logistics implementation into four categories: financial, knowledge, legal and regulatory, and finally, management-related barriers. To investigate these concerns, we found many barriers to the proposed classification that have been described in the literature, which are mentioned below.

Firstly, financial barriers are primarily focused on reverse logistics activities, including a lack of initial capital, an inability to obtain bank loans, increased costs associated with adopting reverse logistics, a lack of funds for product return monitoring systems, a high initial investment with a low return on investment, and finally, expenditure on collecting used products. This is a crucial obstacle for firms seeking rapid advantages from reverse logistic implementation (Waqas et al., 2018).

Secondly, knowledge barriers will include a shortage of reverse logistics professionals, a lack of expertise, training, and experience, and a lack of awareness of reverse logistics practices. Minimal investment in knowledge management and data systems, inaccurate forecasts, a lack of reactivity to reverse logistics, and, finally, an excessively complex implementation. The researcher examines and discusses these barriers to determine the advantage (Tesfaye & Kitaw, 2020).

Thirdly, legal and regulatory barriers include a lack of supportive government policies for reverse logistics, changing regulations as a result of changing political climates, a lack of restrictions imposed, a lack of enforceable laws requiring products to be returned at the end of their useful lives, customers not being informed about returned products, and finally, a lack of political commitment. These impediments are the subject of research by Chileshe et al. (2015).

Finally, management barriers also include a lack of management support, a poor corporate culture, coordination with third-party logistics providers, waste management issues, and management inefficiencies.

2.8 Theoretical Framework

Figure 1 above shows the relationship between barriers to reverse logistics and the implementation of reverse logistics itself. The independent variable of this study consists of Four (4) factors: financial, Knowledge, Law and regulation, and management-related barriers. Therefore, the dependent variable of this study is Reverse Logistics Implementation within the small and medium enterprises of Pahang. This is in line with the statement made by Waqas et al. (2018); Nik Abdullah et al., 2011; Abdulrahman et al., 2014).

3 Research Methodology

The research design looks at the overall strategy to include components of the analysis coherently and logically (Garbarino & Holland, 2009). For this study, the research design is quantitative. As defined by Sundram et al. (2018), this research is focused on a relationship between two variables in which to learn which variables are related to one another.

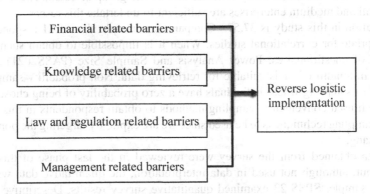

Fig. 1 Conceptual Framework

Descriptive analysis is used to define the characteristics of the population or phenomenon that decides the response to questions about who, what, where, and how (Nik Abdullah et al., 2011; Sundram et al., 2018; Waqas et al., 2018). The present study's objective is to identify barriers to the implementation of reverse logistics in the small and medium enterprise industry and identify types of barriers that primarily affect the industry within Pahang, Malaysia. A survey was used to understand the planned research aims. Questionnaires help researchers discover information from people in a wide variety of fields. (Lau & Wang, 2009).

3.1   Scale and Measures of Research

The first section of the survey assessed the respondent's information within the organisation of the study. The survey question rating scales were based on the Nominal scale, where the respondents will choose what information fits their positions.

The second section of the survey assessed the respondent's attitude about company readiness to reverse logistics implementation. The survey question was, "To what extent do you agree with the following statements for each type of reverse logistic related barriers?" Rating scales were based on a 5-point Likert scale, which one characterised Strongly Disagree, two represented Disagree, three marked either Agree or Disagree, four indicated Agree, and five characterised Strongly Agree. In completing this questionnaire, the respondents were required to think of a situation that they had been involved in or experienced closely.

According to Sundram et al. (2018), population refers to a broad community of people or activities for a researcher. This study's population was all 37,573 small and medium enterprises within the state of Pahang, Malaysia. In January 2019, the number of small and medium enterprise companies in Pahang, Malaysia, was 37,537, at 5.1% of the total number of small and medium enterprises within Malaysia. The sample size is the numerically reflected size of the group selected. The businesses of small and medium enterprises are sufficient to undertake this research. Since the population in this study is 37,573 Companies, a sample size of 81 respondents is appropriate for correlational studies. When it is impossible to obtain such a high return of questionnaires, Power Analysis and Sample Size (PASS2021) suggest that a minimum of 81 is suitable for retrieving data. Non-probability sampling is a kind of sampling where individuals have a zero probability of being chosen. This research used convenience sampling methods to obtain respondents in this survey. This sampling technique is to be used since we are explicitly targeting the population of Pahang.

Data obtained from the survey were reviewed in the last phase of this study. As input, although not used in data interpretation, the quantitative data was used as the sample. SPSS 23 examined quantitative survey results. Descriptive figures and research theories were evaluated for performance. This number of small and medium enterprises will be the base number for the population of this research. These companies' contact details were retrieved from the associated websites of the Ministry of entrepreneur development and cooperatives (MEDAC) (http://dun.medac.gov.my/data-umum/).

4 Data Collection Method

The tool used for this research was a survey. A questionnaire is a research tool, according to Sundram et al. (2018), consisting of a list of knowledge collection questions. A significant number of participants can compile information within a limited period, and the questionnaires can typically be easily measured and rapidly collected. Therefore, the primary means of gathering data for the analysis was this method.

Survey methods gathered data. An initial email was sent to companies within the small and medium enterprises. Participants represented different industry roles, such as manufacturing, distribution, third party, and retail. Survey questions identified the barriers to Reverse Logistics implementation. Data were analysed by SPSS using the multivariate analysis method. Other aspects, such as company size and industry, were tested to see the difference in reverse logistics barriers and implementation.

Survey data was gathered using an online survey. For this method, Goggle Forms was used as the online survey tool. The key participants were workers employed at this company, such as supply chain managers, logistics analysts, industrial engineers, and the like, working closely to revert logistics projects and processes to find obstacles to reverse logistics. The research covers manufacturers, retailers, vendors, and businesses with services related to electronic goods.

4.1 Research Findings

This section examines the survey respondents' replies in detail. To understand the respondent's perspectives, an attempt was made to relate the participant replies to the theoretical backdrop. Since this research is based on a survey, the number of responses is vital to the study's outcome. Many companies were approached to get data on specific reverse logistics concepts. To begin, the survey data were analysed using descriptive statistics, alpha coefficients, and item-total mean. Next, the reverse logistics obstacles in the industries analysed were evaluated and ranked using factor analysis (FA). Reliability analysis confirmed the questionnaire's validity, with Cronbach alpha values exceeding 0.70, as recommended by [51]. Due to many barriers in our study, we employed descriptive statistics to discriminate between the most and least influential. As a result, the most significant barrier has the highest mean score, while the least important has the lowest mean score.

Firstly, for financial related barriers, the lowest mean was the statement "Higher costs of adopting reverse logistic." The highest mean was the statement, "Non-availability of bank loans to encourage green products processes.". Based on uncertain profit forecasts, the respondents were unable to obtain a bank loan. Reverse logistics implementation in Malaysia is unusual due to small and medium companies being cost-conscious to remain competitive locally and prioritising short-term advantages and profits above long-term benefits. Secondly, for knowledge-related barriers, the lowest mean was the statement "Lack of skilled professionals in reverse logistic." The highest mean was the statement, "Wrong forecasting.". Nevertheless, the largest group of respondents felt that wrong forecasting is within the barriers to implementing reverse logistics.

Thirdly, the lowest mean for law and regulation-related barriers was the statement "Lack of political commitment", and the highest mean, "Lack of enforceable laws on products' return of end-of-life.". However, the largest group of respondents was felt that the lack of enforceable laws and on products return of end-of-life effect the most since it will impact the whole industry. Lastly, the lowest mean for management-related barriers was the statement "Lack of understanding by the top management." The highest mean was the statement, "Lack of waste management practices.". Nonetheless, the largest group of respondents felt a lack of waste management practices that can boost the number of recycled materials to be processed.

5 Study Conclusion

Climate change impacts are becoming increasingly apparent as material costs rise, resources become scarce, and global awareness grows. Achieving a competitive edge while meeting evolving local and international environmental regulations is a common goal of proactive firms using reverse logistics strategies. Malaysian firms must participate in global manufacturing sustainability efforts as resources become limited. Malaysian small and medium enterprises do not sufficiently understand reverse logistics. This study looked at 113 Malaysian small and medium enterprises from various demographics. According to the research's data analysis, most firms included in this study participate in reverse logistics activities and processes ranging from product return collection through refurbishing or recycling. But reverse logistics is relatively new in Malaysia. A comprehensive item-by-item examination of reverse logistics finds investments, resources, and overall commitment to being average. The financial hurdles revealed indicate a lack of company commitment to reverse logistics—even management commitment scores below average. The most significant barrier to adoption is a lack of knowledge of reverse logistics.

According to this study, Malaysian firms, particularly Pahang, are still wary about using reverse logistics. This is due to a lack of awareness about reverse logistics. Despite management's recognition of the value of reverse logistics, the company's expenditures, resource allocation, and dedication remain below average. The most common reverse logistics activity is collecting customer returns, whereas refurbishing, repackaging, and even recycling are rare operations in the sector. This study has helped firms in Pahang, Malaysia, understand the current level of reverse logistics usage. It also informs businesses, government agencies, lawmakers, managers, and researchers on essential issues that impede companies' use of reverse logistics. This study is also crucial since it tries to determine which reverse logistics operations this firm has implemented. It also shows the significant benefits gained by companies using reverse logistics. This should encourage companies who haven't embraced reverse logistics to do so since it may help them improve their competitiveness.

Reverse logistics implementation provides a wide range of research options. A future study may examine the state of performance in various companies. Although the breadth of reverse logistics deployment varies with every industry, whether independent or moderating and mediating variables, other variables that may influence a company's decision to participate in reverse logistics should be examined.

Notably, the data and results are from a survey exclusively performed in Pahang. Thus, the results do not reflect all Malaysian small and medium enterprises. The research included a limited time for collecting surveys. The questionnaire is distributed as an electronic survey (e-survey) through Google Forms. The inability to ask for a response on the spot or the same day had become a limitation. The data collection is dependent on the sample having time to answer the survey without any option to receive a quick response. A lack of past research studies on the subject revealed a lack of literature for a strong thesis since the number of studies that give the same independent and dependent variables is limited. Even while most research on reverse logistics implementation recommends additional variables be considered, the results of this study utilising reverse logistic barriers as a significant variable have proved to be highly beneficial.

References

Abbas, H. (2006). Reverse logistics in Indian retail chains.

Abbas, H. (2018). Barriers to reverse logistics practices in pharmaceutical supply chains: An ISM approach. *International Journal of Business Excellence, 16*(1), 47–60.

Abbas, H., & Farooquie, J. A. (2018). Reverse logistics operations in a pharmaceutical retail environment. *International Journal of Logistics Economics and Globalisation, 7*(1), 1.

Abbas, H., & Farooquie, J. A. (2007). Reverse logistics management. 323–339.

Abdulrahman, M. D., Gunasekaran, A., & Subramanian, N. (2014). Critical barriers in implementing reverse logistics in the Chinese manufacturing sectors. *International Journal of Production Economics, 147*(PART B), 460–471.

AlShurideh, M., Alsharari, N. M., & Al Kurdi, B. (2019). Supply chain integration and customer relationship management in the airline logistics. *Theoretical Economics Letters, 9*(2), 392–414.

Aryal, C. (2020). Exploring circularity: A review to assess the opportunities and challenges to close loop in Nepali tourism industry. *J. Tour. Adventure, 3*(1), 142–158.

Badenhorst, A. (2016). Prioritising the implementation of practices to overcome operational barriers in reverse logistics. *Journal of Transport and Supply Chain Management, 10*(1), 1–12.

Bianchini, A., Rossi, J., & Pellegrini, M. (2019). Overcoming the main barriers of circular economy implementation through a new visualization tool for circular business models. *Sustainability, 11*(23).

Chileshe, N., Rameezdeen, R., & Hosseini, M. R. (2015). Barriers to implementing reverse logistics in South Australian construction organisations. *Supply Chain Management, 20*(2), 179–204.

de Campos, E. A. R., Tavana, M., ten Caten, C. S., Bouzon, M., & de Paula, I. C. (2020). *A grey-DEMATEL approach for analyzing factors critical to the implementation of reverse logistics in the pharmaceutical care process.*

Dieckmann, E., Sheldrick, L., Tennant, M., Myers, R., & Cheeseman, C. (2020). Analysis of barriers to transitioning from a linear to a circular economy for end of life materials: A case study for waste feathers. *Sustainability, 12*(5).

Frei, R., Bines, A., Lothian, I., & Jack, L. (2016). Understanding reverse supply chains. *International Journal of Supply Chain and Operations Resilience, 2*(3), 246.

Garbarino, S., & Holland, J. (2009). *Quantitative and Qualitative Methods in Impact Evaluation and Measuring Results Issues Paper.*

Hamadneh, S., Pedersen, O., & Al Kurdi, B. (2021). An investigation of the role of supply chain visibility into the scottish bood supply chain. *Journal of Legal, Ethical and Regulatory Issues, 24*(1), 1–12.

Joghee, S., Alzoubi, H. M., Alshurideh, M., & Al Kurdi, B. (2021). The role of business intelligence systems on green supply chain management: Empirical analysis of FMCG in the UAE. In *The International Conference on Artificial Intelligence and Computer Vision, 2021* (pp. 539–552).

Kaviani, M. A., Tavana, M., Kumar, A., Michnik, J., Niknam, R., & de Campos, E. A. R. (2020b). An integrated framework for evaluating the barriers to successful implementation of reverse logistics in the automotive industry. *Journal of Cleaner Production, 272*, 122714.

Kaviani, M. A., Tavana, M., Kumar, A., Michnik, J., Niknam, R., & de Campos, E. A. R. (2020a). An integrated framework for evaluating the barriers to successful implementation of reverse logistics in the automotive industry. *Journal of Cleaner Production, 272*.

Kirchherr, J., et al. (2018). Barriers to the circular economy: Evidence from the European Union (EU). *Ecological Economics, 150*, 264–272.

Kumar, A. (2017). Extended TPB model to understand consumer 'selling' behaviour: Implications for reverse supply chain design of mobile phones. *Asia Pacific Journal of Marketing and Logistics, 29*(4), 721–742.

Lakshmi, S. (2017). The role of transportation in logistic. *International Journal for Research in Applied Science and Engineering Technology, V*(XI), 1267–1270.

Lau, K. H., & Wang, Y. (2009). Reverse logistics in the electronic industry of China: A case study. *Supply Chain Management, 14*(6), 447–465.

Lototsky, V., Sabitov, R., Smirnova, G., Sirazetdinov, B., Elizarova, N., & Sabitov, S. (2019). Model of the automated warehouse management and forecasting system in the conditions of transition to industry 4.0. *IFAC-PapersOnLine, 52*(13), 78–82.

Meyer, A., Niemann, W., Mackenzie, J., & Lombaard, J. (2017). Drivers and barriers of reverse logistics practices: A study of large grocery retailers in South Africa. *Journal of Transport and Supply Chain Management, 11*, 1–16.

Moktadir, M. A., Rahman, T., Ali, S. M., Nahar, N., & Paul, S. K. (2020). *Examining barriers to reverse logistics practices in the leather footwear industry* (Vol. 293, No. 2). Springer US.

Nik Abdullah, N. A. H., Yaakub, S., & Abdullah, H. H. (2011). Reverse logistics adoption among Malaysian manufacturers. *International Journal of Management, Economics and Social Studies*, 513–517.

Nikmehr, B., Reza Hosseini, M., Rameezdeen, R., Chileshe, N., Ghoddousi, P., & Arashpour, M. (2017). An integrated model for factors affecting construction and demolition waste management in Iran. *Engineering, Construction and Architectural Management, 24*(6), 1246–1268.

Ormazabal, M., Prieto-Sandoval, V., Puga-Leal, R., & Jaca, C. (2018). Circular economy in Spanish SMEs: Challenges and opportunities. *Journal of Cleaner Production, 185*, 157–167.

Pacheco, E. D., Kubota, F. I., Yamakawa, E. K., Paladini, E. P., Campos, L. M. S., & Cauchick-Miguel, P. A. (2018). Reverse logistics: Improvements and benefits when shifting parts exchanging process in a household appliance organization. *Benchmarking, 25*(5), 1447–1460.

Raut, R., Narkhede, B. E., Gardas, B. B., & Luong, H. T. (2018). An ISM approach for the barrier analysis in implementing sustainable practices: The Indian oil and gas sector. *Benchmarking, 25*(4), 1245–1271.

Ravi, V., & Shankar, R. (2015). Survey of reverse logistics practices in manufacturing industries: An Indian context. *Benchmarking, 22*(5), 874–899.

Reddy, D., & Audo, S. (2016). A study on reverse logistics master thesis work (KPP231) master program in product and process development-production &logistics.

Ritzén, S., & Sandström, G. Ö. (2017). Barriers to the circular economy—integration of perspectives and domains. *Procedia CIRP, 64,* 7–12.

Rubio, S., Jiménez-Parra, B., Chamorro-Mera, A., & Miranda, F. J. (2019). Reverse logistics and urban logistics: Making a link. *Sustainability, 11*(20), 1–17.

Sari, D. P., Mujiya Ulkhaq, M., Rinawati, D. I., & Rasyida, D. R. (2018). Barriers of reverse logistics implementation: A case study in a car battery industry in Indonesia. *International Journal of Supply Chain Management, 7*(5), 53–67.

Shao, J., Huang, S., Lemus-Aguilar, I., & Ünal, E. (2019). Circular business models generation for automobile remanufacturing industry in China: Barriers and opportunities. *Journal of Manufacturing Technology Management, 31*(3), 542–571.

Sharma, S. K., Panda, B. N., Mahapatra, S. S., & Sahu, S. (2011). Analysis of barriers for reverse logistics: An Indian perspective. *International Journal of Modeling and Optimization, 1*(2), 101–106.

Shohan, S., Ali, S. M., Kabir, G., Ahmed, S. K. K., Haque, T., & Suhi, S. A. (2020). Building theory of green supply chain management for the chemical industry: An emerging economy context. *Management of Environmental Quality An International Journal, 31*(5), 1285–1308.

Singh, R. K., Rastogi, S., & Aggarwal, M. (2016). Analyzing the factors for implementation of green supply chain management. *Competitiveness Review, 26*(3), 246–264.

Sivakumar, K., Jeyapaul, R., Vimal, K. E. K., & Ravi, P. (2018). A DEMATEL approach for evaluating barriers for sustainable end-of-life practices. *Journal of Manufacturing Technology Management, 29*(6), 1065–1091.

Škapa, R. (2011). Reverse logistics in the Czech Republic: Barriers to development. *Acta Universitatis Agriculturae et Silviculturae Mendelianae Brunensis, 59*(4), 363–370.

Starostka-Patyk, M., Zawada, M., Pabian, A., & Szajt, M. (2014). Reverse logistics barriers in Polish enterprises. *International Journal of Services and Operations Management, 19*(2), 250–264.

Sundram, V. P. K., Bahrin, A. S., Abdul Munir, Z. B., & Zolait, A. H. (2018). The effect of supply chain information management and information system infrastructure: The mediating role of supply chain integration towards manufacturing performance in Malaysia. *Journal of Enterprise Information Management, 31*(5), 751–770.

Tesfaye, W., & Kitaw, D. (2020). Conceptualizing reverse logistics to plastics recycling system. *Social Responsibility Journal.*

Thiyagarajan, G., & Ali, S. (2016). Analysis of reverse logistics implementation barriers in online retail industry. *Indian Journal of Science and Technology, 9*(19).

Vieira, B. O., Guarnieri, P., Nofal, R., & Nofal, B. (2020). Multi-criteria methods applied in the studies of barriers identified in the implementation of reverse logistics of e-waste: A research agenda. *Logistics, 4*(2), 11.

Vieira, B. de. O., Guarnieri, P., e Silva, L. C., & Alfinito, S. (2020). Prioritizing barriers to be solved to the implementation of reverse logistics of e-waste in Brazil under a multicriteria decision aid approach. *Sustain, 12*(10).

Waqas, M., Dong, Q. L., Ahmad, N., Zhu, Y., & Nadeem, M. (2018). Critical barriers to implementation of reverse logistics in the manufacturing industry: A case study of a developing country. *Sustain., 10*(11), 1–25.

Yin, W. (2011). Reverse supply chain management. *Management*.

Younas, G., & Bajwa, F. A. (2019). Barriers and drivers governing implementation of reverse logistics : A case study of Sandvik Coromant.

Does Customer Loyalty Lead to Successful Automotive Industry? a Study of Malaysian Consumer

Mohd Syafiq Ruslan, Nurhanan Syafiah Abdul Razak, Musmuliadi Kamaruding, and Muhammad Alshurideh

Abstract Customer loyalty is the performance index that is being inspired by many companies to ensure their businesses are profitable, sustainable competitive, expandable, and practical in the current era. On a similar note, customer satisfaction plays a big role as indicators to connect the customer to a company in which expressed their experiences and values owning a brand new car from a company. Derived from customer satisfaction leads to customer loyalty which is a customer is likely loyal to a brand and inspired to buy a new car from the same brand. Competitive advantages are the key factors that contribute to customer satisfaction and customer loyalty. One of the challenges is related to market uniqueness in which a company required special strategies to offer a product that suits the demographic of the customer which will have a significant impact on the competitive advantages for a company. The research is to examine the correlation of competitive advantages towards customer loyalty specifically in Malaysian automotive industries. This correlation study will utilize quantitative methodology and the data will be analyzed and interpreted through IBM

M. S. Ruslan · N. S. A. Razak
Faculty of Business and Management, DRB, HICOM University of Automotive Malaysia, 26607 Pahang, Malaysia
e-mail: syafiq.ruslan89@gmail.com

N. S. A. Razak
e-mail: hana.syafiah@gmail.com; nurhanan@meritus.edu.my

N. S. A. Razak
MERITUS University, Kuala Lumpur, Malaysia

M. Kamaruding
School of Civil Engineering, College of Engineering, University of Technology Mara, 40450 Shah Alam, Selangor, Malaysia
e-mail: musmuliadi@uitm.edu.my

M. Alshurideh (✉)
Department of Management, College of Business Administration, University of Sharjah, Sharjah, United Arab Emirates
e-mail: m.alshurideh@ju.edu.jo

Marketing Department, School of Business, The University of Jordan, Amman, Jordan

© The Author(s), under exclusive license to Springer Nature Switzerland AG 2023   2325
M. Alshurideh et al. (eds.), *The Effect of Information Technology on Business and Marketing Intelligence Systems*, Studies in Computational Intelligence 1056,
https://doi.org/10.1007/978-3-031-12382-5_127

SPSS. As a note to research scholars and future research undertaking, this research can be used as guidance to other automotive brands either national or non-national segments.

Keywords Customer loyalty · Competitive advantages · Market uniqueness · Price Tag · Leadership · Product Feature

1 Introduction

In terms of automotive industries, vehicles are essential for everyone nowadays as transportation to commute from one place to another place. Due to the uprising demand for vehicles, automotive industries getting expand globally thus competition between car manufacturers are competitive. The automotive business already passes a century old, "yet it remains one of the most competitive businesses on earth" (MAA, 2020). Edward added that over the past few decades, the competition keeps competitive in the U.S industries where recent market entrants such as car brands from China and Korea making the road into the markets. This situation not only occurred in the U.S, but the whole region already faces the challenge condition which entrance of another foreign brand to the domestic market.

The performance of global automotive industries can be measured based on sales and production of vehicles. Ministry of International Trade and Industry (MITI) summarized and conclude the market situation and economic climate change from 2014 to 2018. Vehicles have been sold in the year 2018 have recorded 7.7% positive growth from 2014 to 2018 which the value from 85.3 million units to 91.9 million units. As for global production statistics, increasing from 87.6 to 95.2 million units from 2014 to 2018. The explanation above support by Fig. 1 shows the global sales and production achievements from the year 2014 to 2018 (MAA, 2020).

2 Background Study

The automotive industries in the ASEAN region growth by 12% for sold units and 10% for production vehicles from 2014 to 2018. This increasing trend is mainly due to ASEAN countries offers investment opportunities for distributors and manufacturers. Japanese-based OEM dominated the market segment in ASEAN. The result can be seen in Fig. 2.

Based on Fig. 3, total sales vehicles production in 2019 is 4,158,983 units in the ASEAN region which Thailand nominates the highest number of production followed by Indonesia and Malaysia.

Indonesia and Thailand are the automotive hub of ASEAN which almost of suppliers localize their parts in both of the countries especially for Japanese car manufacturers such as Toyota, Daihatsu, Honda, Mazda, and others (BOI, 2015).

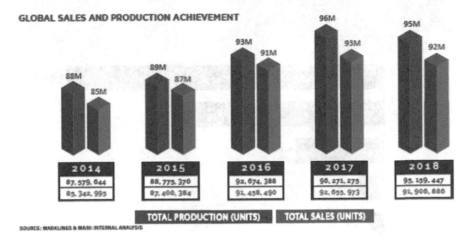

Fig. 1 Global sales and production achievement (MAA, 2020)

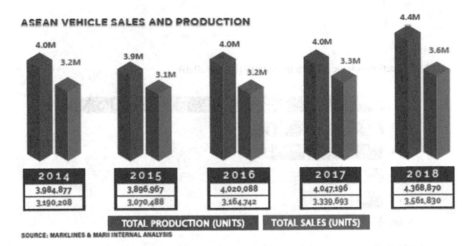

Fig. 2 ASEAN vehicles sales and production (MAA, 2020)

Hence, both of the countries enjoy the benefit of lesser tax imposed, high production volume thus contribute to high revenue. Competition between car manufacturer at Thailand and Indonesia are competitive since both of the countries are having open policies to any brands and don't have the policy to consider the impact towards national car compare to Malaysia (BOI, 2015).

The Malaysian automotive industries are unique compare to major automotive players which are national brands consist of Proton and Perodua. Figure 4 shows that Perodua holds 43.7% while Protons holds 18.2% in the year 2019. Thus, national car nominates more than 50% of the total industry volume (MAA, 2020).

(in 1,000s)

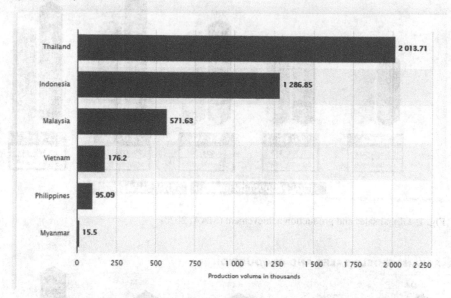

Fig. 3 Production volume in 2019 by country (MAA, 2020)

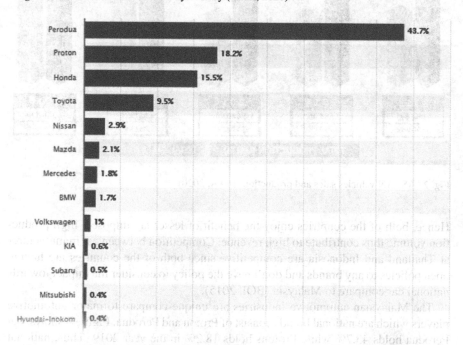

Fig. 4 Market share by brand 2019 (MAA, 2020)

The market share explained above shows that there are huge differences between total sales between national segments which are Proton and Perodua compare to other non-national brands. ABC Company which is a key player in the non-national segment in Malaysian automotive industries by holds a big percentage in market share in total industry volume especially in the non-national brand segment. Several achievements have been recorded for the years of 2020:

(a) ABC Company being nominated as number one for the non-national segment for six consecutive years from 2014 until 2020. (Awarded by MAA)
(b) ABC Company was nominated as first runner-up for Malaysia's 100 Leading Graduates for the automotive segment for the year 2020. (Awarded by GradMalaysia)
(c) ABC Company awarded as the Best Employer Brand Awards for the year 2020 (Awarded by Employer Branding Institute)
(d) ABC Company awarded as Most Attractive Graduate Employer to Work in 2021 (Award by Talentbank).

Despite recognition received by ABC Company, there is a little guarantee on the customer preference, market situation, customer demand, customer perception, and other factors that will determine either ABC Company will keep expanding as the customer loyalty increase or being overtaken by other competitors due to potential of customer brand switching. A previous study conducted by Ariffin and Sahid (2017) on the competitiveness analysis of ASEAN automotive industry, (Wad & Chandran Govindaraju, 2011) on the Malaysian automotive industry development assessment, (Tai & Ku, 2013) on the political-economic analysis of automotive industrial policies in Malaysia and Thailand, and other general study related to competitive advantage and customer loyalty either in Malaysia context or globally. There is a limited study conducted for non-national automotive industries in Malaysia especially in ABC Company on the dimension of competitive advantage that contributes to customer loyalty.

3 Literature Review

3.1 Customer Loyalty

Customer for current era and generation is different compare to previous decades since the customer has sort of information and choices for a product and services which every company tries to penetrate and emerging market over globally. According to Juanamasta et al. (2019), the role of customer service is essential to create customer loyalty by providing quality service to the customer by implementing Customer Relationship Management (CRM). CRM is a helpful tool to have communication with the customer is deemed very effective in delivering services to them. This can be the basis of acquiring new customers, care, maintaining, and developing customers. Many problems faced by a company lead to a bad image and will reduce the level of customer loyalty. Service to the consumers is very important in increasing the satisfaction of them then increase the intention to re-purchase or have deals with the same company as a sign of loyalty (Juanamasta et al., 2019).

Another perception from a profound study conducted by Xhema et al. (2018), on the corporate image, switching cost, and product quality can affect customer loyalty in which companies are highly oriented customers and become a center of attention. Companies invest in the database which required huge amounts of money, so they can focus on customers that bring more profit to the company. Thus, many global companies focus on penetrating the market to gain new customers however new customers cost a lot of time and money due to required information on customer preferences, forecasts, databases, and find solutions. Alternatively, focusing on loyal customers seems to be difficult but can reduce costs and improve profitability. Thus, essential for a company to enhance its competitive advantages so that it can improve on customer loyalty.

In terms of automotive industries in Malaysia, there are relationships between service quality, customer loyalty, and brand image after sales at national automotive service centers in Malaysia (Saidin et al., 2020). According to the study, customer loyalty was affected 87% by brand image and service quality at the service center. Customer satisfaction in terms of service quality will directly correlate with customer loyalty. This will help to deter customers from switching to another brand. There are 27% of consumers reported switched to another brand due to not meet their expectations. This study is conducted on a national carmaker which is Proton. Proton was able to get the highest sales during the early years of introduction among ASEAN countries however, already fell to third place due to stiff competition with other brands.

Various study has been conducted previously prove that the behavior of customers changes drastically. Scherpen et al. explained in the research entitled customer experience management to leverage customer loyalty in the automotive industries, management plays a big role in the current digital age to reach sustainable enterprise and long-term customer loyalty. According to a survey conducted and Germany, France, and the United Kingdom about the relevance of customer experience management, 70% of the sample test believed that this approach was a central success factor for their company strategy in the digital age. Dr. Dieter Zetsche, Head of Mercedes-Benz Car and Chairman of the Board of Management of Daimler AG stated that: "We are transitioning from car manufacturer to networked mobility provider, whereby the focus is always on the individual—as customer and employee" (Daimler, 2016).

Based on the above literature mention that customer loyalty is the major factor that contributes to profitability and enterprise growth. Most of the studies conclude that after-sales quality, customer relations, corporate image, and others contribute to customer loyalty. However, there is a limited study related to customer behavior on purchase intention due to competitive advantages of a company based on product offer to them such as market uniqueness, price tag, leadership, product features, and promotion. Thus, essential to extend the study on this dimension to determine either have a positive correlation with customer loyalty.

3.2 Competitive Advantage

The automotive industry in ASEAN already passed thirty years under different automotive policies, regulations, and approaches leasing to variation of performance among the countries. Ariffin & Sahid, (2017), conducted a study related to comparative analysis of ASEAN automotive industries which compare between Malaysia and Thailand industries. The differences between both Malaysia and Thailand are mainly due to the policies adopted in which Malaysia utilizes independent development while Thailand utilizes dependency development (Wad & Chandran Govindaraju, 2011) as shown in Tables 1 and 2.

The 'Independent development" strategy is applied to establish a national brand while "dependency development" has made its domestic automotive industry as a part of the worldwide supply chain (Tai & Ku, 2013).

In the study utilizing Porter Five Forces analysis (Ariffin & Sahid, 2017), compare: (a) bargaining power of buyer, (b) bargaining power of supplier, (c) threat of substitute product, (d) threat of new entrant, and (d) intensity of rivalry which leads to the conclusion of Malaysian automotive industry is at the turning point which will see a huge potential for foreign investor or MNCs to invest and expand in Malaysia such as the latest condition on 2018 where a strategic partnership occurred between Geely Auto and Proton as China is the largest car manufacturer in Malaysia compare to Thailand which already saturated and dominated by Japanese and American manufacturers. Below Fig. 5 presented the comparison between Thailand and Malaysia.

According to a different researcher, (Wad & Chandran Govindaraju, 2011), argued on the assessment of Malaysian automotive industry development: (a) Why the Malaysian-controlled automotive industries have not become internationally competitive, and (b) whether Malaysian automotive sector can become a regional and global

Table 1 Automotive Policy /Strategy Malaysia vs Thailand (Wad & Chandran Govindaraju, 2011)

Automotive Policy 1 Strategy	Advantages	Disadvantages
Thailand Dependency Development!	• ••Smaller financial burden, on. the government • ••Less Pressure on she opea market	» Hie' automoave industry Ls dominated by foreign capital * More difficult to develop the economic scale of mass production § Cannot bring about the growth of rekted industries
Malay & La Independent Develop mean	• ••Protection brings mass production • ••More likely to support the growth of related industries • ••Can use th.fr home coimtiy's resources	§ Greater financial burden on the government * Over-protection cannot res-pond no pressure from trade liberalization

Table 2 Comparative Automotive Policy of Thailand and Malaysia (Tai & Ku, 2013)

Thailand	Malaysia
• Increasing Productivity by develop Thailand automotive industry to an industry wide lean supply chain and. creating production supply chain network to enable comparison of competitive advantage, thereby mate Thailand automotive industry highly competitive • Expanding domestic and ASEAN markets by developing smaLl passenger car together with maintain Thailand position as the production base for pick-up truck. Also develop infra sanctures to increase efficiency in transportation • Develop design and engineering Technology as a foundation of sustainable and systematic competitioa and value creation, using technology roadmap as an. esseatial tool to enable collaboration, oa research and. development and testing projects • Develop human resource by incus ay-wide development of human resource in management and production. Human, resource development is a key factor in. creating competitive advantage for Thailand automotive industry emphasising on formal education system, training system thai meet the industry demand • Promote domestic and. foreign Investment to promote the industry growth and linking to international Level	**To develop a competitive and capabLe domestic automotive incus ay *To develop Malaysia as the regional automotive hub m Energy Efficient Vehicle (EEV) **To Increase value-added activities in. a sustainable way while continuously developing dom^sac capabilities *To Inc rease exports of vehicle s. auto mo tiv e components, spare partis and related products In the manufacturing and after market sectors *To Increase the participation of competitive Bumiputera companies in the domestic automotive industry, meluding in the after-market sector *To safeguard consumer interests by offering safer and better quality products at c ompe titiv e p tic e *To enhance the ecosystem of the manufacturing and after mark en sectors of the domestic automotive industry

in the export industry? (Wad & Chandran Govindaraju, 2011) criticizes that the automotive industries still lack of product upgrading and technological specifically among parts and component supplier, weak global marketing capabilities and low level of skills among employee. Another criticizes for the stimulus packages offered by the government which targets the national automotive sector only, thus discriminate against the non-national brand and turning them not to invest in Malaysia.

Another perspective study conducted by Mohd Rosli (2012) on the Malaysian small and medium enterprises competitive strategy, there are several findings which are: (a) price setting, marketing, and promotion are the most competitive strategies emphasized by manufacturing SMEs, and (b) Less exposure and adoption of innovation and global orientation in their competitive strategies in which low investment on R&D and least emphasis on the application of internet in business to go global. The study conducted by Mohd Rosli (2012) may reflect on the local automotive supplier in which supplies to the car manufacturer however, the study more focus on food and beverages (F&B) and textile and clothing (T&C) industries.

Fig. 5 The five competitive forces analysis for automotive components and parts industry (Ariffin & Sahid, 2017)

There were limited studies on the factors that contribute to the competitive advantage of automotive industries MNC in Malaysia. One of the researcher's objectives is to add to the existing theories or works of literature and to increase value to the body of knowledge especially in the automotive industry.

Previous studies on competitive advantages, a lot of dimensions have been tested towards customer loyalty in various industries such as policies adopted by the government in Malaysia (Ariffin & Sahid, 2017); 4Ps concept in related to influence customer (Büschken, 2007); market uniqueness and environment in ASEAN (BOI, 2015); inter-relation between price offer by a brand at a store with the perception of brand, perception of store and perception of price that will lead customer to have perceived of quality, thus customer will have perceived on the amount of sacrifice needed to purchase a product (Dodds et al., 1991); strategic alliances through good leadership conceptualized on how the China banking industry to keep competitive (Chia, 2009); assessment of automotive industry development in Malaysia that the combination of high marketing and high technology in the industrial policies will attract and nursing FDI in strategic sectors of economic development (Wad & Chandran Govindaraju, 2011); price setting, marketing and promotion, exposure and adoption of innovation and global orientation lead to low investment on research and development (Tai & Ku, 2013); dependency development or independent development adopted by government either protect national car or open market strategies (Tai & Ku, 2013); utilizing Porter Five Forces analysis to compare each competitors through: (a) bargaining power of buyer, (b) bargaining power of supplier, (c) threat of substitute product, (d) threat of new entrant, and (d) intensity of rivalry which lead to the conclusion of Malaysian (Ariffin & Sahid, 2017).

Based on the literature explained above, the researcher extends this study in dimensions of competitive advantages which relevant to customer loyalty specifically in automotive industries such as: (a) market uniqueness, (b) price tag, (c) leadership, and (d) product features. The literature for the aforementioned dimensions will be elaborated on in the next section.

3.2.1 Market Uniqueness

Over the last three decades, the marketing literature focused on brand studies which are mostly due to brand image. If the company creates strong brand values, the customer will surely seek the value of the brand and reject other alternatives. However, in the current globalization era, consumers have a lot of brand selection with the most identical value, thus consumers have a lot of alternatives. According to Lanza, (2008) , in the automotive industry for the past 30 years in U.S automotive industry, car manufacturers over-manipulated their brands and made the marketplace confusing by offering conflict of brand image, therefore diluting all of the values offer by others and become the most desirable automotive brands.

According to Dodds et al. (1991) on the research related to effects of buyers' product evaluation based on the store information, price, and brand, there are key determinates of brand satisfaction is related to market uniqueness such as: (a) quality, (b) value, (c) equity, (d) brand commitment, (e) interact to discover brand loyalty and (f) repurchase intention. Specifically, (Dodds et al., 1991) point out that superior financial performance, loyalty, and customer retention are linking to brand satisfaction, quality, commitment, and value.

In the 1950s, marketing theories being developed to include the exploitation and manipulation of market demand by winning the battle for customer's choice (Gupta & Wang, 2007). Marketing mix strategies being adopted by MNCs as defined by Büschken (2007) related to influence customer demand. The 4Ps concepts which are: (a) the product or service, (b) the price imposed and the terms associated with the sale, (c) promotional, communication activities, and advertising, and (d) the place where the product was sold including the logistics and distributions process involved.

A profound study conducted by the Thailand Board of Investment, Thailand being nominated as the automotive hub of Asia as become the biggest automotive producer in ASEAN and also being number 12th in the world. Thailand expects to increase its production by 3.5million units in 2020 or definitely increase 80% compared to 2016 (BOI, 2015). Thailand Board of Investment mention the market of Thailand is unique to other countries in ASEAN due to the uniqueness of the market and environment which are:

(a) The geographical advantage which Thailand is located at the center of Indochina Peninsular thus connected with 2 billion people in ASEAN and China by land and sea with world-class infrastructure provided. Furthermore, the third phase extension of Laern Chabang deep-sea port is expected to provide capacity up to 3 million cars being export.

(b) Eastern Economic Corridor (EEC) which Thailand promoted 13,285km^2 of EEC due to the advantage of automotive production is reinforced and developed by supporting facilities and special investment policies. An example of advantage is EEC will provide National Automotive Tire and Testing Centre which is the only testing center in ASEAN and ready for tire and carmakers in 2018.

(c) Abundant skilled labor which Thailand consists of 600,000 workers in automotive industries in 2016 with the advantage of 29 universities and other institutes which provide automotive and mechanical engineering programs that will lead to an adequate supply of graduates and forecasted by 2021, 61% of all workforces in Thailand industry are high-skilled labor.

(d) Research and development (R&D) which most of the car manufacturers in Thailand established their R&D such as: (1) Toyota Motor Asia Pacific Engineering and Manufacturing, (2) Nissan Technical Center South East Asia, (3) Isuzu Technical Center of Asia, (4) Honda R&D Asia Pacific and (5) Mitsubishi Motors Proving Ground.

(e) Investment incentives linked with national development activities which wide range of tax and non-taxable incentives being offered by the Thailand Board of Investment such as corporate income tax exemption for 8 years without a cap for research and development activities.

3.2.2 Price Tag

According to Gupta and Wang (2007), "designing products and services that can be manufactured and delivered at ultra-low prices while yielding satisfactory profit margins". This statement is supported by Knab (2008) on the research study going global by implementing re-invention that requires products and services that can be manufactured and delivered at ultralow prices while the profit margin is at a satisfactory level. Toyota Motor Corp.'s launched efforts to develop and manufacture an ultra-low-cost car for an emerging market such as India, China, and other emerging markets as an approach.

Knab (2008) on the same research found that most of MNCs focus on selling products identical and similar to those in their domestic markets, thus the product offers higher quality and sells at the higher price, therefore, targeted to the wealthy customers because of status associated with owning imports, their designer nature, and inherent quality. The draws by implementing this strategy only allow MNCs to capture a small percentage of the total market while they rather than focus on potential mass-market consumers at the bottom of the economic pyramid.

Based on the profound study by Juanamasta et al. (2019), there are inter-related between the price offered by a brand at a store with the perception of the brand, perception of the store, and perception of the price that will lead the customer to have perceived of quality. Then customers will have perceived the amount of sacrifice needed to purchase a product. Combining both perceived quality and price will

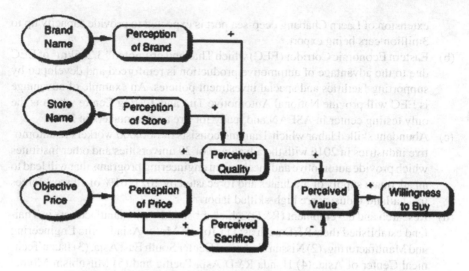

B. Extended Conceptualization to Include Brand Name and Store Name

Fig. 6 Conceptualization of brand name, store name, and objective price (Juanamasta et al., 2019)

determine the perceived value that will turn directly influences willingness to buy. This correlation can be seen in Fig. 6.

According to Büschken (2007), although customers mostly seeking for low-priced, arguably cars do not fall into this category which cars represent almost a major percentage of household's consumption expenses and use for a longer period. In addition, certain aspects of new cars such as durability, maintenance cost, and resale value cannot be determined before purchase, thus can be concluded purchase of a new car is risky. Büschken (2007) conduct research on attributes that lead the customer to have purchase intention which are: (a) reliability, (b) workmanship, (c) price worthiness, and (d) service quality that brand's dealer.

According to Young (2014), there are two views on the price effect towards competitive advantages in study regarding competition between private labels and national brands for fresh milk. As for (a) competitive view, lower-priced for the private label will provide an incentive for a national brand to lower price to be competitive in the market. (b) Market segmentation view which the introduction of private labels leads to higher branded prices. (Young, 2014) explained most retailers have the ability on pricing power, so they have an incentive to raise their brand price and induce customers to switch from the national brand since retailers earn higher margins on their brand.

Ariffin & Sahid (2017) mentioned that interest rates for the car loan and fuel price also contribute to the price offer. Overall interest rate levels remain quite low and also easy to get loan approval to encourage many young graduates to manage to own national cars for the first few years and the experience or other professional workers most likely upgrading from national car to non-national brands.

3.2.3 Leadership

According to Northouse (2012), there are four elements in leadership: (a) leadership is a process that leaders use to affect a group of people, not a trait or a characteristic of leaders nor a one-time event, (b) leadership involves influence to affect followers' behavior and minds (c) leadership occurs in a group situation, not in single person context, and (d) leadership involves the same goal with the team as a common goal.

Based on profound research conduct by Matsui (2013) related to leadership development of western companies in Japan, only 29% of the U.S workforce is properly engaged and almost 71% does not have proper engagement with leaders and management, thus influence organization productivity thus cost incurred huge amount of dollars every year. Besides productivity, employee disengagement leads to increase turnover cost which the cost impacts two times employees' annual salary. Matsui (2013) added that the impact goes worst with the industrial restructuring, series of specific changes in acquisitions and mergers on a global scale resulting in the weakened economic power of corporations. Effective leadership is one of the keys to increasing productivity and employee engagement.

According to Israel (2018), leadership is one of the major factors that contribute to a company which top executives are responsible towards that as well as designing conditions that can guarantee organization survival. Business organizations need leaders who understand how to handle organizational issues and work closely with culturally diverse employees and customers. There is the impact of strategic decisions fail due to mistakes of the top executive which are: (a) their personal interests interfere with their vision for the organization; (b) they are involved in, or accept, unethical behavior; (c) they have disregard for quality, innovation, and productivity, depending too much on intuition to the detriment of logical analysis; or (d) they allocate money and time injudiciously or irresponsibly.

Chia (2009) conceptualizes that in the globalization era and under exaggerate competitive environment, enhancing competitive advantages are the key towards that. Based on the profound research on the China banking industry to keep competitive, three variables contribute: (a) Strategic alliances by linking with a partner to acquire capabilities and resources could enhance competitive advantages, (b) relational asset enhancements by building relational capital with alliances that beneficial for learning, know-how transfer and, (c) linkage for assets exchange between alliances. Chia (2009) added that the foundation of relational capital includes relationship building, respect, friendship, and mutual trust in the alliances. Foreign banks could invest in the domestic bank as competitive advantages in the Chinese market which consists of a unique socio-economic environment and culture.

In terms of Malaysia perspective, (Ariffin & Sahid, 2017) explained in the threats of new entrants for the comparison between Malaysia and Thailand automotive industries, Thailand will continue with the domination of Japanese manufacturers, thus allowing the potential for global Original Equipment Manufacturer (OEM) companies to invest in Malaysia. This correlation is proven by Ariffin and Sahid (2017) on the strategic alliances between Proton and China automotive Geely foresee the new entry of components manufacturer in Malaysia.

Tai & Ku (2013) added that the decision of strategic alliances by leaders will determine the competitive advantages of Malaysian automotive MNCs although the policies regulated is protective for the national brand such as Proton and Perodua. Examples of the strategic alliance such as DRB-Hicom and Honda, UMW and Toyota, Mazda and Bermaz, Naza with Kia, and Inokom with Hyundai. The strategic alliances offer competitive product offer to the local marketplace. The argument can be supported by the decision of manufacturers either to fully import completely build units (CBU) or manufacture in Malaysia through a complete knockdown scheme (CKD) in which the tax and incentive offer has a huge difference. Hence, local assemble CKD will have better advantages (MAA, 2020; MITI, 2014).

3.2.4 Product Feature

Wad & Chandran Govindaraju (2011) explained in the assessment of automotive industry development in Malaysia that the combination of high marketing and technology in the industrial policies will attract and nursing FDI in strategic sectors of economic development. Local manufacturers and firms are unable to excel in both high marketing and technologies except to increase their competitive advantages through import and adopt the technologies from a developed country such as Japan, United Kingdom, Germany, China, and others. Wad & Chandran Govindaraju (2011) added that average around 2% of investments in R&D and technology which are still low in Malaysia durin the period of 2000 to 2005 and spending only 0.14% on other equipment manufacturers.

According to Tai and Ku (2013) in a thesis related to the comparative political-economic analysis of automotive industrial policies between Thailand and Malaysia, criticize the automotive industries in Malaysia is way behind compared to Thailand. From the perspective of the economic point of view, the competitive advantages of Thailand automotive industries lie into following elements: (a) industrial technology, (b) export capabilities, (c) product cost, (d) manufacturing capacity, and other measures of industrial competitiveness. The details in Table 3 show the comparison between Malaysia and Thailand based on research by the Japanese Automotive Manufacturers Association (JAMA).

Tai & Ku (2013) added that the government keeps continuing its protection and intervention although Malaysia set up the industrial scale of automotive industries, Malaysia still not consider as industrial self-reliance. The policy of state protection caused complacency and laziness in operations, and the local management lacks the power to seek for cost optimization, thus lead to low production efficiency. On other hand, Thailand's automotive industries courted MNCs by brought capital and also advanced management and technology methods.

Table 3 JAMA Manufacturing Capacity Ratings for Thailand and Malaysia (Tai & Ku, 2013)

Item	Thailand	Malaysia
Engine	High	Low
Engine parts	Hifih	Low
Electronic systems	Hifth	Standard
Braking system	Standard	High
Interior	High	Low

The current study conducted by MITI (2014) as the review performance of NAP 2014 which is the third version after first introduced in 2006 and second revision in 2009 stated that the key objective of NAP 2014 was to be ASEAN hub for Energy Efficient Vehicles (EEV) through the R&D development, right-hand drive (RHD) vehicles development capabilities and associate technologies such as light material, fuel efficiency, telematics, component design, and tooling. The effort by the government to provide facilities for automotive technology by the establishment of Centers of Excellence (COE) which is the technology centers (MAA, 2020) such as (a) Malaysia Technology Center, (b) National Test Center to implement and educate supply chain and also car maker to utilize Industrial Revolution 4.0, and (c) Automotive design center.

In connection with above mentioned on the key achievement of Ariffin and Sahid (2017) supported that the third version of NAP will see a positive impact for foreign investors to invest in Malaysia after once dominated by the domestic market. This is due to the protective policies have been amended and strategic alliances between national brand, component manufacturers with foreign investor already emphasized towards the fourth revision of NAP which is NAP 2020. The NAP 2020 will continuously cultivate the participation of local car manufacturers and suppliers in both global and domestic supply chains, encourage R&D, and build capacity also capabilities of the local workforce, and enhancing investments, exports, and local production volume.

3.3 Relationship Between Competitive Advantages and Customer Loyalty

The aforementioned literature in the previous section explained the selected dimensions of competitive advantages such as: (a) market uniqueness (BOI, 2015; Gupta & Wang, 2007), (b) price tag (Ariffin & Sahid, 2017; Dodds et al., 1991; Young, 2014) (c) leadership (Israel, 2018; Matsui, 2013; Tai & Ku, 2013), and (d) product features (MAA, 2020; Wad & Chandran Govindaraju, 2011). Four hypotheses have been constructed to verify the correlation between the dimension of competitive advantages and customer loyalty:

H₁ Market uniqueness has a direct positive influence on customer loyalty.

H₂ Price tag has a direct positive influence on customer loyalty.

H₃ Leadership has a direct positive influence on customer loyalty.

H₄ Product features have a direct positive influence on customer loyalty.

4 Theoretical Framework

Customer loyalty is being utilized as an indicator of competitive advantages for a company. There is much literature, modern research on customer behavior in dealing with factors that contribute towards customer loyalty. The theories have been extracted from a theory explained by Shiftan et al. (2015) with research on measuring passenger loyalty to public transport modes. Customer loyalty is derived from three types of product values which are; (a) switching value which reflects the technical effort in switching from one product to another; (b) hedonic value which captures the reflection of emotion experienced by the consumer associated with the products in their mind; and (c) utilitarian value in which represent the functionality of the product to the consumer (Shiftan et al., 2015). The outcome obtained from the model yields the level of satisfaction and repeated choice of products. Satisfaction is the 'consumer fulfillment response … a judgment that a product or service feature, or the product or service itself, provided a pleasurable level of consumption…,' whereas loyalty is a 'deeply held commitment to re-buy or re-patronize a preferred product or service consistently in the future (Oliver, 1996), emphasis added).

Satisfaction is described as a short period judgment of the product, hence loyalty reflects the consumer's commitment and attitude toward the product in the long term. Loyalty creation has been divided into four stages (Oliver, 1996):

(a) Cognitive loyalty is described as knowing—the loyalty created after a short experience with the product, based on the level of satisfaction with the product's physical characteristics.
(b) Affective loyalty is described as an attitude—the creation of an attitude toward the product after a significant period of experience, including a personal commitment toward the product.
(c) Conative loyalty is described as intention—the creation of intention to re-buy the product and an emotional feeling toward the product.
(d) Action loyalty described as re-buy—the highest level of loyalty, involves automatic re-purchasing of the product and blindness to competitors.

Based on literature explained and underpinning theory related to the dimension of competitive advantages in relation to customer loyalty, a research conceptual framework has been constructed to test the direct correlation as below (Fig. 7).

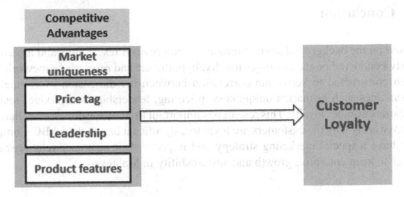

Fig. 7 Research conceptual framework

5 Research Methodology

This research objective is to study dimensions of competitive advantages towards customer loyalty for Malaysian automotive industries in Malaysia. Therefore, a non-experimental design method being implemented which correlation study was conducted to test dimension (independent variables) towards the customer loyalty (dependent variable). Research design is the key structure and strategy to be adopted by the researcher efficiently and effectively throughout the research (Hair et al., 2007).

Based on the previous literature as discussed in the following studies (Mohd Rosli, 2012; Ariffin & Sahid, 2017; Barney, 1995; Knab, 2008; Israel, 2018; Tai & Ku, 2013; Wad & Chandran Govindaraju, 2011), the researchers who had performed similar studies i.e. factors contribute to competitive advantage preferred to adopt the quantitative analysis method. According to Ticehurst (2000), the quantitative analysis method is dependent on numerical results and provides conclusions in systematic ways. Another advantage of utilizing the quantitative analysis technique provides a wide range of research coverage, and it is more efficient, effective, and economical.

All of the data gathered were verified and analyze through Statistical Package for Social Science (SPSS—Ver.22) tools, therefore the outcome such as correlation coefficient, reliability and validity, and multiple-regression are accurate (Ticehurst, 2000).

The methodology of this research is conducted and ends with validity and reliability. The purposive population is consists of 10,000 customers who still choose to maintain and service in ABC Company although their warranty already expired at Melaka region only through convenient sampling method. Thus, recommended sample based on (Krejcie & Morgan, 1970) is 370 respondents to obtain appropriate and accurate data. In the next study, the research design and data analysis method were executed and being performed to interpret the data to collect from the respondents. The data will be translated into figures, tables, and charts, thus easy to analyze and interpret.

6 Conclusion

Based on the background study, literature from previous researchers and a pertinent study conducted on the customer loyalty, hypotheses, and research framework have been constructed to verify the correlation between the dimension of competitive advantages such as market uniqueness, price tag, leadership, and product features towards customer loyalty. This research is important to study the dimensions that lead to answering why the customers are loyal to a significant brand, thus ABC Company can have a special marketing strategy and improve their customer relationship in order to keep enterprise growth and sustainability in Malaysia.

References

(BOI), T. B. of I. (2015). Automotive industry. *Sangyo Igaku.* https://doi.org/10.1539/joh1959. 31.533

Ariffin, A. S., & Sahid, M. L. I. (2017). Competitiveness analysis of ASEAN automotive industry : A comparison between Malaysia and Thailand. *Journal of Science, Technology and Innovation Policy, 3*(2), 23–32.

Barney, J. B. (1995). Looking inside for competitive advantage. *Academy of Management Perspectives, 9*(4), 49–61. https://doi.org/10.5465/ame.1995.9512032192

Büschken, J. (2007). Determinants of brand advertising efficiency evidence from the German car market. *Journal of Advertising, 36*(3). https://doi.org/10.2753/JOA0091-3367360304

Chia, Y. (2009). The relationship between relational assets and competitive advantage of foreign bank in China's banking industry: From the perspective of strategic alliances. *ProQuest Dissertations and Theses, 109.* http://library1.nida.ac.th/termpaper6/sd/2554/19755.pdf

Daimler AG (2016) Daimler AG: Mercedes-Benz presents a new service brand: "Mercedes me"—a new benchmark for service.

Dodds, W. B., Monroe, K. B., & Grewal, D. (1991). Effects of price, brand, and store information on buyers' product evaluations. *Journal of Marketing Research, 28*(3), 307–319. https://doi.org/10.1177/002224379102800305

Gupta, A. K., & Wang, H. (2007). How to get China and India right. *Wall Street Journal, R4.*

Hair, Money, Samuel, & Page, M. (2007). *Research and markets: research methods for business—A skill building approach.* Wiley & Sons, LTD. https://doi.org/10.1108/17506200710779521

Israel, A. (2018). Leadership styles and companies' success in innovation and job satisfaction: A correlational study. *ProQuest LLC, 167.*

Juanamasta, I. G., Wati, N. M. N., Hendrawati, E., Wahyuni, W., Pramudianti, M., Wisnujati, N. S., Setiawati, A. P., Susetyorini, S., Elan, U., Rusdiyanto, R., Muharlisiani, L. T., & Umanailo, M. C. B. (2019). The role of customer service through customer relationship management (CRM) is to increase customer loyalty and good image. *International Journal of Scientific*

Knab, E. F. (2008). Going global: Success factors for penetrating emerging markets. *ProQuest LLC, 9,* 232.

Krejcie, R. V., & Morgan, D. W. (1970). Determining sample size for research activities. *17*(8), 1566–1577. https://doi.org/10.1261/rna.2763111

Lanza, K. M. (2008). The antecedents of automotive brand loyalty and repurchase intentions. Doctoral dissertation, University of Phoenix.

(MAA), M. A. A. (2020). National Automotive Policy (Nap) 2020. In *Ministry of International Trade and Industry.* https://www.miti.gov.my/miti/resources/NAP2020/NAP2020_Booklet.pdf

Matsui, Y. (2013). Key to effective leadership development for western multinational companies in Japan in the eye of the beholder. *ProQuest LLC*.

(MITI), M. of I. T. and I. (2014). National Automotive Policy (NAP). *7*, 219–232.

Mohd Rosli, A. P. (2012). Competitive strategy of Malaysian Small and medium enterprises: An exploratory investigation. *American International Journal of Contemporary Research, 2*(1), 93–105.

Northouse, P. (2012). Leadership: Theory and practice (5th ed.). *Journal of Educational Administration, 50*(3). https://doi.org/10.1108/09578231211232022

Oliver, R. L. (1996). *Satisfaction—A behavioral perspective on the consumer*. Vanderbilt University.

Saidin, Z. H., Abdul Rahman, W. A. W., & Abd Hamid, R. (2020). How the unique industry-specific measures of service quality and brand image can develop customer loyalty in automotive after-sales service. *International Journal of Academic Research in Business and Social Sciences., 10*(11), 847–865.

Shiftan, Y., et al. (2015). Measuring passenger loyalty to public transport modes. *Journal of Public Transportation, 18*(1), 1–16.

Tai, W.-P., & Ku, S. (2013). State and industrial policy: Comparative political economic analysis of automotive industrial policies in Malaysia and Thailand. *JAS (Journal of ASEAN Studies), 1*(1), 52. https://doi.org/10.21512/jas.v1i1.61

Ticehurst, G. W., & Veal, A. J. (2000). Research methods: A managerial approach.

Wad, P., & Chandran Govindaraju, V. G. R. (2011). Automotive industry in Malaysia: An assessment of its development. *International Journal of Automotive Technology and Management, 11*(2), 152–171. https://doi.org/10.1504/IJATM.2011.039542

Xhema, J., Metin, H., Groumpos, P. (2018). Switching-costs, corporate image and product quality effect on customer loyalty: Kosovo retail market. *IFAC-PapersOnLine, 51*(30), 287–292

Young, J. S. (2014). Competition between private labels and national brands: Empirical evidence from homescan data on fluid milk markets. *Purdue University, 9*, 86.

Factors Influencing Car Buyers Purchase Decision During COVID-19 in Malaysia

Mohd Fadzly Suboh, Nurhanan Syafiah Abdul Razak, and Muhammad Alshurideh ⓘ

Abstract Consumer behaviours are largely regulated by multiple factors from need recognition to search for products that would satisfy their requirements. The study in context to car purchase behaviour among the Malaysian consumer highlights increased car buying trends among the Malaysian consumers during the pandemic. Social distancing along with prohibiting the usage of public transport has increased difficulties of consumers thereby resulting in an increased tendency to opt for personal cars. Products features such as fuel consumption, reliability, comfort and performance that provide value and upgraded functionality plays a big role in purchase decision-making. Derived from the the product features leads to brand trust and inspired the decision to buy from the same brand. Affordable and acceptable ownership costs and the consequence purchase power plays important factors in influencing car purchase decision. Higher income people favouring more expensive cars and the subsequent high ownership costs. At the same time, social needs during the pandemic situation has prohibited individuals from close contacts and gatherings that further requires the consumers to purchase cars. COVID-19 pandemic that was known to be airborne and transmittable diseases had truly fundamental impact on purchase decision especially for car buyers. The research examine the correlation of consumer behaviour towards purchase decision specifically in Malaysian context. This research

M. F. Suboh · N. S. A. Razak (✉)
Faculty of Business and Management, DRB-HICOM University of Automotive Malaysia, 26607 Pahang, Malaysia
e-mail: nurhanan@meritus.edu.my; hana.syafiah@gmail.com

M. F. Suboh
e-mail: m_fadzly25@yahoo.com.my

N. S. A. Razak
MERITUS University, Kuala Lumpur, Malaysia

M. Alshurideh
Department of Management, College of Business Administration, University of Sharjah, Sharjah, United Arab Emirates
e-mail: malshurideh@sharjah.ac.ae; m.alshurideh@ju.edu.jo

Marketing Department, School of Business, The University of Jordan, Amman, Jordan

M. Alshurideh et al. (eds.), *The Effect of Information Technology on Business and Marketing Intelligence Systems*, Studies in Computational Intelligence 1056, https://doi.org/10.1007/978-3-031-12382-5_128

2345

can be used as guidance to map the consumer behavior while providing future and practical insights for carmakers towards reviving car industry that being one of the hardest hit by the pandemic.

Keywords Purchase decision · Consumer behaviour · Product features · Ownership costs · Social needs · Purchasing power · Brand trust

1 Introduction

The automobile industry was coping with the downshift in the global demand during COVID-19 pandemic. The lock-down scenarios during this pandemic led to widespread customer confidence loss which had a significant impact on the profitability and revenue of the automakers. The automobile companies in Malaysia also were forced to stop their investment in their continued operations. Funding in the R&D process was also ceased for which technological advancement had not taken place in this industry during this period. The global car sales declined by 20% in 2020 which is depicted in the below figure. In 2018, the automotive industry grew by 3% which was down to 1.9% in 2019 (Deshpande, 2019) (Fig. 1).

Different liquidity issues were also faced by the suppliers and therefore the condition of the market deteriorated. In 2018, around 25 M commercial cars and 70 M cars were made. The wellbeing of the citizens of the world, as well as the economy of the country, is dependent partially on these vehicles. The enhanced rate of car manufacturers increases the rate of employment of a country (Kim, 2020). Due to health concerns, the mobility behaviour changed as most of the customers worked from home while others stayed away from public transportation.

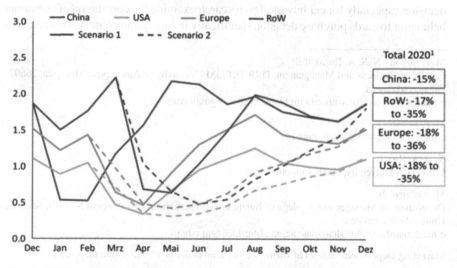

Fig. 1 COVID 19 impact on a global car sale, 2020. *Source* Deshpande (2019)

The car purchasing decisions of the customers are dependent on different factors such as product features, social needs, ownership costs, and brand trust. However, there was a severe and swift impact of COVID 19 on the global automobile industry in 2020. Hence, the large-scale manufacturing process was interrupted all across the world during the outbreak of this pandemic. Consumer behaviour during the pandemic includes lowering the perceived infection threat (Shakibaei et al., 2021). There were also some external factors such as unexpected health emergencies and internal factors such as personal life and attitude that changed the purchasing behaviour of consumers during the pandemic.

2 Background Study

The annual sales of the global automobile industry are predicted to reach 110 M units by 2022. In order to meet the short-term threats, the manufacturer may have to dedicate CAPEX (capital expenditure). It has been seen that the car purchasing rate is higher among the high-income households in Malaysia. As per the view of previous study on current situation, most of the customers are interested in contactless service (Tardivo et al., 2020) (Fig. 2).

The sales of the automobile industry fall to 70 M units in 2021 and it is expected that this industry will grow by 9 Trillion dollars by 2030.

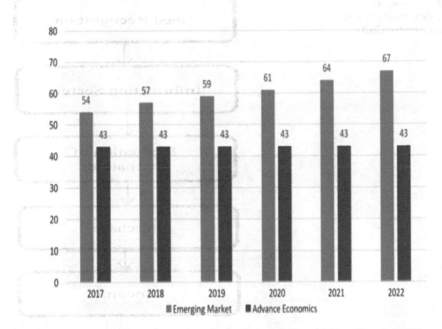

Fig. 2 Sales of global vehicle, medium-term forecast (in million). *Source* Wagner (2021)

Table 1 New motor vehicle registration in Asia before the pandemic

In 2018	In 2017
45,726	46,025
% change 18/17	−0.6
% share 2018	47.0

Source Deshpande (2019)

Number of New Motor Vehicle Registration in Asia Before the Pandemic

In Table 1, the numbers of registered cars in Asia before the pandemic have been highlighted.

Background of Car Purchase Decision During COVID 19

In order to manage the challenging and serious situation, the government of Malaysia had taken different stimulus policies and among them, the most important policy was tax reduction in vehicle purchase (Kumar, 2020).

Figure 3 depicts that the car purchasing decisions of the customers are dependent on need recognition, information search, and alternative evaluation. The purchasing decision during the pandemic was declined as there was a low need for automobiles in the lockdown scenario (Esposti et al., 2021).

Fig. 3 Consumer decision-making model. *Source* Kumar (2020)

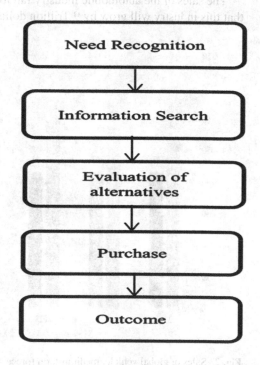

Background of the Car Purchasing Decision Influencing Factors

Consumer behaviour during the pandemic includes lowering the perceived infection threat (Shakibaei et al., 2021). There were also some external factors such as unexpected health emergencies and internal factors such as personal life and attitude that changed the purchasing behaviour of consumers during the pandemic.

As for **product features**, during the pandemic most of the consumers were seeking safety features such as remote vehicle diagnostic and emergency assistance in their vehicles (Awad-Núñez et al., 2021). In the meantime, consumers are attracted by the free and extra product benefits or features of the companies. The features that are in demand in the automobile industry include the availability of spare parts, technical support, and warranty (Bhaduri et al., 2020).

The car **ownership cost** notably includes a monthly repayment but also includes different factors such as petrol cost, road tax, and premium for car insurance (Sheth, 2020). According to recent research, the high car price had affected the purchasing decision of the customers (Garaus & Garaus, 2021).

It has been seen that the major **social need** during the pandemic was social distancing. Different social issues were raised during the pandemic such as lower Gross Domestic Product (GDP) rate as well as the unemployment rate (Abdullah et al., 2020). It has also been analysed that the rate of social media purchasing was increased during the pandemic. The Malaysia's GDP rate in 2020 was −5.59% while the unemployment rate was 4.55% in Malaysia as per (O'Neill, 2021). Thus, it is clear that this lower GDP and high unemployment rate significantly affected the car purchasing power of the customers.

It has been seen that during the COVID-19 pandemic, safety and health became the paramount concern. Thus, changes in the economic condition changed consumer's **purchasing power**. The rate of employees laid-off increased during the pandemic which has affected the purchasing power of the customers (Leow & Husin, 2015). It has been seen that the average income of the Malaysian people during this pandemic was RM 2093. The annual household income of this country fell by 15.5% in the year 2020 (Jabatan Perangkaan Malaysia, 2021). The customers had tuned concisely during the pandemic, it affected their purchasing power (Raza & Masmoudi, 2020).

The availability of service centres, waiting period, brand value as well as dealership reputation are considered as **brand trust** influencing factors (Leow & Husin, 2015). It has been seen that due to improved brand trust in Malaysia, the national carmaker Proton's sales volume was 108.52 thousand units in 2020 despite the outbreak of the COVID-19 pandemic (Müller, 2021).

Since the COVID-19 pandemic, there are little guarantee that factors such as product features, ownership cost, social need, purchasing power and brand trust will continue to have the same influence on car buyers behavior to purchase cars. Previous COVID-19 study focused on travel behavior and preference mode of transport such as Awad-Núñez et al. (2021), Bhaduri et al. (2020), Garaus and Garaus (2021), Kumar (2020), Shakibaei et al. (2021) and Abdullah et al. (2020). Meanwhile, previous researchers like Leow and Husin (2015) had emphasized on car buyers purchase decision during non pandemic period. However, there are some researchers such as

Raza and Masmoudi (2020) who focused on the scope of Bahrain's consumer vehicle purchase decision-making during COVID-19. There is a limited study conducted for car buyers in Malaysia especially during the pandemic on the dimension of consumer behaviour that contributes to purchase decision.

3 Literature Review

3.1 Purchase Decision

Purchase decision in context to availing cars among the Malaysian citizens is largely dependent upon numerous factors for instance Product Features (PF), Ownership Costs (OC), Social Needs (SN), Purchasing Power (PP) and Brand Trust (BT). Dependent variables are the effect whose values change according to the independent variables. In the current study, the purchase decision is the dependent variable as the consumer's intention to avail car depends upon the attractive products features and its utility. Advanced technological usage, multiple utility features, reasonable product prices create a positive impression among the consumers to opt for the products (Shetty, 2021). While on the other hand, decrease purchase power, limited product features, lack of brand trust and lower social requirements creates a negative attitude and belief among the consumers that further prevent them from availing products of their preferences.

3.2 Consumer Behaviour

The decision and action process of the customers for purchasing services and goods for personal consumption is referred to as consumer behavior. There are four types of consumer behaviors such as variety seeking, habitual buying, reducing the dissonance, and a complex behavior. During COVID-19, the behavior of the customers was associated with lowering perceived infection threat. On the other hand, unexpected health emergencies and personal life and attitude changed consumer behavior during the outbreak of the pandemic.

3.2.1 Product Features

Product features are the characteristic attributes that compel consumers to opt for the products of their choice. Product features include artistic components like shape, colour, space, the tone that successfully captures consumer attention (Kowang et al., 2018). Functional aspects, quality and experience associated with the products play a dominating role in determining the purchase behaviour of the consumers (Altaf &

Hashim, 2016). In context to car purchase behaviour of the consumers, product features for instance shape, color, tone of the car adds up to the aesthetic component of the products that result in shaping consumer purchase intention (Alganad et al., 2021). Furthermore, the functional attributes of the products indicate the benefits associated with utilizing a product. The pandemic situation has created trouble for consumers who preferably used public transport for movement. However, increased functionality, comfortable seating arrangements enhance positive experiences of the consumers that further positively enhances consumers buying decisions.

3.2.2 Ownership Costs

Ownership costs highlight the financial resource investment involved in product purchase, operational management and maintenance of the products. In context to availing cars, consumers are required to pay a heavy price for not only availing the products during the initial period but also involves financial investment including the present and future costs while posing the assets. Personal cars require maintenance and servicing costs that help consumers enjoy the services that further adds value to the prices they pay for availing the act of their preferences. As stated by Wadud (2017), often increased ownership costs prevent consumers from availing of cars and thus negatively affecting the purchase decisions of the consumers.

3.2.3 Social Needs

Human beings are social organisms and individuals who affect their purchase behaviour and decisions surround them. Maslow's hierarchical needs highlight that possession of products serve as a basis for the fulfilment of an individual's psychological, safety, belongingness, self-esteem and self-actualisation needs. Since individuals are constantly interacting with other social individuals they prefer to exhibit and perform actions that result in increasing their chances of being socially acceptable (Akar & Dalgic, 2018). However, the current pandemic situation has prohibited individuals from social gatherings evening in public transport that further compels them to adopt alternative strategies to resolve transport difficulties. Social requirements thus positively tend to increase the purchase of cars by consumers.

3.2.4 Purchasing Power

The purchasing power (PP) of the consumer is the ability of the consumer to avail the product of their choices. As opined by Jayaraman et al. (2018), the economy of the country affects the purchasing power of the consumers. Decrease in products prices, increased GDP, decreased inflation rate and increased average annual income of the consumer's results in increased PP of the consumers. Increased PP thus allows consumers to avail products according to their preferences. Thus, interrelationship between PP and Purchase decision indicate higher PP contribute towards positive purchase intension and vice-versa while availing car by consumers.

3.2.5 Brand Trust

In the current competitive ambience, it is essential for the organization to implement innovative strategies to capture consumer attention that further contribute towards providing competitive benefits and increase brand trust. Brand trust (BT) ensures that consumers prefer the selected brand to all other competitors. Product and service quality, experience, perceived risk, brand reputation and security showcased by brand regulate consumer loyalty (Adzharuddin et al., 2017). Superior quality experiences, customer services, enhanced safety and security of the consumers delivered by the brands help consumers had better relate with the organizational goals. Thus, in context to the purchase decision of consumers, BT plays a dominating role in determining consumer preferences while opting for products from particular brands.

3.3 Interrelationship Between Product Features and Purchase Decision

Marketing activities from selling, distribution of goods and promotional strategies cluster around the products being sold by the organization. Product features include discrete areas of upgraded functionality that enhance consumer's value. Both tangible and intangible attributes associated with the products contribute towards determining the purchase behaviour of the consumers (Sahoo & Satpathy, 2020) (Fig. 4).

In the context of purchasing a car, consumers prioritize both the tangible and intangible aspects of the cars. For instance, consumers prefer to look at car size, colour, and material composition while availing a car of their choice. While on the other hand, intangible attributes associated with a car include exchange value, utility benefits. According to the recent report on the sales volume of Proton, the company has reported selling 108,524 cars in 2020 which is comparatively much more than 100,183 cars sold in 2019 (Lim, 2021). The car purchase intention of consumer's increases when the product satisfies their needs and are capable of mitigating the travel difficulties faced by the consumer's number during the pandemic (Peterson & Simkins, 2019). Enhanced utility values constitute increased consumer purchase intention. In context to the increased sales volume of the automobile industries in Malaysia thus sheds light upon the differential product features and highlights the innovative product attributes that are capable of differentiating products from other competitors.

H1: Product features possess a direct positive influence on purchase decision.

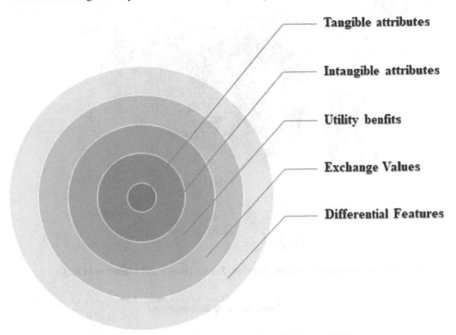

Fig. 4 Products features that affect consumer purchase decisions. *Source* Sahoo and Satpathy (2020)

3.3.1 Interrelationship Between Ownership Costs and Purchase Decision

Ownership cost highlights not only the short-term purchase cost associated with the product but also the long-term operating and maintenance cost associated with the products (Fig. 5).

Ownership cost highlights several factors from petrol price to depreciation cost, road tax and car insurance that must be considered by the consumer while availing the cars. As stated by Havidz and Mahaputra (2020), ownership cost estimation helps individuals obtain a comprehensive view of the products and the values associated with them. Consideration of the maintenance, petrol price, depreciation costs helps individuals to wisely undertake purchase decisions (Fig. 6).

According to the report on the ownership cost associated with the reputed automobile cars, the ownership cost of Perodua Myvi, Perodua Aruz reported to be RM 1579 and RM 2013 respectively. While ownership cost for Honda HRV and Proton is reported to be RM 2658 and RM 2842 respectively (Hasan, 2021). Increased ownership costs create a negative impression in the consumer's mind and adversely affect the purchase decision of the consumers.

H2: Ownership costs possess a direct negative influence on purchase decision.

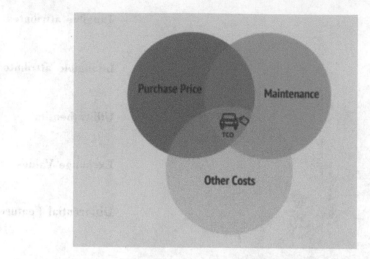

Fig. 5 Ownership costs and purchase decision. *Source* Havidz and Mahaputra (2020)

Fig. 6 Ownership costs and purchase decision. *Source* Hasan (2021)

3.3.2 Interrelationship Between Social Needs and Purchase Decision

Social needs highlight the exhibition of activities by the individuals that increase his or her chances of social acceptance. Lifestyle, income status, family requirements are the crucial factors that determine the purchase intention of the consumers. Product pricing strategy as determined by the marketers depends upon the current market trends, advanced technology usage and innovative product features (Dong et al., 2017). Implementation of the latest technology to enhance product attributes results in higher car prices. The income status of an individual plays a significant role in determining whether the individual can afford a first hand or second-hand car. According to the report on the average monthly salary of Malaysian citizens, the average monthly salary is reported to be 3.2 thousand Malaysian ringgit in 2019 (Jabatan Perangkaan Malaysia, 2021). Furthermore, the salary of the individual may vary depending upon the educational qualification and nature of employment. Individuals with higher disposable income opt for first-hand cars with unique features in comparison to the individuals with lower disposable income who opt for second-hand cars (Du et al., 2018). While on the other hand, households with large families tend to purchase large cars that are spacious and are able to provide comfort to the users (Ali et al., 2019).

H3: Social needs possess a direct positive influence on purchase decision.

3.3.3 Interrelationship Between Purchasing Power and Purchase Decision

Purchasing power indicates the financial resources invested by the consumer while opting for the product of their preferences. The purchasing power of the consumer depends upon the country's GDP, inflation rate, and economic growth (Jannah & Supadmi, 2021). Enhanced economic growth of the country indicates increased GDP and employment within the country. Increased GDP further contributes towards the increased annual income of the citizens of the country thereby strengthening the purchase decisions. The inflation rate on the other hand has an influential role in determining the product price. High inflation rate results in a higher price of cars that can be afforded only by individuals with higher income. However, the reduced inflation rate positively guides consumer purchase intention of the consumers.

H4: Purchasing power possess a direct positive influence on purchase decision.

3.3.4 Interrelationship Between Brand Trust and Purchase Decision

Organizational reputation and culture are significant in the current competitive business ambience that facilitates enhanced corporate revenue. As opined by Novansa and Ali (2017), consumer loyalty is directly interlinked with brand trust. Brand trust allows consumers to avail products from the reputed brands dispensing upon the product and service quality. Provision of superior quality products along with reasonable prices, consideration of consumer safety and security issues plays a significant role in strengthening brand awareness and reputation. Sustainable aspects of products, superior quality consumer experiences and immediately addressing consumer issues generates a positive attitude among the consumer towards the brands thereby contributing towards enhanced purchase decisions among the consumers.

H5: Brand trust possess a direct positive influence on purchase decision.

4 Theoretical Framework

The consumer decision process is the broad aspect that encompasses multiple activities from increased consumer awareness to need recognition, accumulation of information to solve needs, evaluation of alternative options and finally undertaking purchase decisions. *EKB model* highlights the procedures and the factors that shape the consumer purchase behaviours and determine the process followed by the consumers while choosing among the alternatives.

The EKB (Engell–Kollat–Blackwell) model includes information inputs, information processing, decision process and external variables that affect their purchase decisions (Osei & Abenyin, 2016). In context to car purchase behaviours among Malaysian consumers, unique marketing strategies implemented by the marketers creates an urge within the consumers to avail themselves the luxurious cars at a reasonable price. Information processing further highlights memory retention of the attractive product features that further instigate the purchase decision of the consumers (Leljak & Dobrinic, 2019). Furthermore, the necessity of availing cars during pandemic situations would relieve consumers from the difficulties that they face in public transport thereby compelling them to meet their needs by availing personal vehicles that serve to resolve their problem at the earliest. However, evaluation of alternative products is largely influenced by several external variables like consumer attitudes, beliefs, lifestyle, cultural norms, family and group influences.

Consumer purchase intentions are not stable rather it changes according to the market demands, products feature and quality. The current pandemic situation has affected almost every aspect of human lives. Prohibition and limitation on public transportation have resulted in increased trouble for the office goers and has adversely affected their daily lives. Safety protocols in context to social distancing have compelled individuals to prevent social gatherings that threaten social life due to the spread of disease.

The current study has focused on highlighting the increased car purchase in Malaysia among consumers. According to the recent report on sales volume and market growth by Perodua, one of the leading automobile industries in Malaysia, market share has been reported to increase from 39.8% in 2019 to 41.6% in 2020 (Lim, 2021). Increased car sales by the automobile company in Malaysia during the pandemic highlights several factors that played a significant role in regulating the purchase decisions of the consumers. The research framework thus has been formulated keeping in mind the dimension of consumer behavior such as PF, OC, SN, PP and BT that regulates and relation that further affect purchase decisions (Fig. 7).

5 Research Methodology

Quantitative research design has been opted by their searcher to discover the thinking, acting or feeling process of the people in a particular way. It involves huge sample

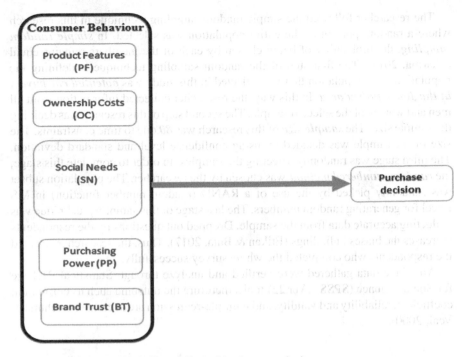

Fig. 7 Research framework (*Source* Created by the researcher)

sizes and concentrates on response quality. There are four main types of quantitative research design and these are correlation, causal-comparative, descriptive and experimental. Among them, the descriptive research design technique was followed by the researcher to conduct a true experiment. A *descriptive research design* enables a researcher to investigate the causal relationship among the variables (Rutberg & Bouikidis, 2018). Therefore, the casual relationship among the dependent variable purchasing decision of the customers and independent variable customers behaviours and Covid-19 has been analysed by the researcher appropriately. The major purpose of this research design is to generate knowledge and creating understanding regarding the social world.

The use of a scientific inquiry is dependent on the observed data for examining questions regarding the sample population (Chih-Pei & Chang, 2017). In this research design, the researcher used the statistical method to analyse the hypotheses. Large subject numbers have been included in this research design that helped the researcher to generate appropriate hypotheses, theories, and models. Imperial data have been collected by the researcher which increased the experimental control, variable manipulation, and improved the measurement methods. In this way, the researcher has gained the benefit to follow a structured method for this research. The outcome of this research is highly reliable and reusable. In this process, the research has gained different numerical outcomes (Rutberg & Bouikidis, 2018).

The researcher followed the simple random sampling technique in this research where a random portion of the entire population was selected. In *simple random sampling*, the probability of being chosen by each of the members remains equal (Sharma, 2017). The first step of the random sampling technique is defining the population. The population that was selected in this study was *potential car buyers in the Johor Bahru area*. In this way, the researcher collected information from all men and women of the selected sample. The second step of this research was deciding the sample size. The *sample size* of this research was *80* due to time constraints. The size of the sample was decided by using confidence level and standard deviation. The third stage was randomly selecting the samples. In order to complete this stage, the *random numbered method* was chosen by the researcher. The population subset was randomly picked by the use of a RAND (random number function) in MS Excel for generating random numbers. The last stage of this sampling technique was collecting accurate data from the sample. Dropped out questions by the respondents increases the biases in findings (Etikan & Bala, 2017). Thus, the researcher selected the respondents who completed the whole survey successfully.

All of the data gathered were verified and analyze through Statistical Package for Social Science (SPSS—Ver.22) tools, therefore the outcome such as correlation coefficient, reliability and validity, and multiple-regression are accurate (Ticehurst & Veal, 2000).

6 Conclusion

In the context of the purchase decision of the consumer opting for cars during Covid-19 highlights several factors that play a significant role in regulating consumer purchase decisions. The study has successfully shed light on the varied factors ranging from Product features, Ownership costs, Social needs, Purchasing power and Brand trust that has a certain impact on the buyer's purchase decisions. Product features from both tangible, intangible attributes along with other factors help consumers to select the products of their preferences. However, individual income along with the purchasing power of the customer is essential in determining the purchase intention of the consumer. Consumers with higher income are more likely to purchase expensive cars in comparison to the ones with lower income who would prefer a second-hand car or public transport. Hence, this study is one that focuses on the car buyers in Malaysia and the significant factors that impacted their purchase decisions at the time of the Coronavirus (COVID-19) outbreak.

References

Abdullah, M., Dias, C., Muley, D., & Shahin, M. (2020). Exploring the impacts of COVID-19 on travel behavior and mode preferences. *Transportation Research Interdisciplinary Perspectives, 8*(November), 100255. https://doi.org/10.1016/j.trip.2020.100255

Adzharuddin, N. A., Moses, I. O., & Yusoff, S. Z. (2017). The influence of brand image of Perodua Axia on consumer's decision making. *International Journal of Academic Research in Business and Social Sciences, 7*(6), 1072–1087. https://doi.org/10.6007/ijarbss/v7-i6/3067

Akar, E., & Dalgic, T. (2018). Understanding online consumers' purchase intentions: A contribution from social network theory. *Behaviour & Information Technology, 37*(5), 473–487. https://doi.org/10.1080/0144929X.2018.1456563

Alganad, A. M. N., Isa, N. M., & Fauzi, W. I. M. (2021). Boosting green cars retail in Malaysia: The influence of conditional value on consumers behaviour. *Journal of Distribution Science, 19*(7), 87–100. https://doi.org/10.15722/jds.19.7.202107.87

Ali, A., Xiaoling, G., Ali, A., Sherwani, M., & Muneeb, F. M. (2019). Customer motivations for sustainable consumption: Investigating the drivers of purchase behavior for a green-luxury car. *Business Strategy and the Environment, 28*(5), 833–846. https://doi.org/10.1002/bse.2284

Altaf, S. N., & Hashim, N. A. (2016). Key factors influencing purchase intentions towards automobiles in Pakistan. *Paradigms, 10*(1), 14–22. https://doi.org/10.24312/paradigms100102

Awad-Núñez, S., Julio, R., Gomez, J., Moya-Gómez, B., & González, J. S. (2021). Post-COVID-19 travel behaviour patterns: Impact on the willingness to pay of users of public transport and shared mobility services in Spain. *European Transport Research Review, 13*(1), 1–44. https://doi.org/10.1186/s12544-021-00476-4

Bhaduri, E., Manoj, B. S., Wadud, Z., Goswami, A. K., & Choudhury, C. F. (2020). Modelling the effects of COVID-19 on travel mode choice behaviour in India. *Transportation Research Interdisciplinary Perspectives, 8*. https://doi.org/10.1016/j.trip.2020.100273

Chih-Pei, H., & Chang, Y. (2017). John W. Creswell, Research design: Qualitative, quantitative, and mixed methods approaches. *Ournal of Social and Administrative Sciences, 4*(June), 3–5.

Deshpande, R. (2019). Global automotive market predictions. *crescendoworldwide.org.*

Dong, X., Suhara, Y., Bozkaya, B., Singh, V. K., Lepri, B., & Pentland, A. (2017). Social bridges in urban purchase behavior. *ACM Transactions on Intelligent Systems and Technology, 9*(3), 1–30. https://doi.org/10.1145/3149409

Du, H., Liu, D., Sovacool, B. K., Wang, Y., Ma, S., & Li, R. Y. M. (2018). Who buys new energy vehicles in China? Assessing social-psychological predictors of purchasing awareness, intention, and policy. *Transportation Research Part F: Traffic Psychology and Behaviour, 58*, 56–69. https://doi.org/10.1016/j.trf.2018.05.008

Esposti, P. D., Mortara, A., & Roberti, G. (2021). Sharing and sustainable consumption in the era of Covid-19. *Sustainability, 13*(4), 1–15. https://doi.org/10.3390/su13041903

Etikan, I., & Bala, K. (2017). Sampling and sampling methods. *Biometrics & Biostatistics International Journal, 5*(6), 215–217.

Garaus, M., & Garaus, C. (2021). The impact of the Covid-19 pandemic on consumers' intention to use shared-mobility services in German cities. *Frontiers in Psychology, 12*, 1–22. https://doi.org/10.3389/fpsyg.2021.646593

Hasan, H. (2021). [Revealed]: The true cost of car ownership in Malaysia. *Balkoni Hijau Blog.* Retrieved September 01, 2021, from https://balkonihijau.com/cost-of-car-ownership/

Havidz, H. B. H., & Mahaputra, M. R. (2020). Brand image and purchasing decision: Analysis of price perception and promotion. *Dinasti International Journal of Economics, Finance & Accounting, 1*(2), 358–372. https://doi.org/10.38035/DIJEFA

Jabatan Perangkaan Malaysia. (2021). *Salaries & wages survey report Malaysia.* Department of Statistics Malaysia Official Portal. Retrieved September 01, 2021, fromhttps://www.dosm.gov.my/v1/index.php?r=column/cthemeByCat&cat=157&bul_id=VDRDc0pGZHpieEUwMDNFWHVHSnpkdz09&menu_id=Tm8zcnRjdVRNWWlpWjRlbmtlaDk1UT09

Jannah, N., & Supadmi, N. L. (2021). The effect of the imposition of value added tax and sales tax on goodsluxury for consumer purchasing power. *American Journal of Humanities and Social Sciences, 4*, 432–437.

Jayaraman, K., Arumugam, S., Mohan Kumar, K., & Kiumarsi, S. (2018). Factors influencing the purchase decision of non-national cars in Malaysia: An empirical study. *International Journal on Global Business Management & Research, 10*(1, Special Issue), 150–162.

Kim, R. Y. (2020). The impact of COVID-19 on consumers: Preparing for digital sales. *IEEE Engineering Management Review, 48*(3), 212–218. https://doi.org/10.1109/EMR.2020.2990115

Kowang, T. O., Samsudin, S. A., Yew, L. K., Hee, O. C., Fei, G. C., & Long, C. S. (2018). Factors affecting car purchase intention among undergraduates in Malaysia. *International Journal of Academic Research in Business and Social Sciences, 8*(8). https://doi.org/10.6007/ijarbss/v8-i8/4437

Kumar, R. (2020). Impact of Covid-19 on Indian economy in terms of consumer buying behavior and sustainability through digitalization. *PalArch's Journal of Archaeology of Egypt, 17*(12), 1190–1204. [Online]. https://www.archives.palarch.nl/index.php/jae/article/view/6713

Leljak, L., & Dobrinic, D. (2019, October). Research of impact factors on behavior of millennials in online buying. In *46th International Scientific Conference on Economic and Social Development—Sustainable Tourist Destinations* (pp. 24–33).

Leow, C. S., & Husin, Z. (2015). Product and price influence on cars purchase intention in Malaysia. *International Journal of Social and Educational Innovation*.

Lim, A. (2021). Vehicle sales performance in Malaysia, 2020 vs. 2019—All the numbers, and how the brands fared last year—paultan.org. *paultan.org*. Retrieved September 01, 2021, from https://paultan.org/2021/01/27/vehicle-sales-performance-in-malaysia-2020-vs-2019-all-the-numbers-and-how-the-brands-fared-last-year/

Müller, J. (2021). Malaysia: Sales volume Proton cars 2020 | Statista. *statista.com*. Retrieved September 02, 2021, from https://www.statista.com/statistics/869527/malaysia-sales-volume-proton-cars/

Novansa, H., & Ali, H. (2017). Purchase decision model: Analysis of brand image, brand awareness and price (Case study SMECO Indonesia SME products). *Saudi Journal of Humanities and Social Sciences, 2*(5), 597–610. https://doi.org/10.21276/sjhss

O'Neill, A. (2021). Malaysia—Unemployment rate 1999–2020 | Statista. *statisca.com*. Retrieved September 02, 2021, https://www.statista.com/statistics/319019/unemployment-rate-in-malaysia/

Osei, B. A., & Abenyin, A. N. (2016). Applying the Engell–Kollat–Blackwell model in understanding international tourists' use of social media for travel decision to Ghana. *Information Technology & Tourism, 16*(3), 265–284. https://doi.org/10.1007/s40558-016-0055-2

Peterson, M., & Simkins, T. (2019). Consumers' processing of mindful commercial car sharing. *Business Strategy and the Environment, 28*(3), 457–465. https://doi.org/10.1002/bse.2221

Raza, S., & Masmoudi, M. (2020). Consumer vehicle purchase decision-making during COVID-19. In *2020 International Conference on Decision Aid Sciences and Application* (pp. 692–696).https://doi.org/10.1109/DASA51403.2020.9317187

Rutberg, S., & Bouikidis, C. D. (2018). Focusing on the fundamentals: A simplistic differentiation between qualitative and quantitative research. *Nephrology Nursing Journal, 45*(2), 209–212.

Sahoo, S. K., & Satpathy, P. B. (2020). Consumer learning leads to purchase intention: A conceptual justification through a proposed model. *Parishodh Journal, IX*(III), 5315–5329.

Shakibaei, S., de Jong, G. C., Alpkökin, P., & Rashidi, T. H. (2021). Impact of the COVID-19 pandemic on travel behavior in Istanbul: A panel data analysis. *Sustainable Cities and Society, 65*, 1–32. https://doi.org/10.1016/j.scs.2020.102619

Sharma, G. (2017). Pros and cons of different sampling techniques. *International Journal of Applied Research, 3*(7), 749–752, 2017. [Online]. www.allresearchjournal.com

Sheth, J. (2020). Impact of Covid-19 on consumer behavior: Will the old habits return or die? *Journal of Business Research, 117*, 280–283. https://doi.org/10.1016/j.jbusres.2020.05.059

Shetty, S. (2021). Factors impacting purchase behaviour of automobiles post Covid-19. *Journal of Educational Psychology, 57*(9), 6154–6161. https://doi.org/10.17762/pae.v57i9.2690

Tardivo, A., Sánchez Martín, C., & Carrillo Zanuy, A. (2020). Covid-19 impact in transport, an essay from the railways' system research perspective. https://doi.org/10.31124/advance.12204836

Ticehurst, G., & Veal, A. (2000, August). Research methods: A managerial approach. *Greg. William.*

Wadud, Z. (2017). Fully automated vehicles: A cost of ownership analysis to inform early adoption. *Transportation Research Part A: Policy and Practic, 101*, 163–176. https://doi.org/10.1016/j.tra.2017.05.005

Wagner, I. (2021). Automotive industry worldwide—Statistics & facts | Statista. *statisca.com.* Retrieved September 02, 2021, https://www.statista.com/topics/1487/automotive-industry/

Environmental Forces Influencing Perceived Acceptance of COVID-19 Vaccination: Social Responsibility as a Moderating Role

Ala'eddin M. Ahmad, Mohammad Abuhashesh⊙, Nawras M. Nusairat, Majd AbedRabbo, Ra'ed Masa'deh⊙, and Mohammad Al Khasawneh

Abstract The main objective of this research is to investigate the environmental forces influencing perceived acceptance of COVID-19 vaccination. Also, to investigate the moderating role of social Responsibility. The research independent variables consist of environmental forces (economic, political/legal, and social), the dependent variable is represented by perceived acceptance of COVID-19 vaccination. A quantitative method was using cross-sectional online research questionnaire to collect the primary data. A sample of (N − 2084) respondents who follow the social media platforms were selected. Therefore, a stratified simple random method was used in this research. The result shows a positive influence of political/legal and social on perceived acceptance of COVID-19 vaccination. In contrast, there is no influence of economic factor on perceived acceptance of COVID-19 vaccination. The research results are beneficial for formulating an effective polices and strategies for vaccination in Jordanian health market. The current research result recommends that Jordanian FDA and MOH, nationally authorized of COVID-19 vaccines, employer

A. M. Ahmad · M. Abuhashesh (✉) · M. AbedRabbo · M. Al Khasawneh
E-Marketing and Social Media Department, Princess Sumaya University for Technology (PSUT), Amman, Jordan
e-mail: m.abuhashesh@psut.edu.jo

A. M. Ahmad
e-mail: a.ahmed@psut.edu.jo

M. AbedRabbo
e-mail: m.abedrabbo@psut.edu.jo

M. Al Khasawneh
e-mail: m.alkhasaawneh@psut.edu.jo

N. M. Nusairat
Department of Marketing, Applied Science Private University (ASU), Amman, Jordan
e-mail: n_nserat@asu.jo

R. Masa'deh
Department of Management Information Systems, School of Business, The University of Jordan, Amman, Jordan
e-mail: r.masadeh@ju.edu.jo

© The Author(s), under exclusive license to Springer Nature Switzerland AG 2023
M. Alshurideh et al. (eds.), *The Effect of Information Technology on Business and Marketing Intelligence Systems*, Studies in Computational Intelligence 1056,
https://doi.org/10.1007/978-3-031-12382-5_129

2363

powerful force and recommendations, friends and family, and positive words of mouth can all play an important role for vaccination acceptance and increase the percentages of Jordanian citizen to be vaccinated which was consistent with previous studies. The current research reflected the positive role of political/legal and social factors on perceived acceptance of COVID-19 vaccination among Jordanian people. Also, the research outcomes are beneficial for formulating an effective polices and strategies for vaccination in Jordanian health market.

Keywords Environmental forces · Economic · Political/Legal · Social · Acceptance of COVID-19 vaccination: social responsibility · Jordan

1 Introduction

Since the emergent of corona virus from China in March 2020 which it quickly spread out into the rest of the world. The World Health Organization has declared that corona virus as pandemic disease. Corona virus has become a sever public health issue (Alameeri et al., 2021; Al-Dmour et al., 2020). Therefore, governments around the world have changed their health strategies to prevent further spread of COVID-19 (Ahmad et al., 2021; Akour et al., 2021). Governments put stricter rules and guidelines such as social distance, avoiding social gatherings, testing every suspected case, staying home, university and school closures, workplace distancing, and lockdowns were all introduced to reduce the impact of corona virus disease. Just like the rest of the world, Jordan has been suffering from the effect of COVID-19 on all aspect of life. Therefore, the Jordanian government took an extraordinary measures and strict emergency law to fight COVID-19 pandemic disease. Governments and their top medical labs, they all were racing to develop and deploy safe and effective vaccines. Vaccines work by training and preparing the body's natural defences—the immune system—to recognize and fight off the viruses and bacteria they target. After vaccination, if the body is later exposed to those disease-causing germs, the body is immediately ready to destroy them, preventing illness (The World Health Organization 2021). Different types of vaccines have been rolled out in many countries. Vaccines are a critical new tool in the battle against COVID-19 and it is hugely encouraging to see so many vaccines proving successful and going into development. Working as quickly as they can, scientists from across the world are collaborating and innovating to bring people tests, treatments, and vaccines that will collectively save lives and end COVID-19 pandemic disease (Abbas et al., 2020; Al Khasawneh et al., 2021a). One year has been passed since the beginning of the global pandemic of COVID-19 (Alshurideh et al., 2021; Nuseir et al., 2021). Today we have several vaccines to choose from to protect ourselves and prevent the further spread of this virus and its newer variations. Research on vaccine production has begun in early 2020 (Jordan et al., 2020). Also, governments around the world have been occupied with vaccinating its populations (Leo et al., 2021; Shah et al., 2021; Taryam et al., 2020). Kwok et al. (2020) Suggest a "a critical (minimum) herd-immunity threshold of 67% to reach safety and go back to living previously accustomed to lifestyles. Also, (Lazarus et al., 2021) attempted a global survey of potential acceptance of a COVID-19 vaccine to measure the likelihood of the public taking the vaccine and whether employer recommendation to take the vaccine would affect this likelihood (Cobb et al., 2021). They state that 71.5% of its participants reported they would take the vaccine while 61.4% said they would take it if their employer requested them to take it (Lazarus et al., 2021).

Vaccine acceptance is a troubling area affected by several factors, such as cultural, trust, fear of future side-effects. Also, scientist plan to measure the effect of vaccine on economic, political/legal, social factors, and all aspect of human lives. Lazarus et al. (2021) Found that greater trust in government significantly affected the acceptance rate positively, as did employer advice. They measured a sample from 19 countries which comprised about 55% of the global population, consisting of 53.5% women, and 63.3% earned more than $32 per day, and 36.3% had a university degree, and 62.4% were between 25 and 54 years old, which is a similar age representation of the majority of Jordan's population. They found a 90% acceptance rate in China, compared to 55% in Russia. In this study, we plan to measure the acceptance rate in Jordan (Gates, 2020). Higher level of income and education populations were found to have more positive acceptance of the vaccine, while higher trust in government information also showed more positive acceptance as long as the information provided was detailed and clear, such as "explaining how vaccines work, as well as how they are developed, from recruitment to regulatory approval based on safety and efficacy. Effective campaigns should also aim to carefully explain a vaccine's level of effectiveness, the time needed for protection (with multiple doses, if required) and the importance of population-wide coverage to achieve public immunity" (Aljumah et al., 2021; Lazarus et al., 2021; Wise et al., 2020). Furthermore, Asian countries had the highest acceptance rate, while middle-income countries "such as Brazil, India and South Africa" also showed significantly high acceptance rates (Lazarus et al., 2021).

Another study conducted on COVID-19 vaccine acceptance in China by Wang et al. (2020) conducted an online survey among 2058 Chinese adults where a majority of 91.3% confirmed they would take the vaccine once available, where the questionnaire attempted to measure sociodemographic characteristics including education, employment status and income, in addition to risk perception and the impact of contractive COVID-19 on their lives, work and studies, as well as vaccination history of the seasonal flu-shot who are considered a higher risk group of infection. The study result revealed that 52% of the targeted population wanted to get the COVID-19 vaccine; however, the rest of the sample which is 48% would like to delay the vaccination until the side effect is safe (Huang et al., 2020).

Wang et al. (2020) also found that being male and married increased the possibility of taking the vaccine since they are perceived as higher risk group and are more likely to follow doctor recommendations, yet less likely to follow employer recommendations to take the vaccine. According to the study result were done by Kreps et al. (2020) showed that 79% of the sample selected a vaccine that has high efficiency and longer duration of protection.

Opel et al. (2020) Mentioned in their study to increase the acceptability of COVID-19 Vaccine, the US official needs to share and transparence data and they should increase informed consent considerations. In addition, Opel et al. (2020) urges frequent and visible communication with public in order to build trust and achieve confidence in COVID-19 Vaccines (Abuhashesh et al., 2021). Graffigna et al. (2020) studied the main factors that affect citizen decision making for taking or not taking COVID-19 vaccine. The study focused on perceived effectiveness, safety and side-effect concerns, vaccine benefits, and where the vaccine was originated. Furthermore, they research studied social factors such as recommendations from family members and health authorities. In addition, individuals also care about vaccination cost, vaccine availability, and accessibility.

A study was done by Tam et al. (2020), measure the differences between older adults and young adults, where older people has been prioritized for corona virus prevention and treatment. On the other hand, young people were less likely to comply with prevention practices and social distance (Fisher et al., 2020). A recent study in the US were done by Reiter et al. (2020) had mentioned the most factors that increase COVID-19 vaccine acceptance. The most important vaccine attribute for selection was increased efficacy, followed by a longer duration of protection and a significantly lower incidence of major adverse events. Also, US citizen were more likely to select a vaccine has an endorsement from the US centers. However, US participants were less likely to select a vaccine was developed outside of the United State, specially from China. Another key finding that US citizens have higher vaccine acceptability rates when endorsed by public health agencies. However, US citizens did not trust politicians' endorsements for public health issues. Opel et al. (2020) urge caution against interpreting the results as further evidence of vaccine hesitancy spreading, although they note that "vaccine hesitancy and a reluctance to accept a COVID-19 vaccine are not completely distinct" therefore, public health authorities should communicate and engage with public in order to encourage people to accept and trust COVID-19 vaccine acceptance (Czeisler et al., 2020; Dror et al., 2020).

In order for Jordanian health official to develop effective corona virus vaccine promotion strategies, it is very important for official to understand what factors would contribute to vaccine acceptance and what factors would result in vaccine rejection. For this reason, Jordanian health official needs to communicate the benefit of vaccines, safety concerns, side effects, vaccine accessibility, and availability in order to change the Jordanian citizens' belief and attitude toward the acceptance of COVID-19 vaccine (Irwin et al., 2017). This study aimed to investigate the factors (economic, political/legal, and social) influencing on perceived acceptance of COVID-19 vaccination among Jordanian people with the moderating role of social responsibility. The research outcomes might be beneficial for health decision makers to establish a concrete strategy for vaccination against epidemics and pandemics (Rosenstock, 1974).

2 Methods

2.1 Research Design, Sample

To test for causal relationships between the research variables, we adopted a quantitative method using cross-sectional online research questionnaire to collect the primary data. A sample of (N = 2084) respondents who follow the social media platforms were selected. Therefore, a stratified simple random method was used in this research.

2.2 Measures

The self-administered questionnaire was designed based on previous studies (Lazarus et al., 2021; Opel et al., 2020; Wang et al., 2020) to examine the influence of environmental forces on perceived acceptance of COVID-19 Vaccination. The research questionnaire consists of (1) socio-demographic characteristics, such as age, gender, occupation, working in medical field, income, and health condition; (2) environmental forces (economic, political/legal, and social) as an independent variable; (3) social responsibility as a moderating variable; (4) perceived acceptance of COVID-19 Vaccination as dependent variable. Both English and Arabic statements were presented in the questionnaire. The questionnaire contained five-point Likert scales 1 = strongly disagree and 5 strongly agree. The original sources of the main scale items are presented in Table 1.

3 Results

3.1 Demographic Examination

Despite the fact that the demographical age and gender factors are of minimal impact on the retrieved data, the researchers managed to collect data from different age groups, both gender, occupation, income level, working in the medical field industry, and health conditions. Table 2 describes the demographic distribution of the research sample.

Table 1 Construct operationalization

Item No	Construct	Source
Perceived economic effect on vaccine acceptance		
PE1	Vaccine price can affect my decision towards being vaccinated	Adopted from Lazarus et al. (2021), Wang et al. (2020)
PE2	The economies of taking the vaccine such as (distance to vaccination sites, waiting time etc.) plays a role in my decision to be vaccinated	
PE3	Costs associated with taking the vaccine influence my decision to take it	
Perceived political and legal effect on vaccine acceptance		
PPL1	FDA & MOH authorization is key to accept taking the vaccine	Adopted from Lazarus et al. (2021), Wang et al. (2020), Opel et al. (2020)
PPL2	Nationally authorized vaccines can provide me with reassurances to accept the vaccine	
PPL3	Employer powerful force and recommendation is an important factor in vaccination acceptance	
PPL4	National laws for vaccinations can play an important role in accepting the vaccine	
Perceived social effect on vaccine acceptance		
PS1	My friends and family play a role in accepting the vaccine	Adopted from Lazarus et al. (2021), Opel et al. (2020)
PS2	I am likely to be influenced by my friends and family's decision to take the vaccine	
PS3	Word of mouth from my social circle may influence my acceptance of the vaccine	
Social responsibility		
SR	Taking the COVID-19 vaccine is part of my social responsibility	Authors
Perceived acceptance of COVID-19 vaccination		

(continued)

3.2 Descriptive Analysis

The following test results are presented as shown in Table 3; descriptive analysis for the answers of all environmental forces dimensions (economic, political/legal, and social) and social responsibility as a moderating factor, and perceived acceptance of COVID-19 vaccination (PAV) as a dependent factor.

The descriptive statistics offered in Table 3 pointed to a positive and negative disposition towards the variables measured.

Table 1 (continued)

Item No	Construct	Source
PAV	How likely will you take a COVID-19 vaccine when it is available? The question was rated using a five-point Likert scale (1 = definitely not take it, 2 = not likely to take it, 3 = I don't know, 4 = likely to take it, and 5 = definitely take it) Based on their responses, participants were divided into three groups including (1) Refusal group, (participants with answers of '1' or '2'); (2) Hesitancy group (participants with answers of '3'); (3) Acceptance group (participants with answers of '4' or '5')	Authors

3.3 Reliability Analysis

Reliability analysis is defined by Wilkinson (1999), p. 598 as a psychometric property of a particular sample's response to a measure monitored under specific scenarios and conditions. Particularly, the recommended scale shall exhibit high extents of internal consistency in reference to Cronbach alpha or composite reliability (Hair et al., 2012). In fact, the accepted and agreed upon percentage of Cronbach alpha is (70%), whilst the composite reliability's advised percentage is (80%) (Streiner, 2003).

The current research shows that Cronbach alpha exceeded (70%) in the five scales variables (e.g., 0.792–0.831), and the composite reliability of each scale was above (80%) as of (0.827–0.907). Moreover, it has been suggested that convergent validity is related to the level of harmony and agreement between a conceptual theory such as the latent construct and its monitoring tool that may be referred to a number of elements that measure the implicit construct) (Hair et al., 2012). The aforementioned validity is reflected through the AVE (average variance extracted) which indicates the shared average variance between a variable and the items related to it.

Table 4 presents the AVE of the five variables which is more than the recommended threshold of (50%). More specifically, the AVE of the five variables was in the range of (0.593–0.636), indicating that the items of each scale are indeed related. Depending on the outcomes of these tests, the instrument of the study is useful, which allows the academics to manage statistical analysis.

Table 2 Demographic characteristics

		Frequency	Percentage
Age	18–30	1110	53.3
	31–40	456	21.9
	41–50	312	15.0
	51–60	150	7.2
	61 +	56	2.7
	Total	2084	100
Gender	Female	1182	56.7
	Male	902	43.3
	Total	2084	100
Occupation	Students	780	37.4
	Unemployed	260	12.5
	Retired	132	6.3
	Part-time employee	114	5.5
	Full-time employee	798	38.3
	Total	2084	100
Working in medical field	No	1818	87.2
	Yes	266	12.8
	Total	2084	100
Income	300 JD and less	822	39.4
	301–700	504	24.2
	701–1100	296	14.2
	1101–1500	162	7.8
	More than 1500	300	14.4
	Total	2084	100
Health condition	No history of known medical conditions	1736	83.3
	I have history with certain medical conditions	256	12.3
	Prefer not to say	92	4.4
	Total	2084	100

Table 3 Descriptive analysis table

Items	Numbers	Mean	SD
Economic	2084	2.21	0.921
Political/legal	2084	3.91	0.897
Social	2084	3.27	0.915
Social responsibility	2084	3.72	0.926
PAV	2084	3.52	0.895

Table 4 Convergent validity and internal consistency of survey

Variables	Cronbach alpha (α)	AVE	CR
Economic	0.831	0.608	0.892
Political/legal	0.825	0.636	0.907
Social	0.810	0.613	0.866
Social responsibility	0.799	0.610	0.827
PAV	0.792	0.593	0.834

Table 5 Discriminant validity

Variables	E	PL	S	SR	PAV
Economic	**0.821**				
Political/legal	0.622	**0.812**			
Social	0.190	0.604	**0.806**		
Social responsibility	0.353	0.459	0.312	**0.801**	
PAV	0.366	0.314	0.290	0.310	**0.813**

3.4 Convergent and Discriminant Validity

The assessment of Discriminant Validity, as suggested by Fornell and Larcker (1981) is possible when comparing and contrasting the correlation between the coefficients of the construct along with the square roots of AVE, which ought to be higher than that of the other constructs, paving the way for the output results of the study model to be disseminated in the study population.

As stated in Table 5, the correlation between the coefficients of the construct is less than the square roots of AVE, reflecting its loadings to their latent construct to be higher than that of other constructs. As a consequence, the discriminant validity of the research model is proven to be satisfactory, enabling the possibility of reflecting the results derived from the study model on the study community. Herein, the discriminant validity is not an issue of concern.

3.5 Structural Model and Hypotheses Testing

With the intention of determining the possibility of accepting the research hypotheses, multiple linear regression tests were set. Besides, significance testing has been undertaken to detect if the hypotheses are accepted or not with a threshold percentage limit not to be exceeding 0.05 to facilitate the substitution of hypotheses and refuse the void ones, particularly due to the fact that the research applied the coefficient of determination to demonstrate the impact of the dependent variable on each responding independent one. In reference to the conceptual theory that declares the value of

Table 6 The results of the first hypothesis

Direct effect	Coefficient (β)	T-value	P-value $P \leq 0.05$	R^2	Decision
Economic → PAV	0.014	1.013	P = 0.925	0.392	Rejected
Political/Legal → PAV	0.342	4.321	P = 0.001		Accept
Social → PAV	0.332	4.278	P = 0.000		Accept
Environmental Forces → PAV	SigF = 0.001		F-Value = 12.119		Accept

R^2: 0.51 = Strong; 0.330 = Moderate; 0.20 = Weak
Effect size: 0.02 = Small effect; 0.15 = Moderate effect; 0.35 = Substantial effect

significance to be less than 0.05 (p < 0.05) for the hypotheses 'H' to be considered accepted, the statistical study applied various linear regressions tests that are presented in Table 6 for the first hypothesis.

Table 6 demonstrates the outcomes of multiple regressions of environmental forces dimensions and the way it can affect *perceived acceptance of COVID-19 vaccination* as follows:

H1: Environmental forces factors and social have a significant impact on perceived acceptance of COVID-19 vaccination at sig. level ≤ 0.05.

The results of the analysis support this hypothesis (Sig. F = 0.001; F = 12.119). Thus, the hypothesis is accepted. In other words, it is confirmed that if Jordanian Government apply theses dimensions, the Jordanian community will accept to take this vaccine.

H1a: Economic factors have a significant impact on perceived acceptance of COVID-19 vaccination at sig. level ≤ 0.05.

The results of the analysis do not support this hypothesis (Sig < 0.05, = 0.925; T-value = 1.013; β Coefficient = 0.014). Therefore, the hypothesis is rejected affirming the negative impact economic factor on the *perceived acceptance of COVID-19 vaccination.*

H1b: Political/legal factors have a significant impact on perceived acceptance of COVID-19 vaccination at sig. level ≤ 0.05.

The results of the analysis support this hypothesis (Sig < 0.05, = 0.001; T-value = 4.321; β Coefficient = 0.342). Accordingly, the hypothesis is accepted indicating that if political and legal factors are applied, then *perceived acceptance of COVID-19 vaccination* will increase.

H1c: Social factors have a significant impact on perceived acceptance of COVID-19 vaccination at sig. level ≤ 0.05.

The results of the analysis support this hypothesis (Sig < 0.05, = 0.000; T-value = 4.278; Coefficient = 0.332). Therefore, the hypothesis is accepted and this demonstrates that if social factor is applied, the *perceived acceptance of COVID-19 vaccination* will increase.

H2: social responsibility would moderate the relationship between environmental forces factors and PAV at sig. level ≤ 0.05.

Moreover, to test for H2 concerning social responsibility moderation between environmental forces factors and perceived acceptance of COVID-19 vaccination, (Hayes, 2017) process macro model was used. The product indicator method was used as it employs all the indicators of the latent predictor and moderator and all the pair combinations possible which serve the interaction term in the structural model. As seen in Table 3 there is a significant influence of social responsibility on the path between environmental forces factors and perceived acceptance of COVID-19 vaccination.

The results show that the significance level p- value is 0.011 (P ≤ 0.05). Thus, the alternative hypothesis test is statistically accepted (supported).

Table 3 also shows that the R^2 rate is (0.421) which points to the precision in *perceived acceptance of COVID-19 vaccination* that relies on environmental forces dimensions (economic, political/ legal, and social), this value is moderated and reliable in the process of interpretation and prediction (Table 7, Fig. 1).

Table 7 The results of the second hypothesis

Indirect effect	Estimated coefficients	S.E	Std. Est. Coef	F2	P-Value	R^2	Hypothesis
Interaction term (Environmental forces factors * social responsibility) PVA	0.120	0.054	0.096	Substantial effect	0.011	0.421	Accepted

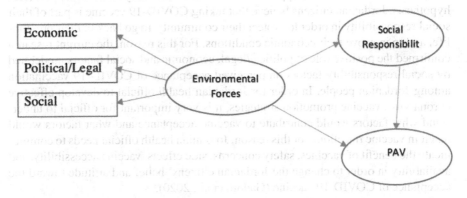

Fig. 1 Structural model

4 Discussions

The current study measures the acceptance rate of Jordanian citizen that would be vaccinated during the COVID-19 pandemic disease. The research stated high acceptance of COVID-19 vaccination among the Jordanian people during the pandemic disease. The study analysis showed that Jordanian with higher level of income and education were found to have more positive acceptance of COVID-19 vaccines. Moreover, the study results illustrate that people show more positive acceptance, when the government provide a trusted information such as the true effectiveness of the vaccine, the origin of the vaccine, the side effect of the vaccine, and how the vaccine works. Two out of three environmental forces affect perceived acceptance of COVID-19 vaccination namely political/legal and social factors. Whereas, economic factor shown no affect perceived acceptance of COVID-19 vaccination. The result of the study analysis supports the main hypothesis (H1), thus, H1 was accepted. Therefore, the study reported high acceptance of COVID-19 vaccination among Jordanian population if some conditions were applied. However, the study result did not support (H1a) hypothesis. The analysis showed that economic factors such as vaccine price, distance to vaccination sites, waiting time, and other related cost did not have an impact on Jordanian citizen to take the vaccine. Thus, the hypothesis (H1a) is rejected. The study analysis showed that political and legal factors have significant impact on vaccine acceptance. Therefore, the hypothesis (H1b) was accepted. Thus; national authorized, FDA & MOH authorization, and Employer powerful force play an important role for COVID-19 vaccine acceptance. The study analysis also showed that social factors have positive impact on citizen acceptance of COVID-19 vaccination. Consequently, family and friends can play a positive role in decision making to accept the vaccine. Moreover, positive word of mouth about the vaccination can influence people to accept the vaccine (Al Khasawneh et al., 2021b). Therefore, the study results accepted (H1c) hypothesis. In addition, the study analysis support (H2) hypothesis. Jordanian citizens believe that taking COVID-19 vaccine is part of their social responsibility in order to protect their community, to go back to their previous life, and to improve their economic conditions. For this reason, the current research confirmed the positive role of political/legal, economic and social factors moderated by social responsibility factors on perceived acceptance of COVID-19 vaccination among Jordanian people. In order for Jordanian health official to develop effective corona virus vaccine promotion strategies, it is very important for official to understand what factors would contribute to vaccine acceptance and what factors would result in vaccine rejection. For this reason, Jordanian health official needs to communicate the benefit of vaccines, safety concerns, side effects, vaccine accessibility, and availability in order to change the Jordanian citizens' belief and attitude toward the acceptance of COVID-19 vaccine (Gadoth et al., 2020).

The current study result recommends that FDA & MOH, nationally authorized of COVID-19 vaccines, employer powerful force and recommendations, friends and family, and positive words of mouth can all play an important role for vaccination acceptance and increase the percentages of Jordanian citizen to be vaccinated which was consistent with previous studies. Moreover, after the high number of deaths among the Jordanian citizen and the hard conditions that people suffer during corona virus, majority of Jordanian citizen changed their attitude toward COVID-19 vaccination. Thus, high percentage of acceptance and positive attitude toward corona virus vaccination reflected the Jordanian people strong demand for COVID-19 vaccination and the high recognition of the importance of vaccines in order to control the pandemics disease (Tian et al., 2020). Furthermore, people started to think seriously about the pros and cons of the vaccinations (Karafillakis & Larson, 2017). Jordanian residents felt the negative impact on their daily life, work, income, and the disease severity on their health and their beloved one. Hence, the majority of the Jordanian citizens thought that COVID-19 vaccination is an effective way to control and prevent corona virus from spreading (Leung et al., 2020). The positive behaviors and attitude toward vaccine, explain the high percentage of vaccine acceptance (Kraemer et al., 2020). Jordanian adults and married citizens perceived the benefit from corona virus vaccine compared with the risk associated on their health and life style (Bone et al., 2010).

5 Conclusions

The current research reflected the positive role of political/legal and social factors on perceived acceptance of COVID-19 vaccination among Jordanian people. The study findings represented a good start in the process of achieving the coverage rate required to ensure herd immunity among the Jordanian residents. Also, the research results are beneficial for formulating an effective polices and strategies for vaccination in Jordanian health market. Firstly, Activate the role of Jordanian FDA & MOH authorization as a critical factor of accept taking the vaccine. Besides, the Jordanian national authorization and laws regarding motivated Jordanian people to accept the vaccination. Secondly, trigger the role of social medial and word of mouth of social circle may influence acceptance of the vaccine. Lastly, the Jordanian government and health bodies should be improving the accessibility of vaccination in terms of the economies of taking the vaccine such as (distance to vaccination sites, waiting time etc.) because of it plays a role in the decision to take the vaccine. Jordanian health official showed design an immunization programs that can remove barriers that hinder people from getting vaccination. As well, health official should communicate with people about any public concern such as vaccine safety (Cheney & John, 2013). Public concern about vaccine safety has frequently been reported as the major obstacle to vaccination decision-making. Some of the Jordanian residents delay their vaccinations until safety of COVID-19 vaccine is confirmed. Therefore, Jordanian health officials need to use all the available tools to persuade people to take the vaccine and to stop the hesitations. Some of the reasons that explain people delay and hesitation are the concerned that the vaccine still under development, no real data and information about vaccine negative side effect and safety. In addition, people concern about the different news and reports about the safety of the vaccine.

Also, negative testimonies from doctors and pharmaceuticals would lower the confidence among public concerns. Besides, our study showed that people who valued convenience tend to delay their vaccination. Thus, Jordanian health officials need to monitor reports that concern vaccine safety with the publics on regular basis. Furthermore, officials can increase doctor and pharmaceutical testimonies regarding safety and future concerns.

To be proactive and respond successfully for any possible epidemic or pandemics in the future, health governing bodies in Jordan (MOH, Royal Medical Services, and Private Sectors) should take in to consideration training programs in terms of vaccination and continuous education for health service providers in general and health providers who working in Epidemiology field in particular.

References

Abbas, K., Procter, S. R., van Zandvoort, K., Clark, A., Funk, S., Mengistu, T., ... & Medley, G. (2020). Routine childhood immunisation during the COVID-19 pandemic in Africa: A benefit–risk analysis of health benefits versus excess risk of SARS-CoV-2 infection. *The Lancet Global Health, 8*(10), e1264–e1272.

Abuhashesh, M. Y., Alshurideh, M. T., Ahmed, A., Sumadi, M., & Masa'deh, R. (2021). The effect of culture on customers' attitudes toward Facebook advertising: The moderating role of gender. *Review of International Business and Strategy*, Vol. ahead-of-print No. ahead-of-print. https://doi.org/10.1108/RIBS-04-2020-0045

Ahmad, A., Alshurideh, M. T., Al Kurdi, B. H., & Salloum, S. A. (2021). Factors impacts organization digital transformation and organization decision making during Covid19 pandemic. In *The effect of coronavirus disease (COVID-19) on business intelligence* (pp. 95–106). Cham.

Akour, I., Alshurideh, M., Al Kurdi, B., Al Ali, A., & Salloum, S. (2021). Using machine learning algorithms to predict people's intention to use mobile learning platforms during the COVID-19 pandemic: Machine learning approach. *JMIR Medical Education, 7*(1), 1–17.

Al Khasawneh, M., Abuhashesh, M., Ahmad, A., Alshurideh, M. T., & Masa'deh, R. (2021a). Determinants of E-word of mouth on social media during COVID-19 outbreaks: An empirical study. *The effect of coronavirus disease (COVID-19) on business intelligence* (p. 347).

Al Khasawneh, M., Abuhashesh, M., Ahmad, A., Masa'deh, R., & Alshurideh, M. T. (2021b). Customers online engagement with social media influencers' content related to COVID 19. *The effect of coronavirus disease (COVID-19) on business intelligence* (p. 385).

Alameeri, K. A., Alshurideh, M. T., & Al Kurdi, B. (2021). The effect of Covid-19 pandemic on business systems' innovation and entrepreneurship and how to cope with it: A theatrical view. *The effect of coronavirus disease (COVID-19) on business intelligence* (Vol. 334, pp. 275–288).

Al-Dmour, H., Salman, A., Abuhashesh, M., & Al-Dmour, R. (2020). Influence of social media platforms on public health protection against the COVID-19 pandemic via the mediating effects of public health awareness and behavioral changes: Integrated model. *Journal of Medical Internet Research, 22*(8), e19996.

Aljumah, A., Nuseir, M. T., & Alshurideh, M. T. (2021). The impact of social media marketing communications on consumer response during the COVID-19: Does the brand equity of a university matter. *The effect of coronavirus disease (COVID-19) on business intelligence* (pp. 367–384).

Alshurideh, M. T., Kurdi, B. A., AlHamad, A. Q., Salloum, S. A., Alkurdi, S., Dehghan, A., Abuhashesh, M., & Masa'deh, R. (2021). Factors affecting the use of smart mobile examination platforms by universities' postgraduate students during the COVID 19 pandemic: An empirical study. *Informatics, 8*(2), 32.

Bone, A., Guthmann, J. P., Nicolau, J., & Lévy-Bruhl, D. (2010). Population and risk group uptake of H1N1 influenza vaccine in mainland France 2009–2010: Results of a national vaccination campaign. *Vaccine, 28*(51), 8157–8161.

Cheney, M. K., & John, R. (2013). Underutilization of influenza vaccine: A test of the health belief model. *SAGE Open, 3*(2), 2158244013484732.

Cobb, N. L., Sathe, N. A., Duan, K. I., Seitz, K. P., Thau, M. R., Sung, C. C., ... & Bhatraju, P. K. (2021). Comparison of clinical features and outcomes in critically ill patients hospitalized with COVID-19 versus influenza. *Annals of the American Thoracic Society, 18*(4), 632–640.

Czeisler, M. É., Tynan, M. A., Howard, M. E., Honeycutt, S., Fulmer, E. B., Kidder, D. P., & Czeisler, C. A. (2020). Public attitudes, behaviors, and beliefs related to COVID-19, stay-at-home orders, nonessential business closures, and public health guidance—United States, New York City, and Los Angeles, May 5–12, 2020. *Morbidity and Mortality Weekly Report, 69*(24), 751.

Dror, A. A., Eisenbach, N., Taiber, S., Morozov, N. G., Mizrachi, M., Zigron, A., ... & Sela, E. (2020). Vaccine hesitancy: The next challenge in the fight against COVID-19. *European Journal of Epidemiology, 35*(8), 775–779.

Fisher, K. A., Bloomstone, S. J., Walder, J., Crawford, S., Fouayzi, H., & Mazor, K. M. (2020). Attitudes toward a potential SARS-CoV-2 vaccine: A survey of US adults. *Annals of Internal Medicine, 173*(12), 964–973.

Fornell, C., & Larcker, D. F. (1981). Structural equation models with unobservable variables and measurement error: Algebra and statistics. *Journal of Marketing Research, 18*(3), 382–388.

Gadoth, A., Halbrook, M., Martin-Blais, R., Gray, A. N., Tobin, N. H., Ferbas, K. G., ... & Rimoin, A. W. (2020). Assessment of COVID-19 vaccine acceptance among healthcare workers in Los Angeles. *medRxiv.*

Gates, B. (2020). Responding to Covid-19—A once-in-a-century pandemic? *New England Journal of Medicine, 382*(18), 1677–1679.

Graffigna, G., Palamenghi, L., Boccia, S., & Barello, S. (2020). Relationship between citizens' health engagement and intention to take the covid-19 vaccine in Italy: A mediation analysis. *Vaccines, 8*(4), 576.

Hair, J. F., Ringle, C. M., & Sarstedt, M. (2012). Partial least squares: The better approach to structural equation modeling? *Long Range Planning, 45*(5–6), 312–319.

Hayes, A. F. (2017). *Introduction to mediation, moderation, and conditional process analysis: A regression-based approach.* Guilford publications.

Huang, C., Wang, Y., Li, X., Ren, L., Zhao, J., Hu, Y., ... & Cao, B. (2020). Clinical features of patients infected with 2019 novel coronavirus in Wuhan, China. *The Lancet, 395*(10223), 497–506.

Irwin, K. L., Jalloh, M. F., Corker, J., Mahmoud, B. A., Robinson, S. J., Li, W., ... & Marston, B. (2017). Attitudes about vaccines to prevent Ebola virus disease in Guinea at the end of a large Ebola epidemic: Results of a national household survey. *Vaccine, 35*(49), 6915–6923.

Jordan, J., Yoeli, E., & Rand, D. (2020). Don't get it or don't spread it? Comparing self-interested versus prosocially framed COVID-19 prevention messaging. PsyArXiv. https://doi.org/10.31234/osf.io/yuq7x

Karafillakis, E., & Larson, H. J. (2017). The benefit of the doubt or doubts over benefits? A systematic literature review of perceived risks of vaccines in European populations. *Vaccine, 35*(37), 4840–4850.

Kraemer, M. U., Yang, C. H., Gutierrez, B., Wu, C. H., Klein, B., Pigott, D. M., ... & Scarpino, S. V. (2020). The effect of human mobility and control measures on the COVID-19 epidemic in China. *Science, 368*(6490), 493–497.

Kreps, S., Prasad, S., Brownstein, J. S., Hswen, Y., Garibaldi, B. T., Zhang, B., & Kriner, D. L. (2020). Factors associated with US adults' likelihood of accepting COVID-19 vaccination. *JAMA Network Open, 3*(10), e2025594–e2025594.

Kwok, K. O., Lai, F., Wei, W. I., Wong, S. Y. S., & Tang, J. W. (2020). Herd immunity–estimating the level required to halt the COVID-19 epidemics in affected countries. *Journal of Infection, 80*(6), e32–e33.

Lazarus, J. V., Ratzan, S. C., Palayew, A., Gostin, L. O., Larson, H. J., Rabin, K., ... & El-Mohandes, A. (2021). A global survey of potential acceptance of a COVID-19 vaccine. *Nature Medicine, 27*(2), 225–228.

Leo, S., Alsharari, N. M., Abbas, J., & Alshurideh, M. T. (2021). From offline to online learning: A qualitative study of challenges and opportunities as a response to the COVID-19 pandemic in the UAE higher education context. In *The effect of coronavirus disease (COVID-19) on business intelligence* (pp. 203–217). Cham.

Leung, K., Wu, J. T., Liu, D., & Leung, G. M. (2020). First-wave COVID-19 transmissibility and severity in China outside Hubei after control measures, and second-wave scenario planning: A modelling impact assessment. *The Lancet, 395*(10233), 1382–1393.

Nuseir, M. T., Aljumah, A., & Alshurideh, M. T. (2021). How the business intelligence in the new startup performance in UAE during COVID-19: The mediating role of innovativeness. *The effect of coronavirus disease (COVID-19) on business intelligence* (Vol. 334, pp. 63–79).

Opel, D. J., Salmon, D. A., & Marcuse, E. K. (2020). Building trust to achieve confidence in COVID-19 vaccines. *JAMA Network Open, 3*(10), e2025672–e2025672.

Reiter, P. L., Pennell, M. L., & Katz, M. L. (2020). Acceptability of a COVID-19 vaccine among adults in the United States: How many people would get vaccinated? *Vaccine, 38*(42), 6500–6507.

Rosenstock, I. M. (1974). The health belief model and preventive health behavior. *Health Education Monographs, 2*(4), 354–386.

Shah, S. F., Alshurideh, M. T., Al-Dmour, A., & Al-Dmour, R. (2021). Understanding the influences of cognitive biases on financial decision making during normal and COVID-19 pandemic situation in the United Arab Emirates. *The effect of coronavirus disease (COVID-19) on business intelligence* (Vol. 334, pp. 257–274).

Streiner, D. (2003). Starting at the beginning: An introduction to coefficient alpha and internal consistency. *Journal of Personality Assessment, 80*(1), 99–103. https://doi.org/10.1207/S15327 752JPA8001_18

Tam, C. C., Qiao, S., & Li, X. (2020). Factors associated with decision making on COVID-19 vaccine acceptance among college students in South Carolina. *medRxiv*.

Taryam, M., Alawadhi, D., Aburayya, A., Albaqa'een, A., Alfarsi, A., Makki, I., ... & Salloum, S. A. (2020). Effectiveness of not quarantining passengers after having a negative COVID-19 PCR test at arrival to Dubai airports. *Systematic Reviews in Pharmacy, 11*(11), 1384–1395.

Tian, H., Liu, Y., Li, Y., Wu, C.H., Chen, B., Kraemer, M.U., ... & Dye, C. (2020). An investigation of transmission control measures during the first 50 days of the COVID-19 epidemic in China. *Science, 368*(6491), 638–642.

Wang, C., Pan, R., Wan, X., Tan, Y., Xu, L., McIntyre, R. S., ... & Ho, C. (2020). A longitudinal study on the mental health of general population during the COVID-19 epidemic in China. *Brain, Behavior, and Immunity, 87*, 40–48.

Wilkinson, L. (1999). Statistical methods in psychology journals: Guidelines and explanations. *American Psychologist, 54*(8), 598.

Wise, T., Zbozinek, T. D., Michelini, G., Hagan, C. C., & Mobbs, D. (2020). Changes in risk perception and self-reported protective behaviour during the first week of the COVID-19 pandemic in the United States. *Royal Society Open Science, 7*(9), 200742.

Ebb and Flow Theory in Tourism, Hospitality, and Event Management

Omar A. Alananzeh ⓘ, Ra'ed Masa'deh ⓘ, and Ibrahim K. Bazazo ⓘ

Abstract This study aims to track the events that affected tourism in Jordan and presents the evolution of tourist inflows and outflows linked to these events. Accordingly, the Ebb and Flow Theory in the service industry is created and discussed. Many terminologies were used in the tourism industry to clarify the incoming and outgoing tourism movement, such as "peak period and recession period", "high–shoulder and low-shoulder", "high-season and low-season", and others. This study sheds light on a new insight that accurately and clearly describes what is happening in the tourism, hospitality, and events industry. It describes the period of prosperity and development in the tourism movement as the Flow, and the period of stagnation and confinement in the global tourist movement as the Ebb. Also, the fluctuation between these two periods is called solstice time. Understanding these periods helps organizations develop their strategies, achieve the desired benefit, improve and innovate in providing optimal service.

Keywords Strategic planning · Ebb and flow theory · Destination · Tourism · Hospitality · Events · Jordan

O. A. Alananzeh (✉)
Department of Hotel Management, Faculty of Tourism and Hotel Management, Yarmouk University, Irbid, Jordan
e-mail: o.ananzeh@yu.edu.jo

R. Masa'deh
Department of Management Information Systems, The University of Jordan, Amman, Jordan
e-mail: r.masadeh@ju.edu.jo

I. K. Bazazo
Department of Travel and Tourism, Faculty of Tourism and Hospitality, The University of Jordan, Amman, Jordan
e-mail: i.bazazo@ju.edu.jo

M. Alshurideh et al. (eds.), *The Effect of Information Technology on Business and Marketing Intelligence Systems*, Studies in Computational Intelligence 1056, https://doi.org/10.1007/978-3-031-12382-5_130

1 Introduction

The economy of many countries depends on a specific industry, but the tourism and hospitality industry are considered one of the basic industries in all countries of the world, regardless if this country is rich in its natural resources or has a scarcity of natural or human resources (Abu Zayyad et al., 2020; Abuhashesh et al., 2019; Alaali et al., 2021; Alananzeh, 2017; Mahmoud et al., 2021). It is undoubtedly one of the largest, most growing, and prosperous industries in the global economy, and it clearly competes with the petroleum or technological industries for some countries (Khan et al., 2020). It is expected that it will maintain its position in the future. Tourism is not a goal but a means to contribute to comprehensive national development, and it is by its nature a multi-lateral and interdependent industry (Mihalic, 2014). Nevertheless, the tourism industry is highly sensitive to various internal and external shocks and crises, such as economic fluctuations, natural disasters such as earthquakes, earthquakes, fires, health epidemics, terrorism, and interstate conflicts (Al-Dmour et al., 2017; Jiang et al., 2019; Sönmez et al., 1999).

It is no longer seen as a secondary sector in the economies of countries, due to its importance in GDP and its role in adjusting the balance of payments, which is usually done through the availability of foreign currencies that enables the country to import its basic needs (Manzoor et al., 2019). This industry witness's rapid growth and development in many countries that have done well to preserve the stability of this sector and immunize it against various natural, political, economic, and health shocks and crises (Breier et al., 2021; Chiu et al., 2019; Obeidat et al., 2017, 2019). Therefore, maintaining a stable and accelerated growth in the tourism sector is not a result of luck and chance in the countries that are successful in tourism, but rather the product of policies, strategies, and plans that are thoughtful and pre-placed within what is called tourism crisis management (Perles-Ribes et al., 2016), which examines the theories, variables, contributors and affected parties (Alshraideh et al., 2017), and procedures that can be taken to avoid or reduce the damage resulting from various crises and disasters before, during and after the crisis for the quick return, recovery, and advancement of tourism after any emergency (Al-Dmour et al., 2021; Alrowwad et al., 2020; Alzoubi et al., 2020, 2022; Masadeh et al., 2019; Masa'deh et al., 2013, 2019; Mashaqi et al., 2020; Tarhini et al., 2015).

That crisis management lies through protecting and preserving the various sectors of this industry and achieving the optimal use of the country's resources during the peak season, or the so-called high-season, as the tourist destination is witnessing a tremendous influx in the number of tourists, which will reflect positively on the prosperity and economic recovery of that tourist destination and at the same time cause some damage or negative effects to the tourist destination (Adamou & Clerides, 2009). And the safe speed of managing crises and risks to which the tourism industry may be exposed in a tourist country or countries of the whole world if that crisis is global as it is now in the Coronavirus pandemic (Covid-19) (Breier et al., 2021). Sometimes we need to manage the regular crisis, which is meant by the natural expectation of a decrease in the number of tourists resulting from natural causes such as high or low temperatures that affect the attractiveness of the tourist destination or what is sometimes called a low-season, which is the decline in the number of tourists to that destination (Aqqad et al., 2019; Breier et al., 2021; Rossi, 2012; Sausmarez, 2007; Uysal & Kılıç, 2020). The impact of crises is also severe on hotel employees, as many companies are laying off a number of employees due to their economic conditions, and this affects the loyalty of employees as well as the orientation of the workforce towards the hospitality sector (Hunaiti et al., 2009; Masa'deh et al., 2017a, 2017b).

2 Ebb and Flow Theory

Ebb and flow (tides) are very long waves that move across the oceans under the influence of the moon and the sun, and when these waves advance towards the coast they appear as a regular rise and fall of sea water, and the reason for these waves is the earth's rotation around itself, and the moon's rotation around the earth. This is as a result of the gravitational pull of the moon water in the oceans towards it, and the tide occurs in the face of the moon, and the opposite side to it (National Ocean Service, 2021).

In other words, Flow is a temporary, gradual rise in the water level of an ocean or sea surface. Ebb is a temporary, gradual decrease in the water level of the surface of the ocean or the sea. The fisherman understands (sectors of the tourism and hospitality industry) from these fish (the tourism movement), they leave their dens (tourists' origin) to eat at the time of the flow (prosperity, prosperity, and peace) and return again at the time of the ebb (disasters and crises). This means that stakeholders of the tourism and hospitality industry must know how to make the most of the flow period and how to deal with strategic planning to manage crises and disasters in the ebb.

In the tourism and hospitality industry, tourism companies rely in particular on the theory of Ebb and Flow in their investments and prosperity. Where the flow indicates the influx of tourists, and therefore a large tourism activity that requires tourism movement and congestion in the means of tourist transport. From here, the tourism companies begin to provide competitive services and coordination with the providers of tourism and travel offices and other tourism companies. The flow is considered joy and prosperity season for tourist transport companies. But every flow must be followed any an ebb, which is a decrease in the number of tourists, or the so-called low season, and here companies must follow strategic plans to deal with the ebb. However, the Corona pandemic period was a long ebb, which required the government to intervene to save the transport sector until the flow came.

The main causes of the ebb in tourist transportation are world wars and the flu epidemic, starting from a year 1918 where there is a comparison between the results of the current pandemic and those disasters in a wide range of consequences. Unfortunately, we can expect other major disturbances in the future (Hendrickson & Rilett, 2020; Jiang et al., 2017).

The tourism industry has gone through many crises over the past 20 years (Ebb). In 2003 during the SARS epidemic, the events of September 11, 2001, and the global economic and financial crisis in 2009. But after these crises, the tourism and hospitality industry will return to its previous state, and perhaps much better, as it flourishes and revives strongly and quickly (Flow).

3 Importance of Ebb and Flow

Ebb and Flow movements are of great importance as they work to cleanse the seas and oceans of all impurities, as well as cleanse estuaries and ports of sediments, as they help ships to enter ports that are located in shallow areas. But the high flow may pose a threat to navigation, especially in the straits.

The phenomenon of Ebb and Flow is one of the natural phenomena that occur in all the seas of the world, and that the ratio of high flow and low ebb varies from the North Pole to the Antarctic through the equator, in some areas of the world it reaches more than 200 cm, and in other areas it does not exceed 30 cm. In other words, in some places, the difference between ebb and flow is not very large, and in others, the difference is huge.

Since ancient times, people have been alerted to the importance of the power of the ebb and flow and have used this in managing their mills to grind grain. The ruins of these mills are still found on the shores of Brittany, in northern France, since the twelfth century AD. The idea by which these mills operate is very simple and is summarized in reserving flow water in a reservoir during high tide, and when the reservoir is filled with water, special gates are closed so that the water level in the reservoir is higher than the sea level when the water begins to recede.

In the tourism, hospitality, and events sector, this happens exactly, as the tourist destination witnesses an active tourism movement, and then this movement begins to decline and shrink. That is, the tourist destinations, in general, go through an ebb and flow time. In a recession period, establishments, tourist destinations, and service facilities can take advantage of this period for strategic planning, training, reconstruction, renewal, and general development in preparation for the period of the return of tourism.

As for the flow time, tourism establishments, service facilities, and tourist destinations must focus on Total Quality Management (TQM) in order to maintain competitiveness, market share, and pay attention to customer satisfaction. The success of tourism destinations and service facilities depends on the ability to strategic planning, which is considered a comprehensive process through which future actions and goals are defined and what steps are necessary to achieve these developmental and competitive goals.

Strategic planning needs to address a number of things during the ebb time for the development of the tourist destination, such as the environment and the impact of tourism on it during the flow time, improving the traffic-related infrastructure, especially in major city centers, and addressing the causes of pressure in some tourist

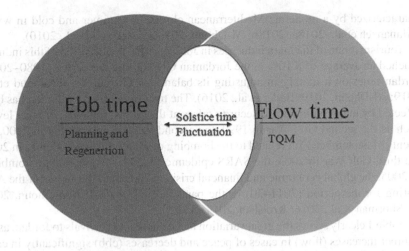

Fig. 1 Ebb and flow theory and its implementation (*Source* Prepared by the researchers)

areas. It makes strategic planning for the flow of tourists easier and does not affect the living conditions of the local population. Strategic planning also helps in maintaining the various service facilities and exploiting them in a healthy way, which can be the main driving force behind the revitalization of tourist destinations and their economic growth, especially during the transition process that destinations go through in the process of transition from ebb time to flow time (Jurdana, 2018) (Fig. 1).

On the other hand, tourism destinations and service establishments such as hotels and restaurants are responsible for total quality management during flow time, which means paying attention to all details and the involvement of all employees in achieving quality required to maintain the level of excellence in providing an ideal competitive service (Mok et al., 2013).

This is achieved by evaluating the policy and procedures followed to develop and implement comprehensive quality requirements and guarantee them that help the company in attracting a large number of tourists regardless of the reasons for travel, whether for business or leisure purposes. The expectations of tourists are high and they demand a certain level of quality; customers are generally willing to pay more for quality.

4   Tourism Industry in Jordan as a Case Study

Jordan is located in the west of the Asian continent, in the heart of the Middle East, and is bordered to the north by Syria, to the east by Iraq and Saudi Arabia, to the south by Saudi Arabia, and to the west by Occupied Palestine, and it has one seaport, which is the Gulf of Aqaba on the Red Sea. Jordan has an area of 89,342 km^2, and its most important natural resources are phosphates, potash, and shale oil. It is characterized by a moderate Mediterranean climate in summer and cold in winter (Alananzeh et al., 2018a, 2018b; Al-Dhoun, 2019; Alsarayreh et al., 2010).

Tourism is one of the main industries in Jordan, as the contribution of this industry reached an average of 8.10% in the Jordanian GDP during the period (1980–2014). Jordan relies on it mainly in adjusting its balance of payments (Abuamoud et al., 2019; Al-Dhoun, 2019; Bader et al., 2016). The tourism industry in Jordan has been affected by many political and security crises at the global, regional, or local levels, such as the second Gulf War in 1990, the second Palestinian uprising in 2000, the events of September 11th related to the bombing of the World Trade Center in 2001, the third Gulf War in 2003, the SARS epidemic IN 2003, Amman hotel bombings in 2005, the global economic and financial crisis in 2009, and the events of the Arab Spring for the period (2011–2014), the pandemic of Covid-19 (Al-Dhoun, 2019; Al-Shorman et al., 2016; Kreishan, 2014).

Table 1 clearly shows the great variation in the numbers of arrivals to Jordan, as the number increases (flow) in cases of peace and decreases (ebb) significantly in cases of crisis whether local, regional, or global. We also note through the statistics that the events that take place in Egypt affect quickly the tourism movement in Jordan for several reasons, including the proximity of Jordan to Egypt and the joint cooperation between the tourism offices in the two countries, as well as the common cultural, religious and civilization factors between Jordan and Egypt.

Figure 2 shows the impact of events and crises on the tourist movement (ebb and flow) coming to Jordan. We notice that there are ebb movements that have negatively affected the tourism movement in Jordan, for example the events of the end of the year 2000, The Second Palestinian Intifada (9-28-2000), which affected tourist areas in the West, but their impact was positive on Jordan. However, some crises and events in the other part of the ball affect positively the tourist, as the flow of tourists begins to the opposite part of the globe (flow movement). The best example of this is the events of September 11, which affected tourist areas in the West, but their impact was positive on Jordan, where the number of arrivals to Jordan increased due to this, and this was confirmed by the study of Al-Shorman et al. (2016), who indicated that Jordan has become a safe haven for tourists in these global crises. This increase was also accompanied by the visit of Pope Paul II to the baptism site, which reflected positively on the movement of tourists to the region (flow).

However, the occurrence of local and regional crises has a negative impact on the flow of tourism to Jordan (the movement of islands), such as the events of the Arab Spring. It leads to the influx of tourists to the other side of the globe (tidal movement).

5 Conclusion

Since the dawn of time, tourist destinations have been affected by the movement of ebb and flow in the movement of tourists. As the movement of flow meant prosperity, development, and economic and social returns for the tourist destination. And the

Table 1 Major events that impacted tourism industry in Jordan

Year	Number of arrivals	The major event	Its impact
1990	0,577.446	Peace and stability	
1991	0,439.481	The second Gulf war (1990-191)	Decrease
1992	0,699.262	Peace and stability	Increase
1993	0,774.871	Peace and stability	Increase
1994	0,857.61	Peace and stability	Increase
1995	1.074.548	Jordan-Israel peace treaty (26-10-1994)	Increase
1996	3, 164,000	Peace and stability	Increase
1997	3,068,000	Local protests linked to soaring bread prices (18-8-1996)	Decrease
1998	3,484,000	The establishment of the Jordan Tourism Board	Increase
1999	3,231,000	Peace and stability	Decrease
2000	2,700,000	The Second Palestinian Intifada (28-9-2000)	Decrease
2001	3,034,000	September 11th (11-9-2001) The visit of Pope Paul II to the Baptism Site	Increase
2002	4,677,000	Peace and stability	Increase
2003	4,600,000	The third Gulf war	Increase
2004	5,387,000	Peace and stability	Increase
2005	5,817,000	Amman Hotel bombing	Increase
2006	6,573,000	War on Hezbollah	Increase
2007	6,529,000	Peace and stability	Decrease
2008	7,100,000	Petra became one of the Seven Wonders of the World in 2007	Increase
2009	7,058,000	The war on Gaza, 27-12-2008	Decrease
2010	8,078,000	The visit of Pope Pope Benedict II to the Baptism Site in 2009	Increase
2011	6,813,000	Civil war in Syria Arab Spring 2010-2011	Decrease
2012	6,314,000	Impact of Arab Spring	Decrease
2013	5,389,000	Egypt-The dismissal of Egyptian President Mohamed Morsi The organization of the Islamic state- Daesh (ISIS)	Decrease
2014	5,327,000	The war on Gaza (2014)	Decrease
2015	4,809,000	The war on Daesh (ISIS)	Decrease
2016	4,236,000	Brussels attacks	Decrease
2017	4,565,000	Peace and stability	Increase
2018	4,922,000	Peace and stability	Increase
2019	5,361,000	Peace and stability	Increase
2020	1, 800,000	Covid-19	Decrease

Sources (1) The World Bank (The World Bank, 2021), (2) Central Bank of Jordan, annual reports (Central Bank of Jordan, 2021)

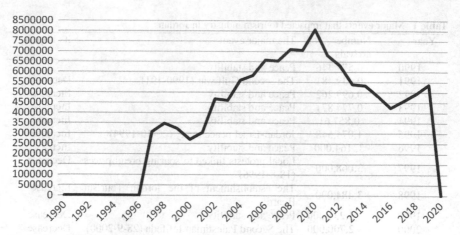

Fig. 2 International inbound tourists (overnight visitors)

flow of traffic may sometimes be above the capacity of the tourist destination, causing the loss of opportunities and negative impacts on the tourist destination and the host community. This is what strongly imposes on those tourist destinations that witness a tremendous turnout of tourists to plan to make the most of this flow movement and reduce the allowance for loss and negative impact.

On the other hand, some tourism destinations are facing at the same time a movement of decline and retreat in the tourism movement (ebb), and this is reflected negatively on the operational process of the facilities and the tourism projects in that tourist destination. Here, too, those countries must follow strategies and contingency plans to manage these crises in order to save the tourist season during the recession.

In the end, ebb and flow times have both benefits and downsides. However, tourist destinations, including tourism facilities and projects, must plan to take advantages of the flow time through the optimal absorption of the increase in the number of tourists, the optimal use of existing resources and resources, and the preservation of the permanence of those resources, focusing on reputation and service quality, and also following the method used in the shores of Brittany, in northern France, in storing water in the flow time to operate the mills in the ebbing time. Alternatively, those tourist destinations that suffer from stagnation in the tourism movement due to seasonality or local, regional, or global events and crises must reduce the risks and save the tourist season by following a developmental strategy and taking advantage of the stock of the peak period (flow time), as well as following the policy of beach cities, which adopts a policy cleaning, arranging, rebuilding and improving the beaches during the ebbing time, because that helps well. Those tourist destinations can follow this policy and strategy in re-developing, cleaning, and expanding the tourist sites in preparation for the flow time.

References

Abu Zayyad, H., Obeidat, Z. M., Alshurideh, M. T., Abuhashesh, M., & Maqableh, M. (2020). Corporate social responsibility and patronage intentions: The mediating effect of brand credibility. *Journal of Marketing Communications*. https://doi.org/10.1080/13527266.2020.1728565

Abuamoud, I., Ibrahim, A., & Hijawi, L. (2019). Estimating the economic impact of tourism in the north of Jordan through the I-O approach. *European Research Studies Journal, XXII*(1), 254–266.

Abuhashesh, M., Al-Khasawneh, M., & Al-Dmour, R. (2019). The impact of facebook on Jordanian consumers' decision process in the hotel selection. *IBIMA Business Review*. https://doi.org/10.5171/2019.928418,2019

Adamou, A., & Clerides, S. (2009). Tourism, development and growth: International evidence and lessons for Cyprus. *Cyprus Economic Policy Review, 3*(2), 3–22.

Alaali, N., Al Marzouqi, A., Albaqaeen, A., Dahabreh, F., Alshurideh, M., Alrwashdh, S., Iyadeh, I., Salloum, S., & Aburayya, A. (2021). The impact of adopting corporate governance strategic performance in the tourism sector: A case study in the Kingdom of Bahrain. *Journal of Legal, Ethical and Regulatory, 24*(Special Issue 1), 1–18.

Alananzeh, O. (2017). The impact of safety issues and hygiene perceptions on customer satisfaction: A case study of four and five star hotels in Aqaba, Jordan. *Journal of Tourism and Hospitality Research, 6*(1), 1–7.

Alananzeh, O., Maaiah, B., Al-Badarneh, M., & Al-Mkhadmeh, A. (2018a). The effect of hotel attributes on length of stay and hotel choices in coastal cities: Aqaba as a case study. *Dirasat, Human and Social Sciences, 45*(4), 275–289.

Alananzeh, O., Jawabreh, O., Al Mahmoud, A., & Hamada, R. (2018b). The impact of customer relationship management on tourist satisfaction: the case of Radisson blue resort in Aqaba city. *Journal of Environmental Management and Tourism, 9*(2), 227–240. https://doi.org/10.14505/jemt.v9.2(26).02

Al-Dhoun, R. M. (2019). The effect of security and political events on incoming package tourism to Jordan for the period (1989–2014). *Dirrasat, Humanities and Social Sciences, 46*(1).

Al-Dmour, R. H., Masa'deh, R., & Obeidat, B. Y. (2017). Factors influencing the adoption and implementation of HRIS applications: Are they similar? *International Journal of Business Innovation and Research, 14*(2), 139–167. https://doi.org/10.1504/IJBIR.2017.086276

Al-Dmour, R., AlShaar, F., Al-Dmour, H., Masa'deh, R., & Alshurideh, M. T. (2021). The effect of service recovery justices strategies on online customer engagement via the role of "customer satisfaction" during the covid-19 pandemic: An empirical study. *The Effect of Coronavirus Disease (COVID-19) on Business Intelligence, 334*, 325–346.

Alrowwad, A., Abualoush, S. H., & Masa'deh, R. (2020). Innovation and intellectual capital as intermediary variables among transformational leadership, transactional leadership, and organizational performance. *Journal of Management Development, 39*(2), 196–222. https://doi.org/10.1108/JMD-02-2019-0062

Alsarayreh, M. N., Jawabreh, O. A., & Helalat, M. S. (2010). The influence of terrorism on the international tourism activities. *European Journal of Social Sciences, 13*(1), 145–160.

Al-Shorman, A., Rawashdih, A., Makhadmih, A., Oudat, A., & Darabsih, F. (2016). Middle Eastern political instability and Jordan's tourism. *Journal of Tourism Research and Hospitality, 5*(1). https://doi.org/10.4172/2324-8807.1000155

Alshraideh, A. T. R., Al-Lozi, M., & Alshurideh, M. T. (2017). The impact of training strategy on organizational loyalty via the mediating variables of organizational satisfaction and organizational performance: An empirical study on Jordanian agricultural credit corporation staff. *Journal of Social Sciences (COESandRJ-JSS), 6*(2), 383–394.

Alzoubi, H., Alshurideh, M., Kurdi, B., Akour, I., & Aziz, R. (2022). Does BLE technology contribute towards improving marketing strategies, customers' satisfaction and loyalty? The role of open innovation. *International Journal of Data and Network Science, 6*(2), 449–460.

Alzoubi, H. M., Alshurideh, M., Al Kurdi, B., & Inairat, M. (2020). Do perceived service value, quality, price fairness and service recovery shape customer satisfaction and delight? A practical study in the service telecommunication context. *Uncertain Supply Chain Management, 8*(3), 579–588.

Aqqad, N., Obeidat, B., & Tarhini, A. (2019). The relationship among emotional intelligence, conflict management styles, and job performance in Jordanian banks. *International Journal of Human Resources Development and Management, 19*(3), 225–265. https://doi.org/10.1504/IJHRDM.2019.100636

Bader, M., Alrousan, R., Abuamoud, I., & Alasal, H. A. (2016). Urban tourism in Jordan: Challenges and opportunities case study: Amman. *British Journal of Economics, Management and Trade, 12*(4), 1–11. https://doi.org/10.9734/BJEMT/2016/24589

Breier, M., Kallmuenzer, A., Clauss, T., Gast, J., Kraus, S., & Tiberius, V. (2021). The role of business model innovation in the hospitality industry during the COVID-19 crisis. *International Journal of Hospitality Management, 92.* https://doi.org/10.1016/j.ijhm.2020.102723

Central Bank of Jordan. (2021). https://www.cbj.gov.jo/

Chiu, L. K., Ting, C. S., Alananzeh, O., & Hua, K. P. (2019). Perceptions of risk and outbound tourism and travel intentions among young working Malaysians. *Dirasat, Human and Social Sciences, 46*(1), 365–379.

Hendrickson, C., & Rilett, L. (2020). The COVID-19 pandemic and transportation engineering. *Journal of Transportation Engineering, Part A: Systems, 146*, 01820001. https://doi.org/10.1061/JTEPBS.0000418

Hunaiti, Z., Masa'deh, R. M., Mansour, M., & Al-Nawafleh, A. (2009). *Electronic commerce adoption barriers in small and medium-sized enterprises (SMEs) in developing countries: The case of Libya.* Paper presented at the Innovation and Knowledge Management in Twin Track Economies Challenges and Solutions—Proceedings of the 11th International Business Information Management Association Conference, IBIMA 2009, (Vol. 1–3, pp. 1375–1383).

Jiang, Y., Ritchie, B. W., & Eichendorff, P. (2017). Bibliometric visualization: An application in tourism crisis and disaster management research. *Current Issues in Tourism.* https://doi.org/10.1080/13683500.2017.1408574

Jiang, Y., Ritchie, B., & Verreynne, M. (2019). Building tourism organizational resilience to crises and disasters: A dynamic capabilities view. *International Journal of Tourism Research, 21.* https://doi.org/10.1002/jtr.2312

Jurdana, D. S. (2018). Strategic planning of tourist development towards sustainability. *Series a, Social Sciences and Humanities, 23*, 239–248. https://doi.org/10.20544/HORIZONS.A.23.2.18.P17

Khan, N., Hassan, U. A., Fahad, S., & Naushad, M. (2020). Factors affecting tourism industry and its impacts on global economy of the world. *SSRN Electronic Journal.* https://ssrn.com/abstract=3559353

Kreishan, F. (2014). The economics of tourism in Jordan: A statistical study during the period 1990–2011. *Arab Economic and Business Journal, 9*, 37–45.

Mahmoud, R., Al-Mkhadmeh, A., Alananzeh, O. A., & Masa'deh, R. (2021). Exploring the relationship between human resource management practices in the hospitality sector and services innovation in Jordan: The mediating role of Human Capital. *GeoJournal of Tourism and Geosites, 35*(2), 507–514. https://doi.org/10.30892/gtg.35231-678

Manzoor, F., Wei, L., Asif, M., Haq, M. Z., & Rehman, H. (2019). The contribution of sustainable tourism to economic growth and employment in Pakistan. *International Journal of Environmental Research and Public Health, 16*(19), 3785.

Masa'deh, R., Alananzeh, O., Algiatheen, N., Ryati, R., Albayyari, R., & Tarhini, A. (2017a). The impact of employee's perception of implementing green supply chain management on hotel's economic and operational performance. *Journal of Hospitality and Tourism Technology, 8*(3), 395–416. https://doi.org/10.1108/JHTT-02-2017-0011

Masa'deh, R., Alananzeh, O., Tarhini, A., & Algudah, O. (2017b). The effect of promotional mix on hotel performance during the political crisis in the Middle East. *Journal of Hospitality and Tourism Technology, 9*(1), 32–47. https://doi.org/10.1108/JHTT-02-2017-0010

Masa'deh, R., Shannak, R. O., & Maqableh, M. (2013). A structural equation modeling approach for determining antecedents and outcomes of students' attitude toward mobile commerce adoption. *Life Science Journal, 10*(4), 2321–2333.

Masa'deh, R., Alananzeh, O., Jawabreh, O., Alhalabi, R., Syam, H., & Keswani, F. (2019). The association among employees' communication skills, image formation and tourist behaviour: Perceptions of hospitality management students in Jordan. *International Journal of Culture, Tourism, and Hospitality Research, 13*(3), 257–272. https://doi.org/10.1108/IJCTHR-02-2018-0028

Masadeh, R., Almajali, D. A., Alrowwad, A., & Obeidat, B. (2019). The role of knowledge management infrastructure in enhancing job satisfaction: A developing country perspective. *Interdisciplinary Journal of Information, Knowledge, and Management, 14*, 1–25. https://doi.org/10.28945/4169

Mashaqi, E., Al-Hajri, S., Alshurideh, M., & Al Kurdi, B. (2020). The impact of E-Service quality, E-Recovery services on E-Loyalty in online shopping: Theoretical foundation and qualitative proof. *PalArch's Journal of Archaeology of Egypt/Egyptology, 17*(10), 2291–2316.

Mihalic, T. (2014). Tourism and economic development issues. In R. Sharpley & D. Telfer, J. (Eds.), *Tourism and development. Concepts and issues* (2nd ed., pp. 77–117). Channel.

Mok, C., Sparks, B., & Kadampully, J. (2013). *Service quality management in hospitality, tourism, and leisure* (1st ed.). Routledge. https://doi.org/10.4324/9780203047965

National Ocean Service. (2021). *How frequent are tides?* https://oceanservice.noaa.gov/facts/tidefrequency.html

Obeidat, B. Y., Al-Hadidi, A., & Tarhini, A. (2017). Factors affecting strategy implementation: A case study of pharmaceutical companies in the Middle East. *Review of International Business and Strategy, 27*(3), 386–408. https://doi.org/10.1108/RIBS-10-2016-0065

Obeidat, Z. M., Alshurideh, M. T., & Al Dweeri, R. (2019). *The influence of online revenge acts on consumers psychological and emotional states: Does revenge taste sweet?* Paper presented at the Proceedings of the 33rd International Business Information Management Association Conference, IBIMA 2019: Education Excellence and Innovation Management through Vision 2020 (pp. 4797–4815).

Perles-Ribes, J. F., Ramón-Rodríguez, A. B., Sevilla-Jiménez, M., & Rubia, A. (2016). The effects of economic crises on tourism success: An integrated model. *Tourism Economics, 22*(2), 417–447.

Rossi, C. (2012). *Tourism security and destination crisis management. Competition and innovation in tourism: New challenges in an uncertain world* (A. Morvillo, Ed.). Enzo Albano Editore.

de Sausmarez, N. (2007). Crisis management, tourism and sustainability: The role of indicators. *Journal of Sustainable Tourism, 15*. https://doi.org/10.2167/jost653.0

Sönmez, S. F., Apostolopoulos, Y., & Tarlow, P. (1999). Tourism in crisis: Managing the effects of terrorism. *Journal of Travel Research, 38*(1), 13–18.

Tarhini, A., Mgbemena, C., & Trab, M. S. A. (2015). User adoption of online banking in Nigeria: a qualitative study. *Journal of Internet Banking and Commerce, 20*(3). https://doi.org/10.4172/1204-5357.1000132

The World Bank. (2021). *International tourism, number of arrivals—Jordan.* Retrieved 2021, from https://data.worldbank.org/indicator/ST.INT.ARVL?locations=JO

Uysal, D., & Kılıç, L. (2020). How well do Turkey-based travel agencies manage the Covid-19 pandemic crisis? *Tourism Academic Journal, 02*, 339–354.

The Impact of Marketing, Technology and Security Orientations on Customer Orientation: A Case Study in Jordan

Omar Jawabreh and Ra'ed Masa'deh

Abstract The aims of this study are multifold: to explore the functionality of hotel websites by performing a content review and to evaluate customer opinions on the value of the design of hotel websites. To perform the study, hotel websites were used to include numerical and visual information. Information providers are assessed by an evaluation of their particular and comprehensive expertise measurements. The research is based on two significant aspects: a report on all 37 five-star hotels in Jordan that examines their website layout in terms of population, and a survey of 100 hotel clients to determine how clients interpret hotel websites. This planned study is the first of its sort, seeking to determine an appraisal method for all hotel measurements provided by hotel websites. The findings are as follows: When value T = 7.519 reflects the results of the independent variable (marketing orientation), the coefficient of simple linear regression has statistical significance (Customer Orientation). The fact that "Aqaba Zone Advertising and Customer Orientation Classified Hotels have a statistically significant influence on their categorization" is acknowledged, as "Aqaba Zone Advertising and Customer Orientation Classified Hotels have a substantial impact on the coefficient of simple linear regression equation (0.605)". As a result, hotel management may utilize a particular predictive evaluation model to analyze the websites of their hotels from a two-sided perspective: the management team and the consumers. This analysis brings together claims from both perspectives for the first time (hotel managers and clients).

Keywords Marketing orientation · Technology orientation · Security orientation · Customer orientation · Jordan

O. Jawabreh (✉)
Department of Hotel Management, Faculty of Tourism and Hospitality, The University of Jordan, Amman, Jordan
e-mail: o.jawabreh@ju.edu.jo

R. Masa'deh
Department of Management Information Systems, School of Business, The University of Jordan, Amman, Jordan
e-mail: r.masadeh@ju.edu.jo

© The Author(s), under exclusive license to Springer Nature Switzerland AG 2023   2391
M. Alshurideh et al. (eds.), *The Effect of Information Technology on Business and Marketing Intelligence Systems*, Studies in Computational Intelligence 1056,
https://doi.org/10.1007/978-3-031-12382-5_131

1 Introduction

The increasing usage of digital technology is providing a solid path for the expansion of the market for services in sectors such as tourism and hospitality, among others (Ahmad et al., 2021; Lee et al., 2022a, 2022b; Tariq et al., 2022b). Use of Information and Communication Technology (ICT) refers to the use of computers, the internet, mobile phones, and other digital devices in the course of corporate activities have drawn attention to the fact that technological advancements have really done their best to define the preferences and prospects of current and future visitors and travelers. Technology, in the form of computers and connected technologies, has made it easier for companies to gather, process, distribute, and store data at a quicker, more accurate, and error-free pace than ever before (Alshurideh, 2022; Alzoubi et al., 2022; Ansah et al., 2012). By modernizing their various company operations, the Indian tourist and hospitality sector has quickly adopted the use of this technology to provide better and more sophisticated services, as well as to explore new commercial possibilities by expanding their existing operations.

Due to the ever-changing marketplace and the rapid growth of information technology, a variety of customers are booking their travel facilities electronically, such as travel, hotel and car rental (Abu Zayyad et al., 2020; Bhuiyan et al., 2020; Chia et al., 2021; Shannak et al., 2010; Tariq et al., 2022a; Yang et al., 2014). In addition, according to a survey of 249 casual travelers (Toh et al., 2011), 80% of travelers searched for hotel information via blogs, with more than half making their bookings via hotel websites or third-party websites. As a result, the Internet plays a crucial function in customer behavior, with more than 50% of the sales of global hotel chains credited to online marketing networks in 2010 (Pan et al., 2013).

This study continues to extend the body of expertise and experience in the online review management assessment of the contents of the hotel website. There have been several academic reports focused on the travel and hospitality industry, which can be classified as either observational or theoretical analysis (Abujamous et al., 2018; Alaali et al., 2021; Daniel et al., 2016; Mahmoud et al., 2021). In the minds of customers, the Internet provides the right to access a hotel directly, without regional or time constraints, to request information and to make transactions. Managers in control of promotions and sales for hotels see lower delivery prices, increased profits, and a greater market share as being great outcomes from hotel websites (Abuhashesh et al., 2021; Al Kurdi & Alshurideh, 2021; Alshurideh et al., 2017; Masa'deh et al., 2018; O'Connor, 2003; Panagopoulos et al., 2014).

An important and well-known marketing resource for hotels is the hotel website (Kim et al., 2017). The job of designing and maintaining websites of the highest standard have been a crucial element in holding the company running. Unfavorable internet experience will limit online purchases, remove future buyers and have a detrimental impact on corporate credit. The analysis described the climate and process efficiency as the two key factors for determining the quality of the electronic services offered by hotel websites. The hotel website acts as a direct connection between the hotel and its guests to meet the needs of the customer (Hahn et al., 2017;

Meng et al., 2020). These days, passengers will use online booking applications to easily pick their lodging. The hotel website presented visitors with high-quality, but cost-effective, details while lowering costs and saving time.

For the evaluation of websites based on studying targets, a number of researchers have identified many variables. For the usage of the fuzzy Delphi screening method, the new study utilizes the present Delphi system, which is composed of many separate variables (Law & Wong, 2010). They explored the effect of blogging on tourism management as a result of their usage of analysis strategies used in other studies. The research discusses a variety of frameworks for website decision-making, including multi-criteria decision-making (MCDM). Promethee is considered to be the most effective promethee model owing to its capacity to accommodate a broad variety of quantitative and qualitative factors, as well as to rank various choices.

There has been a substantial growth in the amount of ways for large hotel chains to capture new market share through their use of the internet in the hospitality industry. As the Internet is still a required networking tool, hotels can make use of it to aid with their marketing (Abuhashesh et al., 2019; Alananzeh et al., 2018; Escobar-Rodrigueza & Carvajal-Trujillo, 2013; Lee & Kozar, 2006; Leung et al., 2016), note that the hotel industry has adapted to this technology and is working hard to create portals from which visitors can book rooms online.

However, only a tiny fraction of the tourism and hospitality literature discusses the relative value of numerous website functionality (Ahmad et al., 2020; Jawabreh, 2020; Yılmaz, 2020). A lot of hypotheses are being made as to what effect the internet has on supporting independent hotels in Jordan. Previous analysis has not researched the independent Jordanian hotel in great depth. Reaction to the dissemination of the Internet is essential for hotel firms in particular, as there is a lack of sufficient data on their opinions on the Internet's importance in their sectors and on how they react to the Internet's impact (Essawy, 2011). As there is a rise in the demand for hospitality, there is a new urgency to introduce innovative ways to advance its place in the industry. Selling and distributing goods and knowledge services on hotel channels to local and global audiences is critical, and the most reliable methods of doing so are in terms of online advertisement and delivery. The Internet is a critical marketing platform for customers.

When consumers are able to do more of their product search and book their travel and accommodation, all at once they are constantly requesting conveniences, specialized, accessible, interactive products, and personal interaction with different hotel providers. This hence puts value on the usage of the Internet for tourism, while also supporting the official websites for the sector (Happ & Horváth, 2020; Jawabreh et al., 2020; Mlambo & Ezeuduji, 2020; Orjuela et al., 2020; Ushakov et al., 2021). Hotel websites can be tested for accuracy as a website marketing platform is built for the current website and consumer perspectives. The aim of this research is therefore to suggest an existing assessment system for assessing the current quality of hotel websites and, by doing so, to help increase the quality of hotel websites for hotel chains (Dai & Lee, 2018).

In order to better understand and measure how various dimensions of hotel websites contribute to different elements of customers' preferences, the aim of the study is to quantitatively analyze the highest current dimension of hotel website evaluations, as well as to recognize distinctions between different hotel website elements and their importance from customers' perspectives.

2   Background of the Study

Many studies have been undertaken to determine the influence of information and communications technology-based marketing tools on company performance and profitability in the tourist and hospitality industry (Jahmani et al., 2020; Law et al., 2014). The tourism and hospitality literature demonstrates an increasing interest in the adoption and use of social media for management and other business operations, notably marketing. Previous studies in the tourism marketing literature have explored and explained various aspects of social media use in this field, including effects of social media on tourist behavior, social media use by tourism organizations by Destination Marketing Organizations (DMOs) and by travel agencies, social media use for co-creation of value and social media use for tourism service personalization and brand management (Kim et al., 2022; Lin et al., 2020). These researches have enhanced knowledge of social media usage in tourist marketing. Tourism marketing is a performance-oriented commercial activity. That is, success or failure of marketing activities is decided by performance results. A performance-oriented activity is described as an activity undertaken by a person or organization that effectively and efficiently induces the behavior and resource allocation required to fulfill a task and to achieve sustained superior performance of the task (Che-Ha et al., 2014). Studies on adoption of service technology systems by enterprises in the hotel sector reveal that the success or failure (Sigala, 2015). Studies of social media usage in tourism-related marketing are quickly developing and several academic works have examined social media use for tourist marketing activities. However, the research reported thus far have utilized a fairly wide viewpoint (Singh & Dhankhar, 2020).

Much of the first research in this area has been undertaken to analyze the quality and functionality of hotel websites (Kang et al., 2021; Law & Wong, 2010; Qandah et al., 2020; Romanazzi et al., 2011; Wan, 2002). A variety of tests have been performed to assess the scope and detail of the knowledge that can be presented to consumers. In light of the findings obtained, these writers recommend that hotel managers provide a "more engaging appearance" on their websites by utilizing the interactive resources accessible on the Internet. The study concludes that cyberhotels must investigate how these features impact or enhance the purpose, mission, and mechanics of their websites. five-star hotel websites have their own mobile app, which might allow travelers to know explicitly that there are any mobile apps accessible to them, the authors add. The study concluded that hotels use their websites to provide details and purchases, but not to communicate with clients. The author concludes that this implies that they ought to make their platforms more immersive.

Hotel portals have been an important aspect of the travel and hospitality sector. Reservation knowledge is considered to be the most important dimension, with nearly all of its features considered to be essential. E-commerce activities are marginal among Jordanian hotels. In reality, none of the hotels offer online reservations.

The construction of websites that offer details in other languages, such as English and Arabic, will support visitors who are interested. Hotel portals have been an imperative aspect of the hospitality business, which is why several independent scholars have researched the actions of online travelers in order to consider their preferences and wishes (Lee & Lee, 2021; Widagdo & Roz, 2021). The measurements and characteristics of the upscale hotel on the websites were more significant than on the websites of the economy hotel. A research that explored the relative value of measurements and quality on hotel websites from the point of view of visitors. Jawabreh et al. (2021), proposed a computational model to improve the customer's intention to revisit hotel websites to book hotel rooms. The chain hotels appear to pay greater attention to the facilities offered by the website. Hotels and advertisers need to discuss their image of hospitality, consumer understanding, aspirations and loyalty (Masa'deh et al., 2019). The name of the hotel website is trustworthy and creates familiar amenities and services. Through putting up a website, hotels will improve their purchasing potential, enhance the reputation of the hotel brand and enhance the maximum impact of their marketing campaigns. The assessment of websites is of vital significance, since it can be checked by the degree to which websites satisfy consumer requirements. It involves studies on the effect of the content of the hotel website on sales and the impact of consumers' expectations of the standard of knowledge and the impact of online marketing campaigns on sales.

Most of the work in this field has been conducted to look at the degree of consumer loyalty and the features of websites. A variety of tests have been performed to assess the scope and detail of the knowledge that can be presented to consumers. Richness exists when the flow of knowledge is broader, wider and quicker than it is in the conventional sector. Bad site design would result in a 50% reduction in future revenue due to consumers being unable to locate what they want, according to one report (Li et al., 2015). The authors suggest that hotel managers are willing to have a "more engaging presence" on their websites using the interactive tools available on the Internet. The findings suggest that hotels use their websites to provide details and purchases, just not to communicate with clients, they claim. Take a look at some of the more popular blogs and other web tools on the topic (Jakovic & Galetic, 2014). Hotel websites that offer details in other languages, such as English and Arabic, may allow visitors who are unfamiliar with Iran to find the information they need. It is proposed that a range of transactional and relationship-building practices in the hotel industry in Jordan and around the world will rise in the immediate future. The study found that web design features are essential for online partnership marketing and web design is a core part of web marketing strategy in the travel industry.

Hotel portals are also at the forefront of the promotion and service of hotels. A well-designed website is the most significant marketing tool for the hotel in today's fast-growing information period. Increasing numbers of researchers are paying attention to website design (Caglione et al., 2009) found that website adoption was positively connected to the success indicator Revenue per Accessible Space. They also noticed that sales growth rates were higher for hotels with their own websites than for hotels with an online presence only via regional portals (Hu, 2009). The "first image" of a hotel has now been moved from the front desk to its website.

The research aims to address the question: What is the strongest dimension in the assessment of the hotel's website? What are the contents? From the customer's point of view, what is the most important feature of the website for the hotel? Are there any gaps in the consistency of new hotel websites and the importance of these contents from the point of view of the customer (Zeki & Alrawadieh, 2017)? Provided a tool for analyzing the websites of international tourist hotels (O'Connor, 2007). Indicated that the assessment of the standard of electronic facilities on the travel platform should be viewed as a multi-criteria decision-making procedure. Hu's requirements are performance, enforcement, device availability, security/privacy, accessibility, reimbursement, communication, and gain. Law and Bai (2007) announced that the effective website of the hotel should be both accessible and functional (Wong & Law, 2005). The following research described the accuracy of details as the most critical aspect of a good hotel website.

Chung and Law (2003) created an information quality measurement model to calculate the efficiency of the hotel website. Law and Cheung (2005) considered the reservation information and website administration to be the measurements with the largest and smallest weights, respectively, the weight meaning representing the perceived significance of the dimension. In comparison to previous reports, the study discussed here focuses solely on a single feature, i.e., the cultural factors of the destination knowledge dimension.

3 Methodology

In this research, the Computational Package for Social Sciences (SPSS) was first used for data entry, and later for data interpretation. Data processing was the initial phase aimed at transforming raw data to a more standardized format that is more suitable for study. Tasks at this point involved data editing, data coding and data entry. Descriptive figures were used to outline the attributes of the respondent, including demographic details such as age, gender; questions about the identity of the participating organization, such as consumer numbers, were also included. Inferential statistics were often used to evaluate theories to assess the association between variables. In specific, the Pearson Correlation was used to check the relationship of the interval amount with the construct, while the Chi-square was used to confirm the association between the nominal variables and the construct. In addition, the Cronbach coefficient alpha was used for the measure of durability. Finally, the Variable Analysis Technique, Variance Analysis (ANOVA) was used to analyze the contents of the official tourist website of the Jordan case study: existence-importance summary (Fig. 1).

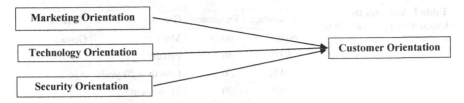

Fig. 1 The model of the study

3.1 Research Hypothesis

H1: There is a statistically relevant effect of Marketing orientation and Customer Orientation of classified hotels in Jordan.

H2: There is a statistically relevant effect of Technology Orientation and Customer Orientation of classified hotels in Jordan.

H3: There is a statistically relevant effect of Security Orientation and Customer Orientation of classified hotels in Jordan.

3.2 Sample Research

The study sample consisted of (100) workers employed in graded Jordanian hotels who were randomly chosen from the study population (Table 1).

3.3 Reliability

To ensure the reliability of the analysis used by Cronbach's Alpha equation to sample the initial analysis, in order to know the stability of the internal consistency for each area of research, Table 2 explains.

Table 2 displays the reliability coefficient for products with high loading where the range of values (0.797–0.950) is adequate.

4 Results

This section comprises the findings of the report, which seeks to analyze the contents of the five-star hotel website for a case study. Jordan: life—significance review; the following is a presentation of the findings, focused on the conclusions of the study:

H1: The marketing orientation and consumer orientation of categorized hotels in the city of Jordan has a statistically relevant effect.

Table 1 Indicates the informative personality trait statistics

Percentage	Frequency	Group	Variable
60.0	60	Male	Gender
40.0	40	Female	
0.00	0	Less than 20 years	Age
20.0	20	21–30 year	
20.0	20	31–40 year	
40.0	40	41–50 year	
20.0	20	51–60 year	
0.00	0	61 years and more	
20.0	20	Less than 1000$	Monthly income
10.0	10	1001$–1500$	
10.0	10	1501–2000	
60.0	60	More than 2001$	
50.0	50	Private sector	Occupation
50.0	50	Public sector	
0.00	0	High school	Education
70.0	70	Bachelor's degree	
20.0	20	Master's degree	
10.0	10	Post graduate	
30.0	30	European	Nationality
30.0	30	Asia	
30.0	30	USA	
10.0	10	Australian	
0.00	0	Canada	
0.00	0	Others	
100.0	100	Total	

Table 2 The values of reliability coefficient by Cronbach's alpha

Factors	Cronbach's alpha
Customer orientation	0.797
Marketing orientation	0.848
Technology orientation	0.805
Security orientation	0.950
Tool overall	0.928

To be confident, we used the Basic Linear Regression Analysis as seen below.

The following is seen in the Table 3. The meaning (F = 56.534) and the statistically meaningful value (0.00) are smaller than the statistically significant amount ($\alpha \leq$ 0.05). The simple linear regression model is therefore suitable for the estimation of the causal relationship between the independent variable (marketing orientation) and the dependent variable (customer orientation).

The value of the correlation coefficient between the independent variable (marketing orientation) and the dependent variable (customer orientation) was (0.605), the value of the coefficient of determination (R^2) (0.366) and the value of the changed coefficient of determination (Adjusted R^2) (0.359) which indicates that the independent variable (marketing orientation) should be explained (35.9).

There is a statistical significance of the coefficient of simple linear regression equation associated with the independent variable (marketing orientation) where the value $(T = 7.519)$ shows the impact of the effect (marketing orientation) on the independent variable (marketing orientation) (Customer Orientation). There is therefore a substantial significance for the coefficient of simple linear regression equation (0.605), which was a positive result, and we therefore accept the statement that 'There is a statistically significant impact on the marketing and customer orientation of classified hotels in the Jordan.

H2: The influence of Technology Orientation and Consumer Orientation of Categorized Hotels in Jordan is statistically important.

To be confident, we used the Basic Linear Regression Analysis as seen below.

The following is seen in the Table 4. The meaning $(F = 148.09)$ and the statistically meaningful value (0.00) are smaller than the statistically significant amount ($\alpha \leq 0.05$). As a consequence, a simple linear regression model is suitable for evaluating the causal relationship between the independent variable (technology orientation) and the dependent variable (technology adoption) (customer orientation).

The coefficient of correlation between the independent variable (technology orientation) and the dependent variable (customer orientation) was (0.776), the coefficient of determination (R^2) was (0.602), and the updated coefficient of determination (Adjusted R^2) was (0.598), both of which mean that the independent variable (technology orientation) should be explained (59).

Table 3 The results for simple linear regression

R	R^2	Adjusted R^2	F	Sig*	Regression coefficients				
					Domain	β	Std error	T	Sig*
0.605	0.366	0.359	56.5	0.00*	Marketing orientation	0.605	0.045	7.519	0.00*

* Statistically significant at the level of statistical significance ($\alpha \leq 0.05$)

Table 4 Effects of basic linear regression

R	R^2	Adjusted R^2	F	Sig*	Regression coefficients				
					Domain	β	Std error	T	Sig*
0.776	0.602	0.598	148.0	0.00*	Technology orientation	0.776	0.041	12.169	0.00*

* Statistically significant at the degree of statistical importance ($\alpha \leq 0.05$)

Table 5 The outcomes of a basic linear regression study

R	R^2	Adjusted R^2	F	Sig*	Regression coefficients				
					Domain	β	Std error	T	Sig*
0.480	0.231	0.223	29.359	0.00*	Security orientation	0.480	0.038	5.418	0.00*

* Statistically significant at the degree of statistical importance ($\alpha \leq 0.05$)

The coefficient of a basic linear regression equation in relation to the independent variable (technology orientation) has statistical significance, with the value (T = 12.169) suggesting the influence (technology orientation) on the independent variable (technology orientation) (Customer Orientation). As a consequence, the coefficient of the simple linear regression equation (0.776), which was a positive influence, has a significant impact, and we accept the statement that "Technology Orientation and Customer Orientation of Classified Hotels in the Aqaba Area have a statistically significant impact."

H3: There is a statistically important effect of the Security Orientation and Consumer Orientation of confidential hotels in Jordan.

To be confident, we used the Basic Linear Regression Analysis as seen below.

The following is shown in the Table 5. The meaning (F = 29.359) and the statistically meaningful value (0.00) are smaller than the statistically significant amount ($\alpha \leq$ 0.05). The association between the independent variable (security orientation) and the corresponding relationship between the dependent variable (correlation) (customer orientation. The correlation coefficient between the independent variable (security orientation) and the dependent variable (customer orientation) was (0.480), with an R^2 of (0.231) and an Updated R^2 of (0.223) indicating that the independent variable (security orientation) could be explained (22.3%).

The coefficient of basic linear regression equation, which is associated with the independent variable (security orientation), has statistical significance (there is a meaningful association between the coefficient and the independent variable) (Customer Orientation). Thus, this implies that the coefficient of the basic linear regression equation (0.480), which was a positive effect, follows the statement that.

5 Conclusion

The statistical relevant number ($\alpha \leq 0.05$) is less than the significance (F = 56.534) but not smaller than the statistically meaningful value (0.00). Since the independent variable (marketing orientation) is the source of the causal interaction between the dependent variable (customer orientation) and the company, it is better to use a simple linear regression model. Correlation coefficient between independent and dependent variables (marketing orientation and consumer orientation) was (0.605), R^2 (0.366) and Updated R^2 (0.359) both indicate that the independent variable (marketing orientation) may be rendered clearer. Pragmatism comes from a belief in our own abilities to work it out as we move along.

Coefficient of basic linear regression has statistical significance where value (T = 7.519) denotes the effects on the independent variable (marketing orientation) (Customer Orientation). Accordingly, there is a considerable impact on the coefficient of simple linear regression equation (0.605), which is a good impact, and we accept the argument that "Jordan Advertisement and Customer Orientation Classified Hotels have a statistically important effect on their categorization." When it comes to the advertising and consumer care aspect of hotels, hotel portals are masters. In today's dynamic environment, the website is the most important marketing asset. The rise in the number of scholars who study website design suggests that increased emphasis on website design is warranted. Website adoption was positively related to revenue per accessible area, according to O'Connor (2007). They observed that revenue growth rates were often higher for hotels with a website to which consumers could only resort from regional portals rather than for hotels with a presence on both the global and regional internet. "First pic" for a hotel is now on the website rather than on the front desk. According to Law and Wong, hotels should make their websites more interesting. According to them, just 20.00% of Croatia's five-star hotel websites have their own mobile app. According to the findings, cyber-hotels should look at how these elements affect or improve the purpose, goal, and mechanics of their websites.

Visitors unfamiliar with Jordan may be able to discover information on hotel websites that provide data in other languages, such as English and Arabic. In today's fast-paced information age, a well-designed website is the most important marketing tool for the hotel. The goal of the study is to find out what the most important factor is when evaluating a hotel's website. What are the contents of the package? What is the most significant element of the hotel's website from the customer's perspective?

Are there any inconsistencies in the consistency of new hotel websites, as well as the significance of these items from the customer's perspective? The research focuses exclusively on a particular characteristic, namely, the cultural variables that influence destination knowledge.

Although the explanation is described as 12.169, it represents the impact of the influence (technology orientation) on the independent variable (technology orientation) (Customer Orientation). To this end, it is necessary to reflect on the coefficient of the simple linear regression equation (0.776), which was a positive influence, and thus lend legitimacy to the finding that "The Technology and Customer Inclination of Classified Hotels in Aqaba Area has a statistically significant impact." The coefficient of basic linear regression associated with the independent variable (security orientation) has statistical significance (meaning: T = 5.418). (Customer Orientation). Also, because of this, we may infer that "The Protection Orientation and Customer Orientation of Classified Hotels in Aqaba Area has a statistically relevant effect".

References

Abu Zayyad, H. M., Obeidat, Z. M., Alshurideh, M. T., Abuhashesh, M., & Maqableh, M. (2020). Corporate social responsibility and patronage intentions: The mediating effect of brand credibility. *Journal of Marketing Communications*. https://doi.org/10.1080/13527266.2020.1728565

Abuhashesh, M. Y., Alshurideh, M. T., & Sumadi, M. (2021). The effect of culture on customers' attitudes toward Facebook advertising: The moderating role of gender. *Review of International Business and Strategy, 31*(3), 416–437.

Abuhashesh, M., Al-Khasawneh, M., & Al-Dmour, R. (2019). The impact of Facebook on Jordanian consumers' decision process in the hotel selection. *IBIMA Business Review, 2019*. https://doi.org/10.5171/2019.928418

Abujamous, I., Jawabreh, O., Jahmani, A., Alsarayreh, M., & Harazneh, A. (2018). Developing tourism through sports events to assist in the rejuvenation of the strategic position of the Aqaba Special Economic Zone Authority (ASEZA). *African Journal of Hospitality, Tourism and Leisure, 8*(4).

Ahmad, A., Madi, Y., Abuhashesh, M., & Nusairat, N. M. (2020). The knowledge, attitude, and practice of the adoption of green fashion innovation. *Journal of Open Innovation: Technology, Market, and Complexity, 6*(4), 1–20. https://doi.org/10.3390/joitmc6040107

Ahmad, A., Alshurideh, M. T., Al Kurdi, B. H., & Salloum, S. A. (2021). Factors impacts organization digital transformation and organization decision making during Covid19 pandemic. In *The Effect of Coronavirus Disease (COVID-19) on Business Intelligence* (pp. 95–106). Springer.

Al Kurdi, B. H., & Alshurideh, M. T. (2021). Facebook advertising as a marketing tool: Examining the influence on female cosmetic purchasing behavior. *International Journal of Online Marketing (IJOM), 11*(2), 52–74.

Alaali, N., Al Marzouqi, A., Albaqaeen, A., Dahabreh, F., Alshurideh, M., Alrwashdh, S., Iyadeh, I., Salloum, S., & Aburayya, A. (2021). The impact of adopting corporate governance strategic performance in the tourism sector: A case study in the Kingdom of Bahrain. *Journal of Legal, Ethical and Regulatory, 24*(Special Issue 1), 1–18.

Alananzeh, O. A., Jawabreh, O., Al Mahmoud, A., & Hamada, R. (2018). The impact of customer relationship management on tourist satisfaction: The case of Radisson blue resort in Aqaba city. *Journal of Environmental Management and Tourism, 9*(2), 227–240. https://doi.org/10.14505/jemt.v9.2(26).02

Alshurideh, M. (2022). Does electronic customer relationship management (E-CRM) affect service quality at private hospitals in Jordan? *Uncertain Supply Chain Management, 10*(2), 1–8.

Alshurideh, M., Al Kurdi, B., Abu Hussien, A., & Alshaar, H. (2017). Determining the main factors affecting consumers' acceptance of ethical advertising: A review of the Jordanian market. *Journal of Marketing Communications, 23*(5), 513–532.

Alzoubi, H., Alshurideh, M., Kurdi, B., Akour, I., & Aziz, R. (2022). Does BLE technology contribute towards improving marketing strategies, customers' satisfaction and loyalty? The role of open innovation. *International Journal of Data and Network Science, 6*(2), 449–460.

Ansah, A. K., Kontoh, M., & Balnkson, V. S. (2012). The use of information and communication technologies (ICTs) in front office operations of chain hotels in Ghana. *International Journal of Advanced Computer Science and Applications, 3*(3), 72–77.

Bhuiyan, M. B., Islam, M. A., Haque, M. Z., & Biswas, C. (2020). Moderating effect of technology readiness on adoption of geotagging technology among social networking sites (SNSS) users for smart tourism. *GeoJournal of Tourism and Geosites, 34*(1), 47–55.

Caglione, M., Schegg, R., & Murphy, J. (2009). Website adoption and sales performance in Valais' hospitality industry. *Technovation, 29*(9), 625–631.

Che-Ha, N., Mavondo, F. T., & Mohd-Said, S. (2014). Performance or learning goal orientation: Implications for business performance. *Journal of Business Research, 67*(1), 2811–2820.

Chia, S. K. S., Lo, M. C., Razak, Z. B., Wang, Y. C., & Mohamad, A. A. (2021). Impact of destination image on tourist satisfaction: The moderating effect of information technology (IT). *GeoJournal of Tourism and Geosites, 34*(1), 88–93.

Chung, T., & Law, R. (2003). Developing a performance indicator for hotel websites. *International Journal of Hospitality Management, 22*(1), 119–125.

Dai, W., & Lee, J.-H. (2018). The effects of website characteristics and delivery service quality on repurchase intention. *Mountain Landscape Research Journal, 9*(5), 17–24.

Daniel, L., Rob, L., & Lee, H. (2016). A modified model for hotel website functionality evaluation. *Journal of Travel & Tourism Marketing, 33*(9), 1268–1285.

Escobar-Rodrigueza, T., & Carvajal-Trujillo, E. (2013). An evaluation of Spanish hotel websites: Informational vs. relational strategies. *International Journal of Hospitality Management, 33*(1), 228–239.

Essawy, M. (2011). Egyptian hotel marketing managers' perceptions of the internet's impact on marketing. *Tourism and Hospitality Research, 11*(3), 207–216.

Hahn, H., Sparks, B., Wilkins, H., & Jin, X. (2017). E-service quality management of a hotel website: A scale and implications for management. *Journal of Hospitality Marketing & Management, 26*(7), 694–716.

Happ, E., & Horváth, Z. (2020). A study of digital marketing tools usage habits among Hungarian tourists. *GeoJournal of Tourism and Geosites, 32*(4), 1283–1289.

Hu, Y. C. (2009). Fuzzy multiple-criteria decision making in the determination of critical criteria for assessing service quality of travel websites. *Expert Systems with Applications, 36*(3), 6439–6445.

Jahmani, A., Bourini, I., & Jawabreh, O. A. (2020). The relationship between service quality, client satisfaction, perceived value and client loyalty: A case study of fly emirates. *Cuadernos De Turismo, 45,* 219–238. https://doi.org/10.6018/turismo.45.426101

Jakovic, B., & Galetic, F. (2014). Marketing and commercial activities offered on Croatian five-star hotel web sites, 24th DAAAM International Symposium on Intelligent Manufacturing and Automation. *Procedia Engineering, 69,* 112–120.

Jawabreh, O. (2020). Innovation management in hotels industry in Aqaba special economic zone authority; hotel classification and administration as a moderator. *GeoJournal of Tourism and Geosites, 32*(4), 1362–1369. https://doi.org/10.30892/gtg.32425-581

Jawabreh, O., Masa'deh, R., Mahmoud, R., & Hamasha, S. A. (2020). Factors influencing the employees service performances in hospitality industry case study Aqaba five stars hotel. *GeoJournal of Tourism and Geosites, 29*(2), 649–661. https://doi.org/10.30892/gtg.29221-496

Jawabreh, O., Abdelrazaq, H., & Jahmani, A. (2021). Business sustainability practice and operational management in hotel industry in aqaba special authority economic zone authority (ASEZA). *GeoJournal of Tourism and Geosites, 38*(4), 1089–1097. https://doi.org/10.30892/gtg.38414-748

Kang, S., Kim, W., & Park, D. (2021). Understanding tourist information search behaviour: The power and insight of social network analysis. *Current Issues in Tourism, 24*(3).

Kim, M. J., Lee, C. K., & Bonn, M. (2017). Obtaining a better understanding about travel-related purchase intentions among senior users of mobile social network sites. *International Journal of Information Management, 37*(5), 484–496.

Kim, T., Jo, H., & Yhee, Y. (2022). Robots, artificial intelligence, and service automation (RAISA) in hospitality: Sentiment analysis of YouTube streaming data. *Electron Markets.* https://doi.org/10.1007/s12525-021-00514-y

Law, R., & Cheung, C. (2005). Weighing of hotel website dimensions and attributes. In A. J. Frew (Eds.), *Information and communication technologies in tourism.* Springer. https://doi.org/10.1007/3-211-27283-6_32.2005

Law, R., & Bai, B. (2007). How do the preferences of online buyers and browsers differ on the design and content of travel websites? *International Journal of Contemporary Hospitality Management, 20*(4), 388–400.

Law, R., & Wong, R. (2010). Analysing room rates and terms and conditions for the online booking of hotel rooms. *Asia Pacific Journal of Tourism Research, 15*(1), 43–56.

Law, R., Buhalis, D., & Cobanoglu, C. (2014). Progress on information and communication technologies in hospitality and tourism. *International Journal of Contemporary Hospitality Management, 26*(5), 727–750. https://ezlibrary.ju.edu.jo:2057/10.1108/IJCHM-08-2013-0367

Lee, Y., & Kozar, K. A. (2006). Investigating the effect of website quality on e-business success: An analytic hierarchy process (Ahp) approach. *Decision Support Systems, 42*(3), 1383–1401.

Lee, J., & Lee, J. H. (2021). A study on effects of repurchase intention of consumer innovativeness and website characteristics: Focused on consumer of overseas direct purchase. *Mountain Landscape Research Journal, 12*(2), 29–40.

Lee, K., Ramiz, P., Hanaysha, J., Alzoubi, H., & Alshurideh, M. (2022a). Investigating the impact of benefits and challenges of IOT adoption on supply chain performance and organizational performance: An empirical study in Malaysia. *Uncertain Supply Chain Management, 10*(2), 1–14.

Lee, K., Azmi, N., Hanaysha, J., Alshurideh, M., & Alzoubi, H. (2022b). The effect of digital supply chain on organizational performance: An empirical study in Malaysia manufacturing industry. *Uncertain Supply Chain Management, 10*(2), 1–16.

Leung, R., Law, R., & Lee, H. (2016). A modified model for hotel website functionality evaluation. *Journal of Travel & Tourism Marketing, 33*(9), 1268–1285.

Li, X., Wang, Y., & Yu, Y. (2015). Present and future hotel website marketing activities: Change propensity analysis. *International Journal of Hospitality Management, 47*(1), 131–139.

Lin, H.-C., Han, X., Lyu, T., Ho, W.-H., Xu, Y., Hsieh, T.-C., Zhu, L., & Zhang, L. (2020). Task-technology fit analysis of social media use for marketing in the tourism and hospitality industry: A systematic literature review. *International Journal of Contemporary Hospitality Management, 32*(8), 2677–2715.

Mahmoud, R., Al-Mkhadmeh, A., & Alananzeh, O. (2021). Exploring the relationship between human resources management practices in the hospitality sector and service innovation in Jordan: The mediating role of human capital. *Geojournal of Tourism and Geosites, 35*(2), 507–514. https://doi.org/10.30892/gtg.35231-678

Masa'deh, R., Alananzeh, O., Tarhini, A., & Algudah, O. (2018). The effect of promotional mix on hotel performance during the political crisis in the Middle East. *Journal of Hospitality and Tourism Technology, 9*(1), 32–47. https://doi.org/10.1108/JHTT-02-2017-0010

Masa'deh, R., Alananzeh, O., Jawabreh, O., Alhalabi, R., Syam, H., & Keswani, F. (2019). The association among employees' communication skills, image formation and tourist behaviour: Perceptions of hospitality management students in Jordan. *International Journal of Culture, Tourism, and Hospitality Research, 13*(3), 257–272. https://doi.org/10.1108/IJCTHR-02-2018-0028

Meng, R., Huy, Q., Gang, L., & Law, R. (2020). Large-scale comparative analyses of hotel photo on the internet. *Cornell Hotel and Restaurant Administration Quarterly, 37*(3), 70–82.

Mlambo, S. S., & Ezeuduji, I. O. (2020). South Africa's Kwazulu-Natal tourism destination brand essence and competitiveness: Tourists' perspectives. *GeoJournal of Tourism and Geosites, 32*(4), 1195–1201.

O'Connor, O. (2003). On-line pricing: An analysis of hotel-company practices. *Cornell Hotel and Restaurant Administration Quarterly, 44*(1), 88–96.

O'Connor, P. (2007). Online consumer privacy an analysis of hotel company behavior. *Cornell Hotel Restaurant Administration Quarterly, 48*(2), 183–200.

Orjuela, A., Escobar, D. A., & Moncada, C. A. (2020). Conditions of territorial accessibility offered by the network of sustainable tourism routes that are part of the coffee cultural landscape—Colombia. *GeoJournal of Tourism and Geosites, 32*(4), 1290–1298.

Pan, B., Zhang, L., & Law, R. (2013). The complex matter of online hotel choice. *Cornell Hotel Restaurant Administration Quarterly, 54*(1), 74–83.

Panagopoulos, A., Kanellopoulos, D., Karachanidis, I., & Konstantinidis, S. (2014). A comprehensive evaluation framework for hotel websites: The case of chain hotel websites operating in Greece. *Journal of Hospitality Marketing & Management, 20*(7), 695–717.

Qandah, R., Suifan, T. S., & Obeidat, B. Y. (2020). The impact of knowledge management capabilities on innovation in entrepreneurial companies in Jordan. *International Journal of Organizational Analysis*. https://doi.org/10.1108/IJOA-06-2020-2246

Romanazzi, S., Petruzzellis, L., & Iannuzzi, E. (2011). The effect of a destination website on tourist choice: Evidence from Italy. *Journal of Hospitality Marketing & Management, 20*(7), 791–813.

Shannak, R. O., Obeidat, B. Y., & Almajali, D. A. (2010). *Information technology investments: A literature review.* Paper presented at the Business Transformation through Innovation and Knowledge Management: An Academic Perspective—Proceedings of the 14th International Business Information Management Association Conference, IBIMA 2010 (Vol. 2, pp. 1356–1368).

Sigala, M. (2015). Social media marketing in tourism and hospitality. *Information Technology & Tourism; Heidelberg, 15*(2), 181–183. https://doi.org/10.1007/s40558-015-0024-1

Singh, L., & Dhankhar, D. (2020). ICT-based marketing and profitability in tourism and hospitality organizations in Indian scenario. In A. Hassan & A. Sharma (Ed.), *The Emerald handbook of ICT in tourism and hospitality* (pp. 311–330). Emerald Publishing Limited.

Tariq, E., Alshurideh, M., Akour, I., & Al-Hawary, S. (2022b). The effect of digital marketing capabilities on organizational ambidexterity of the information technology sector. *International Journal of Data and Network Science, 6*(2), 401–408.

Tariq, E., Alshurideh, M., Akour, E., Al-Hawaryd, S., & Al Kurdi, B. (2022a). The role of digital marketing, CSR policy and green marketing in brand development at UK. *International Journal of Data and Network Science, 6*(3), 1.

Toh, R. S., DeKay, C. F., & Raven, P. (2011). Travel planning searching for and booking hotels on the internet. *Cornell Hotel Restaurant Administration Quarterly, 52*(4), 388–398.

Ushakov, D. S., Kiselev, D. N., Zezyulko, A. V., Imangulova, T. V., Kulakhmetova, G. A., & Kulakhmetova, R. A. (2021). Organization of network basis for transnational tourism activity. *GeoJournal of Tourism and Geosites, 34*(1), 77–87.

Wan, C. S. (2002). The web sites of international tourist hotels and tour wholesalers in Taiwan. *Tourism Management, 23*(2), 155–160.

Widagdo, B., & Roz, K. (2021). Hedonic shopping motivation and impulse buying: The effect of website quality on customer satisfaction. *The Journal of Asian Finance, Economics and Business, 8*(1), 395–405.

Wong, J., & Law, R. (2005). Analysing the intention to purchase on hotel websites: A study of travelers to Hong Kong. *International Journal of Hospitality Management, 24*(3), 311–329.

Yang, C., Guo, X., & Ling, L. (2014). Opening the online marketplace: An examination of hotel pricing and travel agency on-line distribution of rooms. *Tourism Management, 45*(1), 234–243.

Yılmaz, E. (2020). The effects on consumer behavior of hotel related comments on the TripAdvisor website: An Istanbul case. *Advances in Hospitality and Tourism Research, 8*(1), 1–29.

Zeki, M., & Alrawadieh, Z. (2017). Negative word of mouse in the hotel industry: A content analysis of online reviews on luxury hotels in Jordan. *Journal of Hospitality Marketing & Management, 26*(8), 785–804.

The Awareness and Confidence About COVID-19 Vaccines Among Selected Students in Faculty of Health Sciences: Comparison Study Among Malaysia, Pakistan and UAE

Qays Al-Horani⬤, Saher Alsabbah⬤, Saddam Darawsheh⬤,
Anwar Al-Shaar, Muhammad Alshurideh⬤,
Nursyafiq Bin Ali Shibramulisi, Zainorain Natasha Binti Zainal Arifen,
Amina Asif Siddiqui⬤, Anizah Mahmod, Revathi Kathir,
and Siti Shahara Zulfakar

Abstract The COVID-19 pandemic has impacted the world in an unpredictable way, affecting economic, socio-cultural, political, environmental statuses and billions regardless of privilege and wealth. Vaccines were developed quickly, for protection against the Coronavirus, as well as to curb the pandemic. This study aimed to (1) compare awareness about risks and benefits of getting vaccinated against COVID-19,

Q. Al-Horani
Department of Radiography and Medical Imaging, Fatima College of Health Sciences, Abu Dhabi, United Arab Emirates
e-mail: Qays.AlHorani@fchs.ac.ae; P110054@siswa.ukm.edu.my

Q. Al-Horani · N. B. A. Shibramulisi · Z. N. B. Z. Arifen · A. A. Siddiqui · A. Mahmod · R. Kathir · S. S. Zulfakar
Faculty of Health Science, Universiti Kebangsaan Malaysia, Bangi, Selangor, Malaysia
e-mail: P110017@siswa.ukm.edu.my

Z. N. B. Z. Arifen
e-mail: P110185@siswa.ukm.edu.my

A. A. Siddiqui
e-mail: P110507@siswa.ukm.edu.my

A. Mahmod
e-mail: P110101@siswa.ukm.edu.my

R. Kathir
e-mail: P110508@siswa.ukm.edu.my

S. S. Zulfakar
e-mail: sitishahara.zulfakar@ukm.edu.my

S. Alsabbah (✉)
Department of Psychology, Fatima College of Health Sciences, Abu Dhabi, United Arab Emirates
e-mail: saher.alsabbah@fchs.ac.ae

© The Author(s), under exclusive license to Springer Nature Switzerland AG 2023   2407
M. Alshurideh et al. (eds.), *The Effect of Information Technology on Business and Marketing Intelligence Systems*, Studies in Computational Intelligence 1056,
https://doi.org/10.1007/978-3-031-12382-5_132

(2) compare the confidence in usage of vaccines for protection against the COVID-19 (3) determine the association between awareness about risks and benefits of the COVID-19 vaccine and confidence in getting it among selected undergraduate students in the Faculty of Health Sciences in Malaysia, Pakistan and UAE. A quantitative cross-sectional study was carried out using an online validated questionnaire with a 3-point Likert scale, which consisted of three domains, namely sociodemographic, awareness on risks and benefits of the vaccine, and confidence in getting the vaccine. Subjects were Faculty of Health Sciences' undergraduate students from three institutions; one each in Malaysia, Pakistan and UAE, that were recruited through non-probability sampling techniques; convenience and snowball sampling. The subjects' obtained scores on awareness and confidence from each country were averaged, and then compared between the countries using One-Way-ANOVA. Pearson's correlation was used to determine the association between awareness and confidence of the students. All data were analyzed using SPSS version 27.0. A number of 405 students from the three countries; Malaysia (142), Pakistan (124) and UAE (139) participated in this study. One-Way-ANOVA showed that students in Pakistan scored the highest mean awareness with a 1.55 ± 0.23 score, followed by UAE (1.53 ± 0.29) and Malaysia (1.49 ± 0.19). For confidence, students in Malaysia scored the highest mean confidence with a 1.25 ± 0.25 score, followed by Pakistan (1.24 ± 0.31) and UAE (1.20 ± 0.23). Pearson's correlation showed that there was a statistically significant association ($p < 0.001$, $r = 0.442$) between the students' awareness (1.52 ± 0.24) and confidence (1.23 ± 0.26) of getting COVID-19 vaccines. The study found that students in Pakistan had the highest awareness, followed by UAE and Malaysia. Whereas students in Malaysia had the highest confidence followed by Pakistan and UAE. There was a significant association found between the students' awareness regarding the benefits and risks of COVID-19 vaccines and their confidence in getting it.

Keywords COVID-19 · Vaccine · Undergraduate students · Awareness · Confidence

S. Darawsheh
Department of Administrative Sciences, The Applied College, (Imam Abdulrahman Bin Faisal University), Dammam, Saudi Arabia
e-mail: srdarawsehe@iau.edu.sa

A. Al-Shaar
Department of Self Development, Deanship of Preparatory Year and Supporting Studies, (Imam Abdulrahman Bin Faisal University), Dammam, Saudi Arabia
e-mail: anwarshaar5@gmail.com

M. Alshurideh
Department of Management, College of Business Administration, University of Sharjah, Sharjah, UAE
e-mail: malshurideh@sharjah.ac.ae; m.alshurideh@ju.edu.jo

Department of Marketing, School of Business, The University of Jordan, Amman 11942, Jordan

1 Introduction

Coronavirus Disease (CO = Corona; VI = Virus; D = Disease) a.k.a. COVID-19 is defined by the World Health Organization (WHO) as a newly determined virus that spreads through direct contact with respiratory droplets (aerosol) of the infected person in the form of a cough and/or a sneeze. This new virus and disease was unknown before the outbreak began in Wuhan, China, in December 2019 (Al-Sabbah et al., 2021; Lau, 2020; Looi, 2021; Lorini et al., 2018; Lucia et al., 2021; Muqattash et al., 2020; Polit & Beck, 2006; Polit et al., 2007; Regmi et al., 2016; Saied et al., 2021; Sanche et al., 2020; Shafi et al., 2020; Sharma et al., 2021; Sonawane et al., 2021; Syed Alwi et al., 2021; Wadood, et al., 2020; Xu, 2020). The COVID-19 pandemic is being viewed as the most significant worldwide health crisis of the century and one of the biggest challenges since World War II (Aljumah et al., 2021; Alshurideh et al., 2021; Chakraborty & Maity, 2020). It has been named the sixth public health emergency of the last ten years, internationally, by the WHO (Gallegos, 2020). This pandemic has not only led to a large number of fatalities, and health crises but also severe economic difficulties all over the world (Al-Dmour et al., 2021a; Al-Hamad et al., 2021; Chakraborty & Maity, 2020; Leo et al., 2021). The COVID-19 affected people across the world, unsettling and disrupting 'routine' in the lives of many. In almost all countries affected by COVID-19, entire educational, commercial institutions were closed and non-emergency health services were also suspended (Ali & Gatiti, 2020; Al-Maroof et al., 2021; Creese et al., 2021; Shah et al., 2021).

Being very highly contagious, the novel coronavirus virus spread to more than 200 countries. The virus is known to spread through aerosol or droplets emitted from the nose and mouth; hence threatening its transmission among humans through close contact (Alameeri et al., 2021; Al-Dmour et al., 2021b; Nuseir et al., 2021; Sanche et al., 2020). This led to a large majority of countries imposing complete or partial lockdowns since January and February 2020 (Ahmad et al., 2021; Al Khasawneh et al., 2021; Harahsheh et al., 2021; Lau, 2020). Although the symptoms varied from mild to moderate in severity, approximately 1 out of every 6 patients diagnosed with COVID-19 was likely to become seriously ill and develop difficulty breathing, with progression of disease. Older people, and those with underlying medical problems like high blood pressure, heart problems, diabetes and/or individuals with weak immunity systems, are more likely to develop serious illnesses (Sanche et al., 2020).

The second wave of COVID-19 has hit Asia badly (Looi, 2021), and has caused countries to declare a state of emergency in the respective countries. In Malaysia, the government has employed several measures like putting in order the Movement Control Order, stringent screening process at all airports and economic help to the needy citizens, as well as providing COVID-19 vaccines for free. Similar efforts have been observed by countries in Asia such as the United Arab Emirates and Pakistan, where a lockdown was imposed, work from home was enforced and educational institutions initiated online academic instruction (Al-Sabbah et al., 2021).

The COVID-19 vaccine which was well developed on a war footing in the developed countries (Kakar et al., 2021). At least 198 countries have begun the process of inoculating their populations against COVID-19, mobilizing more than a dozen vaccines for emergency use: from the American-manufactured Pfizer-BioNTech and Moderna jabs, to the Chinese Sinopharm and SinoVac. As of the Summer of 2021, more than three billion doses have been administered worldwide. But while several countries—such as the United Arab Emirates (UAE)—have been able to make significant headway towards vaccinating their citizens, others such as Malaysia and Pakistan are gradually stepping up the number of vaccines being administered per day.

Generally, the acceptance and demand of vaccines are complex in nature and are perceived differently by the behavioral nature in the communities (Biasio et al., 2021; Creese et al., 2021; Dutta, 2020). It is more complex when it is needed to be accepted in a pandemic as the nature of vaccines produced are not much known, hence hesitancy from people is very much to be predicted. Hesitancy in receiving the vaccines can be due to various factors like and also due to the already existing anti-vaccine movement which has been misleading and confusing the public, by propoganding mistrust in the system that is delivering the vaccination program (Saied et al., 2021). Therefore, this study will specifically look into examining the awareness and confidence about COVID-19 vaccines among selected students in the Faculty of Health Sciences: comparison study among Malaysia, Pakistan and UAE. In this study, the participants will be only from the Health Sciences Department for two reasons: (a) as healthcare providers, it will be mandatory to receive the vaccine and (b) hesitancy among the healthcare providers will affect the public negatively (Karim et al., 2020; Sharma et al., 2021).

The main objective of this study is to determine the awareness about the risks and benefits of the COVID-19 vaccine among selected undergraduate students in the Faculty of Health Sciences in Malaysia, Pakistan and UAE and their confidence in getting the vaccine. This study aimed to determine the awareness regarding the risks and benefits of the COVID-19 vaccines among selected undergraduate students in the Faculty of Health Sciences in Malaysia, Pakistan and UAE, and their confidence in getting the vaccine.

The on-going global discussion on the safety of the COVID-19 vaccines are potentially leaving communities undecided on whether to trust or not to trust. The concept of 'vaccine hesitancy' can be described as either to prolong or refuse in accepting vaccines that are made available (Gallegos, 2020; Sharma et al., 2021). Generally, vaccine hesitancy which starts with an individual will consequently affect the whole community with the similar hesitancy. And, in order to achieve the herd community, a high coverage rate of the vaccinated population is needed in the efforts to flatten the epidemic curve (Syed Alwi et al., 2021; Wadood, et al., 2020). Vaccine hesitancies are said to be specific to the environment and to be fluctuating in accordance to time and setting. In addition, it is prompted by factors like confidence and awareness (Saied et al., 2021; Sharma et al., 2021). Hence by examining the factors of awareness and confidence in getting COVID-19 vaccines, basic perception of the vaccine can be established.

2 Literature Review

2.1 Awareness Regarding the Risks and Benefits of COVID-19 Vaccines

Awareness denotes trust in vaccination safety, and the knowledge about the vaccines. In this pandemic, people are getting much information and resources of COVID-19 vaccines through various sources including television, radio, newspaper, social media, friends, healthcare providers, governments and NGOs. This information obtained, will shape their awareness level towards the COVID-19 vaccines (Domek et al., 2018), hence it is crucial to disseminate transparent and accurate information about the vaccine safety and efficacy to create a well-informed population (Alkalha et al. 2012; Domek et al., 2018). As advocated by Krejcie and Morgan (1970) people generally trust scientists the most followed by healthcare providers and agencies. Thus, if a country is trying to raise the awareness level of their citizens in this vaccine, all these agencies should team up to create awareness. Without proper awareness, the scene with vaccine hesitancy will continue to grow (Polit et al., 2007).

In a survey of 1068 medical students in India reported that they agree vaccines should be mandatory however they wanted an increase in vaccine advocacy to encourage those who were hesitant in getting it (Hui et al., 2020). This clearly describes low awareness, as they know it is important to receive the vaccine, however they aren't aware of the benefits, safety and efficacy of the vaccines. Similarly, in another study conducted by AlQutob et al. (2020) reported that poor acceptance of COVID-19 vaccines in Jordan is due to the lack of transparency and information dissemination from the national and public health authorities. A survey conducted among the Malaysians, reveals that Malaysians have a high rate of acceptance, but it still remains important to increase their awareness level in vaccine safety and effectiveness (Syed Alwi et al., 2021). Similarly, Pakistan reports have ample understanding on the vaccines and efforts for awareness should be uplifted (Yusoff, 2019a, 2019b).

2.2 Confidence in Getting COVID-19 Vaccines

Confidence denotes trust in vaccination safety, effectiveness and competence of the healthcare system. There is an overlap here with awareness and confidence as we can see with awareness, only then confidence can be built. The overlap factor is vaccination safety, so if the population is aware of the safety, then the confidence in effectiveness and the trust in the healthcare system can be established. Confidence is one of the major factors in vaccine hesitancy as the concerns of the safety, efficacy and the rush in vaccine development and production due to pandemic are lingering in the mind of the population. In addition, many rumors are also being spread which does not help this situation (Lorini et al., 2018; Lucia et al., 2021; Sonawane et al., 2021; Xu, 2020).

As illustrated in the result of a survey of 186 medical university students in Jeddah, the majority of the students were aware of the risks of COVID-19, but 44% of them were hesitant in getting the vaccine. Of these 68.3% were concerned about vaccine side effects. The study recommended vaccine advocacy through educational programs (Chakraborty & Maity, 2020). This shows that there is some kind of awareness level among the participants, however the trust or confidence is not achieved. And to establish confidence, government health departments must ensure that 'good' information is released for public consumption, to ensure their trust towards COVID-19 vaccine (Chakraborty & Maity, 2020; Kakar et al., 2021).

3   Method

The focus in this study is to determine the awareness and confidence level among selected students from the Faculty of Health Sciences on the COVID-19 vaccine. Since COVID-19 is posing to be one of the major threat to the entire population and Asian countries (Lau, 2020), this study would be one of the pioneers in assessing the awareness and confidence level about COVID-19 among undergraduate students in Faculty of Health Sciences in these three countries (Malaysia, Pakistan & UAE), providing a more scientific backup for relevant parties to gear up their efforts in advocating about COVID-19 vaccines in the effort to flatten the pandemic curve.

3.1   Study Design

A quantitative cross-sectional study was conducted for a period of around one week from 29th May to 5th June, 2021.

3.2   Population and Sample

Subjects selected were among undergraduate students in the Faculty of Health Sciences from three universities: Universiti Kebangsaan Malaysia (UKM) in Malaysia, Ziauddin University (ZU) in Pakistan and Fatima College of Health Sciences (FCHS) in UAE. Non-probability sampling strategies through a convenience and snowball sampling technique were used to target all students that consented to fill in an online questionnaire.

The sample size was determined based on Krejcie and Morgan (1970) formula,

$$s = X^2 NP(1 - P) / [d^2(N - 1) + X^2 P(1 - P)]$$

According to the formula, s is the required sample size; X^2 is the table value of chi-square for 1 degree of freedom (3.841) at the desired confidence level of 0.05; N is the population size of 4000, P is the population proportion (assumed to be 0.50 since this would provide the maximum sample size) and d is the degree of accuracy expressed as proportion of 0.05. Based on the formula, the calculated sample size was 351. It was assumed that there would be a 10% drop rate in response. Hence, the required sample size was,

$$s = 351 + (10\% \times 351) = 351 + 35 = 386$$

3.3 Questionnaire's Validity and Reliability

The instrument used was a questionnaire adapted from studies by Al-Sabbah et al. (2021), Muqattash et al. (2020), Biasio et al. (2021). The adapted questionnaire that was named as Student COVID-19 Vaccine (STU-COV) Questionnaire, had three main domains in which were sociodemographic data, awareness on risks and benefits of the vaccine and confidence in getting the vaccine. In this questionnaire, the awareness and confidence domain used a 3-point Likert scale.

Collecting data for a questionnaire based study through traditional techniques (face to face) could be expensive, uncomfortable, unsafe (during the COVID-19 pandemic) and time consuming. The emerging data collection technique based upon internet/e-based modern technologies like online platforms (e.g. SurveyMonkey) was relatively a cost effective approach. These unique data collection techniques could collect a huge number of responses from participants in a short time frame as well as it would be a solution for a situation like COVID-19 pandemic, where no face-to-face contact was recommended (Khan et al., 2020).

Moreover, there was significant evidence that several large cross-country researches had actually been completed making use of online sets of questions to collect different types of data through popular dedicated platforms (e.g. https://www.surveymonkey.co.uk) (Regmi et al., 2016). That's why an anonymous online structured questionnaire was designed using Survey monkey to save time and make it more comfortable for the participants from the three different countries. The link for the questionnaire was exclusively shared by email with the undergraduate students—both individually and through their social media groups. Non-probability sampling strategy through a convenience and snowball sampling technique was used to target all students in the Faculty of Health Sciences consenting and willing to fill the survey. Upon completion of the survey, data was downloaded as a form of SPSS from Survey monkey platform, format and data analysis was conducted using SPSS software version 27.0.

The questionnaire had been tested for validity and reliability, in which two validation protocols i.e. face validation and content validation were conducted for all domains in order to validate the newly developed questionnaire. Whereas, Cronbach's Alpha was used to test for reliability for the awareness and confidence domain. In the pretest of the questionnaire, six experts were involved in the content validation, nine students from the Faculty of Health Sciences were involved in the face validation, and 30 students (10 students from each country) from the Faculty of Health Sciences were recruited to check on the reliability of the STU-COV Questionnaire. The acceptable Face Validity Index (FVI) and Content Validity Index (CVI) values for face validation and content validation are both more 0.83 (Hui et al., 2020).

Meanwhile, the acceptable values for Cronbach's Alpha are 0.70 for 10 items and more (awareness domain), and 0.50 for less than 10 items (confidence domain) (Hui et al., 2020; Looi, 2021). Based on the validity and reliability test conducted, the STU-COV Questionnaire achieved a FVI and CVI values of more than 0.83 for all three domains. Whereas the awareness and confidence domain showed acceptable Cronbach's Alpha value of 0.78 and 0.56.

4 Data Analysis and Findings

A p-value of less than or equal to 0.05 was taken as significant. Data analysis was done using SPSS version 27.0.1.0 and Microsoft Excel. One-Way ANOVA was performed to compare the awareness about the risks and benefits of getting vaccinated against the COVID-19 among selected students in Malaysia, Pakistan and UAE, and also to compare the confidence in the usage of vaccines for protection against the COVID-19 among selected students in the three countries. For determining the association between the awareness about the risks and benefits of the COVID-19 vaccine among selected students and their confidence in getting the vaccine, a Pearson's Correlation test was conducted.

Table 1 showed the sociodemographic characteristics of the subjects. N = 405 students had participated in the study.

4.1 Awareness About the Risks and Benefits of COVID-19 Vaccines

As seen in Table 2 students from Pakistan had the highest awareness (1.55 ± 0.23) compared to UAE (1.53 ± 0.29) and Malaysia (1.49 ± 0.19). One-Way ANOVA was used to compare the awareness about the risks and benefits of getting vaccinated against the COVID-19 among the selected students between the three countries. The test showed that there was no statistically significant difference ($p > 0.05$) on the awareness regarding risk and benefit of COVID-19 vaccines among the students in the three countries.

Table 1 Sociodemographic characteristics

Variable	n (%)	Mean (SD)
Age (year)		21.34 (2.38)
Sex		
Male	47 (11.6)	
Female	358 (88.4)	
Country		
Malaysia	142 (35.1)	
UAE	139 (34.3)	
Pakistan	124 (30.6)	
Year of study		
1st year	95 (23.5)	
2nd year	105 (25.9)	
3rd year	103 (25.4)	
4th year	89 (22.0)	
5th year	13 (3.2)	
Major		
Physical therapy	75 (18.5)	
Occupational therapy	45 (11.1)	
Medical technology	19 (4.7)	
Radiology/medical imaging	76 (18.8)	
Nursing	109 (26.9)	
Emergency health	57 (14.1)	
Speech language therapy	10 (2.5)	
Psychology	13 (3.2)	
Others	1 (0.2)	
Sources of information		
Printed media	62 (15.3)	
Radio/television	147 (36.3)	
Social media	343 (84.7)	
Internet	274 (67.7)	
Scientific journal	57 (14.1)	
Official site	191 (47.2)	
Other	8 (2.0)	

The percentage of respondents who relied on the updates of information from authorized organizations were 95.1% for Malaysia, 87.8% for UAE and 88.7% for

Table 2 The comparison on level awareness of risks and benefits of COVID-19 vaccines among selected students in Malaysia, Pakistan and UAE

Country	Score of awareness (Mean ± SD)	F statistic	*p-value
Pakistan (n = 124)	1.55 ± 0.23	1.890	0.152
UAE (n = 139)	1.53 ± 0.29		
Malaysia (n = 142)	1.49 ± 0.19		

* There is a significant association according to the One-Way ANOVA test ($p < 0.05$)

Pakistan. It was reported in Fridman et al. (2020) that more than half of the respondents who aged less than 25 years old in USA got their information from Communicable Disease Centre (CDC) (64.7%), local health department (63.7%) and WHO (59.2%), which is lower from the findings from this study.

4.2 Confidence in Getting COVID-19 Vaccines

According to Table 3, students from Malaysia had the highest confidence (1.25 ± 0.25) compared to Pakistan (1.24 ± 0.31) and UAE (1.20 ± 0.23). One-Way ANOVA was used to compare the confidence of getting COVID-19 vaccines among selected students in Malaysia, Pakistan and UAE. The test showed that there was no significant difference ($p > 0.05$) on the confidence of getting COVID-19 vaccines among students in the Faculty of Health Sciences in between the three countries.

This study found that 95.1% respondents from Malaysia, 92.1% respondents from UAE and 83.1% of the respondents from Pakistan trusted the healthcare sector in providing accurate and reliable information. In comparison with the results of this study, Imperial College London (2021) reported that confidence in the healthcare sector varies markedly between countries across the globe, ranging from the highest level of confidence on their healthcare sector (70%) in United Kingdom (n = 5005), and the lowest level of confidence (42%) in South Korea (n = 2460). This discrepancy was probably because the targeted population in this study were well-informed undergraduate students from healthcare sectors. Thus, the result could not be generalized into the general population (Saied et al., 2021; Sonawane et al., 2021).

Table 3 The comparison on awareness of getting COVID-19 vaccines among selected students in Malaysia, Pakistan and UAE

Country	Score of confidence (Mean ± SD)	F statistic	*p-value
Malaysia (n = 142)	1.25 ± 0.25	1.493	0.226
Pakistan (n = 124)	1.24 ± 0.31		
UAE (n = 139)	1.20 ± 0.23		

* There is a significant association according to the One-Way ANOVA test ($p < 0.05$)

Table 4 The correlation between awareness about the risks and benefits of the COVID-19 vaccine among selected students and their confidence in getting the vaccine

Item	Mean ± SD	Pearson's correlation (r)	p-value
Awareness	1.52 ± 0.24	0.442	0.001*
Confidence	1.23 ± 0.26		

* There is a significant association according to the One-Way ANOVA test ($p < 0.05$)

4.3 Association Between the Awareness of Benefit and Risk of the COVID-19 Vaccines

Pearson's Correlation was used to determine the correlation between the awareness of benefit and risk of the COVID-19 vaccines among selected students and their confidence in getting the vaccines. Table 4 showed that there was a statistically significant correlation between the awareness of benefit and risk of COVID-19 vaccines among selected students and their confidence in getting the vaccines ($p < 0.001$). Even though the correlation was found to be weak, it was a positive correlation ($r = 0.442$). In their systematic review, Biasio et al. (2021) also found that there was an association between health literacy and vaccination status.

5 Discussion and Conclusion

In conclusion, this study found that there were no significant differences between awareness and confidence level among undergraduates in the Faculty of Health Sciences in Malaysia, Pakistan and UAE. In these three developing countries, the majority of the participating students have awareness on risks and benefits of COVID-19 vaccines. Moreover, most of the students obtained the information from the internet and social media. Hence, government and authorities could use these platforms to educate and create awareness continuously to encourage vaccine uptake among university students, as well as to strengthen overall immunization for COVID-19. It was also found that the confidence level was related to their awareness, thus increasing the awareness of risks and benefits of vaccination would also increase their confidence level in getting it.

6 Recommendations

Questionnaires such as ours could be considered as an important tool that can be applied into a larger population to measure the level of awareness and confidence in getting the vaccine. To ensure increased awareness and confidence among the general population, the elements that define and create awareness should be understood and thus, interventions can be crafted accordingly. A randomized sampling method also can be used to represent the real population. A thematic analysis can be done using a qualitative approach for this study in the future. More participants can be recruited for future study.

7 Limitations

This study recruited university students from the Faculty of Health Sciences only, who can be considered to have more awareness about the vaccine in comparison with the general population of any of the three countries at large. Moreover, this study could not determine the cause-and-effect relationship between the awareness and confidence in getting the COVID-19 vaccine. This study only measured the samples' awareness and confidence at one period of time.

Acknowledgements The authors would like to express their gratitude to everyone who participated in the pilot study and helped validate the STU-COV. The authors also acknowledge all students who have participated in this study. We would also like to acknowledge the Faculty of Health Sciences UKM, ZU and FCHS.

References

Ahmad, A., Alshurideh, M. T., Al Kurdi, B. H., & Salloum, S. A. (2021). *Factors impacts organization digital transformation and organization decision making during Covid19 pandemic* (Vol. 334).

Al Khasawneh, M., Abuhashesh, M., Ahmad, A., Masa'deh, R., & Alshurideh, M. T. (2021). *Customers online engagement with social media influencers' content related to COVID 19* (Vol. 334).

Alameeri, K. A., Alshurideh, M. T., & Al Kurdi, B. (2021). *The effect of Covid-19 pandemic on business systems' innovation and entrepreneurship and how to cope with it: A theatrical view* (Vol. 334).

Al-Dmour, A., Al-Dmour, H., Al-Barghuthi, R., Al-Dmour, R., & Alshurideh, M. T. (2021a). *Factors influencing the adoption of e-payment during pandemic outbreak (COVID-19): Empirical evidence* (Vol. 334).

Al-Dmour, R., AlShaar, F., Al-Dmour, H., Masa'deh, R., & Alshurideh, M. T. (2021b). *The effect of service recovery justices strategies on online customer engagement via the role of "customer satisfaction" during the Covid-19 pandemic: An empirical study* (Vol. 334).

Al-Hamad, M. Q., Mbaidin, H. O., Alhamad, A. Q. M., Alshurideh, M. T., Kurdi, B. H. A., & Al-Hamad, N. Q. (2021). Investigating students' behavioral intention to use mobile learning in higher education in UAE during Coronavirus-19 pandemic. *International Journal of Data and Network Science, 5*(3). https://doi.org/10.5267/j.ijdns.2021.6.001

Ali, M. Y., & Gatiti, P. (2020). The COVID-19 (Coronavirus) pandemic: Reflections on the roles of librarians and information professionals. *Health Information & Libraries Journal, 37*(2), 158–162.

Aljumah, A., Nuseir, M. T., & Alshurideh, M. T. (2021). The impact of social media marketing communications on consumer response during the COVID-19: Does the brand equity of a university matter. *Effects Coronavirus Diseases Business Intelligence, 334*, 367–384.

Alkalha, Z., Al-Zu'bi, Z., Al-Dmour, H., Alshurideh, M., & Masa'deh, R. (2012). Investigating the effects of human resource policies on organizational performance: An empirical study on commercial banks operating in Jordan. *European Journal of Economics, Finance and Administrative Sciences, 51*(1), 44–64.

Al-Maroof, R. S., Alshurideh, M. T., Salloum, S. A., AlHamad, A. Q. M., & Gaber, T. (2021). Acceptance of Google Meet during the spread of Coronavirus by Arab university students. *Informatics, 8*(2), 24.

AlQutob, R., Moonesar, I. A., Tarawneh, M. R., Al Nsour, M., & Khader, Y. (2020). Public health strategies for the gradual lifting of the public sector lockdown in Jordan and the United Arab Emirates during the COVID-19 crisis. *JMIR Public Health and Surveillance, 6*(3), e20478.

Al-Sabbah, S., Darwish, A., Fares, N., Barnes, J., & Almomani, J. A. (2021). Biopsychosocial factors linked with overall well-being of students and educators during the COVID-19 pandemic. *Cogent Psychology, 8*(1), 1875550.

Alshurideh, M. T., Hassanien, A. E., & Masa'deh, R. (2021). *The effect of coronavirus disease (COVID-19) on business intelligence*. Springer.

Biasio, L. R., Bonaccorsi, G., Lorini, C., Mazzini, D., & Pecorelli, S. (2021). Italian adults' likelihood of getting covid-19 vaccine: A second online survey. *Vaccines, 9*(3), 268.

Chakraborty, I., & Maity, P. (2020). COVID-19 outbreak: Migration, effects on society, global environment and prevention. *Science of the Total Environment, 728*, 138882.

Creese, J., Byrne, J.-P., Conway, E., Barrett, E., Prihodova, L., & Humphries, N. (2021). 'We all really need to just take a breath': Composite narratives of hospital doctors' well-being during the COVID-19 pandemic. *International Journal of Environmental Research and Public Health, 18*(4), 2051.

Domek, G. J., et al. (2018). Measuring vaccine hesitancy: Field testing the WHO SAGE Working Group on Vaccine Hesitancy survey tool in Guatemala. *Vaccine, 36*(35), 5273–5281.

Dutta, A. K. (2020). Vaccine against Covid-19 disease–present status of development. *Indian Journal of Pediatrics, 87*(10), 810–816.

Fridman, I., Lucas, N., Henke, D., & Zigler, C. K. (2020). Association between public knowledge about COVID-19, trust in information sources, and adherence to social distancing: Cross-sectional survey. *JMIR Public Health and Surveillance, 6*(3), e22060.

Gallegos, M., et al. (2020). Coping with the coronavirus (Covid-19) pandemic in the Americas: recommendations and guidelines for mental health. *Revista Interamericana de Psicología, 1*–28.

Harahsheh, A. A., Houssien, A. M. A., & Alshurideh, M. T. (2021). The effect of transformational leadership on achieving effective decisions in the presence of psychological capital as an intermediate variable in private Jordanian. In *The effect of coronavirus disease (COVID-19) on business intelligence* (pp. 221–243). Springer Nature.

Hui, D. S., et al. (2020). The continuing 2019-nCoV epidemic threat of novel coronaviruses to global health—The latest 2019 novel coronavirus outbreak in Wuhan, China. *International Journal of Infectious Diseases, 91*, 264–266.

Kakar, A., et al. (2021). COVID vaccines: A step towards ending the pandemic. *Current Medicine Research and Practice, 11*(1), 23.

Karim, W., Haque, A., Anis, Z., & Ulfy, M. A. (2020). The movement control order (mco) for Covid-19 crisis and its impact on tourism and hospitality sector in Malaysia. *International Tourism and Hospitality Journal, 3*(2), 1–7.

Khan, S., Gilani, U. S., Raza, S. M. M., & Hussain, T. (2020). *Evaluation of general awareness among professionals regarding COVID-19: A survey based study from Pakistan.*

Krejcie, R. V., & Morgan, D. W. (1970). Determining sample size for research activities. *Educational and Psychological Measurement, 30*(3), 607–610.

Lau, H., et al. (2020). The positive impact of lockdown in Wuhan on containing the COVID-19 outbreak in China. *Journal of Travel Medicine.*

Leo, S., Alsharari, N. M., Abbas, J., & Alshurideh, M. T. (2021). *From offline to online learning: A qualitative study of challenges and opportunities as a response to the COVID-19 pandemic in the UAE higher education context* (Vol. 334).

Looi. M.-K. (2021). *Covid-19: Japan declares second state of emergency as Asia struggles with virus surge.* British Medical Journal Publishing Group.

Lorini, C., et al. (2018). Health literacy and vaccination: A systematic review. *Human Vaccines & Immunotherapeutics, 14*(2), 478–488.

Lucia, V. C., Kelekar, A., & Afonso, N. M. (2021). COVID-19 vaccine hesitancy among medical students. *Journal of Public Health (bangkok), 43*(3), 445–449.

Muqattash, R., Niankara, I., & Traoret, R. I. (2020). Survey data for COVID-19 vaccine preference analysis in the United Arab Emirates. *Data in Brief, 33*, 106446.

Nuseir, M. T., Aljumah, A., & Alshurideh, M. T. (2021). *How the business intelligence in the new startup performance in UAE during COVID-19: The mediating role of innovativeness* (Vol. 334).

Polit, D. F., & Beck, C. T. (2006). The content validity index: Are you sure you know what's being reported? Critique and recommendations. *Research in Nursing & Health, 29*(5), 489–497.

Polit, D. F., Beck, C. T., & Owen, S. V. (2007). Is the CVI an acceptable indicator of content validity? Appraisal and recommendations. *Research in Nursing & Health, 30*(4), 459–467.

Regmi, P. R., Waithaka, E., Paudyal, A., Simkhada, P., & van Teijlingen, E. (2016). Guide to the design and application of online questionnaire surveys. *Nepal Journal of Epidemiology, 6*(4), 640.

Saied, S. M., Saied, E. M., Kabbash, I. A., & Abdo, S. A. E. (2021). Vaccine hesitancy: Beliefs and barriers associated with COVID-19 vaccination among Egyptian medical students. *Journal of Medical Virology, 93*(7), 4280–4291.

Sanche, S., Lin, Y. T., Xu, C., Romero-Severson, E., Hengartner, N. W., & Ke, R. (2020). *The novel coronavirus, 2019-nCoV, is highly contagious and more infectious than initially estimated.* arXiv: 2002.03268

Shafi, M., Liu, J., & Ren, W. (2020). Impact of COVID-19 pandemic on micro, small, and medium-sized enterprises operating in Pakistan. *Research in Globalization, 2*, 100018.

Shah, S. F., Alshurideh, M. T., Al-Dmour, A., & Al-Dmour, R. (2021). *Understanding the influences of cognitive biases on financial decision making during normal and COVID-19 pandemic situation in the United Arab Emirates* (Vol. 334).

Sharma, M., Davis, R. E., & Wilkerson, A. H. (2021). COVID-19 vaccine acceptance among college Students: A theory-based analysis. *International Journal of Environmental Research and Public Health, 18*(9), 4617.

Sonawane, K., Troisi, C. L., & Deshmukh, A. A. (2021). COVID-19 vaccination in the UK: Addressing vaccine hesitancy. *The Lancet Regional Health, 1.*

Syed Alwi, S. A. R., Rafidah, E., Zurraini, A., Juslina, O., Brohi, I. B., & Lukas, S. (2021). A survey on COVID-19 vaccine acceptance and concern among Malaysians. *BMC Public Health, 21*(1), 1–12.

Wadood, M. A., et al. (2020). *Knowledge, attitude, practice and perception regarding COVID-19 among students in Bangladesh: Survey in Rajshahi University.* Medrxiv.

Xu, X.-W., et al. (2020). Clinical findings in a group of patients infected with the 2019 novel coronavirus (SARS-Cov-2) outside of Wuhan, China: Retrospective case series. *BMJ, 368.*

Yusoff, M. S. B. (2019a). ABC of content validation and content validity index calculation. *Resource, 11*(2), 49–54.

Yusoff, M. S. B. (2019b). ABC of response process validation and face validity index calculation. *Resource, 11*(3).

Predicting Bitcoin Prices Using ANFIS and Haar Model

Jamil J. Jaber⬭, Rami S. Alkhawaldeh⬭, Samar M. Alkhawaldeh⬭,
Ra'ed Masa'deh⬭, and Muhammad Turki Alshurideh⬭

Abstract This study aims to model and enhance the forecasting accuracy of cryptocurrency market data patterns using the daily bitcoin (BTC) close price data with 1535 observations from December 2017 to January 2022. The model employs a nonlinear spectral model of maximum overlapping discrete wavelet transform (MODWT) with Haar mathematical functions in conjunction with an adaptive network-based fuzzy inference system (ANFIS). We have selected the logarithm volume of bitcoin (LV) and logarithm trade count (LCT) as input values according to correlation and multiple regressions. The input and output variables have been collected from the cryptocurrency market. The performance of the proposed model (MODWT-Haar-ANFIS) is compared with traditional models that are the autoregressive integrated moving average (ARIMA) model and the ANFIS model. The obtained results show that the performance of MODWT-Haar-ANFIS is better than that of the traditional models. Therefore, the proposed forecasting model is a promising approach that capable of deploying in the cryptocurrency markets.

Keywords Haar model · Bitcoin · Fuzzy model · Forecasting · ANN

J. J. Jaber
Department of Finance, School of Business, The University of Jordan, Aqaba, Jordan
e-mail: j.jaber@ju.edu.jo

R. S. Alkhawaldeh (✉) · S. M. Alkhawaldeh
Department of Computer Information Systems, School of Information Technology and Systems,
The University of Jordan, Aqaba, Jordan
e-mail: r.alkhawaldeh@ju.edu.jo

R. Masa'deh
Department of Management Information Systems, School of Business, The University of Jordan,
Amman, Jordan
e-mail: r.masadeh@ju.edu.jo

M. T. Alshurideh
Department of Marketing, School of Business, The University of Jordan, Amman 11942, Jordan
e-mail: m.alshurideh@ju.edu.jo

University of Sharjah, Sharjah, UAE

1 Introduction

Financial markets refer to any location or system that allows buyers and sellers to exchange financial instruments such as bonds, shares, foreign traditional currencies, cryptocurrency, and derivatives (Madura, 2020). Financial markets enhance the transfer of surplus cash from savers (surplus cash(to institutions (deficit cash), who subsequently invest them in productive activities. Financial markets encourage players to raise cash and transfer risk through derivatives contracts (forward, future, option) (Madura, 2020). Capital markets are used for long-term financing, whereas money markets are used for short-term finance. The financial market mainly is divided into two parts: primary markets and secondary markets. The primary market deals with the exchange of initial public offering (IPO) of new financial products, whereas the secondary market deals with the exchange of previously issued financial products. There are many types of financial markets such as capital markets, money markets, cryptocurrency markets, derivatives markets, and commodity markets. The capital markets provide long term investment such as stock and Bond. The money markets provide short term debt financing, while the cryptocurrency markets trade digital assets and financial technologies. The derivatives markets give mechanisms for managing financial risk, while commodity markets trade product such as gold, oil, livestock, and wheat (Hull, 2014).

Cryptocurrency is a digital currency that operates without the assistance of a central bank and uses encryption techniques to regulate unit generation and verify payment transfers. It's more accessible, more widely available, and has grown in popularity over time to become one of the most extensively used forms of currency worldwide. Cryptocurrency has been around since 2009, but it wasn't until 2017 that it became popular as a means of investing. The use of encryption to authenticate transactions is referred to as cryptocurrency. This means that cutting-edge code is being utilized to store and transfer cryptocurrency data between wallets and public ledgers. The purpose of encryption is to assure security and safety. Cryptocurrencies are built on blockchain, a decentralized public database that records all transactions that are updated and maintained by currency holders (Fang, 2022). Cryptocurrency units are created through a process known as mining, which entails using computer processing power to solve complicated mathematical problems in order to generate coins. Users can also buy the currencies via brokers, keep them in digital wallets, and spend them.

Advantages of trading cryptocurrency. First, drastic changes refer to the volatility of cryptocurrencies, which is typically likely to attract the curiosity of speculators and investors. The quick volatility of daily prices can give traders with excellent money-making chances, but they also carry a higher level of risk. Second, the cryptocurrency market is open 24 h a day, seven days a week because it is a decentralized market. Third, it offers certain advantages in terms of privacy to investors. Four, peer-to-peer transactions are one of the most significant advantages of cryptocurrencies is that they do not require the involvement of financial institution middlemen. This can minimize transaction costs. Furthermore, this functionality may attract to individuals who are distrust of traditional systems. Some cryptocurrencies can provide holders with additional benefits, such as limited ownership and voting rights (Fang, 2022). Disadvantages of trading cryptocurrency; First, the limitations of technical infrastructure, transaction volume, and transaction speed make it unable to compete with traditional currency trading. Second, cybersecurity issues due to attack from hackers. Finally, regulations issue because it is not organized from some countries (Fang, 2022).

Bitcoin, which was founded in 2009, was the first cryptocurrency and is now the most widely traded (Fang, 2022). Satoshi Nakamoto created the currency, which is commonly assumed to be a codename for an individual or group of individuals whose specific identity is unknown. To improve the privacy of Bitcoin transactions, the currency employs a technique known as "Elliptic Curve Cryptography" (Wang et al., 2020). However, Bitcoin is a decentralized digital currency offering lower fees in transactions using secured decentralized authority in Blockchain architecture.

Artificial neural network (ANN) is a supervised machine-learning model utilized in a variety of business and scientific applications (Khuntia & Hiremath, 2019; Li et al., 2020; Smyth & Narayan, 2018; Xiao et al., 2018). The employment of ANNs has emerged in predicting the stock market. The authors in Lv et al. (2020) used the Haar wavelet and Takagi–Sugeno–Kang (TSK) fuzzy rule-based system to forecast stock prices depending on different technical indicators. The model is trained using the Taiwan Stock Exchange market dataset, which has a performance accuracy of up to 99.1%. The authors in Çatık et al. (2020) proposed a fusion-forecasting model of wavelet as a data preparation technique, fuzzy logic, and ANN. The model was built from a stock market dataset gathered during the period from 2005 to 2010. The hybrid behavior of the model indicates reasonable accuracy instead of employing the models independently. The authors in Munandar (2015) also proposed a fuzzy wavelet neural network (FWNN) model for predicting stock prices. The model is evaluated on a daily stock price dataset of 1000 samples that is 90% for training and 10% for testing. The model achieves remarkable results in comparison with other approaches.

The adaptive network-based fuzzy inference system (ANFIS) is a combination of fuzzy and ANN learning algorithms of five layers (Çatık et al., 2020; Geng & Wang, 2010; Khuntia & Hiremath, 2019; Li et al., 2020; Lv et al., 2020; Smyth & Narayan, 2018; Wang et al., 2020; Xiao et al., 2018) and (Fan et al., 2012) that is used in a variety of applications (Nadimi et al., 2010; Wang et al., 2020). The ANFIS model has achieved superior accuracy results in many applications as a forecasting model in comparison to the ANN model independently (Abiyev & Abiyev, 2012; Chang & Fan, 2008; Homayouni & Amiri, 2011; Honghui & Yongqiang, 2012; Munandar, 2015; Sehgal et al., 2014; Septiarini et al., 2016). Therefore, this paper exploits using the ANFIS model in forecasting the fluctuations in the cryptocurrency stock market. We evaluate the Bitcoin as a recent type of cryptocurrency stock market application. *Thus, our contributions is summarized as follows. First, Gathering a cryptocurrency dataset such as Bitcoin that maintains some features and stock price as a dependent variable. Second, Evaluating the proposed model using evaluation metrics to reveal the effectiveness of the proposed model.*

This study is organized as follows. Materials and methods are explained in Sect. 2. The research design and methodology are discussed in Sect. 3. The empirical results are analyzed in Sect. 4. The conclusions are drawn in Sect. 5.

2 Materials and Methods

This section gives a background of the main concepts used in our study.

2.1 Wavelet Transform Formula

The Wavelet Transform (WT) is a mathematical function that transforms time series data into a time-scale domain. Due to its inherent nature, WT is an appealing solution for non-stationary data, particularly stock market data. Discrete Wavelet Transforms (DWT), Continuous Wavelet Transforms (CWT), and Maximum Overlapping Wavelet Transforms (MOWT) are the three types of WT. In general, these transformations behave the same way. The key difference between DWT and MODWT is that DWT can only be utilized with a certain number of observations (the number of observations should be 2 raised to the power J), but MODWT can handle any size of data. As a result, we'll concentrate on MODWT in this study because of its adaptability (Zhang & Wang, 2012).

WT is essentially an extension of the Fourier Transform (FT) (Adil, 2015), which uses sine and cosine functions. WT satisfies the admission criteria:

$$C_\varphi = \int_0^\infty \varphi(f) \vee \frac{\varphi(t)}{f} df < \infty \tag{1}$$

where $\varphi(f)$ is the FT and a function of frequency f, $\varphi(t)$. The WT is used in many applications such as image analysis and signal processing. It was introduced to overcome the problem of FT, especially when dealing with time, space, or frequency.

There are two types of WT namely the Father wavelet that describes the low-frequency components (smooth data) and the mother wavelet that describes the high-frequency (detailed data) components as shown in Eqs. (2) and (3), with j = 1, 2, 3, ..., J in the J-level wavelet decomposition:

$$\phi j, k = 2^{\left(\frac{-j}{2}\right)} \phi \left(t - \frac{2^j k}{2^j} \right) \tag{2}$$

$$\varphi j, k = 2^{\left(\frac{-j}{2}\right)} \varphi \left(t - \frac{2^j k}{2^j} \right) \tag{3}$$

where J denotes the maximum scale sustainable by the number of data points and the two types of wavelets stated above, namely father wavelets and mother wavelets and satisfies:

$$\int \phi(t)dt = 1 \text{ and } \int \varphi(t)dt = 0 \tag{4}$$

The general mathematical model is presented in Eq. (5)

$$S_{j,k} = \int \phi_{j,k} f(t)dt, \, d_{j,k} = \int \varphi_{j,k} f(t)dt, \tag{5}$$

In more details,

$$F(t) = \sum S_{j,k}\phi_{j,k}(t) + \sum d_{j,k}\varphi_{j,k}(t)$$
$$+ \sum d_{j-1,k}\varphi_{j-1,k}(t) + \cdots + \sum d_{1,k}\varphi_{1,k}(t) \tag{6}$$

$$S_j(t) = \sum S_{j,k}\phi_{j,k}(t) \text{ and } D_j(t) = \sum d_{j,k}\phi_{j,k}(t) \tag{7}$$

The WT is used to calculate the approximation coefficient in Eq. (7) where $S_j(t)$ and $D_j(t)$ are introducing the smooth and detailed coefficients, respectively. The detailed coefficients are used to discover the main variations of the original data, whilst the smooth coefficients contain the most relevant aspects of the original data.

WT has a number of well-known transform functions, including Haar, Daubechies (d4), coiflet (c6), Least Asymmetric (LA8), and the best-localized (bl14). The following are the primary characteristics of these functions: (1) Except for the Haar model, the WT functions are arbitrary regular. (2) Except for the Haar model, WT functions do not have explicit expressions. (3) Real numbers are used in WT functions. (4) The scale function, orthogonal analysis, bio-orthogonal analysis, continuous/discrete transformation, accurate reconstruction, and quick algorithm are all WT functions that are orthogonal, compact, and support an arbitrary number of zero moments. (5) The Haar, LA8, and d4 modes are symmetric, although C6 and Bl14 have near symmetry.

2.2 Autoregressive Integrated Moving-Average Model (ARIMA)

In time series analysis, the auto-regressive moving average (ARMA) models are used to represent stationary time-series data. A moving average (MA) and an auto-regressive (AR) model are combined in the ARMA model. A time series $\{e_t\}$ is called a white noise (WN) process and $\{Y_t\}$ is called Gaussian process iff for all t, e_t is *iid* $N(0, \sigma^2)$. A time series $\{Y_t\}$ is said to follow the ARMA (p, q) model if:

$$Y_t = \mu + \phi_1 Y_{t-1} + \phi_2 Y_{t-2} \cdots + \phi_p Y_{t-p} + e_t - \theta_1 e_{t-1} - \theta_2 e_{t-2} \cdots - \theta_q e_{t-q} \quad (8)$$

where q and p are non-negative integers, p represents order of AR, q is defined as order of the first MA part, and $\{e_t\}$ is the WN process. An extension of the ordinary ARMA model is the auto-regressive integrated moving-average model (ARIMA (p, d, q)) given by:

$$\phi_p(B)(1 - B)^d Y_t = \theta_0 + \theta_q(B)e_t \quad (9)$$

where p, d and q denote orders of auto-regression, integration (differencing) and moving average, respectively. When $d = 0$, the ARIMA model reduces to the ordinary ARMA model.

2.3 ANFIS Model

The ANFIS model combines the fuzzy logic and artificial neural network into a single model. Its operations consist of forwarding and backward learning algorithms. The forward step consists of five layers. The fuzzy inference system under consideration is supposed to have two inputs (x, y) and one output (z) to simplify the explanations. A standard rule base of fuzzy if–then rules for a first order of Sugeno fuzzy model can be expressed as: If x is A_1 and y is B_1 then $f_1 = p_1 x + q_1 y + r_1$, where p, r, and q are linear output parameters. The ANFIS architecture with two inputs and one output is as shown in Fig. 1.

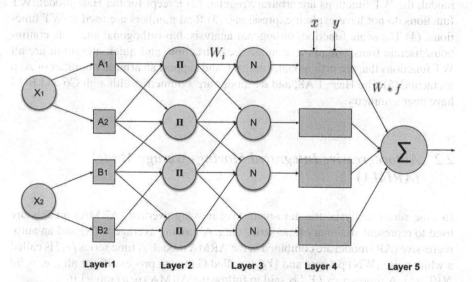

Fig. 1 ANFIS architecture of two inputs and one output with four rules

Layer-1: Every node i in this layer is a square node with a node function.

$$O_{1,i} = \mu_{Ai}(x), \text{ for } i = 1, 2, 3 \text{ and } O_{1,i} = \mu_{Bi-3}(y), \text{ for } i = 4, 5, 6 \qquad (10)$$

where x and y are inputs to node i, and A_i and B_i are linguistic labels for inputs. The $O_{1,i}$ is the membership function of A_i and B_i. Typically, $\mu_{Ai}(x)$ and $\mu_{Bi}(y)$ are selected to be bell-shaped with maximum value of 1 and minimum value of 0, such as $\mu_{Ai}(x), \mu_{Bi-3}(y) = \exp\left(\left(\frac{-(x_i-c_i)}{a_i}\right)^2\right)$, of the set of parameteris a_i, c_i. In this layer, these parameters are referred to as premise parameters. The fuzzification method turns crisp values into linguistic values by utilizing the Gaussian function as the shape of the membership function.

Layer-2: Each node in this layer is a circle node labeled Π that multiplies the incoming signals and sends out the product. For example,

$$O_{2,i} = w_i = \mu_{Ai}(x).\mu_{Bi-3}(y), i = 1, 2, 3, \ldots, 9 \qquad (11)$$

The firing strength of a rule is described by each node output. The inference step uses the t-norm operator (AND operator) in this layer.

Layer-3: Each node in this layer is a circle node called N. The i th node measures the ratio of the i th rule firing strength to the sum of all rules firing strengths:

$$O_{3,i} = \overline{w}_i = \frac{w_i}{(w_1 + w_2 + \cdots + w_9)}, i = 1, 2, 3, \ldots, 9 \qquad (12)$$

In short, the ratio of the strengths of the rules is calculated in this layer.

Layer-4: Each node i in this layer is a square node with a node function:

$$O_{4,i} = \overline{w}_i.f_i = w_i.(p_i x + q_i y + r_i), i = 1, 2, 3, \ldots, 9 \qquad (13)$$

where w_i is the output of layer 3 and $\{p_i, q_i, r_i\}$ is the parameter set. Parameters in this layer are referred to as consequent parameters. In short, the parameters for the consequent parts are measured in this layer.

Layer-5: A circle node called $\sum$ is the single node in this layer that calculates the overall output as the summation of all incoming signals.

$$O_{5,i} = overall\ output = \sum_i \overline{w}_i.f_i = \frac{\sum_i w_i.f_i}{\sum_i w_i} \qquad (14)$$

The backward step is an estimation approach that starts with the membership function parameters and ends with the linear equation coefficients. Because the Gaussian function is employed as the membership function in this method, two parameters of this function are optimized: mean and variance. In this step, the parameter learning is done using the least-squares method.

2.4 Accuracy Criteria

We use several types of accuracy criteria: the Mean absolute percentage error (MAPE), the Mean absolute error (MAE), and the Root means squared error (RMSE).

The MAPE criterion is also known as mean absolute percentage deviation (MAPD) that is a measure of prediction accuracy of a forecasting method in statistics. It usually expresses accuracy as a percentage, and is defined as $MAPE = \frac{1}{n} \sum_{t=1}^{n} \left| \frac{X_t - F_t}{X_t} \right|$, where X_t is the actual value and F_t is the forecasted value. The absolute value in this calculation is summed for every forecasted point in time and divided by the number of fitted points. The MAE is recommended for determining an arithmetic average of the absolute errors to measure comparative accuracy of forecasts. It is estimated by $MAE = \frac{1}{n} \sum_{t=1}^{n} |X_t - F_t|$, where the actual value (X_t) minus the forecasted value (F_t).

The RMSE is also known as root-mean-squared deviation (RMSD) that is a frequently used measure of the estimators differences. It measures the average error produced by the model in predicting the outcome for an observation. It is defined as:

$$RMSE = \sqrt{MSE} = \sqrt{\frac{\sum_{i=1}^{n}(actual\ value - predicted\ value)^2}{n}}$$

where n represents the number of observations.

3 Research Design and Methodology

The objective of this research is to first model the closed price data, and then discuss the fluctuations during the historical data issued from BTC that covers the duration from December 2017 to January 2022. In this regards, we have used ANFIS and MODWT models and considered the Haar WT function. The performance is assessed using the accuracy measure.

We have proposed several models to forecast the closed price BTC data. We used MODWT to convert the original data into a time-scale domain. Then we combined the MODWT with the ANFIS model to improve the forecasting accuracy. We studied Haar WT function of MODWT to evaluate our models. Figure 2 illustrates the steps of the MODWT forecasting process.

When the data pattern is very fluctuated, the wavelet process is repeatedly applied. The aim of pre-processing is to minimize the statistical criteria such as Root Mean Squared Error (RMSE) between the signal before and after transformation. The noise in the original data can thus be removed in this way. Significantly, the adaptive noise in the training pattern may reduce the risk of overfitting in the training phase. Thus, we adopt MODWT twice for the pre-processing of the training data in this study.

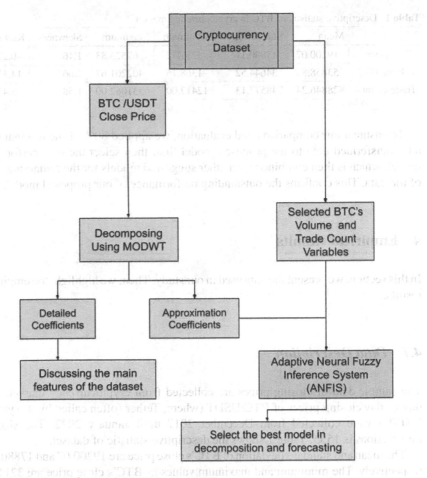

Fig. 2 The flowchart of the MODWT with ANFIS

Further, MODWT converts the data into two sets namely, detail series and approximation series. Since the financial data fluctuates significantly, we have employed these two series because they show good behavior on such data. This makes it easier to anticipate the transformed data with greater accuracy. The MODWT's filtering effect is responsible for these two series' exceptional behavior.

To model the data, we devised the following methodology, which is also shown in Fig. 2. First, the MODWT is used to simulate closed pricing data. The Haar function is then used to dissect the historical return data. Then, using detail series, the variations are recognized and the historical volatility data is examined. The independent variables are chosen in a preprocessing stage based on correlation, causation, and multiple regression. The ANFIS model is then calculated using dependent variables. The ANFIS model is then fitted to the approximation series for forecasting. Finally, using the required criteria, the new technique is compared to the ANFIS model.

Table 1 Descriptive statistic of BTC in cryptocurrency market

	Mean	Std. deviation	Minimum	Maximum	Skewness	Kurtosis
Close price	19200.07	17880.17	3211.72	67525.83	1.16	−0.21
Volume BTC	53858.83	34644.52	1368.15	402201.67	2.66	14.17
Trade-count	828846.24	748577.13	12417.00	6331062.00	1.58	3.41

To ensure a fair comparison and evaluation, we applied 80% of the original data and transformed data to the proposed model first, then select the best performing model, which is then combined with other suggested models for the remaining 20% of the data. This confirms the outstanding performance of our proposed model.

4 Empirical Results

In this section, we present the data used in our study. Then, we highlight the empirical results.

4.1 Data Description

The sample data of closing prices are collected from cryptocurrency market. The day-to-day closing prices of BTC/USDT (where, Tether (often called by its symbol USDT)) were collected from December 2017 until January 2022. The size of observations is 1535. Table 1 shows the descriptive statistic of dataset.

The mean and standard deviation of BTC's close price are 19200.07 and 17880.17, respectively. The minimum and maximum values of BTC's close price are 3211.72 and 67525.83, respectively. The skewness and kurtosis are 1.16 and −0.21. The BTC's volume, mean and standard deviation are 53858.83 and 34644.52 whereas the minimum and maximum values are 1368.15 and 402201.67, respectively. The skewness and kurtosis are 2.66 and 14.17. It should be noted that the mean and standard deviation of trade-count 828846.24 and 748577.13, correspondingly. The minimum and maximum values of trade-count are 12,417 and 6,331,062 whereas the skewness and kurtosis values are 1.58 and 3.41, respectively.

4.2 Endogeneity Issues

In this subsection, we discuss selecting suitable variables by removing multi-collinearity and multiple regression analysis.

Table 2 The correlations between the input variable and the output variables

	LCP	LTC	LV
LCP	1	0.795	−0.852
LTC		1	−0.491
LV			1

4.2.1 Correlation

We deliberately selected independent variables from various other variables, which are eliminated based on certain tests. First, we removed variables as a result of multi-collinearity among independent variables as shown in Table 2. The absence of perfect multi-collinearity, an exact (non-stochastic) linear relationship between two or more independent variables, is generally referred to as no multi-collinearity. We extracted some variables from independent variables according to the strong relation with other independent variables. The correlations between the independent and dependent variables are shown in Table 2. There is a strong correlation between the logarithm of BTC/USDT close price (LCP) with logarithm trade count (LTC) and logarithm volume for BTC (LV) because the percentage of correlation is more than 50%. However, because the correlation between the input variables LV and LTC is less than 50%, there is a weak association.

4.2.2 The Results of Multiple Regressions

Multiple regression is an extension of simple linear regression that is used when the value of a variable must be predicted based on the values of two or more other variables. The dependent variable (output variable (LCP)) is the one that has to be predicted, whereas LTC and LV are the independent variables (input variables) that are utilized to predict LCP. Table 3 shows the results of the multiple regression analysis. The LTC and LV are considerable at 1%. Furthermore, R-square and modified R-square are both around 90%, implying that the independent factors can explain roughly 90% of the dependent variable. The linear regression model is better suited to the results when the F-statistic is 1% significant.

Table 3 The multiple regression

	Estimate	Std. error	t value	Pr(>|t|)
Intercept	−0.028	0.097	−0.289	0.773
LTC	0.457	0.008	54.395	0.000***
LV	−0.709	0.011	−67.06	0.000***
R-square	0.906	F-statistic	7417.445	
Adjusted R-square	0.906	p-value	0.000	

Note "***" 0.001, "**" 0.01, and "*" 0.05

There is a negative relationship between LV and LCP. This indicates that the increase in LV will reduce the LCP in the cryptocurrency market by about 71%. On the other hand, the LTC has a positive relationship with LCP. This implies that the increase in LV will increase the LCP by about 46%. The R-square and adjusted R-square are approximately 90%. Furthermore, F-statistics is significant at a level less than 0.001.

4.3 Results and Discussion

This study examines the LCP of cryptocurrency. It is chosen for different reasons. Cryptocurrency is a digital payment mechanism that does not rely on banks for transaction verification. It's a peer-to-peer payment system that allows anybody, anywhere to send and receive money. Cryptocurrency payments exist purely as digital entries to an online database that identify individual transactions, rather than as physical money that can be carried about and exchanged in the real world. Cryptocurrency is stored in digital wallets. Cryptocurrencies are based on blockchain, a distributed public ledger that keeps track of all transactions that are updated and maintained by currency holders. Furthermore, BTC is an effective indicator of the crypto market in general since it is the largest cryptocurrency by market cap and the rest of the market tends to follow its movements. Therefore, the close price of BTC data is decomposed using MODWT with is Haar function as shown in Fig. 3.

The MODWT based decomposition is an effective approach for revealing variations, magnitudes, and phases of the data. The levels of decomposition can be carried out by the WT using the formula, according to the WT mechanism: $X = TV1 + TW1$ where the original signal is referred to as X. The next component consists of one level of approximation ($TV1$) that shows the plot of the transformed data approximation coefficients. The following parts of $TW1$ reflect the level of detail, whereby $TW1$ is the plot of the first level of the coefficients of detail, so the fluctuation can be explained by this level.

When Bitcoin was first introduced in 2009, it had no value. Its value raised to $0.09 on July 17, 2010. On April 13, 2011, the price of bitcoin climbed again, rising from $1 to a peak of $29.60 on June 7, 2011, a 2.960% rise in three months. Bitcoin's price had fallen below $2.05 by mid-November, following a sharp drop in cryptocurrency markets. By August 15, the next year, the price had risen from $4.85 on May 9 to $13.50. Throughout 2017, Bitcoin's price remained around $1,000 until it crossed $2,000 in mid-May, when it soared to $19,345.49 on December 15. Other institutions began designing cryptocurrencies to compete with Bitcoin as investors, politicians, economists, and scientists took notice. Bitcoin's price meandered laterally during the next two years, with brief bursts of activity. For example, in June 2019, there was a resurgence in price and transaction volume, with prices exceeding $10,000. By mid-December, though, it had decreased to $6,635.84. The economy came to a halt in 2020 owing to the COIVD-19 epidemic, and Bitcoin prices skyrocketed once more. The coin was worth $6,965.72 at the start of the year. Investors' anxieties

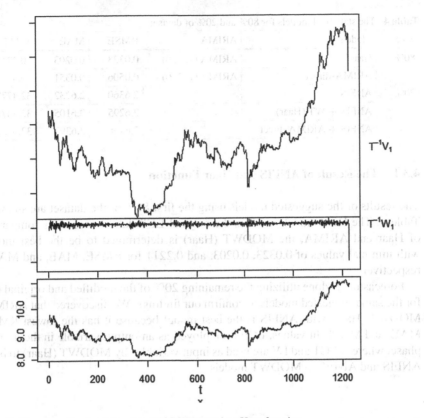

Fig. 3 Decomposing the data using MODWT based on Haar function

about the global economy were increased by the pandemic closure and subsequent government response, hastening Bitcoin's ascent. On November 23, Bitcoin was worth $19,157.16 at the end of the day. Bitcoin's price hit well around $29,000 in December 2020, having risen 416 percent since the beginning of the year.

In 2021, Bitcoin set a new price record in less than a month, reaching $40,000 on January 7, 2021. Bitcoin values achieved new all-time highs of over $60,000 for the first time as Coinbase, a cryptocurrency exchange, went public in mid-April. Bitcoin's price rose even higher as a result of institutional interest, reaching a peak of more than $63,000 on April 12, 2021. Nearly the summer of 2021, prices had plummeted by half, with the lowest being $29,795.55 on July 19. Autumn experienced another bull run in September, with prices reaching $52,693.32 before plunging to around $40,709.59 at the start of 2022.

Table 4 The statistical models for 80% and 20% of dataset

Size	Models	ARIMA	RMSE	MAE	MAPE
80%	Haar	ARIMA (1, 2, 0)	0.0323	0.0203	0.2211
	ARIMA-direct	ARIMA (1, 2, 0)	0.0506	0.0351	0.3823
20%	ANFIS		2.6360	2.6232	32.4777
	ANFIS + WT (Haar)		2.6295	2.6105	32.3071
	ANFIS + ARIMA direct		2.6466	2.6393	32.6862

4.3.1 The Result of ANFIS and Haar Function

The results of the suggested models using the first 80% of the dataset are shown in Table 4. The Haar model provides the original LCP data. Based on a comparison of Haar and ARIMA, the MODWT (Haar) is determined to be the best model, with minimal values of 0.0323, 0.0203, and 0.2211 for RMSE, MAE, and MAPE, respectively.

Forecasting is done utilizing the remaining 20% of the modified and original data for the same suggested models to confirm our findings. We discovered that ARIMA-MODWT (Haar) with ANFIS is the best model because it has the lowest RMSE, MAE, and MAPE-fit values. LCP is employed as an output variable in the training phase, whereas LCT and LV are used as input variables by MODWT (Haar) to build ANFIS and ANFIS + MODWT models.

5 Conclusion

We have proposed a new model called MODWT-Haar-ANFIS. The model is used to forecast the closing price in the cryptocurrency market. We have selected LTC and LV as input values based on correlation and multiple regressions. We found that there is a weak correlation between the input variables ($r = 0.49$). On the other hand, the correlation between LTC and output variable (BTC) is strong ($r = 0.795$). More-over, the correlation between LV and output variable (BTC) is strong ($r = -0.852$). Furthermore, the multiple regression test is used. Therefore, the input variables are significant at level 0.1%. The output and input variables are collected from cryptocurrency market from December 2017 to January 2022 with 1535 observations. The MODWT-Haar mechanism splits output variable into details coefficient and approximation coefficient. On the other hand, the output variable is split into details coefficient (high fluctuated data) and approximation coefficient (which consists of the main features of data). The approximation coefficient data (MODWT-Haar) are used with input variables to build our model MODWT-Haar-ANFIS. Three statistical tests were used to analyze the MODWT-Haar-ANFIS: root mean squared error (RMSE), mean absolute error (MAE), and mean absolute percentage error (MAPE) (MAPE). Traditional models were compared to the MODWT-Haar-ANFIS model (ARIMA and ANFIS models). Traditional models are less accurate than the MODWT-Haar-ANFIS. As a result, the new proposed forecasting methodology can be used to other cryptocurrencies as well. Furthermore, this model is powerful enough to optimize corporate operations for a country's economic progress.

References

Abiyev, R. H., & Abiyev, V. H. (2012). Differential evaluation learning of fuzzy wavelet neural networks for stock price prediction. *Journal of Information and Computing Science, 7*(2), 121–130.

Adil, I. H. (2015). A modified approach for detection of outliers. *Pakistan Journal of Statistics and Operation Research*, 91–102.

Çatık, A. N., Kışla, G. H., & Akdeniz, C. (2020). Time-varying impact of oil prices on sectoral stock returns: Evidence from Turkey. *Resources Policy, 69*, 101845.

Chang, P.-C., & Fan, C.-Y. (2008). A hybrid system integrating a wavelet and TSK fuzzy rules for stock price forecasting. *IEEE Transactions on Systems, Man, and Cybernetics Part C (applications and Reviews), 38*(6), 802–815.

Fan, M.-H., Chen, M.-Y., Huang, H.-F., & Huang, T.-Y. (2012). The case study of adaptive network-based fuzzy inference system modeling for TAIEX prediction. In *2012 Sixth International Conference on Genetic and Evolutionary Computing* (pp. 75–78). IEEE.

Fang, F., et al. (2022). Cryptocurrency trading: A comprehensive survey, *8*(1), 1–59.

Geng, L., & Wang, H. (2010). Grey adaptive-network-based fuzzy inference system for fund volatility forecasting. In *2010 Seventh International Conference on Fuzzy Systems and Knowledge Discovery* (Vol. 3, pp. 1296–1299). IEEE.

Homayouni, N., & Amiri, A. (2011). Stock price prediction using a fusion model of wavelet, fuzzy logic and ANN. *International Conference on E-Business, Management and Economics, 25*, 277–281.

Honghui, Z., & Yongqiang, L. (2012). Application of an adaptive network-based fuzzy inference system using genetic algorithm for short term load forecasting. In *2012 International Conference on Computer Science and Electronics Engineering* (Vol. 2, pp. 314–317). IEEE.

Hull, J. (2014). *Risk management and financial institutions, +Web Site*. Wiley.

Khuntia, S., & Hiremath, G. S. (2019). Monetary policy announcements and stock returns: Some further evidence from India. *Journal of Quantitative Economics, 17*(4), 801–827.

Li, J., Feng, X., & Zhang, Y. (2020). Month-end effect on Chinese stock returns: Explanation of the liquidity hypothesis. *Asia-Pacific Journal of Accounting & Economics*, 1–16.

Lv, X., Lien, D., & Yu, C. (2020). Who affects who? Oil price against the stock return of oil-related companies: Evidence from the US and China. *International Review of Economics & Finance, 67*, 85–100.

Madura, J. (2020). *Financial markets & institutions*. Cengage learning.

Munandar, D. (2015). Optimization weather parameters influencing rainfall prediction using Adaptive Network-Based Fuzzy Inference Systems (ANFIS) and linier regression. In *2015 International Conference on Data and Software Engineering (ICoDSE)* (pp. 1–6). IEEE.

Nadimi, V., Azadeh, A., Pazhoheshfar, P., & Saberi, M. (2010). An adaptive-network-based fuzzy inference system for long-term electric consumption forecasting (2008–2015): A case study of the group of seven (G7) industrialized nations: USA, Canada, Germany, United Kingdom, Japan, France and Italy. In *2010 Fourth UKSim European Symposium on Computer Modeling and Simulation* (pp. 301–305). IEEE.

Sehgal, V., Sahay, R. R., & Chatterjee, C. (2014). Effect of utilization of discrete wavelet components on flood forecasting performance of wavelet based ANFIS models. *Water Resources Management, 28*(6), 1733–1749.

Septiarini, T. W., Abadi, A. M., & Taufik, M. R. (2016). Application of wavelet fuzzy model to forecast the exchange rate IDR of USD. *International Journal of Modeling and Optimization, 6*(1), 66–70.

Smyth, R., & Narayan, P. K. (2018). What do we know about oil prices and stock returns? *International Review of Financial Analysis, 57*, 148–156.

Wang, H., He, D., & Ji, Y. J. F. G. C. S. (2020). Designated-verifier proof of assets for bitcoin exchange using elliptic curve cryptography, *107*, 854–862.

Xiao, J., Zhou, M., Wen, F., & Wen, F. (2018). Asymmetric impacts of oil price uncertainty on Chinese stock returns under different market conditions: Evidence from oil volatility index. *Energy Economics, 74*, 777–786.

Zhang, P., & Wang, H. (2012). Fuzzy wavelet neural networks for city electric energy consumption forecasting. *Energy Procedia, 17*, 1332–1338.

Development of Data Mining Expert System Using Naïve Bayes

Mohammed Salahat, Nidal A. Al-Dmour, Raed A. Said,
Haitham M. Alzoubi⊙, and Muhammad Alshurideh⊙

Abstract The consumer spectrum consists of a wide range, including the affluent, middle-income, and low-income. This consumer shows different behaviors or motivations towards choosing clothes. We want to develop a framework for a Sale Recommendation System. These expert System can be helpful for sale persons, fashion designer, promoter, brand manager as well as sponsor of Recommendation System. The study implemented the Data Science approach and techniques to see how reliable Recommendation Systems are and in our selected dataset we have applied different modelling techniques such as KNN, SVM, Bayes Naïve and Decision Tree and fond the NB as the most suitable and practical method of modelling in regard to the accuracy, recall and runtime.

Keywords Data science · Fashion · Style · Sale · Recommendation system

M. Salahat
College of Engineering and Information Technology, University of Science and Technology of Fujairah, Fujairah, UAE
e-mail: m.salahat@ustf.ac.ae

N. A. Al-Dmour
Department of Computer Engineering, College of Engineering, Mutah University, Mu'tah, Jordan

R. A. Said
Faculty of Management, Canadian University Dubai, Dubai, UAE

H. M. Alzoubi (✉)
School of Business, Skyline University College, Sharjah, UAE
e-mail: haitham.alzubi@skylineuniversity.ac.ae

M. Alshurideh
School of Business, The University of Jordan, Amman, Jordan
e-mail: malshurideh@sharjah.ac.ae

College of Business Administration, University of Sharjah, Sharjah, UAE

2437

1   Introduction

In recent years, fashion and dressing sale have found an abstract meaning in the form of start-ups, branding, design of production and representation. Famous brands have earned large sums of money for their business every year by gaining fame by associating and engaging their customers, and in this case, recommendation system from sale person has not been addressed or paid very little attention (Akhtar et al., 2021; Al Ali, 2021; AlHamad et al., 2021, 2022; Ali et al., 2021). The fashion and clothing industry constitutes high share value of active enterprises worldwide and its amount is increasing every day. New marketing ideas such as fast fashion and entrepreneurship campaigns are emerging (Ali et al., 2022; Alnazer et al., 2017; Alnuaimi et al., 2021; Alsharari, 2021; Alshurideh et al., 2022). In the present work, we want to make a decision support system from the data gained on the sales recommendation of several companies in the fashion and clothing industry, with the help of Data Scientific approaches and compare the reliability of our results in the business world. Are these methodologies suitable in the real world? Based on the findings, is it possible to make suggestions to other activists in the fashion and clothing industry who either do not have a recommendation system or have not made a profit from it, so that they can achieve high profitability and effectiveness? Manufacturers and intermediaries of this category of products generally compete in a win–win or offensive strategy (ex. luxury brands) (Alshurideh et al., 2020; Alwan & Alshurideh, 2022; Alzoubi, 2021a, 2021b; Alzoubi et al., 2020a, 2020b, 2022; Sweiss et al., 2021; Tariq et al., 2022). Clothing goods with customs exemptions are consumer goods and with a small investment can be distributed through the different channel and of course, except for brands that only sell their goods through reputable agencies, specialty stores and branches in hypermarkets and stores (Hanaysha et al., 2021a, 2021b; Joghee et al., 2020; Kashif et al., 2021; Khan, 2021; Lee & Ahmed, 2021). The seller brings his goods mainly to big cities through free trade zones and enters the market and delivers them to the location of his store at different prices at different prices to the consumer. Here, consumer behavior and the tastes of the upper middle class will be considered (Al Kurdi & Alshurideh, 2021; Al Kurdi et al., 2020; Alshurideh, 2019; Alshurideh et al., 2019). Taste or the same behavior of buying fashion and clothing in the traditional middle class mostly follows a single logic in the type of purchase and in the affluent class who have emotional behavior is at very high prices and is sold at the maximum prices in the market (Lee et al., 2022a, 2022b; Mehmood, 2021; Mehmood et al., 2019; Miller, 2021; Mondol, 2021; Obaid, 2021; Radwan & Farouk, 2021; Shamout et al., 2022).

2 Dataset Description

2.1 Preprocessing

First, before examining the size of the our data, we must have a description that refer to UCI repository (Afifi et al., 2020; Al Batayneh et al., 2021; Ali et al., 2021; Alzoubi et al., 2021a, 2021b; Ghazal, 2021; Ghazal et al., 2021a, 2021b, 2021c, 2021d), and because in this study recommendation system is considered independently, we must determine the number of counts and their description to each attribute. From the definition and understanding of this research we have used two filters for data cleansing namely RemovePercentage and AddNoise. Out of the number of samples in the dataset, which was 500, only 200 questionnaires were completely recommended. Therefore, considering dataset as a whole, it can be estimated to generate a 1/1 out of 2/3 ratio among recommendation to constitute 238 versus 238 rebalancing (Fig. 1).

Of the population selected for analysis, rating was with the most skewness in grade between 4 or 5. And price was major in Low and Average (middle class) (Fig. 2).

Fig. 1 Recommendation class

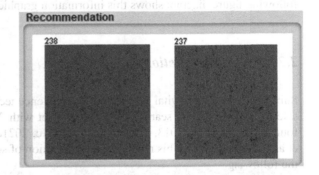

Fig. 2 Price marketing criteria

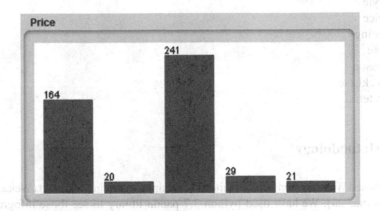

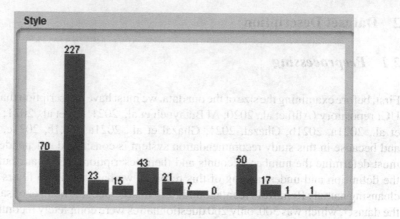

Fig. 3 Style distribution

According to the information on style and size one can guess that population were young aged and female which wear Casual, sexy and party style costume. The following figure diagram shows this information graphically (Fig. 3).

2.2 Attribute Selection

Subset evaluation is a vital step in the Data Science techniques. In our case Weka selected 39 subset and searched for the best set with 74.3% Merit of best subset found (Ghazal et al., 2013, 2021a, 2021b, 2021c, 2021d; Kalra et al., 2020; Khan et al., 2021a, 2021b). This resulted in the selection of seven attributes as shown in the following:

- Style
- Price
- Rating
- Size
- Season
- NeckLine
- Material.

3 Methodology

Java-based Weka version 3.9 was suited for our application in this report (Weka Application manual). We have used python 3.7 panda library to see those infographics. The details of results of coding within the description of all attributes are attached in the Appendix.

```
import pandas as pd
data = pd.read_csv("DressData.csv")
print(data.head(4))
data.info()
data.describe()
data
```

4 Modelling

After the importing the dataset into the Weka, modelling techniques utilised and was calibrated with the usage of hyper-parameter tuning were k-nearest neighbors (kNN), (the number of neighbors is equal to 4) Neural Network (NN), (the batch size is equal to 100) and Naïve Bayes (NB) (Alzoubi & Ahmed, 2019; Alzoubi & Aziz, 2021; Alzoubi & Yanamandra, 2020; Alzoubi et al., 2020a, 2020b, 2021a, 2021b).

5 Results

Table 1 shows the result of accuracy (Recall for the danger of overfitting), F-Measure and Kappa statistics.

Table 1 Data scientific result performed in Weka

Attribute Selection	Classifier	F-measure	Kappa	Recall (%)	Accuracy (%)
12	Bayes Naïve (NB)	60.9%	0.1931	61.0	61.8
	k-Nearest Neighbors (IBk)	50.12%	−0.002	50.8	50.6
	Multilayer Perceptron (NNs)	53.9%	0.0305	53.2	53.2
9	Bayes Naïve (NB)	52.2%	0.0441	52.2	52.21
	k-Nearest Neighbors (IBk)	49.7%	0.0324	51.6	51.57
	Multilayer Perceptron (NNs)	47.7%	−0.044	47.8	47.7
6	Bayes Naïve (NB)	63.6%	0.2486	64.4	64.4
	k-Nearest Neighbors (IBk)	57.2	0.1215	57.2	57.2
	Multilayer Perceptron (NNs)	53.2%	0.0691	58.4	58.4

The Bayes Naïve model estimation is based on the Bayes Probability theory and outperformed other two techniques on all model efficiency and also runtime (Lee et al., 2021; Matloob et al., 2021; Naqvi et al., 2021; Rehman et al., 2021; Suleman et al., 2021). The effect of attribute reduction performed on preprocessing stage can be seen that 3 to 4 attributes are indifferent for our prediction models. Anyway, the precision was not intended to be either/or but rather the practically developed model with more than 60% accuracy. In the following table the confusion matrix is also indicated. The number of correct classified instances for "Yes" 100 and "No" was equal to 222.

=== Confusion Matrix ===

a   b   <-- classified as

100 110 | a = Yes

68 222 | b = No

6 Discussion

Generally, for retailers or clothing stores, this result has less external consequences, but the point of view is very important due to the importance of feasibility of designing of such smart system that can capture the customer sense and interaction with the customer. It is important for any brand as a beneficiary of a recommendation system to take advantage of opportunities and generally use chances and prevent failures in the competitive market, which will be created by differentiating in the mind of the customer (Aziz & Aftab, 2021; Cruz, 2021; Eli, 2021; Farouk, 2021; Ghazal et al., 2021a, 2021b, 2021c, 2021d; Guergov & Radwan, 2021; Hamadneh et al., 2021).

7 Conclusion

We have developed a framework that can model the sale recommendation for better communication between clients. We have firstly extracted the dataset from the UCI and then imported the data for the preprocessing into weka and python. The data was cleansed at the forst step by the filtering tools embedded and was balanced for the danger of overfitting. Then five different machine learning tools was implemented in the dataset to tune the hyper-parameter which resulted into more than 10% improvement in the results. With the information learned from the modeling we understood that with the possibility of neat 65% we can use such system.

Appendix [Python Code with Outputs]

```
import pandas as pd
In [2]:
data = pd.read_csv("DressData.csv")
In [3]:
print(data.head(4))
    Style  Price Rating Size Season NeckLine SleeveLength Waistline \
0   Sexy   Low    4.6   M  Summer  O-neck   Sleeveless   Empire
1   Casual Low    ?     L  Summer  O-neck   Petal        Natural
2   Vintage High  ?     L  Autumn  O-neck   Full         Natural
3   Brief  Average 4.6  L  Spring  O-neck   Full         Natural

   Material  Decoration Pattern Type Recommendation
0      ?     Ruffles    Animal       Yes
1  Microfiber Ruffles   Animal       No
2  Polyster   NaN       Print        No
3      Silk  Embroidary  Print       Yes
In [9]:
data.info()
<class 'pandas.core.frame.DataFrame'>
RangeIndex: 500 entries, 0 to 499
Data columns (total 12 columns):
 #  Column         Non-Null Count  Dtype
--- ------         --------------  -----
 0  Style          500 non-null    object
 1  Price          500 non-null    object
 2  Rating         500 non-null    object
 3  Size           500 non-null    object
 4  Season         500 non-null    object
 5  NeckLine       500 non-null    object
 6  SleeveLength   500 non-null    object
 7  Waistline      500 non-null    object
 8  Material       500 non-null    object
 9  Decoration     262 non-null    object
 10 Pattern Type   500 non-null    object
 11 Recommendation 500 non-null    object
dtypes: object(12)
memory usage: 47.0+ KB
In [11]:
data.describe()
Out[11]:
```

	Style	Price	Rating	Size	Season	NeckLine	SleeveLength	Waistline	Material	Decoration	Pattern Type	Recommendation
count	500	500	500	500	500	500	500	500	500	262	500	500
unique	12	6	17	5	5	16	10	5	20	23	14	2
top	Casual	Average	?	M	Summer	O-neck	Sleeveless	Natural	Cotton	Lace	Solid	No
freq	232	252	120	177	160	271	232	304	152	70	203	290

Data Describe
Out [11]:

	Style	Price	Rating	Size	Season	NeckLine	SleeveLength	Waistline	Material	Decoration	Pattern Type	Recommendation
count	500	500	500	500	500	500	500	500	500	262	500	500
unique	12	6	17	5	5	16	10	5	20	23	14	2
top	Casual	Average	?	M	Summer	O-neck	Sleeveless	Natural	Cotton	Lace	Solid	No
freq	232	252	120	177	160	271	232	304	152	70	203	290

In [21]:
data
Out[21]:

	Style	Price	Rating	Size	Season	NeckLine	SleeveLength	Waistline	Material	Decoration	Pattern Type	Recommendation
0	Sexy	Low	4.6	M	Summer	O-neck	Sleeveless	Empire	?	Ruffles	Animal	Yes
1	Casual	Low	?	L	Summer	O-neck	Petal	Natural	Microfiber	Ruffles	Animal	No
2	Vintage	High	?	L	Autumn	O-neck	Full	Natural	Polyster	NaN	Print	No
3	Brief	Average	4.6	L	Spring	O-neck	Full	Natural	Silk	Embroidary	Print	Yes
4	Cute	Low	4.5	M	Summer	O-neck	Butterfly	Natural	Chiffon	Bow	Dot	No
...	...	...	...	...	...	...	...	...	...	...	...	...
495	Casual	Low	4.7	M	Spring	O-neck	Full	Natural	Polyster	NaN	Solid	Yes
496	Sexy	Low	4.3	Free	Summer	O-neck	Full	Empire	Cotton	NaN	?	No
497	Casual	Average	4.7	M	Summer	V-neck	Full	Empire	Cotton	Lace	Solid	Yes
498	Casual	Average	4.6	L	Winter	Boat-neck	Sleeveless	Empire	Silk	Applique	Print	Yes
499	Casual	Low	4.4	Free	Summer	V-neck	Short	Empire	Cotton	Lace	Solid	No

[1]   500 rows × 12 columns

References

Afifi, M. A. M., Kalra, D., Ghazal, T. M., & Mago, B. (2020). Information technology ethics and professional responsibilities. *International Journal of Advanced Science and Technology, 29*(4), 11336–11343.

Akhtar, A., Akhtar, S., Bakhtawar, B., Kashif, A. A., Aziz, N., & Javeid, M. S. (2021). COVID-19 detection from CBC using machine learning techniques. *International Journal of Technology, Innovation and Management (IJTIM), 1*(2), 65–78. https://doi.org/10.54489/ijtim.v1i2.22

Al Ali, A. (2021). The impact of information sharing and quality assurance on customer service at UAE banking sector. *International Journal of Technology, Innovation and Management (IJTIM)*, *1*(1), 01–17. https://doi.org/10.54489/ijtim.v1i1.10

Al Batayneh, R. M., Taleb, N., Said, R. A., Alshurideh, M. T., Ghazal, T. M., & Alzoubi, H. M. (2021). IT governance framework and smart services integration for future development of Dubai infrastructure utilizing AI and big data, its reflection on the citizens standard of living. *The International Conference on Artificial Intelligence and Computer Vision*, 235–247.

Al Kurdi, B. H., & Alshurideh, M. T. (2021). Facebook advertising as a marketing tool: Examining the influence on female cosmetic purchasing behaviour. *International Journal of Online Marketing (IJOM)*, *11*(2), 52–74.

Al Kurdi, B., Alshurideh, M., Salloum, S., Obeidat, Z., & Al-dweeri, R. (2020). An empirical investigation into examination of factors influencing university students' behavior towards elearning acceptance using SEM approach.

AlHamad, A., Alshurideh, M., Alomari, K., Kurdi, B., Alzoubi, H., Hamouche, S., & Al-Hawary, S. (2022). The effect of electronic human resources management on organizational health of telecommunications companies in Jordan. *International Journal of Data and Network Science*, *6*(2), 429–438.

Alhamad, A. Q. M., Akour, I., Alshurideh, M., Al-Hamad, A. Q., Kurdi, B. A., & Alzoubi, H. (2021). Predicting the intention to use google glass: A comparative approach using machine learning models and PLS-SEM. *International Journal of Data and Network Science*, *5*(3). https://doi.org/10.5267/j.ijdns.2021.6.002

Ali, N., Ahmed, A., Anum, L., Ghazal, T. M., Abbas, S., Khan, M. A., Alzoubi, H. M., & Ahmad, M. (2021). Modelling supply chain information collaboration empowered with machine learning technique. *Intelligent Automation and Soft Computing*, *30*(1), 243–257. https://doi.org/10.32604/iasc.2021.018983

Ali, N., M. Ghazal, T., Ahmed, A., Abbas, S., A. Khan, M., Alzoubi, H., Farooq, U., Ahmad, M., & Adnan Khan, M. (2022). Fusion-based supply chain collaboration using machine learning techniques. *Intelligent Automation & Soft Computing*, *31*(3), 1671–1687. https://doi.org/10.32604/iasc.2022.019892

Alnazer, N. N., Alnuaimi, M. A., & Alzoubi, H. M. (2017). Analysing the appropriate cognitive styles and its effect on strategic innovation in Jordanian universities. *International Journal of Business Excellence*, *13*(1), 127–140. https://doi.org/10.1504/IJBEX.2017.085799

Alnuaimi, M., Alzoubi, H. M., Ajelat, D., & Alzoubi, A. A. (2021). Towards intelligent organisations: An empirical investigation of learning orientation's role in technical innovation. *International Journal of Innovation and Learning*, *29*(2), 207–221. https://doi.org/10.1504/IJIL.2021.112996

Alsharari, N. (2021). Integrating blockchain technology with internet of things to efficiency. *International Journal of Technology, Innovation and Management (IJTIM)*, *1*(2), 1–13.

Alshurideh, M. (2019). Do electronic loyalty programs still drive customer choice and repeat purchase behaviour? *International Journal of Electronic Customer Relationship Management*, *12*(1). https://doi.org/10.1504/IJECRM.2019.098980

Alshurideh, M., Al Kurdi, B., Shaltoni, A. M., & Ghuff, S. S. (2019). Determinants of pro-environmental behaviour in the context of emerging economies. *International Journal of Sustainable Society*, *11*(4). https://doi.org/10.1504/IJSSOC.2019.104563

Alshurideh, M., Gasaymeh, A., Ahmed, G., Alzoubi, H., & Kurd, B. A. (2020). Loyalty program effectiveness: Theoretical reviews and practical proofs. *Uncertain Supply Chain Management*, *8*(3). https://doi.org/10.5267/j.uscm.2020.2.003

Alshurideh, M. T., Al Kurdi, B., Alzoubi, H. M., Ghazal, T. M., Said, R. A., AlHamad, A. Q., Hamadneh, S., Sahawneh, N., & Al-kassem, A. H. (2022). Fuzzy assisted human resource management for supply chain management issues. *Annals of Operations Research*, 1–19.

Alwan, M., & Alshurideh, M. T. (2022, March). *The effect of digital marketing on purchase intention: Moderating effect of brand equity*. https://doi.org/10.5267/j.ijdns.2022.2.012

Alzoubi, A. (2021a). The impact of process quality and quality control on organizational competitiveness at 5-star hotels in Dubai. *International Journal of Technology, Innovation and Management (IJTIM)*, *1*(1), 54–68. https://doi.org/10.54489/ijtim.v1i1.14

Alzoubi, A. (2021b). Renewable green hydrogen energy impact on sustainability performance. *International Journal of Computations, Information and Manufacturing (IJCIM)*, *1*(1), 94–110. https://doi.org/10.54489/ijcim.v1i1.46

Alzoubi, H., & Ahmed, G. (2019). Do TQM practices improve organisational success? A case study of electronics industry in the UAE. *International Journal of Economics and Business Research*, *17*(4), 459–472. https://doi.org/10.1504/IJEBR.2019.099975

Alzoubi, H. M., & Aziz, R. (2021). Does emotional intelligence contribute to quality of strategic decisions? The mediating role of open innovation. *Journal of Open Innovation: Technology, Market, and Complexity*, *7*(2), 130. https://doi.org/10.3390/joitmc7020130

Alzoubi, H. M., & Yanamandra, R. (2020). Investigating the mediating role of information sharing strategy on agile supply chain. *Uncertain Supply Chain Management*, *8*(2), 273–284. https://doi.org/10.5267/j.uscm.2019.12.004

Alzoubi, H., Ahmed, G., Al-Gasaymeh, A., & Kurdi, B. (2020a). Empirical study on sustainable supply chain strategies and its impact on competitive priorities: The mediating role of supply chain collaboration. *Management Science Letters*, *10*(3), 703–708.

Alzoubi, H., Alshurideh, M., Kurdi, B. A., & Inairat, M. (2020b). Do perceived service value, quality, price fairness and service recovery shape customer satisfaction and delight? A practical study in the service telecommunication context. *Uncertain Supply Chain Management*, *8*(3), 579–588. https://doi.org/10.5267/j.uscm.2020.2.005

Alzoubi, H. M., Alshurideh, M., & Ghazal, T. M. (2021a). Integrating BLE beacon technology with intelligent information systems IIS for operations' performance: A managerial perspective. *The International Conference on Artificial Intelligence and Computer Vision*, 527–538.

Alzoubi, H. M., Vij, M., Vij, A., & Hanaysha, J. R. (2021b). What leads guests to satisfaction and loyalty in UAE five-star hotels? AHP analysis to service quality dimensions. *Enlightening Tourism*, *11*(1), 102–135. https://doi.org/10.33776/et.v11i1.5056

Alzoubi, H., Alshurideh, M., Kurdi, B., Akour, I., & Aziz, R. (2022). Does BLE technology contribute towards improving marketing strategies, customers' satisfaction and loyalty? The role of open innovation. *International Journal of Data and Network Science*, *6*(2), 449–460.

Aziz, N., & Aftab, S. (2021). Data mining framework for nutrition ranking: Methodology: SPSS modeller. *International Journal of Technology, Innovation and Management (IJTIM)*, *1*(1), 85–95.

Cruz, A. (2021). Convergence between blockchain and the internet of things. *International Journal of Technology, Innovation and Management (IJTIM)*, *1*(1), 35–56.

Eli, T. (2021). Students perspectives on the use of innovative and interactive teaching methods at the University of Nouakchott Al Aasriya, Mauritania: English Department as a case study. *International Journal of Technology, Innovation and Management (IJTIM)*, *1*(2), 90–104.

Farouk, M. (2021). The universal artificial intelligence efforts to face coronavirus COVID-19. *International Journal of Computations, Information and Manufacturing (IJCIM)*, *1*(1), 77–93. https://doi.org/10.54489/ijcim.v1i1.47

Ghazal, T. M. (2021). Positioning of UAV base stations using 5G and beyond networks for IoMT applications. *Arabian Journal for Science and Engineering*, 1–12.

Ghazal, T., Soomro, T. R., & Shaalan, K. (2013). Integration of Project Management Maturity (PMM) based on Capability Maturity Model Integration (CMMI). *European Journal of Scientific Research*, *99*(3), 418–428.

Ghazal, T. M., Anam, M., Hasan, M. K., Hussain, M., Farooq, M. S., Ali, H. M., Ahmad, M., & Soomro, T. R. (2021a). Hep-pred: Hepatitis C staging prediction using fine Gaussian SVM. *Computers, Materials and Continua*, *69*, 191–203.

Ghazal, T. M., Hasan, M. K., Alshurideh, M. T., Alzoubi, H. M., Ahmad, M., Akbar, S. S., Al Kurdi, B., & Akour, I. A. (2021b). IoT for smart cities: Machine learning approaches in smart healthcare—A review. *Future Internet*, *13*(8), 218. https://doi.org/10.3390/fi13080218

Ghazal, T. M., Hussain, M. Z., Said, R. A., Nadeem, A., Hasan, M. K., Ahmad, M., Khan, M. A., & Naseem, M. T. (2021c). *Performances of K-means clustering algorithm with different distance metrics.*

Ghazal, T. M., Said, R. A., & Taleb, N. (2021d). Internet of vehicles and autonomous systems with AI for medical things. *Soft Computing*, 1–13.

Guergov, S., & Radwan, N. (2021). Blockchain convergence: Analysis of issues affecting IoT, AI and blockchain. *International Journal of Computations, Information and Manufacturing (IJCIM)*, *1*(1), 1–17. https://doi.org/10.54489/ijcim.v1i1.48

Hamadneh, S., Pedersen, O., & Al Kurdi, B. (2021). An investigation of the role of supply chain visibility into the scottish bood supply chain. *Journal of Legal, Ethical and Regulatory Issues*, *24*(Special Issue 1), 1–12.

Hanaysha, J. R., Al-Shaikh, M. E., Joghee, S., & Alzoubi, H. (2021a). Impact of innovation capabilities on business sustainability in small and medium enterprises. *FIIB Business Review*, 1–12. https://doi.org/10.1177/23197145211042232

Hanaysha, J. R., Al Shaikh, M. E., & Alzoubi, H. M. (2021b). Importance of marketing mix elements in determining consumer purchase decision in the retail market. *International Journal of Service Science, Management, Engineering, and Technology (IJSSMET)*, *12*(6), 56–72.

Joghee, S., Alzoubi, H. M., & Dubey, A. R. (2020). Decisions effectiveness of FDI investment biases at real estate industry: Empirical evidence from Dubai smart city projects. *International Journal of Scientific and Technology Research*, *9*(3), 3499–3503.

Kalra, D., Ghazal, T. M., & Afifi, M. A. M. (2020). Integration of collaboration systems in hospitality management as a comprehensive solution. *International Journal of Advanced Science and Technology*, *29*(8s), 3155–3173.

Kashif, A. A., Bakhtawar, B., Akhtar, A., Akhtar, S., Aziz, N., & Javeid, M. S. (2021). Treatment response prediction in hepatitis C patients using machine learning techniques. *International Journal of Technology, Innovation and Management (IJTIM)*, *1*(2), 79–89. https://doi.org/10.54489/ijtim.v1i2.24

Khan, M. A. (2021). Challenges facing the application of IoT in medicine and healthcare. *International Journal of Computations, Information and Manufacturing (IJCIM)*, *1*(1), 39–55. https://doi.org/10.54489/ijcim.v1i1.32

Khan, M. F., Ghazal, T. M., Said, R. A., Fatima, A., Abbas, S., Khan, M. A., Issa, G. F., Ahmad, M., & Khan, M. A. (2021a). An IoMT-enabled smart healthcare model to monitor elderly people using machine learning technique. *Computational Intelligence and Neuroscience, 2021*.

Khan, Q.-T.-A., Ghazal, T., Abbas, S., Khan, W. A., Khan, M. A., Said, Raed, A., Ahmad, M., & Asif, M. (2021b). Modeling habit patterns using conditional reflexes in agency. *Intelligent Automation and Soft Computing*, *30*(2), 539–552.

Lee, C., & Ahmed, G. (2021). Improving IoT privacy, data protection and security concerns. *International Journal of Technology, Innovation and Management (IJTIM)*, *1*(1), 18–33. https://doi.org/10.54489/ijtim.v1i1.12

Lee, S.-W., Hussain, S., Issa, G. F., Abbas, S., Ghazal, T. M., Sohail, T., Ahmad, M., & Khan, M. A. (2021). Multi-dimensional trust quantification by artificial agents through evidential fuzzy multi-criteria decision making. *IEEE Access, 9*, 159399–159412.

Lee, K., Azmi, N., Hanaysha, J., Alzoubi, H., & Alshurideh, M. (2022a). The effect of digital supply chain on organizational performance: An empirical study in Malaysia manufacturing industry. *Uncertain Supply Chain Management, 10*(2), 495–510.

Lee, K., Romzi, P., Hanaysha, J., Alzoubi, H., & Alshurideh, M. (2022b). Investigating the impact of benefits and challenges of IOT adoption on supply chain performance and organizational performance: An empirical study in Malaysia. *Uncertain Supply Chain Management, 10*(2), 537–550.

Matloob, F., Ghazal, T. M., Taleb, N., Aftab, S., Ahmad, M., Khan, M. A., Abbas, S., & Soomro, T. R. (2021). Software defect prediction using ensemble learning: A systematic literature review. *IEEE Access.*

Mehmood, T. (2021). Does information technology competencies and fleet management practices lead to effective service delivery? Empirical evidence from e-commerce industry. *International Journal of Technology, Innovation and Management (IJTIM)*, *1*(2), 14–41.

Mehmood, T., Alzoubi, H. M., & Ahmed, G. (2019). Schumpeterian entrepreneurship theory: Evolution and relevance. *Academy of Entrepreneurship Journal*, *25*(4).

Miller, D. (2021). The best practice of teach computer science students to use paper prototyping. *International Journal of Technology, Innovation and Management (IJTIM)*, *1*(2), 42–63. https://doi.org/10.54489/ijtim.v1i2.17

Mondol, E. P. (2021). The impact of block chain and smart inventory system on supply chain performance at retail industry. *International Journal of Computations, Information and Manufacturing (IJCIM)*, *1*(1), 56–76. https://doi.org/10.54489/ijcim.v1i1.30

Naqvi, R., Soomro, T. R., Alzoubi, H. M., Ghazal, T. M., & Alshurideh, M. T. (2021). The nexus between big data and decision-making: A study of big data techniques and technologies. *The International Conference on Artificial Intelligence and Computer Vision*, 838–853.

Obaid, A. J. (2021). Assessment of smart home assistants as an IoT. *International Journal of Computations, Information and Manufacturing (IJCIM)*, *1*(1), 18–36. https://doi.org/10.54489/ijcim.v1i1.34

Radwan, N., & Farouk, M. (2021). The growth of Internet of Things (IoT) in the management of healthcare issues and healthcare policy development. *International Journal of Technology, Innovation and Management (IJTIM)*, *1*(1), 69–84. https://doi.org/10.54489/ijtim.v1i1.8

Rehman, E., Khan, M. A., Soomro, T. R., Taleb, N., Afifi, M. A., & Ghazal, T. M. (2021). Using blockchain to ensure trust between donor agencies and ngos in under-developed countries. *Computers*, *10*(8), 98.

Shamout, M., Ben-Abdallah, B., Alshurideh, M., Alzoubi, H., Al Kurdi, B., & Hamadneh, S. (2022). A conceptual model for the adoption of autonomous robots in supply chain and logistics industry. *Uncertain Supply Chain Management*, *10*, 1–16.

Suleman, M., Soomro, T. R., Ghazal, T. M., & Alshurideh, M. (2021). Combating against potentially harmful mobile apps. *The International Conference on Artificial Intelligence and Computer Vision*, 154–173.

Sweiss, N., Obeidat, Z. M., Al-Dweeri, R. M., Mohammad Khalaf Ahmad, A., Obeidat, M. A., & Alshurideh, M. (2021). The moderating role of perceived company effort in mitigating customer misconduct within Online Brand Communities (OBC). *Journal of Marketing Communications*, 1–24. https://doi.org/10.1080/13527266.2021.1931942

Tariq, E., Alshurideh, M., Akour, I., Al-Hawary, S., & Al, B. (2022). The role of digital marketing, CSR policy and green marketing in brand development. *International Journal of Data and Network Science*, *6*(3), 1–10.

Does Strategic Change Mediate the Relationship Between Total Quality Management and Organizational Culture

Enass Khalil Alquqa, Bader Obeidat, and Muhammad Turki Alshurideh

Abstract The study conducts an empirical analysis of the relationship between total quality management (TQM) practices, with the organizational culture (OC) Cameron and Quinn (Diagnosing and changing organizational culture: Based on the competing values framework, John Wiley & Sons, 2011) model which was divided into four dimensions: leadership supportive, customer focus, strategic planning, and workforce focus. In addition, the study evaluates the mediating role of strategic change (SCII), between organizational culture (OC) and total quality management (TQM). The study uses data obtained from a survey of 341 respondents by using google form online from the employees of the Emirates Post Group (EPG), which is considered as one of the government-owned agencies that controls the logistics business in the UAE, Dubai. The result of the study revealed that strategic change has a direct impact on the TQM practices. In addition to that, the strategic change can be explained as a partial mediator in the relationship between organizational culture variables and TQM practices. It also found that there is a positive and significant effect of organizational culture on strategic change. This study makes a novel contribution by offering a detailed and deeper understanding of the relationship between Organizational culture, strategic change, and total quality management. According to the results, the study recommended that future studies might investigate if other possible mediating variables may also mediate whatever totally or partially the effect of the relationship between TQM and organizational culture.

E. K. Alquqa (✉) · B. Obeidat
Faculty of Business and Law, The British University in Dubai, Dubai, United Arab Emirates
e-mail: 20191464@student.buid.ac.ae

B. Obeidat
e-mail: bader.obeidat@buid.ac.ae

M. T. Alshurideh
Department of Management, College of Business, University of Sharjah, 27272 Sharjah, United Arab Emirates
e-mail: m.alshurideh@ju.edu.jo

Department of Marketing, School of Business, The University of Jordan, Amman 11942, Jordan

M. Alshurideh et al. (eds.), *The Effect of Information Technology on Business and Marketing Intelligence Systems*, Studies in Computational Intelligence 1056,
https://doi.org/10.1007/978-3-031-12382-5_135

Keywords Total Quality Management · Organizational culture · Organizational change

1 Introduction

In recent years, the TQM has directly contributed to the increased competition potential of most institutions and remains at the forefront of expanding changes in the economic, political, and administrative systems (Almaraz, 1994; Cao et al., 2000; Duck, 1993). The current global pandemic of the COVID-19 problem is the most visible illustration of the catastrophe which has unexpectedly struck the world and produced catastrophic change at the economic, political, and social level (Baldwin & Mauro, 2020). Numerous changes also occur at the local and global level to systems and legislation. Such changes demand a flexible, quick response to all the factors, including those at the strategic level, which encompass all the administrative procedures of the organization, that have altered on many levels (Balogun & Johnson, 2005). To encourage change, the organization's culture must be considered, based on the strong conviction of the employees at this institution that change is important and that they feel they are part of the organization, given its role in achieving directional changes requiring speed, preciseness, and quality of employment (Cameron & Quinn, 2011; Denison, 1990). Global rivalry encouraged companies to concentrate more on their clients (Lindquist, 2006). Consumers want items of the highest quality. The focus of the firm is on continual improvement and quality of production and product shipment for customers, with most companies modifying the whole quality management system for the organization and production structure. Total management of quality is crucial to the corporate culture and impacts all areas of employee performance. Every part of the performance of its employees must be carried out to achieve the objective of staff who are able to satisfy excellent customer demands (Prajogo & Sohal, 2006). Through distinguishing its goods and developing a competitive market position, companies have adopted TQM to enhance their company's success. by developing an adequate climate, corporate culture may truly influence company and business success (Cadden et al., 2013). Therefore, the study examined the relationship between the three elements—OC, TQM, and SCH, which are crucial to gain competitive advantage in any company. Additionally, the study has been investigated the relationship between the dimensions of (OC) in the Emirates Post Group and the extent to which (TQM) practices are used, as well as the role of strategic change (SCH) in mediating the relationship between (OC) and (TQM).

2 Importance of the Research

The study investigated the relationship between strategic change (cultural, hierarchical, and human resources); organizational culture (in terms of clan and hierarchy;

external environment: adhocracy and the market) (Cameron & Quinn, 2011). In addition, TQM (customer focus, strategic planning, and workforce focus on terms of leadership support). The point of consideration is that the relationships between these three groups of variables must be investigated, since there are discrepancies in recognizing such relations, as the literature has shown the connection between TQM with the culture and performance of organizations (Baird et al., 2011; Cheng & Liu, 2007; Valmohammadi & Roshanzamir, 2015). Furthermore, few studies have addressed TQM's relationship to strategic change (Huq, 2005; Leonard & McAdam, 2001). Thus, it is crucial to examine the impact of TQM practices on the relationship between strategic change and organizational culture.

3 Problem Statement

The role of management in institutions emerges increasing the extent of awareness of employees to change (Shin et al., 2012). Given the paucity of studies in the Middle East that are concerned with the organizational culture of the institution and the indicators of comprehensive quality and aspects of strategic change. The researchers first discussed the theoretical framework for those aspects that relied on previous studies in this field, then targeted a field framework to analyze the impact and relationship between the three variables: organizational culture, strategy change, and Total quality management (Cameron & Quinn, 2011; Haffar et al., 2013). Therefore, the study investigates the nature of the relationship between the three variables: total quality management, organizational culture, and strategic change management, using which some of the assumptions have been statistically tested in the Emirates Post Group (EPG), which is considered as one of the government-owned agencies that controls the logistics business and has been profitable for over 45 years. They lived through a period of tranquility during which the Emirate's post was safe. When compared to the previous year, in 2019, they lost 41% of their profits. As a result, if they are to recover and keep their market position, they must halt and make the required modifications. There are numerous barriers to overcome, a lot of misused resources, and a heavy burden on the organization. Rejoining the market will not be easy, as there are many logistic competitors where only the strong survive. In this case, we will discuss empirically the impact of the organizational culture of the organization and TQM on the mediation of strategic change.

The study focused on the independent variable total quality management with its indicators and its impact on the change strategy, as well as quality management as an independent variable and its impact on change strategies. It also focused on quality management indicators as an intermediate variable between organizational culture and strategic change.

4 Goals and Objectives

Accordingly, the importance of the study is in determining the relationship between organizational culture, strategic change, and total quality management when it comes to defining the research problem. Moreover, determine whether strategy change plays as a mediator between organizational culture and total quality management. To achieve the following goals:

1. To specify the level of TQM indicators, Strategic Change, and Organizational Change.
2. To investigate the relationship between Organization Culture (OC), Strategic Change (SCH and total quality management (TQM).
3. To examine strategic change as a mediator between (OC) and (TQM).

5 Purpose of This Research

Using empirical data collected from the emirates post group, this study attempts to achieve two primary goals. First, it aims to discuss the association between OC, strategic change, and total quality management. Secondly, to analyze the appropriateness of strategic change a mediation of the relationship between TQM and OC.

The study consists of four parts. The first focused on the general framework and the theoretical side and previous studies that dealt with all research variables and through which the research hypotheses were defined, which some considered on a relationship between organizational culture and strategic change (Abuhashesh et al., 2021; Al-bawaia et al., 2022; Alshurideh et al., 2021a; Choi & Ruona, 2011; Hamadneh et al., 2021; Heracleous & Langham, 1996; Johnson, 1992; Odeh et al., 2021). While the other considerations on the relationship between SCH, TQM and OC (Fares, 2017). The second part focused on the method of data collection and the mechanism of data analysis and determining the study population, while the fourth part dealt with several aspects. Moreover, the study laid the groundwork for future research to explore and extend the connection between Total quality management, organizational culture, and strategic change.

6 Research Questions

There are valuable explanations to investigate the relationship between (TQM), organizational culture (OC), and strategic change of the organization. The study aims to answer the following questions considering defining the relationship between organizational culture, strategic change, and total quality management:

1. What is the impact of Organizational Cultures (OC) on Strategic Change (SCH) for the employees at the emirates post group?
2. What is the impact of strategic change on total quality management for the employees at the emirates post group?
3. What is the impact of Organizational Culture (OC) on total quality management for the employees at the emirates post group?
4. Does the strategic change (SCH) mediate the relationship between the (OC) and (TQM) for the employees at the emirates post group?

7 Literature Review and Hypotheses Development

Total Quality Management Concepts

Different definitions have led to various interpretations of TQM, Persico (Persico, 1989) argue that TQM is an organizational culture approach which includes employees in numerous business areas as well as the continuous improvement in output through cooperation to achieve the organization's goals. The credit for TQM since 1950 is given to scientists in the United States, such as Edward Goran and Philip Crosby. Juran and Deming prefer to employ TQM measurement by using statistical techniques. While Crosby and Free (1979) did not accept statistical quality, 14 actions have been taken to increase quality (Deming & Edwards, 1982). TQM has also been the main reason for the achievement of profitability and sustainability for small and big businesses in recent years. Ciampa (1992) described TQM as an organizational effort to build a sustainable environment where employees continuously enhance their customer satisfaction delivery capacity. In addition, an orderly manner to ensure the fluidity of planned operations is an efficient way to prevent problems, improve good conduct and apply management strategies effectively. Furthermore, Ciampa noted that an organization maintaining the flow of an organization's planned operations is an efficient approach to averting problems, improving good conduct, and making optimal use of management strategies. TQM also offers an organizational approach to building and developing a productive atmosphere in which employees consistently increase their ability to supply value-added products and services on demand. It is a commitment to meet client expectations at the best price, through improved performance.

Organizational Culture (OC): Organizational culture is defined as the general way that people share behaviour, standards, and attitudes and influences the organization's actions, practices, and other values that can readily be absorbed. Therefore, culture is distinguished factors between one organization and another (Schein, 1985). Organizational culture is also a process that aims to improve ways and techniques for reducing workplace conflict and barriers (Moorhead & Griffin, 1989). Furthermore, it is seen as a set of shared fundamental assumptions discovered and formed by a group of individuals in the same context.

Strategic Change (SCH): The study investigates some of the theoretical groundwork for change and, as a result, analyzes how academics are approaching the topic of organizational culture and total quality management in their studies. In the public and private sectors, the pace of change has recently increased considerably. And it is expected to become much more extreme in a highly competitive world where all businesses will be forced to compete for survival in a new era of "intense competition" (Aupperle & Karimalis, 2001). Researchers propose many elements of strategic change concepts and apply them in various ways, therefore there is no precise or single notion for each one. In terms of dealing with change, the current societal situation is unparalleled. Organizations have accepted the theory that if they want to survive, they must adapt for many years since the industrial revolution (Beer & Nohria, 2000). In general, if people have different interests and values, the change will not be properly implemented (Jackson, 1995). To attain the aim of greater effectiveness based on intellectual ability rather than physical assets, organizations must modify the way they engage with their people (Jarrar & Aspinwall, 1999). Furthermore, the strategy lays out defined objectives, outlines activities to attain them, and allocates resources to carry them out. In addition, when the firm adapts or competes in its environment, a strategy is created as an operational pattern (Alkhafaji & Nelson, 2013). Whereas the operations of the strategy include strategic thinking and strategic planning (Pfeffer & Salancik, 2003). Also, from the perspective of Max Mckeown (2011) the strategy is to draw the future of the organizations and ensure success for the long term (Hill et al., 2014). Strategic change is considered essential for maintaining competitive advantage and long-term organizational sustainability. However, over the last 5 years, Google Scholar has published approximately (791,000) papers on "strategic change" address.

Organizational culture (OC) on strategic change (SCH) context

Cameron and Quinn (2011), Johnson (1992), Heracleous and Langham (1996), Choi and Ruona (2011), Henry (2008), Ashal et al. (2021) studies consider that there is a relationship between organizational culture and strategic change Table 1 summarizes the view of researcher on organizational culture effects on strategic change based on the previous discussion. The study Hypothesize that *H1* organizational culture positively impacts strategic change.

Strategic change (SCH) impacts on total quality management (TQM)

Total quality management seeks methods to enhance quality on a continuous basis. There are three criteria or questions that a company should examine when determining whether a change is required: Will the change improve overall consumer satisfaction? Is the change allowing the business and the client to get outcomes more quickly? Also, does the concept help the firm make better use of its resources? If you responded yes to any of these questions, you should adopt the concept to enhance overall organizational and product quality. Several studies, such as (Dutton & Duncan, 1987; Zeffane, 1996), have underlined the importance of culture within the framework of strategic change.

Thus, the second hypotheses are focused on investigating the relationship between strategic change and total quality management TQM.

H2. Strategic change positively impacts total quality management.

Table 1 Organizational culture effect on strategic change

Research	Effects
Henry (2008)	Culture is more powerful than anything else in business, thus every suggestion for change, whatever its nature, must always fit into the unique kind of culture that exists in the company if it is to succeed
Jones et al. (2005)	The employees of organizations that deal with the group's values and norms feel that by participating in a successful change process, they will gain advantages such as awards or promotions. Therefore, it will improve change awareness
Zammuto et al. (2000)	Organizations that follow a centralized structure with a bureaucracy culture and hierarchy, on the other hand, do not foster creativity and innovation; instead, they oppose change. As a result, the participants develop a negative attitude towards the change process
Cameron and Quinn (2011)	Leaders of companies are more likely to be positive about recommended improvements since they expect their firm to not begin or enforce any changes that may negatively impact their employees. This is due to the company's focus on internal employees, which emphasizes the long-term benefits of human capital and places a high value on boosting morale

The relationship between TQM, OC, and SCH

Various studies have been conducted on the relationship between Total Quality Management and organizational culture, with the conclusion that some businesses have failed to adopt because the application process is culturally neglected (Dale & Cooper, 1992). Furthermore, numerous research has claimed the connection or effect of organizational culture (OC) and total quality management (TQM), (Heracleous & Langham, 1996; Rad, 2006). While few have looked at TQM's empirical mediation role in the relationship between change methods and organizational culture. Table 2 highlighted the literature on the factors' mediating function in total quality management, as well as their relationships with other variables.

Therefore, according to the previous discussion, the hypotheses below has been formulated:

- *H3.* organizational culture positively impacts TQM.
- *H4.* Strategic change (SCH) partially mediates the relationship between OC and TQM.

8 Research Methodology

When this study aims to analyze the relationship between three variables, total quality management (TQM), organizational culture (OC), and strategic change (SCH), the analysis is descriptive, organization culture in the empirical analytics model is

Table 2 The factors' mediating function in total quality management

Research	Effects
Cao et al. (2000)	refers to the assessment of multiple functional areas of TQM as well as strategic change in response to changes in organizational procedures, changes in hierarchical structure and frameworks, and changes in beliefs, norms, and actions related to the distribution of authority and personal rules and procedures
Cameron and Quinn (2011)	Leaders of companies are more likely to be positive about recommended improvements since they expect their firm to not begin or enforce any changes that may negatively impact their employees. This is due to the company's focus on internal employees, which emphasizes the long-term benefits of human capital and places a high value on boosting morale
Haffar et al. (2013)	Investigated the mediation role of total quality management as a possible mechanism through which organizational culture influences change tactics empirically. The findings of this study supported the idea that IRFC plays a mediating role in the relationship between OC and TQM. This contributes to elucidating the connections between OC and TQM implementation
Heracleous and Langham (1996)	Developed a set of three factors to identify the strategic change mediation concept between knowledge and TQM, the study discovered that strategic change mediates the relationship between knowledge management procedures and total quality management indicators
Dutton and Duncan (1987)	Created a framework for how leaders understand strategic concerns for the creation of effective change that helps managers to forecast how members will respond to a changing environment by emphasizing the systematic influence of two variables: the organizational structure and its resources
Demirbag et al. (2006)	Found a positive link between TQM and non-financial SMEs' performance, and a weak relationship between TQM and SMEs' financial performance. Furthermore, there is a mediating effect of SME non-financial performance on TQM practices
Mahmud and Hilmi (2014)	Found that TQM supported both SMEs' performance and organizational learning. Furthermore, culture has an indirect positive impact on total quality management. While (Cao et al., 2000) refers to the assessment of multiple functional areas of TQM as well as strategic change in response to changes in organizational procedures, changes in hierarchical structure and frameworks, and changes in beliefs, norms, and actions related to the distribution of authority and personal rules and procedures

known as the independent variable, while the strategic change as mediate and second independent variable, whereas TQM considered as the dependent variable.

In this study, a questionnaire-based tool was used to collect data, by using online google Microsoft form. 18 questions for evaluating TQM, 18 questions for organizational culture, and 16 questions for strategic change were considered on a Likert scale of 5 points. From which the criteria for evaluating TQM and organizational culture are provided by Cameron and Quinn (2011), Valmohammadi and Roshanzamir (2015) which OC based on the internal and external environment including clan culture, hierarchy, adhocracy, and consumer culture. TQM was also identified based on Cameron and Quinn (2011), contingent on his model, which was divided into four dimensions: leadership supportive, customer focus, strategic planning, and workforce focus. For the evaluation of strategic change (SCH), based on the questionnaire in Fares' research thesis, (Fares, 2017) which included three dimensions: organizational culture, hierarchy, and human resource.

The scope of the research

Emirates post group—Dubai plans for long-term preparation for its two divisions of E documents services and its branches, and Wall Street exchange centers. Currently, they provide high-quality services in the postal division starting from domestic mailing to global shipping, with routes of express mailing. They are continuously developing their services and seeking new solutions for their deliveries being motivated by the growing number of customers and public expectations from the corporation. One of the ideas they came up with is the one-stop-shop for post offices. Additionally, the corporation facilitates financial solutions for both government and private sectors for services like payment of bills, banking, and travelling.

Measurement and Tools

Validity and Reliability

The study used three reliable and valid tools from previous research to develop the questionnaire and methods to achieve the main goals and objectives of this article. The valid and reliable tools developed by Cameron and Quinn (2011) were constructed to determine the level of TQM and organizational culture in the Emirates Group Post (EPG). While measuring the valid and reliable of strategic change in the EPG via the tools developed by the study of Fares (2017). The study has used Cronbach alpha to measure and estimate the degree of the internal consistency of each construct. Table 3 shows that all scales have high reliability and values of Cronbach's alpha derived for the constructs ranging from 64.0 to 92.4. We assessed the construct validity of each construct by using SPSS factor analyses (Samson & Terziovski, 1999). The study result shows in Table 3 that the Cronbach alpha for 16 items of strategic change (SCH) ($\alpha = 91.6$) highly reliable, while the Cronbach alpha for 18 Total Quality Management (TQM) ($\alpha = 92.3$) highly reliable, and the Organizational Culture for 18 Items (OC) ($\alpha = 88.3$) are also highly reliable. Moreover, the study used Pearson correlation, see Table 4. In addition to that, the experts assessed the construct validity. The population includes all the staff in the Emirates post in

Table 3 The variables of the study and Test of reliability-results of factor and reliability analysis

Variables	Items	No. item	Cronbach's alpha
Strategic change	Organizational culture	5	80.2
	Organizational hierarchy	4	79.4
	Human resource	7	91.2
	Total	16	91.6
Total quality management	Leadership supportive	5	86.5
	Customer focus	4	68.1
	Strategic planning	4	92.4
	Workforce focus	5	83.9
	Total	18	92.3
Organizational culture	Clan	5	80.8
	Hierarchy culture	4	64.0
	Adhocracy culture	5	76.4
	Market culture	4	76.8
	Total	18	88.3

Dubai-2020, the total number of employees in person is 2000. The questionnaire was circulated online by using 1000 people in stratified random sampling. 341(198 male, 143 female) Questionnaires were returned. Were chosen and statistically analyzed. Different statistical methods were used for testing the conceptual model such as the Pearson correlation for testing the relations between variables and structural equation modelling for testing the fitness of the model (Table 4).

Testing the study Hypotheses

A series of regression analyses were performed to investigate the proposed relationships among TQM practices, OC, and strategic change (SCH).

From the correlation between the four factor's scores, there is evidence to support convergent validity.

Convergent validity was demonstrated by the positive significant correlation between the four factors (p-value $= 0.01$). In addition, inspection of the Table 4 reveals the significant correlation between the three factors which r ranging between (940**–732**) hence no item was deleted. As well as the Table 4 reveals the significant correlation between four factors which r ranging between (0.868**–0.456**).

In addition to that, the validity indicators of the tools were extracted by checking the correlation of each item with the dimension, to which it belongs, as well as the

Table 4 Internal context validity by using person correlation

	SCH3	SCH1	SCH2	SCH	
SCH3	1				
SCH1	0.816**	1			
SCH2	0.732**	0.795**	1		
SCH	0.940**	0.937**	0.891**	1	
	TQM1	TQM2	TQM3	TQM4	TQM
TQM1	1				
TQM2	0.863**	1			
TQM3	0.868**	0.865**	1		
TQM4	0.484**	0.456**	0.479**	1	
TQM	0.855**	0.834**	0.852**	0.840**	1
	OC1	OC2	OC3	OC4	OC
OC1	1				
OC2	0.543**	1			
OC3	0.613**	0.506**	1		
OC4	722**	450**	0.720**	1	
OC	886**	752**	0.839**	0.851**	1

correlation of each dimension with the total degree of the scale and the correlation of dimensions with each other.

Descriptive Statistics

Table 5 shows the mean score of organizational culture (OC), TQM, and strategic change. According to the value of mean and standard deviation SD, it is clear in a natural value.

Direct effect

Multiple regression analysis was used to test the direct effect of organizational culture (OC) on Strategic change (SCH) practices as Table 6 (Model1) shows, Clan ($\beta = 0.823$, P < 0.001), adhocracy culture ($\beta = 0.849$, P < 0.001), Hierarchy culture ($\beta = 0.573$, P < 0.001), Market culture ($\beta = 0.748$, P < 0.001) have positive and significant effects on SCH. in addition to that Table 6 shows that organizational culture (OC) ($\beta = 0.902$, P < 0.001) have positive and significant effects on strategic change (SCH). The result strongly supports the study hypotheses *H1*.

Multiple regression analysis was conducted to evaluate the influence of strategic change on TQM. The result in Table 6 (Model2) show that strategic change ($\beta = 0.784$, P < 0.001, $R^2 = 0.614$), organizational culture ($\beta = 0.663$, P < 0.001, $R^2 = 0.465$), and human resource ($\beta = 0.822$, P < 0.001, $R^2 = 0.676$) have a positive and significant effect on total quality management TQM, whereas the effect of hierarchy ($\beta = 0.525$, P < 0.001, $R^2 = 0.276$) affect less than other factors on TQM. In general, the result in Model2 supports *H2*.

Table 5 Mean score of SCH, TQM, and OC

Variables		No. item	MEAN	SD
Strategic change	Organizational culture	5	19.53	3.14
	Organizational hierarchy	4	15.72	2.46
	Human resource	7	26.83	4.69
	Total	16	62.09	9.31
Total quality management	Leadership supportive	5	18.92	4.27
	Customer focus	4	15.17	3.58
	Strategic planning	4	15.02	4.37
	Workforce focus	5	18.91	5.18
	Total	18	68.05	15.96
Organizational culture	Clan	5	19.37	3.84
	Hierarchy culture	4	14.79	2.98
	Adhocracy culture	5	19.49	2.89
	Market culture	4	15.52	2.42
	Total	18	69.18	10.13

The result in model 4 shows that there is a strong impact of OC on TQM that supports *H3*. On the other hand, the Table 6 shows that the culture ($\beta = 0.809$, P < 0.001, $R^2 = 0.654$). while the adhocracy culture ($\beta = 0.612$, P < 0.001, $R^2 = 0.375$), market culture ($\beta = 0.694$, P < 0.001, $R^2 = 0.480$) and hierarchy culture ($\beta = 0.461$, P < 0.001).

Mediating Effect: The main idea of the mediator between three variable impacts is to examine when Z leading the relationship between X and Y. in another word Z change the relationship between X and Y, it may increase or decrease (Sampedro-Matarín et al., 2006). Although to evaluate the mediating role of SCH strategic change between organizational culture (OC) and total quality management (TQM), the study follows Woodworth's model 1928 and (Baron & Kenny, 1986). The first step to test the direct relationship between the organizational culture-independent variable and SCH. the dependent variable to establish that there is an impact that may be mediated. The previous analysis in Table 5 Model1. Shows that the variables of OC significantly indirect impact on SCH see also (Fig. 1 and Table 7). Which the criteria for the first one was fit (MacKinnon, 2008). The second step is to ensure that the mediator variable affects the dependent variable. When we look back to Table 5. Which shows there is a positive impact on TQM. That is mean the criteria of the second step were met. The third step is to examine if the mediator factors affect the dependent variables. Table 5 and Fig. 1 show a positive and significant impact on TQM. Therefore, criteria three were met. In the fourth step, the study conducted a regression analysis to evaluate the mediating role of SCH strategic change in the relationship between OC variables and TQM practices. which the result of Table 6 Model 3 & 4, Figs. 1 and 2, and Table 7, indicates that the strategic change (SCH) lead to a decrease in the effect size such as (the strength of clan variable of organizational

Table 6 summary of the result of the multiple linear regression models

Model	Unstandardized coefficients		Standardized coefficients	T		Sig
	B	Std. error	Beta		R^2	
MODEL1						
Dependent Variable: SCH — Clan	1.790	0.075	0.823	26.716	0.678	0.000
Hierarchy culture	1.995	0.139	0.573	12.861	0.328	0.000
Adhocracy culture	2.734	0.092	0.849	29.584	82.1	0.000
Market culture	2.872	0.139	0.748	20.736	0.559	0.000
MODEL2						
Dependent Variable: TQM — Organizational culture	3.467	0.201	0.663	17.206	0.465	0.000
Organizational hierarchy	3.408	0.300	0.525	11.364	0.276	0.000
Human resource	2.796	0.105	0.822	26.579	0.676	0.000
Model3						
Dependent Variable: TQM — Clan	3.644	0.108	0.877	33.655	0.770	0.000
Hierarchy culture	2.472	0.258	0.461	9.566	0.213	0.000
Adhocracy culture	3.380	0.237	0.612	14.249	0.375	0.010
Market culture	4.568	0.258	0.694	17.737	0.480	0.000
Model4						
Dependent Variable: TQM WITH SCH Mediator — Clan	2.994	0.186	0.721	16.078	0.781	0.000
Hierarchy culture	1.235	0.221	0.018	0.446	0.614	0.000
Adhocracy culture	-1.053	0.348	-0.191	-3.022	0.624	0.003
Market culture	1.611	0.324	0.245	4.797	0.640	0.000

culture change with mediator SCH from ($\beta = 0.877$ to $\beta = 0.772$), market culture change from ($\beta = 0.694$ to $\beta = 0.245$). while the significant remains the same. The result of the study agrees with other studies such as (Haffar et al., 2013). Thus the study SCH strategic change can be explained as a partial mediator in the relationship between organizational culture variables and TQM practices. Therefore, the result indicates that OC indirectly affects TQM practices via the impact on strategic change SCH. That result support *H4* (Figs. 3, 4, 5 and Tables 8 and 9).

Table 7 (Multiple linear regression analysis) (H1–H4)

Models	Hypotheses	R	R square	Adjusted R square	Std. an error of the estimate	Change statistics				
						R square change	F change	df1	df2	Sig. F change
1	H1	0.902[a]	0.813	0.813	4.03098	0.813	1476.157	1	339	0.000
2	H2	0.784[a]	0.614	0.613	9.93561	0.614	539.266	0.784[a]	0.614	0.000
3	H3	0.809[a]	0.654	0.653	9.40236	0.654	641.713	1	339	0.000
4	H4 (mediating testing)	0.819[a]	0.670	0.668	9.20019	0.670	343.144	2	338	0.000

Table 8 Multiple linear regression analysis

Dependent	Variables	B	SD Error	Beta	t	P-value
Total Quality Management (TQM)	Constant	−31.859	3.855		−8.265	0.000
	OC	0.862	0.109	0.547	7.932	0.000
	Gender	6.666	1.037	0.206	6.430	0.000
	SCH	0.438	0.119	0.255	3.667	0.000
	Work experience	1.464	0.531	0.087	2.759	0.006
	$R = 0.841$ $R^2 = 0.707$ Std. Error = 8.70			$F = 202.270$ P-value $= 0.000$ df. $= 4$		

9　Results and Discussion

The aim of the study was to clarify of the mediating role of strategic change on the relationship between total quality management, organizational culture the result of the study has shown that the level of strategic change has a direct impact on the process of TQM practice. These findings support by other studies such as Haffar et al. (2013). Additionally, the strategic change can be explained as a partial mediator in the relationship between organizational culture variables and TQM practices. As well as the study emphasized the role of organizational culture to effectively transition across the various layers of organizational structure. On the other hand, the finding of this study indicates that organizational culture is highly affected by the strategic change. As a result, the organization believes how important a culture of belief and change is to be created. The organization often involves its members to make a change, and the organizational structure is distinguished by a high degree of cooperation between the various units to achieve strategic change. Therefore, the findings of the study indicate that the organizational culture (OC) is an impact on strategic change. In addition, some scholars, such as jones (2005) have found that clan/adhocracy and culture forms are correlated with higher levels of change. While (Iljins et al., 2015) have found that the organizational factors influencing organizational climate. In the current research, similar results were found where the degree of change is positively affected by the clan (that people get things in common. deep links of commitment, custom, and shared experience typically form) and styles of the culture of adhocracy. Furthermore, it is obvious from the findings of this analysis that a high degree of change is correlated with consumer culture. Therefore, the findings of this study suggest that organizations governed by forms of clan/adhocracy and consumer culture promote enhanced applicability of change. Hence, if they consider their work environment to have the factors associated with either a clan or adhocracy culture forms, members of an organization are more likely to have higher levels of change. In addition to that, the leadership of such organizations also used to appreciate employee engagement and empower their workers by setting up a fair system of benefits and rewarding the members who share in the progress of the implementation process. This contributes to improving the emotions and confidence of the participants in honesty and personal benefits that will be achieved because of involvement in the introduction

Table 9 Estimate—regression weight

	Estimate	S. E	C.R	P-VALUE	Standardized regression weights	Direct effect	Decision
SCH <-- OC H1	0.829	***	38.477	***	0.902	0.829	Supported H1
TQM <-- SCH H2	0.497	0.124	4.020	***	0.290	0.497	Supported H2
TQM <-- OC H3	0.863	0.114	7.596	***	0.548	0.863	Supported H3

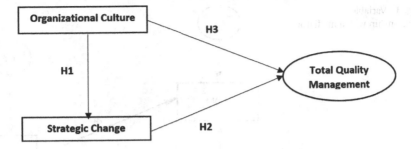

Fig. 1 Explained the study conceptual framework

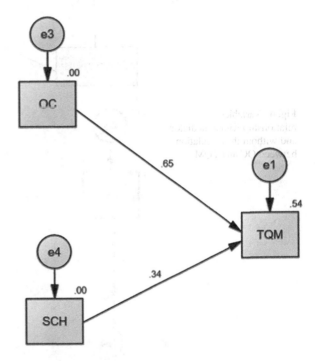

Fig. 2 Three variables relationship without mediator and without the correlation between OC–SCH

of new intrusive change initiatives. This confirms the results of several scholars, such as (Latta, 2009). who found that if members of organizational fairness in processes of change, they would have more personally prepared to adopt readiness of change? Moreover, the findings of this research show that the most positive culture groups for TQM are compared to cultural forms, community, and consumer culture. While the Cultural forms of hierarchy and adhocracy are the least encouraging to TQM. This is consistent with the results of various research studies, such as (Gimenez-Espin et al., 2013).

Fig. 3 Variables
relationship with a mediator

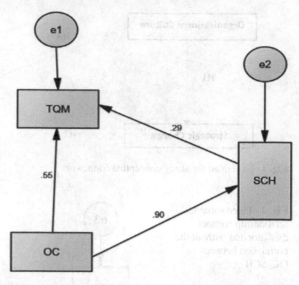

Fig. 4 Variables
relationship with a mediator
and without the correlation
between OC and TQM

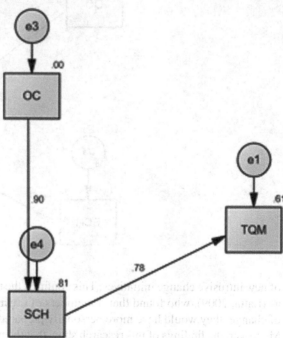

Fig. 5 Variables correlation
without the correlation
between TQM–SCH

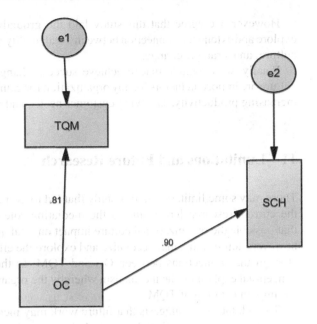

According to the effective implementation of the TQM includes long-term strategic planning. Authors stress strongly on the value of strategic planning (Hitchcock & Willard, 2002; Sinclair & Zairi, 2001). Involvement and dedication of managers at various levels is important for the implementation of Juran and Godfrey (1999), Crosby (1989). Between OC and supportive leadership in TQM is consistent with the results of previous studies such as (Mustafa & Bon, 2012; Waldman, 1993; Zhang et al., 1999). Furthermore, the findings of this research demonstrate the lack of relevance for total quality management (TQM) of the hierarchical society. In addition, the finding of this study shows that the culture of adhocracy has less of a noticeable effect on TQM. Help for this performance with the study of Yong and Pheng (2008).

10 Conclusion

The result of the study revealed that strategic change has a direct impact on the TQM practices. In addition to that, the strategic change can be explained as a partial mediator in the relationship between organizational culture variables and TQM practices. The result of the study also found that there is a positive and significant effect of organizational culture on strategic change. On the other hand, the results analysis for TQM and strategic change revealed that the strategic change positively effects on total quality management in various areas.

However, we agree that this study laid the groundwork for future research to explore and extend the connection between Total quality management, organizational culture, and strategic change.

Finally, we conclude that to achieve success change TQM and organizational culture are important factors for any organization for gaining competitive advantages, increasing productivity, satisfying customer needs, and survive.

11 Limitations and Future Research

There may some limitation in this study that led to avenues in future research. First, the current research has evaluated the mediating role of strategic change (SCH) that possible the organizational culture impact on total quality management (TQM), however, future study, may recognize and explore the effect of other potential mediators in the connections between OC and TQM, as they will support extend our conceptual explaining the mechanism whereby the organizational culture influences the implementation of TQM.

Second, the study suggests that future work may include samples from a greater geographic area in UAE or be expanded to a multi-country analysis. Thus, the current study findings contribute to the global picture form the basis for future, more integrated studies.

According to the results, the study recommended that future research might investigate if other possible mediating variables may also mediate whatever totally or partially the effect of the relationship between TQM and organizational culture. Second thing, future studies can expand the respondents more than this study. Additionally, future studies may use the (SEM) Structure Equation Modeling that enhances the result by building a more convenient model of the study. Moreover, the researcher may use the result of this study to examine more specifically the relationship between two variables such as organizational culture with strategic change, TQM and organizational culture, or organizational culture and strategic change.

Finally, in current research, as in most recent research, regression analysis tests for hypotheses have been used. Although structural equation modelling (SEM) can be a useful technique for the analysis of large data, such as Kline (2015; Mueller & Hancock, 2008). considering that SEM is a large sample technique that needs to be used but the study couldn't implement because of the number of participants in the research is less than 400. For future research, it is suggested to use SEM with greater sample size and a higher number of participants' expectations.

12 Implication and Recommendations

The current study conducted within the COVID-19 pandemic. As most organizations are making a change in different fields of their organization to face the crisis and survive (Ahmad et al., 2021; Alameeri et al., 2021; Alshurideh et al., 2021b, 2021c). It is hoped that the findings of the analysis will be useful to managers, even after the present crisis has ended. Having taken the study techniques and organizational culture variables into consideration to apply total quality management.

In addition to that, creating a positive environment will also promote the willingness of employees for the change process and encourage their participation in the implementation of TQM. In turn, this increases the chance of a successful TQM implementation in organizations, especially in the current situation with the COVID 19 pandemic, which gives the managers new experience to come up with innovative ideas to keep up and maintain the TQM success.

What Emirate Post Group have to do to improve their business?

- Foster high performing organization culture by defining the organization ethically, enabling two-way communication, rewarding employee, and empowering the team.
- Redesign the performance evaluation process to ensure that the individual goals align with the department goals.
- Develop learning manuals for new joiners.
- Document job descriptions.
- Update the HR process.
- Establish a cultural periodical review and continuous improvement.

References

Abuhashesh, M. Y., Alshurideh, M. T., & Sumadi, M. (2021). The effect of culture on customers' attitudes toward Facebook advertising: The moderating role of gender. *Review of International Business and Strategy, 31*(3), 416–437.

Ahmad, A., Alshurideh, M. T., Al Kurdi, B. H., & Salloum, S. A. (2021). Factors impacts organization digital transformation and organization decision making during Covid19 pandemic. In *The Effect of Coronavirus Disease (COVID-19) on Business Intelligence* (pp. 95–106). Springer, Cham.

Alameeri, K. A., Alshurideh, M. T., & Al Kurdi, B. (2021). The effect of Covid-19 pandemic on business systems' innovation and entrepreneurship and how to cope with it: A theatrical view. In *The effect of coronavirus disease (COVID-19) on business intelligence* (Vol. 334, pp. 275–288).

Al-bawaia, E., Alshurideh, M., Obeidat, B., & Masa'deh, R. (2022) The impact of corporate culture and employee motivation on organization effectiveness in Jordanian banking sector. *Academy of Strategic Management Journal, 21*(Special Issue 2), 1–18.

Alkhafaji, A., & Nelson, R. A. (2013). *Strategic management: Formulation, implementation, and control in a dynamic environment.* Routledge.

Almaraz, J. (1994). Quality management and the process of change. *Journal of Organizational Change Management.*

Alshurideh, M. T., Al Kurdi, B., & Salloum, S. A. (2021a). The moderation effect of gender on accepting electronic payment technology: A study on United Arab Emirates consumers. *Review of International Business and Strategy, 31*(3), 375–396.

Alshurideh, M. T., Hassanien, A. E., & Ra'ed Masa'deh. (2021b). *The effect of coronavirus disease (COVID-19) on business intelligence.* Springer.

Alshurideh, M. T., Kurdi, B. A., AlHamad, A. Q., Salloum, S. A., Alkurdi, S., Dehghan, A., ... & Masa'deh, R. E. (2021c). Factors affecting the use of smart mobile examination platforms by universities' postgraduate students during the COVID 19 pandemic: An empirical study. In *Informatics* (Vol. 8, No. 2, pp. 1–21). Multidisciplinary Digital Publishing Institute.

Ashal, N., Alshurideh, M., Obeidat, B., & Masa'deh, R. (2021). The impact of strategic orientation on organizational performance: Examining the mediating role of learning culture in Jordanian telecommunication companies. *Academy of Strategic Management Journal, 21*(Special Issue 6), 1–29.

Aupperle, K., & Karimalis, G. (2001). Using metaphors to facilitate cooperation and resolve conflict: Examining the case of Disneyland Paris. *Journal of Change Management, 2*(1), 23–32.

Baird, K., Hu, K. J., & Reeve, R. (2011). The relationships between organizational culture, total quality management practices and operational performance. *International Journal of Operations & Production Management*.

Baldwin, R., & Di Mauro, B. W. (2020). Economics in the time of COVID-19: A new eBook. *VOX CEPR Policy Portal*, 2–3.

Balogun, J., & Johnson, G. (2005). From intended strategies to unintended outcomes: The impact of change recipient sensemaking. *Organization Studies, 26*(11), 1573–1601.

Baron, R. M., & Kenny, D. A. (1986). The moderator–mediator variable distinction in social psychological research: Conceptual, strategic, and statistical considerations. *Journal of Personality and Social Psychology, 51*(6), 1173.

Beer, M., & Nohria, N. (2000). Cracking the code of change. *HBR's 10 Must Reads on Change, 78*(3), 133–141.

Cadden, T., Marshall, D., & Cao, G. (2013). Opposites attract organisational culture and supply chain performance. *Supply Chain Management: An International Journal*

Cameron, K. S., & Quinn, R. E. (2011). *Diagnosing and changing organizational culture: Based on the competing values framework*. John Wiley & Sons.

Cao, G., Clarke, S., & Lehaney, B. (2000). A systemic view of organisational change and TQM. *The TQM Magazine*.

Cheng, C. W., & Liu, A. M. (2007). The relationship of organizational culture and the implementation of total quality management in construction firms. *Surveying & Built Environment*.

Choi, M., & Ruona, W. E. (2011). Individual readiness for organizational change and its implications for human resource and organization development. *Human Resource Development Review, 10*(1), 46–73.

Ciampa, D. (1992). *Total quality: A users' guide for implementation*. Addison Wesley Publishing Company.

Crosby, P. B. (1989). *Let's talk quality: 96 questions you always wanted to ask Phil Crosby*. McGraw-Hill Publishing Company.

Crosby, P. B., & Free, Q. I. (1979). The art of making quality certain. *New York: New American Library, 17*, 174–183.

Dale, B. G., & Cooper, C. L. (1992). *Total quality and human resources: An executive guide*. Blackwell Business.

Deming, W. E., & Edwards, D. W. (1982). *Quality, productivity, and competitive position* (Vol. 183). Massachusetts Institute of Technology, Center for advanced engineering study.

Demirbag, M., Tatoglu, E., Tekinkus, M., & Zaim, S. (2006). An analysis of the relationship between TQM implementation and organizational performance: Evidence from Turkish SMEs. *Journal of Manufacturing Technology Management*.

Denison, D. R. (1990). *Corporate culture and organizational effectiveness*. John Wiley & Sons.

Duck, J. D. (1993). Managing change: The art of balancing. *Harvard Business Review, 71*(6), 109–118.

Dutton, J. E., & Duncan, R. B. (1987). The influence of the strategic planning process on strategic change. *Strategic Management Journal, 8*(2), 103–116.

Fares. (2017). The mediatory role of domains of strategic change in the relationship between knowledge management processes and indicators of total quality management at Salah Uddin University—Colleges-Erbil. http://repository.sustech.edu/handle/123456789/20265

Gimenez-Espin, J. A., Jiménez-Jiménez, D., & Martínez-Costa, M. (2013). Organizational culture for total quality management. *Total Quality Management & Business Excellence, 24*(5–6), 678–692.

Haffar, M., Al-Karaghouli, W., & Ghoneim, A. (2013). The mediating effect of individual readiness for change in the relationship between organisational culture and TQM implementation. *Total Quality Management & Business Excellence, 24*(5–6), 693–706.

Hamadneh, S., Hassan, J., Alshurideh, M., Al Kurdi, B., & Aburayya, A. (2021). The effect of brand personality on consumer self-identity: The moderation effect of cultural orientations among British and Chinese consumers. *Journal of Legal, Ethical and Regulatory Issues, 24*, 1–14.

Henry, A. (2008). *Understanding strategic management*. Oxford University Press.

Heracleous, L., & Langham, B. (1996). Strategic change and organizational culture at Hay Management Consultants. *Long Range Planning, 29*(4), 485–494.

Hill, C. W., Jones, G. R., & Schilling, M. A. (2014). *Strategic management: Theory & cases: An integrated approach*. Cengage Learning.

Hitchcock, D., & Willard, M. (2002). Sustainability: Enlarging quality's mission. *Quality Progress, 35*(2), 43.

Huq, Z. (2005). Managing change: a barrier to TQM implementation in service industries. *Managing Service Quality: An International Journal*.

Iljins, J., Skvarciany, V., & Gaile-Sarkane, E. (2015). Impact of organizational culture on organizational climate during the process of change. *Procedia-Social and Behavioral Sciences, 213*, 944–950.

Jackson, S. (1995). Love: Complicity, resistance and change. *Romance Revisited*, 49.

Jarrar, Y. F., & Aspinwall, E. M. (1999). Integrating total quality management and business process re-engineering: Is it enough? *Total Quality Management, 10*(4–5), 584–593.

Johnson, G. (1992). Managing strategic change—Strategy, culture and action. *Long Range Planning, 25*(1), 28–36.

Jones, R. A., Jimmieson, N. L., & Griffiths, A. (2005). The impact of organizational culture and reshaping capabilities on change implementation success: The mediating role of readiness for change. *Journal of Management Studies, 42*(2), 361–386.

Juran, J., & Godfrey, A. B. (1999). Quality handbook. *Republished McGraw-Hill, 173*(8).

Kline, R. B. (2015). *Principles and practice of structural equation modeling*. Guilford Publications.

Latta, G. F. (2009). A process model of organizational change in cultural context (OC3 Model) The impact of organizational culture on leading change. *Journal of Leadership & Organizational Studies, 16*(1), 19–37.

Leonard, D., & McAdam, R. (2001). The relationship between total quality management (TQM) and corporate strategy: The strategic impact of TQM. *Strategic Change, 10*(8), 439–448.

Lindquist, M. (2006). For better business results, focus on your customer, not your competition. *Cost Engineering, 48*(3), 10.

MacKinnon, D. P. (2008). *Introduction to statistical mediation analysis*. Routledge.

Mahmud, N., & Hilmi, M. F. (2014). TQM and Malaysian SMEs performance: The mediating roles of organization learning. *Procedia-Social and Behavioral Sciences, 130*, 216–225.

Max Mckeown, B. (2011). Business performance measures and alignment impact on strategy. *International Journal of Operations & Production Management, 2*(1), 972–996.

Moorhead, G., & Griffin, R. W. (1989). Individual differences. *Organizational Behavior*

Mueller, R. O., & Hancock, G. R. (2008). Best practices in structural equation modeling. *Best Practices in Quantitative Methods, 488508*.

Mustafa, E., & Bon, A. T. (2012). Role of top management leadership and commitment in total quality management in service organization in Malaysia: A review and conceptual framework. *Elixir Human Resource Management, 51*, 11029–11033.

Odeh, R. B. M., Obeidat, B. Y., Jaradat, M. O., & Alshurideh, M. T. (2021). The transformational leadership role in achieving organizational resilience through adaptive cultures: the case of Dubai service sector. *International Journal of Productivity and Performance Management, ahead-of-print*(ahead-of-print). https://doi.org/10.1108/IJPPM-02-2021-0093

Persico, J. (1989). Team up for quality improvement. *Quality Progress, 22*(1), 33–37.

Pfeffer, J., & Salancik, G. R. (2003). *The external control of organizations: A resource dependence perspective.* Stanford University Press.

Prajogo, D. I., & Sohal, A. S. (2006). The relationship between organization strategy, total quality management (TQM), and organization performance—The mediating role of TQM. *European Journal of Operational Research, 168*(1), 35–50.

Rad, A. M. M. (2006). The impact of organizational culture on the successful implementation of total quality management. *The TQM Magazine.*

Sampedro-Matarín, C., Gámiz, F., Godoy, A., & Ruiz, F. G. (2006). The multivalley effective conduction band-edge method for Monte Carlo simulation of nanoscale structures. *IEEE Transactions on Electron Devices, 53*(11), 2703–2710.

Samson, D., & Terziovski, M. (1999). The relationship between total quality management practices and operational performance. *Journal of Operations Management, 17*(4), 393–409.

Schein, E. H. (1985). Defining organizational culture. *Classics of Organization Theory, 3*(1), 490–502.

Shin, J., Taylor, M. S., & Seo, M. G. (2012). Resources for change: The relationships of organizational inducements and psychological resilience to employees' attitudes and behaviors toward organizational change. *Academy of Management Journal, 55*(3), 727–748.

Sinclair, D., & Zairi, M. (2001). An empirical study of key elements of total quality-based performance measurement systems: A case study approach in the service industry sector. *Total Quality Management, 12*(4), 535–550.

Valmohammadi, C., & Roshanzamir, S. (2015). The guidelines of improvement: Relations among organizational culture, TQM and performance. *International Journal of Production Economics, 164*, 167–178.

Waldman, D. A. (1993). A theoretical consideration of leadership and total quality management. *The Leadership Quarterly, 4*(1), 65–79.

Yong, K. T., & Pheng, L. S. (2008). Organizational culture and TQM implementation in construction firms in Singapore. *Construction Management and Economics, 26*(3), 237–248.

Zammuto, R. F., Gifford, B., Goodman, E. A. (2000). Managerial ideologies, organization culture and the outcomes of innovation: A competing values perspective.

Zeffane, R. (1996). Dynamics of strategic change: Critical issues in fostering positive organizational change. *Leadership & Organization Development Journal.*

Zhang, Z., Waszink, A., & Wijngaard, J. (1999). *Developing an instrument for measuring TQM implementation in a Chinese context.* Graduate School/Research Institute Systems, Organisation and Management.

Educational Supply Chain Management: A View on Professional Development Success in Malaysia

K. L. Lee(iD), **G. Nawanir**(iD), **J. K. Cheng**(iD), **Haitham M. Alzoubi**(iD), and **Muhammad Alshurideh**(iD)

Abstract The industrial environment is unpredictably shifting. Technology advancements and pandemics force the corporate environment to evolve even faster in order to stay competitive and thrive. As a result, in today's fast-paced business environment, supply chain management has grown in popularity. The study's major goal is to apply the concept of industrial supply chain management to educational management. The empirical study will be conducted separately and do not cover in this paper. In dynamic and uncertain business operations, the educational model is more appropriate to switch from mass production delivery to industrial experiences in building individual talent excellence. This minimizes the educator's knowledge and experiences delivery gap to the learner in the educational supply chain. From the issues, This study identifies the education attributes and how it affects professional development success. This study further investigates the moderating effect of industrial experiences on the relationship between education attributes and professional development success. Besides, this study aims to propose an educational supply chain model (EduSCM) for 21st-century professional development success in

K. L. Lee · G. Nawanir · J. K. Cheng
Faculty of Industrial Management, Universiti Malaysia Pahang, Lebuhraya Tun Razak, 26300 Gambang, Kuantan, Pahang, Malaysia
e-mail: leekhailoon@ump.edu.my

G. Nawanir
e-mail: gusman@ump.edu.my

J. K. Cheng
e-mail: jackkie@ump.edu.my

H. M. Alzoubi (✉)
School of Business, Skyline University College, Sharjah, UAE
e-mail: haitham.alzubi@skylineuniversity.ac.ae

M. Alshurideh
Department of Marketing, School of Business, University of Jordan, Amman, Jordan
e-mail: m.alshurideh@ju.edu.jo; malshurideh@sharjah.ac.ae

Department of Management, College of Business Administration, University of Sharjah, Sharjah, UAE

Malaysia higher education to gain competitive advantages. To achieve the objectives, mixed methods research design is proposed. The qualitative data collection will be collected after the completion of quantitative data analysis. The data collected from the interview use to explain, interpret, understand, and triangulate the results of quantitative analysis to produce a potent conclusion. Lastly, this study will provide a unique conceptual diagram expected to aid researchers and practitioners to create a more comprehensive understanding of the linkages between learning competencies, industrial experiences, and professional development.

Keywords Educational supply chain management · Professional development success · Industry experience · Malaysia

1 Introduction

The initial concept of logistics began in 1950 and the concept matured in 1970 (Habib & Jungthirapanich, 2008a, 2008b). The industry noticed that the logistics concept plays an essential role in the business environment and gets critical, and the logistics elements also get more comprehensive. Therefore, in 1980, the focus of the logistics concept is switched, and the supply chain management concept is initiated. Due to the power of the supply chain concept, researchers and academicians start researching the concept, and the publication was available in 1982. In 1985, the manufacturing industry realized the usefulness of the supply chain management concept in managing their operation. The manufacturer applies the supply chain management concepts as strategic and systematic coordination of business activities to reduce costs, improve customer service, enhance quality, and increase competitiveness (Tan et al., 2002). In 1990, more and more researchers started exploring the concept and the area that can be researched in the supply chain (Huan et al., 2004; Cooper, 2006).

The power of supply chain management concept and the result of research activities from academician influenced the service industry as one of the main sections in a country start to implement supply chain management concept in-service operation in 1995. After five years, in 2000, it gains incredible attention from academics and practitioners (Chan & Qi, 2003). The concept of educational supply chain management started to emerge in 2007 as it is one of the service industries and aims to produce the right talent that can meet the industry needs (Habib & Jungthirapanich, 2008a, 2008b). In 2010, the market no longer competed between the organizations but between the supply chain to gain competitiveness (Gunasekaran et al., 2008). Technology advancement is moving faster than ever in the current decades. Therefore, the digital supply chain starts to emerge in 2017, focusing on integrating the information system to increase the efficiency, reliability, and transparency of the overall supply chain (Büyüközkan & Göçer, 2018; Korpela et al., 2021). In short, the supply chain management concept becomes more and more important in the current

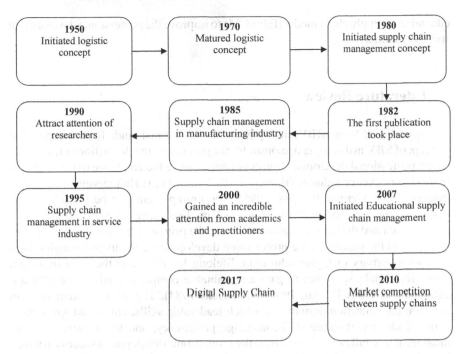

Fig. 1 The Evolution of Supply Chain Management. Source. Adapted from Lee et al. (2018)

industries, not only in the manufacturing but also in the service sectors, including the education sector. The evolution of supply chain management is illustrated in Fig. 1.

The graduates' statistics 2019 reported by DOSM (DOSM, 2018) shows that the number of graduates in Malaysia was increased from 4.94 million to 5.29 million in 2019. Based on the statistics, the number of graduates was increased. However, the employment rate was slightly decreased from 83.6 to 83.5%. The statistical number shows the employment rate and whether the graduates produce and are employed in the right field based on the program graduate is another issue. The mismatch in the supply and demand of graduates between higher institutions and industry needs remains a critical discussion topic until today. Employers claimed graduates lacked the necessary knowledge, skills, and attitudes. Computing skills, teamwork, English language proficiency (Markussen-Brown et al., 2017), previous training, and the graduate's personality are the five most important employability qualities, according to a recent research (Belwal et al., 2017). According to McKinsey report in education to employment, only 40% of graduates are completely prepared by the university to work in the real industry ield (Coffield, 2012). Obviously, Malaysian education needs to have a new professional development success model in handling contemporary issues.

Therefore, this paper investigates the relationship between education attributes and professional development success in Malaysia and how the moderating effect of industrial experiences in the relationship. Ultimately, this paper aims to propose an educational supply chain model (EduSCM) to improve the professional development success in Malaysia.

2 Literature Review

Resource-Based View (RBV) theory becomes an essential guide for this paper. The concept of RBV in this paper demonstrates the process of transformation of learners to be the professional development success which can be the employee or entrepreneur in the job market or industry (Barney, 2001; Grant, 1991; Halldorsson et al., 2007; Peteraf, 1993; Wernerfelt, 1984, 1995). The entrepreneur can be the job creator that offers the job opportunity to the job seeker. This paper believed that the parent participation and the learning environment are the primary resources for the learning process and the success of the professional development need to go through school education, tertiary or higher education, lifelong learning, and the most important is the industrial experience to gain a sustainable competitive advantage (Barney, 2001; Grant, 1991; Peteraf, 1993; Wernerfelt, 1984). The six measurements will measure the education attributes in which leadership skills, ethics and spirituality, national identity, thinking skills, language proficiency, and knowledge. The four measurements will be used to measure the professional development success in which holistic, entrepreneurial and balanced graduates, talent excellence, lifelong learner, and quality technical and vocational graduates. The application of RBV theory in educational supply chain management is shown in Fig. 2.

There are 672 higher learning institutions in Malaysia, consisting of 70 private universities, 20 public universities, 34 private university colleges, 410 private colleges, 33 polytechnics, 91 community colleges, and 14 higher institutions centres of excellence (HICoE), as shown in Table 1. In addition, there are 67,747 academic staff and 1,157,000 students in higher learning education in Malaysia, as shown in Table 2. The Ministry of Higher Education (2014) research noted that only 75% of

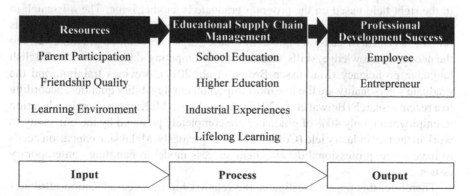

Fig. 2 Resource-based view theory in educational supply chain management. *Note* Adapted from Lee et al. (2018)

graduates are employed. Moreover, McKinsey (2012) noticed that only 40% of graduates are entirely prepared to work in the real industry. The International Standard Classification of Education (ISCED) level five encompasses Diploma, Advanced Diploma, Postgraduate Diploma, Professional Certificate, Bachelors, Master's, and Ph.D. programmes.

The phenomenal expansion of Malaysia's higher education industry over the previous two decades has aided in the country's internationalisation. The primary goals, according to the Ministry of Higher Education (MOHE), are to guarantee that all Malaysians have better access to excellent higher education by maximising the resources of current public and private higher education institutions. Malaysia's well-structured higher education system is critical to accomplishing professional development goals. Ethics and spirituality (Frich et al., 2015; Puchalski et al., 2014) leadership skills, national identity, language competence (Murray, 2011; Noom-Ura, 2013), thinking skills (Petersen, 2007) and knowledge are six main qualities for individual pupils presented in a recent study. Industrial experience is recommended to play a moderating role in enabling higher education to work closely with industry to guarantee that supply and demand are met demand. As a result, modern educational trends place a greater emphasis on delivery, accountability, openness, and outcomes rather than input.

Table 1 Higher learning institution in Malaysia

Higher learning institution in Malaysia	Number of institution
Public Universities	20
Private Universities	70
Private University Colleges	34
Private Colleges	410
Community Colleges	91
Polytechnics	33
Higher Institutions Centres of Excellence (HICoE)	14
Total	672

Source Adopted from Malaysia Education Blueprint 2015–2025

Table 2 Students and academic staff in higher learning institution in Malaysia

Higher learning institution in Malaysia	Number of academic staff	Number of students
Public Universities	33,199	560,000
Private Universities, Private University Colleges, and Private Colleges	24,476	485,000
Community Colleges	2,816	22,000
Polytechnics	7,256	90,000
Total	67,747	1,157,000

Source Adopted from Malaysia Education Blueprint 2015–2025

It's recommended that a graduate be considered outstanding academic performance is comprised of holistic, entrepreneurial, talent excellence, lifelong learner,
and quality technical and vocational. Education entrepreneurs must improve professional development success (Ruskovaara et al., 2015) through curricula that earned
the country international recognition. Lifelong learning can be achieved through the
current trend of globalized online learning (Fine & Zygouris-Coe, 2011). This is
supported by the fact that Malaysia presently has a 67% internet penetration rate,
which is the seventh-highest penetration rate in Asia, indicating significant potential. Besides, In dynamic and uncertain business operations, the educational model is
more appropriate to switch from mass production delivery to personalized learning
experiences in building individual talent excellence (Alshurideh, 2012; Fraser-Seeto
et al., 2014; Nijs et al., 2014). The conceptual framework of educational supply chain
management is illustrated in Fig. 3.

By focusing on supply chain reliability, responsiveness, agility, and costs, today's
supply chains are more likely to respond quickly, efficiently, and effectively to
changes, volatility, and turbulence in the marketplace to achieve, succeed, and sustain
competitive advantage in the new edge of business environment that is more global,
customer-driven, and technologically oriented. In Malaysia, several supply chain
management studies have been done. Supply chain management, according to the
trends, has offered several benefits to the supply chain or company. Firms have used
the supply chain to globalise their business, particularly in recent years, in order to

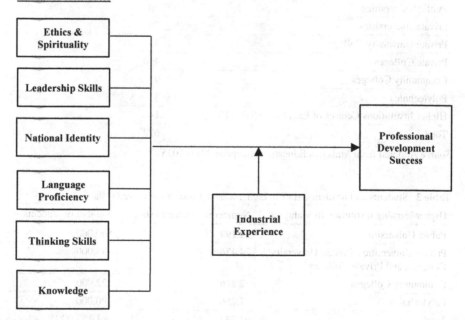

Fig. 3 Conceptual framework of professional development success

satisfy customers with high-quality products and cost-effective delivery (Al Dmour et al., 2014; Crinis, 2012; Lam & Postle, 2006). However, there are just a few studies that have used supply chain management theory to manage educational operations.

Educational supply chain management is based on the same principles as existing industrial supply chain management methods Kurdi et al. (2021). The supply chain also has upstream and downstream components. The upstream educational supply chain refers to the movement of students, information, and finances from early child-hood education (kindergarten and pre-school) to primary, secondary, higher educa-tion, and higher education. However, parental involvement and the quality of chil-dren's friendships are crucial, and they can have a considerable influence on the quality of graduates produced. This stage may be viewed as laying the groundwork for producing high-quality graduates in the future. This is due to the fact that both elements are considered raw material providers. With low-quality raw materials and inefficient manufacturing methods, it is difficult to generate a high-quality product. The downstream educational supply chain, on the other hand, refers to the delivery of graduates to industry as employees or the business world as entrepreneurs. According to the newest Malaysia Education Blueprint 2015–2025, graduates are not need to be job seekers in order to be employed, instead, they may be job creators in order to be entrepreneurs who give work possibilities to job seekers.

Parents' involvement and the quality of their friendships are seen as requirements for generating high-quality graduates (Alshurideh, Salloum, Al Kurdi, & Al-Emran, 2019; Pittman & Richmond, 2008; Salloum et al., 2020). Since their children are born, parents may have a big influence on them. Before sending their children to school, they spend a long time together. Even at school, the school day is shorter than the time spent with parents. According to statistics, parent involvement in the educational supply chain is important (Alshurideh, 2022; Alzoubi et al., 2022; Joghee et al., 2021; Shamout et al., 2022). Furthermore, one of the most important factors in creating high-quality graduates is friendship quality (AlHamad et al., 2022; Amarneh et al., 2021; Lee et al., 2022a, 2022b). Although friendship quality did not have a direct impact on academic achievement, it was found to be strongly connected to the reduction of problem behaviour in individuals (Bettayeb et al., 2020; Lee et al., 2022a, 2022b). Problem behaviour is thought to be the cause of poor academic achievement (Al Kurdi et al., 2021; Alshurideh et al., 2022). To make the registration process go well, however, coordination between parents, friends, and the school is necessary.

Collaboration between higher education and schools is essential. The partnership can guarantee that the appropriate resources are developed and available for higher education. This not only enhances the student's competence and ability for further schooling, but it also raises the student's potential for future grooming. A high-quality input is thought to provide a high-quality output in supply chain management (Al Ali, 2021; Alshurideh et al., 2020a, 2020b; Kashif et al., 2021). Some of the issues that arise in higher education stem from students' academic performance. One of the most important factors in a student's ability to succeed in higher education is the school's academic success (Akour et al., 2021; Alzoubi, 2021a, 2021b; Lee & Ahmed, 2021). A shaky foundation creates a gap in the learning process for the pupil. This had an impact on pupils who had a weak foundation, as well as the entire teaching and learning process. It might have an impact on how a lecturer teaches and how other students in the same class learn.

Furthermore, collaboration between higher education and industry cannot be ended at any point throughout the educational process. For the transition from student to employee or entrepreneur, both sides must communicate effectively (Cruz, 2021; Kurdi, Alshurideh, & Salloum, 2020a, 2020b, 2020c; Radwan & Farouk, 2021). This is because the industry's demands can only be met if the industry's higher education requirements are shared, as the industry's primary source of graduates (Alshurideh, Al Kurdi, et al., 2019a, 2019b, 2019c; Alzoubi, 2021b; Aziz & Aftab, 2021). All of the grads are not necessarily searching for full-time employment. Graduates, on the other hand, can become entrepreneurs who generate jobs and give them to others. This seeks to lower the unemployment rate among graduates, as stated in the Malaysia Education Blueprint 2015–2025 (Ali et al., 2022).

According to the explanation above, data, information, and knowledge exchange are essential throughout the educational supply chain process (Alzoubi and Yana-mandra 2020; Mehmood et al., 2019; Mehrez et al., 2021). As illustrated in Fig. 2, the sharing must be evenly balanced between curriculum design and industry demands (Al-Maroof et al., 2021a, 2021b; Alshurideh et al., 2019a, 2019b, 2019c; Farouk, 2021; Obaid, 2021). This is to guarantee that the information shared from industry to higher education is successful in developing curriculum that produces high-quality graduates who meet real-world industry requirements (Akhtar et al., 2021; Eli, 2021). This method maintains higher education curriculum design current while also lowering the jobless rate. Industry is continuously on the lookout for people who have the necessary knowledge and abilities for the job (Al Kurdi et al., 2021; Alshurideh et al., 2020a, 2020b; Alzoubi et al., 2020a, 2020b; Kurdi et al., 2020a, 2020b, 2020c). This indicates that the recruited person is capable of doing the job without any basic training, therefore meeting the ultimate expectation from industry to higher education. Prior to that, all programmes need an industrial internship for students to learn about real-world industrial functions. Consistent measurement is used and discussed to achieve consistent performance at all levels, including school, higher education, and industry (Alshurideh, 2014; Alshurideh et al., 2015, 2021a, 2021b; Al-Maroof et al., 2021a, 2021b; Alnazer et al., 2017; Ghazal et al., 2021; Kurdi et al., 2020a, 2020b; Nazir et al., 2021).

In educational supply chain management, student performance should be evaluated in all elements of the learning process, including ethics and morals, as well as knowledge and abilities. (1) ethics and spirituality, (2) leadership skills, and (3) national identity make up ethics and morals, whereas (1) language proficiency, (2) thinking skills, and (3) knowledge make up knowledge and skills. However, as shown in Table 3, the measures may be divided into two stages: school level, higher education, and industrial level (Alshamsi et al., 2020; Alsharari, 2021; Alsharari & Alshurideh, 2021; Alshurideh et al., 2021a, 2021b; Leo et al., 2021; Mehmood, 2021; Sultan et al., 2021).

In responding to these needs, this study summarised the concept of educational supply chain management as shown in Fig. 4. is expected to be fundamental to provide a comprehensive, integrated view on all aspects and facets of learning competencies,

Table 3 Definition of measurements used in conceptual framework

Measurements	Definition	
	School level	Higher education and industry level
Ethics and spirituality	Possesses solid moral foundation and courage to make right decisions	Ethically and morally upright, spiritually grounded, compassionate and caring; appreciates sustainable development and a healthy lifestyle
Leadership skills	Has strong communication skills, is entrepreneurial, resilient, can lead and work in teams	An effective communicator, emotionally intelligent and able to work across cultures; is socially responsible, competitive, resilient, and confident
National identity	Proudly identifies as Malaysian and embraces diversity	Pride in Malaysia and an understanding of Malaysia in relation to the world
Language proficiency	Operationally proficient in at least "Bahasa Melayu" and English	Proficient in "Bahasa Melayu" and English, and encouraged to learn one additional global language
Thinking skills	Inquisitive and innovative can apply, create, and connect knowledge to provide solutions	Appreciates diverse views, can think critically and innovative, as a problem-solving initiative, and has an entrepreneurial mindset
Knowledge	Mastery of core subjects and general knowledge about the world	Has mastery of own disciplines, can harness, connect and apply knowledge learnt, and has an appreciation of culture, arts, and Science, Technology, Engineering and Mathematics (STEM)

Source Adapted from Puchalski et al. (2014)

personalized learning experience, and professional development success (Guergov & Radwan, 2021; Khan, 2021; Miller, 2021; Mondol, 2021).

3 Methodology

There are three primary research paradigms in social science research: quantitative, qualitative, and mixed-method research design (Hanaysha et al., 2021a, 2021b). Convergent parallel, explanatory sequential, exploratory sequential, embedding, transformational, and multiphase designs are the six types of mixed-method study designs (Alhamad et al., 2021). This study concentrates on professional development success through educational supply chain management in Malaysia higher education. Therefore, the population in this study is Malaysian higher education graduates

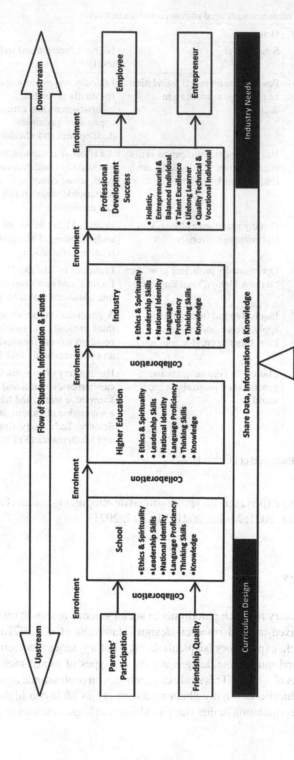

Fig. 4 Conceptual diagram of educational supply chain management

(Puchalski et al., 2014) to provide information for this research. Thus, the unit of measure of this study would be individual. The population of this study will be 504,000 individuals, and the sample size will be 384 individuals (Alzoubi & Aziz, 2021). A convenient sampling technique will be applied to determine the sample.

In a mixed method research design, survey questionnaires will be the instrument used to collect quantitative data, while the case study will be used to collect qualitative data. The respondents will be asked to answer the simple and convenient 5-point Likert scale questions (Hamadneh et al., 2021; Hanaysha et al., 2021a, 2021b). The survey questionnaire will distribute to 384 individuals, and three samples will be chosen from quantitative respondents for the case study. After the data collection, the collected data will be analyzed using the structural equation modeling (SEM) technique using SmartPLS 3.0 software. After completing quantitative data analysis, the next step is to prepare an interview protocol or semi-structured questionnaire for the case study, a face-to-face interview to collect qualitative data (Alzoubi et al., 2021). According to Ali et al. (2021) and Alzoubi et al. (2020a, 2020b), there is no rule of thumb to identify the sample in the qualitative study. The study is proposed to have three samples for the qualitative phase, and the sample will be selected from the quantitative respondents based on the result of quantitative data analysis. After the interview with the individuals, the software ATLAS.ti will analyze the data collected from the interview session. The combination of quantitative and qualitative data will produce a potent conclusion for this study (Alnuaimi et al., 2021; Alzoubi & Ahmed, 2019; Joghee et al., 2020).

3.1 Expected Implications

An appropriate combination of existing education attributes, industrial experience, and professional development success can help firms gain a more comprehensive picture of operational effect across their value chain. Institutions along the education supply chain would make more use of clear and consistent education attributes. In addition, by highlighting the effects of Malaysian educational attributes and industrial experience on the country's professional development success consequences of higher education institutions. Academicians of these institutions can benefit from the six education attributes that can be developed in addressing the issue. The model is expected to help the academicians to recognize and control the effects of education attributes.

It is strongly believed that the six education attributes and industrial experience initiated from the model developed will mitigate school, university, or company's social impact. This impact will shape not only the quality of graduates but also make professional development success and business operation more effective and efficient. In the long term, these will impact the economy of Malaysia, which is robust in advance professional development and ultimately contribute to the Gross Domestic

Product (GDP) of Malaysia. Finally, contribute to Malaysia's reputation and international standards by professional development through an educational supply chain model well-known globally.

Acknowledgements The authors are grateful to the Ministry of Higher Education (KPT), who has provided generous financial support through the MyGrant, Fundamental Research Grant Scheme, FRGS/1/2018/SS03/UMP/02/3.

References

Akhtar, A., Akhtar, S., Bakhtawar, B., Kashif, A. A., Aziz, N., & Javeid, M. S. (2021). COVID-19 detection from CBC using machine learning techniques. *International Journal of Technology, Innovation and Management (IJTIM), 1*(2), 65–78.

Akour, I., Alshurideh, M., Al Kurdi, B., Al Ali, A., & Salloum, S. (2021). Using machine learning algorithms to predict people's intention to use mobile learning platforms during the COVID-19 pandemic: Machine learning approach. *JMIR Medical Education, 7*(1), 1–17. https://doi.org/10.2196/24032

Al-Maroof, R., Ayoubi, K., Alhumaid, K., Aburayya, A., Alshurideh, M., Alfaisal, R., & Salloum, S. (2021a). The acceptance of social media video for knowledge acquisition, sharing and application: A comparative study among YouYube users and TikTok users' for medical purposes. *International Journal of Data and Network Science, 5*(3). https://doi.org/10.5267/j.ijdns.2021a.6.013

Al-Maroof, R. S., Alshurideh, M. T., Salloum, S. A., AlHamad, A. Q. M., & Gaber, T. (2021b). Acceptance of Google Meet during the spread of Coronavirus by Arab university students. *Informatics, 8*(2), 24.

Al Ali, A. (2021). The impact of information sharing and quality assurance on customer service at UAE banking sector. *International Journal of Technology, Innovation and Management (IJTIM), 1*(1), 01–17. https://doi.org/10.54489/ijtim.v1i1.10

Al Dmour, H., Alshurideh, M., & Shishan, F. (2014). The influence of mobile application quality and attributes on the continuance intention of mobile shopping. *Life Science Journal, 11*(10).

Al Kurdi, B., Alshurideh, M., Nuseir, M., Aburayya, A., & Salloum, S. A. (2021). The Effects of subjective norm on the intention to use social media networks: An exploratory study using PLS-SEM and machine learning approach. In *Advances in Intelligent Systems and Computing* (Vol. 1339). https://doi.org/10.1007/978-3-030-69717-4_55

AlHamad, A., Alshurideh, M., Alomari, K., Kurdi, B. A., Alzoubi, H., Hamouche, S., & Al-Hawary, S. (2022). The effect of electronic human resources management on organizational health of telecommuni-cations companies in Jordan. *International Journal of Data and Network Science, 6*(2), 429–438. https://doi.org/10.5267/j.ijdns.2021.12.011

Alhamad, A. Q. M., Akour, I., Alshurideh, M., Al-Hamad, A. Q., Kurdi, B. A., & Alzoubi, H. (2021). Predicting the intention to use google glass: A comparative approach using machine learning models and PLS-SEM. *International Journal of Data and Network Science, 5*(3). https://doi.org/10.5267/j.ijdns.2021.6.002

Ali, N., Ahmed, A., Anum, L., Ghazal, T. M., Abbas, S., Khan, M. A., Alzoubi, H. M., & Ahmad, M. (2021). Modelling supply chain information collaboration empowered with machine learning technique. *Intelligent Automation and Soft Computing, 30*(1), 243–257. https://doi.org/10.32604/iasc.2021.018983

Ali, N., M. Ghazal, T., Ahmed, A., Abbas, S., A. Khan, M., Alzoubi, H., Farooq, U., Ahmad, M., & Adnan Khan, M. (2022). Fusion-based supply chain collaboration using machine learning techniques. *Intelligent Automation & Soft Computing, 31*(3), 1671–1687. https://doi.org/10.32604/iasc.2022.019892

Alnazer, N. N., Alnuaimi, M. A., & Alzoubi, H. M. (2017). Analysing the appropriate cognitive styles and its effect on strategic innovation in Jordanian universities. *International Journal of Business Excellence, 13*(1), 127–140. https://doi.org/10.1504/IJBEX.2017.085799

Alnuaimi, M., Alzoubi, H. M., Ajelat, D., & Alzoubi, A. A. (2021). Towards intelligent organisations: An empirical investigation of learning orientation's role in technical innovation. *International Journal of Innovation and Learning, 29*(2), 207–221. https://doi.org/10.1504/IJIL.2021.112996

Alshamsi, A., Alshurideh, M., Al Kurdi, B., & Salloum, S. A. (2020). The influence of service quality on customer retention: A systematic review in the higher education. *International Conference on Advanced Intelligent Systems and Informatics*, 404–416.

Alsharari, N. (2021). Integrating blockchain technology with internet of things to efficiency. *International Journal of Technology, Innovation and Management (IJTIM), 1*(2), 1–13.

Alsharari, N. M., & Alshurideh, M. T. (2021). Student retention in higher education: The role of creativity, emotional intelligence and learner autonomy. *International Journal of Educational Management, 35*(1). https://doi.org/10.1108/IJEM-12-2019-0421

Alshurideh. (2012). The effect of previous experience on mobile subscribers' repeat purchase behaviour. *European Journal of Social Sciences, 30*(3).

Alshurideh, M. (2022). Does electronic customer relationship management (E-CRM) affect service quality at private hospitals in Jordan? *Uncertain Supply Chain Management, 10*(2), 1–8.

Alshurideh, M., Al Kurdi, B., Salloum, S. A., Arpaci, I., & Al-Emran, M. (2020a). Predicting the actual use of m-learning systems: A comparative approach using PLS-SEM and machine learning algorithms. *Interactive Learning Environments*. https://doi.org/10.1080/10494820.2020.1826982

Alshurideh, M., Gasaymeh, A., Ahmed, G., Alzoubi, H., & Kurd, B. A. (2020b). Loyalty program effectiveness: Theoretical reviews and practical proofs. *Uncertain Supply Chain Management, 8*(3). https://doi.org/10.5267/j.uscm.2020b.2.003

Alshurideh, M.T., Al Kurdi, B., AlHamad, A. Q., Salloum, S. A., Alkurdi, S., Dehghan, A., Abuhashesh, M., & Masa'deh, R. (2021a). Factors affecting the use of smart mobile examination platforms by universities' postgraduate students during the COVID-19 pandemic: An empirical study. *Informatics, 8*(2). https://doi.org/10.3390/informatics8020032

Alshurideh, M., Al Kurdi, B., & Salloum, S. (2021b). The moderation effect of gender on accepting electronic payment technology: A study on United Arab Emirates consumers. *Review of International Business and Strategy, 31*(3), 375–396.

Alshurideh, M., Bataineh, A., Al kurdi, B., & Alasmr, N. (2015). Factors affect mobile phone brand choices—Studying the case of Jordan universities students. *International Business Research, 8*(3), 141–155.

Alshurideh, M. (2014). The factors predicting students' satisfaction with universities' healthcare clinics' services. *Dirasat. Administrative Sciences, 41*(2), 451–464.

Alshurideh, M., Al Kurdi, B., & Salloum, S. A. (2019a). Examining the main mobile learning system drivers' effects: A mix empirical examination of both the Expectation-Confirmation Model (ECM) and the Technology Acceptance Model (TAM). In *International Conference on Advanced Intelligent Systems and Informatics* (pp. 406–417).

Alshurideh, M., Salloum, S. A., Al Kurdi, B., & Al-Emran, M. (2019b). Factors affecting the social networks acceptance: an empirical study using PLS-SEM approach. In *Proceedings of the 2019b 8th International Conference on Software and Computer Applications* (pp. 414–418).

Alshurideh, M., Salloum, S. A., Al Kurdi, B., Monem, A. A., & Shaalan, K. (2019c). Understanding the quality determinants that influence the intention to use the mobile learning platforms: A practical study. *International Journal of Interactive Mobile Technologies, 13*(11).

Alshurideh, M. T., Al Kurdi, B., Alzoubi, H. M., Ghazal, T. M., Said, R. A., AlHamad, A. Q., Hamadneh, S., Sahawneh, N., & Al-kassem, A. H. (2022). Fuzzy assisted human resource management for supply chain management issues. *Annals of Operations Research*, 1–19.

Alzoubi, A. (2021a). The impact of process quality and quality control on organizational competitiveness at 5-star hotels in Dubai. *International Journal of Technology, Innovation and Management (IJTIM), 1*(1), 54–68.

Alzoubi, A. (2021b). Renewable Green hydrogen energy impact on sustainability performance. *International Journal of Computations, Information and Manufacturing (IJCIM), 1*(1), 94–110. https://doi.org/10.54489/ijcim.v1i1.46

Alzoubi, H., & Yanamandra, R. (2020). Investigating the mediating role of information sharing strategy on agile supply chain in supply chain performance. *Uncertain Supply Chain Management, 8*(2), 273–284.

Alzoubi, H. M., Vij, M., Vij, A., & Hanaysha, J. R. (2021). What leads guests to satisfaction and loyalty in UAE five-star hotels? AHP analysis to service quality dimensions. *Enlightening Tourism. A Pathmaking Journal, 11*(1), 102–135.

Alzoubi, H., & Ahmed, G. (2019). Do TQM practices improve organisational success? A case study of electronics industry in the UAE. *International Journal of Economics and Business Research, 17*(4), 459–472. https://doi.org/10.1504/IJEBR.2019.099975

Alzoubi, H., Alshurideh, M., Kurdi, B. A., Akour, I., & Azi, R. (2022). Does BLE technology contribute towards improving marketing strategies, customers' satisfaction and loyalty? The role of open innovation. *International Journal of Data and Network Science, 6*(2), 449–460. https://doi.org/10.5267/j.ijdns.2021.12.009

Alzoubi, H., Alshurideh, M., Kurdi, B. A., & Inairat, M. (2020a). Do perceived service value, quality, price fairness and service recovery shape customer satisfaction and delight? A practical study in the service telecommunication context. *Uncertain Supply Chain Management, 8*(3), 579–588. https://doi.org/10.5267/j.uscm.2020.2.005

Alzoubi, H. M., Ahmed, G., Al-Gasaymeh, A., & Al Kurdi, B. (2020b). Empirical study on sustainable supply chain strategies and its impact on competitive priorities: The mediating role of supply chain collaboration. *Management Science Letters, 10*(3), 703–708. https://doi.org/10.5267/j.msl.2019.9.008

Alzoubi, H. M., & Aziz, R. (2021). Does emotional intelligence contribute to quality of strategic decisions? The mediating role of open innovation. *Journal of Open Innovation: Technology, Market, and Complexity, 7*(2), 130. https://doi.org/10.3390/joitmc7020130

Amarneh, B. M., Alshurideh, M. T., Al Kurdi, B. H., & Obeidat, Z. (2021). The impact of COVID-19 on E-learning: Advantages and challenges. In *The International Conference on Artificial Intelligence and Computer Vision* (pp. 75–89).

Aziz, N., & Aftab, S. (2021). Data Mining Framework for Nutrition Ranking: Methodology: SPSS Modeller. *International Journal of Technology, Innovation and Management (IJTIM), 1*(1), 85–95.

Barney, J. (2001). Is the resource based view a useful perspective for strategic management research? Yes. *Academy of Management Review, 26*(1), 41–56.

Belwal, R., Priyadarshi, P., Al Fazari, M. H. (2017). Graduate attributes and employability skills: Graduates' perspectives on employers' expectations in Oman. *International Journal of Educational Management, 31*(6), 814–827. https://doi.org/10.1108/IJEM-05-2016-0122

Bettayeb, H., Alshurideh, M. T., & Al Kurdi, B. (2020). The effectiveness of mobile learning in UAE universities: A systematic review of motivation, self-efficacy, usability and usefulness. *Int. J. Control Autom, 13*(2), 1558–1579.

Büyüközkan, G., & Göçer, F. (2018). Digital supply chain: Literature review and a proposed framework for future research. *Computers in Industry, 97*, 157–177. https://doi.org/10.1016/j.compind.2018.02.010

Chan, F. T. S., & Qi, H. J. (2003). An innovative performance measurement method for supply chain management. *Supply Chain Management: An International Journal, 8*(3), 209–223. https://doi.org/10.1108/13598540310484618

Coffield, F. (2012). Why the McKinsey reports will not improve school systems. *Journal of Education Policy, 27*(1), 131–149. https://doi.org/10.1080/02680939.2011.623243

Cooper, W. D. (2006). Textile and apparel supply chain management technology adoption. *Journal of Textile and Apparel, Technology and Management, 5*(2), 1–22.

Crinis, V. (2012). Global commodity chains in crisis: The garment industry in Malaysia. *Institutions and Economies, 4*(3), 61–82.

Cruz, A. (2021). Convergence between Blockchain and the Internet of Things. *International Journal of Technology, Innovation and Management (IJTIM), 1*(1), 34–53.

DOSM. (2018). *Department of Statistics Malaysia Official Portal.*

Eli, T. (2021). Students perspectives on the use of innovative and interactive teaching methods at the University of Nouakchott Al Aasriya, Mauritania: English Department as a Case Study. *International Journal of Technology, Innovation and Management (IJTIM), 1*(2), 90–104.

Farouk, M. (2021). The Universal Artificial intelligence efforts to face coronavirus COVID-19. *International Journal of Computations, Information and Manufacturing (IJCIM), 1*(1), 77–93. https://doi.org/10.54489/ijcim.v1i1.47

Fine, J., & Zygouris-Coe, V. (2011). Secondary teachers' knowledge, beliefs, and self-efficacy to teach reading in the content areas: Voices following professional development. *P,* 24–28.

Fraser-Seeto, K. T., Howard, S. J., & Woodcock, S. (2014). An investigation of teachers' awareness and willingness to engage with a self-directed professional development package on gifted and talented education. *Australian Journal of Teacher Education, 40*(1), 1–14. https://doi.org/10.14221/ajte.2015v40n1.1

Frich, J. C., Brewster, A. L., Cherlin, E. J., & Bradley, E. H. (2015). Leadership development programs for physicians: A systematic review. *Journal of General Internal Medicine, 30*(5), 656–674. https://doi.org/10.1007/s11606-014-3141-1

Ghazal, T. M., Hasan, M. K., Alshurideh, M. T., Alzoubi, H. M., Ahmad, M., Akbar, S. S., Al Kurdi, B., & Akour, I. A. (2021). IoT for smart cities: Machine learning approaches in smart healthcare—A review. *Future Internet, 13*(8), 218. https://doi.org/10.3390/fi13080218

Grant, R. M. (1991). The resource-based theory of competitive advantage: Implications for strategy formulation. *California Management Review, 33*(33), 114–135.

Guergov, S., & Radwan, N. (2021). Blockchain convergence: analysis of issues affecting IoT, AI and blockchain. *International Journal of Computations, Information and Manufacturing (IJCIM), 1*(1), 1–17. https://doi.org/10.54489/ijcim.v1i1.48

Gunasekaran, A., Lai, K. H., & Cheng, T. C. E. (2008). Responsive supply chain: A competitive strategy in a networked economy. *The International Journal of Management Science, 36*(4), 549–564. https://doi.org/10.1016/j.omega.2006.12.002

Habib, M., & Jungthirapanich, C. (2008a). Integrated Educational Supply Chain Management (IESCM) for the Universities. In *Sixth AIMS International Conference on Management.*

Habib, M. M., & Jungthirapanich, C. (2008b). An integrated framework for research and education supply chain for the universities. In *4th IEEE International Conference on Management of Innovation and Technology* (pp. 1027–1032). https://doi.org/10.1109/ICMIT.2008b.4654509

Halldorsson, A., Kotzab, H., Mikkola, J. H., & Skjott-Larsen, T. (2007). Complementary theories to supply chain management. *Supply Chain Management: An International Journal, 12*(4), 284–296. https://doi.org/10.1108/13598540710759808

Hamadneh, S., Pedersen, O., & Al Kurdi, B. (2021). An investigation of the role of supply chain visibility into the scottish bood supply chain. *Journal of Legal, Ethical and Regulatory Issues, 24*(Special Issue 1), 1–12.

Hanaysha, J. R., Al-Shaikh, M. E., Joghee, S., & Alzoubi, H. (2021a). Impact of innovation capabilities on business sustainability in small and medium enterprises. *FIIB Business Review, 1–12.* https://doi.org/10.1177/23197145211042232

Hanaysha, J. R., Al Shaikh, M. E., & Alzoubi, H. M. (2021b). Importance of marketing mix elements in determining consumer purchase decision in the retail market. *International Journal of Service Science, Management, Engineering, and Technology (IJSSMET), 12*(6), 56–72.

Huan, S. H., Sheoran, S. K., & Wang, G. (2004). A review and analysis of supply chain operations reference (SCOR) model. *Supply Chain Management: An International Journal, 9*(1), 23–29. https://doi.org/10.1108/13598540410517557

Joghee, S., Alzoubi, H. M., Alshurideh, M., & Al Kurdi, B. (2021). The role of business intelligence systems on green supply chain management: Empirical analysis of FMCG in the UAE. In *The International Conference on Artificial Intelligence and Computer Vision* (pp. 539–552).

Joghee, S., Alzoubi, H. M., & Dubey, A. R. (2020). Decisions effectiveness of FDI investment biases at real estate industry: Empirical evidence from Dubai smart city projects. *International Journal of Scientific and Technology Research, 9*(3), 3499–3503.

Kashif, A. A., Bakhtawar, B., Akhtar, A., Akhtar, S., Aziz, N., & Javeid, M. S. (2021). Treatment response prediction in hepatitis C patients using machine learning techniques. *International Journal of Technology, Innovation and Management (IJTIM), 1*(2), 79–89. https://doi.org/10.54489/ijtim.v1i2.24

Khan, M. A. (2021). Challenges facing the application of IoT in medicine and healthcare. *International Journal of Computations, Information and Manufacturing (IJCIM), 1*(1), 39–55. https://doi.org/10.54489/ijcim.v1i1.32

Korpela, K., Hallikas, J., & Dahlberg, T. (2021). Digital supply chain transformation toward blockchain integration. In *Proceedings of the 50th Hawaii International Conference on System Sciences* (pp 4182–4191).

Kurdi, B. A., Elrehail, H., Alzoubi, H. M., Alshurideh, M., & Al-adaileh, R. (2021). The interplay among HRM practices. *Job Satisfaction and Intention to Leave: An Empirical Investigation, 24*(1), 1–14.

Kurdi, B. A., Alshurideh, M., & Alnaser, A. (2020a). The impact of employee satisfaction on customer satisfaction: Theoretical and empirical underpinning. *Management Science Letters, 10*(15). https://doi.org/10.5267/j.msl.2020a.6.038

Kurdi, B. A., Alshurideh, M., & Salloum, S. A. (2020b). Investigating a theoretical framework for e-learning technology acceptance. *International Journal of Electrical and Computer Engineering, 10*(6). https://doi.org/10.11591/IJECE.V10I6.PP6484-6496

Kurdi, B. A., Alshurideh, M., Salloum, S. A., Obeidat, Z. M., & Al-dweeri, R. M. (2020c). An empirical investigation into examination of factors influencing university students' behavior towards elearning acceptance using SEM approach. *International Journal of Interactive Mobile Technologies, 14*(2). https://doi.org/10.3991/ijim.v14i02.11115

Lam, J. K. C., & Postle, R. (2006). Textile and apparel supply chain management in Hong Kong. *International Journal of Clothing Science and Technology, 18*(4), 265–277. https://doi.org/10.1108/09556220610668491

Lee, C., & Ahmed, G. (2021). Improving IoT privacy, data protection and security concerns. *International Journal of Technology, Innovation and Management (IJTIM), 1*(1), 18–33.

Lee, K L, Bakar, Z. A., Hanaysha, J. R., & Deraman, S. (2018). A review of educational supply chain management In Malaysia tertiary education. *IEBMC, 2017*(8), 249–260. https://doi.org/10.15405/epsbs.2018.07.02.26

Lee, K. L., Azmi, N. A. N., Hanaysha, J. R., Alzoubi, H. M., & Alshurideh, M. T. (2022a). The effect of digital supply chain on organizational performance: An empirical study in Malaysia manufacturing industry. *Uncertain Supply Chain Management, 10*(2), 495–510. https://doi.org/10.5267/j.uscm.2021.12.002

Lee, K. L., Romzi, P. N., Hanaysha, J. R., Alzoubi, H. M., & Alshurideh, M. (2022b). Investigating the impact of benefits and challenges of IOT adoption on supply chain performance and organizational performance: An empirical study in Malaysia. *Uncertain Supply Chain Management, 10*(2), 537–550. https://doi.org/10.5267/j.uscm.2021.11.009

Leo, S., Alsharari, N. M., Abbas, J., & Alshurideh, M. T. (2021). From offline to online learning: A qualitative study of challenges and opportunities as a response to the COVID-19 pandemic in the UAE higher education context. *The Effect of Coronavirus Disease (COVID-19) on Business Intelligence, 334*, 203–217.

Markussen-Brown, J., Juhl, C. B., Piasta, S. B., Bleses, D., Højen, A., & Justice, L. M. (2017). The effects of language- and literacy-focused professional development on early educators and children: A best-evidence meta-analysis. *Early Childhood Research Quarterly, 38*, 97–115. https://doi.org/10.1016/j.ecresq.2016.07.002

Mehmood, T. (2021). Does information technology competencies and fleet management practices lead to effective service delivery? *Empirical Evidence from E-Commerce Industry, 1*, 14–41.

Mehmood, T., Alzoubi, H. M., Alshurideh, M., Al-Gasaymeh, A., & Ahmed, G. (2019). Schumpeterian entrepreneurship theory: Evolution and relevance. *Academy of Entrepreneurship Journal*, *25*(4), 1–10.

Mehrez, A. A. A., Alshurideh, M., Kurdi, B. A., & Salloum, S. A. (2021). Internal factors affect knowledge management and firm performance: A systematic review. In *Advances in intelligent systems and computing: Vol. 1261 AISC*. https://doi.org/10.1007/978-3-030-58669-0_57

Miller, D. (2021). The best practice of teach computer science students to use paper prototyping. *International Journal of Technology, Innovation and Management (IJTIM)*, *1*(2), 42–63.

Mondol, E. P. (2021). The Impact of block chain and smart inventory system on supply chain performance at retail industry. *International Journal of Computations, Information and Manufacturing (IJCIM)*, *1*(1), 56–76. https://doi.org/10.54489/ijcim.v1i1.30

Murray, N. (2011). Widening participation and English language proficiency: A convergence with implications for assessment practices in higher education. *No. March, 2015*, 37–41. https://doi.org/10.1080/03075079.2011.580838

Nazir, M. I. J., Rahaman, S., Chunawala, S., & AlHamad, A. Q. M. (2021). Perceived factors affecting students academic performance; Nazir, J., Rahaman, S., Chunawala, S., Ahmed, G., Alzoubi, H., Alshurideh, M., AlHamad, A. (2022) Perceived factors affecting students academic performance. *Academy of Strategic Management Journal*, *21*(Special Issue 4), 1–15, *21*(Special Issue 4), 1–15.

Nijs, S., Gallardo-Gallardo, E., Dries, N., & Sels, L. (2014). Talent: A multi disciplinary review. *Journal of World Business*, *49*(2), 180–191.

Noom-Ura, S. (2013). English-teaching problems in Thailand and Thai teachers' professional development needs. *English Language Teaching*, *6*(11), 139–147. https://doi.org/10.5539/elt.v6n11p139

Obaid, A. J. (2021). Assessment of smart home assistants as an IoT. *International Journal of Computations, Information and Manufacturing (IJCIM)*, *1*(1), 18–36. https://doi.org/10.54489/ijcim.v1i1.34

Peteraf, M. A. (1993). The cornerstones of competitive advantage: A resource based view. *Strategic Management Journal*, *14*(3), 179–191.

Petersen, N. (2007). The professional development of teachers in terms of their understanding, experience and application of indigenous knowledge and cooperative learning in life sciences Josef de Beer. P, 536–545.

Pittman, L. D., & Richmond, A. (2008). University belonging, friendship quality, and psychological adjustment during the transition to college. *The Journal of Experimental Education*, *76*(4), 343–362. https://doi.org/10.3200/JEXE.76.4.343-362

Puchalski, C. M., Vitillo, R., Hull, S. K., & Reller, N. (2014). Improving the spiritual dimension of whole person care: Reaching national and international consensus. *Journal of Palliative Medicine*, *17*(6), 642–656. https://doi.org/10.1089/jpm.2014.9427

Radwan, N., & Farouk, M. (2021). The growth of Internet of Things (IoT) In the management of healthcare issues and healthcare policy development. *International Journal of Technology, Innovation and Management (IJTIM)*, *1*(1), 69–84. https://doi.org/10.54489/ijtim.v1i1.8

Ruskovaara, E., Pihkala, T., Seikkula-Leino, J., & J"arvinen, M. R. (2015). Broadening the resource base for entrepreneurship education through teachers' networking activities. *Teaching and Teacher Education*, *47*, 62–70. https://doi.org/10.1016/j.tate.2014.12.008

Salloum, S. A., Alshurideh, M., Elnagar, A., & Shaalan, K. (2020). Machine learning and deep learning techniques for cybersecurity: A review. In *Advances in Intelligent Systems and Computing: Vol. 1153 AISC*. https://doi.org/10.1007/978-3-030-44289-7_5

Shamout, M., Ben-Abdallah, R., Alshurideh, M., Alzoubi, H., Kurdi, B. A., & Hamadneh, S. (2022). A conceptual model for the adoption of autonomous robots in supply chain and logistics industry. *Uncertain Supply Chain Management*, *10*(2), 577–592. https://doi.org/10.5267/j.uscm.2021.11.006

Sultan, R. A., Alqallaf, A. K., Alzarooni, S. A., Alrahma, N. H., AlAli, M. A., & Alshurideh, . M. T. (2021). How students influence faculty satisfaction with online courses and do the age of faculty matter. In *The International Conference on Artificial Intelligence and Computer Vision* (pp. 823–837).

Tan, K. C., Lyman, S. B., & Wisner, J. D. (2002). Supply chain management: A strategic perspective. *International Journal of Operations & Production Management, 22*(6), 614–631. https://doi.org/10.1108/01443570210427659

Wernerfelt, B. (1984). A resource based view of the firm. *Strategic Management Journal, 5*(2), 171–180.

Wernerfelt, B. (1995). The resource based view of the firm: Ten years after. *Strategic Management Journal, 16*(3), 171–174.

Linear Discrimination Analysis Using Image Processing Optimization

Raed A. Said, Nidal A. Al-Dmour, Liaqat Ali, Haitham M. Alzoubi⬤, Muhammad Alshurideh⬤, and Mohammed Salahat

Abstract When we talk about Machinery Vision and Deep Learning, we often talk about algorithms. In fact, mathematical models with computer knowledge are the basis of how we deal with graphical data to process the Image and make decision. Machine learning can play an important role in determining agricultural plant type in order to optimize the harvesting steps in an automated way. How to process and introduce the products to the market often requires detailed information about the stages of planting and harvesting. In addition, by using this method, sophisticated research can be designed in plant genetics and effect of environmental variables on the end product. The ultimate goal of this work is to use Linear Discrimination Analysis for the Image Processing and classification of harvested wheat grain which are belonged to different types of grain namely Rosa, Kama and Canadian. The above discovery has proved with the statistics to have with more than 94% of accuracy.

R. A. Said
Faculty of Management, Canadian University Dubai, Dubai, UAE

N. A. Al-Dmour
Department of Computer Engineering, College of Engineering, Mutah University, Mu'tah, Jordan

L. Ali · M. Salahat
College of Engineering and Information Technology, University of Science and Technology of Fujairah, Fujairah, UAE
e-mail: l.ali@ustf.ac.ae

M. Salahat
e-mail: m.salahat@ustf.ac.ae

H. M. Alzoubi (✉)
School of Business, Skyline University College, Sharjah, UAE
e-mail: haitham.alzubi@skylineuniversity.ac.ae

M. Alshurideh
Department of Marketing, School of Business, The University of Jordan, Amman, Jordan
e-mail: malshurideh@sharjah.ac.ae

Department of Management, College of Business, University of Sharjah, Sharjah, UAE

M. Alshurideh et al. (eds.), *The Effect of Information Technology on Business and Marketing Intelligence Systems*, Studies in Computational Intelligence 1056, https://doi.org/10.1007/978-3-031-12382-5_137

Keywords Data science · Agricultural products · Clustering · Harvesting expert system

1 Introduction

Wheat grain with the highest nutritional value is the main resource of starch, fibers, amino acids, phenolics and polycarbonates. The wheat grains can be used as the supplemental ingredient of foods with low amino acids, phytases and antioxidants. Lactic fermentation also significantly improved protein digestibility and a satisfactory reduction in glycemic index compared to bread made from wheat flour alone (Akhtar et al., 2021; Al Ali, 2021; AlHamad et al., 2022). Lactic acid bacteria have a great effect on improving the nutritional and functional properties of the product and the enzymatic activity of these bacteria is responsible for increasing phenolic compounds and soluble fibers in the final product and reducing the risk of cardiovascular disease, type 2 diabetes, some types Cancers and obesity (Alhamad et al., 2021; Ali et al., 2021, 2022; Alnazer et al., 2017).

The wheat grains are damaged in their harvesting or post harvesting stages due to mechanical and thermal shocks (Alnuaimi et al., 2021; Alsharari, 2021; Alshurideh et al., 2022). Wheat Kernel damage as the main driver for the detrimental effects affect their biological and physical properties. One of the most important factors causing damages in the wheat kernel is its moisture absorption, which can be measured from prepared seed sample with weighting after maintaining in several hours in distilled water at certain temperature (Alshurideh et al., 2020; Alzoubi, 2021a, 2021b). For the kernels with higher potency i.e., high tendency for damage, a significant difference can be measured by the electrical conductivity of the solution obtained from seeds with different potency. The electrical conductivity of seeds with higher potency can be significantly lower than that of seeds with lower potency. The laboratorial methods required for the measurement of physical properties need a controlled environment induce higher cost. Other methods such as Magnetic Resonance Imaging, laser or scanning electron microscopy is also very cost intensive (Alzoubi & Aziz, 2021; Alzoubi & Yanamandra, 2020; Alzoubi et al., 2020a, 2022).

As an optimal alternative, Soft X-ray Image processing measuring the geometrical and statistical parameters of kernels are well posed as the way of quality inspection and assessment most cereal internal structure where the naked eye cannot separate the wheat grain type. The main reason is that X-ray can be differently exposed in the damaged and undamaged regions of internal kernel (Alzoubi et al., 2021; Alzoubi et al., 2020b; Alzoubi & Ahmed, 2019; Aziz & Aftab, 2021). The main steps for the determination of the damage index of wheat kernels are.

Companies are now rapidly developing agricultural performance-enhancing algorithms that can show farmers what the best crop method is (Alhashmi et al., 2020; Cruz, 2021; Eli, 2021; Farouk, 2021; Ghazal et al., 2021a, 2021b). Despite some concerns about the difficulty of doing this kind of analysis in nature, farmers and others can make great strides in maximizing yields simply by using intelligent algorithms and generators designed to help mimic computers. In simpler terms, the work of artificial intelligence has supported the idea that machines can provide deep insight into farmers and take many human arguments out (Al AlShamsi et al., 2021; Guergov & Radwan, 2021; Hamadneh et al., 2021; Hanaysha et al., 2021a; Shebli et al., 2021; Yousuf et al., 2021). Finally, it should be noted that population growth and climate change will be major challenges for the future of world agriculture, but the establishment and use of artificial intelligence can mitigate the impact of these challenges and other challenges and make smart agriculture in the face of problems. What farmers are facing is more resilient (Hanaysha et al., 2021b; Joghee et al., 2020; Kashif et al., 2021; Khan, 2021; Nuseir et al., 2021).

Linear Discrimination Analysis (LDA) is the method used in Machine learning for the recognition of two or more objects. In LDA, which is very similar to variance and regression analysis dependent variable is linearly relate to the other variables (Lee & Ahmed, 2021; Lee et al., 2022a, 2022a, 2022b; Mehmood, 2021). The difference between these methods is that in LDA, instead of distance relationship, dependent variable is solved by the ranking or ordinal outranking (Siddiqui et al., 2021; Zitar, 2021; Zitar et al., 2021). The LDA is also closely related to Principal Component Analysis (PCA) core component analysis. Because both methods look for a linear combination of variables that best describe the data. The LDA also attempts to model differences between different data classes. The LDA is used when the observation sizes are continuous values (Mehmood et al., 2019; Miller, 2021; Mondol, 2021).

In this paper, the essential data preprocessing stages are described in Chap. 2 then in Chap. 3 the configuration of the Weka Program for the LDA analysis are explained. Modelling methods is the subject of Chap. 4 and their tuning parameters are described in Chap. 5. The results are shown for the classification accuracy are discussed and the confusion matrix is represented. At last, the results are summarized and the usage of the Machine Deep Learning in wheat classification are discussed in the conclusion.

2 Dataset Description

A. Preprocessing stage

The dataset consisted of 210 instances corresponding to the three equal batches of grain for each class with 7 attributed. Wheat grain dataset can be extracted from the UCI repository dataset (Obaid, 2021; Radwan & Farouk, 2021; Shamout et al., 2022). The file was in the form of txt file and attributes a compaigned only with numerical values, then we used microsoft excel to name each column and export the result as a csv file. Since the ultimate goal of this work is to measure the misclassification of the randomly ordered classes with linear discrimination modelling, it is needed to use two filters for data preparation in this step (Al-Dmour, 2016; Al Hamadi et al., 2017; Al Neaimi et al., 2020). These are namely NumerictoNominal for object class

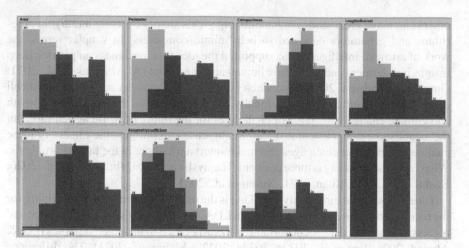

Fig. 1 Attribute distribution versus multi-nominal Class

and Normalize for all others. The normalization process transform all attributes to range [0,1] with mean 1 and variance 0. This can be used to control the bias error and reduce the risk of overfitting. As it is shown in Fig. 1, it is not possible to cover all the differentiae each of types with a single variable. That's evident from the fact that a coverage of the some of the area is existed where make the classification in that region impossible. For instance, Rosa Grain (red color) class can fairly be explained by the Area, Perimeter and length of kernel groove (Bibi et al., 2021; Ghazal, 2021; Maasmi et al., 2021).

Canadian and Kama Grain centroid are neat which is therefore difficult to separately classified by one single attribute (Al-Dmour & Teahan, 2005; Al-Hamadi et al., 2015, 2021). The advantage of the Linear Discrimination Analysis is where it is not possible to differentiate the class by a single attribute. LDA algorithm makes a linear relationship between the independent variable and classes to make this happen. The effect of each variable will be on the whole classification. The classification problem is to find a predictor for each class from the same distribution (not necessarily from the training set) given by the observation set x. LDA is assumed a conditional probability density function exist which are both have normal distributions with mean and covariance parameters.

B. Dimension Reduction

Linear Discrimination Analysis is a method similar to Principal Component Analysis and factor analysis; Both of these statistical methods are used to linearly combine variables in a way that best describes the data (Ali & Dmour, 2021; Al-Naymat et al., 2021; Aslam et al., 2021). A major application of both of these methods is to reduce the number of dimensions of the data. However, there are major differences between these methods: in Linear Discrimination Analysis, class differences are modeled, while in principal component analysis, class differences are ignored. The reduced dataset was exported to reduced.arff in order to further analysis the reduced result by Linear Discrimination Analysis. In Fig. 2 the new variables are represented:

As it can be seen from the PCA analysis in Fig. 2, the first and second are posed well as a Principal Component Analysis. Figure 3 shows the graph of second versus first dimensions and the region inside the ovals are for each type of grain kernel.

(0.717Assymetrycoefficient-0.529Compactness
+0.377lengthofkernelgroove+0.206Lengthofkernel-0.117Widthofkernel...)   versus
(0.444Area+
0.442Perimeter+0.433Widthofkernel+0.424Lengthofkernel+0.387lengthofkernelgroove...)

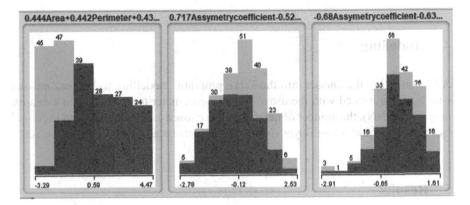

Fig. 2 Dimension reduction by PCA

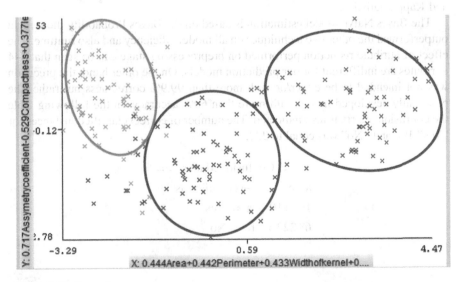

Fig. 3 PCA dimension

3 Methodology

We have used python 3.6 to import the dataset into Jupyter Notebook. The details of results of coding within the description of every attributes are attached in the Appendix; For machine learning the Weka Program V 3.9 was applied.

import pandas as pd
data = pd.read_csv("wheat.csv")
data.info()
data.describe()

4 Modelling

After importing the dataset into the Weka program, modelling techniques utilized which was calibrated with the usage of hyper-parameter tuning where in k-nearest neighbors (kNN), the number of neighbors was chosen to be equal to 4 and in Neural Network (NN) and Naïve Bayes (NB), the batch size was with the volume of 100.

5 Results

Table 1 shows the result of accuracy (Recall for the danger of overfitting), F-Measure and Kappa statistics.

The Bayes Naïve model estimation is based on the Bayes Probability theory and outperformed the other two techniques on all model efficiency and also runtime. The effect of attribute reduction performed on preprocessing stage can be seen that 3–4 attributes are indifferent for our prediction models. On the other hand, the precision was not intended to be either/or with more than 99.9% correctness but rather the practically developed model with more than 60% accuracy. In the following table the confusion matrix is also indicated. The number of correct classified instances for "Yes" 100 and "No" was equal to 222.

$$=== \text{Confusion Matrix} ===$$
$$\text{a} \quad \text{b} \quad < -- \text{classified as}$$
$$100\,110 \mid \quad \text{a} = \text{Yes}$$
$$68\,222 \mid \quad \text{b} = \text{No}$$

Table 1 Data Scientific result performed in Weka

Attribute selection	Classifier	F-measure (%)	Kappa	Recall (%)	Accuracy (%)
12	Bayes Naïve (NB)	60.9	0.1931	61.0	61.8
	k-Nearest Neighbors (IBk)	50.12	−0.002	50.8	50.6
	Multilayer Perceptron (NNs)	53.9	0.0305	53.2	53.2
9	Bayes Naïve (NB)	52.2	0.0441	52.2	52.21
	k-Nearest Neighbors (IBk)	49.7	0.0324	51.6	51.57
	Multilayer Perceptron (NNs)	47.7	−0.044	47.8	47.7
6	Bayes Naïve (NB)	63.6	0.2486	64.4	64.4
	k-Nearest Neighbors (IBk)	57.2	0.1215	57.2	57.2
	Multilayer Perceptron (NNs)	53.2	0.0691	58.4	58.4

6 Discussion

Generally, for retailers or clothing stores, this result has fewer external consequences, though the point of view is very important due to the importance of feasibility of designing of such a smart system that can capture the customer sense and recommend a cloth for the customer. It is important for any brand as a beneficiary of a recommendation system to take advantage of opportunities and generally use chances and prevent failures in the competitive market, which will be created by differentiating in the mind of the customer. On the other word, this study could be the first evaluation of application of such framework in the stores which can be developed by the help of expert in fashion design and apparel entrepreneurship.

7 Conclusion

We have developed a framework that can model the sale recommendation for better communication between clients. We have firstly extracted the dataset from the UCI and then imported the data for the preprocessing into weka and python. The data was cleansed at the first step by the filtering tools embedded in Weka Application. Then the dataset was balanced to 50/50 distribution of target class for the danger of overfitting. Atlast, five different machine learning tools was implemented in the dataset to tune the hyper-parameter which resulted into more than 10% improvement in the results. With the specific learning from the modeling, we understood that with the possibility of near 65% of success such a recommendation system can be applied.

Appendix [Python Code with Outputs]

```
<class 'pandas.core.frame.DataFrame'>
RangeIndex: 210 entries, 0 to 209
Data columns (total 8 columns):
 #   Column                Non-Null Count  Dtype
---  ------                --------------  -----
 0   Area                  210 non-null    float64
 1   Perimeter             210 non-null    float64
 2   Compactness           210 non-null    float64
 3   Lengthofkernel        210 non-null    float64
 4   Widthofkernel         210 non-null    float64
 5   Assymetrycoefficient  210 non-null    float64
 6   lengthofkernelgroove  210 non-null    float64
 7   type                  210 non-null    int64
dtypes: float64(7), int64(1)
memory usage: 13.2 KB
Out[3]:
```

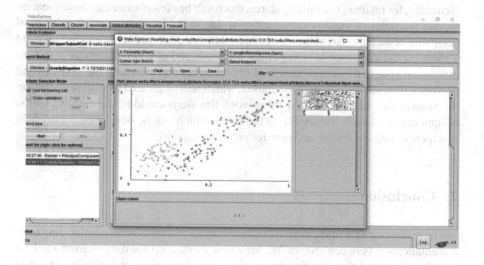

References

Akhtar, A., Akhtar, S., Bakhtawar, B., Kashif, A. A., Aziz, N., & Javeid, M. S. (2021). COVID-19 detection from CBC using machine learning techniques. *International Journal of Technology, Innovation and Management (IJTIM), 1*(2), 65–78. https://doi.org/10.54489/ijtim.v1i2.22

Al-Dmour, N. (2016). Using unstructured search algorithms for data collection in IoT-based WSN. *International Journal of Engineering Research and Technology.* ISSN, 974–3154.

Al-Dmour, N. A., & Teahan, W. J. (2005). Peer-to-Peer protocols for resource discovery in the grid. *Parallel and Distributed Computing and Networks*, 319–324.

Al-Hamadi, H., Gawanmeh, A., & Al-Qutayri, M. (2015). An automatic ECG generator for testing and evaluating ECG sensor algorithms. In *2015 10th International Design & Test Symposium (IDT)* (pp. 78–83).

Al-Hamadi, H., Nasir, N., Yeun, C. Y., & Damiani, E. (2021). A verified protocol for secure autonomous and cooperative public transportation in smart cities. In *IEEE International Conference on Communications Workshops (ICC Workshops)* (pp. 1–6).

Al-Naymat, G., Hussain, H., Al-Kasassbeh, M., & Al-Dmour, N. (2021). Accurate detection of network anomalies within SNMP-MIB data set using deep learning. *International Journal of Computer Applications in Technology, 66*(1), 74–85.

Al Ali, A. (2021). the impact of information sharing and quality assurance on customer service at UAE banking sector. *International Journal of Technology, Innovation and Management (IJTIM), 1*(1), 01–17. https://doi.org/10.54489/ijtim.v1i1.10

Al Hamadi, H., Gawanmeh, A., & Al-Qutayri, M. (2017). Guided test case generation for enhanced ECG bio-sensors functional verification. *International Journal of E-Health and Medical Communications (IJEHMC), 8*(4), 1–20.

Al Neaimi, M., Al Hamadi, H., Yeun, C. Y., & Zemerly, M. J. (2020). Digital forensic analysis of files using deep learning. In *2020 3rd International Conference on Signal Processing and Information Security (ICSPIS)* (pp. 1–4).

Al Shebli, K., Said, R. A., Taleb, N., Ghazal, T. M., Alshurideh, M. T., & Alzoubi, H. M. (2021). RTA's employees' perceptions toward the efficiency of artificial intelligence and big data utilization in providing smart services to the residents of Dubai. In *The International Conference on Artificial Intelligence and Computer Vision* (pp. 573–585).

AlHamad, A., Alshurideh, M., Alomari, K., Kurdi, B., Alzoubi, H., Hamouche, S., & Al-Hawary, S. (2022). The effect of electronic human resources management on organizational health of telecommuni-cations companies in Jordan. *International Journal of Data and Network Science, 6*(2), 429–438.

Alhamad, A. Q. M., Akour, I., Alshurideh, M., Al-Hamad, A. Q., Kurdi, B. A., & Alzoubi, H. (2021). Predicting the intention to use google glass: A comparative approach using machine learning models and PLS-SEM. *International Journal of Data and Network Science, 5*(3). https://doi.org/10.5267/j.ijdns.2021.6.002

Alhashmi, S. F. S., Alshurideh, M., Al Kurdi, B., & Salloum, S. A. (2020). A systematic review of the factors affecting the artificial intelligence implementation in the health care sector. In *Advances in Intelligent Systems and Computing: Vol. 1153 AISC.* https://doi.org/10.1007/978-3-030-44289-7_4

Ali, L., & Dmour, N. (2021). The shift to online assessment due to COVID-19: An empirical study of university students, behaviour and performance, in the region of UAE. *International Journal of Information and Education Technology, 11*(5), 220–228.

Ali, N., Ahmed, A., Anum, L., Ghazal, T. M., Abbas, S., Khan, M. A., Alzoubi, H. M., & Ahmad, M. (2021). Modelling supply chain information collaboration empowered with machine learning technique. *Intelligent Automation and Soft Computing, 30*(1), 243–257. https://doi.org/10.32604/iasc.2021.018983

Ali, N., M. Ghazal, T., Ahmed, A., Abbas, S., A. Khan, M., Alzoubi, H., Farooq, U., Ahmad, M., & Adnan Khan, M. (2022). Fusion-based supply chain collaboration using machine learning techniques. *Intelligent Automation & Soft Computing, 31*(3), 1671–1687. https://doi.org/10.32604/iasc.2022.019892

Alnazer, N. N., Alnuaimi, M. A., & Alzoubi, H. M. (2017). Analysing the appropriate cognitive styles and its effect on strategic innovation in Jordanian universities. *International Journal of Business Excellence, 13*(1), 127–140. https://doi.org/10.1504/IJBEX.2017.085799

Alnuaimi, M., Alzoubi, H. M., Ajelat, D., & Alzoubi, A. A. (2021). Towards intelligent organisations: An empirical investigation of learning orientation's role in technical innovation. *International Journal of Innovation and Learning, 29*(2), 207–221. https://doi.org/10.1504/IJIL.2021.112996

AlShamsi, M., Salloum, S. A., Alshurideh, M., & Abdallah, S. (2021). Artificial intelligence and blockchain for transparency in governance. In *Artificial Intelligence for Sustainable Development: Theory, Practice and Future Applications* (pp. 219–230). Springer.

Alsharari, N. (2021). Integrating blockchain technology with internet of things to efficiency. *International Journal of Technology, Innovation and Management (IJTIM), 1*(2), 1–13.

Alshurideh, M., Gasaymeh, A., Ahmed, G., Alzoubi, H., & Kurd, B. A. (2020). Loyalty program effectiveness: Theoretical reviews and practical proofs. *Uncertain Supply Chain Management, 8*(3). https://doi.org/10.5267/j.uscm.2020.2.003

Alshurideh, M. T., Al Kurdi, B., Alzoubi, H. M., Ghazal, T. M., Said, R. A., AlHamad, A. Q., Hamadneh, S., Sahawneh, N., & Al-kassem, A. H. (2022). Fuzzy assisted human resource management for supply chain management issues. *Annals of Operations Research*, 1–19.

Alzoubi, A. (2021a). The Impact of Process Quality and Quality Control on Organizational Competitiveness at 5-star hotels in Dubai. *International Journal of Technology, Innovation and Management (IJTIM), 1*(1), 54–68. https://doi.org/10.54489/ijtim.v1i1.14

Alzoubi, A. (2021b). Renewable Green hydrogen energy impact on sustainability performance. *International Journal of Computations, Information and Manufacturing (IJCIM), 1*(1), 94–110. https://doi.org/10.54489/ijcim.v1i1.46

Alzoubi, H. M., & Aziz, R. (2021). Does emotional intelligence contribute to quality of strategic decisions? The mediating role of open innovation. *Journal of Open Innovation: Technology, Market, and Complexity, 7*(2), 130. https://doi.org/10.3390/joitmc7020130

Alzoubi, H. M., Vij, M., Vij, A., & Hanaysha, J. R. (2021). What leads guests to satisfaction and loyalty in UAE five-star hotels? AHP analysis to service quality dimensions. *Enlightening Tourism, 11*(1), 102–135. https://doi.org/10.33776/et.v11i1.5056

Alzoubi, H. M., & Yanamandra, R. (2020). Investigating the mediating role of information sharing strategy on agile supply chain. *Uncertain Supply Chain Management, 8*(2), 273–284. https://doi.org/10.5267/j.uscm.2019.12.004

Alzoubi, H., Ahmed, G., Al-Gasaymeh, A., & Kurdi, B. (2020a). Empirical study on sustainable supply chain strategies and its impact on competitive priorities: The mediating role of supply chain collaboration. *Management Science Letters, 10*(3), 703–708.

Alzoubi, H., Alshurideh, M., Kurdi, B. A., & Inairat, M. (2020b). Do perceived service value, quality, price fairness and service recovery shape customer satisfaction and delight? A practical study in the service telecommunication context. *Uncertain Supply Chain Management, 8*(3), 579–588. https://doi.org/10.5267/j.uscm.2020.2.005

Alzoubi, H., Alshurideh, M., Kurdi, B., Akour, I., & Aziz, R. (2022). Does BLE technology contribute towards improving marketing strategies, customers' satisfaction and loyalty? The role of open innovation. *International Journal of Data and Network Science, 6*(2), 449–460.

Alzoubi, H., & Ahmed, G. (2019). Do TQM practices improve organisational success? A case study of electronics industry in the UAE. *International Journal of Economics and Business Research, 17*(4), 459–472. https://doi.org/10.1504/IJEBR.2019.099975

Aslam, M. S., Ghazal, T. M., Fatima, A., Said, R. A., Abbas, S., Khan, M. A., Siddiqui, S. Y., & Ahmad, M. (2021). *Energy-efficiency model for residential buildings using supervised machine learning algorithm.*

Aziz, N., & Aftab, S. (2021). Data mining framework for nutrition ranking: Methodology: SPSS modeller. *International Journal of Technology, Innovation and Management (IJTIM)*, *1*(1), 85–95.

Bibi, R., Saeed, Y., Zeb, A., Ghazal, T. M., Rahman, T., Said, R. A., Abbas, S., Ahmad, M., & Khan, M. A. (2021). Edge AI-based automated detection and classification of road anomalies in VANET using deep learning. *Computational Intelligence and Neuroscience.*

Cruz, A. (2021). Convergence between blockchain and the Internet of Things. *International Journal of Technology, Innovation and Management (IJTIM)*, *1*(1), 35–56.

Eli, T. (2021). Students perspectives on the use of innovative and interactive teaching methods at the University of Nouakchott Al Aasriya, Mauritania: English department as a case study. *International Journal of Technology, Innovation and Management (IJTIM)*, *1*(2), 90–104.

Farouk, M. (2021). The universal artificial intelligence efforts to face coronavirus COVID-19. *International Journal of Computations, Information and Manufacturing (IJCIM)*, *1*(1), 77–93. https://doi.org/10.54489/ijcim.v1i1.47

Ghazal, T., Alshurideh, M., & Alzoubi, H. (2021a). Blockchain-Enabled Internet of Things (IoT) Platforms for pharmaceutical and biomedical research. *The International Conference on Artificial Intelligence and Computer Vision* (pp. 589–600).

Ghazal, T. M., Hasan, M. K., Alshurideh, M. T., Alzoubi, H. M., Ahmad, M., Akbar, S. S., Al Kurdi, B., & Akour, I. A. (2021b). IoT for smart cities: Machine learning approaches in smart healthcare—A review. *Future Internet, 13*(8), 218. https://doi.org/10.3390/fi13080218

Ghazal, T. M. (2021). Internet of Things with artificial intelligence for health care security. *Arabian Journal for Science and Engineering, 1–12.*

Guergov, S., & Radwan, N. (2021). Blockchain Convergence: Analysis of issues affecting IoT, AI and blockchain. *International Journal of Computations, Information and Manufacturing (IJCIM)*, *1*(1), 1–17. https://doi.org/10.54489/ijcim.v1i1.48

Hamadneh, S., Pedersen, O., & Al Kurdi, B. (2021). An Investigation of the role of supply chain visibility into the scottish bood supply chain. *Journal of Legal, Ethical and Regulatory Issues, 24*(Special Issue 1), 1–12.

Hanaysha, J. R., Al-Shaikh, M. E., Joghee, S., & Alzoubi, H. (2021a). Impact of innovation capabilities on business sustainability in small and medium enterprises. *FIIB Business Review, 1–12.* https://doi.org/10.1177/23197145211042232

Hanaysha, J. R., Al Shaikh, M. E., & Alzoubi, H. M. (2021b). Importance of marketing mix elements in determining consumer purchase decision in the retail market. *International Journal of Service Science, Management, Engineering, and Technology (IJSSMET), 12*(6), 56–72.

Joghee, S., Alzoubi, H. M., & Dubey, A. R. (2020). Decisions effectiveness of FDI investment biases at real estate industry: Empirical evidence from Dubai smart city projects. *International Journal of Scientific and Technology Research, 9*(3), 3499–3503.

Kashif, A. A., Bakhtawar, B., Akhtar, A., Akhtar, S., Aziz, N., & Javeid, M. S. (2021). Treatment response prediction in hepatitis C patients using machine learning techniques. *International Journal of Technology, Innovation and Management (IJTIM)*, *1*(2), 79–89. https://doi.org/10.54489/ijtim.v1i2.24

Khan, M. A. (2021). Challenges facing the application of IoT in medicine and healthcare. *International Journal of Computations, Information and Manufacturing (IJCIM)*, *1*(1), 39–55. https://doi.org/10.54489/ijcim.v1i1.32

Lee, C., & Ahmed, G. (2021). Improving IoT privacy, data protection and security concerns. *International Journal of Technology, Innovation and Management (IJTIM)*, *1*(1), 18–33. https://doi.org/10.54489/ijtim.v1i1.12

Lee, K., Azmi, N., Hanaysha, J., Alzoubi, H., & Alshurideh, M. (2022a). The effect of digital supply chain on organizational performance: An empirical study in Malaysia manufacturing industry. *Uncertain Supply Chain Management, 10*(2), 495–510.

Lee, K., Romzi, P., Hanaysha, J., Alzoubi, H., & Alshurideh, M. (2022b). Investigating the impact of benefits and challenges of IOT adoption on supply chain performance and organizational performance: An empirical study in Malaysia. *Uncertain Supply Chain Management, 10*(2), 537–550.

Maasmi, F., Morcos, M., Al Hamadi, H., & Damiani, E. (2021). Identifying applications' state via system calls activity: A pipeline approach. In *28th IEEE International Conference on Electronics, Circuits, and Systems (ICECS)* (pp. 1–6).

Mehmood, T. (2021). Does information technology competencies and fleet management practices lead to effective service delivery? Empirical evidence from E-commerce industry. *International Journal of Technology, Innovation and Management (IJTIM)*, *1*(2), 14–41.

Mehmood, T., Alzoubi, H. M., & Ahmed, G. (2019). Schumpeterian entrepreneurship theory: evolution and relevance. *Academy of Entrepreneurship Journal*, *25*(4).

Miller, D. (2021). The best practice of teach computer science students to use paper prototyping. *International Journal of Technology, Innovation and Management (IJTIM)*, *1*(2), 42–63. https://doi.org/10.54489/ijtim.v1i2.17

Mondol, E. P. (2021). The impact of block chain and smart inventory system on supply chain performance at retail industry. *International Journal of Computations, Information and Manufacturing (IJCIM)*, *1*(1), 56–76. https://doi.org/10.54489/ijcim.v1i1.30

Nuseir, M. T., Al Kurdi, B. H., Alshurideh, M. T., & Alzoubi, H. M. (2021). Gender discrimination at workplace: Do Artificial Intelligence (AI) and Machine Learning (ML) have opinions about it. In *The International Conference on Artificial Intelligence and Computer Vision* (pp. 301–316).

Obaid, A. J. (2021). Assessment of smart home assistants as an IoT. *International Journal of Computations, Information and Manufacturing (IJCIM)*, *1*(1), 18–36. https://doi.org/10.54489/ijcim.v1i1.34

Radwan, N., & Farouk, M. (2021). The growth of Internet of Things (IoT) in the management of healthcare issues and healthcare policy development. *International Journal of Technology, Innovation and Management (IJTIM)*, *1*(1), 69–84. https://doi.org/10.54489/ijtim.v1i1.8

Shamout, M., Ben-Abdallah, B., Alshurideh, M., Alzoubi, H., Al Kurdi, B., & Hamadneh, S. (2022). A conceptual model for the adoption of autonomous robots in supply chain and logistics industry. *Uncertain Supply Chain Management, 10*, 1–16.

Siddiqui, S. Y., Haider, A., Ghazal, T. M., Khan, M. A., Naseer, I., Abbas, S., Rahman, M., Khan, J. A., Ahmad, M., & Hasan, M. K. (2021). IoMT Cloud-based intelligent prediction of breast cancer stages empowered with deep learning. *IEEE Access, 9*, 146478–146491.

Yousuf, H., Zainal, A. Y., Alshurideh, M., & Salloum, S. A. (2021). Artificial intelligence models in power system analysis. In *Studies in Computational Intelligence* (Vol. 912). https://doi.org/10.1007/978-3-030-51920-9_12

Zitar, R. A. (2021). A review for the Genetic Algorithm and the Red Deer Algorithm applications. In *2021 14th International Congress on Image and Signal Processing, BioMedical Engineering and Informatics (CISP-BMEI)* (pp. 1–6).

Zitar, R. A., Abualigah, L., & Al-Dmour, N. A. (2021). Review and analysis for the Red Deer Algorithm. *Journal of Ambient Intelligence and Humanized Computing*, 1–11.

Pricing Policies and Their Impact on Customer Satisfaction

Omar Jawabreh [ID], Ra'ed Masa'deh, Yaqoub Bqaa'een, and Muhammad Alshurideh [ID]

Abstract This research aims to focus on pricing, policies and their impact on customer satisfaction in the five-star hotels in Aqaba. The technique that was used in the present research is described in this section. There are many components of this document: the research model, operational definitions of the study's independent and dependent variables, research hypotheses, as well as a data collection instrument, research population, and research sample. The components of this study are developed on the basis of prior literature, which may be either theoretical or empirical in nature. Indeed, the factors utilized in this research are often found in the literature on tourist development. The data analysis results have shown that services in restaurants in Aqaba hotels are applied at a very high level, which the mean score is 4.38. Also, customer satisfaction and pricing cognition (pricing) existed on a high base with mean of 3.86 and 3.74 respectively. Specifically, the research seeks to develop price policies for all categories and to work on the design of all pricing criteria to fit the quality of service and the degree of customer satisfaction, which shows that a high level of customer satisfaction may be achieved.

Keywords Services in restaurant · And pricing cognition (Pricing) · Customer satisfaction

O. Jawabreh (✉) · Y. Bqaa'een
Faculty of Tourism and Hospitality, Department of Hotel Management, The University of Jordan, Amman, Jordan
e-mail: o.jawabreh@ju.edu.jo

R. Masa'deh
School of Business, Department of Management Information Systems, The University of Jordan, Amman, Jordan
e-mail: r.masadeh@ju.edu.jo

M. Alshurideh
School of Business, Department of Marketing, The University of Jordan, Amman, Jordan
e-mail: m.alshurideh@ju.edu.jo

University of Sharjah, Sharjah, UAE

© The Author(s), under exclusive license to Springer Nature Switzerland AG 2023   2503
M. Alshurideh et al. (eds.), *The Effect of Information Technology on Business and Marketing Intelligence Systems*, Studies in Computational Intelligence 1056,
https://doi.org/10.1007/978-3-031-12382-5_138

1 Introduction

Being customer focused nowadays has to be accepted as a bare necessity to conduct business. Globalization and value-driven business imperatives will therefore mean that mistakes are not going to be tolerated; substandard products and services will ensure that the weakest are not going to continue competing; and fragile practices and poorly defined and managed processes will not be accepted (Jawabreh, 2020). Apart from anything else, true competitive advantage will only be established through excellence in customer value and the ensuing relentless care and attention provided. Customer satisfaction (CS) is a very often misused and abused expression. Many organizations use it casually in order to state that their customers are happy and satisfied with the levels of service rendered and the products and services purchased, but they actually have never tried to measure that satisfies. But if a company takes its customers seriously, it should not behave like this because the results of customer satisfaction measurements provide significant information for modern management processes and a warning signal about future business results. This enables an understanding of how customers perceive the organization, i.e., whether its performance meets their expectations, identifies priorities for improvement, benchmarks the performance of the organization against other organizations, and increase profits through improved customer loyalty. Through the process of creating a customer and supplier chain at all levels, a better focus can be achieved and, ultimately, all the work carried out will be of value. This customer–supplier communication will help to ensure quality and thus the customer's satisfaction (Al Khasawneh et al., 2021; Jawabreh, 2017; Jawabreh & Al Sarayreh, 2017; Lepkova & Žukaitė-Jefimovienė, 2012).

For decades, the conception of trust has achieved appreciable implications in the field of marketing, not only for products as well as for services. When we have to make a choice between different products in the same category, trust is involved. It is an essential asset of a person, product, organization, institution, and skill. The buyer will go for the trustworthy asset mentioned above. Customer trust is an important variable that enhances customer commitment. He also stated that the quality of services enhances customer trust. Loyalty has, for years, been an important and fundamental ingredient of marketing. It has been a burning subject matter in marketing, not only for commerce academics, but also for industry managers (Qandah et al., 2020; Sarwar et al., 2012). For long-term success, organizations work very hard. The organizations' success is determined by numerous factors. Those organizations which earn lots of profit are not thought to be powerful organizations, because profit generation is one of the tasks of the organization. So, not just organizations now have to focus on shareholders, but also for customers, because now the customers are the king. In the existent world, maintaining the old ones and drawing the attention of new customers is considered to be a victory. Now the telecommunication sector is growing to be a vital sector of the world. Voice communication, graphics, data, and delivery of video at a very high speed are provided by the telecommunication sector. The world economy is being influenced by the telecommunication sector. Competition is going to be more intense. For the purpose of achieving a maintained competitive edge, the telecommunication industries are compelled to pursue innovation and to execute the best thing for the satisfaction of their customers. So, the marketing relationship plays a vital part in the industry of telecommunication (Alrowwad et al., 2020; Mahmoud et al., 2021; Malik et al., 2012; Obeidat et al., 2019).

Much emphasis has been placed on revenue from the oil and gas industry, thereby leading to the neglect of other vital sectors of the economy, such as tourism, power, etc. It is based on this fact that the hospitality industry has really remained under-explored in Nigeria for too long as an alternative source of generating revenue. The operations of hotels and other tourist centers gave rise to the function of what we see today as the hospitality industry. The hospitality industry is an important variable in the affairs of a nation and, most recently, it has sprung up to be a viable fund-generating medium in Jordan. Tourist centers, to a great extent, tend to bring investors and tourists, both local and foreign, to these countries, which, in turn, helps to improve the productivity level of the nation. The hotels in an environment serve as residing and accommodating places for people. These tourists, investors, and leisure seekers stay in hotels for different reasons, ranging from business meetings, conferences, leisure, etc. The government, on the other hand, derives revenue from the taxes imposed on these hospitality centers, which they usefully utilize in other productive ventures (Abuhashesh et al., 2019; Akaegbu, 2013).

While customer relationship marketing may be a firm's important driver towards the satisfaction of its customers, switching barriers and price perceptions are also likely to influence retention. Recent research on switching barriers and pricing perceptions may have interaction effects on customer retention. If they are high, service firms may continue to retain customers even if they are not highly satisfied (Kyriazopoulos & Rounti, 2007). Customers' satisfaction with their purchase is a significant factor that leads business to success. In recent times, customer satisfaction has gained new attention within the context of the paradigm shift from trans-actional marketing to relationship marketing. Customers who are satisfied with a purchased product will buy the same product again, more often, and will also recommend it to others (Masadeh et al., 2019). Customer satisfaction is commonly related to two fundamental properties, including the customer's judgment of the quality of the product and his evaluation of the interaction experience he or she has had with the product provider. He sums this up when he states: "The key to customer retention is customer satisfaction." Even though it is agreed in the literature that price and quality have a high effect on customers' satisfaction, there is little empirical evidence exploring this relationship. Almost nothing is known about the human neural mechanisms through which they affect the decisions made by individuals (Dapkevicius & Melnikas, 2009).

Traders usually use various pricing tools and promotion policies to attract new customers and increase customer loyalty. It is surprising that pricing strategies have proved to be a fruitful research area for marketers. Marketing scientists have offered both theoretical and empirical estimates showing that for all types of consumers, different price policies are likely to attract. Although there are well-known reasons why a consumer selects a store, we know relatively little about how retailers choose pricing strategies because of the complexity of such a decision should take into account both consumer preferences and current and anticipated actions of rivals (Marinescu et al., 2010).

One important factor that has been considered in many exchange relationships is price, which is the financial value that is given out in exchange for a product. The literature on consumer studies regarding prices has argued that customers hold an internal reference price which serves as a standard against which newly encoded prices are compared. In consumer markets. Price and satisfaction, so far, only measured aspects of the financing. Although price satisfaction has a direct impact on consumer behavior and loyalty. Obviously, this increased loyalty can reduce the cost of switching customers to other brands and the use of marketing tools such as word of mouth. In today's competitive and turbulent environment, according to a new marketing philosophy, customer-oriented organizations focus on customers and look at things from the perspective of the customer. Thus, it is clear what the customer wants and then begins searching for their research (Hortamani et al., 2013). With increasing globalization, local retailers find themselves having to compete with large foreign players by targeting niche markets. To excel and flaunt as a market leader in an ultramodern era and a globalized world, organizations must strive to harvest from their marketing strategies, benchmarking and company quality policies. Ranking and selecting projects is a relatively common, yet often difficult, task. It is complicated because there is usually more than one dimension for measuring the impact of each criterion and more than one decision maker.

It's impossible for anyone to neglect the impact of the air transport business on the world economy. Sir Colin Marshal, the former CEO of British Airways, has described it as "the flywheel for the engines of the world's industry". According to the Air Transportation Action Group, the industry at the beginning of this decade produced more than €1.5 trillion worth of economic activities and employed directly and indirectly 57 million people. It is, furthermore, an essential element in the world transport network, as it is currently used by more than two and a half billion passengers per year, and ships around 35% of the world's manufactured products (by value). The industry has, therefore, turned from a sign of military and national power to a key element of business and investment. Technologically driven innovations have boosted the industry, but economic developments have had even more influence on the market (Jahmani et al., 2020; Jawabreh, 2021). Despite being faced with rising fuel prices, environmental concerns, and small margins, the airline business will still face an intensely competitive market (Al-Jabaly & Khraim, 2014).

The study aims to provide pricing policies for all categories and work on the design of all pricing requirements to suit the quality of service and customer satisfaction, which indicates reaching a high level of customer satisfaction.

2 Literature Review

Trust is the name of confidence and belief which a customer attaches to some organization and considers what he or she thinks should be delivered. Actually, trust is a relationship that attaches the customer to the company. Trust is also involved between the employees of an organization. The high level of trust between each other in multinational and multicultural organizations creates productive relationships, which, in the end, generates long-term benefits for the organizations (Alananzeh et al., 2019; Jawabreh et al., 2020). Basically, trust plays an important role in e-business. Because in the e-market, privacy and security are key elements to developing trust. Trust development is more suitable for trade when considering the business to consumer market. Even in store salesperson behavior influences building trustworthy relationships. Sales effectiveness ultimately increases the trust of customers. Salesperson behavior plays a vital role in trust building (Alshurideh et al., 2021; Masa'deh et al., 2018; Sarwar et al., 2012).

The restaurant service practitioner provides high-quality food and drink service to guests. A food service practitioner generally works in the commercial sector, offering a range of services to customers. There is a direct relationship between the nature and quality of the service required, and the payment made by the guest. Therefore, the practitioner has a continuing responsibility to work professionally and interactively with the guest in order to give satisfaction and thus maintain and grow the business. The Marketing Accountability Standards Board (MASB) endorses the definitions, purposes, and constructs of classes of measures that appear in Marketing Metrics as part of its ongoing Common Language in Marketing Project. In a survey of nearly 200 senior marketing managers, 71% responded that they found a customer satisfaction metric very useful in managing and monitoring their businesses. The pricing/marketing strategy is based on the theory that certain prices have a psychological impact. Retail prices are often expressed as "odd prices": a little less than a round number, e.g., $19.99 or £2.98. There's evidence that consumers tend to perceive "odd prices" as being significantly lower than they actually are, tending to round to the next lowest monetary unit (Jawabreh, 2020). Thus, prices such as $1.99 are associated with spending $1 rather than $2. The theory that drives this is that lower pricing such as this institutes greater demand than if consumers were perfectly rational. Psychological pricing is one cause of price points. In this context, customer retention is a burning issue. In this context, trustworthy customers facilitate the organization to achieve its core objective, even if the organization is a profit-making concern or a nonprofit organization. Retaining more customers and delighting them leads the customer to retention with the organization's products or services. Since mid-1990, customer retention has been a very beloved topic for researchers. Those companies who know about the value of customer retention invest in it even from their profit, because customer retention gives long-term benefits to the organization in return. That's why not only big organizations, but small firms care about retaining customers (Sarwar et al., 2012).

The building of brand image is also an important matter for service firms as it is a powerful determinant for product marketing. Brand image as "the brand relations retained in consumers' minds causes assumptions about a brand." According to Nel et al., (2018), the product brand image is negatively affected by brand expansion, but the negative effect is being reduced by the relationship between the original and expansion brands. The great fall in the brand image will occur due to the brand expansion if the supposed image and quality of the original brand is superior. There have been positive effects on the apparent quality of the brand, consumers' brand awareness, fit perceived by the consumer, and consumers' attitudes towards the expansion of the product brand image after the expansion. The behavioral study reveals that there is a positive effect of brand image on customer loyalty and dedication towards market offerings. The supposed quality of the expansion is being affected by the degree of the relationship between the corporate brand and the service extension, which, as a result, affects the corporate image, particularly for those corporate brands which have very high-rated images (Malik et al., 2012).

A marketing concept that encompasses a customer's impression, awareness, and/or consciousness about a company or its offerings. Perception is the process by which we select, organize, and interpret information inputs to create a meaningful picture of the world. Perception is very important because it affects customers' actual behavior. This is a concept that constantly arises when trying to evaluate customers' satisfaction. This is in view of checking how the customer perceives or rates the organization in terms of product, quality, pricing, and service delivery. It is a term frequently used to measure how well products/services supplied by a company meet or surpass customers' expectations. Evaluating customer perception regularly helps in managing and monitoring business progress. When a particular firm has loyal customers, it gains positive word-of-mouth marketing, which is both free and highly effective, but very disastrous if it is negative and disloyal. Customer perception is how your target market sees your product or service, and this can go a long way towards determining the success or failure of any business. For example, if a particular hotel has a reputation for having the best customer service as well as internet facilities, the general customer perception would be that you should go there if you need assured services. This sentiment could influence the hotel's ability to make a profit. When the customer makes his buying decision, he evaluates the benefits to be derived from a particular service/product and compares them with the cost (Akaegbu, 2013).

Companies win or lose based on what percentage of their customers they can keep. Success is largely about the retention of customers, which again depends on the CS level. It would be a great help to be able to comprehensively measure the quality of products and services by relating the measures of quality to real customer behavior. Some companies get feedback about CS through the percentage of complaints and some through non-systematic surveys, but some do not measure CS at all, because the system would not add anything useful and is very time-consuming. Due to the small size of the market, there is intense rivalry among different airlines companies in Jordan. This rivalry has taken the form of severe price competition, with airlines ruthlessly undercutting each other with fare promotions. The service the airlines sell (air transport) is pretty homogenous, and there is not much service differentiation.

Airlines strive to create customer loyalty by offering frequent flyer programs, but this competitive advantage is quickly eroded by almost all airlines offering such programs. At present, it is much easier for travelers to search around for the best price. In a small market like Jordan, intense competition can lead to excess seat capacity due to declining demand because of macro-economic factors, and the high fixed costs and low marginal costs make the airline industry very price competitive. This study will yield a better understanding of the factors that lead to customer satisfaction and the value they are looking for, and ultimately, perceived price fairness for the overall service obtained by them (Al-Jabaly & Khraim, 2014).

3   Research Methodology

This section provides the methodology applied in the current study. It consists of the research model, operational definitions of the study's independent and dependent variables, research hypotheses, besides data collection tool and research population and sample.

3.1   Research Model

The elements of this research are established based on preceding literature, either theoretically or empirically. Indeed, this study used variables that are common in tourism development literature. Figure 1 represents a model for the study that shows the independent variables, the dependent variable, and the proposed relationship between them.

Fig. 1 Research model

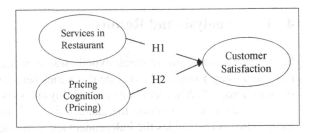

3.2   Operational Definitions

Adapted from (Lepkova & Žūkaitė-Jefimovienė, 2012), the current research considers two independent variables (i.e., Services in Restaurant, and Pricing Cognition (Pricing)) which were measured in the research questionnaire through five items for each variable; one dependent variable (i.e., Customer Satisfaction) which is adapted from (Lepkova & Žūkaitė-Jefimovienė, 2012) and measured through five items.

3.3   Research Hypotheses

In order to test the research model, the study is hypothesized as follows:
   H1: There is a significant relationship between Services in Restaurant and Customer Satisfaction.
   H2: There is a significant relationship between Pricing Cognition (Pricing) and Customer Satisfaction.

3.4   Population and Sampling

The study population consisted of all spectrums of the community interested in tourism in Aqaba city includes all age levels (from 18 to 40 years and above). 1000 questionnaires were distributed and considered 600 of them for statistical analysis, thus the response rate was 60% which is high for the community of the city of Aqaba.

4   Data Analysis and Results

In order to examine the research hypotheses, in which these variables have been measured using 5-points Likert scale that varies between strongly disagree $= 1$ and strongly agree $= 5$. Also, reliability and validity analyses were conducted; descriptive analysis was used to describe the characteristic of sample and the respondent to the questionnaires besides the independent and dependent variables. Also, multiple regression analysis was employed to test the research hypotheses.

Table 1 The Cronbach's alpha coefficients of study variables

Variables	Number of items	Cronbach alpha
Services in restaurant	5	0.763
Pricing cognition (Pricing)	5	0.799
Customer satisfaction	5	0.812

4.1 Validity and Reliability

Validity and reliability are two important measures to determine the quality and usefulness of the primary data. Validity is about accuracy and whether the instrument measures what it is intended to measure while reliability is about precision; it is used to check the consistency and stability of the questionnaire. Indeed, the researchers depended on scales and items that were previously developed and used by other researchers with similar interest. Also, a draft of the questionnaire was formulated, and then it was reviewed by four academic lecturers—who have a sufficient knowledge and experience in this scope—to ensure that each item is measuring what is intended to be measured, and to avoid the ambiguity and complexity in the phrasing of questions. The reliability of the instrument was measured by the Cronbach's alpha coefficient. Further, some scholars (e.g., Bagozzi & Yi, 1988) suggested that the values of all indicators or dimensional scales should be above the recommended value of 0.60. Table 1 represents the results of Cranach's alpha for the independent and dependent variables. Cronbach's alpha coefficients of all the tested variables are above 0.60 which suggesting the composite measure is reliable.

4.2 Respondents Demographic Profile

As indicated in Table 2, the demographic profile of the respondents for this study showed that they are typically females, most of them between 20 and 30 years old, and the majorities' experiences between 3 and less than 10 years.

4.3 Descriptive Analysis

In order to describe the responses and thus the attitude of the respondents toward each question they were asked in the survey (Poh et al., 2012), the mean and the standard deviation were estimated. While the mean shows the central tendency of the data, the standard deviation measures the dispersion which offers an index of the spread or variability in the data (Sekaran & Bougie, 2013). In other words, a small standard deviation for a set of values reveals that these values are clustered closely about the mean or located close to it; a large standard deviation indicates the opposite. The level of each item was determined by the following formula: (highest point in

Table 2 Description of the respondents' demographic profiles

Category	Category	Frequency	Percentage (%)
Gender	Males	18	18.0
	Females	82	82.0
	Total	100	100
Age	20 years–less than 25	37	37.0
	25 years–less than 30	42	42.0
	30 years–less than 35	13	13.0
	More than 35 years old	8	8.0
	Total	100	100.0
Experience	1 years–less than 3 years	18	18.0
	3 years–less than 6 years	38	38.0
	6 years–less than 10 years	36	36.0
	Without experience	8	8.0
	Total	100	100

Likert scale – lowest point in Likert scale)/the number of the levels used = (5–1)/5 = 0.80, where 1–1.80 reflected by "very low", 1.81–2.60 reflected by "low", 2.61–3.40 reflected by "moderate", 3.41–4.20 reflected by "high", and 4.21–5 reflected by "very high". Then the items were being ordered based on their means. Tables 3 and 4 show the results.

As presented in Table 3, data analysis results have shown that services in restaurant in Aqaba hotels are applied to a very high level in which the mean score is 4.38. Also, customer satisfaction and pricing cognition (pricing) were existed in high bases with means of 3.86 and 3.74 respectively. Table 4 demonstrates the mean, standard

Table 3 Overall mean and standard deviation of the study's variables

Type of variable	Variables	Mean	Standard deviation	Level	Order
Independent variables	Services in restaurant	4.3860	0.58690	Very high	1
	Pricing cognition (Pricing)	3.7420	0.73816	High	2
Dependent variable	Customer satisfaction	3.8620	0.71220	High	

Table 4 Mean and standard deviation of the study's variables

Services in restaurant	Mean	SD	Level	Order
The food level is good in restaurants	4.59	0.698	Very high	1
Staff apply the rules of public safety in cooking	4.40	0.667	Very high	2
Use the latest cooking tools	4.34	0.819	Very high	3
Work as a team to reach a high level of service	4.33	0.817	Very high	4
Follow-up by officials on performance and productivity	4.27	1.043	Very high	5
Pricing cognition (Pricing)				
Pricing for marginal profit per unit or at the aggregate level	3.64	1.078	High	5
Pricing to achieve large volume sales with market spread	3.73	0.941	High	3
Pricing in order to stabilize the market and establish a stable mental image of the market for the products provided by the institution	3.68	0.952	High	4
Increase customer purchase rate for products on the market	3.78	0.960	High	2
Increase demand from non-users and win new customers	3.88	1.018	High	1
Customer satisfaction				
The food level fits the customer's satisfaction	4.00	0.953	High	1
The staff treat fits the customer's satisfaction	3.86	0.899	High	3
The prices are commensurate with the quality of service	3.82	0.936	High	4
Restaurants reach a high level of customer satisfaction	3.75	0.957	High	5
The meal speed is suitable for the customer	3.88	0.967	High	2

deviations, level, and order scores for services in restaurant and pricing cognition (pricing) toward customer satisfaction.

4.4 Hypotheses Testing Results

The current research is mainly seeking to investigate the impact of services in restaurant, and pricing cognition (pricing) toward customer satisfaction in Aqaba city in Jordan. Consequently, in order to test the hypotheses developed for this study, multiple regression technique was used. Further, the level of significance (α-level) was chosen to be 0.05 and the probability value (p-value) obtained from the statistical hypotheses test is considered to be the decision rule for rejecting the null hypotheses (Creswell, 2009). If the p-value is less than or equal to α-level, the null hypothesis will be rejected and the alternative hypothesis will be supported. However, if the p-value is greater than the α-level, the null hypothesis cannot be rejected and the alternative hypothesis will not be supported. In addition, normality of the independent variables and the absence of multi co-linearity problem (a case of multiple regression in which the independent variables are themselves highly correlated) were checked. According to (Pallant, 2005), most of the values should be inside the adequate ranges

Table 5 Skewness and VIF for the independent variables

Variables	Tolerance	VIF	Skewness
Services in restaurant	0.987	1.013	0.525
Pricing cognition (Pricing)	0.987	1.013	0.721

Table 6 Result for the study model[b]

Variable	r	R²	F	Sig (f)	β	t	Sig (t)
Services in restaurant	0.536	0.287	19.561	0.000[a]	0.124	1.433	0.155
Pricing cognition (Pricing)					0.508	5.884	0.000

[a] Predictors: (Constant), Services in Restaurant, Pricing Cognition (Pricing)
[b] Dependent variable: Customer Satisfaction

for normality (i.e., −1.0 to +1.0). For this purpose, skewness and Variance Inflation Factor (VIF) were investigated; Table 5 includes the results.

As can be noticed from Table 5, the skewness values were within the normal values (−1.0 to +1.0) suggesting that the data of the independent variables is normal. The VIF values were less than the critical value (10) which is most common among the most studies, suggesting no multi co-linearity problem among the independent variables. However, the results of testing the two hypotheses on the impact of services in restaurant and pricing cognition (pricing) on customer satisfaction are demonstrated in Table 6.

Refer to Table 6 the multiple correlation coefficient R = 0.536 indicates that there is a positive correlation between the independent variables (services in the restaurant, and pricing cognition (pricing)) and customer satisfaction. The R2 indicated the generalizability of the model. It allows us to generalize the results taken from the respondents to the whole population. In this case it equals 0.287. The results showed that the F-ratio for these data is equal to 19.561, which is statistically significant at p < 0.05. Therefore, we conclude that there is a statistically significant impact of the independent variables on customer satisfaction.

The β indicates the individual contribution of each predictor (independent variable) to the model, if other predictors are held constant. Table 6 shows the standardized coefficients for each independent variable. The value of β for services in restaurant is 0.124 which has a small value. While for pricing cognition (pricing); the value of β is 0.508, which has a good value compared with another predictor. The level of effect of these variables depends on the β value, the higher β value, the higher effect on dependent variable. We can infer from the values of beta that the variable that has the highest contribution in the model is merely pricing cognition (pricing).

5   Discussion and Conclusions

In light of the results of the study answers questions and test hypotheses has been reached following results; the study results showed H1 was rejected whereas H2 was accepted by reference to the Table 6. In this study we aim to identify quality methods in restaurants, Focus on the core aspects of customer satisfaction and to develop cooking methods and staff development and give awareness sessions in all cooking styles and focus on all segments of society, because in general this leads to a high level of quality of pricing, service and development, which leads to increased customer satisfaction.

The findings of the data analysis revealed that services at restaurants in Aqaba hotels are provided at a very high level, with a mean score of 4.38 out of 5. Customer satisfaction and price cognition (pricing) were both found to be at high levels, with mean scores of 3.86 and 3.74, respectively, indicating high levels of both. Table 4 shows the mean, standard deviations, level, and order scores for services in a restaurant and price cognition (pricing) in relation to customer satisfaction in the restaurant industry. The multiple correlation coefficient $R = 0.536$ shows that there is a positive relationship between the independent variables (restaurant services and price cognition (pricing)) and customer satisfaction. The model's generalizability was shown by the R2 value. It enables us to extrapolate the findings obtained from the respondents to the whole population as a whole. In this particular instance, it equals 0.287. The findings revealed that the F-ratio for these data points is equal to 19.561, which is statistically significant at a 0.05 level of significance. As a result, we reach the conclusion that the independent factors have a statistically significant effect on customer satisfaction.

Customer retention is a hot topic globally, and trustworthy customers help organizations accomplish their goals, whether for profit or not. Retaining more consumers and making them happy contributes to customer retention. Since mid-1990, client retention has been a hot study subject. Because customer retention provides long-term advantages to the company, not just large organizations, but even small businesses worry about keeping consumers (Sarwar et al., 2012). Building a brand image is essential for service companies since it influences product marketing. This is because the brand image is "the brand relations maintained in customers' minds." According to (Nel et al., 2018), brand, expansion adversely impacts product brand image, although the impact is mitigated by the relationship between the original and expansion brands. Brand expansion will cause a significant loss in brand image if the original brand's image and quality are better. The product brand image has improved due to increased customer knowledge, perceived fit and consumer attitudes towards the expansion. The behavioral research shows that brand image has a beneficial impact on consumer loyalty and commitment to market offerings. The degree of connection between the corporate brand and service extension affects the perceived quality of the expansion and therefore the corporate image, especially for those corporate brands with high rated images (Abu Zayyad et al., 2021; Aljumah et al., 2021; Alshurideh, 2022; Alshurideh et al., 2015; Alwan & Alshurideh, 2022;

ELSamen & Alshurideh, 2012; Hamadneh et al., 2021; Malik et al., 2012; Sweiss et al., 2021; Tariq et al., 2022).

Each predictor (independent variable specific) contribution to the model is represented by the symbol, assuming that all other predictors are kept constant. The standardized coefficients for each independent variable are shown in Table 6. Table 6 showed a little amount of money is spent on services at a restaurant, as shown by the figure of 0.124. Whereas, for pricing cognition (price), the value of the predictor is 0.508, which is an excellent value when compared to other predictors. The magnitude of the impact of these variables is determined by the value; the greater the value, the greater the effect on the dependent variable. Based on the values of beta, we may conclude that the variable that makes the most contribution to the model is simply pricing cognition (price discrimination) (pricing) (Alzoubi et al., 2020).

References

Abu Zayyad, H. M., Obeidat, Z. M., Alshurideh, M. T., Abuhashesh, M., Maqableh, M., & Masa'deh, R. E. (2021). Corporate social responsibility and patronage intentions: the mediating effect of brand credibility. *Journal of Marketing Communications, 27*(5), 510–533.

Abuhashesh, M., Al-Khasawneh, M., & Al-Dmour, R. (2019). The impact of Facebook on Jordanian consumers' decision process in the hotel selection. *IBIMA Business Review.* https://doi.org/10.5171/2019.928418

Akaegbu, J. B. (2013). An exploratory study of customers' perception of pricing of hotel service offerings in Calabar Metropolis, Cross River State, Nigeria. *International Journal of Business and Social Science, 4*(11).

Al Khasawneh, M., Abuhashesh, M., Ahmad, A., & Alshurideh, M. T. (2021). Customers online engagement with social media influencers' content related to COVID 19. https://doi.org/10.1007/978-3-030-67151-8_22

Alananzeh, O., Al-Badarneh, M., Al-Mkhadmeh, A., & Jawabreh, O. (2019). Factors influencing MICE tourism stakeholders' decision making: The case of Aqaba in Jordan. *Journal of Convention & Event Tourism, 20*(1), 24–43. https://doi.org/10.1080/15470148.2018.1526152

Al-Jabaly, S. M., & Khraim, A. S. (2014). The effect of perceived value and customer satisfaction on perceived price fairness of airline travelers in Jordan. *Universal Journal of Management, 2*(5), 186–196.

Aljumah, A., Nuseir, M. T., & Alshurideh, M. T. (2021). The impact of social media marketing communications on consumer response during the COVID-19: Does the brand equity of a university matter. *The effect of coronavirus disease (COVID-19) on business intelligence* (pp. 367–384).

Alrowwad, A., Abualoush, S. H., & Masa'deh, R. (2020). Innovation and intellectual capital as intermediary variables among transformational leadership, transactional leadership, and organizational performance. *Journal of Management Development, 39*(2), 196–222. https://doi.org/10.1108/JMD-02-2019-0062

Alshurideh, M. (2022). Does electronic customer relationship management (E-CRM) affect service quality at private hospitals in Jordan? *Uncertain Supply Chain Management, 10*(2), 325–332.

Alshurideh, M., Bataineh, A., Alkurdi, B., & Alasmr, N. (2015). Factors affect mobile phone brand choices—Studying the case of Jordan universities students. *International Business Research, 8*(3), 141–155.

Alshurideh, M. T., Al Kurdi, B., AlHamad, A. Q., Salloum, S. A., Alkurdi, S., & Dehghan, A. (2021). Factors affecting the use of smart mobile examination platforms by universities' postgraduate students during the COVID-19 pandemic: An empirical study. *Informatics, 8*(2). https://doi.org/10.3390/informatics8020032

Alwan, M., & Alshurideh, M. (2022). The effect of digital marketing on purchase intention: Moderating effect of brand equity. *International Journal of Data and Network Science, 10*(3), 1–12.

Alzoubi, H. M., Alshurideh, M., Al Kurdi, B., & Inairat, M. (2020). Do perceived service value, quality, price fairness and service recovery shape customer satisfaction and delight? A practical study in the service telecommunication context. *Uncertain Supply Chain Management, 8*(3), 579–588.

Bagozzi, R., & Yi, Y. (1988). On the evaluation of structural evaluation models. *Journal of the Academy of Marketing Science, 16*(1), 74–94.

Creswell, J. (2009). *Research design: Qualitative, quantitative, and mixed methods approaches* (3rd ed.). Sage Publications.

Dapkevicius, A., & Melnikas, B. (2009). Influence of price and quality to customer satisfaction: neuromarketing approach. *Mokslas: LietuvosAteitis, 1*(3), 17.

ELSamen, A., & Alshurideh, M. (2012). The impact of internal marketing on internal service quality: A case study in a Jordanian pharmaceutical company. *International Journal of Business and Management, 7*(19), 84–95.

Hamadneh, S., Hassan, J., Alshurideh, M., Al Kurdi, B., & Aburayya, A. (2021). The effect of brand personality on consumer self-identity: The moderation effect of cultural orientations among British and Chinese consumers. *Journal of Legal, Ethical and Regulatory Issues, 24*, 1–14.

Hortamani, A., Ansari, A., & Akbari, M. (2013). Studying impact of price satisfaction on loyalty: A case study in Electric Generating Plant Snowa. *International Journal of Academic Research in Accounting, Finance and Management Sciences, 3*(4), 46–50.

Jahmani, A., Bourini, I., & Jawabreh, O. A. (2020). The relationship between service quality, client satisfaction, perceived value and client loyalty: A case study of fly emirates. *Cuadernos De Turismo,* (45), 219–238. https://doi.org/10.6018/turismo.45.426101

Jawabreh, O. A. (2017). An exploratory study of the motives of Jordanian out bound tourism and its impact on the development of tourism in Jordan. *International Journal of Applied Business and Economic Research, 15*(19 Part-II), 443–467.

Jawabreh, O. A. (2020). Innovation management in hotels industry in aqaba special economic zone authority: Hotel classification and administration as a moderator. *GeoJournal of Tourism and Geosites, 32*(4), 1362–1369. https://doi.org/10.30892/gtg.32425-581

Jawabreh, O. A. (2021). Tourists and local community of the case study Aqaba special economic zone authority (ASEZA). *GeoJournal of Tourism and Geosites, 35*(2), 490–498. https://doi.org/10.30892/gtg.35229-676

Jawabreh, O., & Al Sarayreh, M. (2017). Analysis of job satisfaction in the hotel industry: A study of hotels five-Stars in Aqaba special economic zone authority (AZEZA). *International Journal of Applied Business and Economic Research, 15*(26), 389–407.

Jawabreh, O., Masa'deh, R., Mahmoud, R., & Hamasha, S. A. (2020). Factors influencing the employees service performances in hospitality industry case study Aqba five stars hotel. *GeoJournal of Tourism and Geosites, 29*(2), 649–661. https://doi.org/10.30892/gtg.29221-496

Kyriazopoulos, P., & Rounti, I. S. (2007). Can price perceptions influence customer satisfaction? *Revista de Administração FACES Journal, 6*(1).

Lepkova, N., & Žūkaitė-Jefimovienė, G. (2012). Study on customer satisfaction with facilities management services in Lithuania. *Slovak Journal of Civil Engineering, 20*(4), 1–16.

Mahmoud, R., Al-Mkhadmeh, A., & Alananzeh, O. (2021). Exploring the relationship between human resources management practices in the hospitality sector and service innovation in Jordan: The mediating role of human capital. *Geojournal of Tourism and Geosites, 35*(2), 507–514. https://doi.org/10.30892/gtg.35231-678

Malik, M. E., Ghafoor, M. M., & Hafiz, K. I. (2012). Impact of brand image, service quality and price on customer satisfaction in Pakistan telecommunication sector. *International Journal of Business and Social Science, 3*(23).

Marinescu, P., Mihai, N. S., & Toma, S. (2010). Pricing strategy used as a tool for building customer satisfaction in the retail sector. *Science, 16*(1), 60–80.

Masa'deh, R., Alananzeh, O., Tarhini, A., & Algudah, O. (2018). The effect of promotional mix on hotel performance during the political crisis in the Middle East. *Journal of Hospitality and Tourism Technology, 9*(1), 32-47. https://doi.org/10.1108/JHTT-02-2017-0010

Masadeh, R., Alananzeh, O., Jawabreh, O., Alhalabi, R., Syam, H., & Keswani, F. (2019). The association among employee's communication skills, image formation and tourist behaviour: Perceptions of hospitality management students in Jordan. *International Journal of Culture, Tourism and Hospitality Research, 13*(3), 257–272. https://doi.org/10.1108/ijcthr-02-2018-0028

Nel, J., Williams, A., Steyn, R., & Hind, C. (2018). Customer experiences and expectations of sit-down restaurants: An auto-ethnographic perspective on restaurant failure. *African Journal of Hospitality, Tourism and Leisure, 7*(3).

Obeidat, Z. M., Alshurideh, M. T., & Al Dweeri, R. (2019). The influence of online revenge acts on consumers psychological and emotional states: Does revenge taste sweet? Paper presented at the Proceedings of the 33rd International Business Information Management Association Conference, IBIMA 2019: Education Excellence and Innovation Management through Vision 2020, 4797–4815.

Pallant, J. (2005). *SPSS survival manual: A step guide to data analysis using SPSS for windows version 12.* Open University Press.

Poh, A., Saludin, M., & Mukaidono, M. (2012). Deriving consensus rankings via multicriteria decision making methodology. *Business Strategy Series, 13*(1), 3–12.

Qandah, R., Suifan, T. S., & Obeidat, B. Y. (2020). The impact of knowledge management capabilities on innovation in entrepreneurial companies in Jordan. *International Journal of Organizational Analysis.* https://doi.org/10.1108/IJOA-06-2020-2246

Sarwar, M. Z., Abbasi, K. S., & Pervaiz, S. (2012). The effect of customer trust on customer loyalty and customer retention: A moderating role of cause related marketing. *Global Journal of Management and Business Research, 12*(6).

Sekaran, U., & Bougie, R. (2013). *Research methods for business: A skill-building approach* (6th ed.). Wiley.

Sweiss, N., Obeidat, Z. M., Al-Dweeri, R. M., Mohammad Khalaf Ahmad, A., M. Obeidat, A., & Alshurideh, M. (2021). The moderating role of perceived company effort in mitigating customer misconduct within Online Brand Communities (OBC). *Journal of Marketing Communications,* 1–24.

Tariq, E., Alshurideh, M., Akour, E., Al-Hawaryd, S., & Al Kurdi, B. (2022). The role of digital marketing, CSR policy and green marketing in brand development at UK. *International Journal of Data and Network Science, 6*(3), 1–10.

The Impact of Changes in Oil Prices on the Global and Saudi Arabia Economy

Mona Halim(iD)**, Walaa Rezk**(iD)**, Saddam Darawsheh**(iD)**, Anwar Al-Shaar, and Muhammad Alshurideh**(iD)

Abstract This study aimed to identify the impact of COVID-19 on oil sector which faced the significant challenges, globally and in Saudi Arabia. The study sheds light on the unprecedented crisis that occurred in the decline in demand for oil and the collapse of its prices, and its effects and consequences on world economies and Arab economies, and the extent to which the Saudi economy has been affected by them in light of what markets have witnessed global oil, and opportunities to benefit from this crisis and the study relied on the descriptive and analytical methodology. The study paper will use the NARDL to explain the asymmetric effect of changes in oil price on Government Budget for Saudi Arabia. The study reached that this situation provides a further opportunity for Saudi Arabia to diversify the economy away from a reliance on oil, Government policies seek to facing these challenges by diversifying the economic base and developing and diversifying revenues non-oil products to reduce financial volatility, in addition to enhancing the role of the private

M. Halim (✉) · W. Rezk
Financial Sciences Department, The Applied College, Imam Abdulrahman Bin Faisal University, P.O. BOX 1982, Dammam 31441, Saudi Arabia
e-mail: mahalim@iau.edu.sa

W. Rezk
e-mail: wmrezk@iau.edu.sa

S. Darawsheh
Department of Administrative Sciences, The Applied College, Imam Abdulrahman Bin Faisal University, P.O. Box 1982, Dammam, Saudi Arabia
e-mail: srdarawsehe@iau.edu.sa

A. Al-Shaar
Deanship of Preparatory Year and Supporting Studies, Department of Self Development, Imam Abdulrahman Bin Faisal University, P.O. Box 1982, Dammam 43212, Saudi Arabia

M. Alshurideh
Department of Management, College of Business Administration, University of Sharjah, Sharjah, UAE
e-mail: malshurideh@sharjah.ac.ae; m.alshurideh@ju.edu.jo

Department of Marketing, School of Business, The University of Jordan, Amman 11942, Jordan

© The Author(s), under exclusive license to Springer Nature Switzerland AG 2023   2519
M. Alshurideh et al. (eds.), *The Effect of Information Technology on Business and Marketing Intelligence Systems*, Studies in Computational Intelligence 1056,
https://doi.org/10.1007/978-3-031-12382-5_139

sector focusing on supporting non-oil GDP growth rates to enhance the resilience of the economy.

Keywords Oil prices · Oil and gas sector crude oil futures · Global economy · Saudi Arabia

1 Introduction

The oil markets witnessed in April 2020 a historic event, where the prices of American light crude collapsed below zero, in an unprecedented event, and the world will remember them for a long time, which raises questions about how this happens and its meaning, those affected by it and the potential or expected effects after That and such prices. The Corona pandemic was not the main reason for the decline in oil demand and the collapse of its prices in the manner that occurred, but the last actions by oil producers was the reason for increasing pressure in a market that was previously suffering.

This decrease in oil prices (American crude) came in the context of the global downturn in the COVID -19 pandemic that struck the world's economies in a recession, causing the demand for points to drop by nearly a third, in addition to the disturbances that dominated the relations of OPEC members, After Russia withdrew from the Alliance of Oil Producers, which was aiming to balance the market, in early March 2020.

And where there are producing countries that produce oil, there are other countries that store only, and based on the agreement of the OPEC countries, the producing countries must maintain the equivalent of 90 days of oil reserves, to face any crisis, especially in times of war. However, it did not happen that the OPEC member countries kept this amount of reserves, until the COVID-19 pandemic came and caused the demand for oil to drop, and the rates of over storage increased dramatically—due to the oversupply increase to 35 million barrels per day. Exceeds the actual storage capacity in countries that do so.

In this context, this research paper attempts to shed light on this unprecedented crisis and its effects and consequences on the global economy, the extent to which the Saudi economy was affected by it and the opportunities to benefit from this crisis.

2 Literature Review

The oil and gas industry are experiencing its third price collapse in 12 years. After the first two shocks, the industry rebounded, and business as usual continued. This time is different. The current context combines a supply shock with an unprecedented demand drop and a global humanitarian crisis. Additionally, the sector's financial and structural health is worse than in previous crises. The advent of shale, excessive supply, and generous financial markets that overlooked the limited capital discipline have all contributed to poor returns. Today, with prices touching 30-year lows, and accelerating societal pressure, executives' sense that change is inevitable. The COVID-19 crisis accelerates what was already shaping up to be one of the industry's most transformative moments.

The pandemic is first and foremost a humanitarian challenge, as well as an unprecedented economic one. The industry has responded with a Herculean effort to successfully and safely operate essential assets in this challenging time. The current crisis will have a profound impact on the industry, both short and long term. How radically the oil and gas ecosystem will reconfigure, and when, will depend on potential supply–demand outcomes and the actions of other stakeholders, such as governments, regulators, and investors. In any scenario, however, we argue that the unprecedented crisis will be a catalytic moment and accelerate permanent shifts in the industry's ecosystem, with new future opportunities.

Taking bold action during the crisis to secure resilience and accelerated repositioning Hard questions, indeed. In the meantime, winners will accept the crisis for what it is: a chance to form their own views of the future and to lead to capture new opportunities. Leaders will adopt tailored strategies that fit within their specific environment and markets in which they choose to compete, and the capabilities they bring (such as low-cost production, regional-gas or downstream-oil market leadership, value-chain integration, and specialized strengths in for example retail, trading, and distribution). In our view, all companies should act boldly on five themes, consistent with their chosen strategy: Reshape the portfolio, and radically reallocate capital to the highest-return opportunities, Take bold M&A moves, Unlock a step-change in performance and cost competitiveness through re-imagining the operating model, Ensure supply-chain resilience through redefining strategic partnership approaches, Create the Organization of the Future, in both talent and structure (Al-Jarrah et al., 2012; Assad & Alshurideh, 2020; Barbosa et al., 2020).

Aloui et al. (2020) indicated that on 12 March 2020, the sharp fell of U.S. crude oil price to 30 dollars was explained by the outspreads of coronavirus pandemic and the OPEC's inability to reach a production quota agreement. We employ the structural VAR model with time-varying coefficients and stochastic volatility (TVP-SVAR model) developed by (Primiceri, 2005) to assess the impact of COVID-19 shocks on the energy futures markets, particularly on crude oil and natural gas S&P GS Indexes. The findings confirm that energy commodities S&P GS Indexes respond to COVID-19 shock that varying over time due to fundamentals factors as well as behavioral and psychological factors.

Using a TVP-SVAR with a stochastic volatility model, they investigate the impact of COVID-19 on the commodity futures market. First, we show the importance of investors' expectations in future commodity price dynamics. Speculators in the futures markets anticipate price movements which caused high stochastic volatility during the period that precedes the huge increase in the number of deaths. Then, we show that commodity responses to COVID-19 shock varied from one period to another. The determinants likely to explain the dynamics of the SP GSCI Crude oil index and SP GSCI natural gas index face the COVID-19 include fundamental factors as well as financial factors. In fact, the fall in oil futures prices is partly due to a structural imbalance between supply and demand, particularly a strong decreasing of energy demand following the COVID 19 in large emerging countries such as China and an oil shock caused by disagreements between the Organization of the Petroleum Exporting Countries (OPEC) and Russia, triggering the drop in oil spot prices. In addition, the spread of the virus in the USA negatively impacted the commodity futures market. On the other side, based on behavioral finance theorist, the commodity indices fluctuations are linked to the investors' expectations. In fact, during the crisis, speculators operating on a purely financial logic, turned to the commodity markets to increase their returns which lead to a slight increase in commodity futures index during stock market crash. The short-term outlook for the commodity futures market will depend on the efficiency of government policy to contain the COVID-19 outbreak, and to weaken the impact of the global health crisis has on economic activity.

Recently many studies investigated COVID-19's impact on global economic activity (Ahmad et al., 2021; Al-Dmour et al., 2021; Alameeri et al., 2021; Aljumah et al., 2021; Alshurideh et al., 2021; Harahsheh et al., 2021; Shah et al., 2021). Yilmazkuday (2022) by using daily data on the global coronavirus disease 2019 (COVID-19) deaths, shows that the effect of COVID-19 deaths is insignificant, and explained the plunge of oil price in March 2020 by the OPEC disagreement. In the same line of work, Musa et al. (2020) investigates the impact of coronavirus outbreak on the global energy demand by using daily data on China's population, currency Chinese exchange rate, and international crude oil prices, from 23rd January to 8th February 2020. The result revealed that the total population has a positive and significant impact on total coronavirus infected cases while the crude oil price is negative and significantly related to the coronavirus infected cases.

3   Oil Production (American Crude) and OPEC Countries and Oil Prices in the Period 2011–2020

Oil prices ranged 2012–2020 between $110 in 2011 and below the dollar per barrel in 2020, decreased in 2015 and ranged to $80 and below $30 per barrel until 2020. It is worth noting that there are two types of American crude Both are used as a benchmark for setting world oil prices:

- **The first type**: West Texas Intermediate (WTI) oil. It is extracted from oil fields in the United States, especially from Texas, Louisiana and North Dakota, and then transported via a pipeline to Cushing, Oklahoma, for delivery. WTI is used as the benchmark for determining the price of oil in North America.
- **The second type**: Brent Crude, extracted from oil fields in the North Sea, which is a mixture of four crude oils, but collectively referred to as "Brent Crude Oil (Wingfield et al., 2020).

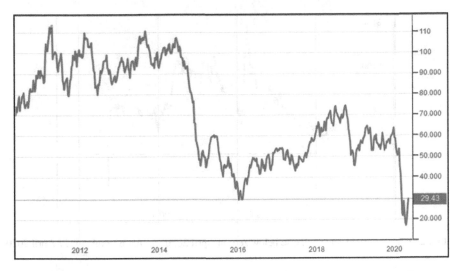

Fig. 1 Oil prices (Brent crude) in the period 2011–2020 (dollars per barrel). *Source* tradingecono mies.com (2020)

- Brent crude is the most traded globally and is used as a standard for global pricing and measurement. Brent crude and West Texas are classified as relatively light oils (Energy, 2020) (Fig. 1).

These prices mentioned in the previous figure refer to the prices of American crude (Brent), as it is considered the main reference for oil prices and purchases in the world Major Benchmark Price for purchases of oil worldwide. Brent prices have fallen to nearly $20, compared to previous years, when they reached the ceiling of $100 a barrel. As for the oil whose price has reached below zero (37-), it is WTI, as shown in the following figure (Amadeo, 2021) (Fig. 2).

West Texas Intermediate was trading at a higher price than Brent crude, and due to the revolution in shale oil production in the United States in early 2000, along with more imports to the United States from Canada, the price of WTI decreased, and it is now usually trading at a price Low on Brent. But in general, there are no major differences between the prices of Brent crude and West Texas crude, as shown in the following figure in the last five months. It also indicates the decrease of WTI prices to below zero last April and then returning to rise again, but below the levels recorded in the months Previous (Sharma, 2020) (Fig. 3).

The total oil production in the world is between 100 and 105 million barrels per day, of which 41–43 million barrels are produced by OPEC countries daily. OPEC controls approximately 75% of the total crude oil reserves in the world and produces 42% of the total crude oil production in the world.

Before the Corona pandemic, the supply of oil was greater than the demand for it (global demand = 90 million barrels per day), equivalent to 12–15 million barrels per day. Since the United States is the largest producer and consumer, as well as an importer of oil in the world, producing close to 12 million barrels per day, until

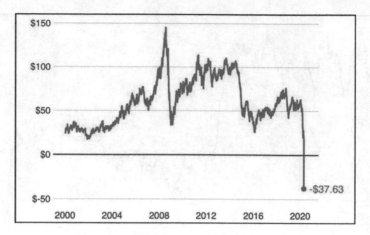

Fig. 2 US WTI crude price collapse below zero in April 2020. *Source* Bloomberg, 27 April 2020

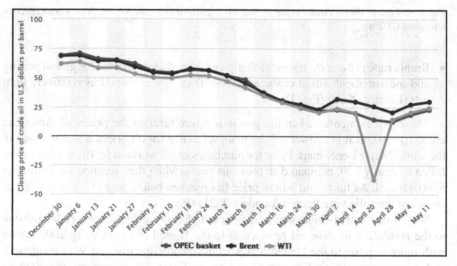

Fig. 3 Oil prices for: Brent, OPEC basket, and WTI from December 2019 to May 2020. *Source* www.statista.com

2012, it used to import about 5 million barrels per day from all over the world. but as of that year, the United States began to rely on "shale gas and shale oil", and paid attention to everything related to extractive industry to not fall under the pressure of the oil producing countries. Since that time, American oil production increased to reach 13 million barrels per day in early 2020, after it was only 8 million barrels in 2014, as shown in the following figure (Sönnichsen, 2020) (Fig. 4).

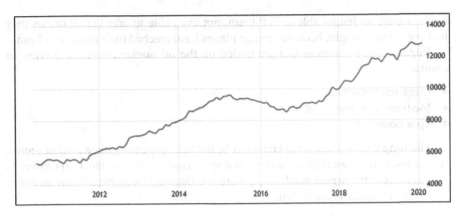

Fig. 4 The development of American crude production in the period 2011–2020 (thousand barrels). *Source* tradingeconomies.com 2020, U.S. Energy Information administration

4   Reasons for the Recent Sudden Collapse of Oil Prices in Late April 2020?

When the world invaded the newly emerging epidemic of 19-COVID, countries did not think that a comprehensive ban would be imposed in all sectors and at all levels in the manner we are witnessing today, but the United States did not expect the pandemic to extend to it in this way and cause economic life to stop in all Countries of the world, and as a result, the demand for oil decreased significantly.

If we look at Kuwait and the Kingdom of Saudi Arabia as two countries that depend on oil and its exports at 90% of their national product for each of them, and assuming that Saudi Arabia exports 3 million barrels per day, their value before 2016 was approximately $400 million, and today their value is only $120 million, meaning that A loss of $300 million per day and $9 billion per month that would harm the Saudi economy, likewise in Arab economies dependent on oil.

The value of futures contracts for US crude fell unprecedentedly, as the price of a barrel fell to minus $37, the lowest level since the beginning of the sale of futures contracts in 1983, which are obligated to sell a specific product at a specific date at specific prices, which has increased the economic pressures that it suffers from during Corona pandemic.

In a deeper view of the American crude market, the buying and selling in the oil market is not done on quantities on time. Rather, there is a circulation of buying and selling contracts between speculators who do not obtain oil in kind, but rather seek to achieve profits only through trading in contracts.

Because of the state of closure and stagnation that prevailed in the world due to the Corona pandemic, a crisis of surplus oil production was generated, while Saudi Arabia and Russia continued to pump large quantities to the market, and with the decrease in demand to stop the movement of air, land and sea navigation, contract holders were no longer able to Sell them, not even able to take the oil drums they had previously bought, because storage places have reached their maximum limits.

Regarding the contracts that are traded on the oil market, there are 3 types of contracts:

- Long-term contracts
- Medium-term contracts
- Spot contracts.

The long-term contracts are known to be the least expensive, then medium-term, where prices increase slightly, and the last spot contracts are the most expensive, as they represent the urgent need. Also, there are three oil exchanges in the world in which these contracts are traded:

- West Texas Stock Exchange
- Brent Stock Exchange
- Gulf Stock Exchange.

In the first type (West Texas) contracts were concluded in May 2020 until 20 April 2020. On the Brent Stock Exchange, June contracts were concluded at $25 a barrel, and May contracts closed at $23, with an outlook for July of $30 a barrel. the Gulf Stock Exchange, May, June and July contracts were concluded without any storage problems.

The problem of storage appeared in the United States only, and it became difficult to receive any additional quantities from oil producers, and the cost of storage until June became equivalent to $40 a barrel per month, while the cost of producing a barrel is only $25, Thus, the cost of storage in two months is $80 per barrel.

To get out of this deadlock, which is called the loss-lose situation, which is the worst situation the world has reached, the offer to sell for $0 would be better, in order to avoid the cost of storage for two months ($80 per barrel), which happened on Monday, 20 April 2020, but It was required that the sale be negative, meaning that each barrel added 37 dollars, which is better than the product costing 80 dollars storage per barrel/two months. Thus, this solution was to discard WTI stock.

Not all the Brent and Gulf stock exchanges were exposed to this problem, as Gulf oil is directly to the refining markets in India, Korea and Japan, and these markets was far from the oil market turmoil in North America.

Oil prices often change, and prices are affected by a wide range of factors, the most important of which are:

- Increase or decrease in supply by oil producers.
- Increase or decrease in demand by oil users and importers.
- Subsidies to oil companies or other energy companies.
- International policy (agreements concluded between countries).
- The internal policy of oil producers.
- Oil supply worldwide.
- Competition from other energy sources.
- Geopolitical disturbances, pressures, and insecurity (leads to an increase in prices).

the global oil price curve will rise in light of:

- The tendency to reduce production due to lower demand, and hence a decrease in the amount of surplus or stock.
- The return of economic and air to work in most countries of the world, as the weakness of the global economy has an adverse effect that would reduce the demand for oil.

The price of a barrel of oil will rise to $40–$60, which is the fair price of oil after adding variable costs to the production cost, which is estimated at $23 a barrel, while variable costs range between $15 and $25 a barrel and selling at any price below that constitutes Loss of the product. Also, the lower price than the fair price pushes small producers (whose production ranges between 7 and 10 million barrels per day) out of the market (Aloui et al., 2020).

5 The Impact of the 19-COVID Pandemic on the world's Economics

Before examining the economic effects of changes in the oil market and its prices on the global economies and the Saudi economy in particular, it is necessary to identify the most important effects of the Corona pandemic through exposure to the most important economic indicators summarized in the following table to know the impact of the pandemic on economic performance in those countries (Soñnichsen, 2021) (Table 1).

When looking at the indicators it is necessary to look at them somewhat conservatively, especially indicators of Saudi, because these indicators were calculated by the end of the fiscal year in June 2019, and not in December 2019 as in other countries. That is, the impact of the COVID-19 pandemic has not yet appeared in economic indicators, Noting that there is an estimate of some indicators for some countries that were calculated in the first quarter of 2020, which makes them appear somewhat deteriorated, as shown in the unemployment rate in the United States (14.7%), the growth rate in China (−6.8%), The growth rate in the euro area (−3.2%), in addition to the increase in the ratio of public debt to gross domestic product in countries that achieved negative growth at the end of the first quarter of this year (*Trading Economies*, n.d.).

If we look at, for example, the rate of unemployment in Saudi Arabia and its development in the period July 2017–January 2020, we find that it has made remarkable progress at the last estimate of 5.7%, as shown in the following figure (Fig. 5):

Table 1 Some economic indicators in some economies in 2018/2019

The indicator Country	The growth rate economic (%)	The interest rate (%)	The rate of inflation (%)	The unemployment rate (%)	Budget deficit (%)	Public debt/GDP (%)
United State	0.3	0.25	0.3	14.7	−4.60	106.90
China	−6.8	3.85	3.3	6	−2.80	50.50
Euro-zone	−3.2	0.0	0.3	7.4	−0.60	84.10
Japan	−2	−0.10	0.1	2.5	−3.80	238.20
United Kingdom	−1.6	0.10	0.8	3.9	−1.80	80.80
India	4.7	4	5.84	7.8	−3.34	69.62
Brazil	1.7	3	2.4	12.2	−5.90	75.79
Russia	1.6	5.5	3.1	4.7	1.80	14.60
Canada	1.5	0.25	−0.2	13	−0.70	89.70
South Korea	1.3	0.75	0.1	3.8	−1.60	36.60
Saudi	**−0.3**	**1.0**	**1.3**	**5.7**	**−9.20**	**19.10**
UAE	1.6	1.5	−1.9	2.64	−0.80	18.60
Egypt	5.6	9.25	5.9	7.7	−8.00	90.00
Singapore	−2.2	0.09	0.0	2.4	−0.30	126.30
Iran	1.8	18	19.8	10.4	−4.40	44.20
South Africa	−0.5	3.75	4.1	29.1	−6.30	62.20%

Source TRADINGECONOMIES (2020) https://tradingeconomics.com/

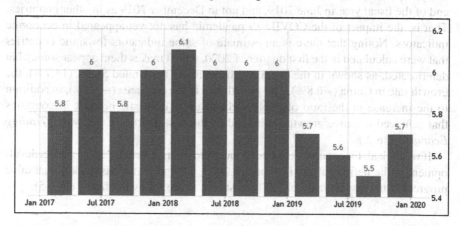

Fig. 5 The unemployment rate in Saudi Arabia (January 2017–January 2020). *Source* tradingeconomies.com 2020, CAPMAS, Saudi Arabia

6 The Expected Effects of Changes in the Oil Market and Its Prices on Arab and Saudi Economies During Corona Pandemic

It is noted in recent years that the world is on the way to reducing dependence on oil as an energy source, and therefore has a direct negative impact on oil prices. But at the same time, oil remains an irreplaceable source of energy in many areas with other sources of energy. The natural resources that exist in the world, such as minerals, timber, and textiles, will not be enough for the inhabitants of the earth at one time, so the alternative is "the petrochemical industry".

Therefore, we find the policy of Arab countries, including the Gulf states, especially Saudi Arabia, to establish a number of advanced refineries to encourage the manufacturing industry by relying on oil, and then export those products and move as far as possible from the export of oil as crude.

we can say that the Arab common market and the creation of an integrative market among the countries of the region is a real opportunity, and Arab economies should work in this direction, because it will achieve a great added value that is not comparable to that generated from the export of crude oil.

Demand for oil declined due to the emerging Corona Virus "Covid–19", which pushed the price of oil down, and consequently, the Kingdom's oil revenues decreased to increase the fiscal deficit with it.

Demand for oil has fallen historically due to the emerging Corona Virus "Covid-19", which caused the oil price crash to an unprecedented degree. Government oil revenues decreased in the first quarter of 2020 by 24.0% from the same quarter of the previous year, to record 129.0 billion Saudi riyals. This was accompanied by increased expenditures, bringing the budget deficit in the first quarter of 2020 to 34.1 billion Saudi. With the weak expectations of oil demand continuing until the end of 2020, due to the decline in economic activities, high financial spending and decreased financial revenues, we expect the Kingdom to record this year the largest historical deficit in its budget. Early estimates are that the Kingdom's budget deficit will range between 420 and 370 billion Saudi riyals. The longer the epidemic spreads, the greater the impact on the economy.

7 Model and Methodology

The study tries to investigate the short-run and long-run asymmetric effect of oil price changes on the aspects of the government budget in Saudi Arabia. The study uses the nonlinear ARDL approach.

The asymmetric integration interrelationship can be written in the following form:

$$y_t = \gamma^+ z_t^+ + \gamma^- z_t^- + \varepsilon_t \tag{1}$$

where y_t is the dependent series, z_t^+ and z_t^- are the partial summation of negative and positive developments in z_t, and ε_t is the error component. In addition, γ^+ and γ^- measure the long run asymmetric impact for positive and negative changes respectively.

$$z_t^- = \sum_{i=1}^{t} \Delta z_i^- = \sum_{i=1}^{t} \max(\Delta z_i, 0) \tag{2}$$

$$z_t^+ = \sum_{i=1}^{t} \Delta z_i^+ = \sum_{i=1}^{t} \max(\Delta z_i, 0)) \tag{3}$$

Therefore, the asymmetric effect of oil price changes on the different aspects of government budget can be expressed as follows:

$$y_t = \rho y_{t-1} + \sum_{k=0}^{p-1} \vartheta_k^+ OP_t^+ + \sum_{k=0}^{p-1} \vartheta_k^- OP_t^- + \sum_{k=1}^{p-1} \varphi_{yk} \Delta y_{t-k}$$

$$+ \sum_{k=1}^{q} \varphi_k^+ \Delta OP_t^+ + \sum_{k=1}^{q} \varphi_k^- \Delta OP_t^- + \varepsilon_t \tag{4}$$

where, 'y_t represents the different aspects of public budget, ε_t reflects the error component, $\gamma^+ = -\vartheta^+/\rho$ and $\gamma^+ = -\vartheta^+/\rho$ reflect the asymmetric long run coefficients.

In order to check for the existence of the cointegration relation in (4), we will follow (Shin et al., 2014) and (Sims et al., 1982) by utilizing the Pesaran (2001)'s cointegration bound (F_{pss}) test (Pesaran et al., 2001). Which depends on F test. F_{pss} based on the null hypothesis that there is no cointegration in the underlying relationship; it means the joint summation for the long run coefficients are equal to zero or $\rho = \vartheta^+ = \vartheta^- = 0$. Therefore, refusing the null hypothesis means that the existence of the long run cointegration in the underlying relationship. Moreover, usual Wald test can by utilized to test for the asymmetry in short and long run relationship.

Data

The paper explores the hypothesis that changes in the crude oil price would have a considerable impact on government budget, total government expenditure, current government expenditure, capital government expenditure, total government revenue, and the ratio of surplus or deficit to the GDP in Saudi Arabia. The data covers annual data for the years 1990 to 2020 for West Texas Intermediate crude oil price, total government expenditure, current government expenditure, capital government expenditure, total government revenue, and the ratio of surplus or deficit to the GDP. The dataset is captured from IFS and the General Authority for Statistics in Saudi Arabia (Ministry of Finance, 2020b).

Table 2 shows the descriptive statistics for the raw data of the utilized variables where Table 3 depicts unit root tests for the variables. The ARDL model has one condition is that employed series are integrated of order I(0) or I(1) and it is not allowed to be integrated of order I(2). We depend on Augmented Dickey Fuller test to check for the unit root. As it is shown in Table 3, all variables are stationary in the level of the first difference and there is no any series is stationary in the second difference; which means it is appropriate to apply ARDL model.

Table 4 depicts the results of the linear ARDL for the effect of oil price changes on a public budget. The first and the second column show that over the long run changes in total government revenue is significantly affected by its first lag, the level and the lag of oil price. Over the short run, a change in total government revenue is significantly affected by changes in its lag and changes in the level oil price changes. In addition, the cointegration term is negative and significant which means the relationship is stable.

Table 4 shows estimates for current government expenditure, capital government expenditure, and total government expenditure. Over the long run, current government expenditure is considerably affected by its lag and the level of oil price where over the long run it is significantly determined by changes in its lag and changes in

Table 2 Descriptive statistics for variables

	OP	TGR	TGX	CGX	CPGX	DCGDP
Mean	33.13043	335,243.2	315,143.7	227,375.8	87,767.87	1.610638
Median	20.82000	205,500.0	221,272.0	162,350.0	54,652.00	−2.600000
Maximum	110.2200	1,247,398	1,109,903	739,658.0	370,245.0	43.20000
Minimum	1.300000	7940.000	6293.000	3989.000	2304.000	−25.30000
Std. Dev	29.88969	332,748.2	278,764.6	196,192.7	91,699.93	14.65213
Skewness	1.410193	1.470927	1.416078	1.217659	1.404911	0.802962
Kurtosis	3.904577	4.091775	4.026157	3.551291	4.148236	3.699688
Jarque–Bera	17.18013	19.28269	17.77012	12.20960	18.04320	6.009254
Probability	0.000186	0.000065	0.000138	0.002232	0.000121	0.049557
Observations	47	47	47	47	47	47

Table 3 Unit root results

	I(0)		I(1)	
	Z(t)	P-value	Z(t)	P-value
OP	−1.566	0.4913	−5.936	0.000
TGR	−1.758	0.3957	−7.576	0.000
CEXP	−0.431857	0.8948	−8.698	0.000
CPEXP	−1.019	0.7387	−5.972	0.000
CDGDP	−2.564	0.107	−6.897	0.000

Table 4 Results of the linear ARDL of oil price on the public budget

Total government revenue D(TGR)		Current government expenditure (CGE_t)		Capital government expenditure (PGE_t)		Total government expenditure TGE		The ratio of surplus or deficit to GDP (RGDP)	
C	4.521 (0.001)***	C	2.658 (0.000)***	C	7.879 (0.000)***	C	4.8718 (0.006)***	C	-1.783 (0.717)
TGR_{t-1}	0.5076 (0.001)***	CGE_{t-1}	0.7086 (0.000)***	PGE_{t-1}	0.489 (0.028)**	TGE_{t-1}	0.525 (0.000)	$RGDP_{t-1}$	0.511 (0.000)
OP_t	0.995967 (0.000)***	OP_t	0.971(0.000)***	OP_t	0.995 (0.013)**	OP_t	0.290 (0.0251)	OP_t	23.14 (0.000)***
OP_{t-1}	0.5059 (0.002)***	OP_{t-1}	0.104 (0.484)	OP_{t-1}	0.09 (0.877)	OP_{t-1}	0.316 (0.008)***	OP_{t-1}	-19.88 (0.000)***
$D(TGR_{t-1})$	-0.232 (0.2984)	$D(CGE_{t-1})$	-0.1772 (0.541)	$D(PGE_{t-1})$	0.0183 (0.034)**	$D(TGE_{t-1})$	0.484 (0.009)	$D(RGDP_{t-1})$	-0.0111 (0.9513)
$D(OP_t)$	1.021 (0.000)***	$D(OP_t)$	0.283 (0.000)***	$D(OP_t)$	0.378 (0.07)*	$D(OP_t)$	0.290 (0.025)**	$D(OP_t)$	23.14 (0.000)***
$D(OP_{t-1})$	0.1692 (0.4662)	$D(OP_{t-1})$	0.181 (0.227)	$D(OP_{t-1})$	0.163 (0.675)	$D(OP_{t-1})$	-0.04 (0.81)	$D(OP_{t-1})$	3.173 (0.55)
ECT_{t-1}	-0.492 (0.001)	ECT_{t-1}	-0.291)0.000)***	ECT_{t-1}	-0.38 (0.010)**	ECT_{t-1}	-0.504 (0.000)***	ECT_{t-1}	-0.488 (0.000)***
Adjusted R^2	0.945	Adjusted R^2	0.954	Adjusted R^2	0.628	Adjusted R^2	0.956	Adjusted R^2	0.709
F-statistic	260.08 (0.000)***	F-statistic	472.25 (0.000)***	F-statistic	19.17 (0.000)	F-statistic	328.1 (0.00)	F-statistic	27.90 (0.000)***
D.W	2.19	D.W	2.181	D.W	1.855	D.W	1.98	D.W	2.06
Q-statistic	0.5750 (0.448)	Q-statistic	0.1563 (0.693)	Q-statistic	0.0270 (0.987)	Q-statistic	0.012 (0.913)	Q-statistic	0.057 (0.810)

(continued)

Table 4 (continued)

	Total government revenue D(TGR)		Current government expenditure (CGE$_t$)		Capital government expenditure (PGE$_t$)		Total government expenditure TGE		The ratio of surplus or deficit to GDP (RGDP)	
Heteroscedasticity	1.612441	(0.2008)	Heteroscedasticity	1.789 (0.150)	Heteroscedasticity	1.143 (0.374)	Heteroscedasticity	0.775 (0.514)	Heteroscedasticity	0.775 (0.514)
Serial correlation	0.986	(0.3817)	Serial correlation	1.595 (0.216)	Serial correlation	0.2844 (0.755)	Serial correlation	0.32 (0.72)	Serial correlation	0.320 (0.72)
Ramsey reset	1.309	(0.2591)	Ramsey reset	2.273 (0.1398)	Ramsey reset	0.397 (0.695)	Ramsey reset	1.15 (0.28)	Ramsey reset	1.154 (0.2889)

*** Reflects the coefficient is significant with probabilities 1%, 5%, 10%; ** the coefficient is significant with probabilities 5%, 10%; * the coefficient is significant with probability 10%

oil price. Further, Table 4 shows that capital government expenditure is significantly affected by its lag and the leg of oil price over the long run. However, over the short run, capital government expenditure is significantly determined by changes in its lag and changes in the level of oil price. Furthermore, Table 4 reveals that total government expenditure is significantly affected by its lag, level of oil price and the first lag of oil price. On another hand, over the short run, total government expenditure is significantly affected by changes in its lag and changes in the level of oil price.

Table 4 presents the results for the deficit or the surplus of government budget as a ratio of GDP. It can be observed that over the long run RGDP is significantly determined by its lag, the level of oil price and the lag of oil price. However, over the short run, RGDP is significantly determined by changes in oil price.

Table 5 depicts the estimates for the asymmetric effect of oil price changes on the government budget. The first two columns show factors that effect on total government revenues, we can observe that over the long run total revenues is significantly affected by its lag, positive and also negative shocks of oil price. However, the effect of positive socks is stronger than negative shocks. Over the short run, the total revenue is considerably only affected by changes in negative oil price shocks. Moreover, the cointegration term is negative and significant which reflects the stability of cointegration relationship.

Table 5 shows the current government expenditure is significantly affected by its lag and the negative oil price shocks. Further, current government expenditure is significantly affected by positive and negative oil price shocks. However, the effect of positive shocks is more obvious than negative shocks. On the other hand, capital government expenditure is significantly affected by its lag and positive oil price shocks over the long run. In addition, capital government expenditure is mainly determined by changes in its lag, changes in the level of positive oil price shocks and the changes in negative oil price shocks. Furthermore, the total government expenditure is mainly determined by its lag, the level of positive oil price shocks and the lag of positive oil price shocks. Contrary, the total government expenditure is significantly affected by changes in its lag, changes in the level and in the lag of positive oil price shocks.

Table 5 presents the results of the asymmetric effect of oil price changes on the ratio of surplus or deficit in the public budget to the GDP. It can be observed that this ratio is considerably affected by its lag, the level of negative and positive shocks of oil price and the lag of positive shocks. However, the ratio of the surplus or deficit to GDP is significantly affected by changes in positive shocks and changes in the lag of changes in negative shocks.

In sum, our results support the asymmetric effect of oil price shocks on different aspects of public budget in Saudi Arabia. This is confirmed by the results of Wald test which refer to rejecting the null hypothesis over the short run and long run for the majority of the underlying models.

Table 5 Results of the non-linear ARDL of oil price on the public budget

Total government revenue D(TGR)		Current government expenditure (CGE_t)		Capital government expenditure (PGE_t)		Total government expenditure (ge)		The ratio of surplus or deficit to GDP (RGDP)	
C	2.758 (0.25)	C	3.805 (0.000)***	C	1.27 (0.52)	C	6.06 (0.00)***	C	2.46 (0.58)
ρ	0.559 (0.00)***	ρ	0.597 (0.000)***	ρ	0.623 (0.000)***	ρ	0.704 (0.00)***	ρ	0.337 (0.016)**
ϑ_0^+	1.50 (0.00)**	ϑ^+	0.156 (0.39)	ϑ^+	1.148 (0.039)**	ϑ^+	0.821 (0.00)***	ϑ^+	35.82 (0.00)***
ϑ_0^-	0.67 (0.00)**	ϑ^-	0.228 (0.02)**	ϑ^-	0.131 (0.719)	ϑ^-	-0.148 (0.484)	ϑ^-	19.198 (0.00)***
ϑ_1^+	-0.875 (0.11)	ϑ_1^+	0.171 (0.41)	ϑ_1^+	0.081 (0.91)	ϑ_1^+	-0.474 (0.04)**	ϑ_1^+	-41.96 (0.00)***
ϑ_1^-	-0.262 (0.41)	ϑ_1^-	-0.004 (0.988)	ϑ_1^-	0.774 (0.422)	ϑ_1^-	-0.082 (0.76)	ϑ_1^-	7.25 (0.34)
φ_{y1}	-0.206 (0.35)	φ_{y1}	-0.092 (0.60)	φ_{y1}	-0.298 (0.03)**	φ_{y1}	-0.25 (0.074)*	φ_{y1}	-0.08 (0.58)
φ_0^+	1.503 (0.00)	φ_0^+	2.658 (0.000)***	φ_0^+	1.148 (0.039)**	φ_0^+	0.821 (0.00)***	φ_0^+	35.8 (0.00)***
φ_0^-	0.673 (0.00)***	φ_0^-	0.283 (0.00)***	φ_0^-	0.131 (0.719)	φ_0^-	-0.148 (0.48)	φ_0^-	-1.78 (0.75)
φ_1^+	-0.021 (0.95)	φ_1^+	-0.065 (0.81)	φ_1^+	1.195 (0.024)**	φ_1^+	0.474 (0.0441)**	φ_1^+	-2.2 (0.74)
φ_1^-	-0.045 (0.89)	φ_1^-	0.069 (0.83)	φ_1^-	-1.15 (0.29)	φ_1^-	-0.482 (0.109)	φ_1^-	30.40 (0.00)***
$\vartheta_k^+ = \vartheta_k^-$	6.49 (0.015)**	$\vartheta_k^+ = \vartheta_k^-$	8.46 (0.006)***	$\vartheta_k^+ = \vartheta_k^-$	4.865 (0.03)**	$\vartheta_k^+ = \vartheta_k^-$	4.689 (0.0401)**	$\vartheta_k^+ = \vartheta_k^-$	7.90 (0.008)***

(continued)

Table 5 (continued)

	Total government revenue $D(TGR)$	Current government expenditure (CGE_t)	Capital government expenditure (PGE_t)	Total government expenditure tge	The ratio of surplus or deficit to GDP (RGDP)
$\varphi_k^+ = \varphi_k^-$	15.13 (0.00)***	4.29 (0.045)	4.97 (0.036)**	17.41 (0.00)***	3.27 (0.051)*
ECT_{t-1}	−0.306 (0.022)**	−0.29 (0.002)***	−0.078 (0.071)*	−0.043 (0.021)**	−0.407 (0.00)***
Adjusted R^2	0.945	0.955	0.671	0.92	0.84
F-statistic	41.73 (0.00)***	241.19 (0.000)***	6.71 (0.000)***	31.47 (0.00)***	21.2 (0.00)***
D.W	2.01	2.02	2.02	1.86	2.1
Q-statistic	6.471 (0.89)	6.75 (0.87)	10.329 (0.568)	11.14 (0.431)	16.147 (0.185)
Heteroscedasticity	0.596 (0.856)	0.525 (0.71)	1.73 (0.12)	1.243 (0.305)	0.505 (0.92)
Serial correlation	0.34 (0.715)	0.51 (0.60)	1.60 (0.216)	0.655 (0.528)	0.622 (0.544)
Ramsey reset	0.520 (0.478)	0.420 (0.52)	0.647 (0.426)	2.30 (0.142)	2.19 (0.3)

*** Reflects the coefficient is significant with probabilities 1%, 5%, 10%; ** the coefficient is significant with probabilities 5%, 10%; * the coefficient is significant with probability 10%

8 How Did Saudi Arabia Face This Price Crisis in the Oil Market?

The Kingdom and the oil-exporting countries are experiencing many challenges, as restricting oil production and developments on the supply side are other negative effects added to the slowdown in the growth of the global economy, which affects price stability and thus affect the ability of financial planning for countries. Oil markets witnessed great fluctuations during the year 2019, as Brent crude prices achieved their lowest levels in January at 2.53 dollars a barrel, and its highest levels in April, to reach $9.74 a barrel.

The Kingdom, through its oil policy, seeks to stabilize the global market. Crude Oil, which balances the interests of producers and consumers together and provides the right environment for sustainable growth in the global economy through its active leadership role in commitment in quotas agreed in accordance with the OPEC agreement. Government policies also seek to facing these challenges through diversification of the economic base and developing and diversifying revenues from mon-oil resources to reduce the volatility of financial flows, in addition to strengthening the role of the private sector Emphasis is placed on supporting non-oil GDP growth rates to enhance the economy's resilience (Ministry of Finance, 2020a).

The following are important points that can be taken into account when such crises occur and minimize their negative effects on the Saudi market.

- Encouraging the manufacturing industry and stopping the export of oil or gas in its raw form.
- Relying on alternative sources of electric power generation, which is a global trend sought by different countries of the world, and has many positive consequences and effects, the most important of which is reducing pollution rates. Evidence for this is that the low dependence on petroleum products and the consequent stoppages of flights and land and the stopping of many factories worldwide have had the greatest impact in reducing pollution rates in the world by rates that reached 25%.
- Working to form a fully supply chain in the Arab region, and to use the crude oil produced in one country to produce finished goods in another country and then export them abroad(Barbosa et al., 2020).
- Forming a moderate axis by the oil-producing Arab countries, and not aligned with the Chinese or American poles, considering the global power imbalance. Power is not only a military force, but there is an economic force that will not be achieved in the Arab region except by integration between the countries of the region to form one economic force that produces what is needed and exports what others need (Fig. 6).

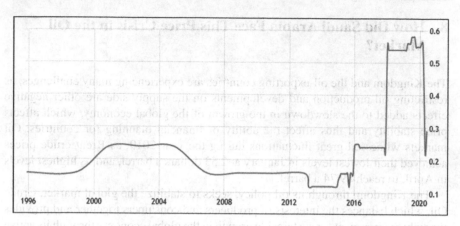

Fig. 6 Saudi   Arabia   Gasoline   Prices.   *Source*   https://tradingeconomics.com/
Saudi Arabia/gasoline-prices

9   Saudi Arabia Reforms of Energy Products Prices (Increasing the Efficiency of Government Support)

- The Kingdom has been providing energy products to local consumers at a subsidized price and much less. Of their prices when exporting. The difference between the price of a product domestically and its price when exporting represents the cost of an alternative opportunity per barrel of oil or the benefits that the domestic consumer acquires from the subsidized price.
- The value of government support for energy and water products in 2015 was approximately (300) billion Saudi riyals, based on the export price of energy products at that time. The value of subsidies can increase significantly over time due to the rapid growth of domestic consumption of energy and water products and changes in the export prices of energy products. Subsidized prices also contribute to stimulating rapid growth in the consumption of state resources.
- The Kingdom relies on oil and non-renewable natural resources as a main source of income, and therefore it is not possible in the long run to continue consuming these resources locally in this unfair form and at subsidized prices.

The Kingdom aims to adjust the prices of energy products as the cost of the alternative opportunity wastes on the Kingdom the opportunity to collect the optimum price for its natural resources, and therefore to use those lost revenue in investment in other sectors to achieve economic diversification and provide various benefits for the residential and non-residential sectors.

10 Conclusion

In this study we presented and shed light on unprecedented crisis for the oil industry which generated from the COVID-19 pandemic demand disruption, and a supply surplus. The oil sector is witnessed its third price collapse in 12 years. After the first two shocks, but this time is different. This one combines a supply shock with an unprecedented demand drop and a global humanitarian crisis. The depth and duration of this crisis are uncertain, this crisis will have a deep impact on the industry, both short and long term. We think, it will be difficult to return to the attractive industry performance without fundamental change. In the Short-term scenarios for supply, demand, and prices we expected oil prices could recover in 2021 or 2022 to pre-crisis levels of $50/bbl to $60/bbl. We think it will be two years before demand recovers, with the outlook for jet fuel particularly bleak. The immediate effects are already staggering companies must figure out how to deal with full storage, prices falling below cash costs for some operators. We think in the long run the macro-environment will become more challenging and we predict growth in demand for oil, to peak in the 2030s, and then begin a slow decline.

The challenge of the energy transition will continue. Saudi Arabia must begin a new age and change in its strategy, where the industry will be entering an era defined by intense competition, technology-led rapid supply response, flat to declining demand. The unprecedented crisis will be a catalytic moment and accelerate permanent shifts in the industry's ecosystem, with new future opportunities.

References

Ahmad, A., Alshurideh, M. T., Al Kurdi, B. H., & Salloum, S. A. (2021). Factors impacts organization digital transformation and organization decision making during Covid19 pandemic. In *Studies in systems, decision and control* (Vol. 334). https://doi.org/10.1007/978-3-030-67151-8_6

Al-Dmour, A., Al-Dmour, H., Al-Barghuthi, R., Al-Dmour, R., & Alshurideh, M. T. (2021). Factors influencing the adoption of E-payment during pandemic outbreak (COVID-19): Empirical evidence. In *Studies in systems, decision and control* (Vol. 334). https://doi.org/10.1007/978-3-030-67151-8_8

Al-Jarrah, I. M., Al-Zu'bi, Z. M. F., Jaara, O. O., & Alshurideh, M. (2012). Evaluating the impact of financial development on economic growth in Jordan. *International Research Journal of Finance and Economics, 94*, 123–139.

Alameeri, K. A., Alshurideh, M. T., & Al Kurdi, B. (2021). The effect of Covid-19 pandemic on business systems' innovation and entrepreneurship and how to cope with it: A theatrical view. In *Studies in systems, decision and control* (Vol. 334). https://doi.org/10.1007/978-3-030-67151-8_16

Aljumah, A., Nuseir, M. T., & Alshurideh, M. T. (2021). The impact of social media marketing communications on consumer response during the COVID-19: Does the brand equity of a university matter. *The effect of coronavirus disease (COVID-19) on business intelligence* (Vol. 334, pp. 384–367).

Aloui, D., Goutte, S., Guesmi, K., & Hchaichi, R. (2020). *COVID 19's impact on crude oil and natural gas S&P GS Indexes.*

Alshurideh, M., Al Kurdi, B., Shaltoni, A. M., & Ghuff, S. S. (2019a). Determinants of pro-environmental behaviour in the context of emerging economies. *International Journal of Sustainable Society, 11*(4). https://doi.org/10.1504/IJSSOC.2019.104563

Alshurideh, M., Alsharari, N., & Al Kurdi, B. (2019b). Supply chain integration and customer relationship management in the airline logistics. *Theoretical Economics Letters, 9*(02), 392–414.

Amadeo, K. (2021). *Oil price forecast 2022–2050 current and future crude oil prices.*

Assad, N. F., & Alshurideh, M. T. (2020). Financial reporting quality, audit quality, and investment efficiency: Evidence from GCC economies. *WAFFEN-UND Kostumkd. J, 11*(3), 194–208.

Barbosa, F., Bresciani, G., Graham, P., Nyquist, S., & Yanosek, K. (2020). Oil and gas after COVID-19: The day of reckoning or a new age of opportunity. *McKinsey & Company, May, 15,* 2020.

Energy, I. A. (2020). *Global oil demand to decline in 2020 as coronavirus weighs heavily on markets.*

Harahsheh, A. A., Houssien, A. M. A., & Alshurideh, M. T. (2021). The effect of transformational leadership on achieving effective decisions in the presence of psychological capital as an intermediate variable in private Jordanian. In *The effect of coronavirus disease (COVID-19) on business intelligence* (pp. 243–221). Springer Nature.

Ministry of Finance. (2020a). *State general budget statement, Saudi Arabia.*

Ministry of Finance. (2020b). *The program of achieving the Saudi financial balance.*

Musa, K. S., Maijama'a, R., Mohammed, N., & Yakubu, M. (2020). Covid-19 pandemic, oil price slump and food crisis nexus: An application of ARDL approach. *Open Access Library Journal, 7*(6), 1–19.

Pesaran, M. H., Shin, Y., & Smith, R. J. (2001). Bounds testing approaches to the analysis of level relationships. *Journal of Applied Econometrics (chichester, England), 16*(3), 289–326.

Primiceri, G. E. (2005). Time varying structural vector autoregressions and monetary policy. *The Review of Economic Studies, 72*(3), 821–852.

Shah, S. F., Alshurideh, M. T., Al-Dmour, A., & Al-Dmour, R. (2021). Understanding the influences of cognitive biases on financial decision making during normal and COVID-19 pandemic situation in the United Arab Emirates. In *Studies in systems, decision and control* (Vol. 334). https://doi.org/10.1007/978-3-030-67151-8_15

Sharma, R. (2020). *OPEC vs. the US: Who controls oil prices?*

Shin, Y., Yu, B., & Greenwood-Nimmo, M. (2014). Modelling asymmetric cointegration and dynamic multipliers in a nonlinear ARDL framework. In *Festschrift in honor of Peter Schmidt* (pp. 281–314). Springer.

Sims, C. A., Stephen, M. G., & Jeffrey, D. S. (1982). Policy analysis with econometrics models. *Brooking Papers on Economic Activity, 13*(1), 107–164. https://doi.org/10.2307/2534318

Soñnichsen, N. (2021). Daily demand for crude oil worldwide from 2006 to 2020, with a forecast until 2026. *Statista.* Retrieved September, 22, 2021.

Sönnichsen, N. (2020). Weekly Brent, OPEC Basket, and WTI crude oil prices from December 30, 2019 to October 12, 2020. *Statista,* October, 16.

Trading Economics. (n.d.).

Wingfield, B., Dodge, S., & Pogkas, D. (2020). New Decade, New OPEC Oil Curbs. Same Mixed Results.

Yilmazkuday, H. (2022). Coronavirus disease 2019 and the global economy. *Transport Policy, 120,* 40–46.

IoT Applications in Business and Marketing During the Coronavirus Pandemic

Muhammad Turki Alshurideh⬤, Barween Al Kurdi⬤, Salaheddin Saleh, Karim Massoud, and Abdullah Osama

Abstract The internet of things (IoT) is a widely used technology in day-to-day life and in many business fields. the applications of IoT are used in many businesses and in marketing as main part of business activities, but during the coronavirus pandemic, IoT was intensively used, as a practical solution to many challenges. In this paper, the use of IoT is discussed generally in business such as in supply chain and logistics, human resources management, and in organizational performance. The paper will also tackle the use of IoT in both marketing and business activities and finally discussing some of the limitations and challenges businesses face when implementing IoT.

1 Introduction

The coronavirus pandemic has necessitated changes in organizational management. Thus, the pandemic has revealed the aspects pertaining to the management of companies and its efficiency. Furthermore, its effect has echoed the need for greater engagement between businesses and their consumers. As a result of the pandemic, community changes have become the key factor necessitating provision of new organizational management models. One of the key areas that organizational management

M. T. Alshurideh (✉)
Department of Management, College of Business, University of Sharjah, 27272 Sharjah, United Arab Emirates
e-mail: malshurideh@sharjah.ac.ae; m.alshurideh@ju.edu.jo

Department of Marketing, School of Business, The University of Jordan, Amman 11942, Jordan

B. Al Kurdi
Department of Marketing, Faculty of Economics and Administrative Sciences, The Hashemite University, Zarqa, Jordan
e-mail: barween@hu.edu.jo

S. Saleh · K. Massoud · A. Osama
University of Sharjah, 27272 Sharjah, United Arab Emirates

© The Author(s), under exclusive license to Springer Nature Switzerland AG 2023
M. Alshurideh et al. (eds.), *The Effect of Information Technology on Business and Marketing Intelligence Systems*, Studies in Computational Intelligence 1056,
https://doi.org/10.1007/978-3-031-12382-5_140

has had a chance to integrate is using the internet of things (IoT) (Friedow et al., 2018a).

The Internet of Things (IoT) devices and applications are utilized in several fields (Nasajpour et al., Dec. 2020) as it offers a great chance for businesses to have an influential model for tackling and managing several aspects of business such as monitoring customers' needs, measuring business performance, marketing, and business processes; thus, making IoT the key tool in, defining management roles and potential improvement reaching down to the level of monitoring physical engagement of employees. Notably, the coronavirus pandemic has provided an insight into using IoT applications in the management of organizations.

2   What is IoT or IoE (Internet of Thing or Internet of Everything)

The Internet of Things (IoT), also known as the Internet of Everything or the Industrial Internet, is basically the interconnection of machines and devices through a network or internet that enables their communication (Lee & Lee, 2015). The communication happens through sensors by sending and receiving information without the direct involvement of a human being (Ranger, 2020). This communication uses Radio Frequency Identification (RFID), in which information systems will be invisibly embedded in the environment around us (Gubbi et al., 2013). The history of this machine-to-machine communication started in 1843 when a sensor was put on the clouds transmitting data every half-hour to predict the weather patterns (J.-F. (JF) Fava-Verde and S. Forster 2017). Another example used nowadays is a smartwatch. As it monitors the movement of the person wearing it, it automatically collects information such as steps walked, calories burnt, and even sleeping patterns, without the direct input by the person.

3   The Applications of IoT in Business

IoT has a significant impact on several aspects of life and behavior of individuals and on businesses. From the business perspective, using IoT has greatly impacted automation, business processes and logistic (Atzori et al., Oct. 2010). IoT devices and applications include robots, drones, smartphone applications, and wearables. They were significantly utilized during the coronavirus pandemic (Nasajpour et al., Dec. 2020) as they were used for data collection and generation. The application of IoT in the form of sensors in different locations of a smart building for example, are used for collecting and managing data from different parts of the smart building. The data produced from each sensor would be collected and saved through a cloud server. Then, it will be managed through a remote device such as a mobile, functioning through a mesh of network by IoT technology. The purpose of collecting these data in smart building, is to promote energy-friendly smart buildings which would also be called Green Smart Buildings (Plageras et al., May 2018). In addition, data collection was extensively used during the coronavirus pandemic. Drones were to collect data for crowd monitoring and smartphone applications were used to capture data of people of close contact such as the duration of contact (Nasajpour et al., Dec. 2020).

Notably, during the use of IoT technology, it was seen to reduce maintenance costs. For instance, a study was done in a hospital in the UAE that presented a Predictive Maintenance (PdM) approach in which it works to monitor the equipment, predict, and classify healthy vs faulty parts thus enhance their performance. Moreover, they have proved how IoT technology is important in the economic sector as their approach caused up to 25% repair cost savings and functioned as an investment payback period of a year. In addition, deploying IoT along with other technologies was found to have a chance in providing the e-health system at high availability, accessibility, performance, with low maintenance cost (Shamayleh et al., 2020).

3.1 IoT in Supply Chain

Businesses use IoT to help managing supply chain and saving on inventory management activities. The application of IoT in companies helps create relevance in dealing with the supply and demands. Hence, the use of IoT in companies pertains to the designed operations and functionality elements (Kumar et al., 2020). In this case, IoT helps companies to maintain costs through tracking the use of smart devices. The application of smart solutions for printers, is another example of operating cost reduction (Nasajpour et al., Dec. 2020). In addition, in its application in the supply chain, IoT supports smart inventory management by collecting data and helps companies use right steps in assigning suppliers, retailers, and even managing employee activities. Inventory management of companies using IoT is less tedious to use, as it assigns roles for every party and increases the chances of attaining higher efficiency level. Similarly, the inventory management role of the IoT simplifies conducting logistic modeling for the companies (Nasajpour et al., Dec. 2020). Moreover, the use of IoT marks a chance to align and integrate the community's functional segments to specific organizational objectives. Hence, IoT enables companies have better management of their supply chain framework, logistics, and job assignment platforms. Therefore, integration of these functional areas marks the chance to use the IoT in managing organizational functions and increasing effectiveness of company operations (Nasajpour et al., Dec. 2020). It also offers, for example, a chance for companies to use smart tagging solutions to create a relevant model of tracking location of their products. IoT, in this sense, brings in an easier alternative for business to respond to the needs of the consumers when delivering goods. Nonetheless, it also increases the company's chances for saving on costs and marking the channel to grow. These advantages increase the demand for IOT in companies, particularly, in the times of pandemic. Hence, IoT contributes to ensuring efficiency of management and identification of proper avenues to target the organizational growth (Kumar et al., 2020).

3.2 Organizational Performance

IoT application will increase the chances and capacity to attain better performance for companies. The use of IoT in companies provides market analysis and information from consumers. Thus, IoT real-time data will help companies gain an increased role over the consumer demands. Therefore, the IoT will help business institutions to pinpoint the best activities and implement relevant changes to cater for their customers' needs. Similarly, IoT application will appease the required level of addressing the needs of the consumers and target all parties' demands. Likewise, the use of IoT in conducting logistic management and customer service will increase efficiency of company operations (Friedow et al., 2018b). Thus, IoT will ensure an increased capacity and ability of the company to target the consumers' demands as well as to structure the company operations to attend them. Managing the logistics will bring out a high level of engagement with the consumers. Moreover, the use of the customer service model will ensure the company has an adequate level of connection with the consumers as well as can focus on appreciating demands of the stakeholders within the community (Nasajpour et al., 2020).

Another application of IoT is truck telematics, where it was utilized to better understand, and improve driving behaviors. Remote control centers monitor live sensor data from the company's fleet of vehicles, capturing speed, location, braking, and engine data. The gathered data is used to enhance future training programs. A combination of truck telematics and geo-information are being used to enable proactive alerts to be sent to drivers regarding possible upcoming hazards. Camera-based technologies have been adopted to improve driver safety, and fatigue management, capturing evidence of important driving events, improve truck routing, recommend optimal fuel purchasing times/locations, and forecasting predictive and proactive maintenance schedules. Improvements in utilization and routing have the potential to reduce traffic congestion, which is responsible for losses in productivity, increases in fuel consumption, air pollution and noise, and can incite stress, aggression, anger, and unsafe behaviors in drivers. Predictive analytics, which generate refueling and maintenance schedules, have the potential to be adopted by all vehicle manufacturers, and could generate reductions in customer fuel costs, whilst improving the performance, efficiency, and life expectancy of future motor all vehicles (Hopkins & Hawking, 2018) (Jianli, 2012). Furthermore, automated tracking integrates the communication of IoT devices with blockchain, which play a role in monitoring products and automating the tracking and clearance processes (Augusto et al., 2019).

3.3 IoT and Customer Service Management

Similarly, IoT can be applied in companies when executing customer service. The use of customer service bots in companies reduces the need for hiring physical experts to handle the customers. Nonetheless, this application also relates to the chance and capacity to integrate 24–7-h service for customers. In its turn, the use of the IoT bots to instruct a better customer service model encourages identifying and highlighting the steps to handle through the IoT advancement within the company. In effect, the IoT customer service bots ensure efficiency of the customers' engagement and proper organizational connection with the market. Hence, the use of IoT in the customer service segment boosts company image (Friedow et al., 2018b).

3.4 IoT and Human Resources Management

IoT helps companies save on the employment costs by eliminating the need to hire more employees or performing the roles that were carried by individual employees. For example, bigger companies employ baristas in their canteens and cleaners. The application of IoT technology provides smart solutions to these roles. Thus, cleaning activities can be replaced with smart vacuums that would activate as pre-programmed. Consequently, IoT will reduce the costs of paying the employees to perform cleaning services and increase the efficiency of cleaning services (Friedow et al., 2018b). In their turn, smart coffee-machines will help increase work efficiency of the employees and save servicing time as they may offer pre-programmed solutions to every employee's drink preference. Smart coffee makers are, therefore, a beneficial solution to increase the level of productivity among employees. Thus, the use of IoT will ensure a better level of managing the company's activities (Friedow et al., 2018b).

3.5 Environmental and Energy Conservation

IoT helps companies to ensure environmental conservation through temperature control within the premises. The use of IoT smart technology helps to set the optimal temperatures for the office thermostats and can be used in creating office heat map. Therefore, IoT helps in reducing energy costs in companies through introduction of new effective ways of handling environmental and energy needs. Another example is the use of intelligence lighting as it contributes to energy conservation. As intelligent lighting is programmed to light only at specific hours, reduces the maintenance costs that the company must conduct on its energy sources (Hussain, 2017).

3.6 Data Security Management

IoT maintains the data management function of companies in various models, such as data learning and data integration. It also provides smarter security options for configuring appropriate steps and channels to handle security breaches in companies. The use of IoT is instrumental in identifying the core levels of handling the security needs of databases and server rooms as well as detecting real-time threats and mitigating them before causing harm to anyone. Integration of the smart security system on a company platform ensures increased relevance and potential to engage the company's best segments in safety management (Kumar et al., 2020). Therefore, the IoT security management system is a key factor in ensuring safety in technological networks and platforms within the companies. To this end, the IoT helps in appealing to a company's core segments by providing security and handling information provided to all company members. Real-time data from IoT helps define the channels of threat, deploy solutions to mitigate the threat, achieve best outcomes and manage provision of instructions to the company employees.(Nasajpour et al., Dec. 2020).

3.7 Organizational Management and Decision Making

IoT assists in increasing the organizational capacity to make informed decisions. The IoT plays a key role in designation of values and provision of decisions to meet the company's demands. In using IoT, companies look into data and have new insights into carrying out their activities (Nasajpour et al., Dec. 2020). The role designates an increased chance to affiliate and works with the elements of valuing community changes. IoT is considered a major factor in informing the companies on better ways to carry out their activities, as it emphasizes the weak points of the companies and defines the ways they can achieve efficiency (Friedow et al., 2018b). To enumerate, IoT looks into industry trends and ways that the company's activities suit the trends (Kumar et al., 2020). In cases, where the application of IoT does not lead to better handling of the trends, the institutions can challenge their models and decide on better ways that would elevate their performance. Besides, the IoT helps select the best practices within companies and analyze at the statistics of dealing with the organizational needs to achieve the optimal outcome in every operation carried out (Hussain, 2017). Hence, IoT helps companies to attend to core demands and requirements through improved management model.

In addition, IoT helps managers to establish the best channel of working towards addressing pertinent issues in times of pandemic. By obtaining information from the market segments, the customer behavior models, and industry trends, managers would be directed towards data-driven decision -making. Notably, real-time information will also increase efficiency of the managers. For instance, the use of realistic levels of attending to employee demands in the pandemic and instructing remote working capacities will increase the overall company performance (Friedow et al., 2018b). Thus, regarding the employees work from home, the company's management will be able to efficiently designate duties and ensure the employees can meet the set deadlines. Equally important, IoT can help company management to address the pertinent trends and issues that offer value to their activities. In this sense, a company will relate to industry trends and begin working, for example, on e-commerce platforms. Most companies applied e-commerce as an avenue to maintain contact with their customers during the pandemic. Restriction of movement and reduced working hours has rendered e-commerce platform more appealing to most customers. Thus, IoT application and the use of insights from the industry helped to ensure an effective and impactful integration of activities (Friedow et al., 2018b).

3.8　Workplace Sanitation

The coronavirus pandemic has indicated the increased need for cleanliness and sanitation within workplaces and public environments. The use of IoT was able to assist in the automation of these functions to ensure exclusion of the employees' services, thus ensure health security of the latter. Automation of the cleaning and hygiene functions increased the role of IoT and the structural engagement of companies to confer with the community's demands and the pandemic sanitation requirements (Kumar et al., 2020). To demonstrate, IoT automation can be used to operate workplace fumigation and sanitizer dispensers. These machines will fumigate the workplace to ensure the place is free from bacteria and viruses. Increased application of the automated service will ensure dispensation of liquid sanitizers to employers, when they get in and out of the workplace, as the machine will be measuring their temperature. Application of IoT will create a sufficient level of ease of use and relevance for the employees. Thus, the use of IoT in companies will ensure compliance with health and sanitation regulations and reduce the risk of employees becoming sick (Kumar et al., 2020).

4　IoT Applications in Marketing and Advertising

There are many applications of marking using IoT. Some examples of these applications are contextual advertising: With the wide spread of smartphones, targeted advertising using IoT applications could be easily done by sends pop-up ads to the customers and informing them with promotions or offers. Another application is in-store advertising which is done by using an electronic device called "beacon" to send ads and notifications to customers' smartphones when they are in or nearby the store. In addition, location-based advertisement by using GPS on smartphones, ads or activities within a certain range could be sent to customers. Another commonly used application is targeting customers by collecting data about a customer's preference or behavior then sending smart ads based on the collected data every time the customer goes online. IoT is also used in interactive marketing with high potential of opportunities, such as using interactive smart mirrors that enable customers to try on clothes only by looking at the mirror and without wearing them. Another example is using QR codes to open a menu or use them for a discount (Hashem, 2021).

4.1 Customer Behavior Analysis

Another area of IoT application in businesses is ensuring proper level of attending to customer needs. The use of IoT attracts insights into customer behavior and changes within the customer segment (Hussain, 2017). In this sense, IoT facilitates identification and definition of optimal steps to handle customers' requirements. The application of IoT provides an opportunity to integrate company values and mitigate the challenges related to satisfying customer needs across all segments. Thus, the use of IoT ensures an increasing application of value in a company's operations. IoT management of consumer data is a key factor in depicting organizational actions and activities in the market segment. IoT establishes a sustainable channel of advancing the needs of the company by defining the best activities to conduct at any given moment.(Kumar et al., 2020).

4.2 IoT and Customer Relationship Management (CRM)

Data-driven marketing is a success factor in customer relationship management (CRM), thus utilizing all sorts of data collected from different sources including ones from IoT, help in predicting customers' needs and behavior, design marketing plans that fit predictions, personalizing marketing campaigns and advertisements, and increase the accuracy of decision-making process and data driven marketing strategies. The relationship between IoT and CRM could be summarized by: IoT provides decision making capabilities, data collection and exchange, marketing management, and by providing customers with rich experiences and high level of personalized interaction. These factors are connected directly to better CRM leading to better marketing management. (Hashem, 2021).

4.3 IoT Applications in Customer Service and Support

Manufacturing companies operate in an environment characterized as increasingly volatile, uncertain, complex, and ambiguous. At the same time, customers' orientation makes it increasingly important to ensure high delivery reliability. A concept to connect supply network and manufacturing more closely by integrating events from the supply network into manufacturing control's decisions. This approach enables an effective and efficient response to events from the supply network through smarter manufacturing control, and thus more resilient manufacturing (Bauer & Bauernhansl, 2021).

4.4 *IoT and Product/Service Support*

IoT can be helpful in the products support when it comes to supply and demand management. In the supply management, operations at the manufacturing companies can be a bit of a mess if not handled properly. The more complex the operation, the more attention and monitoring it requires. IoT comes in hand in introducing concepts that would connect supply network with manufacturing decisions (Bauer et al., Mar. 2021).

5 Limitations and Challenges

The use of IoT has several limitations and challenges that needs to be taken into consideration: Privacy and secrecy concerns: Since IoT is all about gathering all sorts of data which could be accessed by hackers, thus could be causing security threats and breaches depending on the type of data and information collected. This requires the adoption of technology that is designed to prevent information leakage and the implementation of network guards and firewalls ((Hashem, 2021; Razzak, 2012)). Data interpretation: It is equally important to note that the companies should consider the bulk of data generated and to design proper processes to handle it terms of processing, storage, and interpreting it in a useful manner such as in the field of marketing (Hashem, 2021).Cyber threats: due to the interconnection of IoT devices, the network could be facing a security attack by malware or worm.

Unsolicited intrusions: Leaked data could take place while transferring it through IoT, thus security protocols should be adopted to protect private. Loss of jobs: Since IoT is considered an efficient and effective automation process, it could replace many jobs due to its capabilities and accuracy (Razzak, 2012).

6 Conclusion

To conclude, IoT technology has a powerful impact on day-to-day life and on businesses in many aspects (Ghazal et al., 2021a)-(Lee et al., 2022). As seen in this paper, the applications of IoT varies from minor daily tasks to major strategic decisions and from simple process automation to transformative innovations which were all evident during the coronavirus pandemic. Since the adoption of this technology along with other technologies is inevitable, businesses are required to benefit from its applications in achieving goals and competitive advantage (Al Kurdi et al., 2020)-(Tariq et al., 2022). Therefore, setting the right implementation methodology is a must especially with consideration to challenges such as privacy and social impact.

References

Al Kurdi, B., Alshurideh, M., & Salloum, S. A. (2020). Investigating a theoretical framework for e-learning technology acceptance. *International Journal of Electrical and Computer Engineering (IJECE), 10*(6), 6484–6496.

Alshurideh, M. (2022). Does electronic customer relationship management (E-CRM) affect service quality at private hospitals in Jordan? *Uncertain Supply Chain Management, 10*(2), 325–332.

Alshurideh, M. T., Al Kurdi, B., & Salloum, S. A. (2021). The moderation effect of gender on accepting electronic payment technology: A study on United Arab Emirates consumers. *Review of International Business and Strategy, 31*(3), 375–396.

Alshurideh, M., Al Kurdi, B., & Salloum, S. A. (2019, October). Examining the main mobile learning system drivers' effects: A mix empirical examination of both the Expectation-Confirmation Model (ECM) and the technology acceptance Model (TAM). In *International Conference on Advanced Intelligent Systems and Informatics* (pp. 406–417). Springer, Cham.

Alzoubi, H., Alshurideh, M., Kurdi, B., Akour, I., & Aziz, R. (2022). Does BLE technology contribute towards improving marketing strategies, customers' satisfaction and loyalty? The role of open innovation. *International Journal of Data and Network Science, 6*(2), 449–460.

Atzori, L., Iera, A., & Morabito, G. (2010, October). The internet of things: a survey, *Computer Networks, 54*(15), pp. 2787–2805. https://doi.org/10.1016/j.comnet.2010.05.010.

Augusto, L., Costa, R., Ferreira, J., & Jardim-Gonçalves, R. (2019). An application of Ethereum smart contracts and IoT to logistics, *IEEE*.

Bauer, D., & Bauernhansl T. (2021) Applied sciences Improvement of delivery reliability by an intelligent control loop between supply network and manufacturing.

Bauer, D., Bauernhansl, T., & Sauer, A. (2021, March) Improvement of delivery reliability by an intelligent control loop between supply network and manufacturing, *Applied Sciences (Switzerland), 11*(5), pp. 1–22, https://doi.org/10.3390/app11052205.

Friedow, C., Völker, M., & Hewelt, M. (2018a). Integrating IoT devices into business processes. In: *International Conference on Advanced Information Systems Engineering*, pp. 265–277.

Friedow, C., Völker, M., & Hewelt, M. (2018b). Integrating IoT devices into business processes chimera: Fragment-based case management approach view project process information and guidance systems view project integrating IoT devices into business processes, https://www.bosch-iot-suite.com/things/

Ghazal, T. M., Alshurideh, M. T., & Alzoubi, H. M. (2021a, June). Blockchain-Enabled internet of things (IoT) platforms for pharmaceutical and biomedical research. In *The International Conference on Artificial Intelligence and Computer Vision* (pp. 589–600). Springer, Cham.

Ghazal, T. M., Hasan, M. K., Alshurideh, M. T., Alzoubi, H. M., Ahmad, M., Akbar, S. S., ... & Akour, I. A. (2021b). IoT for smart cities: Machine learning approaches in smart healthcare—A review. *Future Internet, 13*(8), 218, 1–19.

Gubbi, J., Buyya, R., Marusic, S., & Palaniswami, M. (2013). Internet of Things (IoT): A vision, architectural elements, and future directions. *Future Generation Computer Systems, 29*(7), 1645–1660. https://doi.org/10.1016/j.future.2013.01.010

Hasan, O., McColl, J., Pfefferkorn, T., Hamadneh, S., Alshurideh, M., & Kurdi, B. (2022). Consumer attitudes towards the use of autonomous vehicles: Evidence from United Kingdom taxi services. *International Journal of Data and Network Science, 6*(2), 537–550.

Hashem, T. (2021). The reality of internet of things (IoT) in creating a data-driven marketing opportunity: mediating role of customer relationship management (CRM), *Journal of Theoretical and Applied Information, 31*(2), https://www.researchgate.net/publication/348922750

Hopkins, J., & Hawking, P. (2018). Big data analytics and IoT in logistics: a case study, *The International Journal of Logistics Management, 29*(2). https://doi.org/10.1108/IJLM-05-2017-0109.

Hussain, F. (2017). *Internet of Things Building Blocks and Business Models*. Springer.

Fava-Verde, J.-F. (JF)., & Forster, S. (2017). The history of Internet of Things (IoT), *Innovate UK Blog*.

Jianli, S. (2012). Design and implementation of IOT-based logistics management system, *IEEE*.

Kamaruddeen, A., Rui, L., Lee, SK., Alzoubi, H., Alshurideh, M. (2022) Determinants of emerging technology adoption for safety among construction businesses. *Academy of Strategic Management Journal, 21*(Special Issue 4), 1–20.

Kumar, S., Raut, R. D., & Narkhede, B. E. (2020, October). A proposed collaborative framework by using artificial intelligence-internet of things (AI-IoT) in COVID-19 pandemic situation for healthcare workers, *International Journal of Healthcare Management*, Taylor and Francis Ltd., *13*(4). pp. 337–345, https://doi.org/10.1080/20479700.2020.1810453.

Lee, K., Ramiz, P., Hanaysha, J., Alzoubi, H., & Alshurideh, M. (2022). Investigating the impact of benefits and challenges of IOT adoption on supply chain performance and organizational performance: An empirical study in Malaysia. *Uncertain Supply Chain Management, 10*(2), 1–14.

Lee, I., & Lee, K. (2015). The internet of things (IoT): Applications, investments, and challenges for enterprises. *Business Horizons, 58*(4), pp. 431–440. https://doi.org/10.1016/j.bushor.2015.03.008.

Nasajpour, M., Pouriyeh, S., Parizi, R. M., Dorodchi, M., Valero, M., & Arabnia H. R. (2020, December). Internet of things for current COVID-19 and future pandemics: an exploratory study, *Journal of Healthcare Informatics Research, 4*(4), pp. 325–364. https://doi.org/10.1007/s41666-020-00080-6.

Plageras, A. P., Psannis, K. E., Stergiou, C., Wang, H., & Gupta, B. B. (2018, May). Efficient IoT-based sensor BIG data collection–processing and analysis in smart buildings, *Future Generation Computer Systems, 82*, pp. 349–357, https://doi.org/10.1016/j.future.2017.09.082.

Ranger, S. (2020). What is the IoT? Everything you need to know about the Internet of Things right now, *ZDNet*.

Razzak, F. (2012). Spamming the Internet of Things: A possibility and its probable solution. *Procedia Computer Science, 10*, 658–665. https://doi.org/10.1016/j.procs.2012.06.084

Shamayleh, A., Awad, M., & Farhat, J. (2020, April). IoT Based Predictive Maintenance Management of Medical Equipment, *Journal of Medical Systems, 44*(4), https://doi.org/10.1007/s10916-020-1534-8.

Tariq, E., Alshurideh, M., Akour, I., & Al-Hawary, S. (2022). The effect of digital marketing capabilities on organizational ambidexterity of the information technology sector. *International Journal of Data and Network Science, 6*(2), 401–408.

Impact of Workplace Happiness on the Employee Creativity: A Systematic Review

Shamsa Alketbi and Muhammad Alshurideh

Abstract This study examines the antecedents of workplace happiness and how the later impact the employees' performances and their levels of creativity. Based on our literature review, researcher identified four main potential drivers of workplace happiness, namely, leadership style, work environment, compensation and opportunities for learning and personal development.

Keywords Workplace happiness · Work environment · Leadership · Compensation · Opportunities for learning and personal development · Employees' performance · Employees' creativity

1 Introduction

Systematic reviews are types of literature reviews that collect and critically analyze multiple research studies or papers, using methods that are selected before one or me research questions are formulated, and then finding and analyzing studies that relate to and answer those questions in a structured methodology.

S. Alketbi (✉)
University of Sharjah, 27272 Sharjah, United Arab Emirates
e-mail: u19106217@sharjah.ac.ae

M. Alshurideh
Department of Management, College of Business, University of Sharjah, 27272 Sharjah, United Arab Emirates
e-mail: malshurideh@sharjah.ac.ae

© The Author(s), under exclusive license to Springer Nature Switzerland AG 2023
M. Alshurideh et al. (eds.), *The Effect of Information Technology on Business and Marketing Intelligence Systems*, Studies in Computational Intelligence 1056,
https://doi.org/10.1007/978-3-031-12382-5_141

2   Problem Statement

Happiness is very important nowadays, especially in United Arab Emirates, as we know UAE is a rapid and growth country, which contains a various number of nationalities. Who came from different cultures, so we need to keep in mind the importance of keeping those nationalities happy with us in order to give us a wide range of creative ideas as well as perform in best way? The competition in the business environment has been increasing over time. At present, organisations have a large number of rivals, in every market and industry (Chandrasekar, 2011). This rivalry and competition have made it impossible for organisations to afford to waste the potential of their human resources and workforce. Several factors come together to determine the workplace environment suitability for employees. For instance, organisations, in the modern era, are promoting new programs for employees which address issues such as health, fitness, work-life balance, lifestyle changes, and other benefits so that the employees perceive their work environment as favourable (Chandrasekar, 2011).

Even some organisations have introduced policies such as performance-based salaries, motivation programmers, and employment security agreements. The main reason behind such practices is that organisations, practitioners, and experts have realised that workplace happiness has an important role in determining the performance and success of the organisation. Workplace happiness can ensure motivation among employees, which can, in turn, further promote the necessary skills and ability to perform their jobs effectively.

Being happy holds significant importance to many people. More importantly, most of the societies around the world, value happiness to a great extent-it causes it to become the basis of the human emotional framework. Fisher (2010) reported that feeling happy is a fundamental experience for human beings and people must remain happy to be emotionally and psychologically well. Fisher (2010) defined happiness as a person's satisfaction with their relationships, work, leisure, and health. It also defined as the set of positive feelings which arise due to the positive experiences in life (Ashkanasy & Dorris, 2017). Based on this definition of happiness, the same feeling at the workplace can be viewed to have constructed of job satisfaction and motivation level of an employee (Fisher, 2010). According to Fisher (2010), there are a large number of happiness-related constructs which affect the creativity and performance of an organisation. These include job solicitation, a typical mood at work, dispositional effect, job involvement, and personal career development. Oswald et al. (2015) present evidence from their findings to support the hypothesis that happiness at workplace is positively related with the performance of an organization and employee performance (Oswald et al., 2015).

There are several factors which can affect the happiness, performance, and creativity of personnel in an organisation. Career development, workload, organisational culture, salary, monetary incentives, and infrastructure, are some of the most important indicators of workplace happiness (Mazuki et al., 2017).

Career development is one of the functions of HRM of any organisation. It focuses on issues such as upward mobility of employees and promotions in the hierarchy of an organisation. According to Mazurka et al. (2017), the major concern of career development is to ensure that personnel performance increased over the period. In addition to this, career development also aims to promote the overall growth of employees and their skill acquisition. Organisations that promote career development opportunities for their employees have better workplace happiness than the organisations that do not have any programs for career development. The workload is also an important factor which affects the workplace happiness of organisations. Organisations, where people have an adequate workload, have better culture. More importantly, employees' satisfaction with their jobs needs to ensure before the personnel can perform at their maximum potential (Liu et al., 2016).

Better infrastructure has also been found to be associated with better workplace happiness. Several large organisations have worked to create new designs of office buildings, equipment, and facilities which lead to better creativity, happiness, and performance among employees (Mazuki et al., 2017). The ultimate goal of such efforts is to maximise the creativity and performance of employees by promoting workplace happiness (Chandrasekar, 2011).

The culture of an organisation is also important to consider when it comes down to workplace happiness. For instance, the work ecosystem or organisational culture refers to the relationships and values inside an organisation or workplace (Mazuki et al., 2017). Better interpersonal relationships at workplaces are associated with better happiness at work. It has been often found to lead to better performance and creativity among employees as open communication is promoted (Sageer et al., 2012).

Monetary incentives and raises in salaries are also an important factor which determines workplace happiness. A reasonable and high salary can ensure that employees feel more connected to their workplace (Mazuki et al., 2017). Further, it has been found that employees, receiving high salaries, are more productive and creative because they feel motivated. In addition to this, prizes and incentives can also promote workplace happiness and performance among employees.

Leadership style at the workplace is also important in determining the performance and creativity of employees. Yang (2014) reported that leadership style shapes the culture of an organisation, which further influences the workplace happiness. Ethical leadership found to associated with a better focus on the well-being of employees. Any culture which gives adequate attention to the well-being of employees is more likely to have a better index of workplace happiness since the employees feel a sense of connection with their workplace (Yang, 2014). An adequate leadership style in an organisation can ensure that proper team-building skills promoted among the employees. It has a positive influence on workplace happiness, which further positively affects the performance and creativity of employees. Overall, the performance of the organisation begins to rise (Ashkanasy & Dorris, 2017).

Since workplace happiness connected with the level of satisfaction that a person has about their work environment, leadership and organisational culture directly tied to their performance and creativity (Joo & Lee, 2017). Even more important is the fact that the creativity and performance of employees are factors which govern the performance of an organisation. Happy employees are highly engaged in their work which heightens their performance as well as creativity (Chandrasekar, 2011).

Several research studies have found various methods which can be implemented at workplaces to maximise the potential of the workforce. For instance, goal-setting is a process which can be enforced at the workplace, so the meaning given to the work that an employee does. In this way, the employee is assigned a supervisor and becomes a part of the management processing maximises their job satisfaction since performance measures (Chandrasekar, 2011). Similar to the goal-setting, performance feedback can be enforced at workplaces so that the upper management and leadership know the problems and issues of the employees.

Additionally, the employees feel a sense of belonging to their organisation when the management takes relevant steps regarding their problems (Chandrasekar, 2011). Furthermore; the culture of an organisation can have a significant impact on how employees interact with each other and with the top management. Therefore, it can be said that that leadership style and management of an organisation could determine the workplace happiness of an organisation. This level of happiness, further, determines the performance and level of creativity among employees. In the long run, all of this process affects the performance of an organisation.

Large companies have been becoming more and more concerned about their employees' happiness and well-being. Proto (2016) has found that the workplace environment can be very unhealthy and unsafe for workers in several industries. Unsafe unhealthy, the workplace environment causes the workers to become unhappy regarding their jobs if the problems not dealt with, accordingly (Proto, 2016). It has also found that the majority of offices and workplaces have unsuitable furniture, excessive noise, poorly designed workstations, lack of proper ventilation, and lack of security agreements for employees.

In addition to this, there are many other factors which can affect the workplace happiness among employees, such as an increase in pays or wages, growth in career, and personal development opportunities (Chandrasekar, 2011). In several organisations, the problems mentioned above not adequately addressed which renders the employees unhappy. With the employees growing unhappy, the performance and motivation level drops considerably, as has been noted by Proto (2016). It becomes highly challenging for organisations to keep up their performance or to improve their performance. The overall effect of such problem results in lower performance and creativity among employees. The performance of any organisation is related to the performance and creativity of employees (Proto, 2016).

It has further noted that industry giants such as Google have better performance because their employees are happier than other organisations (Han, 2015). Therefore, the happiness of employees becomes a major problem to be dealt with. Only by promoting happiness at the workplace can an organisation provide the right environment for performance and creativity among employees. Proto (2016) had also found that psychological well-being was of primary concern when happiness at the workplace discussed. For instance, it has found that the motivation level of an employee is directly related to their psychological well-being at their workplace (Han, 2015).

In any organisation, the performance of employees is reliant on workplace happiness. It is defined as the level of satisfaction that the workers have, regarding their work environment (Mazuki et al., 2017). In addition to this, a happy workforce has more motivation to get a job done as compared to the unhappy staff. Therefore, Mazurka et al. (2017) found that unhappy staffs are usually unable to execute their operations productively as compared to a happy workforce. To maintain sustainability performance offices and workplaces, organisations need to maintain long-term happiness for their employees, as well. Happiness is another factor, in terms of which, the performance of an organisation can be measured, apart from monetary units (Chandrasekar, 2011). Therefore, this study will attempt to investigate the antecedents and consequences/outcomes of workplace happiness.

3 Research Purpose

As has been mentioned before, workplace happiness is something that needs to be addressed in all organisations to ensure performance and success. The concept, at present, is relatively newer than other concepts of management and HRM (Mazuki et al., 2017). Therefore, the specialists of the subjects are adding knowledge to the literature, regarding happiness at work. This concept has to deal with the psychological aspects of human nature, the leadership style of the organisation, the culture of an organisation and the policies of an organisation. Therefore, workplace happiness is a vast topic to discuss (Oswald et al., 2015). The emotional dimension is one of the most important aspects of workplace happiness. By studying the workplace happiness, the various factors can be uncovered which are into play when the performance of an organisation is to be determined.

More importantly, the impact of workplace happiness on employee performance and employee creativity can also be found out. The major reason to study employee performance and creativity is that both of these are key determinants of organisations performances. Unless the performance and creativity among employees of an organisation maximised, it is not possible to achieve the desired results or aims of an organisation. It is also important to mention here that the research study will aid in revealing the role of leadership in promoting happiness at workplace Yang (2014), noted that ethical leadership at workplaces could help in maximising the performance of workers by promoting workplace happiness. Not only workplace happiness is beneficial for the organisations in achieving the organisational goals, but the employees exposed to career development and skills acquisition opportunities (Yang, 2014).

The overall effect of such culture is that the employees have a greater sense of motivation and feel happier about working in their workplace. Enables the employees to learn a wider range of skills and helps them in improving their career. As for the organisation, the benefit is simple and straight-forward, as mentioned before—growth in performance and performance (Chandrasekar, 2011). This importance of workplace happiness makes it a very valuable concept to be studied and investigated. Furthermore, it becomes important to explore the various underlying factors which can either promote or demote happiness among employees, at the workplace. By understanding the reasons and factors affecting workplace happiness, the performance and creativity of employees can improve.

4   Research Questions

The objective of this study is to examine the main antecedents of work happiness and how the later may impact employees' performance and creativity in context.

RQ1: What are the main determinants of workplace happiness?

RQ2: What is the impact of workplace happiness on employees' performance?

RQ3: What is the effect of employees' performance on employees' creativity?

RQ4: What are the main research methods and research outcomes addressed in the collected studies?

RQ5: What is the main effect of payment criteria in employee happiness?

RQ6: What is the effect of leadership style in employee happiness and creativity?

RQ7: What are the results of employees' performance in creativity?

RQ8: what is the effect of opportunity of learning and personal development in happiness and creativity?

5   Research Hypotheses

Based on our literature review, we propose to test the following:

H1: Leadership style influences positively employees' workplace happiness.

H2: Work environment influences positively employees' workplace happiness.

H3: Flexible pay influences positively employees' workplace happiness.

H4: Opportunities for learning and personal development influences positively employees' workplace happiness.

H5: Employees' workplace happiness influences positively employees' performance.

H6: Employees' workplace happiness influences positively employees' level of creativity.

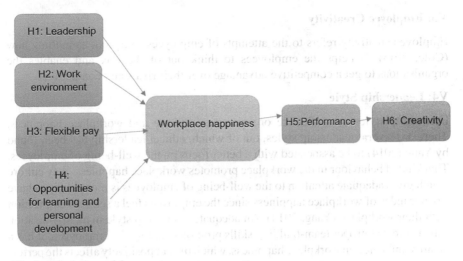

Fig. 1 The study model

6 The Study Proposed Model

See Fig. 1.

7 The Study Terms Definitions

In the following, we define the variables included in the conceptual model.

V1: Workplace Happiness

Workplace happiness is how satisfied and excited the employees feel about their workplace and job. It can make sure that motivation among employees promoted, which can, in turn, further increase the necessary skills and ability to perform their jobs effectively. It has also been found to be associated with the level of satisfaction that a person has about their work environment.

V2: Employee Performance

Employee performance refers to the efficiency and efficacy of a worker to carry out the job or task assigned to them (Bratnicka, 2015). This variable enables an organisation to assess their workers' output in terms of the task they did and the time duration they took to complete it.

V3: Employee Creativity

Employee creativity refers to the attempts of employees to try out something new (Cole, 2015). It helps the employees to think out of the box and enables the organisations to get a competitive advantage over their rivals and competitors.

V4: Leadership Style

Leadership styles can influence organisational culture and workplace happiness. There are several leadership styles, out of which, ethical leadership has been found by Yang (2014) to be associated with a better focus on the well-being of employees. This kind of behaviour in the workplace promotes workplace happiness. Any culture which gives adequate attention to the well-being of employees is more likely to have a better index of workplace happiness since the employees feel a sense of connection with their workplace (Yang, 2014). An adequate leadership style in an organisation can ensure that proper team-building skills promoted among the employees. It has a positive influence on workplace happiness, which further positively affects the performance and creativity of employees. Overall, the performance of the organisation begins to rise if the leadership resonates with the expectations of employees.

V5: Work Environment

Work environment refers to the surroundings of employees while at work. It has a direct influence on employees' performance and their involvement with work (Chancellor, 2015).

V6: Flexible Pay

Pay and financial benefits are important factors when it comes to employee performance and creativity. Employees who receive better pay and benefits are more motivated to work effectively (Davis, 2015).

V7: Opportunities for learning and personal development

Opportunities for learning and personal development refer to special training workshops and sessions that focus on improving the skills and performance of employees. Through such opportunities, an organization can help handle change and promote productivity (Bratnicka, 2015).

8 Literature Review

It is a general concept that happy employees tend to be productive employees. The happiness of the employees at their workplace exerts not only a positive influence on the professional but also the personal lives of the employees. The happiness of the employees at the workplace depends on several factors like the environment of the workplace, leadership at the workplace, availability of the career development opportunities, characteristics and nature of the job and the compatibility of the person with the environment of the workplace (Joo, 2016).

8.1 Workplace Happiness

Boehm and Lyubomirsky (2008) preliminarily define a happy person as someone who frequently experiences positive emotions like joy, satisfaction, contentment, enthusiasm and interest. Fisher (2010) suggested that happiness can be explained in several ways. Happiness constitutes the pleasant feelings and sensations. Extensive research work has done on the antecedents of happiness at the workplace and the consequences of happiness at the workplace. The definition of happiness at the work-place extends far beyond the mere job- satisfaction. In the modern age of technology and business, there is high competition and rivalry in the business environment. Every industry is, at the moment, facing a large number of rivals. The current situation of competition has made it impossible for the organisations to afford to waste any useful potential of their human resources and workforce. Therefore, the need for maximising performance and creativity arises. Several factors come together to determine the workplace environment suitability for employees. All of these factors can have significant on the performance of any organisation.

Many organisations, nowadays, are promoting new policies for employees which address issues such as health, fitness, work-life balance, lifestyle changes, and other benefits so that the employees perceive their work environment as favourable. All of these efforts are to promote happiness among the workforce.

Based on the above discussion, this research study will attempt to map the patterns of happiness in various organisations across the world. Also, it will try to observe the case of the UAE, so that it can determine the behaviour of employees and their happiness, at their workplaces. More importantly, the study will observe the impact of happiness on employee performance and employee creativity. Research on employee happiness, at present, is quite limited. Therefore, the paper will attempt to bring together knowledge, based on primary sources, to give a valuable contribution to this area.

In most of the societies, the happiness is thought to be the most valued goal of the individuals, groups and organisations. It is the basic human emotion expressed in the form of joy. For the human experience, happiness is the fundamental need, and most of the people are mildly happy at most of the times in their lives (Cole, 2015).

Happiness in the workplace has attracted the attention of several researchers. Over the last past years, several constructs are overlapping the concept of happiness. An important construct of workplace happiness is job satisfaction. The happiness at the workplace related constructs influences the organisational research in many ways like the level of the constructs, duration and stability of a certain construct over time and the content of each construct. The happiness related constructs are usually measured and defined at the personal level like the job satisfaction, mood at work and effective commitment of the employee (Sharifzadeh, 2014).

Its concluded that happiness is the experience of the frequent positive feeling, infrequent negative feeling and the sense of satisfaction with life. To win satisfaction customers, the organisation needs enthusiastic employees who are capable of transferring the loyalty and enthusiasm to the customers, since the success of a business relies upon the satisfaction of the customers.

Work is a significant aspect of the lives of people. Work was done in exchange to intrinsic and extrinsic rewards. The nature and the consequences of work have been changing rapidly over the past few years. The changing environment of the organisations and work like the development of new technological products, the rapidly increasing globalisation and the employment of new organisational practices led to the changing nature of the work (Wesarat, 2015).

Certain practices and strategies adopted by the organisations tend to reduce the happiness and satisfaction of the employees like corporate restructuring and downsizing. Unhappy and unsatisfied employees feel less valued and perceive low job security. The organisations expect performance from their employees and employ managerial tools and techniques to enhance employee performance. By maintaining a higher level of employee happiness and satisfaction, organisations can achieve higher levels of employee performance.

Rego (2008) also suggested that happy employees are productive employees. There is a strong correlation between employee happiness and job satisfaction. In addition to this, job satisfaction and life satisfaction are also influenced by each other. The happiness at the workplace refers to the happiness of an individual at the workplace and life (Rego, 2008).

The significance of workplace happiness is important to understand because the performance of the organisations heavily relies on the happiness and satisfaction of the employees. Happy employees tend to add value towards the organisational goals and set them at their priority as compared to their counterparts who merely work for the earning and do not enjoy the content of their job.

8.2 Employee Performance

The performance of the employees can be perceived as a driver to the organisational performance. The employees who deliver help the organisation move a step forward towards the attainment of its goals and objectives.

Hanaysha (2016) found that one of the most important objectives of a business organisation is to enhance the level of the performance of the employees. The enhanced performance of the employees entitles an organisation and its employees with a wide range of advantages. Higher rate of performance of the employees leads to a higher rate of economic growth, the higher rate of profitability and faster progress of the society. The employees who are more productive and perform well in an organisation get higher wages, benefits, salary packages, better working conditions and other favourable employment opportunities. Better performance of the employees enhances the degree of output and reduces the costs associated with production and operations thus tending to help the organisation gain a competitive advantage. Employee performance is essential to gain long term success of the organisation (Hanaysha, 2016).

Previous researchers have found that the companies must seek to invest in employee engagement to gain higher levels of employee performance. There exists a positive relationship between employee engagement and performance. The engaged employees are more involved in the work and tend to be more productive. They are more focused than disengaged employees. The engaged employees work more efficiently since they place the organisational success and goal achievement on the priority position in their minds. Employee engagement drives the performance of the employees and results in higher levels of organisational outcomes (Pradhan, 2016).

As highlighted by Pradhan (2016), nowadays, enhancing the level of employee performance are of major concern to every together business organisation. It can be defined as the measure of the efficiency of an employee or a group of employees. Performance, as defined in the objective terms, is the effort or the measures taken that enhance the profitability of the organisation. Performance can be gauged in terms of the output delivered by an employee over a specific period. The performance of an employee can be measured relative to the average work performed by the other employees doing a similar task. In addition to this, the performance of the employees can also be measured in terms of the units produced or managed by an employee in the defined period. The success of an organisation is defined by the level of the performance of the employees, thus making it an important objective for the businesses (Pradhan, 2017).

The performance of the employees can be assessed in several ways. However, there is a need for the development of some standardised criteria to measure the performance of the employees. Employee performance defined as the amount of the time that an employee spends on the job while being physically and mentally present and working efficiently. Time can effectively be as a measure of employee performance. The time that an employee spends at the job or while doing the task that he or she was hired to perform help in evaluating the performance of the employees (Muda, 2014).

The study of the past researches regarding employee performance is related to our study because the more productive the employees of an organisation would be, the more the company would have to deliver to its customers. Attaining customer satisfaction and loyalty is the ultimate goal of every business which can only be achieved when the employees are productive.

8.3 Employee Creativity

Creativity defined by several researchers in different ways. Some researchers defined it as a personal characteristic, and some define creativity as a process. The researchers suggest that a creative proposition can be a product, service or idea if it fulfils the following conditions: it must be original and made for the very first time and it must be relevant to one's job or beneficial for the organisation towards the attainment of its goals. The researchers also determined the basic personality traits that are associated with the degree of creativity one can depict (Mubarak, 2018).

Hassan (2013) addressed the role of the personal attributes in determining the level of creativity of an employee. Unsworth's matrix of creativity that consists of four types of creativities that vary based on two dimensions I the driver of employee engagement that can be an internal or external stimulus and the problem that the employee is confronting that can be open or close. When the driver of the engagement it external, the employee is engaged to be creative by some external driver of stimulus, and when there is some internal engagement, the employee urges to be creative rather than being forced by some other or external stimulus. The open idea is those ideas that come from the individual employees, and the closed ideas are those that presented to the individuals.

The four types of creativity include the responsive creativity that arises to respond to the problems arising due to the external drivers, the expected creativity that helps in discovering the problems due to external drivers, contributory creativity responds to the problems arising from the internal drivers' proactive creativity that discovers the problems that arise due to the internal drivers. The creativity of the employees is enhanced by the personal as well as by the contextual factors. The contextual factors include the goals and expectations for the pursuance of the creative activities, social influence on the creativity of an individual, relationship with the employer, co-workers and the organisational work setting.

As found by Qureshi (2016), an employee who comes up with new and innovative ideas or suggestions is thought to be creative employee. A creative employee proposes the new ideas for the products or the services to be constructed, influences the communication and the perception which ultimately affects the way done during his or her labour hours. Creativity is the design in which the employee brings an innovation that seeks to resolve the work related problems following the step by step procedure. It can also be explained as the ability of the individuals to develop resourceful solutions to meet the challenges that the employees face while being at a job working individually and in groups.

The creative employee is aware of the organisation and the workplace setting. He or she can be flexible and sensitive to the problems and come up with innovative solutions. The creative employee tends to have a strong memory and remembers the task for a longer period. The creative employee is adaptive and easily copes up with the changing environment and readily accepts technological advancements. Thus, an individual who accepts all the changes or advancements like the change in the organisational culture of the technological changes which are essential to keep pace with the competition termed as a creative employee (Shahzadi, 2014).

The large organisations are more likely to have creative employees. In the current era, the employees constantly exposed to the challenging environment where it is essential to learn and adapt quickly to keep up with the pace of the competition. To achieve the goals promptly, the employees need to be creative. Creativity can also explain as the process of introducing new techniques by individuals and groups to achieve the goals efficiently. The firms that want to compete in the market hire creative and extrovert employees. The creative employees are expected to bring positive and constructive change in the organisation. The creative employees are flexible and adapt to the technological advancements and cope up with the changes in the external environment of the business. The creative employee has strong interpersonal and analytical skills that enable him or her to devise the ways to resolve the business processes on time (Dul, 2010).

The studies conducted by Nawaz (2018) suggested that empowered employees tend to be more creative since they have a say in the matters of the organisation. Creativity may depend on the employee or the situation that the employee is subject to. The employee may put in all his or her efforts and capabilities to get the best outcome and develop resourceful and creative solutions (Nawaz, 2018).

The past researches suggest that creativity among the employees help the organisations find an efficient means of resolving the work and operational issues. The creative employees not only help the companies find resourceful solutions for the problems but also assist in enhancing the efficiency of the processes.

Fisher, Happiness at Work (2010) Found that happiness mediates the relationship between authentic leadership and employee creativity.

8.4 Antecedents of Employee Happiness

This section will discuss the factors that influence the degree of the happiness of the employees at the workplace. According to Herzberg's Theory of Employee motivations, the factors that influence the job satisfaction of an employee can categorise as the hygiene factors and the motivational factors. The motivational factors are those factors that are concerned with the content of the job and the factors like recognition, acknowledgement, growth opportunities, career advancement, responsibilities and the tasks assigned to him or her. On the other hand, the hygiene factors are the factors that are not related to the content of the job the employee is performing like the salary, job security, workplace settings, company policies and strategies and the interpersonal links at the workplace. As explained by Herzberg's Theory, the motivational factors being employee happiness and satisfaction while the absence of the hygiene factors brings dissatisfaction of the employees. The extrinsic factors are the events that arise from the efforts of the employees as well as the other factors that are not directly involved with the job of the employee. The compensation, benefits, interpersonal ties with the colleagues and the workplace setting influence the outcomes of the employees but they are not a fundamental part of the job of the employee (Pauceanu, 2016).

8.5   The Extrinsic Factors of an Employee's Happiness at the Workplace

The factors affecting the happiness of the employees at the workplace can categorise as the extrinsic and intrinsic factors (Shu Zhou, 2013). The extrinsic factors of an employee's happiness at the workplace are.

8.5.1   Workplace Environment

The level of the employees' happiness and their satisfaction gauged by the environment of the work placement. Since the employees spend the major part of their day at the workplace, so it is very important for the organisation to ensure the availability of a clean and suitable workplace for them. Proper working conditions promoting a professional culture must provide to the employees. Proper working conditions will promote the performance of the employees, and they will work in a relaxed manner, thus feeling happier with the workplace (Štreimikienė, 2009).

The organisation must ensure to provide all the necessary resources to the employee that are necessary to perform the particular task or the job that he assigned. If the work conditions are not appropriate, the employees might lose their interest in their work and will not enjoy it. The dimensions that constitute the work environment satisfaction are: the natural environment of the workplace like humidity, noise, brightness, cold, smell and other natural environmental factors, the second dimension of the workplace conditions is the availability of the required tools and equipment. Are the tools and equipment necessary to perform a job readily available to the employee?

The third dimensions are the working timings and hours and the overtime hours. The fourth factor is the safety of the workplace. Are the health and safety legislation properly being followed and the employees are safe?

8.5.2   Compensation

The pay and compensation include whatever the employee paid against the tasks that he performs while being at the job. Pay defined by the researchers in several ways like the basic salary and the other fringe benefits like the housing rent, medical allowance, transport allowance and the overtime salary. Pay or compensation includes the financial as well as the non-financial awards that the employee gets against his services for the organisation. If the organisation hinders the pay of the employees, their happiness and satisfaction levels will greatly decline. When the employees are compensated rightly and justifiably for the task that they are performing, and they can easily manage the workload assigned to them, they will feel happy with the job. Salary is one of the hygiene factors described by Herzberg's Theory, and it greatly accounts for the employee's happiness and satisfaction at the workplace. The compensation plan of the organisation greatly influences the level of the happiness of employees at the workplace. The better the compensation the company offers, the more satisfied and happier the employees would be. Compensation and benefits are very important factors influencing employee happiness in the workplace.

8.5.3　Organisational Commitment

The strongly committed the employees are to the organisation, the higher will be their loyalty and dependence with the organisation. When the employees are committed to the organisation, they have a stronger sense of ownership and thus feel happier.

8.5.4　Self-realisation

The self-realisation comes at the top of Maslow's Hierarchy of needs and is an important determinant of the employee happiness at the workplace. The feeling of self-fulfilment had a strong influence on the level of employee happiness.

8.5.5　Health Concerns

The organisations focus on employee health and well-being. The organisations invest in employee health maintenance programs and policies. The organisations that impart significance to the employee well-being and health tend to have satisfied and happier employees.

8.5.6　Nature of the Job

In addition to the hygiene factors, the motivational factor that is the nature of the job also greatly influences the happiness of the employees at the workplace. Different aspects of the job like the salaries of the growth opportunities and terms with the supervisors. Influence the happiness level of the employees. When employees perceive that they are performing a productive task, they feel happier and more satisfied. The challenges associated with the tasks that are worthwhile in an organisation, drive the motivation and confidence of the employees and they develop an interest in their work, thus feeling happier with the workplace.

8.6 The Intrinsic Factors Influencing the Level of Employee Happiness at the Workplace

The intrinsic factors influencing the level of employee happiness at the workplace are.

8.6.1 Degree of Autonomy Imparted to the Employee

The employees want to feel empowered in the organisation where they work. The employees need the authority to perform the tasks assigned to them properly. There is a positive relationship between employee happiness and autonomy in the workplace. When the employees empowered in their jobs, they take pride in the work that they do.

8.6.2 Recognition

Recognition can be defined as the situation in which the organisation acknowledges the efforts of the employees and appreciate them by rewarding them some status in the organisation. Among the recognition of the intrinsic and non-financial reward is the most important one. It makes the employees feel valued and appreciated, and they fee motivated thus doing all the tasks with more interest and focus. The organisation can gain a competitive advantage by awarding employees with recognition. When the employees rewarded for their efforts, they become more loyal and happier with the organisation and go the extra mile for the attainment of the organisational goals and objectives.

8.6.3 Meaningful Work

The meaningful work is a very important factor shaping up the perceived happiness of the employees. The employees want to be engaged in productive tasks and qualitative work. Meaningful work is very crucial for the success of the organisation as well as the employee. Apart from money, there are some factors that an employee wishes to share with the community and his circle. Sense of achievement and responsibility are among the most important such factors. The employees who are engaged in meaningful work feel happier and more satisfied because they perceive themselves responsible.

A large number of factors act as the antecedents of employee happiness and well-being. The wages and compensation do not always motivate the employees, but some hygienic factors as well that include the organisational work setting.

In this study, we decide to focus only on the work environment, leadership style and compensation as antecedents of workplace happiness.

8.7 Role of Workplace Happiness

The author says that it is necessary for an organization to understand what the employees feel and what they want. In this way, the organization can shape up its policies and strategies and can devise the ways to enhance the loyalty of the employees. Enhancing the happiness and satisfaction of the employees is crucial for the efficient performance of the organization (Malaolu, 2013).

Workplace happiness means how often the employee feels joy in relation to his or her job. The feelings of happiness are constructive and positive. There is a positive relation between employee happiness at the workplace and the organizational performance. The workplace happiness provides input for the better performance of the organization. Workplace happiness creates positive feelings among the employees regarding their jobs, and they tend to perform better resulting in greater outcomes for the organization.

The outputs are the obvious features of the performance of an organization. Most of the organizations measure their performance in relation to the degree to which they are achieving their mission. The organizations that are moving towards the attainment of their mission are said to be performing well. When the employees are happy and satisfied, they serve the customers of the organization as per plan and thus helps the organization increase the cash inflows (Nur, 2015).

9. Organizations can dramatically be changed by leveraging creativity and innovation. Creativity is the starting point of bringing innovation in an organization. The creativity is contagion across the organization. The individual creativity has an effect on the creativity of the group. The creativity of the group has an effect on the creativity of the organization and the creativity of the organization ultimately drives the innovation across the entire organization (Han, 2015). The creativity of the employees helps

9 Data Collection and Analysis

9.1 Keyword Search

The researcher has been used some key words in order to find similar factors, see the Table 1.

9.2 Journal Databases

The researcher has been found many results in the databases that has been used, Table 2.

Refer to search scores, the researcher found that 307 studies were gotten based on the search terms that are explained in Table 2 and Fig. 2. While presenting data analysis, the researcher removed some studies such as low-quality studies, and 27

Table 1 Keyword search

Keyword search
"Workplace" and "Happiness" and "Employee" and "Performance"
"Happiness" and "Work placement"
"Workplace happiness" and "Creativity" and "Performance"
"Happiness" and "Performance" and "Innovation"
"Happiness" and "Employee" systematic review
"Workplace happiness" and "Employees"
"Workplace happiness" and "Employees" and "Systematic review"

Table 2 Initial search results across the database

Journal databases	Frequency
Emerald	41
Google scholar	104
ScienceDirect	53
Epsco	33
ProQuest	76
Total	307

duplicate results were excluded. The researcher also removed 231 articles from the remaining after screening inclusion and exclusion. Furthermore, qualitative studies that did not give specific information that help us in finding needed information were removed.

Fig. 2 Initial search results across the database

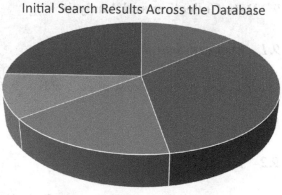

Initial Search Results Across the Database

■ Emerald　■ Google Scholar　■ ScienceDirect　■ Epsco　■ ProQuest

9.3   Flowchart for the Selected Studies

The researcher has been searching in five databases to find the external factors frequently used and joined the results together by using below criteria in order to give the connection between the papers Table 3. In around 80 studies there was more than 210 external factors were identified, but we chose only six external factors that have a relationship which described in Table 3 such as (Work Placement Happiness, leadership, Payment Methods, Work Environment, Creativity and Performance) (Figs. 3, 4 and 5).

Table 3 The most commonly used external factors across databases

External factors	Database					
	emerald	Google scholar	ProQuest	ScienceDirect	Epsco	Total
Work placement Happiness	5	7	9	4	2	27
Leadership style	4	5	0	10	3	22
Flexible pay	5	8	8	11	9	41
Work environment	3	8	9	2	7	29
Creativity	4	5	1	6	0	16
Performance	3	0	6	1	8	18

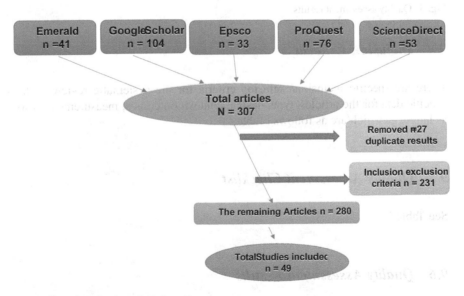

Fig. 3 Flowchart for the selected studies

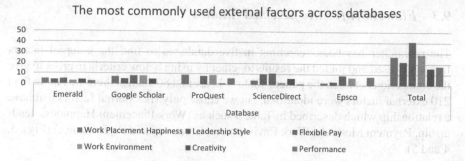

Fig. 4 The most commonly used external factors across databases

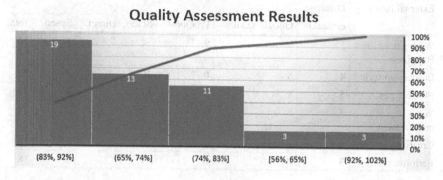

Fig. 5 Quality assessment results

9.4 *Inclusion Criteria*

There are specific inclusion/exclusion criteria for the systematic review such as specific data for the articles, type, design, context, outcomes, measurement as well as language which are as follows (Table 4).

9.5 *Quality Assessment Checklist*

See Table 5.

9.6 *Quality Assessment Results*

See Table 6.

Table 4 Inclusion and exclusion criteria for customer retention studies

Inclusion criteria	Exclusion criteria
– Context: should be in workplace happiness impact – Measurement: articles in which the impact of the workplace happiness on the employee performance and level of creativity – Language of articles should be In English – Date: Published in the period 2010 to 2020 – Type of Studies: Qualitative, Peer-reviewed, empirical studies – Outcome: employee performance and creativity – Study design: systematic, randomized and meta-analysis – Could include articles, reports, or theses	– Knowledge management process or factors that are used in context other than impact of workplace happiness – Information in knowledge management which not related and not support to Workplace happiness and impact on employee – Articles in not English Language

Table 5 Quality assessment checklist

Numbers	Questions
1.	Are the research aims clearly specified?
2.	Was the study designed to achieve these aims?
3.	Are the variables considered by the study clearly specified?
4.	Is the study context/discipline clearly specified?
5.	Are the data collection methods adequately detailed?
6.	Does the study explain the reliability/validity of the measures?
7.	Are the statistical techniques used to analyze the data adequately described?
8.	Do the results add to the literature?
9.	Does the study add to your knowledge or understanding?

Table 6 Quality assessment results

Study	Q1	Q2	Q3	Q4	Q5	Q6	Q7	Q8	Q9	Total	Percentage (%)
1	1	1	0.5	1	1	1	0.5	0.5	1	7.5	83
2	1	1	1	1	1	1	1	1	1	9	100
3	1	1	1	1	0.5	0.5	1	1	1	8	89
4	1	1	0.5	0.5	1	1	0.5	1	1	7.5	83
5	1	0.5	1	1	1	1	1	0.5	1	8	89
6	1	1	1	0.5	0.5	0.5	1	1	0.5	7	78
7	0.5	0.5	1	1	0.5	1	0.5	0.5	0.5	6	67
8	1	0.5	0.5	1	1	1	0.5	1	1	7.5	83
9	1	1	1	0.5	0.5	1	0.5	0.5	1	7	78
10	0.5	0.5	0.5	0.5	1	1	0.5	1	0.5	6	67
11	1	1	1	0.5	0.5	0.5	0.5	0.5	1	6.5	72

(continued)

Table 6 (continued)

Study	Q1	Q2	Q3	Q4	Q5	Q6	Q7	Q8	Q9	Total	Percentage (%)
12	1	1	1	0.5	1	1	1	0.5	1	8	89
13	1	0.5	0.5	0.5	1	1	1	1	1	7.5	83
14	0	0.5	1	0.5	1	0.5	0.5	1	0.5	5.5	61
15	1	1	0.5	1	0.5	0.5	0.5	1	0.5	6.5	72
16	0.5	0.5	1	1	1	0.5	0.5	1	1	7	78
17	1	0.5	1	0.5	1	0.5	1	0.5	0.5	6.5	72
18	0.5	1	1	0.5	0.5	0.5	0.5	0.5	0.5	5.5	61
19	1	0.5	1	1	1	0.5	0.5	0.5	1	7	78
20	1	1	0.5	1	1	1	0.5	0.5	1	7.5	83
21	1	1	1	1	0.5	1	0.5	1	0.5	7.5	83
22	1	1	1	1	1	0.5	1	0.5	1	8	89
23	1	1	0.5	0.5	0.5	1	0.5	1	1	7	78
24	1	0.5	1	1	0.5	1	1	0.5	1	7.5	83
25	1	1	0.5	0.5	0.5	0.5	1	1	0.5	6.5	72
26	0.5	0.5	0.5	1	0.5	1	1	0.5	0.5	6	67
27	1	0.5	0.5	1	1	1	0.5	1	1	7.5	83
28	1	1	0.5	0.5	1	1	1	0.5	1	7.5	83
29	0.5	0.5	1	1	1	0.5	0.5	1	0.5	6.5	72
30	1	1	1	1	0.5	0.5	1	0.5	1	7.5	83
31	1	1	0.5	0.5	1	0.5	1	0.5	1	7	78
32	1	0.5	0.5	1	1	1	1	0.5	1	7.5	83
33	0	0.5	1	1	1	0.5	0.5	1	0.5	6	67
34	1	1	1	1	0.5	0.5	1	1	0.5	7.5	83
35	0.5	0.5	1	0.5	1	1	0.5	1	1	7	78
36	1	1	1	0.5	1	0.5	1	0.5	1	7.5	83
37	0.5	1	1	0.5	1	0.5	0.5	0.5	0.5	6	67
38	1	0.5	0.5	1	1	0.5	0.5	0.5	1	6.5	72
39	1	1	0.5	0.5	1	1	0.5	0.5	1	7	78
40	1	1	1	1	1	1	1	1	1	9	100
41	1	1	0.5	1	0.5	0.5	1	1	1	7.5	83
42	1	1	0.5	0.5	1	0.5	0.5	1	1	7	78
43	1	0.5	1	1	0.5	1	1	1	1	8	89
44	1	1	1	0.5	0.5	0.5	1	1	0.5	7	78
45	0.5	0.5	1	0.5	0.5	0.5	0.5	0.5	0.5	5	56
46	1	1	1	1	1	1	1	1	1	9	100
47	1	0.5	1	0.5	0.5	1	0.5	0.5	1	6.5	72
48	0.5	0.5	0.5	0.5	1	1	1	1	0.5	6.5	72
49	1	1	1	0.5	0.5	1	0.5	0.5	1	7	78

9.7   Classification of the Study Process and Previous Studies

Author	Year	Study title	Variables	Location & sample size	Findings
NGUYEN NGOC AN	2015	"Actors affecting employee creativity & ITS impact On organizational innovation capability in state Owned banks in Hanoi, Vietnam"	Employee creativity and organizational innovation	Vietnam 24	This study helped us identify the relationship between employee creativity and organizational innovation. We found positive relation between the two variables
Malikeh Beheshtifar and Elaheh Zare	2013	"Employee creativity: a compulsory factor in organizations"	Employee creativity	Iran Secondary research	The results indicated that the organizations can improve creativity by hiring the candidates who are potentially creative and are energetic to support the positive change in organization
Katarzyna Bratnicka	2015	"Creativity and effectiveness in organizations"	A framework was established to show the relationship between the variables—Creativity and effectiveness	Katowice Secondary Research	The study depicted that the organizational creativity not only adds towards the efficiency and effectiveness of the organization, rather it also influences the emergence
Larry E. Cole and Michael S. Cole	2014	"Employee satisfaction and organizational performance"	Employee satisfaction and organizational performance are the variables of this study. The study used secondary data	USA 200	This study concluded that the performance of employees is related to the performance of organization and this finding can be generalized across all the firms and industries
Jan Dul and Canan Ceylan	2014	"Work environments for employee creativity"	Work environment and employee creativity	Netherlands 409	This study investigated the effect of the social, personal and psychological factors on the creativity of the individuals. The conceptual model depicted this relationship

(continued)

(continued)

Author	Year	Study title	Variables	Location & sample size	Findings
Jae Hoon Han, Jin Taek Jung and Hyung Kun Joo	2015	"A study on effects of creativity to organizational innovation"	Individual creativity, group creativity and organizational performance	Korea 260	The results reflected that the individual creativity and creativity in groups have no significant impact on the organizational creativity
Mobrezi, H., & Khoshtinat, B	2016	"Improving employee productivity through work engagement: Evidence from higher education sector"	Employee engagement and employee creativity	Malaysia 242	The results indicated that the work engagement positively influences the employee creativity. The study also proved that all the dimensions of work engagement influences the productivity of the employees
Masood UL Hassan, Ahmad Mehmood Malik and Amna Hasnain	2013	"Measuring employee creativity and its impact on organization innovation capability and performance in the Banking Sector of Pakistan"	Employee creativity, organizational performance and organizational innovation	Multan 164	The results showed that the complexity of the job and relationship with the manager directly influence the creativity of the employees. Moreover it showed positive relationship between employee creativity and organizational performance
Baek-Kyoo Joo and Insuk Lee	2016	"Workplace happiness: work engagement, career satisfaction, and subjective well-being"	Workplace happiness and employee well being	Korea 550	The study indicated that the employees who are dedicated to their work and enjoy their jobs tend to have greater level of workplace happiness
Andrew J. Oswald	2009	Happiness and productivity	Happiness—productivity—performance and organization	Bonn 180 male and female	Positive effect on productivity and happiness as this study found happiness in a workplace carries with it a return in productivity

(continued)

(continued)

Author	Year	Study title	Variables	Location & sample size	Findings
David Sweetman, Fred Luthans, James and Brett	2010	Relationship between positive psychological capital and creative performance	Positive organizational behavior—organizational—performance creative—employee	Working adults 459 males & 417 females N = 899	The study examine the significant relationship between working adults specific positive psychological resources with their performance in creative exercise. Employee creativity is a widely recognized ingredient of innovation in organizations
Micu Nngela & Necula Ramona	2017	Pursuing happiness in the workplace, purpose and challenge for human resources management	Job—happiness—satisfaction	N = 50 employee	study examine to be truly happy at work we need to contribute to the happiness of others who work with us and who also want their stat of well being-happiness at work cannot be overlooked but consciously assessed if it is to achieve organizational performance and profit
Abdul Ghani Kanesan	2016	Understanding wokplace happiness in schools: the relationship between workplace happiness, teachers' sense of efficacy, affective commitment and innovative behavior	Innovative behavior—self efficacy and trust	Teachers in Malaysia N = 835	This study show that workplace happiness has a relationship with teacher's innovative behavior and examine that, there is a direct Relationship between happiness at work on affective commitment

10　Research Limitations

The research limitations refer to the issues and problems which have been faced during the course of design and analysis. The research has been focused on workplace happiness and how it impacts employee creativity and performances. The overall research has been conducted in terms of primary data analysis. This method has posed different issues within the course of research. These issues include diverse activities which are related to the data analysis and collection of data as well. The limitations which have been mentioned were tackled on the basis of using ethical guidelines and standards. These were also based on achieving the complete aims and analysis of the research.

11 Conclusion

The research has been compiled on the basis of workplace happiness and its impacts on employee creativity and performance as well. Human Resource Management practices revolve around the factors of implementing change and techniques. The implementation of these changes defines how the organizational structure is also balanced. These services and products also define the company values towards the people and the community. Workplace happiness is something that needs to be addressed in all organizations to ensure performance and success. The focus of this research is to develop different hypotheses about employee's happiness and his performance. Different factors will be tested to see their correlation with employees' happiness, and then this factor of happiness is used to evaluate its effect on employees' performance. It is concluded that leadership style, flexible pay, effective working environment positively affects employees' happiness at the workplace and in turn employees' happiness effects their performance at the organization.

References

1nguyen Ngoc An, 2. T. (2015). Factors affecting employee creativity & its impact on organizational innovation capability in state owned banks in Hanoi, Vietnam.
Ashkanasy, N. M., & Dorris, A. D. (2017). Emotions in the workplace. *Annual Review of Organizational Psychology and Organizational Behavior,* 67–90. Retrieved from http://sci-hub.tw/10.1146/annurev-orgpsych-032516-113231.
Beheshtifar, M. (2013). Employee creativity: A compulsory factor in organizations.
Bratnicka, K. (2015). Creativity and effectiveness in organizations. *Management.*
Chancellor, J. (2015). Recalling positive events at work makes employees feel happier, move more, but interact less: A 6-week randomized controlled intervention at a Japanese workplace. *Journal of Happiness Studies,* 871–887.
Chandrasekar, K. (2011). Workplace environment and its impact on organisational performance in public sector organisations. *International Journal of Enterprise Computing and Business Systems.* Retrieved from http://citeseerx.ist.psu.edu/viewdoc/download?doi=10.1.1.300.8598&rep=rep1&type=pdf.
Cole, L. E. (2015). Employee satisfaction and organizational performance: A summary of key findings from applied psychology.
Davis, S. (2015). Increasing happiness in the workplace. *The kabod,* 1–11.
Dul, J. (2010). Work environments for employee creativity.
Fisher, C. D. (2010). Happiness at work. *International Journal of Management Reviews,* 384–412. Retrieved from http://sci-hub.tw/10.1111/j.1468-2370.2009.00270.x.
Han, J. H. (2015). A study on effects of creativity to organizational innovation.

Hanaysha, J. (2016). Improving employee productivity through work engagement: Empirical evidence from higher. *International Journal of Industrial Engineering Computations*.

Hassan, M. U. (2013). Measuring employee creativity and its impact on organization innovation capability and performance in the banking sector of Pakistan. *World Applied Sciences Journal*.

Joo, B.-K. (2016). Workplace happiness: Work engagement, career satisfaction, and subjective well-being. *Evidence-Based HRM: A Global Forum for Empirical Scholarship*.

Joo, B.-K., & Lee, I. (2017). Workplace happiness: Work engagement, career satisfaction, and subjective well-being. *Evidence-Based HRM: A Global Forum for Empirical Scholarship*, 206–221. Retrieved from http://sci-hub.tw/10.1108/EBHRM-04-2015-0011.

Liu, D., Gong, Y., Zhou, J., & Huang, J.-C. (2016). Human resource systems, employee creativity, and firm innovation: The moderating role of firm ownership. *Academy of Management Journal*, 1164–1188. Retrieved from http://sci-hub.tw/10.5465/amj.2015.0230.

Malaolu, V. A. (2013). Training and manpower development, employee productivity and organizational performance in Nigeria: An empirical investigation.

Mazuki, R., Samah, A. A., Omar, S. Z., Zawawi, D., Idris, K., D'Silva, J. L., et al. (2017). Towards developing a framework on happiness and productivity among personnel in institution of higher learning. *International Journal of Academic Research in Business and Social Sciences*, 269–278. Retrieved from http://hrmars.com/hrmars_papers/Towards_developing_a_framework_on_happiness_and_productivity_among_personnel_in_Institution_of_Higher_Learning.pdf.

Mubarak, F. (2018). Effect of authentic leadership on employee creativity in project-based organizations with the mediating roles of work engagement and psychological empowerment.

Muda, I. (2014). Factors influencing employees' performance: A study on the Islamic Banks in Indonesia.

Nawaz, K. (2018). Impact of person-organization fit on employee creativity: a research on 200 employees working in private banks in Sargodha Punjab, Pakistan. *Arabian Journal of Business and Management Review*.

Nur, A. H. (2015). Mployee job satisfaction and organizational performance: Empirical study from higher education centers in Mogadishu-Somalia. *International Journal In Commerce, It & Social Sciences*.

Oswald, A. J., Proto, E., & Sgroi, D. (2015). Happiness and productivity. *Journal of Labor Economics*, 789–822. Retrieved from http://sci-hub.tw/https://www.journals.uchicago.edu/doi/abs/10.1086/681096.

Pauceanu, A. M. (2016). Variables affecting employee satisfaction—Evidence from sultanate of Oman.

Pradhan, R. K. (2016). Employee performance at workplace: Conceptual model and empirical validation.

Pradhan, R. K. (2017). Employee performance at workplace: Conceptual model and empirical.

Proto, E. (2016). Are happy workers more productive? *IZA World of Labor*. Retrieved from https://wol.iza.org/uploads/articles/315/pdfs/are-happy-workers-more-productive.pdf.

Qureshi, H. S. (2016). The impact of employees' creativity on the performance of the firm. *Research Issues in Social Sciences*.

Rego, A. (2008). Workplace spirituality and organizational commitment: An empirical study. *Journal of Organizational Change*.

Sageer, A., Rafat, S., & Agarwal, P. (2012). Identification of variables affecting employee satisfaction and their impact on the organization. *IOSR Journal of Business and Management*, 32–39. Retrieved from https://s3.amazonaws.com/academia.edu.documents/34377954/E0513239_shakeela.pdf?AWSAccessKeyId=AKIAIWOWYYGZ2Y53UL3A&Expires=1546867392&Signature=8kCmqHfSMvN9%2FSh%2F79jAdbys%2F8s%3D&response-content-disposition=inline%3B%20filename%3DIdentification_of_Var.

Shahzadi, I. (2014). Impact of employee motivation on employee performance.

Sharifzadeh, M. (2014). Happiness and productivity in the workplace.

Shu Zhou, L. Q. (2013). The differences of factors influencing employees' Happiness. *American Journal of Industrial and Business Management*.

Štreimikienė, D. (2009). Life satisfaction and happiness—The factors in work performance.

Wesarat, P.-O. (2015). A conceptual framework of happiness at the workplace. *Asian Social Science.*

Yang, C. (2014). Does ethical leadership lead to happy workers? A study on the impact of ethical leadership, subjective well-being, and life happiness in the Chinese Culture. *Journal of Business Ethics,* 513–525. Retrieved from http://sci-hub.tw/10.1007/s10551-013-1852-6.